11th International Scientific Symposium on Electric Power Engineering (ELEKTROENERGETIKA 2022)

Stará Lesná, Slovakia
12-14 September 2022

Editors:

Michal Kolcun
Iraida Kolcunová
Juraj Kurimský

ISBN: 978-1-7138-5555-2

Printed from e-media with permission by:

Curran Associates, Inc.
57 Morehouse Lane
Red Hook, NY 12571

Some format issues inherent in the e-media version may also appear in this print version.

Technical University of Kosice – Department of Electrical Power Engineering
Letna 9
042 00 Kosice
Slovak Republic

Phone: 421 55 602 1111

http://www.tuke.sk

Additional copies of this publication are available from:

Curran Associates, Inc.
57 Morehouse Lane
Red Hook, NY 12571 USA
Phone: 845-758-0400
Fax: 845-758-2633
Email: curran@proceedings.com
Web: www.proceedings.com

TECHNICAL UNIVERSITY OF KOŠICE

Faculty of Electrical Engineering and Informatics

Department of Electrical Power Engineering

Proceedings of the 11th International Scientific Symposium on Electrical Power Engineering

ELEKTROENERGETIKA 2022

September 12–14, 2022, Stará Lesná, Slovak Republic

GENERAL SPONSOR

VSD

Východoslovenská distribučná a.s.

SPONSORS AND SUPPORTERS

seps
Slovenská elektrizačná prenosová sústava

Slovenská elektrizačná prenosová sústava, a. s.

SLOVENSKÉ ELEKTRÁRNE

Slovenské elektrárne, a.s.

HASMA

MEDIA AND OTHER PARTNERS

ENERGETIKA

Časopis ENERGETIKA

World Energy Council
Slovenský výbor svetovej energetickej rady

AUTOMA

časopis pro automatizační techniku

TECHNICAL UNIVERSITY
OF KOŠICE

Faculty of Electrical Engineering and Informatics
Department of Electrical Power Engineering

Proceedings of the 11th International Scientific Symposium on Electrical Power Engineering

ELEKTROENERGETIKA 2022

September 12–14, 2022, Stará Lesná, Slovak Republic

Name: Proceedings of the 11th International Scientific Symposium on Electrical Power Engineering ELEKTROENERGETIKA 2022

Publisher: Technical University of Košice
Faculty of Electrical Engineering and Informatics
Department of Electrical Power Engineering

Published: September 2022

Editors: Michal Kolcun, Iraida Kolcunová, Juraj Kurimský

Manuscripts of papers submitted to Symposium Proceedings are peer reviewed by two reviewers. The review results are compiled according following procedure: the paper is accepted or rejected according match of recommendations of both reviewers. If there is conflict of review results, the paper are reviewed by third reviewer and his result is conclusive.

The Authors are responsible for contentual and lingual accuracy of their papers and the materials presented.

Publisher address

Technical University of Košice
Letná 9
Košice
Slovak Republic

ELEKTROENERGETIKA 2022 – General Informations

Symposium Guarantee

Dr.h.c. prof. Ing. Michal Kolcun, PhD., Technical University of Košice, SK

Members of International Scientific Committee

Symposium Steering Committee

Reviewers for Proceedings of the 11th International Scientific Symposium on Electrical Power Engineering ELEKTROENERGETIKA 2022

Many experts have given generously of their time and expertise to review manuscripts for Proceedings International Scientific Symposium on Electrical Power Engineering ELEKTROENERGETIKA during years. We are extremely grateful to each of these reviewers for their commitment to the Proceedings. — Editors.

- Beláň Anton
- Beňa Ľubomír
- Braciník Peter
- Bucko Samuel
- Cimbala Roman
- Čonka Zsolt
- Dolník Bystrík
- Dudiak Jozef
- Dvorský Emil
- Diahovchenko Illia
- Džmura Jaroslav
- Georrgi Georgiev
- Goňo Radomir
- Gutten Miroslav
- Gyurov Valentin
- Knápek Jaroslav
- Kiraly Jozef
- Kolcunová Iraida
- Kot Sebastian
- Krbal Michal
- Krištof Vladimír
- Krokavec Dušan
- Kurimský Juraj
- Liptai Pavol
- Lomane Tatjana
- Marci Martin
- Martínek Zbyněk
- Máslo Karel
- Medved Dusan
- Mešter Marián
- Mišák Stanislav
- Müller Zdeněk
- Orságová Jaroslava
- Otčenášová Alena
- Pavlík Marek
- Pavlov Ľuboš
- Perduková Daniela
- Petráš Jaroslav
- Rusek Stanislav
- Rusnák Jozef
- Semančík Peter
- Skurčák Ľuboš
- Vojtek Martin
- Zbojovský Jan
- Zicmane Inga

Table of Contents

Index of Authors

A: Generation of Electricity, Transmission, Distribution and Consumption of Electricity

Operability Satisfaction of Energy Systems with Nuclear Power Plants at Different Photovoltaic Penetration Levels

Illia Diahovchenko, Ihor Yevtushenko
Department of Electric Power Engineering
Sumy State University
Sumy, Ukraine
i.diahovchenko@etech.sumdu.edu.ua

Michal Kolcun, Zsolt Čonka, Dušan Medveď, Ľuboš Šárpataky
Department of Electric Power Engineering
Technical university of Košice
Košice, Slovakia
michal.kolcun@tuke.sk, zsolt.conka@tuke.sk, dusan.medved@tuke.sk, lubos.sarpataky@tuke.sk

Abstract—**Power output of photovoltaic (PV) generation can vary periodically and irregularly, depending on weather conditions. At high PV penetration levels this peculiarity of the technology can not only cause voltage and power flow fluctuations in local distribution grids, but also violate demand-supply balance of a whole energy system, resulting into issues with frequency control and difficulty of demand supply management. This study is primarily focused on demand and supply balancing of an energy system with high PV penetration levels, assuming a significant share of nuclear power plants (NPPs), as well as thermal power plants (TPPs), and a strong transmission system. The potential benefits of flexible nuclear operations in an energy system are analyzed. It is demonstrated that nuclear flexibility can reduce the share of environmentally unfriendly thermal power units and substantially reduce curtailment of renewable energy.**

Keywords: nuclear energy; flexible operation; renewable energy integration; demand-supply balancing; transmission power system

I. Introduction

A. Balancing Operation and Generation Dispatch

Energy plays a crucial role in the global economic development. Therefore, the improvement of generation, transmission and distribution systems is a priority for experts in the field. In order to maintain the power balance in a system, it is necessary to schedule the generation of each dispatchable generation unit.

The total power demand, which reflects all the changes of all individual demands, varies hourly, daily, weekly, depending on a season and a year. In a power system the balance of demand and supply is ensured by controlling the output of the dispatchable generators in a demand-and-supply balancing area. When the instantaneous balance is insufficient or lost, the power system's frequency (nominally 50 Hz or 60 Hz) or voltage will fluctuate, reducing the quality of supply [1]. In the worst case multiple devices, including power plants with rotating machines, which are designed to operate within a specified range of frequency deviation, are disconnected from the power system, leading to a blackout (see Fig. 1).

The balancing between demand and supply is realized through sophisticated generation schedule to make the best use of the features of each generation unit and system: the hourly balancing through generation unit's starts and stops, the balancing in minutes through centralized automatic generation control specifying the production of each unit, the balancing in seconds through independent governor control of each unit, and the remaining mismatch is transformed into a fluctuation of the system frequency [2]. Because the balancing requirements vary by time, by day, by season, reflecting the variation of the demand and supply structure including the share of renewable energy sources (RESs) (e.g., photovoltaic), the key concept to accommodate large amount of variable generation is the flexibility of a power system to cope with the balancing.

B. Integration of Variable Renewable Energy

At present, world energy adheres to the path of decarbonization, development and dissemination of energy-saving technologies and renewables. Ukraine has also joined the "green course". Recently in Ukraine there is a steady trend towards reducing electricity production at pulverized coal power units and increasing the share of carbon-neutral electricity from RESs [3]. One of Ukraine's commitments under the Association Agreement with the European Union is the development and maintenance of RESs, taking into account the principles of economic feasibility and environmental protection. Renewable energy is recognized as one of the main priorities of energy reform, and the promotion of energy production from alternative energy sources is the task of state policy in the energy sector. This is enshrined in the adopted Energy Strategy of Ukraine for the period up to 2035 and in the provisions of the legal framework. Renewable energy guarantees environmental security and energy independence to the entities that implement it [4].

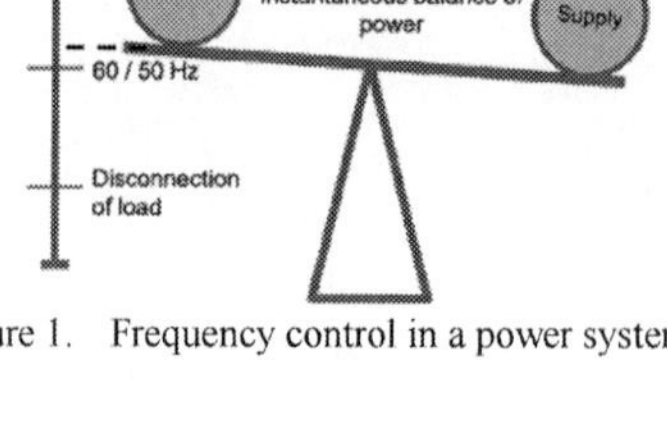

Figure 1. Frequency control in a power system [1]

The United Energy System (UES) of Ukraine is a set of power plants, electrical and thermal networks operating in the general mode of generation, transmission and distribution of electrical and thermal energy. NPPs, TPPs and hydroelectric power plants (HPPs), combined heat and power (CHP) plants, as well as power plants running on alternative energy resources (i.e., solar, wind, biofuel and others) operate in parallel in the UES of Ukraine. All of them are connected by main electric networks. As of mid-2021, the share of electricity generated by NPPs is 58.7%, HPPs and pumped storage hydropower – 10.2%, RESs – 9.1%, TPPs and CHPs – 20.5% [3].

Solar energy has received the greatest development among RESs in Ukraine [7]. It should be noted that the operation of solar power plants (SPPs) has a number of peculiarities, such as a fairly narrow period of time during the day when the PV installations can generate electricity, poor predictability of their operating capacity in the short term, and significant fluctuations in energy production periods due to weather conditions. The SPPs significantly affect the schedule of operation of TPPs, which are the second main source of electricity, along with NPPs. This results in additional costs spent for operation of TPPs. As the installed capacity of SPPs increases, these costs will rise further, and will reduce the efficiency of TPPs in the long run.

NPPs, which account for the largest share of generation, operate on a uniform schedule and create an energy base during the day. NPPs increase or decrease power very slowly, so abrupt maneuvers are simply dangerous. As a result, nuclear power plants cannot increase production during the evening peaks and decrease it at night, when there is a "night dip" in electricity consumption. At the same time, energy share from RESs is growing rapidly, which is a cause for concern. Unlike NPPs, which emit energy constantly and in the same amount, RESs are variable and dependent on weather conditions (i.e., availability of sun or wind) [5], [6]. These are two extremes, and both need to be balanced by those energy sources that can be more easily turned on and off as needed.

The daily and seasonal variability of wind and solar generation present a challenge to their efficient integration into existing electrical grids. In general, the variability and uncertainty introduced by renewable energy calls for a higher level of system flexibility. The amount of flexibility needed to accommodate the introduction of new RESs depends on their capacity and the existing flexibility in the system's infrastructure and operation [2]. This article considers one of the possible ways to overcome this problem – the theoretical possibility of increasing the flexibility of NPPs.

The main tasks of the study are as follows.

- Assess the possibility of integrating high levels of variable solar energy without compromising reliability with operational and institutional changes.
- Analyze the impact of unregulated use of SPPs on the modes of electricity generation.
- Estimate the possibility of regulating the operation of the UES with different shares of SPPs in the energy balance.
- Assess the feasibility of flexible regulation of NPPs in terms of power losses' and PV curtailment reduction and lowering of carbon emissions.

Of practical importance is the possible introduction of a mechanism to maintain the balance of demand and supply in the energy system with a significant share of RESs, using the maneuverability of NPPs and TPPs.

II. Maintaining the Necessary Balance of Demand and Supply of Electricity

A. Covering Consumers Demand of Electric Energy in the UES of Ukraine

The balance of production and consumption of electric energy in the UES of Ukraine is mainly maintained jointly by power units of TPPs and HPPs [8]. At the same time, the maneuverability of HPP power units is usually used during the morning and evening maximums of the power demand. Therefore, the uneven consumers' demand in the UES of Ukraine is predominantly covered with the maneuverability of power units of TPPs. It is impossible to completely exclude at least a part of thermal power units from the daily schedule of load covering in order to reduce their daily starts-stops, because without their participation it is difficult to cover uneven demand through the day, including the maximal demand periods. In addition, TPP power units often remain the only means of regulating the modes of electricity generation at night [9].

Obviously, the forced use of TPP power units as maneuverable generating capacities of the UES is associated with significant additional costs for their daily start-ups, as well as with costs for the energy generation by these power units [10]. Moreover, this mode of operation is not provided by the design of thermal power units, which leads to the increased wear of equipment, reduced reliability of its operation, as well as increased costs for scheduled and post-emergency repairs [8].

The greatest difficulties in managing the modes of electricity generation in the UES occur during the night demand minimum [11]. These complications are especially noticeable on summer weekends, when consumers' demand slightly exceeds the base (almost unregulated) capacity of operating NPP units. In this situation, even relatively small fluctuations in consumers' demand are extremely difficult for the power system. In the event of reduced power demand it may be necessary to quickly shut down one of the NPP units, which is not only undesirable for economic reasons, but also impossible for technical reasons, as well as in terms of NPP safety [12].

Another problem is the growing share of RESs in the country's energy balance. If there is an excess of unregulated generation from SPPs in the UAS, while the power demand remains unchanged, the need for electricity production by the thermal power units will decrease. Thus, there will be a need to reduce the output of TPP units or even reserve part of them. It is clear that this will increase the uneven loading of TPP units, significantly worsen their operation modes, leading to rise in costs for their operation and maintenance [8].

Given that, the uncontrolled use of SPPs can gradually exacerbate one of the biggest problems of the UES of Ukraine, which is the uneven daily load schedules of traditional generation, including TPPs. Combined with the shortage of maneuverable generating capacities, this will inevitably decrease the reliability and efficiency.

B. Applications of NPP's Flexibility for Maintaining the Balance of Demand and Supply

Increasing the flexibility of generation sources is one of the mechanisms to address the generation variability and ensure the balance of demand and supply. Power systems with increasing penetrations of variable renewable energy sources (i.e., wind and solar power) require greater system flexibility, including operating reserves and ramping capability to ensure that the supply-demand balance is maintained at all times [13]-[15].

This paper considers the viability of the flexibility enhancing of existing NPPs to allow faster response to demand changes. It should be noted that flexible operation of conventional generators often results in increased fuel, maintenance and capital costs that must be balanced against the benefits of increased levels of renewable energy in the system [2].

NPPs are commonly operated in a "baseload" mode, producing their maximum rated capacity whenever online, while the electricity demand varies during the day and year. In the power system generation must constantly correspond to consumption, and if there are too many NPPs in the system, they will have to change their capacity in response to changes in demand [16]. NPPs are technically capable of more flexible operation, changing their power output over time (i.e., ramping or load following) and contributing to power system reliability needs, including frequency regulation and operating reserves [17]. Flexible operation can help manage daily and seasonal variability in demand or renewable energy output or respond dynamically to hourly market prices or system operator dispatch.

For the UES of Ukraine and for similar power systems with NPPs supplying a substantial portion of the net load and/or with a significant share of variable energy sources, the flexible capabilities of NPPs are essential for maximizing revenues for reactor owners, ensuring system's reliability, reducing system's operational costs, integrating renewable energy, and reducing greenhouse gas emissions [17]. However, in literature nuclear units are typically represented as inflexible "must-run" (baseload) resources [15], [16], [18]. These traditional representations do not accurately capture the flexible capabilities of NPPs or the peculiar operational constraints arising from nuclear reactor dynamics and fuel irradiation cycles [17].

For example, pressurized water reactors (PWR), which are common in Ukraine and throughout the world, are capable of flexible operation by adjusting power output primarily by withdrawing neutron absorbing control roads into the core to increase power and inserting control roads to reduce power [17]. The PWR reactor operates in a double-circuit nuclear power system, and ordinary non-boiling water is a neutron moderator and coolant and is under high pressure (~ 16 MPa) [18]. Inserting or withdrawing control rods is an effective way of modulating power output for flexible operation, but the maximum rate at which reactors can adjust electricity production, or "ramp," is constrained by limitations on the thermal and mechanical stresses incurred by nuclear fuel assemblies.

Depending on the design, French and German reactors can safely operate with ramp rates of up to 2–5.2% of rated power capacity per minute, without increasing the rate of fuel cladding failure [19], [20]. However, in practice ramp maneuvers performed by operators typically proceed at a more conservative pace (e.g., at < 0.5% per minute) to limit stress on reactor components [17]. Existing nuclear plants in France and Germany contribute up to 5% of their maximum rated power to frequency regulation [19], [20].

III. Regulation and Constraints on Flexible Operation of Nuclear Power Plants

A. Power Output of Solar and Nuclear Units

The solar irradiation data and the ambient temperature data were extracted from the Photovoltaic Geographical Information System (PVGIS) [21] for the geographical location of Southern Ukraine, latitude: 50°89', longitude: 34°8', as an example. It is assumed that the slope of the PV modules (i.e., the angle with the horizontal plane) is 37° for a fixed (non-solar) mounting type, the azimuth is –1°. The crystalline silicon photoelectric technology is considered. The PV electric output was estimated in accordance with [22]:

$$P_{PV} = \eta_{PV} \cdot \eta_{PC} \cdot A_{PV} \cdot G_{PV} \cdot \left(1 + \gamma + \left(T_{PV} - T_{ref}\right)\right), \quad (1)$$

where, P_{PV} is power produced from the PV system in kW; η_{PV} is the power conversion efficiency of the PV module in p.u.; η_{PC} is the efficiency of the power converter in p.u.; A_{PV} is the area of the PV array in m^2; G_{PV} is the solar irradiance incident on the plane of the PV array in kW/m^2; γ is the temperature coefficient of the PV module; T_{PV} is the PV module temperature in °C; and T_{ref} is the reference temperature in °C.

The output of the nuclear power units is regulated with regard to the ramp rate of the reactors, total power demand and amount of power output from PV installations:

$$P_i^{NPP} = \begin{cases} P_{i-1}^{NPP}\left(1+K_{ramp}\right), & \dfrac{P_i^{\Sigma cons}\cdot\psi}{P_{i-1}^{NPP}+P_i^{SPP}} \geq 1+K_{ramp} \\ P_{i-1}^{NPP}\left(\dfrac{P_i^{\Sigma cons}\cdot\psi}{P_{i-1}^{NPP}+P_i^{SPP}}\right), & \left[1 \leq \dfrac{P_i^{\Sigma cons}\cdot\psi}{P_{i-1}^{NPP}+P_i^{SPP}} < 1+K_{ramp}\right] \vee \ldots \\ & \ldots \vee \left[1-K_{ramp} < \dfrac{P_i^{\Sigma cons}\cdot\psi}{P_{i-1}^{NPP}+P_i^{SPP}} < 1\right] \\ P_{i-1}^{NPP}\left(1-K_{ramp}\right), & \dfrac{P_i^{\Sigma cons}\cdot\psi}{P_{i-1}^{NPP}+P_i^{SPP}} \leq 1+K_{ramp} \end{cases} \quad (2)$$

In (2) P_i^{NPP} is the instant active power output of the NPP at the i-th time step; P_i^{SPP} is the instant active power output of the SPP at the i-th time step; $P_i^{\Sigma cons}$ is the total instant active power consumption at the i-th time step; ψ is the maximum possible share of NPPs in total energy production in the UES; K_{ramp} is the ramping rate of the NPP in p.u. per a time step. The coefficient ψ depends on the available capacity of nuclear power units and balancing capabilities of the UES, and is assumed to be 0.6.

B. Operational Constraints

During the operation of the UES it is important to ensure that system's constraints are satisfied. At each step of the modelling the power system must meet operational constraints, such as power balance, voltage limit at each node, and transmission lines' current capacity limits. In this work the demand and generation values change in hourly steps. To simulate different operating states of the UES, power flow should be recalculated for all the snapshots of load demand values. The constraints are determined as follows [23].

Voltage limits constraints:

$$\left|V_n^{\min}\right| \leq \left|V_n\right| \leq \left|V_n^{\max}\right| \quad (3)$$

Branch capacity constraints:

$$\begin{gathered} \left|I_{fn}\right| \leq \left|I_{fn}^{\max}\right| \\ \left|I_{rn}\right| \leq \left|I_{rn}^{\max}\right| \end{gathered} \quad (4)$$

SPP real and reactive power constraints:

$$\begin{gathered} P_{\min}^{SPP} \leq P^{SPP} \leq P_{\max}^{SPP} \\ Q_{\min}^{SPP} \leq Q^{SPP} \leq Q_{\max}^{SPP} \end{gathered} \quad (5)$$

In the equations (3)-(5) V_i is the voltage at the n-th bus, I_{fn} is the forward flow capacity of the n-th branch of the UES; I_{rn} is the reverse flow capacity of the n-th branch of the UES; P^{SPP} and Q^{SPP} are the available real and reactive power capacities of the SPP. The superscripts and subscripts *min* and *max* represent the maximum and minimum allowable limits of the corresponding values. It is accepted in this work that voltage deviations should not exceed ±0.1 p.u. threshold [24].

IV. Case Study

The object under study is an energy system with a significant share of NPPs, TPPs and SPPs, which in some approximation reminds the UES of Ukraine and some European countries. The IEEE 9-bus test system with the base 100 MWA was chosen to simulate such a UES. The diagram of the system is shown in Fig. 2.

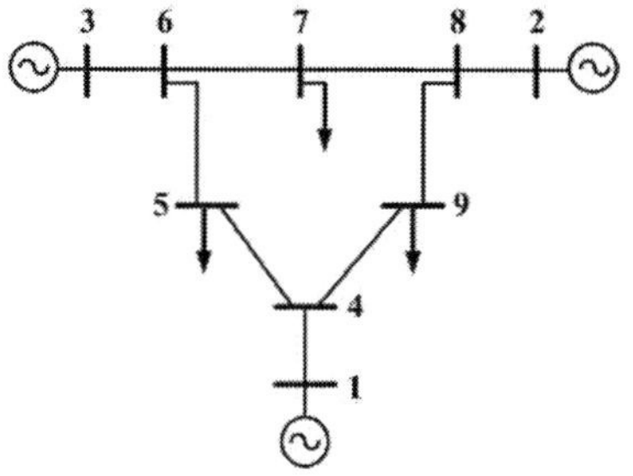

Figure 2. Single-line diagram of the IEEE 9-bus test system

The modified IEEE 9 bus system was modelled in PowerWorld simulator, and load flow studies were performed to assess the system's performance under different PV penetration levels, using Newton-Raphson method. The rated bus voltages, load values and transformers' impedances are in accordance with [25]. The overhead transmission lines are of 50 km length and have AC-240/32 wires with 605 A current capacity and 217 MW power capacity. The bus data and the rated load data is shown in Table 1, and the branch data is given in Table 2.

A single-line diagram of the UES modelled in PowerWorld simulator is shown in Fig. 3.

Table 1. Bus data for the power system under study

Bus No.	Bus type	Voltage, p.u.	Voltage, kV	Load	
				MW	Mvar
1	TPP (Slack)	1.04	16.5	0	0
2	NPP	1.025	18.0	0	0
3	SPP	1.025	13.8	0	0
4	PQ	1	230	0	0
5	PQ	1	230	125	50
6	PQ	1	230	90	30
7	PQ	1	230	0	0
8	PQ	1	230	100	35
9	PQ	1	230	0	0

Table 2. Branch data for the power system under study

Line from	Line to	Resistance, R, Ω/km	Reactance, X, Ω/km	Conductivity, B, ℧/km
1	4	0	0.0576	0
4	5	0.121	0.435	$2.6 \cdot 10^{-6}$
4	6	0.121	0.435	$2.6 \cdot 10^{-6}$
5	7	0.121	0.435	$2.6 \cdot 10^{-6}$
6	9	0.121	0.435	$2.6 \cdot 10^{-6}$
7	2	0	0.0625	0
7	8	0.121	0.435	$2.6 \cdot 10^{-6}$
9	3	0	0.0586	0
9	8	0.121	0.435	$2.6 \cdot 10^{-6}$

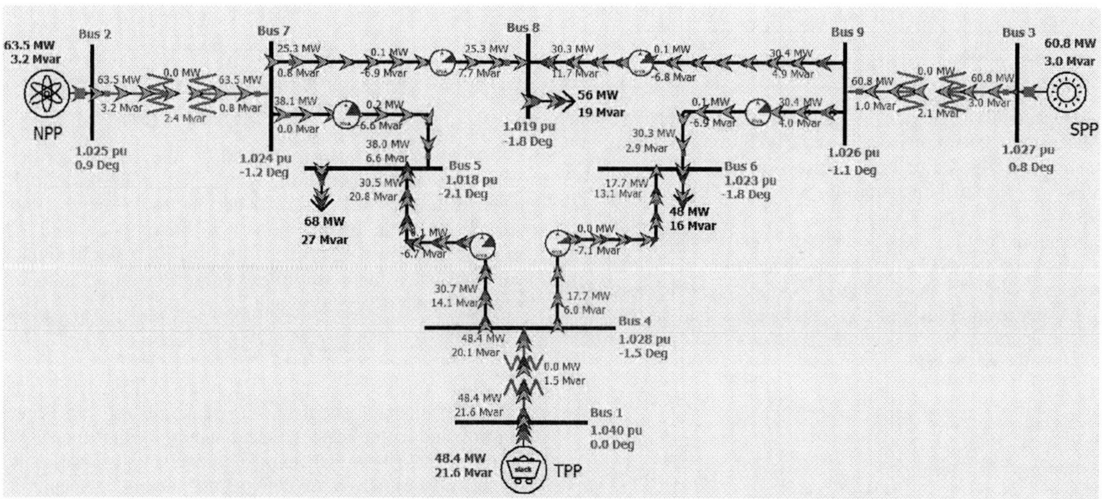

Figure 3. Single-line diagram of the UES in PowerWorld (the simulation is ongoing)

SPPs are the most unstable sources of the UES with a variable output. Scenarios with the SPPs' installed capacity of 10%, 20%, 30%, 40%, 50%, 75% and 100% of the total system's capacity are considered. The installed capacity of the solar generation units with regard to the total system's capacity is further referred as the penetration level.

TPP power units are maneuverable generating capacities of the UES, and, therefore, they are located at the slack bus, which can adjust its output in wide limits to match the demand and supply.

NPPs are less maneuverable. In the UES of Ukraine the nuclear units usually operate in a "baseload" mode. For the first seven scenarios it is considered that the NPP operates with the ramp rate up to 2% of rated power per a time step, while the installed capacity of SPPs moderately changes from 10% to 100% of the power system's capacity. For the next seven scenarios the NPP power units are considered to have more flexibility so that they are able to operate with the ramp rate of up to 20% per a time step. In this case the NPPs will better contribute to frequency regulation and provision of an operating reserve.

In total, there are fourteen scenarios that differ in the installed capacity of the SPPs and the NPPs' flexibility. Their summary is given in Table 3.

V. Results and Discussion

The simulated scenarios are operationally feasible, i.e. power flow converged successfully, and none of the established constraints were violated. The results of the simulation are given in Table 4.

Table 3. Scenarios for the case study

Scenarios	Description
1-7	The installed capacity of the SPPs is 10%, 20%, 30%, 40%, 50%, 75%, 100% of the total power system's capacity. The NPPs have low flexibility and operate with the ramp rate up to 2% of the rated power per a time step.
8-14	The installed capacity of the SPP is 10%, 20%, 30%, 40%, 50%, 75%, 100% of the total power system's capacity. The NPPs have higher flexibility and operate with the ramp rate up to 20% of the rated power per a time step.

Table 4. Simulation results for different scenarios

Scenario	TPP Share, p.u.	NPP Share, p.u.	SPP Share, p.u.	Power Loss Share, p.u.	PV curtailment necessary?	PV curtailment, %
1	41.35	56.29	2.37	0.67	No	0
2	41.70	53.56	4.74	0.65	No	0
3	40.94	51.95	7.11	0.64	Yes	$9.02 \cdot 10^{-5}$
4	39.58	50.95	9.48	0.64	Yes	$6.9 \cdot 10^{-2}$
5	38.05	50.11	11.84	0.66	Yes	0.47
6	33.44	48.81	17.75	0.76	Yes	3.21
7	28.64	47.75	23.61	0.96	Yes	7.06
8	41.35	56.29	2.37	0.67	No	0
9	41.70	53.56	4.74	0.64	No	0
10	40.94	51.95	7.11	0.63	No	0
11	39.58	50.95	9.48	0.64	No	0
12	38.05	50.11	11.84	0.66	Yes	$13.32 \cdot 10^{-4}$
13	33.44	48.81	17.75	0.77	Yes	0.64
14	28.64	47.75	23.61	0.94	Yes	3.23

Dependences of the share of thermal and nuclear power units in the energy generation balance of the UES from PV penetration are shown in Fig. 4. For the scenarios with mild ramp rate of the NPPs (i.e., baseload operation) the shares of TPPs and NPPs in energy generation mix both decrease with the increase of the PV penetration. For the scenarios with additional flexibility of NPPs the share of the NPPs in energy generation mix constantly decreases, while the share of the TPPs moderately increases until the PV penetration level of 50%. Further the share of the TPPs starts to decrease, but its declining proceeds notably slower than the declining of the NPPs.

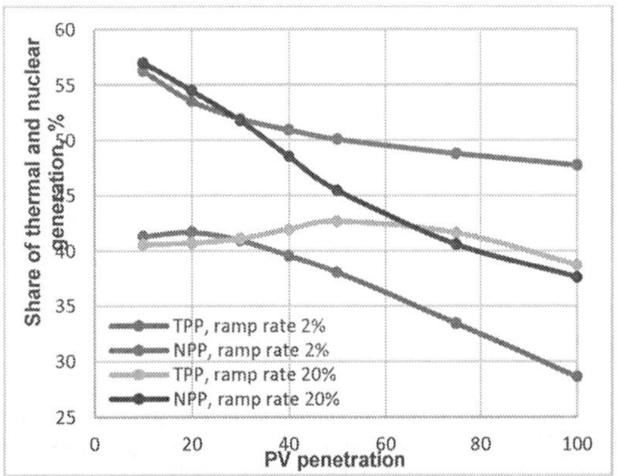

Figure 4. Dependences of the share of TPPs and NPPs in total energy generation

Comparing flexible and inflexible operation of the nuclear, the share of NPPs in the total generation mix is higher with the NPPs' ramp rate of 20% than with the NPPs' ramp rate of 2% until the PV penetration reaches 30%. At the same time, the share of TPPs in the total energy generation mix is lower with the NPPs' ramp rate of 20% than with the NPP's ramp rate of 2% until the PV penetration reaches 30%. This means that flexible operation of nuclear units allows to better engage them in the demand-supply balancing and to reduce the share of environmentally unfriendly thermal power units. However, when the PV penetration exceeds 30%, the situation changes to the opposite. This can be explained by the peculiarities of NPP regulation under high penetration of RESs. Thus, higher PV penetrations reduce the value of nuclear flexibility and enhance demand for reserves. Forecasting of solar generation and scheduling of NPPs with regard to the expected SPPs output can be a helpful measure to increase the share of cleaner energy technologies.

Comparison of power losses in the energy system with different ramp rates of the NPPs is shown in Fig. 5. As it can be seen, additional flexibility of the NPPs has negligible influence on the power losses.

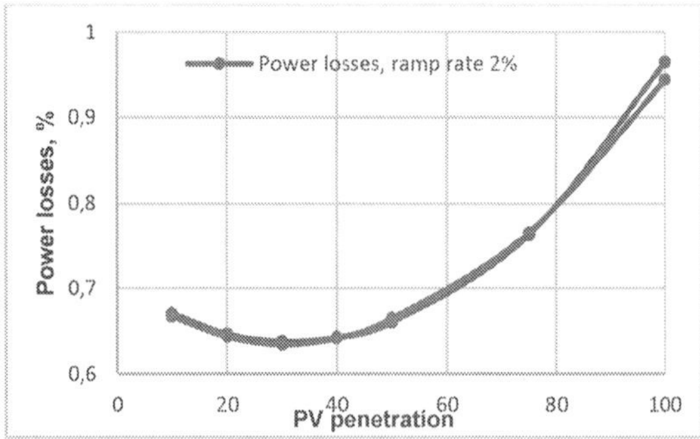

Figure 5. Comparison of power losses in the energy system with different ramp rates of the NPP

Additionally, nuclear flexibility allows to significantly reduce PV curtailment in the system. For example, at high PV penetrations of 75% and 100% (scenarios 13 and 14) the PV curtailment with flexible nuclear power is 0.64% and 3.23% accordingly, while 3.21% and 7.06% of PV generation must be curtailed when there are inflexible nuclear power units in the energy system (scenarios 6 and 7).

VI. Conclusions

High penetration levels of PV in the united energy systems can violate demand-supply balance, cause issues with frequency control and complicate demand-supply management. While TPPs have the best maneuverable generating capacities, NPPs can also contribute to the demand and supply balancing in the UES.

In this paper the potential impacts of flexible nuclear operations in the energy system with a significant share of NPPs, high solar penetration, and a strong transmission system has been investigated. The results have shown that flexible nuclear operation can substantially reduce curtailment of renewables and, in some cases, reduce the share of environmentally unfriendly thermal energy. Higher than 30% PV penetration reduces the value of nuclear flexibility and enhances demand for reserves. Forecasting of solar generation and scheduling of NPPs with regard to the expected SPPs output can be a helpful measure to increase the share of cleaner energy technologies, such as nuclear and renewable energy. Although the numerical results are case specific, flexible operation of nuclear power stations is likely to yield similar benefits in power systems with comparable shares of RESs and NPPs. Rigorous sensitivity analysis exploring which factors are most crucial for economic benefits of nuclear flexibility would be a productive avenue for future work.

ACKNOWLEDGMENT

This work was supported by the Ministry of Education, Science, Research and Sport of the Slovak Republic and the Slovak Academy of Sciences under the contract VEGA 1/0757/21. This work was supported by the Slovak Research and Development Agency under the contract No. APVV-19-0576.

REFERENCES

[1] F. Gafaro. (2017). "Planning the operability of power systems – Overcoming technical and operational bottlenecks." *International Renewable Energy Agency (IRENA).* [On-line]. pp. 1-27, Available: www.irena.org/-/media/Files/IRENA/Agency/Events/2017/Jan/17/IRENA-WFES-Scaling-up-VRE---Operation-Planning---Final.pdf. [Apr. 30, 2022].

[2] K. Ogimoto, (Nov. 2014). "Power System Operation and Augmentation Planning with PV Integration." *IEA-PVPS.* Report IEA PVPS T14-04:2014. [On-line]. Available: https://iea-pvps.org/wp-content/uploads/2020/01/Power_System_Operation_Planning_with_PV_Integration_T14_05_2015_LR_2.pdf [Apr. 30, 2022]. ISBN: 978-3-906042-26-8.

[3] Ministerstvo energie Ukrajiny. "The share of thermal generation in total electricity production has decreased by 5% since 2016." Internet: http://mpe.kmu.gov.ua/minugol/control/publish/article?art_id=245538724, May 26, 2021 [Apr. 30, 2022].

[4] U. Terletska. "Renewable energy in Ukraine (RES) is reality." Internet: http://www.institute.lviv.ua/index.php/uk/2018-02-16-08-36-56/126-2019-11-15-11-44-25, 2010. [Apr. 30, 2022].

[5] V. V. Volokhin, I. M. Diahovchenko, B. V. Derevyanko. "Electric energy accounting and power quality in electric networks with photovoltaic power stations," in *2017 IEEE International Young Scientists Forum on Applied Physics and Engineering (YSF)*, 2017, pp. 36-39. DOI: 10.1109/YSF.2017.8126588.

[6] S. Yu. Shevchenko, V. V. Volokhin, I. M. Diahovchenko. "Power quality issues in smart grids with photovoltaic power stations," in *Energetika*, 63(4), pp. 146-153, 2017. DOI: 10.6001/energetika.v63i4.3623.

[7] GreenDea. (2021) "The capacity of RESs in Ukraine in 2021 increased by almost a third, SPP – by less than 5%." [Online]. Available: https://greendeal.org.ua/potuzhnist-ves-v-ukrayini-za-pidsumkamy-2021-roku-zrosla-majzhe-na-tretynu-ses-mensh-nizh-na-5/ [Apr. 30, 2022].

[8] O. Chernousenko. "The state of energy of Ukraine and the results of modernization of TPP power units." Problems of general energy, 39(4), pp. 20-28, 2014.

[9] M. P. Kuznetsov, O. V. Lysenko, O. A. Melnyk, "Optimal regulation of a local power system with renewable energy sources," in *Bulletin of the National Technical University (KhPI)*, 2(1), pp. 52-61, 2021.

[10] V. F. Nakhodov. "Management of consumption regimes and efficiency of electricity use in energy systems." Doctoral thesis, Kuiv, 2018.

[11] G. Morva, I. Diahovchenko, "Effects of COVID-19 on the electricity sectors of Ukraine and Hungary: challenges of energy demand and renewables integration," in *2020 IEEE 3rd International Conference and Workshop in Óbuda on Electrical and Power Engineering (CANDO-EPE)*, 2020, pp. 000041-000046. DOI: 10.1109/CANDO-EPE51100.2020.9337785.

[12] V. P. Severin et. al., "The problem of maneuverability of the NPP power unit and development of the models of its control systems," in *Bulletin of the National Technical University (KhPI)*, 1103(61), pp. 24-29, 2014.

[13] P. D. Lund, J. Lindgren, J. Mikkola, J. Salpakari. "Review of energy system flexibility measures to enable high levels of variable renewable electricity," in *Renewable and Sustainable Energy Reviews*, vol. 45, pp. 785-807, 2015. DOI: 10.1016/j.rser.2015.01.057.

[14] A. Pfeifer, L. Herc, I. B. Bjelić, N. Duić, "Flexibility index and decreasing the costs in energy systems with high share of renewable energy," in *Energy Conversion and Management*, vol. 240, pp. 114258, 2021. DOI: 10.1016/j.enconman.2021.114258.

[15] S. Impram, S. Varbak Nese, B.T. Oral, "Challenges of renewable energy penetration on power system flexibility: A survey," in *Energy Strategy Reviews*, vol. 31, pp. 100539, 2020, DOI: 10.1016/j.esr.2020.100539.

[16] Ekodia. (2020). "What to know about the "conflict" between nuclear and renewable energy," *Ecoaction.org.ua* [On-line]. Available: https://ecoaction.org.ua/renewables-vs-nuke.html. [Apr. 30, 2022].

[17] J. D. Jenkins, Z. Zhou, R. Ponciroli, R.B. Vilim, F. Ganda, F. de Sisternes, A. Botterud. "The benefits of nuclear flexibility in power system operations with renewable energy," in *Applied Energy*, vol. 222, pp. 872-884, 2018. DOI: 10.1016/j.apenergy.2018.03.002.

[18] K.B. Denisevich, Yu.A. Landau, V.A. Neiman, V.N. Suleimanov, B.A. Shyliiaiev. (2013) "Power Engineering: History, Present and Future. Development of Nuclear Power Engineering and Integrated Power Systems," [On-line]. Available: http://energetika.in.ua/ua/books/book-4. [Apr. 30, 2022].

[19] A. Lokhov. (2011) "Technical and Economic Aspects of Load Following with Nuclear Power Plants," in *Paris: Nuclear Energy Agency.* [Online]. Available: www.oecd-nea.org/ndd/reports/2011/load-following-npp.pdf. [Apr. 30, 2022].

[20] H. Ludwig, T. Salnikova, A. Stockman, U. Waas, "Load cycling capabilities of German Nuclear Power Plants (NPP)," in *International Journal for Nucl Power*, 55(8/9), pp. 2-8, 2010.

[21] European commission (2022). "Photovoltaic Geographical Information System (PVGIS)," Internet: https://re.jrc.ec.europa.eu/pvg_tools/en/#MR [Apr. 30, 2022].

[22] L. Petrichenko et. al., "Energy losses in power distribution systems with prosumers owning individual photovoltaic installations and shares in solar energy communities," unpublished.

[23] I. M. Diahovchenk, G. Kandaperumal, A. K. Srivastava, Z.I. maslova, S.M. Lebedka. "Resiliency-driven strategies for power distribution system development," in *Electric Power Systems Research*, vol. 197, pp. 107327, 2021. DOI: 10.1016/j.epsr.2021.107327.

[24] I. Diahovchenko, N. Sushchenko, A. Shulumei, O. Strokin, "Influence of supply voltage and frequency variations on the electrical equipment and power consumption in LV and MV distribution networks," in *Energetika*, 65(4), pp. 172-187, 2019. DOI: 10.6001/energetika.v65i4.4246.

[25] M. Anderson. "Power System Control and Stability," in *IEEE Press Power Engineering Series*, 2003, pp. 651. ISBN: 0-471-23862-7.

Statistical Analysis of Photovoltaic Power Plant Production in Second Resolution

Jozef Novotný
Innovation Department
joint stock company sféra, a.s
Bratislava, Slovakia
jozef.novotny@sfera.sk

Abstract—**Nowadays, measured data at 15-minute resolution are used in the electric power industry mainly for commercial balancing or in various other business and technical processes. However, the temporal resolution of these data does not allow to study in detail the variability of generation, the rapid changes and fluctuations of the generated power of photovoltaic plants caused mainly by the passage of clouds and fog in a given location. However, the statistical analysis of measured high-frequency PV plant production data at a second resolution allows to investigate and analyze power output variations down to the level of transients.**

Keywords: high-frequency data; renewable energy; photovoltaic power plant output variability; statistical data analysis; high-frequency fluctuations; statistical software R

I. Introduction

The production of electricity from renewable energy sources has increased significantly over the last decade, mainly due to energy policy and environmental factors. In the conditions of the Slovak Republic, priority has been given to the construction of photovoltaic power plants, whose aggregate installed capacity reached 535 MW in 2020, according to the annual report of the short-term electricity market operator OKTE [1].

Renewable energy sources are generally known for their seasonality and fluctuations in performance on different time scales. Therefore, one of the biggest challenges is the technical management of these fluctuations. Power system analysis, modelling and research are in the vast majority of cases carried out on 15, 30 or 60 minute time scales. Therefore, this paper focuses on very short time resolution by analyzing high-frequency second-by-second PV generation data. The aim is to represent the importance of variability in modeling and simulations of power system operation, while providing a statistical basis for second-based analyses. Understanding the characteristics of renewable resources using high-frequency data is an important aspect in the design of power systems, for example microgrids in islanded operation, because the high variability of the generated power leads to voltage fluctuations as a local parameter and frequency fluctuations as a global parameter. For this reason, microgrids are challenging not only to control but also to size the electrical protections and to maintain the dynamic and static stability of their islanded operation.

In general, high-frequency fluctuations are minimally studied in academia, mainly because of the unavailability of such data and also because of the natural mutual damping of fluctuations that arises from the distribution of renewable resources in the geographic relief of the landscape. This distribution reduces variability, but does not eliminate it completely. Some research studies in the past have used various statistical techniques to transform from low frequency (common commercial measurements) to high frequency data or tools to generate synthetic high frequency data [2].

This paper presents numerical and graphical results of a time series analysis in the form of measured values of the active power output of a 16.2 kWp photovoltaic power plant connected to the electricity distribution network. The intention of the paper is to provide insight into the second order dynamics of PV generation and to justify its rationale in power system modelling and design.

II. High-Frequency Data

The database of measured values contains high-frequency data in second resolution. All measured parameters are listed in Table 1. The time period from 1.7.2021 to 31.7.2021 of the total generated power P is analyzed. The day 4.7.2021, 5.7.2021 and 6.7.2021 are omitted from the analysis due to the malfunction of the metering system. In total there are 28 days analyzed. The location is Bratislava, Slovakia. The measurement system produces one text file in .txt format on a daily basis. Data processing and analysis is performed in the statistical software R. The total number of measured parameters is 19.

Table 1. Header of the processed .txt file

Parameter	Label
Timestamp	–
Voltage [V]	$V_{RMS\,1}$, $V_{RMS\,2}$, $V_{RMS\,3}$
Total Harmonic Voltage Distortion [%]	THD_U
Current [A]	I_1, I_2, I_3
Total Harmonic Current Distortion [%]	THD_I
Active power [W]	P_1, P_2, P_3, P
Power factor [–]	Cos φ_1, Cos φ_2, Cos φ_3, Cos φ
Electricity consumption [Wh]	E_{in}
Cumulative electricity produced [Wh]	E_{out}

The photovoltaic power plant contains 69 polycrystalline panels with an output of 235 Wp in three strings. The analysis is based on a comparison of the two days with the lowest and highest variability of production.

III. Methodology for Selecting the Two Days with the Lowest and Highest Production Variability and Their Analysis

The selection methodology is based on quantifying the variability of the time series of power production using the sum of the absolute values of the first differences according to the formula

$$VAR_{Pout} = \sum_{2}^{86400} \left| \frac{P_{(t)} - P_{(t-\Delta t)}}{3600} \right| [Wh] \quad (1)$$

$P_{(t)}$ is the power at time t, $P_{(t-1)}$ is the power at time $t - 1$. Both quantities are measured in watts. The differentials are divided by the number of seconds in an hour so that the given value corresponds to the energy expressed in Wh. The value 86 400 corresponds to the number of seconds per day. Since this is the first differential, the time step in this case is $\Delta t = 1$. From the above formula, it follows that the higher the magnitude of the differentials and the more frequent the step changes, the higher the variability becomes. Table 2 shows the VAR_{Pout} for the days with the lowest and highest variability of electricity generation.

Table 2. Days with the lowest and highest production variability

Day	VAR_{Pout} [Wh]
16.7.2021	126.33
14.7.2021	814,56

For comparison, Figure 1 shows the pattern of second values of power production on the days with the lowest and days with the highest variability. A significant difference in production variability is evident. The timeline itself is expressed in hourly resolution for clarity.

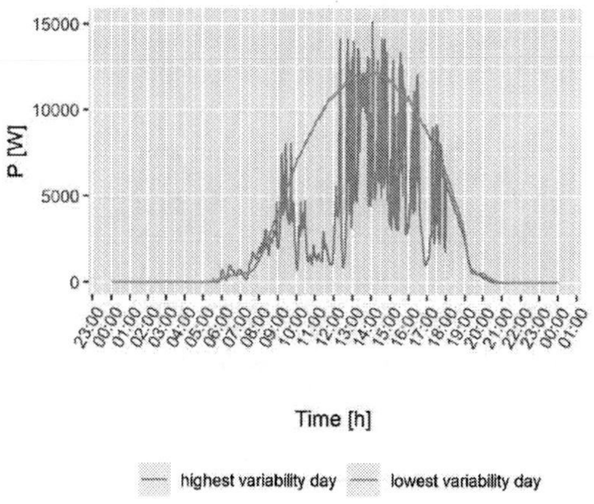

Figure 1. Profile comparison of days with the lowest and highest variability

Since the most commonly used data in analytical and simulation practice is 15-minute time resolution data, which involves expressing one quarter-hour with a single numerical value, Figure 2 is derived from Figure 1 with the time period from 14:00 to 14:15. Figure 2 thus provides a detailed view of the variability of production within a quarter-hour. The time axis in this case is expressed in minute resolution.

For interest, Table 3 shows the statistical characteristics of the selected quarter-hour. Means, medians and standard deviations are given to describe the width of the distribution of values in the set of measured data. Conventional measurement systems calculate the average value within quarter-hourly readings at predefined time slices within a quarter-hour. The resulting measured value is the average of these values. Thus, information about the intra-quarter-hour variability is not retained.

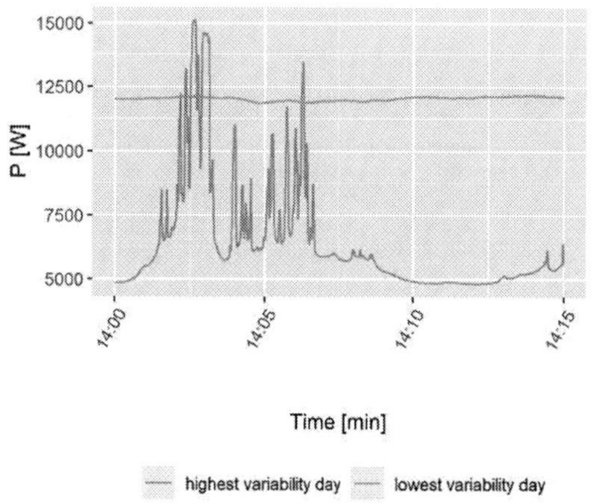

Figure 2. Time zoom of the selected quarter-hour in second resolution

Table 3. Statistical characteristics of quarter-hours for selected days

	The day with the lowest variability	The day with the highest variability
Minimum value	11 860	4 824
Median	12 071	5 722
Mean	12 051	6 398
Maximum value	12 184	15 130
Standard deviation	82,02	2 214,23

It is also useful to visualize the intra-quarter-hour power variability using a histogram, which interprets the measured power values by indicating the number of measured values that lie within a certain range. The higher the frequency of values within a given range, the higher the bar in the histogram. Figure 3 visualizes the histogram for selected quarters. The figure shows a significant difference in the distribution of values between the high and low variability quarters. The quarter-hour with low variability is represented by one high bar and one very low bar. Hence, in this quarter-hour, the concentration of measured values oscillating very close to the mean value of 12 051 W, which is shown in Table 2. On the contrary, at high variability, a concentration of values is evident especially in the range from 5 000 to 7 500 W. Values above 7 500 W are also present, but their frequency is low. The histogram also gives a rough indication of the underlying probability distribution of the values.

The table also shows another indicator of variability, namely the percentage difference between the mean and the median. Both characteristics belong to measures of central tendency. A quarter with low variability has a percentage difference of 0.17%, while a quarter with high variability has 11.81%. The difference can be explained by the non-uniform asymmetric distribution of the high variability values, with the mean, unlike the median, as an indicator of the central value being sensitive to outliers.

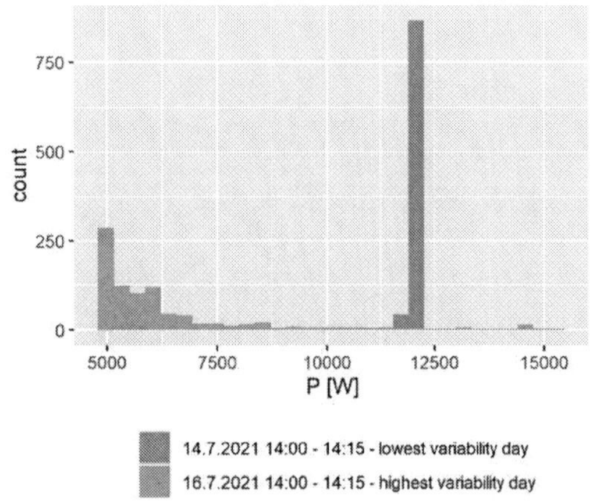

Figure 3. Visualization of measured second data in selected quarters by histogram

IV. Statistical Analysis of Fluctuations

The power fluctuation $\Delta P_{\Delta t}(t)$ is defined at time t for a given time resolution period. The parameter Δt is the difference between two consecutive values of the produced power $P(t)$. The values are converted to the installed power P* [3]:

$$\Delta P_{\Delta t}(t)=\frac{[P_{(t)}-P_{(t-\Delta t)}]}{P^{*}}.100\ [\%] \quad (2)$$

The statistical analysis is performed for the measured values of 02.07.2021, when the largest positive and negative change in the differenced time series occurred simultaneously for $\Delta t = 1$. Figure 4 shows the high variability and irregularity of the produced power.

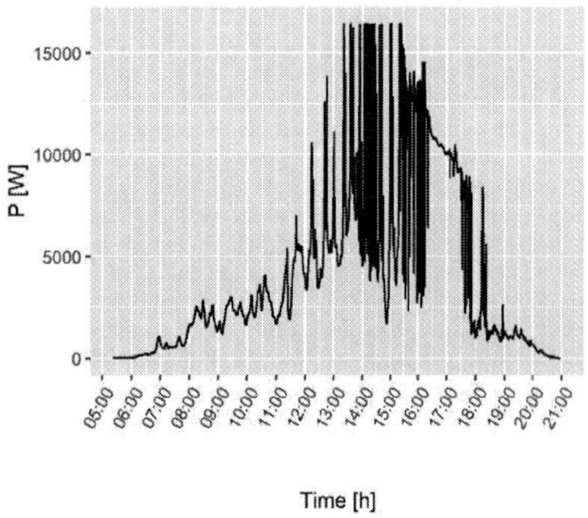

Figure 4. Visualisation of the active power for the day 02.07.2021

The percentage changes in performance according to (2) are also shown graphically in the time domain in Figure 5. Figure 6 shows a histogram of the percentage changes. The values when production was zero are removed from the analysis. The histogram clearly declares that the largest concentration of percentage changes is around zero. This fact is also expressed in Table 4, which shows that the maximum value of the negative fluctuation and therefore the decrease in output was more than 50% between two consecutive seconds. The increase in this case was 31.49%.

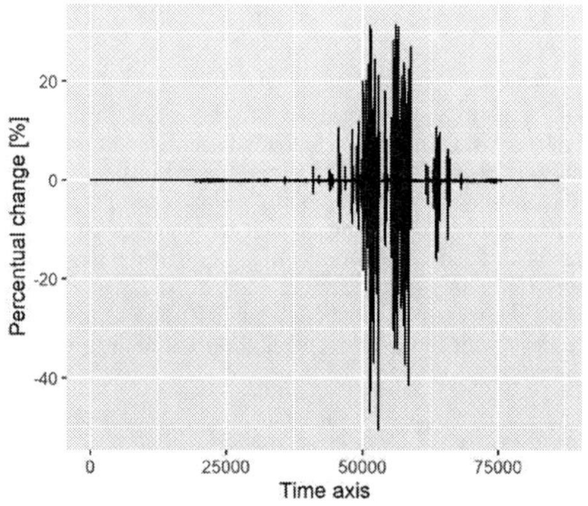

Figure 5. Visualisation of percentage changes in power on 02.07.2021 in the time domain

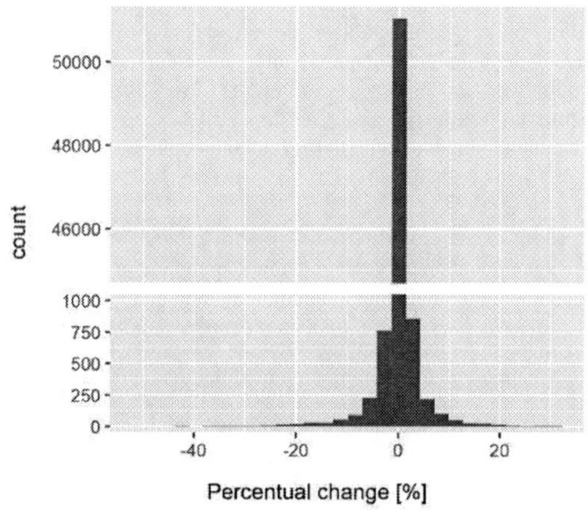

Figure 6. Visualisation of percentage changes in power on 02.07.2021 using histogram

Table 4. Statistical characteristics of fluctuations expressed as a percentage

	$\Delta P_{\Delta t}$[%]
Minimum value	-50.50
Median	0.00
Mean	0.00
Maximum value	31.49
Standard deviation	1.59

Since the percentage of fluctuations is burdened by a dominant number of very low values oscillating around zero, all values falling in the interval (-5%, 5%) are removed for further analysis, thus creating two data sets with positive and negative values. The statistical characteristics of these data sets are expressed in Table 5.

Table 5. Statistical characteristics of the two datasets of fluctuations expressed as a percentage

	$(\min(\Delta P_{\Delta t}), -5\ \%)[\%]$	$(5\ \%, \max(\Delta P_{\Delta t}))[\%]$
Minimum value	-50.50	5.01
Median	-8.62	8.34
Mean	-11.22	10.25
Maximum value	-5.02	31.46
Standard deviation	7.13	5.54

V. Statistical Analysis of the Rate of Decline and Increase in Performance

The above analyses show that a step change in power drop can reach up to 50% of the installed power within one second, which in the case of microgrids or other isolated arrays with high PV penetration can have a negative impact on the quality of the delivered electricity.

From a power management perspective, it is therefore important to quantify the rate of decline and increase in power output given in MW/min in common practice, especially for distribution and transmission system operators. Since second values are available, this chapter works with W/s. The calculation is made according to [4]:

$$R_{RATEPV} = \frac{P_{(t)} - P_{(t-\Delta t)}}{\Delta t} \quad [W/s] (3)$$

Where R_{RATEPV} is the rate of change of power, $P_{(t)}$ is the power over time and Δt is the time step. The analysis is performed for a moving window with time difference $\Delta t = 1$, 2, 3 and 4, corresponding to time window widths of 2, 3, 4 and 5 seconds, respectively. Table 6 and Table 7 show the statistical characteristics for the power decrease and power increase, respectively. The values are given in W/s.

The tables show that as the width of the window used increases, the absolute value of the characteristics decreases and a smoothing effect occurs.

Both datasets for all time differences were tested for normality by the Shapiro-Wilk test. In practice, it is often assumed that the data come from a normal distribution. However, testing rejected the hypothesis that the data sets fit a normal distribution [5]. Hence, it is more appropriate to consider the median as a measure of central tendency. The tables also show a slight difference between the absolute values of the medians for the decrease and increase in power. The values suggest the conclusion that the decline in performance is generally faster. For example, for $\Delta t = 2$, the median decline is 10.13% faster than the increase. Based on this fact, the hypothesis was formulated that the difference in the rate of decline and increase is greater than 0 and hence statistically significant. Since the data did not come from a normal distribution, the Wilcoxon test was chosen for testing [6]. The test decides based on the calculated p-value whether the median difference between the two data sets is statistically significant or not. The results from the testing are presented in Table 8. In general, if the p-value is greater than 0.05, the hypothesis that the difference is not statistically significant is accepted. If the value is less, the alternative hypothesis is accepted and the difference of the two medians is statistically significant.

Table 6. Statistical characteristics of the decline in active power

	$\Delta t = 1$	$\Delta t = 2$	$\Delta t = 3$	$\Delta t = 4$
Minimum value	-8 182	-4 934	-3 635	-2 771
Median	-1 396	-1 413	-1 321	-1 272
Mean	1 818	-1 673	-1 506	-1 409
Maximum value	-813	-812	-810	-811
Standard deviation	1 155	847	641	497

Table 7. Statistical characteristics of the increase in active power

	$\Delta t = 1$	$\Delta t = 2$	$\Delta t = 3$	$\Delta t = 4$
Minimum value	812	813	812	811
Median	1 351	1 283	1 205	1 147
Mean	1 660	1 493	1 368	1 268
Maximum value	5 101	4 520	3 548	2 844
Standard deviation	898	677	517	405

Table 8. Evaluation of testing the statistical hypothesis of the difference in medians

time step	p-value	conclusion
$\Delta t = 1$	0.189	The difference is not statistically significant
$\Delta t = 2$	0.005	The difference is statistically significant
$\Delta t = 3$	0.014	The difference is not statistically significant
$\Delta t = 4$	0.000	The difference is statistically significant

VI. Conclusion

The analyses show a high degree of variability in the power produced by PV plants, which needs to be considered when designing microgrid control and understanding the dynamics of renewable energy sources. The maximum decrease between two consecutive seconds reached a magnitude of -50.50 %, which is 8 181 W, while the installed capacity is 16 200 W. The maximum increase in power between two consecutive seconds reached 31.49 %, which is 5 101 W.

Based on the median differences in the rate of decline and increase in power for different time windows, the hypothesis was formulated that there is a statistically significant difference between the rate and decline in generation. This was demonstrated at $\Delta t = 2$ and $\Delta t = 4$. Numerically, the rate of decline is generally slightly greater than the rate of power production ramp-up after the clouds depart.

Acknowledgment

This article has been produced thanks to the support under the Operational Programme Integrated Infrastructure for the project International Centre of Excellence for Research on Intelligent and Secure Information and Communication Technologies and Systems - Phase II, ITMS code: 313021W404, co-funded by the European Regional Development Fund.

References

[1] OKTE. "Annual reports." Internet: www.okte.sk/en/information/annual-reports/, 2011-2021 [Mar. 30, 2022].

[2] M. Lave, M.J. RENO, R.J. Broderick. "Characterizing local high-frequency solar variability and its impact to distribution studies," in *Solar Energy*, vol .118, 2015, pp. 327-337. ISSN: 0038092X, DOI: 10.1016/j.solener.2015.05.028.

[3] J. Marcos, I. de la Parra, M. Garcia, L. Marroyo. "Control Strategies to Smooth Short-Term Power Fluctuations in Large Photovoltaic Plants Using Battery Storage Systems," in *Energies*. 2014, 7(10), pp. 6593-6619. ISSN: 1996-1073, DOI: 10.3390/en7106593.

[4] S. Maleki, M.T. Hagh. "Review of PV power ramp rate control methods and their requirements," in *SVU-International Journal of Engineering Sciences and Applications*, 2021, 2(2), pp. 14-25. ISSN: 2735-4571, DOI: 10.21608/svusrc.2021.83458.1011.

[5] GeeksforGeeks. "Shapiro–Wilk Test in R Programming." Internet: https://www.geeksforgeeks.org/shapiro-wilk-test-in-r-programming/, Jul. 16. 2020 [Mar. 31, 2022].

[6] STHDA. "Unpaired Two-Samples Wilcoxon Test." Internet: www.sthda.com/english/wiki/unpaired-two-samples-wilcoxon-test-in-r, [Mar. 31, 2022].

Utilisation of Energy Storage System in Power System

Dušan Medveď, Michal Kolcun, Jozef Király
Department of Electrical Power Engineering
Electrical Engineering and Informatics
Technical University of Košice
Mäsiarska 74
041 20 Košice, Slovak Republic
Dusan.Medved@tuke.sk , Michal.Kolcun@tuke.sk , Jozef.Kiraly@tuke.sk

Olexandr Shavolkin, Iryna Shvedchykova
Department of Computer Engineering and Electromechanics
Kyiv National University of Technologies and Design
Kyiv, Ukraine
shavolkin@gmail.com , shvedchykova.io@knutd.edu.ua

Abstract — **This paper deals with the contribution of energy storage system (ESS) to power system. According to a distribution network parameters there were implemented elements that are very popular and common in power system configuration. For that reason a power network consisted of sources such as wind farm, photovoltaic farm and in specific undesirable condition the distribution network connection as a backup was considered as a power source. Load, in this power system, was considered of constant power. During the simulation, there was implemented energy storage system that compensate the missing power until the specific state of discharge, when the connection to distribution system (DS) was implemented. Simulations were realized in Matlab/Simulink environment where 48-hours interval of various states of above mentioned elements cooperation was investigated. Results were analyzed and the recommendations were mentioned in conclusion.**

Keywords: energy storage system; photovoltaic power plant; wind power plant; energy accumulation

I. Introduction

Consumers in the field of renewable energy sources (RES) are households or organizations that produce excess fuel or energy and supply it to the national (or local) distribution network at times, while consuming the same fuel or energy from the network at other times (when their fuel or energy requirements exceed their own production). Households achieve this by installing PV panels on their rooftops that generate power. These homes can also employ battery storage to enhance their portion of their own PV power consumption, a practice known as prosumer in the literature. Companies that create biogas and feed it to the gas network can also do this while utilizing gas from the same network at different times or in different areas.

If prosumers participate in benefit programs and so supply free capacity for the network operator's objectives, the power network can already employ such an element to manage the power in the network. The study described in this article will be employed for this purpose, in which the prosumer (Energy Storage System (ESS)) is used as a conventional storage device and, in the event of a power outage in the network, the prosumer acts as a producer of energy. The influence of such a prosumer's operation was investigated using the Simulink program, which was used to model numerous situations that may occur in the electrical network.

II. Case Study Simulation

The given case study is the connection of the on-grid electrical network, which is connected in the 22 kV node (marked M1 in Figure 1) with the distribution system (DS) for the purpose of delivery, respectively electricity sales and vice versa. In case of insufficiency energy to cover the load, the required power is taken from the DS. This model also contains other sources, such as a photovoltaic power plant (PVP) and a wind farm (WPP).

The M1 node establishes the connection between the prosumer and the distributor in the distribution system. The prosumer is made up of an electricity storage system (ESS) and a photovoltaic power plant (PVP) together with wind power plant (WPP) at node M5 and M6. A load measuring at node M2 also serves as the measurement site for a load measurement system.

The model provides real-time measurements of crucial parameters. Voltage monitoring in each node, frequency monitoring, and condition monitoring of the SoC (State of Charge) ESS are incorporated in addition to current and active power flow.

A comparison of the active power of production and non-production units is also provided by the model. These systems were constructed using the same components as the prosumer system shown in Figure 1.

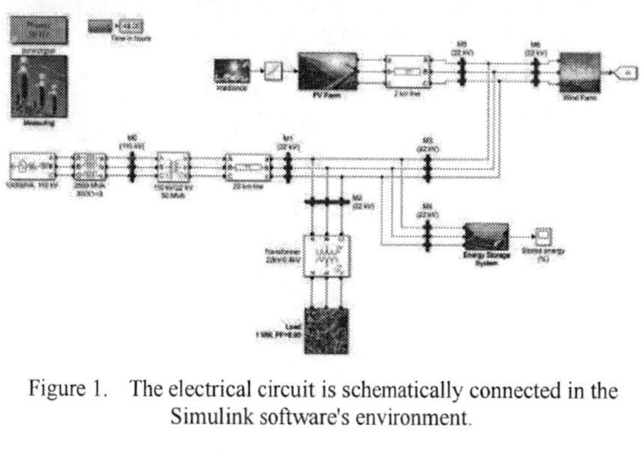

Figure 1. The electrical circuit is schematically connected in the Simulink software's environment.

A. The Default Configuration Settings are as Follows:

- connected photovoltaic power plant (PVP) with a total installed capacity of $P_i = 1800$ kWp, usable area of solar panels 12,000 m^2 with an efficiency of 15 %,
- characteristics of solar radiation was based on a 2-day sun diagram (Figure 2),

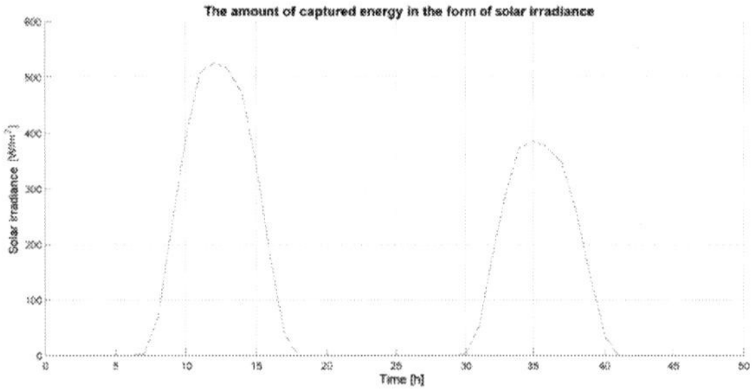

Figure 2. Solar radiation characteristics

- load with the peak power of 1 MW (in case of excess power from RES, possibility to connect other technological equipment as a load with a total power of 3×50 kW),
- load characteristics is based on Figure. 3,

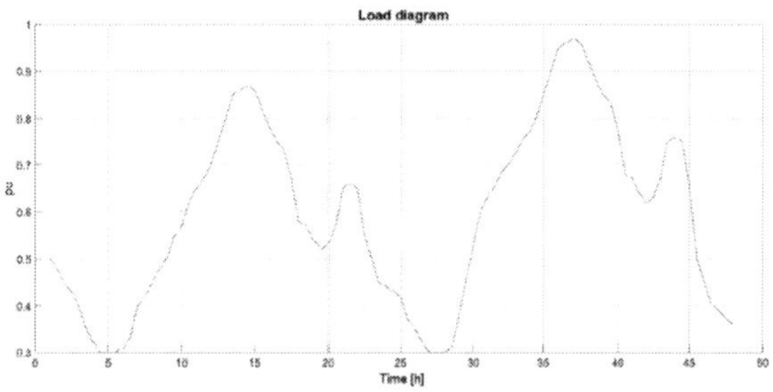

Figure 3. Diagram showing the load during a 48-hour simulation

- ESS (Energy Storage System) - can represent several possible ways of energy storage such as flywheels, compressed air, supercapacitors, etc.,
- the set ESS parameters are according to Figure 4, except parameter no. 5, i.e. the maximum permissible power taken from the DS has been changed from the original 400 kW to 250 kW,
- connected wind power plant with $P_i = 400$ kW at nominal wind speed of $v_{nom} = 9$ m/s,

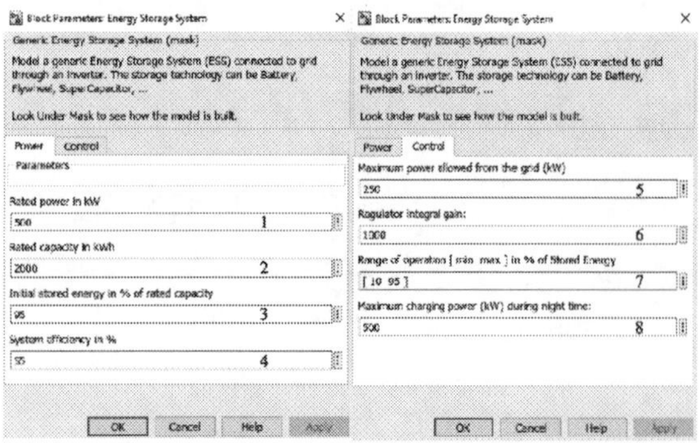

Figure 4. Basic ESS settings

Legend (parameters) to Figure. 4:

- 1 - rated power in kW,
- 2 - nominal ESS capacity in kWh,
- 3 - initial ESS capacity in%,
- 4 - system efficiency in%,
- 5 - the maximum permitted power consumption in kW from the distribution network after which the ESS will start supplying power,
- 6 - constant of the integration regulator,
- 7 - range of usable battery capacity in%,
- 8 - maximum charging power in kW.

III. Description of Transients

In describing the transient states of this case simulation, we considered the initial conditions of the model configuration, and the observed period (48 hours) was divided into certain intervals, where we described the course of power, voltage, frequency and other important characteristics.

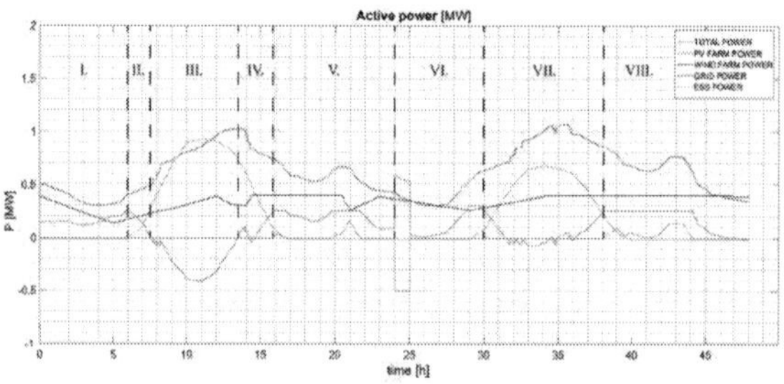

Figure 5. Active power characteristics

A description of the active power characteristics from Fig. 5:

In the *I. interval* in the time from 0 o'clock to 6 o'clock, the power from the wind power plant (WPP) and from the DS is used to cover the load. ESS is fully charged.

At the beginning of *II. interval*, it starts supplying PVP power to the network, which causes a decrease in the power taken from the DS.

In *III. interval*, the excess power from the RES is supplied to the DS for the purpose of its sale. If the power supplied to the DS from RES exceeds the value of 50 kW, the technological equipment is connected as a load in 3 states with a total power of 3×50 kW, with the shutdown of technological equipment being coupled not only to the value of power supplied to the DS from RES, but also for a specified amount of time, when they take power from the network. This means that these technological devices can take power from the network even if the power from the RES is not supplied to the DS.

The *IV. interval* is characterized by the disconnection of technological equipment, because the conditions for their further operation are not met.

At the beginning of the *V. interval*, it will start supplying ESS power to the network, because the condition of the maximum allowed power taken from the DS, specifically 250 kW, has been met.

In *VI. interval* from 24 hours to 30 hours, the time is reserved for charging the ESS by switching on a low tariff. In this time zone, even if the ESS is charged to the built-in rated capacity, the ESS does not supply any power to the network.

The *VII. interval* is characterized by a reduced supply of power from PVP compared to III. interval, the wind farm delivers a rated output of 400 kW from approximately 34 o'clock almost until the end of the simulation. Again similar to III. interval, in the event of excess power from RES, technological equipment is connected as a load.

In the last *VIII. interval* at a time when it is getting dark and the power supplied by the PVP is rapidly declining, it is again put into operation by the ESS, which stabilizes the power taken from the DS at 250 kW. As can be seen from the course of the SOC ESS in Figure 14, ESS capacity stabilized at 60% SOC. As we can see from Table 1, most electricity was supplied to the network during the entire monitored period WPP, namely 16.17 MWh, followed by PVP (9.85 MWh). The power taken by the prosumer from the DS represented 19.93% of the total power to cover the load.

Table 1. Electrical energy consumption for case simulation

W [MWh]	LOAD	TOTAL	ESS	GRID	PV FARM	WIND FARM
W_{IN_A}*	29,94	0,00	-0,71	-1,54	-0,51	0,0
W_{OUT_A}**	0,00	29,94	0,48	6,19	9,85	16,17

* WIN_A is the electricity consumed by the load, accordingly in the form of losses.

** WOUT_A is electricity delivered by production units.

The following graphs show the courses of the relevant quantities in nodes M0 to M6.

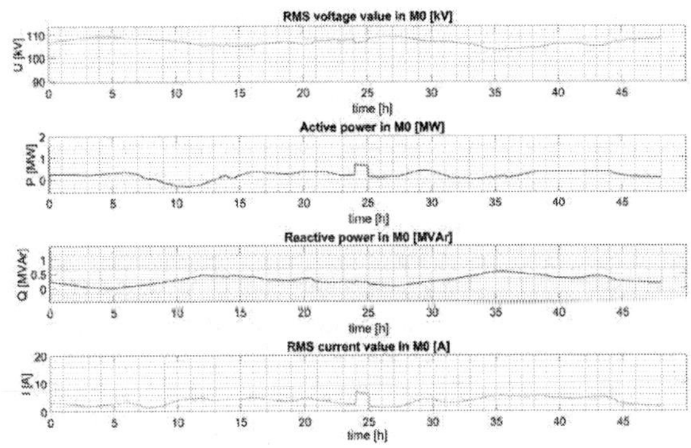

Figure 6. Measured electrical quantities in the M0 node

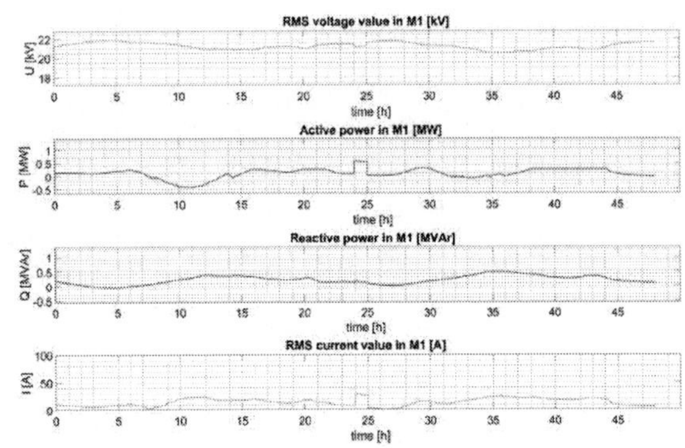

Figure 7. Measured electrical quantities in the M1 node

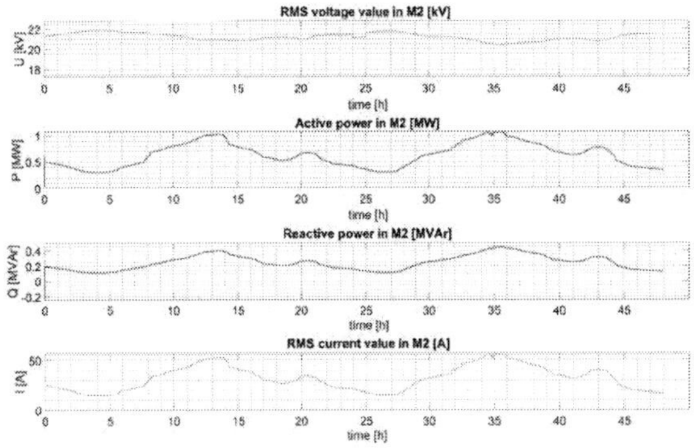

Figure 8. Measured electrical quantities in the M2 node

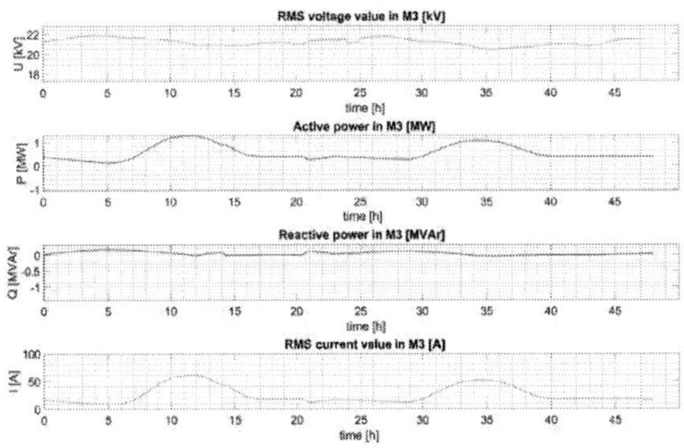

Figure 9. Measured electrical quantities in the M3 node

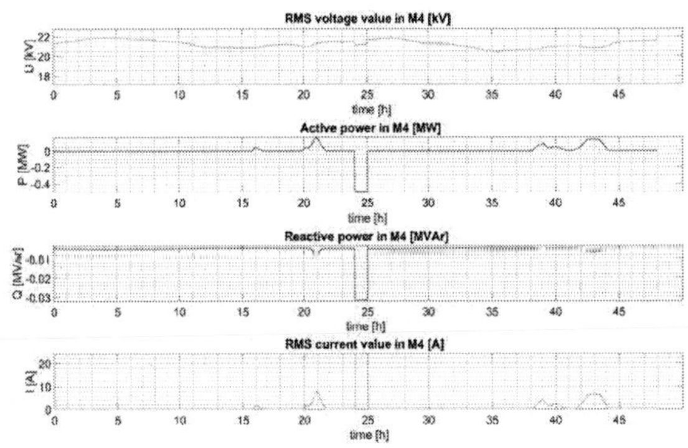

Figure 10. Measured electrical quantities in the M4 node

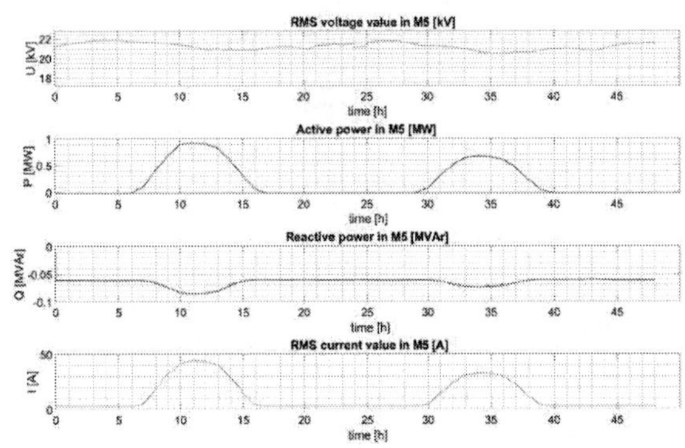

Figure 11. Measured electrical quantities in the M5 node

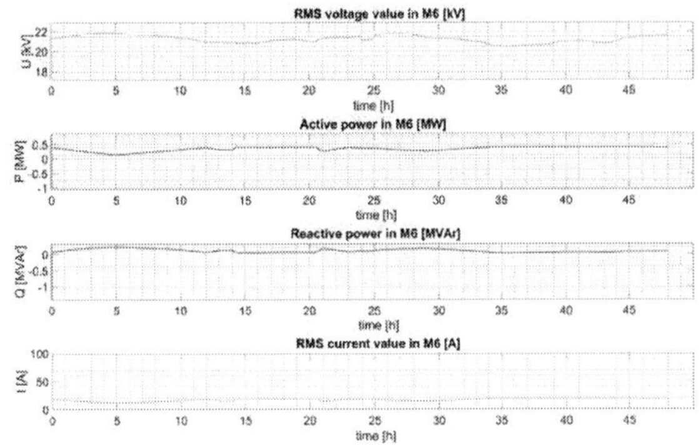

Figure 12. Measured electrical quantities in the M6 node

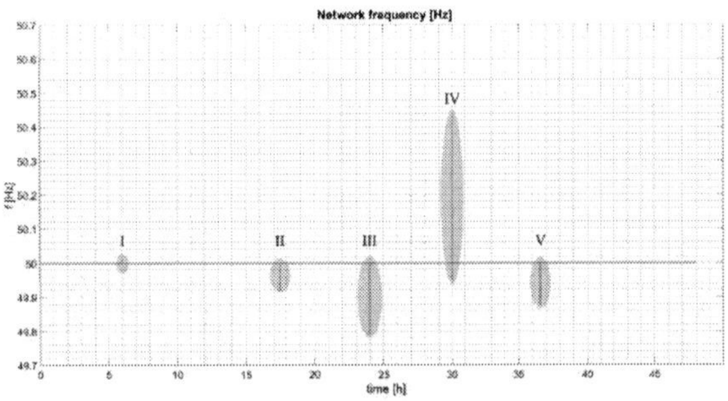

Figure 13. Frequency course in the network in the monitored interval

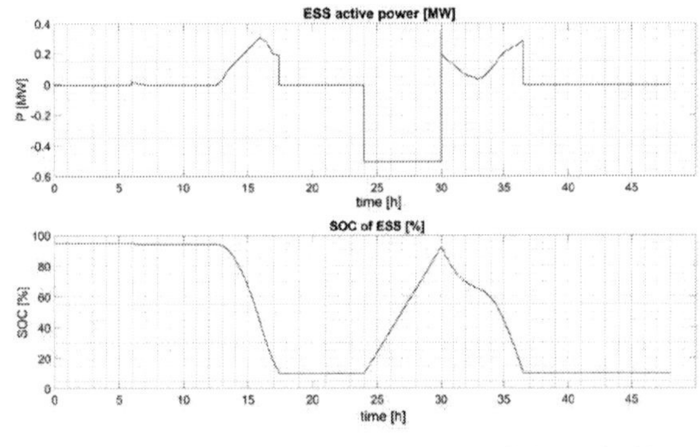

Figure 14. Course of active power and state of charge of ESS

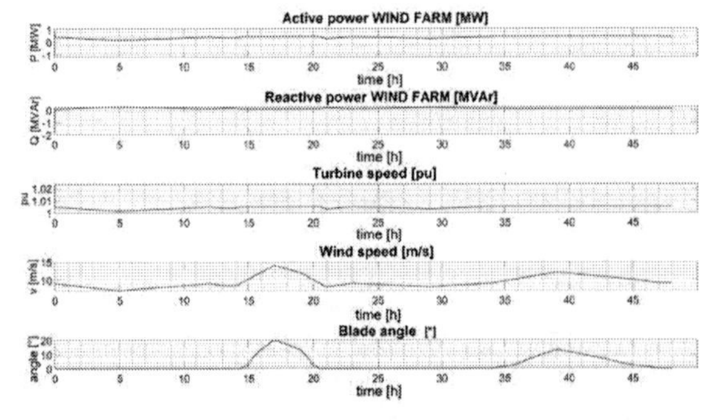

Figure 15. Observed characteristic values of wind power plant (blade rotation, wind speed, rotor speed)

A. Description of frequency characteristics in the network (Figure 13):

I. - the decrease of the frequency below the nominal value f_n is caused by the gradual connection of the load, respectively technological equipment in three stages 3×50 kW,

II. - the increase of the frequency above the nominal value f_n is caused by the gradual disconnection of the load, respectively technological equipment,

III. - the decrease of the frequency below the nominal value f_n is caused by the charging of the ESS and the increase of the frequency by the disconnection of the ESS (end of charging, SOC = 95%),

IV. - the increase of the frequency above the nominal value f_n is caused by the connection of the ESS due to the exceeding of the maximum allowed power consumption from the DS by the prosumer, i.e. 250 kW,

V. – decrease (resp. increase) of the frequency below, (resp. above) the nominal value is again due to connection, (resp. disconnection) the load, where the regulation was reduced, (resp. increased) by the supply of power from DS.

IV. Conclusion

In the given case simulation, the configuration of the prosumer's electrical circuit was supplemented by a WPP with an installed power $P_i = 400$ kW at a wind nominal speed $v_{nom} = 9$ m/s. The installed capacity of the PVP plant was increased to $P_i = 1800$ kWp (the usable area of the solar panels was 12 000 m^2 with an efficiency of 15%). In the case of the ESS, the parameter that determined the maximum allowable power consumption from the DS was changed from the original 400 kW to 250 kW.

Within the case simulation, 29.94 MWh of electricity was consumed in the given time period, which is caused by connecting an additional active power load in the form of technological equipment (e.g. electric boiler) in three stages of 3×50 kW. Electricity from RES, which was used directly to cover the load, accounted for 80.07% (23.97 MWh). In addition, there was a total volume of 1.54 MWh within the excess power from RES supplied to the DS.

In terms of the evaluation of electricity quality according to STN 50160 for LV and MV, we can evaluate that in the entire interconnected electricity network when connecting, respectively the disconnection of the load or the production unit did not cause any dangerous phenomena and in terms of the most important quality indicators, i.e. voltages and frequencies were kept within acceptable limits. From the frequency characteristics we see that in the case of the on-grid system, the frequency deviations in the power control were in the order of tenths and did not exceed ± 0.5 Hz.

Acknowledgment

This paper was supported by the Slovak Research and Development Agency, under the contracts APVV-19-0576, APVV-21-0312 and SK-UA-21-0024 and the Slovak Academy of Sciences, under the contracts VEGA 1/0757/21.

This work was also supported by Ministry of Education and Science of Ukraine within the framework of joint Ukrainian-Slovak scientific and research project "*Energy management improvement of hybrid photovoltaic systems of local objects with storage batteries*" (2022/2023) between Kyiv National University of Technologies and Design (Ukraine) and Technical University of Kosice (Slovakia).

References

[1] J. Dang, J. Seuss, L. Suneja, R.G. Harley, "SOC feedback control for wind and ESS hybrid power system frequency regulation", In *Proceedings of the IEEE Power Electronics and Machines in Wind Applications*, Denver, CO, USA, 16-18 July 2012; pp. 1-7, doi: 10.1109/PEMWA.2012.6316384.

[2] G. S. Lakshmi, O. Rubanenko, A. Hunko, "Control of the Sectioned Electrical Network Modes with Renewable Energy Sources", In *Proc. of the 2021 International Conference on Sustainable Energy and Future Electric Transportation (SEFET)*, Hyderabad, India, 21-23 January 2021; pp. 1-6, doi: 10.1109/SeFet48154.2021. 9375781.

[3] G. Ren, "MATLAB/Simulink-Based Simulation and Experimental Validation of a Novel Energy Storage System to a New Type of Linear Engine for Alt. Energy Vehicle Applications", *IEEE Trans. Power Electron.* 2018, 33, 8683-8694, doi: 10.1109/TPEL.2017. 2784563.

[4] M. Makhlouf, F. Messai, K. Nabti, H. Benalla, "Modeling and simulation of grid-connected photovoltaic distributed generation system", In *Proceedings of the 2012 First International Conference on Renewable Energies and Vehicular Technology*, Nabeul, Tunisia, 26-28 March 2012; pp. 187-193, doi: 10.1109/REVET.2012.6195269.

[5] N.K. Roy, H.R. Pota, "Current status and issues of concern for the integration of distributed generation into electricity networks", *IEEE Syst.* J. 2014, 9, 933-944, doi: 10.1109/JSYST.2014.2305282.

[6] B. Dolník, et.al., "The Response of a Magnetic Fluid to Radio Frequency Electromagnetic Field". In: *Acta Physica Polonica A*. doi: 10.12693/APhysPolA.131.946.

[7] I. Kolcunová, et.al., *"Experimental observation of negative differential characteristic of corona discharge in ultraviolet spectrum"*. In: *Journal of Electrostatics*. doi: 10.1016/j.elstat.2016.12.016.

[8] J. Zbojovský, L. Kruželák, M. Pavlík, "Impedance Spectroscopy of Liquid Insulating Materials". In: *IEEE CANDO-EPE 2018, IEEE International Conference and Workshop in Óbuda on Electrical and Power Engineering*, Nov. 20-21, 2018, Budapest, Hungary. ISBN: 978-1-7281-1154-4.

[9] M. Pavlík, I. Kolcunová, L. Lisoň, "Measuring the shielding effectiveness and reflection of electromagnetic field of building material". In: *2015 16th International Scientific Conference on Electric Power Engineering (EPE)*, 2015, pp. 56-59, doi: 10.1109/EPE.2015.7161186.

[10] I. Kolcunová, J. Zbojovský, M. Pavlík, S. Bucko, J. Labun, M. Hegedus, M. Vavra, R. Cimbala, J. Kurimský, B. Dolník, J. Petráš, J. Džmura, "Shielding Effectiveness of Electromagnetic Field by Specially Developed Shielding Coating". In: *Acta Physica Polonica A. Warsaw (Poland): Instytut Fizyki*. Vol. 137, No. 5 (2020), pp. 711-713. ISSN 0587-4246. doi: 10.12693/APhysPolA.137.711.

[11] J. Jirinec, D. Rot, R. Velev, "The Energy Efficient Solution for Intelligent Lighting", In: *2019 20th International Scientific Conference on Electric Power Engineering (EPE)*, 2019, pp. 1-4, doi: 10.1109/EPE.2019.8777938.

[12] K. Noháč, L. Raková, V. Mužik, K. Máslo, "Open Source Platforms for Dynamic Stability Assessment", In: *2019 20th International Scientific Conference on Electric Power Engineering (EPE)*, 2019, pp. 1-6, doi: 10.1109/EPE.2019.8778087.

[13] D. Kapral, P. Bracinik, M. Roch, M. Hoger, "Optimization of distribution network operation based on data from smart metering systems", In: *Electrical Engineering*, Vol. 99, Issue 4, Springer, New York, USA, 2017, December, pp: 1417-1428, ISSN 0948-7921. doi: 10.1007/s00202-017-0628-x.

[14] Z. Eleschova, B. Cintula, V. Volcko, A. Belan, J. Bendik, M. Cenky, "The Influence of Smart Grids on a Large Synchronous Generators Operation", In: *Proceedings of the 2018 19th International Scientific Conference on Electric Power Engineering (EPE)*, pp 43-48. doi: 10.1109/EPE.2018.8395948.

[15] K. Nohac, M. Tesarova, L. Nohacova, J. Veleba, V. Majer, "Utilization of Events Measured by WAMS-BIOZE-Detector for System Voltage Stability Evaluation", In: *IFAC Workshop on Control of Transmission and Distribution Smart Grids (CTDSG)*, Volume: 49, Issue: 27, pp 364-369, ISSN 2405-8963, doi: 10.1115/1.4043103.

[16] D. Rot, J. Kozeny, S. Jirinec, J. Jirinec, A. Podhrazky, I. Poznyak, "Induction melting of aluminium oxide in the cold crucible", In: *18th International Scientific Conference on Electric Power Engineering (EPE)* 2017, pp 1-4, doi: 10.1109/EPE.2017.7967281.

[17] D. Pál, Ľ. Beňa, J. Urbanský, M. Oliinyk, "The impact of reconfiguration on power losses in smart networks", In: *2020 21st International Scientific Conference on Electric Power Engineering (EPE)*, 2020, pp. 1-5, doi: 10.1109/EPE51172.2020.9269189.

[18] M. Sulír, J. Porubän, "Augmenting Source Code Lines with Sample Variable Values", In: *2018 IEEE/ACM 26th International Conference on Program Comprehension (ICPC)*, 2018, pp. 344-347. doi: 10.1145/3196321.3196364.

[19] M. Tesařová, R. Vykuka, "Impact of Distributed Generation on Power Flows Along Parallelly Operated MV Feeders", In: *Int. Conf. on Env. and El. Eng. and 2018 IEEE Ind. and Com. Power Syst. Europe, EEEIC/I and CPS Europe* 2018, doi: 10.1109/EEEIC.2018.8493640.

[20] P. Girovský, J. Žilková, J. Kaňuch, "Optimization of Vehicle Braking Distance Using a Fuzzy Controller". In: *Energies*. 2020; 13(11):3022. doi: 10.3390/en13113022.

[21] M. Bereš, D. Kováč, T. Vince, I. Kováčová, J. Molnár, I. Tomčíková, J. Dziak, P. Jacko, B. Fecko, Š. Gans, "Efficiency Enhancement of Non-Isolated DC-DC Interleaved Buck Converter for Renewable Energy Sources". In: *Energies*. 2021; 14(14):4127. doi: 10.3390/en14144127.

[22] B. Cintula, Ž. Eleschová, M. Cenký, P. Janiga, J. Bendík, A. Beláň, "Three-Phase and Single-Phase Measurement of Overhead Power Line Impedance Evaluation". In: *Energies*. 2021; 14(19):6314. doi: 10.3390/en14196314.

[23] J. Kanuch, P. Girovsky, "The device to measuring of the load angle for salient-pole synchronous machine in education laboratory", In: *Measurement*, Volume 116, 2018, pp 49-55, ISSN 0263-2241, doi: 10.1016/j.measurement.2017.10.043.

[24] M. Cenký, J. Bendík, Ž. Eleschová, A. Beláň, B. Cintula, P. Janiga, "Probabilistic Model of Electric Vehicle Charging Demand to Distribution Network Impact Analyses", In: *2019 20th International Scientific Conference on Electric Power Engineering (EPE)*, 2019, pp. 1 6, doi: 10.1109/EPE.2019.8777973.

[25] B. C. Sánchez, M. Váry, M. Perný, F. Janíček, V. Šály, J. Packa, "Prediction and production of small PV power plant", In: *2017 18th International Scientific Conference on Electric Power Engineering (EPE)*, 2017, pp. 1-5, doi: 10.1109/EPE.2017.7967276.

[26] M. Pavlik, "Compare of shielding effectiveness for building materials," *Przeglad Elektrotechniczny*, 95(5), 137-140, doi: 10.15199/48.2019.05.33.

[27] S. Bucko, M. Krchnak, R. Cimbala, L. Kruzelak, J. Zbojovsky, "Abrasive properties of transformer oil-based magnetic nanofluid," In: *19th International Scientific Conference on Electric Power Engineering (EPE)*, 2018, pp. 1-4, doi: 10.1109/EPE.2018.8396038.

[28] A. Beláň, B. Cintula, M. Cenký, P. Janiga, J. Bendík, Ž. Eleschová, A. Šimurka, "Measurement of Static Frequency Characteristics of Home Appliances in Smart Grid Systems". *Energies* 2021, 14, 1739. doi: 10.3390/en14061739.

[29] M. Kajanova, P. Bracinik, "Social welfare-based charging of electric vehicles in the microgrids fed by renewables" In: *International Journal of Electrical Power & Energy Systems*, Vol. 138, doi: 10.1016/j.ijepes.2022.107974.

Control of Modular Multilevel Converter for Compensation of Unsymmetrical Loads

Martin Bejvl, Petr Šimek, Viktor Valouch
Department of Electrical Engineering and Electrophysics
Czech Academy of Sciences
Prague, Czech Republic
bejvl@it.cas.cz, simek@it.cas.cz, valouch@it.cas.cz

Abstract— **The paper introduces principles of an active compensation of a general single-phase load connected to the three-phase network. The aim of the compensation is a symmetrical and only active load of the three-phase network. A modular multilevel voltage converter is considered as a compensation device. Relations among the single-phase load current and desired current references for the multilevel converter are formulated. In order to reach savings of devices, a compensation converter with only two branches is considered. A concept of the way to reach desired currents of the three-phase network and simulation results are presented.**

Keywords: three-phase grid, single-phase load, unbalance compensation, multi-level voltage converter, current control, simulation

I. Introduction

If we have a single-phase load connected to a three-phase network, this creates an asymmetrical load that adversely affects the operation of this three-phase network. Therefore, single-phase loads are normally distributed between phases so that the network is loaded evenly in all three phases as much as possible. This can be done with smaller loads, of which there are a large number. The problem is if the load is a large appliance that cannot be divided in this way. A typical example of this is electric traction. Partial layout is possible by feeding multiple sections of the trolley from different phases. The trolley is thus divided into several sections assigned to different phases. This will reduce the asymmetry. However, if we want to have a trolley without transitions, it is necessary to ensure the symmetrization of a relatively large and, in addition, variable load.

Several articles and producers proposed compensation equipments for unsymmetrical loads in three-phase grids, which consist of three branches of a multilevel voltage converter [1]-[5]. These three branches were connected to delta and every branch was connected between the phases *ab*, *bc*, *ca* of the grid. Every branch contains series of H-bridges and every H-bridge has a *DC* intermediate circuit with a capacitor. Every branch also contains a reactor in series with H-bridges.

The company Siemens produced the system Sitras RAB, which is an example of 3-branches balancing equipment. Some solutions with STATCOMs (Static Compensators) were suggested as well, specifically for a single-phase traction loads.

The mentioned connections have an essential disadvantage. The converter with three branches under conditions of high voltage is too expensive. In order to reach savings of devices, we want to develop a compensation converter with only two branches. If we want to operate with only compensation converter with two branches, we have to ensure the active power exchange between branches of the converter.

In the contribution we will introduce basic scheme and basic circuit equations of the system. Then we compare several variants of solutions. The aim of the compensation is a symmetrical and only active load of the three-phase network. A modular multilevel voltage converter is considered as a compensation device. Relations among the single-phase load current and desired current references for the multilevel converter are formulated. A concept of the way to reach desired currents of the three-phase network and simulation results are presented.

II. Principle of Symmetrization

The classic solution of symmetrization of a single-phase load consists in adding elements *C* and *L* of suitable size to the other phases. Sizes *C* and *L* are given by the value of the single-phase load current. In addition, the solution depends on the phase sequence. From the point of view of the source, then, at suitably selected values, such a system appears to be three-phase purely active and symmetrical, even if the currents of the triangular branches are not symmetrical in such an arrangement.

The connection is applicable to a load whose size does not change. For a variable load (which is also the case considered for electric traction), the values of inductance *L* and capacitance *C* of the balancing element must also be variable.

Multi-level pulse converters will be used for this purpose as a variable capacitor and inductance. Because no active power is required, *DC* power may not be present. This supply would greatly increase the complexity of the converter, as this supply must be floating and, in the case of a modular converter, needs many floating sources.

The simulated system thus consists of a single-phase load connected between the phases *A* and *B* and a three-phase modular converter connected in a delta. The individual phases of the converter are indicated only symbolically in Fig. 1, and the *AB* phase is drawn in dashed lines because it will in fact be omitted. It is in the picture to maintain generality.

III. Calculation of Compenzating Currents

In order for the unbalance to be compensated, the converter must generate currents that make all currents in this balance state. It is desirable that the value of the consumed active power is maintained.

Let us analyze the conditions in the circuit in a complex plane. This procedure is relatively simple and straightforward.

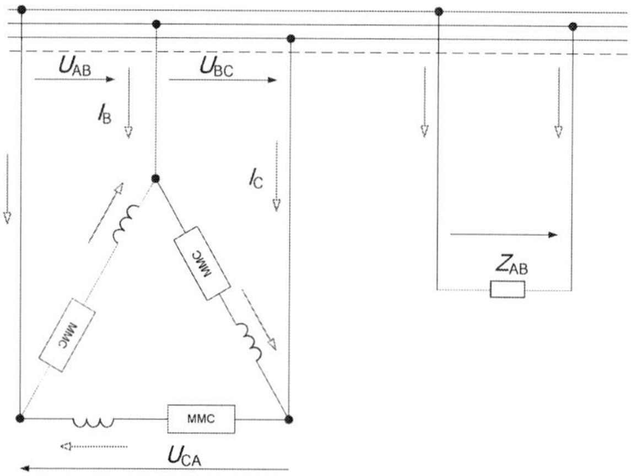

Figure 1. Scheme of converter and load

Let us assume that we have the measured phase-to-phase voltages of the network u_{ab}, u_{bc} and u_{ca} (one of them is redundant, so only two are sufficient) and the measured load currents i_{Lab}, i_{Lbc}, i_{Lca} (in general case of a three-phase load). After conversion to the complex plane we have the vectors $\bar{U}_{ab}$, $\bar{U}_{bc}$, $\bar{U}_{ca}$ and $\bar{I}_{Lab}$, $\bar{I}_{Lbc}$, $\bar{I}_{Lca}$. The currents of the converter that works as a compensator of asymmetry are denoted by the index M, the load currents by the index L, and the network currents (= load + converter) by the index S.

Consider that all currents are oriented according to the principle of the consumer system. The currents in the triangle are marked according to the direction of rotation of the used reference system (a→c, b→c, c→a).

After the conversion to vectors, we have the associated network voltages or the phase-to- phase voltages

$$\overline{U_f} = \begin{pmatrix} \overrightarrow{U_a} \\ \overrightarrow{U_b} \\ \overrightarrow{U_c} \end{pmatrix}, \quad \overline{U_s} = \begin{pmatrix} \overrightarrow{U_{ab}} \\ \overrightarrow{U_{bc}} \\ \overrightarrow{U_{ca}} \end{pmatrix} \tag{1}$$

In the three phase system the following relations are valid

$$\begin{gathered} \overrightarrow{U_a} + \overrightarrow{U_b} + \overrightarrow{U_c} = 0 \\ \overrightarrow{U_{ab}} + \overrightarrow{U_{bc}} + \overrightarrow{U_{ca}} = 0 \end{gathered} \tag{2}$$

and

$$\begin{gathered} \overline{I_L} = \begin{pmatrix} \overrightarrow{I_{Lab}} \\ \overrightarrow{I_{Lbc}} \\ \overrightarrow{I_{Lca}} \end{pmatrix} \quad \overline{I_L} = \begin{pmatrix} \overrightarrow{I_{La}} \\ \overrightarrow{I_{Lb}} \\ \overrightarrow{I_{Lc}} \end{pmatrix} \\ \overline{I_M} = \begin{pmatrix} \overrightarrow{I_{Mab}} \\ \overrightarrow{I_{Mbc}} \\ \overrightarrow{I_{Mca}} \end{pmatrix} \quad \overline{I_M} = \begin{pmatrix} \overrightarrow{I_{Ma}} \\ \overrightarrow{I_{Mb}} \\ \overrightarrow{I_{Mc}} \end{pmatrix} \end{gathered} \tag{3}$$

The currents withdrawn from the network are

$$\overline{I_S} = \begin{pmatrix} \overrightarrow{I_{Sa}} \\ \overrightarrow{I_{Sb}} \\ \overrightarrow{I_{Sc}} \end{pmatrix} = \overline{I_L} + \overline{I_M} = \begin{pmatrix} \overrightarrow{I_{La}} \\ \overrightarrow{I_{Lb}} \\ \overrightarrow{I_{Lc}} \end{pmatrix} + \begin{pmatrix} \overrightarrow{I_{Ma}} \\ \overrightarrow{I_{Mb}} \\ \overrightarrow{I_{Mc}} \end{pmatrix} \tag{4}$$

The converter can produce a circular current $\bar{I}_0$ of any value that does not influence the currents of the network

$$\overrightarrow{I_{ab}} + \overrightarrow{I_{bc}} + \overrightarrow{I_{ca}} = \overrightarrow{I_o} \tag{5}$$

The active power is the sum of the phase powers or the sum of the respective branch powers

$$P = P_{ab} + P_{bc} + P_{ca} = P_a + P_b + P_c \tag{6}$$

Partial powers are given by the real part of the product of voltage and complex conjugate current of the respective part. I we assume vectors of the voltages and currents expressed in the scale of magnitudes, we must add the coefficient ½. The symbol * denotes a complex conjugate number, x replaces the relevant voltage and current index from the equation (6)

$$P_x = \frac{1}{2}\Re\left\{\overrightarrow{U_x}\overrightarrow{I_x^*}\right\} \tag{7}$$

Let the active power of the full three phase load is

$$P_L = \frac{1}{2}\Re\left\{\overrightarrow{U_a}\cdot\overrightarrow{I_{La}}^* + \overrightarrow{U_b}\cdot\overrightarrow{I_{Lb}}^* + \overrightarrow{U_c}\cdot\overrightarrow{I_{Lc}}^*\right\} \tag{8}$$

The active power of the considered one phase load is

$$P_L = \frac{1}{2}\Re\left\{\overrightarrow{U_{ab}}\cdot\overrightarrow{I_{Lab}}^*\right\} \tag{9}$$

This power must be maintained even after balancing (compensator losses are not considered now). The total active power of each branch (from the source's point of view) will therefore be

$$P_x = \frac{1}{3}P_L \tag{10}$$

The respective currents are

$$\overrightarrow{I_x}^* = 2\frac{P_x}{\overrightarrow{U_x}} \tag{11}$$

$$\overrightarrow{I_x} = \left(2\frac{P_x}{\overrightarrow{U_x}}\right)^* \tag{12}$$

where $x=a,b,c$.

If we subtract the load current (i.e. the measured current) from the current that should flow for a purely active symmetrical load from the previous relation (12), we get the current that is missing for this state, in other words, we obtain the references for currents of individual phases of the converter. If the converter implements these currents, it will generate a current that will replenish the load current to the required symmetrical consumption.

$$\overline{I_M} = \begin{pmatrix} \overrightarrow{I_{Ma}} \\ \overrightarrow{I_{Mb}} \\ \overrightarrow{I_{Mc}} \end{pmatrix} = \overrightarrow{I_{Sideal}} - \overline{I_L} = \begin{pmatrix} \overrightarrow{I_{Sa}} \\ \overrightarrow{I_{Sb}} \\ \overrightarrow{I_{Sc}} \end{pmatrix} - \begin{pmatrix} \overrightarrow{I_{La}} \\ \overrightarrow{I_{Lb}} \\ \overrightarrow{I_{Lc}} \end{pmatrix} \tag{13}$$

We consider the converter connected in a triangle (for now we will not complicate the derivation the impact of absence of one branch). Therefore, two branches participate in each of the currents. Thus, it is necessary to convert the network current to the internal currents of the inverter.

So far, everything could be applied to any load. Next we will focus on single phase load.

In a situation of single-phase active load, the current $\overline{I}_{Mab}$ should be zero and the remaining two pure reactive, namely capacitive in the branch *BC* and inductive in the branch *CA* (in accordance with the basic principle of symmetrization described earlier).

The currents of the individual branches of the triangle are

$$\overrightarrow{I_{M\Delta}} = \begin{bmatrix} \overrightarrow{I_{Mab}} \\ \overrightarrow{I_{Mbc}} \\ \overrightarrow{I_{Mca}} \end{bmatrix} = \frac{1}{3}\begin{bmatrix} 1 & -1 & 0 \\ 0 & 1 & -1 \\ -1 & 0 & 1 \end{bmatrix} \begin{bmatrix} \overrightarrow{I_{Ma}} \\ \overrightarrow{I_{Mb}} \\ \overrightarrow{I_{Mc}} \end{bmatrix} + \begin{bmatrix} \overrightarrow{I_0} \\ \overrightarrow{I_0} \\ \overrightarrow{I_0} \end{bmatrix} \tag{14}$$

The circular current $\overline{I}_0$ does not contribute to the inverter current from the network and its magnitude is arbitrary. In the case of a complete triangle, it represents the degree of freedom applicable to the control needs (for example, the possibility of voltage balancing even at zero external current). In the case of an incomplete triangle, its magnitude is clearly based on the condition of zero current in the third non-existent phase.

It can be seen that a non-zero current emerges in the *AB* branch, so we must choose a circuit current equal to its magnitude with the opposite sign and add it to all currents. This circuit current will be only virtual and it has no trace to close, but it must be considered because the $\overline{I}_{ab}$ must be equal to zero. Therefore, the magnitude of the required circuit current is

$$\overrightarrow{I_0} = -\overrightarrow{I_{Mab}}^{\,r} \tag{15}$$

Thus, the search current value is

$$\overrightarrow{I_0} = \frac{1}{3}\left(\overrightarrow{I_{Mb}} - \overrightarrow{I_{Ma}}\right) \tag{16}$$

This value is given by the power circuit topology (only 2 branches) and cannot be changed.

Substituting (16) into (14) we get a formula for the required current values

$$\overrightarrow{I_{M\Delta}} = \begin{bmatrix} \overrightarrow{I_{Mab}} \\ \overrightarrow{I_{Mbc}} \\ \overrightarrow{I_{Mca}} \end{bmatrix} = \frac{1}{3}\left(\begin{bmatrix} 1 & -1 & 0 \\ 0 & 1 & -1 \\ -1 & 0 & 1 \end{bmatrix} \begin{bmatrix} \overrightarrow{I_{Ma}} \\ \overrightarrow{I_{Mb}} \\ \overrightarrow{I_{Mc}} \end{bmatrix} + \left(\overrightarrow{I_{Mb}} - \overrightarrow{I_{Ma}}\right) \right)$$

(17)

Thus, the currents of the individual branches of the triangle are

$$\overrightarrow{I_{M\Delta}} = \begin{bmatrix} \overrightarrow{I_{Mab}} \\ \overrightarrow{I_{Mbc}} \\ \overrightarrow{I_{Mca}} \end{bmatrix} = \frac{1}{3}\begin{bmatrix} 0 \\ -\overrightarrow{I_{Ma}} + 2\overrightarrow{I_{Mb}} - \overrightarrow{I_{Mc}} \\ -2\overrightarrow{I_{Ma}} + \overrightarrow{I_{Mb}} + \overrightarrow{I_{Mc}} \end{bmatrix} \tag{18}$$

and considering $\overline{I}_a = \overline{I}_b = \overline{I}_c = 0$ the final equation is

$$\overrightarrow{I_{M\Delta}} = \begin{bmatrix} \overrightarrow{I_{Mab}} \\ \overrightarrow{I_{Mbc}} \\ \overrightarrow{I_{Mca}} \end{bmatrix} = \frac{1}{3}\begin{bmatrix} 0 \\ -\overrightarrow{I_{Ma}} + 2\overrightarrow{I_{Mb}} + \overrightarrow{I_{Ma}} + \overrightarrow{I_{Mb}} \\ -2\overrightarrow{I_{Ma}} + \overrightarrow{I_{Mb}} - \overrightarrow{I_{Ma}} - \overrightarrow{I_{Mb}} \end{bmatrix} = \tag{19}$$

$$= \frac{1}{3}\begin{bmatrix} 0 \\ 3\overrightarrow{I_{Mb}} \\ -3\overrightarrow{I_{Ma}} \end{bmatrix} = \begin{bmatrix} 0 \\ \overrightarrow{I_{Mb}} \\ -\overrightarrow{I_{Ma}} \end{bmatrix}$$

IV. Model in Matlab/Simulink environment

The model assembled in the Simulink® environment consists of two separate parts. The first part is a model of a controlled system containing a source, a single-phase load and a converter. The second is the control system model.

The first part, which simulates a real system, is simulated as continuous, while the control system is simulated as discontinuous. A trigger subsystem was chosen for its modeling, which is a technique that ensures that the included blocks are executed only when a signal arrives at the trigger (sampling) input. Thus, the sampling period can be arbitrary, changed on the fly (possibly linked to some asynchronous event) and does not even have to be constant.

V. Model of Power Part

Several models have been created for the simulation. The system is relatively complex and its simulation is time consuming. However, it is not necessary to simulate in detail all the events taking place in the converter to tune the control unit. For this purpose, a replacement model is sufficient, which behaves outwards in the same way as the considered converter. Thus, several substitutes were made, namely:

a) Replacement by one voltage source switched to the reference at the sampling half-life (actually a two-level converter model),

b) Inverter voltage replacement in the meantime by step voltage, switching between reference values takes place in the selected number of intermediate stages (it simulates the response of a multi-level converter in the mean value),

c) The PWM voltage is considered (faithfully simulates MMC response except dead times, the simulation of which was not included in this model).

The concept of the model presupposes the possibility of exchanging individual substitutions of the power part and thus a relatively easy transition between individual variants.

The full involvement model tries to get as close as possible to the real system. This model contains N basic cells, each of which contains its own modulator. This modulator also has a mechanism for balancing the voltage of the capacitors. The modulation index is adjusted according to the deviation of the voltage from the average (block input) so that this deviation decreases. The inverter consists of a cascade of these cells, which are floating and powered from the *AC* side.

The basic scheme of the power part is in Fig. 2. In this figure are shown also signals for a control system.

A. Load Model

The load model contains two blocks that can be connected or disabled (control is independent, so both can be connected at the same time, as well as both disabled).

The first block consists of several resistors and switches, thus allowing the simulation of jumps. The jumps are secured by switching the resistance sections. In addition to the two resistors, switch resistors (R_{on}) are also used.

The second block consists of a resistor simulated by a controlled current source (the source is controlled so that its current corresponds to the resistor current of the considered value). It can be used for simulation of a programmable load course.

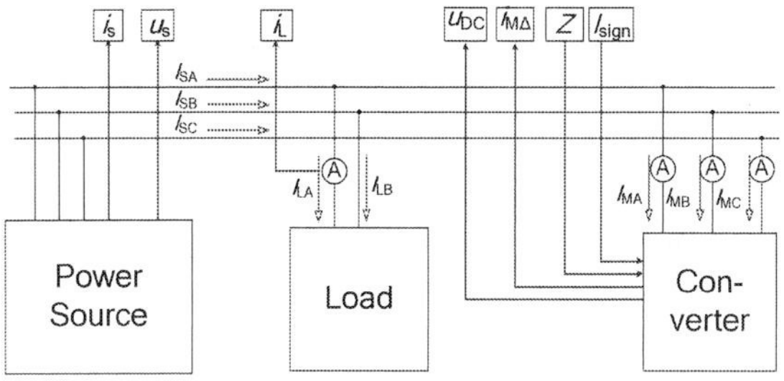

Figure 2. Basic scheme of power part

B. *Converter Model*

The main component is the converter model, which ensures the balancing of the consumption. It contains two single-phase cell converters. The third phase is unconnected.

C. *Source Model*

The power supply model simulates the power supply system. It consists of three voltage sources and network impedance. If a simulation other than pure and symmetrical voltage is needed, it can be replaced by a more complex source.

D. *Measurement of Variables*

The measured values of *ac* voltages and currents are converted into a complex plane into vectors. The next step is the transfer to the synchronous rotating system.

The parameters of the simulated system are in TABLE I.

Table 1. Parameters of simulated system

Variable, parameter	**Values for HV model**
Number of cells	10
Cell voltage	1050 V
Cell capacitance C_{DC}	5.5 mF
Nominal *AC* current [1]	1000/√3 A
AC phase voltage	$\frac{25}{4\sqrt{3}}$ kV
AC phase-to-phase voltage	$\frac{25}{4}$ kV
Filter inductance L_f	2 mH
Filter resistance R_f	0.1 Ω
Active load current	1000 A
Sampling frequency	2 000 Hz
Network frequency	50 Hz

[1]It is based on the considered load current.

VI. Modulator

Two types of the pulse width modulation were considered.

As the first the was tested PSC-PWM (Phase Shifted Carrier PWM) modulation with FBCK-BAL balancing, where balancing is performed by means of feedback P-regulation of voltage on cell capacitors. The sampling period of the discontinuous part is given by the clock generator.

The second was SA-PWM (Sampled Average PWM) with voltage balancing on SORT-BAL cells, where cells are sorted according to their voltage. The whole model was therefore externally continuous and contains a part that is discontinuous. Here, the sampling period of the continuous part is synchronized with the modulation, all cells has a common synchronization (clock) signal.

A. *Phase Shifted Carier PWM*

For the first simulations the PSM (Phase Shifted Modulation) was used. The asymmetric PWM modulation is used. Each cell has its own modulator. Modulation is generated based on the required switching ratio; this value is modified according to the deviation of the cell voltage from the average in accordance with the current direction and the desired influence. (Individual cells don't produce exactly required ratio, but in sum, the average switch ratio agrees.)

The PWMs of the individual cells are evenly distributed over the time of the sampling period, so that the composition of the waveforms creates a multi-level voltage. The modulation factor is updated once a period.

The implementation of this strategy is relatively simple. It uses repetitive base modules, only with different timings. Balancing is also easy for it. But there are many drawbacks of them. The basic ones are unnecessarily frequent switching and a more complex response (by the sum of the responses of many cells distributed over time). The main drawback is switching of all cells for every regulation period (for large amount of cells it is not necessary).

B. *Sampled Average PWM (SA-PWM)*

Unlike previous strategy, individual cells are timed synchronously, but do not switch simultaneously. Only one cell switches, the others are switched on at all times to full voltage of one or the other polarity or to zero.

The composition of the waveforms thus results in a multi-level stepped voltage with a small number of switching (theoretically always only one changeover during the switching period).

The modulation factor is updated once a period, but for the whole block of cells at one time due to cell synchronization. This makes signal transmission delays better predicted.

First step is to transpose the duty cycle to 0...N range. Fix part of the number says how much cells should be fully on. The fractional part of the duty cycle determinates duty cycle of last cell. Other cells are in neutral state.

In order to keep the voltages of individual cells balanced, there is used permutation of duty cycles sort in order to influence cell voltages by right way (to decrease its difference).

VII. Converter Current Responses

The current of each of the converter phases is determined by the inductance of the filter and the difference of the total voltage of all inverter cells (i.e. the inverter voltage) against the network voltage. These two voltage sources are connected via the impedance, the magnitude of which is known. The situation can be illustrated in Fig. 3.

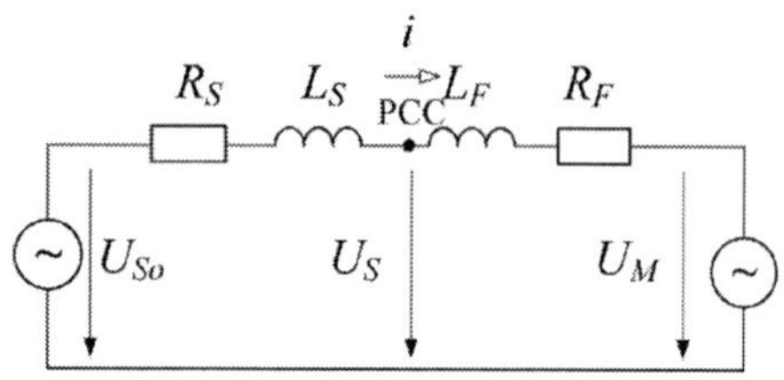

Figure 3. Scheme of controlled system

Thanks to the delta connection, the legs (phases) are independent. For testing and tuning the control algorithm there is satisfactory to simulate only one phase. In part II the required current was determinate. This principle was checked by simulation. Now the aim of simulation is not to verify the request. The aim is to reach this current with given topology.

Figure 4 shows the topology of the one phase. There are chosen number of full H-Bridges, every with own floating DC capacitor. The capacitor voltages U_{DC1} … U_{DCn} are measured. Every H-bridge has own PWM modulator. Depending on used modulation strategy, there are distributed or synchronized.

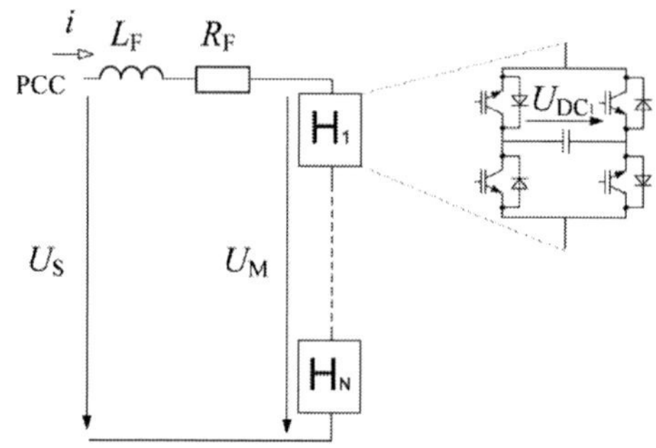

Figure 4. Diagram of one phase of power part

VIII. Control Part

In the part II the required current was determined. Beside this calculation, the control part must ensure the implementation of this current. When using SA-PWM, the control unit must also actively balance the capacitors of the DC circuits.

It means in particular:

- processing of measured voltages (transformation or conversion to the complex plane)
- calculation of current reference values
- calculation of the required converter voltage - current and voltage regulators

Control strategy is shown in Fig. 5. From the grid voltage is derived direct and signal. Since the sampling frequency is very small, the sampling delay must be compensated by rotation of appropriate angle.

Voltage magnitude given by

$$U_{max} = \sqrt{u_d^2 + u_q^2} \tag{20}$$

After normalization, these signals are used as a base for generation the active and reactive current.

The active and reactive power are used to determine active and reactive current

$$I_d = \frac{2P^*}{U_{max}} \qquad I_q = \frac{2Q^*}{U_{max}} \tag{21}$$

The components of converter voltage drop

$$U_{md} = I_d R_f - I_q \omega L_f$$

$$U_{mq} = I_q R_f + I_d \omega L_f \tag{22}$$

Finally the number is converted to the sinusoidal shape by using multiplication by the base voltages u_{sdn} u_{sqn}. These two components in sum present the desired voltage drop. Converter voltage is given by sum of voltage drop with voltage in advance (estimated voltage in the instant of its realization). From this voltage the duty cycle ratio is calculated. The duty cycle ratio is distributed to the individual cells in according to modulation strategy.

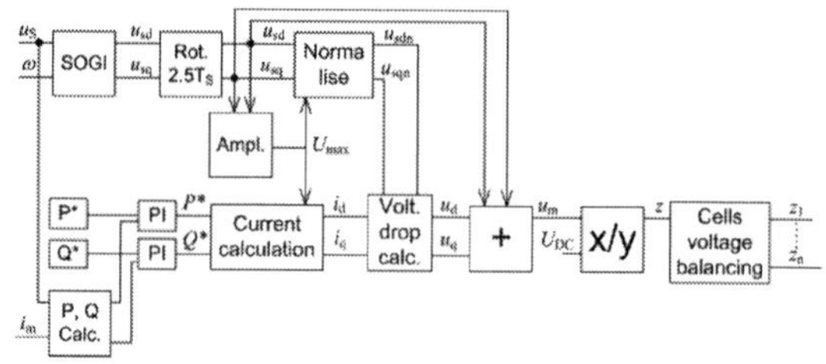

Figure 5. Diagram of control algorithm

IX. Simulation Results

A lot of simulations are performed. Simulation results presented here are acquired from single phase full model (i.e. the cells are modeled in detail) with six cells. Scale of this model is in low voltage, connection to the phase-to-phase voltage 400 V is supposed. The SA-PWM is used.

Figure 6 shows active and reactive power and sampled voltage and current. Figure 7 shows converter voltage in comparison with grid voltage and converter current. The grid current and voltage are the same quantities as shown in Fig. 6, here there are not sampled.

There are two transient responses in the figures. First is change from capacitive load to the inductive one, second is increase the required power. Active power is required to zero.

It is evident a good agreement with required value. There is only a little difference to suppress by the controller.

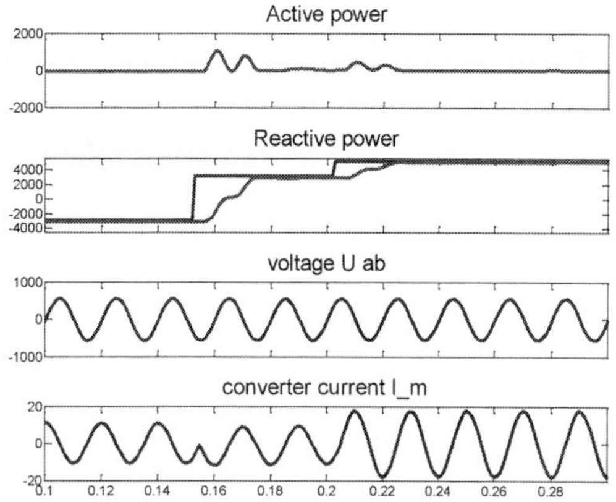

Figure 6. Response to the required reactive power change. From -3000 VAr to 3000 VAr in the time 0.152s and increase to 5000 VAr in time 0.202 s.

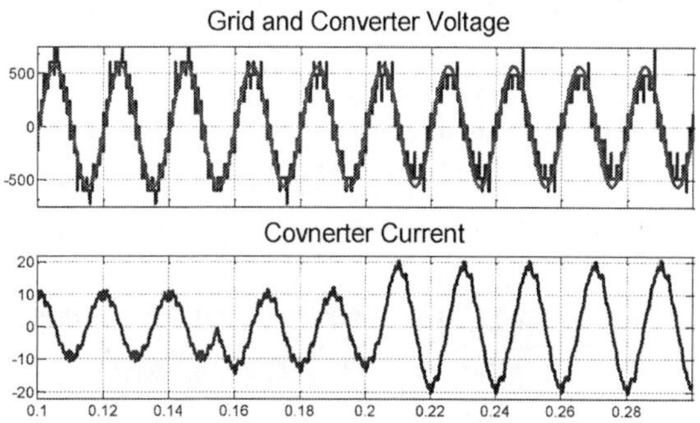

Figure 7. Voltage and current of the converter.

X. Conclusion

The paper introduces principles of an active compensation of a general single-phase load connected to the three-phase network. The aim of the compensation is a symmetrical and only active load of the three-phase network.

Control of the symmetrization device based on delta connected multilevel converter is presented in this contribution. Basic principles, way of calculation and simulated responses of compensating currents are shown.

The model assembled in the Simulink® environment consists of two parts. The first part is a model of a controlled system containing a source, a single-phase load and a converter. The second part is the control system model.

Selected transient responses of the converter current, active and reactive power following changes in the references for powers are presented. The responses confirm functionality of the developed control strategy.

Acknowledgment

The financial support from the project RVO//:61388998, Institute of Thermomechanics of the Czech Academy of Sciences is highly acknowledged.

References

[1] Sitras RAB plus. "Rail active balancer for AC traction power supply." © Siemens AG2018. Product information/Version 1.0.2, Nr. A6Z00036124075.

[2] W. Baoan, S. Jiao, Ch. Hao, D. Ningyi. "Steinmetz Theory Applied to Fundamental Phase Reference Current Calculation for a STATCOM with Delta Configuration," *TENCON 2015 - 2015 IEEE Region 10 Conference*, Macao, 2015, pp. 1-5. doi: 10.1109/TENCON.2015.7372845.

[3] J. Shang, N. Y. Dai, B. Wang, H. Chen, "Railway Power Conditioner Based on Deltaconnected Modular Multilevel Converter," *2016 IEEE Energy Conversion Congress and Exposition (ECCE)*, Milwaukee, WI, 2016, pp. 1-7.doi: 10.1109/ECCE.2016.7855376.

[4] P. H. Wu, P. T. Cheng, "Control of Delta-Connected Cascaded Converter in the Railway Power Conditioner Application", *2018 IEEE Energy Conversion Congress and Exposition (ECCE)*, Portland, OR, 2018, pp. 2790-2797.doi: 10.1109/ECCE.2018.8558458.

[5] A. Zafeiropoulos. "An MMC- based topology with Dual-Active-Bridge Power Channels for load balancing in 50Hz Railway Applications". *MSc Thesis, KTH Royal Institute of Technology*, Sweden, Nov 2016

[6] Z. He, F. Ma, Q. Xu, Y. Chen, C. Li, M. Li, J. M. Guerrero, A. Luo. "Reactive Power Strategy of Cascaded Deltaconnected STATCOM Under Asymmetrical Voltage Conditions.! *I E E E Journal of Emerging and Selected Topics in Power Electronics*, 5(2), 784 - 795 2017. https://doi.org/10.1109/JESTPE.2017.2649683.

Examination and Analysis of the Energy Efficiency of Pesa Swing 122NaSF2 Tram Types

Georgi Pavlov, Lyubomir Sekulov
Department of Power Supply and Electrical Equipment of Transport
Todor Kableshkov University of Transport
Sofia, Bulgaria
gpavlov@vtu.bg, lsekulov@vtu.bg

***Abstract* — proposed in the present paper is a comparative analysis of the measured energy regained through recuperation and a simulated model of a specific section of the tram route system in Sofia. The relations between the tram car traffic scheduling and the recuperated energy used have been accounted for.**

***Keywords* — trams, recuperation, urban transport, system, energy efficiency, traffic mode**

I. Introduction

The paper focuses exclusively on Pesa Swing 122NaSF2 tram types. Two models of PESA SWING trams operate in the city of Sofia, with the first type having been brought into service since 2013, and the second one - since 2016. On the whole, the total number of trams amounts to 24. Technically speaking, they provide public transportation services along two main routes in Sofia, tram lines 7 and 18. The lines differ in their route lengths, stop spacing, surface slopes and vehicle stoppages and downtime, as determined by the distinctive features of the urban infrastructure. Presented in the paper are some concrete details as to the parameters of the two tram routes, obtained as a result of an accomplished experimental research study.

The surveyed tramcars use alternating current asynchronous motors, composed of four individual traction motors within the range of 105 kW. The traction motors employ two double traction inverters 2xFT105-600, one for each bogie construction. The trams under consideration run at a maximum speed of 70 km/h, have the capability to overcome an inclined surface, the maximum gradient of which is 60 ‰, with a track gauge of 1009 mm and permissible kerb (without passengers) weight of 20 000kg.

The air conditioning system in the lounge area consists of two pieces of HVAC 6406 equipment, each of which has a rated capacity of up to 24 kW and 17 1000W heaters. As for the tram driver's cabin, there is a separate HVAC 3405 3-6kW unit and a 5400W heater. The power for their own consumption is provided by two PSM36-TI static converters with 3x400V/25kW and 24VDC/10kW output voltage range.

The experimental studies were performed under varied operating conditions and across all main modes of operation (traction and electric braking) of the two tram models hereto discussed. Considered carefully are also the parameters indicative of their energy efficiency – specific and relative energy consumption in the mode of traction, recuperative braking mode efficiency, as well as the power consumption for their own needs, heating and air conditioning. The values obtained from the performance measurements and their subsequent processing were closely examined to put forward a set of specific recommendations with regard to some practical opportunities for improved energy efficiency of the surveyed passenger vehicles.

II. Brief Overview of the Service Route and Technical Characteristics of the Surveyed Tramway

The examined tramway is served by tram line №18 and is part of the transportation system connecting Zhurnalist Sq and Orlandovtsi Quarter running through the central part of the city of Sofia. The route is clearly distinguished by two distinct sections with restricted motor vehicle traffic, which, in turn, empowers the trains to accelerate at higher speeds and promotes regenerative braking efficiency (in the mode of recuperation). Represented schematically in Fig.1 is the route of the tram line №18 and the corresponding feeding points of power supply.

Figure 1 The route of Sofia tram line №18 and its power supply feeding points

The length of the tramway from Orlandovtsi Quarter to Zhurnalst Sq. is 5551m, with the opposite direction being 5541m. The positive denivelation in the direction to Zhurnalist Sq. is 49m, and the negative one is 28m. Depicted in Fig. 1 is the tram line's power supply, accomplished by 4 rectifying or converting stations (TIS), RS Veslets,

RS Perlovets, RS Vazrazhdane, RS Yunak, with discernible outline of the numbers of the feeders for each section and the numbers of the sectional disconnectors. The study takes due consideration of the parameters of the feeder cables, as well as their length. The power supply of the traction network is ensured by the four converting stations (TISa), and is divided into 9 sections. Examined further in the experimental study are also the parameters of the traction contact DC network (CN), which within the section itself can be of two types – simple and chain.

The four converting stations supplying power to the surveyed route were constructed upon a three-phase six-pulse rectifier circuit with its well-proven lower power-conversion efficiency relative to the new and modernized RS with a twelve-pulse rectifier circuit. Hence, the energy consumed is of low quality and exerts a negative impact upon the traction currents of the other consumers. The measurement taken during the experimental research conducted in RS Veslets and Dimitrovska substantiates conclusively the poor energy behavior of the six-pulse rectifiers. The schematic diagram of such a rectifier station (RS), in operation in Bulgarian older substations for continuous power supply is depicted in Fig. 2.

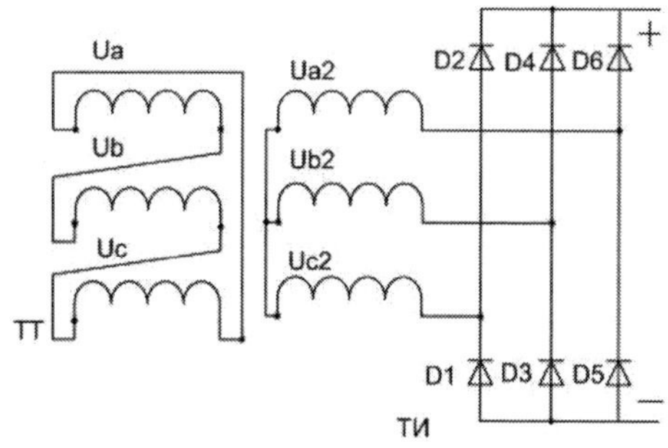

Figure 2 Three-phase six-pulse circuit of a rectifier station

The three-phase RS comprises a three-phase power transformer and a rectifier with six diodes D1÷D6 (Larionov's circuit). The traction transformer primary winding is connected by a „triangle", and the secondary one by a „star". Each terminal of the secondary winding is connected to one diode of the anode group and one diode of the cathode group. The diodes in the traction rectifiers are arranged into anode and cathode groups, with the cathode group represented by the plus symbol of the rectifier, and the anode group – by the minus. One of the most effective ways for the RS to improve the technical and economic performance indicators of the DC traction power supply system is the use of more complex rectifier circuits with larger numbers of the pulses in the rectified voltage curves [2].

Illustrated in Fig.3 is the tramway profile and the speed of movement along the route in the direction to Orlandovtsi Quarter with the speed values being approximated and carefully defined by a substantial number of measurements. The tramway profile is shown in green and the speed-in blue. In outline, the average speed of movement in the direction to Orlandovtsi Quarter is 12.73km/h with the maximum one reaching 51.72km/h, and the completed instances of stoppages equal to 19.

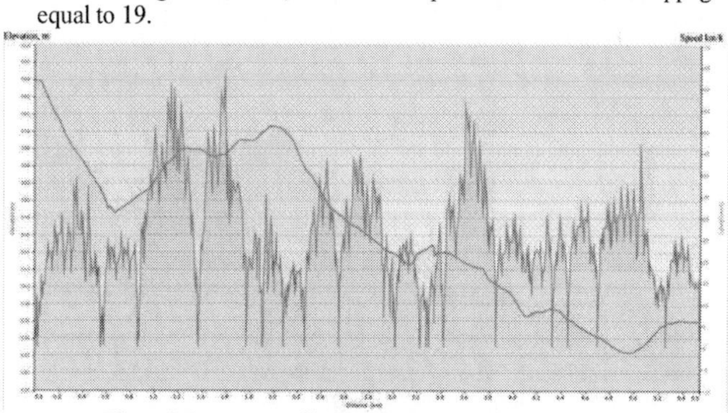

Figure 3 Tramway profile and tram speeds of movement towards Orlandovtsi Quarter.

Presented in Fig.4 is the movement speed and the tramway profile towards Zhurnalist Sq.

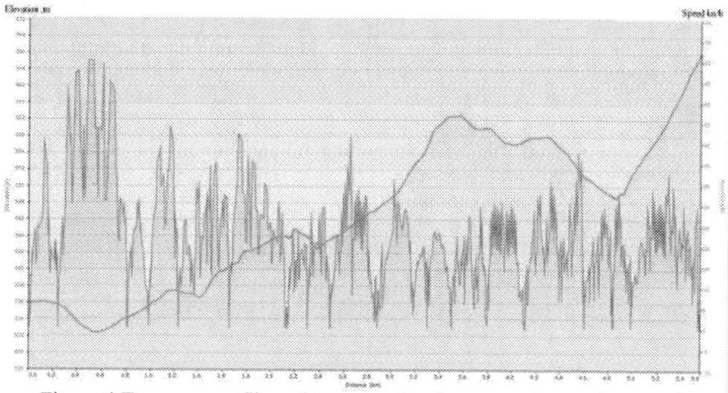

Figure 4 Tramway profile and tram speeds of movement towards towards Zhurnalist Sq.

The number of the tram stops in both directions inclusive of the last ones is 14 altogether. The average speed of movement in the direction to Zhurnalist Sq. is 11.49km/h, with the maximum one being 48.78 км/h , and the number of completed stoppages - 18.

The main parameters determining the mode of traction and ensuring minimal energy consumption are conditioned by the effective increase in speed V_p, dependent on the length and profile of the tramway, the number and type of the intersections, the stops, traffic light modes of control, the pedestrian crossings, and others. This parameter is computed by means of some traction calculations and by the formula:

$$V_p = -137{,}88.\, l_s^2 + 158{,}76.\, l_s + 6{,}1 \qquad (1)$$

Where l_s- the length of movement, km;

V_p – rational acceleration speed, km/h.

If $l_s > 0{,}5\ km$, then the assumed $V_p \approx 55\ km/h$.

Further optimization estimates were made in order to select rational speeds of acceleration for a definite length of the section between those two stops where the lowest electricity consumption is obtained. Recorded, accordingly, in Table 1 are some statistical acceleration speeds, determined experimentally, on the basis of the length of the tramway, for a movement length of $l_s \leq 0{,}5\ km$ and an equivalent gradient of ($i_e \leq 5‰$).

Table 1 Statistical acceleration speeds

V_p, km/h	19.9	27.3	32.7	37.4	37.8	45	47.5	48.6	51.1
, km	0.1	0.15	0.2	0.25	0.3	0.35	0.4	0.45	0.5

In the ground-based urban electric transportation (GUET) when vast numbers of electric vehicles (EV) are connected to a single power distribution network, there is a strong likelihood for their modes of traction and recuperation to coincide. It becomes quite possible at that moment for the energy released in the CN by the EV in recuperation to be consumed by those in mode of traction. Nowadays, considering the level of the emerging technologies in transport engineering and their possibilities for recuperation, the efficiency of instantaneous utilisation of regenerative energy should be sought in the other subsystems. Such a system, for instance, is the power supply system, whose load tolerance is principally determined by the location of all EVs within a single feeder or substation zone, the power they consume, etc. Optimal solution might, therefore, be reached, if the research study proceeds from real processes, taking notice of all decisive factors of substantial impact rather than from idealized processes occurring under the influence of constant loads. [5]

If this is the case, then the mathematical description of the EV movement is obtained by incorporating the fixed and random factors as well, as constituent parts of the process of movement. Formulated, in that framework, is an equation for EV movement, of the type:

$$F_{tr} = (1+\gamma).\, V.\frac{dV}{ds} \qquad (2)$$

where: F_{tr}- force acting upon the EV; V –speed; $(1+\gamma)$ – the coefficient of inertia, S – the distance travelled.[4]

The work performed by the EV while in motion in mode of traction is established as the sum of the amount of work undertaken to overcome the resistance to movement and the accumulation of kinetic and sometimes potential energy in the EV, which can be expressed along the lines of:

$$A_{tr} = \int_0^{S_i}. G.(w_0 - w_i).dS + \frac{102.G.(1+\gamma).v_b^2}{2.3,6^2} \quad (3)$$

where S_i is the distance travelled by the vehicle during traction; G – the weight of the vehicle, kN; w_0, w_i– respectively, the specific intrinsic and additional resistance to movement, N/ kN;Vb – the speed at the very moment of switching off the traction motors, km/h.

The movement, however, continues even with the shutting down of traction motors, in view of the already accumulated kinetic energy, consumed on overcoming the resistance to movement in motion of inertia or extinguished by the EV 's braking systems during braking. The EV 's mode of movement is graphically depicted in Fig. 5.

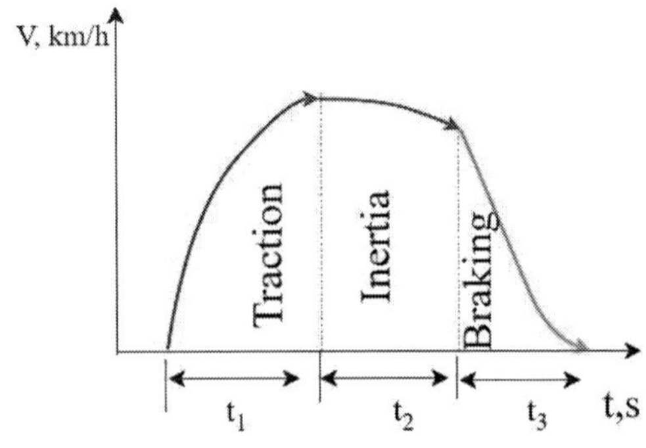

Figure 5 Electric vehicles mode of movement

The braking energy in the mode of recuperation is computed in kWh according to the following dependency:

$$E_r' = 1{,}073.G.(V_s^2 - V_f^2).(1+\gamma).10^{-5}.\left(1 - \frac{\omega_0 - i_{cp}}{102(1+\gamma)\alpha}\right).\eta \quad (4)$$

Where:

V_s and V_f are the starting and finishing speeds when braking;

G mass of the EV, t;

α EV's delay, m/s2;

η – the EV efficiency during recuperation, approximately 0,85.

The EV's delay is calculated through:

$$\alpha = \frac{\xi}{1+\gamma}.(i_m - \omega_0 - b) \quad (5)$$

where: ξ = 0,00981 is transfer coefficient; b is a specific value of the regenerative braking torque, relative to the EV's mass N/kN, i_m mean inclination, ‰.[4]

The tram intrinsic resistance to movement was obtained experimentally by the following dependency:

$$\omega_0' = 1{,}5 + \frac{8+0.038.v^2}{G} \quad (6)$$

The loss of energy in the braking resistors can also be determined by the kinetic energy gained at the end of the process of braking, provided that account is taken of the energy expenditures for overcoming the resistance to movement, and including the power for own consumption as well.

The simulations of the mathematical model comprise concrete mechanical and electrical parameters, the specific characteristics of the surveyed tram trains, the profile of the tramway and the speed of movement, the tramway length, the CN's resistance and the individual parts of the sections. Adopted are the parameters of those tram trains which operate exclusively on these lines and run through the particular section under study. Namely, these are Pesa Swing 122NaSF2 tram type models. [3]

The parts of the sections, pertaining to the surveyed tramway, with the most likely occurrence of effective recuperation, are determined through formula 1 and Table 1, accounting for the speeds of movement and the parameters of the tramway itself. Travelling in the direction to Orlandovtsi Quarter, these are mostly inter-stop sections: Zhurnalist Sq. stop –UACEG stop (400m), UACEG stop - Vassil Levski Blvd stop (400m), Vassil Levski Blvd stop- Slaveykov Sq. stop (700m), Kozloduy Str. stop - Dunav Str. stop (400m), Dunav Str. stop - Hladilen Zavod stop (700m), Hladilen Zavod stop – CSC stop (500m), CSC stop - Catholic Cemetery Park stop (400m), Catholic Cemetery Park stop – Orlandovtsi Quarter stop. (400m). The respective streets forming the above-named inter-stop sections are: Hristo Smirnenski Blvd., Graf Ignatiev Str. and Kamenodelska Str. Sequentially, the first section along Hristo Smirnenski Blvd. is powered by RS Yunak, the second section along Graf Ignatiev Str. - by RS Perlovets, and the third one along Kamenodelska Str. - by RS Veslets.

The inter-stop section Zhurnalist Sq. stop – UACEG stop (400m) has an equivalent gradient of $i_e = -23‰$, which defines the section as being effective for efficient recuperation in this particular direction of movement where the rational acceleration speed V_p might be different and lower than the one reported in Table 1.

Alternatively, the sections for efficient recuperation in the opposite direction, that is towards Zhurnalist Sq. are between the stops Orlandovtsi Quarter - Catholic Cemetery Park stop (400m), Catholic Cemetery Park stop - CSC stop (400m), CSC stop –Hladilen Zavod stop (500m), which are located along Kamenodelska Str. and are powered by RS Veslets.

MATLAB-based simulation modelling of the tram train movement along the route of line №18 is constructed through the use of the mathematical model hereto discussed, and the results obtained were duly analysed but do not come into the focus of the present report.

III. Principal Results from the Conducted Experimental Research Studies

Taken were the readings from 13 tram trains of Pesa Swing 122NaSF2 type for the period from March 11th, 2022 to March 13th, 2022, with inventory № within the range of 2326-2338, keeping in mind that not all of them were servicing line №18. The average daily temperature for the period under discussion is – 4.4 C⁰. The surveyed tram trains had different length of movement, different numbers of completed courses of service and different business mileage rates prior to their actual service over the line in consideration , and even operated by different tram-drivers.

Represented in a tabular form are the major results from the performance measurements of the 13-th tram trains for the three days from March 10th , 2022 to March 13th , 2022. It should be also pointed out that on March 13th (Saturday) the line №18 was serviced by tram trains with inv. №№2326, 2331, 2332, 2337. The measurements themselves refer to the length of movement, the total energy consumption, power for own consumption, energy for traction needs and recuperated to traction power as recorded in Table 2.

Table 2 Measurement results for Pesa Swing 122NaSF2 tram trains

10.03.2022 r.	2326	2327	2328	2329	2330	2331	2332	2333	2334	2335	2336	2337	2338
Length of movement,km	168	37	0	0	96	277	103	151	0	167	0	187	277
Total energy consumption,kWh	1036	283	0	0	646	1692	645	1045	0	826	0	1127	1707
Power for own consumption,kWh	692	192	0	0	443	1068	431	669	0	428	0	717	934
Traction power,kWh	344	72	0	0	202	623	215	377	0	398	0	411	573
Recuperated power,kWh	54	9	0	0	26	111	26	82	0	113	0	74	111

11.03.2022 r.	2326	2327	2328	2329	2330	2331	2332	2333	2334	2335	2336	2337	2338
Length of movement,km	101	164	169	0	260	170	183	0	0	99	0	284	159
Total energy consumption,kWh	698	1105	1148	0	1782	1059	1250	0	0	534	0	1790	1084
Power for own consumption,kWh	484	727	742	0	1177	654	782	0	0	289	0	1184	693
Traction power,kWh	215	379	406	0	605	405	469	0	0	245	0	607	391
Recuperated power,kWh	32	70	78	0	111	78	120	0	0	69	0	93	67

12.03.2022 r.	2326	2327	2328	2329	2330	2331	2332	2333	2334	2335	2336	2337	2338
Length of movement,km	280	29	1	173	0	186	289	1	0	99	0	190	1
Total energy consumption,kWh	1680	150	4	1264	2	1176	1883	2	4	459	231	1166	1
Power for own consumption,kWh	794	72	3	801	1	721	1188	0	3	233	232	736	1
Traction power,kWh	586	79	1	463	1	455	693	1	0	226	0	430	1
Recuperated power,kWh	101	18	0	97	1	107	144	0	0	60	0	84	0

In accordance with the measurement results obtained, performed are further calculations as to the relative consumption of the total energy consumed and the ratio of the recuperated to traction power, with the findings being reported in Table 3[6].

Table 3 Calculation results

	2326	2327	2328	2329	2330	2331	2332	2333	2334	2335	2336	2337	2338
10.03.2022 г.													
Relative consumption kWh/km	6.1667	7.6486	0	0	6.729	6.108	6.2621	6.9205	0	4.946	0	6.0267	6.362
Relative traction consumption kWh/km	2.0476	1.9459	0	0	2.104	2.249	2.0874	2.4967	0	2.383	0	2.1979	2.069
Recuperated to traction power	0.157	0.125	0	0	0.129	0.178	0.1209	0.2175	0	0.284	0	0.18	0.194
11.03.2022 г.													
Relative consumption kWh/km	6.9109	6.7378	6.793	0	6.854	6.229	6.8306	0	0	5.394	0	6.3028	6.818
Relative traction consumption kWh/km	2.1287	2.311	2.402	0	2.327	2.382	2.5628	0	0	2.475	0	2.1171	2.459
Recuperated to traction power	0.1488	0.1847	0.192	0	0.183	0.193	0.2559	0	0	0.282	0	0.1532	0.171
12.03.2022 г.													
Relative consumption kWh/km	6	5.1724	0	7.306	0	6.323	6.5156	0	0	4.656	0	6.1368	0
Relative traction consumption kWh/km	2.0929	2.7241	0	2.676	0	2.446	2.3979	0	0	2.283	0	2.2632	0
Recuperated to traction power	0.1724	0.2278	0	0.21	0	0.235	0.2078	0	0	0.265	0	0.1953	0

The trend of the ratio of the relative consumption of the total energy with respect to the percentage of the recuperated power is outlined in Fig. 6.

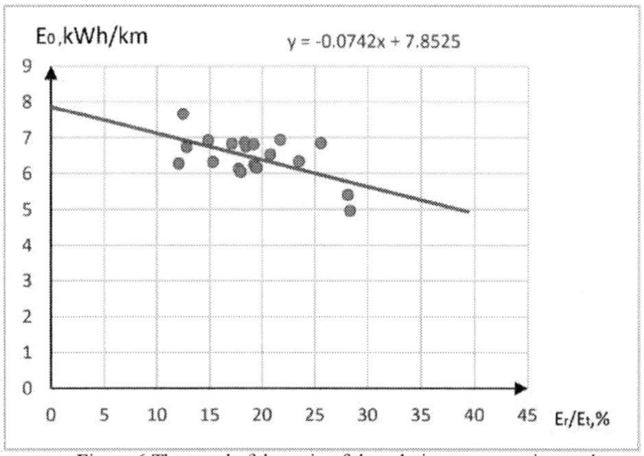

Figure 6 The trend of the ratio of the relative consumption to the percentage of the recuperated power

The dependency between the relative consumption of the total energy and the ratio of the recuperated to traction power for the Pesa Swing 122NaSF2 tram trains along the route of Sofia №18 tram line, as of today, is computed via the formula:

$$E_o = -0.0742.\frac{E_R}{E_t} + 7.8525\,,\frac{\text{kWh}}{\text{km}} \qquad (7)$$

, where E_o is the relative consumption of the total EV energy, E_R recuperated power, E_t is the traction power [1].

IV. Conclusion

It follows, from the measurements performed, that there seems to be a dependency between the total energy consumption and the recuperative braking energy. As for the tram trains, the underlying conclusion is that the lower the relative energy consumption, the higher the ratio of the recuperated to traction power. The energy efficiency of the vehicles may be considerably increased, provided that the tram-driver's style of driving is fully consistent with the parameters of the route itself. As clearly indicated in Table 6 and Fig.6, the percentage of recuperative braking is likely to reach up to 28% with a more efficient control of the respective means of transportation.The recuperative power relative to that of traction averages to 17.6% on weekdays and 20.26% at weekends and public holidays, which is assumed from the higher tram trains V_P values, the greater acceleration when starting from rest, due to the lack of heavy traffic along that route. It should be noted, however, that the weekend/holiday percentage would be considerably higher, if the tram trains didn't run on special infrequent time schedules.

The relative consumption of the consumed total tram train energy, at minimum efficiency with no recuperation, approximates 7.8 kWh/km, while that with 40% efficiently recuperated power amounts to an average of 4.8 kWh/km. Hence, the highest relative consumption of 7.65kWh/km corresponds to 13% recuperated to traction power, and that of the lowest 4.95 kWh/km values to 28%. The deviations from the extreme values reported herein, subject to the complex factors affecting the performance efficiency as discussed throughout the paper, might, in practice, reach up to 25% for the tramway under consideration. It is also worth mentioning that the results were obtained at daily ambient temperature mean of – 4.4C^0.

The average relative consumption of traction power is 2.19 kWh/km.

The existent methodology prescribing the correct experimental study procedure in assessing the personal qualities of the individual tram drivers requires a more prolonged and extensive research, which is currently in progress.

Indeed, improving the energy efficiency of urban EVs by a certain percentage is certain to bring about a similar reduction in the levels of environmental pollution. This is quite a powerful incentive propelling forwards our research activities towards that goal.

Acknowledgment

This work was supported by by the Bulgarian Scientific Fund, Operational Programme "Fundamental Research", Session 2019 under contract №06-37/28 Date: 16.12.2019. „Research on Electricity Consumption Regimes in Electric Power Supply Systems for Urban Electric Transport with Two-way Power Flow".

References

[1] I. Petrov, G. Dimitrov, T. Lalev. "Experimental study of the energy efficiency of the metro trains of the capital metro in real operating conditions" *Yearbook of the Technical University-Sofia*, volume 67, book 1, p.113, 2017.

[2] M. Tomcheva. "Research and analysis of the parameters of the types of rectifier circuits, operating in rectifier stations." *Scientific Journal of Transport Communications*, volume 16, issue 3/2, 2018, vol № 1730, ISSN 1312-3823.

[3] PESA 120 tram train technical documentation. Unpublished.

[4] D. V. Ustimenko. „Rational modes of recovery of electric locomotives UDC 621.337.522." *Unpublished Report. Ukraine, Dnepropetrovsk, National University of Railway Transport named after Academician V. Lazaryan.*

[5] N.I. Shchurov. „Methods of increasing the efficiency of the use of electric energy in the subsystem of electric transport." *Bulletin of the Krasnoyarsk State Technical University. Transport / Ed. by V.N. Katargin. Krasnoyarsk: KSTU.* 2002. Issue 29.-C.101 – 105.

[6] V. Gyurov, N. Bezhanov. „Study on electric consumption regimes in rectifier stations of trolleybus transport." *Proceedings of the 11th Electrical Engineering Faculty Conference (BulEF)*, 2019, Varna, pp.1-4, ISBN: 978-1-7281-2698-2.

[7] V. Gyurov, N. Bezhanov. „Possibilities for energy planning in electric power supply systems of urban electric transport." *Proceedings of the 11th Electrical engineering faculty conference (BulEF)*, 2019, Varna, pp.1-6, ISBN: 978-1-7281-2698-2.

[8] M. Tomcheva, I. Tarpov. „Study of the Power Modes of Skoda Solaris 26Tr and 27Tr Trolleybuses." *2021 17th Conference on Electrical Machines, Drives and Power Systems (ELMA), IEEE Xplore*, 2021, doi: 10.1109/ELMA52514.2021.9503018.

Degradation of Monocrystalline PV Panels Differences between Ukrainian and Czech Conditions

Milan Belik
Department of Electrical Power Engineering
University of West Bohemia
Plzen, Czech republic
belik4@fel.zcu.cz

Olena Rubanenko
Department of Electric Stations and Systems
Vinnytsia National Technical University
Vinnytsya, Ukraine
olenarubanenko@ukr.net
mailto:olenarubanenko@ukr.net

Abstract — **The purpose of our research is to compare the influence of meteorological factors and technical conditions of PV plants on power generation. The issue is studied on PV plants in Ukraine and the Czech Republic. A method and model for assessing the technical condition of PV plants during their operation in the balancing group have been developed. The Fault tree analysis (FTA) method developed a photovoltaic module damage tree. The main defects of photovoltaic modules operated at PV plants are determined. A method for determining the coefficient of the total residual resource of PV to clarify the actual generation of PV plants taking into account their degradation is proposed. A model of the software complex for determining the technical condition of PV plants equipment has been developed. Indicators coefficients anomalous work of PV modules are offered. Principal results are measurements on monocrystalline Si PV panels used in commercial plants in Ukraine and Czech Republic. Measurements show significant differences in operation behavior and structural conditions and prove that it is necessary to assess the technical condition of PV plants, in particular the degradation of used photovoltaic modules, because during operation there is a gradual decrease in the actual electricity generation of plants compared to their initial value and depending on local conditions. It may be related to the operation of the technical resource of the PV modules, their technological shortage, as well as other operational reasons. Failure to take into account the gradual reduction of capacity leads to errors in assessing their energy efficiency, profitability, payback time, power reserve to compensate the unstable RES generation, and, very important today, hourly forecasting of electricity generation the next period.**

Keywords: technical conditions; PV plant; fault tree analysis

I. Introduction

This paper is shown that it is necessary to assess the technical condition of PV plants, in particular the degradation of their photovoltaic modules. During operation, there is a gradual decrease in the actual electricity generation of PV plants compared to their initial value. It may be related to the operation of the technical resource of the PV modules, their technological shortage, as well as other operational reasons. Failure to take into account the gradual reduction of PV modules capacity leads to errors in assessing their energy efficiency, profitability, payback time, power reserve to compensate for unstable RES generation, and, most importantly today, hourly forecasting of electricity generation for the next day in the power balance. Therefore, in order to more accurately determine the actual generation of PV plants, it is necessary to develop and improve appropriate methods, mathematical models, and tools. The first step for this is the model of the software and hardware complex developed in this paper, which serves to form a database to identify the dependences of the actual generation of electricity PV plants on their technical condition, meteorological conditions, and operating conditions [1, 2].

Peculiarities of RES functioning in modern conditions of decarbonization strategy implementation are investigated. The problems of balancing power and electricity in the power system are studied, the basic technologies of ensuring balance are offered. The influence of meteorological factors and technical conditions on electricity generation and participation of PV plants in balancing power plants modes is studied. A method and model for assessing the technical condition of PV plants during their work in the balancing group have been developed. The main defects of PV modules are operated on PV plants, are determined. A method for determining the coefficient of the total residual resource of PV plants to clarify the actual generation of PV power taking into account their degradation is proposed. A model of the software and hardware complex for determining the technical condition of the PV module has been developed. [1, 3]

As a result of the introduction of methods and tools for assessing the degradation of PV plants, their projected and actual generation in the balancing system of PS regimes will be clarified, which will allow more accurate planning of reserve capacity in the system and reduce penalties for PV plants investors [2, 3].

II. Degradadion And Faults Of Photovoltaic System

Every photovoltaic system fault is an undesirable condition when the power plant or its part cannot operate optimally, in the extreme case it cannot supply energy at all. Faults and the degradation as well need to be identified as soon as possible, because when the power plant is out of power, economical profit is decreased and the payback period is extended. It is ideal to prevent faults by designing the system correctly, selecting quality components, installing them correctly and using advanced protective and monitoring elements. In most significant cases, personals may be injured, or also the fire can be initiated [4, 5].

The influence of particular degradation or fault is non-equal. Final influence can best be assessed by its impact on overall energy supply. For example, a damaged inverter is more serious fault than a malfunction of particular panel in affected string. Fig. 1 demonstrates contradictory directions of degradation influence and the degradation detection [4].

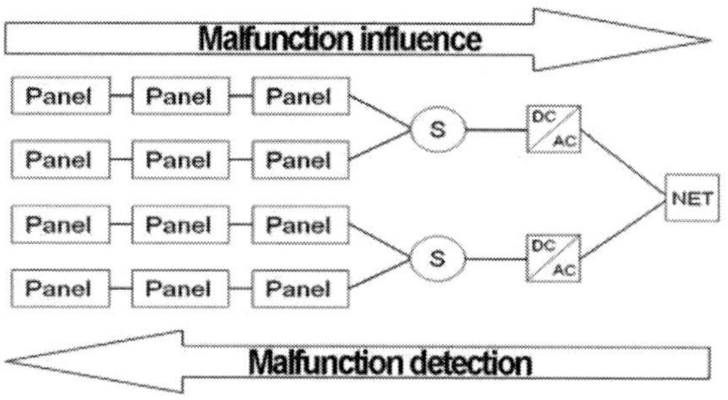

Figure 1. Influence of the fault on PV system

Table 1 summarizes particular types of degradation or fault accompanying panels. This table originates from fundamental research made between 2009 and 2010 at the end of solar boom in Czech Republic. That time, large amount of photovoltaic installations was rapidly commissioned and many investors were forced to use improper components with discussable parameters and lifetime and also many projects were designed non-professionally. [3, 5].

Table 1. Basic degradation and faults

	detection/behaviour	reason	remedy
shading	changes in VA characteristic, low power, visual check	obstacles in surroundings	remove obstacles
delamination	changes in VA characteristic, low power	insufficient material or fabrication	replacement
"bubbles" in module	changes in VA characteristic, low power	material problem	replacement
low transparency of protecting glass	visual check, low power	material problem	replacement
humidity in panel	visual check, low power, high Rs	insufficient material or fabrication	replacement
aerial pollution on the panel	low power, visual check	dust, snow, pollen	cleaning
moss, races, birds excrements	changes in VA characteristic, low power, visual check		cleaning
broken glass	visual check	hailstorm, improper operation	replacement
broken cells	changes in VA characteristic, low power	mechanical damage	replacement
improper module connection	high Rs, low power	improper assembly	replacement
reversed or broken bypass diode	power of string decreased for the power of particular panel	improper assembly	repair/replacement of the diode
missing or defective bypass diode	power of string decreased for the power of particular panel	improper assembly	repair/replacement of the diode
module mismatch	low power of module	improper assembly	replacement
shortcut between cells	low power, low open circuit voltage	improper assembly	replacement

Similar dramatic situation has been in recent time in Ukraine. The energy network is very obsolete, many generating stations and substations are at the end or even over their effective lifetime. The power industry suffers with huge lack of energy coal since beginning of summer 2021. Around 20 coal burning power stations have been out of operation during last 8 months. The country tries to solve their energy problems with wider usage of particular renewable energy sources. Many new photovoltaic installations of various quality are being commissioned and correspond with the same problems like during Czech solar boom. This research was made before the war in Ukraine and it could be expected that after the war the situation will have similar progress [2, 5].

Reliability of particular PV panels can be calculated from reliabilities of particular components according to equation 1. Using results from concerned Czech and Ukrainian PV plants we can expect final monocrystalline panel reliability 0,996 [3].

$$R_p = R_{ps} \cdot R_{pr} \cdot R_{pv} \cdot R_{pdl} = 0{,}996 \quad (1)$$

where:

$$R_{ps} = 1 - Q_{ps} = 0{,}999 \quad (2)$$

$$R_{pr} = 1 - Q_{pr} = 0{,}999 \quad (3)$$

$$R_{pv} = 1 - Q_{pv} = 0{,}999 \quad (4)$$

$$R_{pdl} = 1 - Q_{pdl} = 0{,}999 \quad (5)$$

O_{ps} means probability of broken glass, O_{pr} probability of damaged frame, O_{pv} probability of decreased power and O_{pdl} probability of delamination of particular panel [4].

In real situation, the reliability of particular panels can be retrospectively influenced by degradation or defects of surrounding components. Fig. 2 demonstrates serious damage of panels caused by malfunction of mounting system. Loosen mounting profiles caused during uncontrolled fall significant scratches in the covering surface leading to replacement of all affected panels [2, 4].

Figure 2. Panels damage from mounting system malfunction

Table 2 shows malfunctions of related components that influence the operation of entire PV plant and particular panels as well. As evident from Fig. 2 also failures of non-electrical components can have in result serious influence on electrical parameters of electrical components [1].

Table 2. Malfunctions of related components

	2009	2010
group „inverter“	39	21
group „distribution box“	3	3
group „protections“	2	1
group „monitoring“	4	3
group „mountings“	16	6
group „cables“	9	3

Fig. 3 illustrates serious degradation of mounting system used in 2 MW PV Czech plant. This example shows earlier version of wooden frames suffering with huge putrefaction. It is the result of improper relative position of particular panels, respectively position of the panel joint, above horizontal and vertical girder. That position caused increased leaking during rains and snow melting. This issue was identified after 1,5 year operation of Stage 1 and led to design modification of the mounting system used in further stages. Position shift for 12 cm solved this problem. The original construction was not redesigned and requires regular maintenance leading to relative high operational costs [3].

Figure 3. Result of improper mounting

III. Comparison of PV Plants Operated in Czech Republic and Ukraine

During the last years of Czech solar boom, selected new commissioned power plants were investigated for initial degradation and faults. In year 2009 in total 308 operated PV plants with installed power between 0,004 to 8 MW and during 2010 another 44 plants were analyzed. Results are displayed on Fig. 4.

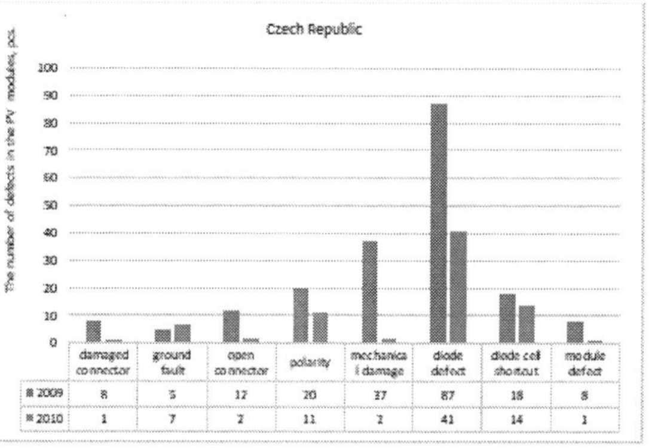

Figure 4. Particular defects (Czech Republic, 2009-2010)

Fig.5 presents the same analysis made during 2020 on 98 PV plants and during 2021 on another 111 stations installed in Ukraine.

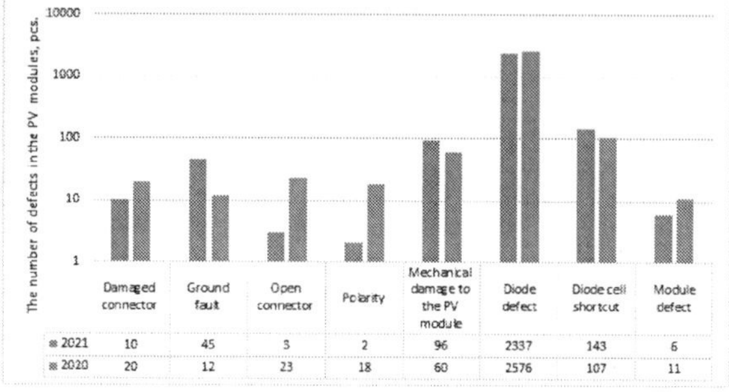

Figure 5. Particular defects (Ukraine, 2020-2021)

It is evident, that relative amount of particular defects and degradations in both events has differs. It is the result of wider usage of modern cheap panels mostly originating from China production. Although these panels have usually better nominal values than older panels produced between 2009 and 2010, their reliability is lower. Table 3 shows typical structure of malfunction for one producer. In total 374246 Jinko Solar panels were tested. 561 various problems were identified in 2020 and 410 during 2021 analysis.

In accordance with the Law of Ukraine on the Electricity Market, the permissible deviation of the projected generation schedule from the actual generation for photovoltaic plants is 5 %. If the production differs for more than 5 % the producer must pay penalties. After studying the forecast of meteorological parameters and improving the interpretation of the obtained data of the generation schedule, there is a significant discrepancy between the forecast and the actual schedule of electricity generation. This phenomenon can also be explained by a large number of technological violations due to the failure of the main electrical equipment of the PV plants, namely power transformers, inverters, cable lines, and so on.

During the operation of the studied PV plants, a detailed analysis of technological violations and the reasons for the deviation of the projected schedule of electricity generation from the actual. During the period of six months at one of the studied photovoltaic power plants (PV plants) in Ukraine, there were four technological violations due to the failure of power

TABLE 3. MALFUNCTIONS OF SAMPLE PRODUCER (JINKO SOLAR)

№	Type	Amount	2020					2021				
			Defect of the diode	Short circuit of cells of the PV module	Mechanical damage to PV module	Melting of a connecting box	Defect of the module	Defect of the diode	Short circuit of cells of the PV module (cells)	Mechanical damage to PV module	Melting of a connecting box	Defect of the module
PV 1	JKM265PP-60	31042	46					26				
PV 2	JKM265PP-60	37048	185	2	5			103				
PV 3	JKM265PP-60	37048	45	4				47	1		2	
PV 4	JKM270PP-60	22220	6		1			13		3		
PV 5	JKM270PP-60	18876	8					7				
PV 6	JKM280M-60	65340	16	13				20	2			
PV 7	JKM340M-72	38232	5					1				
PV 8	JKM270PP-60	23056	119					121		1	2	
PV 9	JKM330PP-72-V	22088	2					2				
PV 10	JKM320M-60-V	19320	81				1	6	1			
PV 11	JKM375M-72-V	34272	21					50				
PV 12	JKM-335M-72-V	25704	1					1	1			
	Sum	374246	535	19	6	0	1	397	5	4	4	0

transformers with a capacity of 2800 kVA. Each technological disruption occurred at different times and under different network conditions, which are in no way related to a single event. Due to a large number of gradual damage to the same type of equipment within one facility, will be necessary analysis of the network during damage, testing, and measurement of electrical equipment in the data and adjacent circuits, as well as comprehensive testing of grounding device and testing of transformers.

IV. Fault Tree Analysis

The functioning of the PV plant as a complex system is influenced by a number of factors explained earlier. FTA of a common PV power plant is designed to take into account the data from Fig. 4, Fig. 5 and Table 3. Structure of this FTA is presented on Fig. 6 [6, 8].

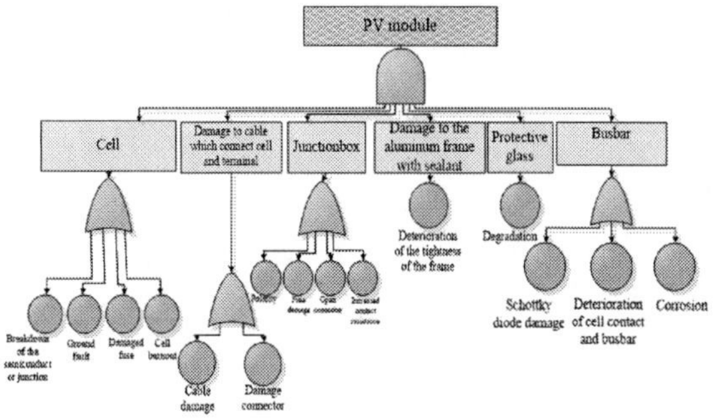

Figure 6. PV plant fault tree analysis

That involves determining the technical condition of the PV module, electrical components and other parts. Detected features allow neuro-fuzzy modeling and expert evaluation in particular and to develop a neuro-fuzzy model of the residual life of the PV module. This model can be used to identify the PV module possible "emergency", "pre-emergency", "with minor deviations of parameters", "serviceable" statement and to predict future generation of the plant [6, 7].

Firstly, we need to define residual resources coefficients for each diagnostic parameter for every PVM element. It is necessary to follow from known values of diagnostic parameters to the corresponding values of RRI for each diagnostic parameter. Therefore, RRI for each diagnostic parameter can be mathematically defined as follows (6):

$$k_{i1} = \left| \frac{x_{i1,lim} - x_{i1,current}}{x_{i1,lim} - x_{i1,start}} \right| \qquad (6)$$

where $x_{i1,lim}$ is limit value of the i_{th} diagnostic parameter; $x_{i1,current}$ is the value of the i_{th} diagnostic parameter at the time of detection; $x_{i1,start}$ is the start value of the i_{th} diagnostic parameter at the time of commissioning of new unit or after major or general repair [1].

If we take into the account, that the failure or degradation of particular elements from Table 3 results in a failure of the PV module, then the total RRI of PV modules is calculated using equation (7):

$$k_{TRR} = \prod_{i=1}^{j} k_i^{P_i}. \qquad (7)$$

where the total RRI of the PV module is evaluated on the i_{th} diagnostic parameter. Many other ambient parameter (temperature, insulation, resistance, current, etc.) affects the RRI of the PV module elements as evident in the equation (8). Coefficients k_1-k_6 are RRI of every element that influences the total RRI of the PV module. Number i is i_{th} diagnostic parameter of elements of PV module, j is the number of diagnostic parameters and p_i is the probability of deviations of the controlled parameter from the allowable limit value [8].

$$p_i = \frac{y_i}{m}, \qquad (8)$$

where y_i is the number of deviations of the controlled parameter from the allowable boundary of normalized value of this parameter. These deviations are detected by controlling the ith diagnostic parameter from the total number of detected deviations of controlled parameters from the allowable boundary of normalized values, while m is the total number of detected deviations of controlled diagnostic parameters allowable boundary of normalized values. Expanding the expression (7), total PV module RRI becomes (9) [8]:

$$k_{tot.res} = k_1^{P_{k1}} \cdot k_2^{P_{k2}} \cdot k_3^{P_{k3}} \cdot k_4^{P_{k4}} \cdot k_5^{P_{k5}} \cdot k_6^{P_{k6}} \qquad (9)$$

where k_1, k_2, k_3, k_4, k_5, k_6 are known at the time of calculation of the values of the RRI for the relevant diagnostic parameters, and pk_1, pk_2, pk_3, pk_4, pk_5, pk_6 are the PV module node malfunction probabilities detected by measured of the diagnostic parameters taking into account total malfunction. These probabilities can be found in Table 3.

Calculated probabilities of particular malfunction or degradation issue are presented on Fig. 7 for PV plants analyzed in Czech Republic and on Fig. 8 for installations in Ukraine.

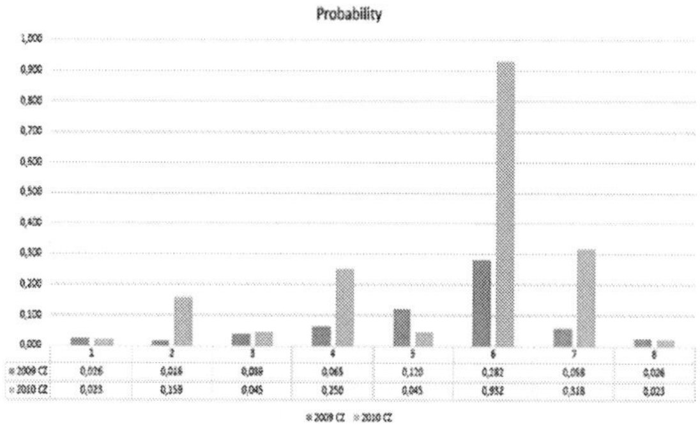

Figure 7. Malfunction probability (Czech PV plants)

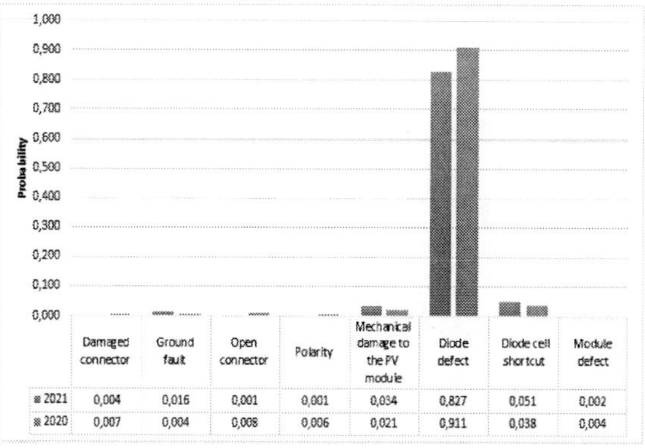

Figure 8. Malfunction probability (Ukrainian PV plants)

V. Results and Conclusions

Fault tree analysis based on neuro-fuzzy networks can be used for quick or online detection of general issue since its elementary stage. This can either prolong the lifecycle of all affected components or decrease eventual economical losses. Another interesting result can be found in better accuracy of power prediction and thus better prediction of conditions in power grid. This has been becoming more important because of still more strict conditions defining proper operation of power sources including RES.

Main problem of these analysis is the correct and accurate mathematic description of particular components. This paper is dealing only with monocrystalline Si panels whose represent the most utilized technology in nowadays PV plants.

Results of presented analyses are the base for additional further analysis. Described method will be verified on identical small experimental PV plants installed in Czech Republic and Ukraine. These PV systems are already installed, but nowadays it is not possible to verify and analyze all data from Ukrainian system. From this reason, analyses dealing with these 2 systems are not taken into account.

ACKNOWLEDGMENT

THE PAPER WAS SUPPORTED BY SGS-2021-18.

REFERENCES

[1] Fault Tree Analysis (FTA), December 2006. INTERNATIONAL IEC STANDARD 61025 Second edition 2006-12.
[2] Fault Tree Handbook, January 1981, NUREG-0492.
[3] M. Belik, O. Rubanenko. "Determination PV Module Technical Condition," Proc. 19 th International Conference on Renewable Energies and Power Quality (ICREPQ'21), RE&PQJ, Volume No.19, September 2021, pp.604–608.
[4] F. Chiacchio, F. Famoso, D. D'Urso, S. Brusca, J.I. Aizpurua, L. Cedola. "Dynamic Performance Evaluation of Photovoltaic Power Plant by Stochastic Hybrid Fault Tree Automaton Model". *Energies* 2018, *11*, 306. DOI: 10.3390/en11020306.
[5] M. Belik. "Evaluation of long term degradation process of monocrystalline Si photovoltaic panels", Proc. 18 th International Conference on Renewable Energies and Power Quality (ICREPQ'20), RE&PQJ, Volume No.18, June 2020, pp.551–555.
[6] A. Garrod, F. Barrara. "Reliability Analysis of Residential Photovoltaic Systems", Proc. International Conference on Renewable Energies and Power Quality (ICREPQ'11) Las Palmas de Gran Canaria (Spain), 13th to 15th April, 2011, pp.1197–1202.
[7] A. Sayed, M. El-Shimy, M. El-Metwally, M. Elshahed. "Reliability, Availability and Maintainability Analysis for Grid-Connected Solar Photovoltaic Systems". *Energies* 2019, *12*, 1213. DOI: 10.3390/en12071213.
[8] A. Haque, K.V.S. Bharath, M.A. Khan, I. Khan, Z.A. Jaffery. "Fault diagnosis of photovoltaic modules". Energy Sci Eng. 2019;7:622-644. DOI:10.1002/ese3.255.

A Probabilistic Model for Assessment of the PV Hosting Capacity in LV Grid

Dmitriy Kim
High School of Digital Technologies
Narxoz University,
Almaty, Kazakhstan
E-mail: dmitriy.kim@narxoz.kz

Georgi Georgiev
Plethora Ltd, Varna, Bulgaria
Email: georgidg@gmail.com

Abstract — **This work studies the distributed photovoltaic (PV) hosting capacity in low voltage (LV) grids by using the stochastic analysis approach. Energy consumption fluctuations together with solar generation variability in such grids have a significant impact on the voltage profile compared to traditional grids. We consider the parameters of the system as random variables and study probability that voltage magnitudes belong to admissible set of values. We study the maximum number of PV units on the roofs of buildings by selecting parameters at which an acceptable probability of grid voltages being within the permissible limits is achieved.**

Keywords: PV hosting capacity, distribution generation; distribution grid; voltage constraints; optimization problem; probabilistic approach.

I. Introduction

Mass distributed energy resources integration into the distribution grid for example rooftop photovoltaic (PV), energy storage and charging stations for electrical vehicles leads to its serious transformation [1]. Traditionally, the flow of power, generated by large generation plants, was from the higher to the lower voltage levels through distribution networks. With the introduction of the PV units, the power flows become bidirectional and the consumer becomes an active element of the energy system or prosumer (producer and consumer) which can consume, reduce and/or shift consumption over time, store and generate electricity in the grid.

At the same time, new problems arise. On the one hand, it is necessary to provide the largest number of consumers to connect their PV units to the grid. On the other hand, it is necessary to ensure the quality of electricity [2]. This problem is connected with assessment of PV hosting capacity. "*Hosting capacity is defined as the total PV capacity that can be accommodated on a given feeder without adversely impacting voltage, protection and power quality and with no feeder upgrades or modifications*" (see [3]). Our work investigates the stochastic approach for assessment of PV hosting capacity.

Random fluctuations in the power of distributed generation caused by solar generation and consumption in grid nodes lead to higher stochastic voltage variations than in traditional networks. Too low or high voltages lead to unacceptable deterioration in power quality. Therefore, the high volatility of PV generation has a severe impact on the voltage profile in distribution networks and this problem is known as one of the challenges for increasing the integration of PV modules in low voltage grid.

The aim of our work is to assess PV hosting capacity with the constraints related to the voltage profile.

Traditional approaches of managing this voltage problem are becoming too restrictive as they are based on worst-case scenarios. Such analysis for determining the maximal voltage of a LV grid usually considers that each PV unit injects power equal to its rated value and each node consumes its lowest expected load. To determine the minimal voltage, the evening peak loads are considered. At the same time, the consumption and generation of each user of the network, as well as the degree of dependence between them, are constantly changing, so based on such extreme cases, the voltage values are clearly overestimated compared to reality.

Moreover, in the traditional case, the nodal voltages are always monotonically decreased at a distance from the transformer, but when PV units are connected, the daily voltages of their nodes can easily be higher than that of the transformer. As a result of the traditional approach, the possibilities to integrate the PV units are greatly underestimated.

The study of voltage variability under new conditions requires more in-depth mathematical methods. In this work we will generalize this deterministic approach using probabilistic tools, which will allow us to take a new look at the problem. We use probabilistic approach to analyze voltages in nodes of a LV network with distributed generation of PV units.

The number of works devoted to the study of voltage in a distribution network is quite large (see last review [4]); and many of them use a probabilistic approach [4]-[5]. In most of them, a functional is selected for study, then either historical data is analyzed, or a simulation model of a network is developed and, after its repeated modeling by the Monte-Carlo method, the probabilistic characteristics of the selected functional are estimated (i.e. [5] - [8]). At the same time, due to the large number of simulations required for convergence, this method is usually time-consuming, in contrast to analytical methods. Moreover, this approach is not very convenient for studying the dependence of the results on input parameters. Analytical methods, as a rule, are devoid of these shortcomings, therefore, the main emphasis in this work will be placed on the development of analytical methods for the analysis of distribution networks.

In addition to the already mentioned random fluctuations in the power of distributed generation it is also important to note that in distribution LV grids the number of available real-time measurements is usually very limited [9]. Therefore, pseudo-measurements with large margins of error are used as additional measurements.

We consider parameters as random variables or random processes with some probabilistic distribution. We use the fact that the principal concern of the distribution network operator is to keep the state of the network within its constraints. It means that it is not necessary for the network operator to know the exact values of the estimated parameters, but rather if these parameters remain within their respective constraints. According to this logic instead of finding of exact values of parameters we investigate the probability that the sought parameters belong to admissible intervals. Moreover, if we can find or assess such probability the next step might be to find such values of some input parameters that maximize probability or other aim function, connected with this probability.

In this work we consider distribution network with random voltages which depend on renewable generation and load of customer. We define PV hosting capacity as the maximum number of some notional PV units that might be placed in the grid. In accordance with standards EN 50160 [10], shortly, the voltage profile of each network node must be within the interval $U_{nom} \pm 10\,\%$ (U_{nom} – nominal voltage) for at least 95% of the time, which might be reformulated as probability that nodes voltages in this interval has to be more than 95%. Then the problem of assessment of PV hosting capacity is reduced to an optimization problem with a linear objective function with probabilistic constraints on the voltages.

II. Problem formulation for linear grid

Figure 1 shows one of the feeders from the MV/LV (medium/low voltage) transformer to the consumers. The circuit is three-phase with a neutral line. Not all consumers use a three-phase supply. Many are connected to a single phase so that the resultant loads of the three phases are almost equal. Then the current in the neutral line will be as minimal as the difference between the phase voltages. This allows only a single phase to be investigated and the results can be applied to the other two phases. For this purpose it is sufficient to have a single-line diagram of the network.

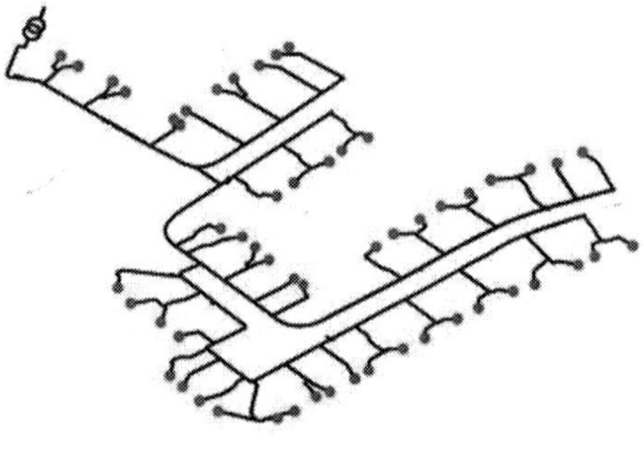

Figure 1. *Example of LV grid.*

We consider linear distribution grid with $m+1$ nodes and their voltages for $t \in [0,T]$ (during the day)

$$U_0(t), U_1(t), U_2(t), \dots, U_m(t).$$

Put that initial node is balancing, i.e. $U_0(t) = U_0$ is known constant. Define the vector of voltages for each node except balancing

$$\boldsymbol{U(t)} = (U_1(t), U_2(t), \dots, U_m(t))^T,$$

the vector of currents

$$\boldsymbol{I} = (I_1(t), I_2(t), \dots, I_m(t))^T$$

and active resistances between nodes of the network

$$R_1, R_2, \dots, R_m.$$

Assumption 1. We can neglect reactive parameters of the LV grid.

The relationship between the vector of voltages and the vector of currents might be expressed as

$$\boldsymbol{U}(t) = \boldsymbol{V} + \boldsymbol{H} \times \boldsymbol{I}(t),\ t \in [0,T], \tag{1}$$

where $\boldsymbol{H}$ – a matrix of nodal resistances and $\boldsymbol{V} = (U_0, U_0, \dots, U_0)^T$ – vector of known values.

We suppose that values of $\boldsymbol{I}$ could be expressed as follows:

$$I_i(t) = -IS_i(t) + J_i(t),\ \ i = 1,2,\dots,m, \tag{2}$$

where $IS_i(t) \geq 0$ – the current from PV unit of the grid and $J_i(t) \geq 0$ – the current from the node i to customer, $t \in [0,T]$.

Assumption 2. We suppose that we can treat $IS_i(t)$ and $J_i(t),\ i = 1,2,\dots,m$, as random processes.

Assumption 3. Suppose that at a certain time of the day (evening consumption peak) the maximum current consumption is reached and solar generation is zero at the same time and for all nodes, i.e. for some $e \in [0,T]$ and every $x > 0$

$$\max_t P(J_i(t) > x) \leq P(J_i(e) > x),\ \ i = 1,2,\dots,m, \tag{3}$$

$$IS_i(e) = 0.$$

In other words, $J_i(e), i = 1,2,\dots,m$, are the random consumption currents during the evening peak consumption and the minimum (zero) solar generation.

Assumption 4. Suppose that at a certain time of the day the minimum current consumption is reached and solar generation is maximal at the same time and for all nodes, i.e. for some $d \in [0,T]$ and every $x > 0$

$$\min_t P(J_i(t) > x) \geq P(J_i(d) > x),\ \ i = 1,2,\dots,m, \tag{4}$$

$$\max_t P(IS_i(t) > x) \leq P(IS_i(d) > x).$$

In other words, $J_i(d),\ i = 1,2,\dots,m$, are the random consumption currents during the minimal consumption and the maximal solar generation.

Assumption 5. Assume that maximal solar generation of every node $IS_i(d),\ i = 1,2,\dots,m$, might be represented as

$$IS_i(d) = q_i \times IS, \quad (5)$$

where IS – the current from the notional PV unit, q_i – coefficient, which characterize technical details of PV module of the node i.

The essence of this coefficient q_i is that if all the roofs of houses are equally located in relation to the sun (or can be reduced to this form by transformations) and have an identical shape, it shows the area of the roof covered by solar panels or the number of notional PV units on the roof.

Assumption 6. Assume that $J_i(e)$ and $J_i(d), i = 1,2,\dots,m$, are independent random variables.

In accordance with standards EN 50160 [10], the voltage profile of each grid's node must be within the interval $\boldsymbol{U_{nom}} \pm$ **10** % for at least 95% of the time, which might be reformulated as

$$\boldsymbol{P}(0.9\boldsymbol{U}_{nom} < \boldsymbol{U}(t) < 1.1\boldsymbol{U}_{nom}) \geq 0.95, \ t \in [0,T], \quad (6)$$

where $\boldsymbol{U}_{nom} = (U_{nom}, U_{nom}, \dots, U_{nom})$ – the vector of nominal voltages.

It means that we can study PV hosting capacity with nonlinear constraint (6). We may find such parameters of the system under which PV hosting capacity is achieved and (6) is satisfied.

III. Main Results

During the evening peak ($t = e$) consumption voltages at the nodes will be the lowest, so that they do not fall below the permissible limit ($0.9\ U_{nom}$) it is necessary to raise the voltage U_0 at the beginning of the line. At the same time, during maximum solar generation and minimum consumption ($t = d$), the voltages in the nodes may exceed the upper voltage limit $1.1\ U_{nom}$, so the voltage at the beginning of the line U_0 must not be very high.

Firstly, choose the minimum U_0 at which during the evening peak the voltages will be above the lower voltage limit with a large probability.

A. *Lower limit for the voltage U_0 during the evening peak.*

Assumption 7. Assume that currents $J_i(e)$ and $J_i(d), i = 1,2,\dots,m$, have Gaussian (normal) probabilistic distribution with means α_e, α_d and variations σ_e^2, σ_d^2:

$$J_i(e) \in \Phi_{\alpha_e,\sigma_e^2},\ J_i(d) \in \Phi_{\alpha_d,\sigma_d^2}. \quad (7)$$

Remark 1. This assumption is made to simplify calculations and is unnecessary for grids with a large number of nodes (see [11]).

It is known that for normal (or Gaussian) distribution the currents at evening peak ($t = e$) might be represented as

$$I_i(e) = a_i(e) + \sigma_i(e) \times \xi_i(e), \ i = 1,2,\dots,m, \quad (8)$$

where $\xi_i(e)$, $i = 1,2,\dots,m$, are independent and have standard normal distributions.

Then we can write

$$\boldsymbol{U}(e) = \boldsymbol{\Gamma e} + \boldsymbol{Be} \times \boldsymbol{\Xi}(e), \quad (9)$$

where

$$\boldsymbol{\Gamma e} = \boldsymbol{V} + \boldsymbol{H} \times \boldsymbol{Ae}, \quad \boldsymbol{Be} = \boldsymbol{H} \times \boldsymbol{W}(e),$$

$$\boldsymbol{Ae} = \begin{pmatrix} \alpha_1(e) \\ \alpha_2(e) \\ \dots \\ \alpha_m(e) \end{pmatrix}, \ \Xi(e) = \begin{pmatrix} \xi_1(e) \\ \xi_2(e) \\ \dots \\ \xi_m(e) \end{pmatrix},$$

$$\boldsymbol{W}(e) = \begin{pmatrix} \sigma_1(e) & 0 & 0 & \dots & 0 \\ 0 & \sigma_2(e) & 0 & \dots & 0 \\ & & \dots & & \\ 0 & 0 & 0 & \dots & \sigma_m(e) \end{pmatrix}.$$

We find the minimum voltage U_0 at which the probability of voltages crossing the lower voltage limit ($0.9\ U_{nom}$) will be higher than some fixed probability.

Let us formulate the optimization problem:

$$U_0 \to min, \quad (10)$$

under nonlinear constraints: for some $\varepsilon > 0$

$$\boldsymbol{P}(0.9\ \boldsymbol{U}_{nom} < \ \boldsymbol{U}(e) \ < 1.1\ \boldsymbol{U}_{nom}) \geq 1 - \varepsilon. \quad (11)$$

Example 1. Let

$$m = 25,\ \varepsilon = 0.03,$$
$$U_{nom} = 230\ V,$$
$$\alpha_i(e) = 9\ A,\ i = 1,2,\dots,m,$$
$$\sigma_i(e) = 3\ A,$$
$$R_i = 0.01\ Om.$$

Then the solution of optimization problem (10)

$$U_0 = 240.5\ V.$$

Remark 2. In the conditions of Example 1, we assumed that each node consumes a random current value of

$$9 \pm 9\ A$$

with probability 0.997. If we assume that each node consumes a constant current value rather than a random current value, e.g., 12 A, we get that U_0 should be more than 246 V.

B. *Daily minimum consumption and maximum solar radiation*

The currents at time $t = d$ might be represented as

$$I_i(d) = \ -q_i \times IS + a_i(d) + \ \sigma_i(d) \times \xi_i(d), \quad (12)$$

where $\xi_i(d)$, $i = 1,2,\dots,m$, are independent and have standard normal distributions.

Then we can write

$$\boldsymbol{U}(d) = \boldsymbol{\Gamma d} + \boldsymbol{Bd} \times \boldsymbol{\Xi}(d), \quad (13)$$

where

$$\boldsymbol{\Gamma d} = \boldsymbol{V} + \boldsymbol{H} \times \boldsymbol{Ad}, \quad \boldsymbol{Be} = \boldsymbol{H} \times \boldsymbol{W}(d),$$

$$\boldsymbol{Ad} = \begin{pmatrix} -q_1 \times IS + \alpha_1(d) \\ -q_2 \times IS + \alpha_2(d) \\ \dots \\ -q_m \times IS + \alpha_m(d) \end{pmatrix}, \ \Xi(d) = \begin{pmatrix} \xi_1(d) \\ \xi_2(d) \\ \dots \\ \xi_m(d) \end{pmatrix},$$

$$W(d) = \begin{pmatrix} \sigma_1(e) & 0 & 0 & \dots & 0 \\ 0 & \sigma_2(e) & 0 & \dots & 0 \\ & & \dots & & \\ 0 & 0 & 0 & \dots & \sigma_m(e) \end{pmatrix}.$$

Definition 1. By maximum PV hosting capacity

$$q_1 + q_2 + \dots + q_m \quad (14)$$

we will understand the number (or their area) of notional PV units that can be deployed in the grid *without adversely impacting voltage and with no feeder upgrades or modifications*.

Let us formulate the optimization problem:

$$q_1 + q_2 + \dots + q_m \rightarrow max \quad (15)$$

under nonlinear constraints: for some $\varepsilon > 0$

$$\boldsymbol{P}(0.9\,\boldsymbol{U}_{nom} < \boldsymbol{U}(e) < 1.1\,\boldsymbol{U}_{nom}) \geq 1 - \varepsilon, \quad (16)$$

$$\underline{q_i} \leq q_i \leq \overline{q_i},\ i = 1,2,\dots,m,$$

where constants $\underline{q_i}, \overline{q_i}$ are due to technical restrictions.

Example 2. Let

$$m = 25,\ \varepsilon = 0.03,$$
$$U_0 = 240.5\ V,$$
$$U_{nom} = 230\ V,$$
$$\alpha_i(e) = 9\ A,\ i = 1,2,\dots,m,$$
$$\sigma_i(e) = 3\ A,$$
$$\alpha_i(d) = 4\ A,$$
$$\sigma_i(d) = 1.5\ A,$$
$$IS = 5\ A,$$
$$R_i = 0.01\ Om,$$
$$\underline{q_i} = 0,\ \overline{q_i} = \infty.$$

Then

$$q_1 + q_2 + \dots + q_m = 146,$$

$$q_1 = 130, q_2 = 16, q_3 = q_4 = \dots = q_m = 0.$$

Example 3. Let in the conditions of Example 2 put

$$\underline{q_i} = 0,\ \overline{q_i} = 2.$$

Then

$$q_1 + q_2 + \dots + q_m = 41.9$$

and $q_i, i = 1,2,\dots,m,$ values are shown in the figure.

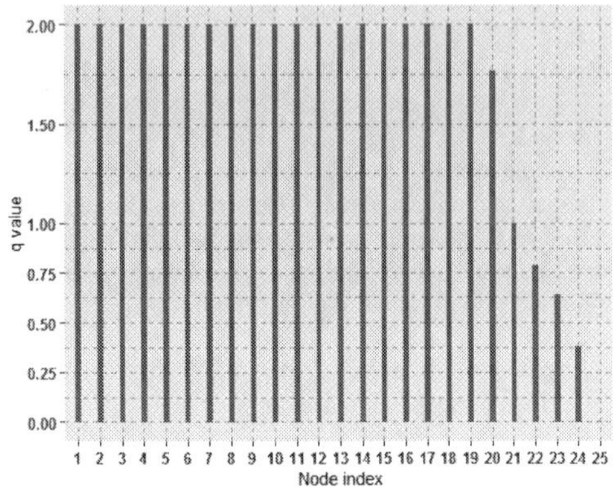

Figure 2. $q_i, i = 1,2,\dots,m,$ values.

IV. Conclusion

In this work we consider the problem of assessment of PV hosting capacity of LV grid based on probabilistic approach. Traditional approaches of managing this voltage problem usually consider that each PV unit injects power equal to its rated value and each node consumes its lowest expected load. To determine the minimal voltage, the evening peak loads are considered. At the same time, the consumption and generation of each node of the grid, as well as the degree of dependence between them, are constantly changing, so based on such extreme cases, the voltage values are clearly overestimated compared to reality. Thus the traditional deterministic approach to voltage analysis focuses on the worst-case scenario, which is very rare happening or which can be called extremely unlikely. As a result of the traditional approach, the possibilities to integrate the PV units are greatly underestimated (see Remark 2).

The study of voltage variability under new conditions requires more in-depth mathematical methods that, in addition to extreme cases, will take into account the most expected network scenarios. In this work we use probabilistic approach to analyze voltages in nodes of a LV grid with distributed generation of PV units. It is also important to note that many technical parameters of most LV networks (line length, resistance of ground loops, etc.) are known with some error, which again suggests in favor of probabilistic methods.

Therefore, we consider the system parameters as random variables and apply probabilistic methods to analyze them. We introduced assumptions and build strict mathematical model for low voltage grid with PV units. The currents and voltages at each node we treated like random processes. We defined the notion PV hosting capacity and gave an algorithm for finding it. A few examples for a linear single-phase grid were given.

PV hosting capacity is defined as the maximum number of some notional PV units that might be deployed in the grid without adversely impacting voltage. We formulate the optimization problem of finding the maximum number of notional PV units at which the voltage is within acceptable limits with a probability close to unity. This means that with a probability near unity during the evening peak the voltage drop does not violate the lower limit $0.9\boldsymbol{U}_{nom}$, and during the maximum solar generation the voltage values do not exceed the upper limit $1.1\boldsymbol{U}_{nom}$. If we find an analytical representation for this probability, we can first find the minimum U_0 (or transformer ratio) at the beginning of the feeder, at which the node voltages during the evening peak are within acceptable limits (see example 1). Then we find the maximum number of notional PV units at which the nodal voltages are within acceptable limits with a probability close to unity (see example 2 and 3).

The general aim of the work is to develop mathematical methods for reliable maintenance of voltage levels in the LV grid corresponding to the standard ($U\_nom \pm 10\,\%$), under deep penetration of PV units.

References

[1] M. Eisenreich, G. Balzer, J. Backes and B. Maurer, "Integration of PV systems into low voltage networks using standard load profiles," *International Conference on Renewable Energies and Power Quality (ICREPQ'10)*, 2010, pp. 972-976.

[2] A.P. Kenneth, K. Folly, "Voltage rise issue with high penetration of grid connected pv." *IFAC Proceedings*, vol. 47, pp. 4959–4966, 2014.

[3] F. Ding, B. Mather and P. Gotseff, "Technologies to increase PV hosting capacity in distribution feeders," *2016 IEEE Power and Energy Society General Meeting (PESGM)*, 2016, pp. 1-5.

[4] K.A. Alboaouh, S. Mohagheghi, "Impact of Rooftop Photovoltaics on the Distribution System.", *Journal of Renewable Energy*, pp. 1-23, 2020.

[5] V. Klonari, "Probabilistic analysis of low voltage distribution networks with distributed generation." PhD Thesis, Université Libre de Bruxelles, pp. 1-314, 2016.

[6] S. Conti, S. Raiti, "Probabilistic load flow using Monte Carlo techniques for distribution networks with photovoltaic generators." *Sol. Energy*, vol. 81, pp. 1473–1481, 2007.

[7] S. Heslop, I. MacGill, J. Fletcher, S. Lewis, "Method for determining a PV generation limit on low voltage feeders for evenly distributed PV and Load." *Energy Procedia*, vol. 57, pp. 207–216, 2014.

[8] R. Torquato, Q. Shi, W. Xu, W. Freitas, "A Monte Carlo simulation platform for studying low voltage residential networks." *IEEE Trans. Smart Grid*, vol. 5, pp. 2766–2776, 2014.

[9] A. Abur and A. G. Expósito, *Power System State Estimation Theory and Implementation*, New York: Marcel Dekker, Inc., 2004, pp. 327.

[10] "Voltage characteristics of electricity supplied by public electricity networks," EN 50160 Standard, Dec. 21, 2012.

[11] G. Georgiev, D. Kim and M. Negnevitsky, "Probabilistic approximation of voltage values for power distribution network with large number of nodes and solar generation." *Proceedings of the 10th International Scientific Symposium on Electrical Power Engineering*, 2019, pp. 53-57.

On Energy Storage Options for Small Modular Nuclear Reactors

Branislav Vrban, Štefan Čerba, Jakub Lüley, Vendula Filová, Vladimír Nečas
Institute of Nuclear and Physical Engineering
Slovak University of Technology in Bratislava, FEI STU
Bratislava, Slovakia
branislav.vrban@stuba.sk, stefan.cerba@stuba.sk, jakub.luley@stuba.sk, vendula.filova@stuba.sk, vladimir.necas@stuba.sk

Abstract — **The electricity grid comprising supply, distribution, and demand is continuously developing and utilizing the advantages of modern technologies. For the last decades, the emphasis has been placed on renewable sources posing higher operational requirements on existing sources. As a result, current and future nuclear power plants are being called to operate more flexibly than ever before in the evolving electricity grid that looks very different than for which they were originally designed. In response to new requirements, intensive research is focused on energy storage technologies able to reduce costs and increase revenue via more stable operation and a combination of ancillary services. These technologies may be categorized into three forms of energy: chemical, mechanical, and thermal. This paper presents the results of a literature survey trying to assess the deployment capabilities for the future electricity grid in Slovakia. The assessment of these energy storage systems starts from the requirements of the current fleet of reactors, such as VVER-440, but is focusing also on the aspects of small modular reactors (SMR).**

Keywords: energy storage, nuclear, electricity grid

I. Introduction

The baseload electricity generation in most countries is provided by hydro, fossil fuels, and nuclear [1]. Currently, the strong movement toward renewable energy imposes higher requirements on nuclear reactors including significantly variable power demands. Contrary, new devices such as electric vehicles could back-export electricity to the grid, and thus may reduce the dependency of variations in demand on the weather. Anyhow, these variable power demands might influence the capacity factor of reactors and thus negatively impact their economy. The missing link in the energy system could be the symbiotic cooperation of small and medium nuclear reactors [2] with various energy storage systems. This cooperation has the potential to allow the operation of the reactor at a constant power level by storing cheaper off-peak electricity, which can then be sold at a higher price in the event of high demand. In addition, to achieve the climate goals stipulated in the Paris Agreement [3], nuclear power seems to be one of the primary alternatives to renewables for reducing carbon emissions in the electric sector [4]. The necessity to reduce carbon emissions is emphasized by the high-level nature of the Paris Agreement, which is a legally binding international treaty on climate change adopted by 196 Parties at COP 21 in Paris, on 12 December 2015 and entered into force on 4 November 2016.

II. Energy Mix of Slovakia

A. General Overview

Slovakia generates electricity in nuclear power stations, hydro-power stations, natural gas, and coal-fired power stations as well as from renewable energy sources. In 2020, Slovakia generated 29,010 GWh of electricity while the total consumption amounted to 29,328 GWh. Thus, after years, the electricity demand was almost covered by production. As shown in Fig. 1, this well-balanced production and consumption were mainly achieved thanks to the decreasing electricity demand noticeable in the last 4 years [5].

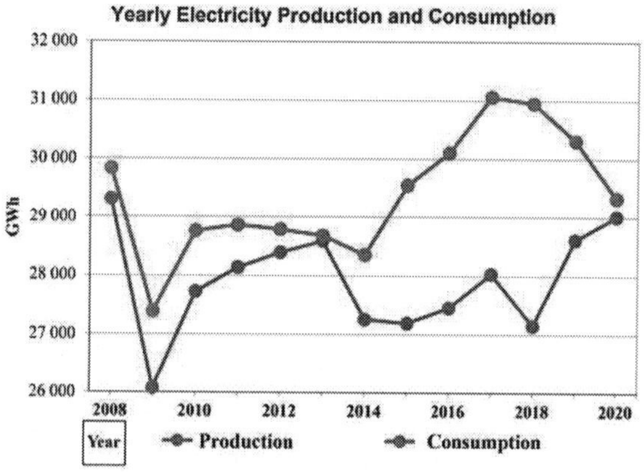

Figure 1. Yearly electricity production and consumption [5]

The Slovak Republic already has a low-carbon mix of electricity sources; the share of carbon-free electricity production in 2020 was almost 80%. Nuclear power plants had the highest share of electricity generation in 2020 (52.66%). They were followed by thermal power plants (21.31%), hydroelectric power plants (16.61%), renewables (7.96%), and other power plants (0.38%) [5]. To give a reader a better understanding of the shares, the load-curve coverage on the day of the yearly peak in 2020 is shown in Fig. 2. From a long-term perspective, the production of electricity in thermal power plants in the Slovak Republic is gradually declining and the importance of nuclear energy and energy from renewable sources is growing. This can be demonstrated by the plans of the Slovak government to close the Novaky thermal plant till 2023 due to its economic, social, climate, health, and environmental impacts. Units no. 3 and 4 (2 x 110 MWe) of ENO B were already shut down as of 31 December 2015.

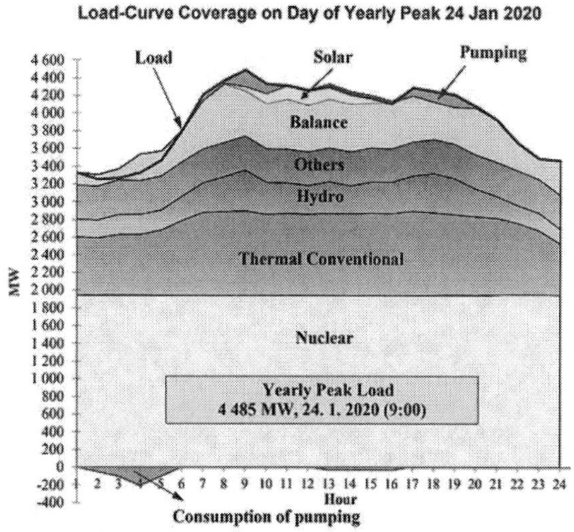

Figure 2. Load-curve coverage on the day of yearly peak in 2020 [5]

B. Nuclear Power Specifics

Slovakia needed to import electricity after the shutdown of two nuclear power reactors in Jaslovské Bohunice in 2006 and 2008. The country is expected to again become self-sufficient, and even become a net exporter, after the third and four nuclear power station units in Mochovce (EMO) are put into operation. The IAEA usually defines small and medium-sized or modular reactors (SMRs) as reactors producing up to 300 MW(e) (small-sized or small modular) and reactors producing 300–700 MW(e) (medium-sized). The reactors currently operated and constructed in Slovakia (VVER-440) might be thus already judged as medium-sized reactors. A new generation of SMRs would benefit from flexible power generation options, a wide range of applications, enhanced safety resulting from inherent passive safety features, reduced upfront capital investment, and possibilities for cogeneration and non-electrical applications. The higher awaited share of nuclear energy in the grid with a significant portion of renewables will enlarge the need for load-following operation for baseload energy sources and will open the possibility for storing the off-peak electricity in energy storing systems. The expected connection to the grid for EMO 3 is planned this year and the first criticality of EMO 4 is planned in the next 2 years. Due to the volatility of the political situation and the actual war in Ukraine, serious concerns exist about the completion of the last unit.

Despite the nuclear power is the strategic electric source for Slovakia, there exists continuous political discussion about the future and the tax conditions of the nuclear industry which brings investor distrust and effectively prevents the capital-intensive build of new possible sources. As an example, in 2022 Slovakia as a response to increasing electricity prices has agreed to drop previously announced plans to impose a 50% tax on profits from the sale of nuclear energy in exchange for a fixed electricity price, under a memorandum of understanding signed with Slovak NPP provider Slovenské elektrárne a.s. [6].

Worldwide and in Slovakia as well, NPPs had difficulty adapting to the shifting electricity grid dynamics due to regulatory constraints and limited experience ramping electricity production and operating flexibly. Slovak VVERs have a limited capacity to adjust the electric output. Limitations are mainly given by the performance of nuclear fuel and associated turbogenerator. There is no limit to the number of power changes in the core below 5% of rated power (RP). Within the 5% and 10% RP there are no limits for changes slower than 3% RP per minute, but limitations of approx. 2500 cycles are given for the faster changes [7]. For bigger RP changes, the allowed numbers are decreasing dramatically. In VVER-440 modernized designs, there are 2 turbines per reactor with approx. 250 MW_e with an allowed change of 3% of rated power per minute. The current version of European Utilities Requirements expects that an NPP must at least be capable of daily load cycling operation between 50% and 100% of its rated power, with a rate of change of electric output from 3-5% of rated power per minute [8]. The request for load-following operation of NPPs naturally opens questions about the economics, safety (limited by the fuel performance), and impact on the aging of equipment due to load following. The fuel performance needs to be studied and the new generation of fuel needs to be developed with the support of the new simulation tools and experimental data [9]. According to the literature [10], no significant impact is estimated with the aging of equipment components. It is understandable that load-following decreases the utilization factor and thus negatively impacts the overall economy.

As can be seen, the load-following operation of NPPs is economically and technically difficult and more viable is to store the excess or off-peak electricity using energy storage technologies. The significant use of these technologies has the potential to improve the economics of NPPs and might provide cost-effective grid services such as backup power, frequency regulation, and fast response reserves. As shown in previous paragraphs, an increase in the share of nuclear energy and renewables in Slovakia in the future will set workable conditions for incorporating energy storage options into the grid.

III. Energy Storage Systems

To allow load-following operation and storing of off-peak electricity, several large-scale energy storage systems can be assumed. The literature is replete with schematic descriptions, but many of them are just in the conceptual stage of development. These energy storage systems can be categorized by the forms of energy they are utilizing.

A. Chemical energy

Chemical energy storage works by storing energy in chemical bonds that can be later broken to release energy. The most common representatives of this category are super-capacitors (electrolytic) and batteries (electrochemical). Supercapacitors store energy in an electric field between two electrodes separated by an ion-permeable membrane. Supercapacitors are usable just for short-term small capacity storage. On the other hand, several battery types are in various stages of development worldwide and are already used to provide ancillary services to the grids. Batteries can be categorized into conventional batteries and flow batteries. A conventional battery is one in which the movement of chemical species between anode and cathode occurs within the battery compartment. Flow batteries are larger in size and offer many advantages over conventional batteries. Flow batteries store energy in the electrolyte of the battery, rather than the electrodes, which allows a flow battery to be quickly recharged by simply replacing the electrolyte tanks in the battery system. The tank size depends on the required battery capacity. When in operation, the electrolyte from the tanks flows into the battery to produce an electrochemical reaction either to store or retrieve the electricity. The common conventional batteries are lithium-ion, sodium-sulfur, lead-acid and nickel-cadmium batteries. The zinc-bromine and vanadium redox technologies batteries are the most promising representatives of flow batteries.

In a lithium-ion battery cell, energy is stored by causing positively charged lithium ions to travel through a liquid electrolyte as electrons are transferred through an external circuit. In the past couple of years, lithium-ion batteries have dominated the market for stationary grid-scale energy storage applications [11]. The lithium-ion batteries are characterized by high energy densities, very high cycle efficiency, and quick response times. However, lithium is also highly reactive when exposed to oxygen or water and must be packaged carefully. Another issue is a $LiFP_6$ organic carbonate solution, which is highly toxic and flammable [12].

The main characteristics of sodium-sulphur (NaS) battery systems are long discharge periods, high pulse power capacity, and extreme operating conditions. In a NaS battery, molten sodium (with possible exothermic reaction with water) and sulfur act as the battery's two electrodes, with beta-alumina acting as the solid electrolyte. Therefore, the operating temperature of the battery must be kept between 300 and 350°C. Due to the nature of materials used in the battery, strict safety measures need to be applied. Despite the safety challenges, NaS batteries offer superior energy densities and show promise for use in applications that require short and powerful bursts of energy [13].

Lead-acid batteries are the oldest form of battery storage available. However, the cycle life of these batteries and their poor performance at lower temperatures do not favor them for several energy management services [11]. Nickel-cadmium batteries offer similar performance to lead-acid batteries, but with superior cycling abilities, energy density, and memory effect. In a NiCd battery, nickel oxide hydroxide is used to make the cathode and the anode is made from metallic cadmium [14]. These batteries do not require thermal management to maintain a specific operating temperature. However, if the battery is not completely recharged immediately after discharging, the maximum capacity of the battery system will begin to decrease rapidly. The limitations of a large deployment of these batteries are the numerous environmental hazards related to cadmium and nickel. Both are toxic heavy metals; cadmium is classified as a carcinogenic substance for humans and nickel can induce lung and gastrointestinal cancer. Zinc-bromide flow batteries are hybrid flow batteries where the part of the energy is stored in the electrolyte and another part on another by plating it with zinc during charging [15]. This type of battery excels in a long lifetime, modularity, and almost no energy loss. However, the technology complexness yields higher costs and stricter operating conditions. Zinc-bromide batteries are well-suited for grid-scale energy management due to sub-millisecond response times suitable to provide fast ancillary services and because they are not restricted by power capacity, depth-of-discharge, and self-discharge rate of the battery. From the other view, bromic is an environmental contaminant and also attacks seals, and induces corrosion.

The next representative of flow batteries is the vanadium redox battery (VRB) which is the most mature type of flow battery available today. This batter stores energy with vanadium redox couples that are kept in two separate electrolyte tanks. As the electrolyte flows through the battery during charging, vanadium ions accept electrons at the anode and deposit electrons at the cathode. The reactions run in the reverse direction when the battery is discharging. A significant advantage of VRB systems is that the two electrolyte solutions are chemically identical, which makes the operation of the battery much simpler and less expensive. However, still face some technical challenges, including low electrolyte stability and solubility which can lead to decreased energy densities which are comparatively lower than other battery storage systems [13]. However, the capital cost of these systems remains far too high for deep market penetration [16].

The battery energy storage systems (BEES) comprise mainly batteries, control, and power conditioning systems. As shown in the previous paragraphs, some of these systems involve rare materials which negatively impacts the capital cost and reduces the sustainability of such solutions. Conventional batteries also suffer from the low number of charge-discharge cycles which implies additional costs needed for replacing batteries. With respect to large-scale energy storage requirements, an important issue is the relatively smaller storage capacity caused by very low specific energy and low energy density. Thus, the need for a higher number of battery packs may be compensated by distributed battery storage utilizing electric cars connected to the grid when are not in operation. This is indicated as a vehicle-to-grid operation [17]. The Tehachapi Energy Storage Project [18] can serve as an example of the real utilization of BEES systems in the grid. The Tehachapi Energy Storage Project (2014-2021) was the 8 MW/ 32 MWh lithium-ion battery-based grid energy storage system at the Monolith Substation of Southern California Edison in Tehachapi, California, sufficient to power between 1 600 and 2 400 homes for four hours. One important technology is hydrogen energy storage. In this form of energy storage, the electricity that has been generated by a power plant can be used to produce hydrogen mainly through high-temperature (>800 °C) [19] steam electrolysis which reduces the amount of electrical energy compared with low-temperature methods. Another possibility is to utilize thermos-chemical processes that split water using a series of chemical reactions at varying temperatures. Both sulfur-iodine thermochemical cycle and copper-chloride cycle require relatively high temperatures (approx. 900 °C and 500 °C). All methods mentioned above require a significant amount of electricity from NPP with just a small fraction of process heat.

After the hydrogen is produced, it must be stored in a specific form. In the solid phase, it is stored as sodium, magnesium, or lithium hydrides [20], [21]. In the gas phase, hydrogen is stored in large quantities underground in salt caverns, depleted oil and gas reservoirs, and aquifers [1]. Due to reasons of material properties and operating costs, large amounts of gaseous hydrogen are usually not stored at pressures exceeding 100 bar in aboveground vessels and 200 bar in underground storages. As the storage pressures are limited, so are the achievable hydrogen storage densities: at 100 bar and 20 °C, the density of hydrogen gas is approximately 7.8 kg/m^3 [22]. The low hydrogen density leads to large storage-specific volumes, and, thus, high investment costs. Large amounts of hydrogen are already stored underground: e.g. salt cavity storages in Teeside, UK, and Texas, USA, have proven the applicability of the approach [23]. As an alternative to salt caverns, the most promising option for large-scale storage of hydrogen in metallic vessels is pipe storage. The pipe storages have been already applied for the store of natural gas, mainly to manage peaks in demand for storage facilities [24]. In addition to compression, the density of pure hydrogen may also be increased via its liquefaction. Liquefaction has the advantage that very high hydrogen storage densities can be attained already at atmospheric pressure: the density of saturated liquid hydrogen at 1 bar is 70 kg/m^3 [25]. Liquid hydrogen requires significant amounts of energy because it must be cooled below its critical point of 33 K. The storage vessels are most commonly double walls with spherical shapes to minimize the boil-off rates and are specially insulated to minimize the heat transfer via radiation. As a result of the high degree of insulation and the low surface-to-volume ratio, boil-off rates are very low for larger spherical tanks, commonly below 0.1% per day [26]. Another alternative is chemically bond hydrogen by chemical hydrides such as methanol, ammonia, or formic acid. However, it should be noted that the utility of these chemicals clearly goes beyond the storage of hydrogen.

The chemical energy stored in the hydrogen bonds can then be used later in a fuel cell or combustion turbine to produce useful work. Due to high costs, hydrogen production is not typically considered to be well-suited for the daily management of fluctuations in electricity markets caused by variable electricity generation from renewable energy sources. Moreover, none of the mentioned advanced production processes have been demonstrated beyond the laboratory scale. As a result of intensive research, the hydrogen fuel cell costs are expected to drop significantly in the next decade [27]. Hydrogen production systems can also be easily expanded for grid-scale applications. In addition, hydrogen energy storage systems are compact and can be easily integrated into a power plant's generation cycle.

B. *Mechanical energy*

Mechanical energy storage accounts for most of the installed energy storage capacity in the world. There exist several mechanical energy-based technologies such as compressed air energy storage (CAES), pumped storage hydropower (PSH), and flywheels energy storage (FES). CAES stores electricity as potential energy due to differences in gas pressure reservoirs. Only two full-scale CAES systems are in operation in the world today: one in Germany and one in the U.S. state of Alabama [11]. Although CAES plants offer quick ramp rates like PSH facilities, the efficiency of the energy storage and conversion process is relatively low compared to other energy storage technologies. Underground CAES storage systems are the most cost-effective with storage capacities of up to 400 MW and with discharge times of 8 (above-ground) to 26 (underground installation) hours [28]. Flywheels store kinetic energy due to the internal mass of the spinning rotor inside the frictionless vacuum chamber. In modern flywheel systems, the spinning rotor is contained in a thick, steel vessel that protects the rotor and the motor-generator used to convert electrical energy into mechanical energy and vice versa. FES systems have the highest energy transfer efficiencies but only have very limited energy capacity just in tens of kWh [28]. Pumped storage hydropower is the most developed and well-proofed energy storage technology in the world. Pumped storage hydropower plants store energy by pumping water from a lower reservoir to a higher reservoir using electricity generated during off-peak periods. During peak demand periods, the water is allowed to flow back down to the lower reservoir, generating electricity in the same way a conventional hydropower plant would. The International Energy Agency estimates that PSH installations account for 99% of the energy storage capacity worldwide [29]. Čierny Váh pumped storage power plant is Slovakia's largest pumped storage power plant and with its installed capacity also the largest hydroelectric power plant. The upper reservoir located 1160 meters above sea level has no inflow of its own. The installed capacity of Čierny Váh is 735.16 MW_e [30]. Its utilization in the Slovak electricity grid can be seen in Fig. 2. Major issues with PSH include siting (geographic conditions), permitting processes, public acceptance, and associated environmental impact processes that can take many years to complete. Anyhow, is crystal clear that PSH is the most ideal option for large-energy storage of off-peak electricity.

C. *Thermal energy*

Thermal energy storage (TES) technologies store energy in the form of heat. They are mostly used at power plants to store thermal energy before it is converted to electric energy or for other utilizations like district heating. TES technologies are usually considered in conjunction with concentrated solar power renewable energy sources [31] but could be also directly incorporated into the NPP designs [31]. There exist two main mechanisms to store energy in an inert material, either as sensible heat or as latent heat. For sensible heat storage, either solid or liquid materials are used based on the operating temperature. The solid heat storage materials include concrete, rocks, and ceramics. Molten salts, water, or synthetic oils are usually chosen as liquid storage media [32]. In the case of latent heat storage, a phase change material stores heat as the heat of fusion or the heat of vaporization. One of the interesting thermal energy storage systems is the well-established underground thermal energy system in which energy is stored in boreholes, aquifers, and caverns. Typically, the storage medium used is water [14]. The economical effectiveness of these systems is largely dependent on surrounding geographic conditions. Another usable technology is the hot and cold water storage system. These systems are used primarily to shift the energy demand for the heating and cooling of residential and commercial buildings to off-peak periods to reduce costs. In a typical hot and cold water storage system, the device chills or transfers heat to the water, which is then stored in an insulated tank. The performance of these systems is limited by the output temperate which is dependent on the freezing and boiling temperatures of the water.

Molten salts are a phase change material that is commonly used for thermal energy storage. Molten salts are solid under room conditions but are liquid when heated. In the majority of heat storage systems, the molten salt energy storage systems, the molten salt is maintained as a liquid throughout the energy storage process. The interesting proposal of coupling an Advanced High-Temperature Reactor with a lithium chloride (utilizing latent heat) molten salt thermal energy block can be found in the literature [31]. In the system block diagram, the LiCl thermal energy storage with the capacity to store 150 MWd of thermal energy is located between the traditional primary and secondary loop of NPP. It should be noted that this concept is also applicable to other TES. According to the paper authors, the load following characteristics of the nuclear power reactor are significantly enhanced and the additional safety margin is introduced in case of an unplanned shutdown.

IV. Regulatory Aspects

The coupling of a large energy storage system to a nuclear reactor brings several challenges to guaranteeing the safety of the combined system. It can be assumed that thermal large energy storage will improve safety due to the additional thermal inertia coupled to the reactor, no temperature transient on the side of the core, and the capability to absorb decay heat if needed. Some of the energy storage systems may provide emergency backup power useful for cooling, starting after the grid black-out or other purposes. However, one can imagine concerns about the reliability and integrity of such systems, fires in case of BEES systems, chemical interaction of sodium in NaS, floodings caused by failures of hot and cold water storage systems, molten salts frozen and blocking heat exchange from reactor, hydrogen explosions and other. Finally, the regulatory challenges posed by the coupling of a large energy storage system to a nuclear reactor are significant, where the whole methodology of safety assessment including new simulation codes supported by experimental work needs to be developed. Based on these finding a proper internationally accepted licensing framework need to be established and adopted by the national regulators.

V. Conclusion

The strong movement toward renewable energy and the construction of new NPPs in Slovakia brings challenges to the effective and economical operation of baseload power sources. On the other hand, these challenges may be judged as possibilities for new technologies. Undoubtedly energy storage systems are of great interest to the scientific and industrial community and their possible interaction with nuclear reactors seems to be mutually beneficial. This viable cooperation may even help to decarbonize the electric sector and thus achieve the set climate goals. There is a variety of potential large energy storage systems with different levels of maturity that could be considered for the application. A big advantage for the future is that in terms of power level, the nuclear units located in Slovakia can be logically replaced by the new generation of small modular reactors without compromising the already deployed standalone energy storage systems. In addition to the overview performed, this paper lays out the preliminary steps for Slovak grid modeling utilizing the European Sodium Fast Reactor (as the SMRs representative) in cooperation with standalone or integrated energy storage technologies in close future.

ACKNOWLEDGMENT

This publication was created thanks to support under the Operational Program Integrated Infrastructure for the project: International Center of Excellence for Research on Intelligent and Secure Information and Communication Technologies and Systems – 2. stage, ITMS code: 313021W404, co-financed by the European Regional Development Fund, by the Scientific Grant Agency of the Ministry of Education of Slovak Republic No. VEGA 1/0615/21 and partially by the ESFR-SIMPLE project funded by the European Union's Horizon Europe research and innovation program.

REFERENCES

[1] N. Gnanapragasam, D. Ryland and S. Suppiah, "Status of energy storage options for electricity from nuclear power plants," *IEEE International Conference on Smart Energy Grid Engineering (SEGE)*, 2013, pp. 1-11.

[2] IAEA, *Benefits and challenges of small modular fast reactors*, Proceedings of a technical meeting, IAEA-TECDOC-1972, Vienna, IAEA, 2021, pp. 360.

[3] Paris Agreement to the United Nations Framework Convention on Climate Change, Dec. 12, 2015, T.I.A.S. No. 16-1104.

[4] IAEA, "Climate change and nuclear power 2020," Vienna, IAEA, 2020, pp. 97.

[5] SEPS, "Individual and Consolidated Annual Report 2020," Bratislava, SEPS as., 2021, pp. 102.

[6] J. Stojaspal, "Slovakia to drop tax in return for fixed electricity price," Internet: https://news.bloombergtax.com/daily-tax-report-international/slovakia-to-drop-tax-in-return-for-fixed-electricity-price [Feb. 15, 2022].

[7] Slovenské elektrárne a.s., "Prevádzkový predpis - Bezpečnostná správa JE V-2 (6-BSP-001)," SE a.s., Bohunice, 2012.

[8] NEA, "Technical and economic aspects of load following with nuclear power plants," OECD NEA, 2011, pp. 53.

[9] H.C. Kim, S.K. Seo, S.U. Lee, Y.S. Yang, "Development of NUFORM3D module with FRAPCON3.4 for simulation of pellet-cladding mechanical interaction," *Nuclear Engineering and Design*, vol. 318, pp. 61-71, 2017.

[10] C. Bruynooghe, A. Eriksson, and G. Full, "Load-following operating mode at nuclear power plants (NPPs) and incidence on operation and maintenance (O&M) costs – compatibility with wind power variability," EUR 24583 by JRC scientific and technical reports, 2010.

[11] A.A. Akhil, G. Huff, A.B. Currier, B.C. Kaun, D.M. Rastler, S. B. Chen, A.K. Cotter, D.T. Bradshaw, and W.D. Gauntlett, *Electricity Storage Handbook in Collaboration with NRECA*, DOE/EPRI 2013, Sandia National Laboratory, Albuquerque, NM, 2013, pp. 340.

[12] Z. Zhou, T. Levin, and G. Conzelmann, "Survey of U.S. Ancillary Markets," Argonne National Laboratory, Argonne, IL, 2016, pp. 59.

[13] X. Luo, J. Wang, M. Dooner, J. Clarke, "Overview of current development in electrical energy storage technologies and the application potential in power system operation," *Applied Energy*, Vol. 137, pp. 511-536, 2015.

[14] J. Coleman, S. Bragg-Sitton, E. Dufek, S. Johnson, J. Rhodes, T. Davidson, and M.E. Webber, "An evaluation of energy storage options for nuclear power", INL/EXT-17-42420 Rev. 0, INL, USA, June 2017, pp. 148.

[15] J. Intrator, E. Elkind, A. Abele, S. Wiessman, M. Sawchuk, and E. Barklett, "2020 Strategic Analysis of Energy Storage in California," Public Interest Energy Research (PIER) Program, 2011.

[16] Z. Abdin, K. R. Khalilpour, "Chapter 4 - Single and Polystorage Technologies for Renewable-Based Hybrid Energy Systems," Polygeneration with Polystorage for Chemical and Energy Hubs, Academic Press, 2019, pp. 77-131.

[17] K. Clement-Nyns, E. Haesen, J. Driesen, "The impact of vehicle-to-grid on the distribution grid," *Electric Power Systems Research*, vol. 81, pp. 185-192, Jan. 2011.

[18] N. Pinsky, L. O'Neill, *Tehachapi Wind Energy Storage Project*, Technology performance report #3, Southern California Edison, USA, 2017, pp. 221.

[19] J. E. O'Brien, M.G. McKellar, E.A. Harvego, and C.M Stoots, "High-temperature electrolysis for large-scale hydrogen and syngas production from nuclear energy – summary of system simulation and economic analyses," *International Journal of Hydrogen Energy*, vol. 35, pp. 4808- 4819, Oct. 2009.

[20] B. Sakintuna, F. Lamari-Darkrim, M. Hirscher, "Metal hydride materials for solid hydrogen storage: A review," *International Journal of Hydrogen Energy*, vol. 32, pp. 1121-1140, Jan. 2007.

[21] D. D. Macdonald, J. McLafferty, J. Tokash, A. Saleg, R. Sharna, S. Sharifi-Asl, S.K. Lee, K. Williams, and M. Taylor, "Electrochemical hydrogen storage systems," Final report by Department of Materials Science and Engineering, Pennsylvania State University, June 2010.

[22] J. Andresson, S. Grönkvist, "Large-scale storage of hydrogen," *International Journal of Hydrogen Energy*, vol. 44, pp. 11901-11919, May 2019.

[23] E. Wolf, "Large-scale hydrogen energy storage," In Garche J, editor. Electrochemical energy storage for renewable sources and grid balancing. Amsterdam: Elsevier; 2015. pp. 129-42.

[24] O. Kruck, F. Crotogino, R. Prelicz, T. Rudolph,"Overview on all known underground storage technologies for hydrogen," HyUnder; 2013 August. Deliverable, 2013.

[25] A. Witkowski, A. Rusin, M. Majkut, K. Stolecka, "Comprehensive analysis of hydrogen compression and pipeline transportation from thermodynamics and safety aspects," *Energy*, vol. 141, pp. 2508-2518, 2017.

[26] W. A. Amos, "Costs of storing and transporting hydrogen," Golden, CO (US): National Renewable Energy Lab., 1999, pp. 216.

[27] M. Beaudin, H. Zareipour, A. Schellenberglabe, W. Rosehart, "Energy storage for mitigating the variability of renewable electricity sources: An updated review," Energy for Sustainable Development, vol. 14, pp. 302-314, 2010.

[28] D. Rastler, "Electricity energy storage technology options – A white paper primer on applications, costs, and benefits," Technical report by Electric Power Research Institute, 2010, pp. 170.

[29] International Energy Agency, "Technology Roadmap: Energy Storage," Paris, France, 2014.

[30] Slovenské elektrárne, "The Čierny Váh pumped storage power plant." Internet: https://procurement.seas.sk/cierny-vah-hpp, [Apr. 07, 2022].

[31] S.A. Alameri and J.C. King., "A coupled nuclear reactor thermal energy storage system for enhanced load following operation," *Proc. International Nuclear Atlantic Conference – INAC 2013*, 2013. pp. 1-12.

[32] A. Gil, M. Medrano, I. Martorell, A. Lázaro, P. Dolado, B. Zalba, L.F. Cabeza, "State of the art on high-temperature thermal energy storage for power generation. Part 1–Concepts, materials, and modelization," *Renewable and Sustainable Energy Reviews*, vol. 14, pp. 31–55, 2010.

Small Modular Nuclear Reactors for Future Electricity Grids

Jakub Lüley, Branislav Vrban, Štefan Čerba, Vendula Filová, Vladimír Nečas
Institute of Nuclear and Physical Engineering
Slovak University of Technology in Bratislava, FEI STU
Bratislava, Slovakia
jakub.luley@stuba.sk, branislav.vrban@stuba.sk, jakub.luley@stuba.sk, vendula.filova@stuba.sk, vladimir.necas@stuba.sk

Abstract — **The electricity grid comprising supply, distribution, and demand is continuously developing and utilizing the advantages of modern technologies. For the last decades the emphasis has been placed to renewable sources posing higher operational requirements on existing or newly deployed sources. As a result, current and future nuclear power plants are being called to operate more flexibly than ever before in the evolving electricity grid that looks very different or in the complicated grids where big units cannot be installed. In response to new requirements, intensive research is focused on nuclear technologies able to reduce investments costs and increase revenue via more stable operation and a combination of ancillary services. This paper presents the results of a comprehensive literature survey trying to assess the deployment capabilities of small modular reactors (SMR) for the future electricity grids also in consideration of Slovakia.**

Keywords: SMR; Smart grid; microgrid; hybrid power system

I. Introduction

Implementation of the digital solution to the management of electric power grids is driven by the worldwide move towards carbon neutrality. Establishing so called smart grids is a key component for future and sustainable energy systems [1]. The Smart Grid will be an inherent part of the electricity power system which will intelligently integrate the actions of all users connected to the grid like generators, consumers and those that do both, in order to efficiently deliver sustainable, economic and secure electricity supplies. The development of the smart grids will be accelerated by the electrification or hydrogenation of the transportation, industry, and district heating. Other motivation behind the transformation of the electric power grids is not based only on the cost reduction, which can be realized by the automatization, but also establishing an involvement of the costumers [2]. Therefore, the electricity grids are approaching a new era and could undergo a fundamental architectural change [3].

Special part of the Smart grid, or its subpart, is the microgrid, where diverse energy sources can be integrated. The microgrid is able to combine multiple sources, loads and energy storages to a self-sustaining energy system that can be operated also independently to a large-scale utility grid [4,5]. The microgrid can eliminate power transmission losses and provide increased resiliency especially in highly industrialized or rural areas. Thus are managed locally and are considered as a single entity in the electricity grid [6].

A potential solution as a power source suitable for deployment in the microgrids is a Small Modular Reactor (SMR) as a dispatchable energy source [7]. The SMR can provide low-carbon power at a cost comparable or below that is anticipated for diesel generation. The size of considered SMRs is competitive to the renewable energy sources planned for microgrid. The main advantages of the SMRs, compare to renewable sources, are power generation flexibility, which can be combine with local industry within non-electrical application, and long-term stability due to long refueling cycles [6]. In special cases, developers are going a step further and are looking to address yet another market niche for smaller sources of reliable power supply in remote places, such as a mine where the only source of reliable power comes from local off-grid combustion generation [8].

II. Small Modular Reactor

Development of the SMR technology can be considered as an evolution and an adaptation of the nuclear industry to the global power market. Limitations of grid capacity and slow growth in power demand led to exploring smaller and scalable nuclear power technologies, which can be used either to replacement of large nuclear power plants or to supplement existing installed capacity. Moreover, the growing utilization of intermittent renewable capacity like wind, solar, tidal and geothermal generation requires introducing small baseload nuclear plants with enhanced load following capabilities to stabilize the supply to the grid. Generally, new nuclear reactor, or any other power unit in the electric power grid should not represent more than 10 % of the installed capacity [8]. The main reason is to ensure that a single reactor trip does not involve unacceptable grid fluctuations in the system voltage and frequency. At the same time, stability of the grid should be ensured to continue to provide reliable supply of electricity.

In 2022, 51 nuclear power reactors were under construction under various stages with total net installed capacity of 53 644 MWe [9]. All of these power reactors are evolutionary based on small prototype or demonstration reactors where the former designs ranged from less than 100 MWe to no more than 300 MWe [8]. Small number of these demonstration units are safely operated till today. Therefore, large experience exists from construction and operation of such small units. The main interest lies now to enhance safety features, integrate primary components to a compact design, minimize force or external power regulation and mitigate proliferation risks. Several concepts of the reactor designs exist and are based on variety of large scale units. The most expanded concept is derived from pressurized water reactors (PWR), but Generation IV concepts offer variants in the scale of SMR in all their alternatives:

- Supercritical water-cooled reactor (SCWR);
- Very-high-temperature reactor (VHTR);

- Molten salt reactor (MSR);
- Sodium-cooled fast reactor (SFR);
- Lead-cooled fast reactor (LFR);
- Gas-cooled fast reactor (GFR).

The nuclear industry is a quite conservative market, which means that while the reactor or NPP is not constructed and connected to the grid, it is still just a paper concept. To demonstrate viability of all design features, three demonstration SMR units are in an advance stage of the deployment with scheduled begin of the commercial operation between 2020 and 2023:

- Akademik Lomonosov in the Russian Federation;
- CAREM in Argentina;
- HTR-PM in China.

More commercial fleets of SMRs are expected to start operation in 2025–2035.

CAREM, see Fig. 1, was one of the first concepts to fully integrate the pressurizer, control rod drive mechanisms, and 12 mini-helical, once-through steam generators inside the reactor pressure vessel. Because of its innovative integration of major PWR components into a single unit, this particular concept has been referred to as an integral pressurized water reactor (iPWR) and is now the basis for several designs around the world. A 32 MW(e) prototype plant is currently under construction to validate the innovative technologies, prove the licensing approach and study reliability issues. CNEA envisions that the prototype will be followed by construction of a larger version, providing an electrical power output of between 100 and 200 MW(e). [8]

Figure 1. Construction of the CAREM reactor [10]

China started its R&D program on high temperature gas cooled reactors (HTGRs) in the 1970s. In 1992, the Chinese Government approved the construction of a small high temperature gas cooled test reactor (HTR-10) and in January 2003 the reactor reached full power (10 MWth) operation. Since that time, Tsinghua University's Institute of Nuclear and New Energy Technology (INET) has used the HTR-10 to conduct experiments and verify crucial inherent safety features associated with modular HTRs, see Fig. 2. China's next step in HTGR development began in 2001 when the High Temperature Gas Cooled Reactor–Pebble Bed Module (HTR-PM) project was launched. The HTR-PM will be a commercial demonstration plant for electricity production. Compared with the HTR-10, the components for the HTR-PM have been scaled up, although the design bases are similar. In the HTR-PM demonstration plant two reactor modules (each with their own SG) are connected to a single turbine generator set. This configuration will allow the developer to fully demonstrate the feasibility of connecting multiple modules to a single turbine. Future commercial deployment of the HTR-PM is likely to connect as many as six modules together to a single turbine. [8]

Figure 2. HTR-10 site location in China [11]

The KLT-40S is a PWR developed for a barge mounted nuclear power plant and provides 35 MW(e) per module (Table A–3). The design is based on the commercial KLT-40 marine propulsion plant and is an advanced version of the reactor, which historically provided power to icebreakers under more severe conditions when compared with stationary nuclear power plants. The floating nuclear power plant, see Fig. 3, is being manufactured in shipyards and can be delivered to customers as fully assembled, tested and ready for operation. This approach eliminates the need to build special transportation links and preparatory infrastructure, which are required for land based nuclear plants. The KLT-40S is designed with proven safety solutions such as a compact structure of the SG unit with short nozzles connecting the main equipment, small diameter primary circuit pipelines, and proven reactor emergency shutdown actuators based on different operation principles. [8]

Figure 3. Akademik Lomonosov transported to permanent location [12]

III. SMR APPLICATIONS

The most promising deployment areas which have been identified for SMR technology are as follows: (1) remote or isolated communities; (2) heavy industry; (3) on-grid applications [6]. The compact and modular designs allow SMRs to be deployed scalable in a wide variety of environments and applications, including those that require both electrical and thermal power sources [13]. Several modules can be incorporated into microgrid to meet specific load demand as a reliable low-carbon electric source or thermal energy provider. The SMR powered microgrid is able to manage residential, commercial, and industrial loads also in combination with solar photovoltaics panels, wind turbines and energy storage systems (ESS) [6].

ESS can be classified based on various forms of energy used to store electric power. Generally, ESS can be characterized as follows: (1) mechanical; (2) electrochemical; (3) electrical; (4) thermal, (5) hybrid, and (6) chemical [14]. The most commonly utilized ESS for microgrid application is Battery-based Energy Storage System (BESS) [15], Compressed Air-based Energy Storage System [16], Flywheel-based Energy Storage System [17], SuperCapacitor-based Energy Storage System [18], Superconducting Magnet-based Energy Storage System [19], Hydrogen Fuel Cell-based Energy Storage System [20], and Hybrid Energy Storage System [21].

The idealized concept considers operating the SMR in the load follow mode. The microgrid energy management system is than responsible for selecting when and how much to change the SMR output power level. The optimal operation strategy for the whole microgrid have to deal with multiple objectives and constraints, which include fuel costs, thermal stress on heat transport systems, efficiency, reserve capacity, reliability, and equipment conditions. For example, predictions of the overnight load demand can be used to schedule output power level adjustments for the SMR over that period, while the ESS can be used to compensate for any short-term discrepancy between the available power and the load demand.

An alternative approach to load follow operation of the SMR is to hold the reactor at constant power level and alternate the power to ESS or redirect a portion of the thermal output to another industrial thermal process [22]. Thus, effectively maximize the use of the generated thermal energy while still meeting the electricity demand. Some industrial processes can utilize both thermal and electrical energy. Several combinations of this approach have been proposed, also considering a combination of electrical and thermal outputs for hydrogen production using a high-temperature electrolysis process, which is more efficient than conventional electrolysis [23]. The hydrogen is an alternative reservoir for storing non-utilized reactor output. The hydrogen production can be very flexible due to that the electrolysis process can be restarted on demand, while the quality of the hydrogen production is maintained. The hydrogen serves as an energy carrier, which can be subsequently used in the transportation or reverted back to the electricity grid using fuel cell modules. Taking into account this approach a faster return on investment can be achieved than pure nuclear to electricity applications [6]. Related approaches have been proposed how the redundant or non-utilized heat energy can be used. The list of suitable processes and SMR candidates are shown in Tab. 1. The most simple solution which can effectively applied in coastal or island regions is to operate desalination plant [24]. Potable as well as technical water can be easily stored and transported to regions where needed. The use of hybrid systems has direct consequences to the microgrid operating strategy and dynamic models should be considered [6].

From the point of view of Slovakia, the eastern part of the republic can ideally serve for the installation of SMR. There is no significant source of electricity in this locality, but the region covers metallurgical production and several larger cities. There are several hydroelectric power plants, solar power plants, but also thermal power plants in the region. At the same time, the interconnection to the nationwide electricity system is not ideal. The deployment of at least one 100 - 300 MWe unit would allow to reduce the region's dependence on fossil fuels, mostly imported from abroad, make more efficient use of hydropower plants and accelerate deployment of carbon neutral sources. Stable production of the base band and guaranteed energy prices prove to support the development of industry in the region. Although the network connection with the surrounding countries could be characterized as basic, it is nevertheless sufficient for the export of electricity. In this way, the network would not be pushed to the west, and at the same time a self-sufficient region would be created, which could be characterized from the microgrid view. Strategic location of the SMRs in the region would allow also hybrid utilization. Connection of the SMR with district heating system is the most natural and most effective exploitation of SMR potential. Alternative could be also currently developing hydrogen industry for mobility applications.

Table 1. Industrial heat processes and SMR candidates [25]

Selected processes	*Process temperature (°C)*	*Candidate reactors for heat input*
District heating	80-200	PWR or waste heat from higher temperature reactor
Seawater desalination		
Feedstock drying, minerals concentration		
Petroleum refining	250–550	PWR, MSR, FR
Oil shale and oil sands processing	300–600	MSR, FR
Cogeneration of electricity and steam	350–800	MSR, FR, VHTR, PWR with heat augmentation
Steam reforming of natural gas or coal (methanol production); synthetic gas production	500–900	MSR, FR, VHTR, PWR with heat augmentation
Hydrogen production via HTSE[1]	800-1000	FR, VHTR, MSR, or PWR with heat augmentation
Hydrogen production via steam methane reforming		
Calcination		
Coal gasification		

[1]High Temperature Steam Electrolysis

IV. CONCLUSION

Implementation of the SMR technologies is the most promising alternative for expansion of the nuclear energy market relying on stability in the energy production to the grid in strategic locations, remote territories or countries with small or non-interconnected electricity grid where large nuclear units would not be suitable. Moreover, due to adaptation of the environmental policies and growing effort to combat CO2 emissions, older fossil fuel plants need to be phased out. These localities have already developed electricity grid infrastructure and can accommodate new power plant with comparable power output. The main advantage, which should be highlighted, of the SMR utilization is the hybrid application. Standalone or fleet of SMRs can work as energy island and backbone of the microgrids almost universally for all options. Highly industrialized areas, remote locations or as support of expanding, but intermittent, renewable sources, SMRs are the solutions to meet the energy demand and fulfil low-carbon policy. The biggest drawback of the SMR is the current unit prize. Although, the term 'modular' is often related to an approach where the design, components, assemblies, or structures are subdivided into modules that can be more easily assembled and manufactured, which should primarily decrease construction and maintenance prices, the first deployed SMRs will be constructed as on-site facilities with perhaps the testing of prototypical small volume module manufacturing lines to acquire manufacturing experience.

Acknowledgment

This publication was created thanks to support under the Operational Program Integrated Infrastructure for the project: International Center of Excellence for Research on Intelligent and Secure Information and Communication Technologies and Systems – 2. stage, ITMS code: 313021W404, co-financed by the European Regional Development Fund, by the Scientific Grant Agency of the Ministry of Education of Slovak Republic No. VEGA 1/0615/21 and partially by the ESFR-SIMPLE project funded by the European Union's Horizon Europe research and innovation program.

References

[1] ENTSOG and ENTSO-E, TYNDP Scenario report, June 2020.

[2] European Commission, Joint Research Center (JRC): Global Energy and Climate Outlook 2019: electrification for the low-carbon transition, 2020.

[3] S. Borenius, H. Hämmäinen, M. Lehtonen, P. Ahokangas, "Smart grid evolution and mobile communications-Scenarios on the Finnish power grid," *Electric Power Systems Research*, vol. 199, pp. 1-13, October 2021.

[4] S. Parhizi, H. Lotfi, A. Khodaei, S. Bahramirad, "State of the art in research on microgrids: a review," *IEEE Access*, vol. 3, pp. 890-925, 2015.

[5] N. Hatziargyriou, H. Asano, R. Iravani, C. Marnay "Microgrids," *IEEE Power and Energy Magazine*, vol. 5, pp. 78-94, 2007.

[6] D. Michaelson, J. Jiang, "Review of integration of small modular reactors in renewable energy microgrids," *Renewable and Sustainable Energy Reviews*, vol. 152, pp. 1-9, December 2021.

[7] J. Vujić, R.M. Bergmann, R. Škoda, M. Miletić, "Small modular reactors: simpler, safer, cheaper?," *Energy*, vol. 45, pp. 288-295, 2012.

[8] IAEA, "Technology Roadmap for Small Modular Reactor Deployment," IAEA nuclear energy series, No. NR-T-1.18, August 2021.

[9] PRIS, "Power Reactor Information System. World Statistics. Net Electrical Capacity and Number of Reactors." Internet: https://pris.iaea.org/PRIS/WorldStatistics/UnderConstructionReactorsByCountry.aspx, [Apr. 01, 2022].

[10] World Nuclear News, "Nucleoeléctrica contracted to complete CAREM-25." Internet: https://world-nuclear-news.org/Articles/Nucleoelectrica-contracted-to-complete-CAREM-25, [March 26, 2022].

[11] Tsinghua University News, "World's first HTR-PM nuclear power plant connects to grid." Internet: https://www.tsinghua.edu.cn/en/info/1244/10921.htm, [Apr. 01, 2022].

[12] Power technology, "Is floating nuclear power a good idea?" Internet: https://www.power-technology.com/analysis/floating-nuclear-power/, [Apr. 01, 2022].

[13] G. Iyer, N. Hultman, S. Fetter, S. H. Kim, "Implications of small modular reactors for climate change mitigation," *Energy Economics*, vol. 45, pp. 144-154, September 2014.

[14] S. Choudhury, "Review of energy storage system technologies integration to microgrid: Types, control strategies, issues, and future prospects," *Journal of Energy Storage*, vol. 48, pp. 1-19, April 2022.

[15] I. Alsaidan, A. Khodaei, W. Gao, "A comprehensive battery energy storage optimal sizing model for microgrid applications," *IEEE Trans. Power Syst.*, vol. 33, pp. 3968-3980, April 2017.

[16] H. Ibrahim, K. Belmokhtar, M. Ghandour, "Investigation of usage of compressed air energy storage for power generation system improving-application in a microgrid integrating wind energy," *Energy Procedia*, vol. 73, pp. 305-316, 2015.

[17] A.K. Arani, H. Karami, G. Gharehpetian, M. Hejazi, "Review of flywheel energy storage systems structures and applications in power systems and microgrids," *Renew. Sustain. Energy Rev.*, vol. 69, pp. 9-18, 2017.

[18] F.A. Inthamoussou, J. Pegueroles-Queralt, F.D. Bianchi, "Control of a supercapacitor energy storage system for microgrid applications," *IEEE Trans. Energy Convers.*, vol. 28, pp. 690-697, March 2013.

[19] T.T. Nguyen, H.J. Yoo, H.M. Kim, "Applying model predictive control to SMES system in microgrids for eddy current losses reduction," *IEEE Trans. Appl. Supercond.*, vol. 26, pp. 1-5, April 2016.

[20] S.A. Konstantinopoulos, A.G. Anastasiadis, G.A. Vokas, G.P. Kondylis, A. Polyzakis, "Optimal management of hydrogen storage in stochastic smart microgrid operation," *Int. J. Hydrog. Energy*, vol. 43, pp. 490-499, January 2018.

[21] M.S. Guney, Y. Tepe, "Classification and assessment of energy storage systems," *Renew. Sustain. Energy Rev.*, vol. 75, pp. 1187-1197, 2017.

[22] M.F. Ruth, O.R. Zinaman, M. Antkowiak, R.D. Boardman, R.S. Cherry, M.D. Bazilian, "Nuclear-renewable hybrid energy systems: opportunities, interconnections, and needs," *Energy Convers. Manag.*, vol. 78, pp. 684-694, 2014.

[23] D. Ingersoll, Z. Houghton, R. Bromm, C. Desportes, M. McKellar and R. Boardman, "Extending nuclear energy to non-electrical applications," *The 19th pacific basin nuclear conference (PBNC 2014)*, 2014, pp. 1-15.

[24] M. Rath, M.G. Morgan, "Assessment of a hybrid system that uses small modular reactors (SMRs) to back up intermittent renewables and desalinate water", *Progress in Nuclear Energy*, vol. 122, pp. 1-7, April 2020.

[25] S.M. Bragg-Sitton, *Hybrid energy systems using small modular nuclear reactors (SMRs)*, in Handbook of Small Modular Nuclear Reactors (Second Edition), D.T. Ingersoll and M. D. Carelli, Woodhead Publishing, 2020, pp 34.

Analysis and Evaluation of Photovoltaic Cell Defects and Their Impact on Electricity Generation

Marek Pavlík, Michal Kolcun, Samuel Bucko
Department of Electric Power Engineering
Technical University of Košice
Košice, Slovak republic
marek.pavlik@tuke.sk, michal.kolcun@tuke.sk, samuel.bucko@tuke.sk

Abstract — **Renewable sources are currently a widely used source of electricity. They also supported within the European Union. Defects in photovoltaic panels often occur during the implementation and operation of photovoltaic panels. Some defects are not visible. In practice, it happens that no attention to defects that occur during the operation of the photovoltaic power plant. These defects cause a decrease in electricity production, which reduces the efficiency of this renewable source.**

This paper describes defects that occur during the implementation and operation of a photovoltaic panel. Based on the performed simulations in program Protheus, the authors describe how defects affect the performance of the photovoltaic panel. Finally, the measurement of the V-A characteristic of a photovoltaic cell without a defect and with a defect were realize. Based on these measurements, simulations were create to determine the parameters of the mathematical model of the photovoltaic cell. By changing these parameters, it is possible to find out how the production of electricity will change in practice.

Keywords: photovoltaics panel; electricity produce; P-N junction; MPP point; Photovoltaic cell parameters

I. Introduction

Photovoltaics (PV) is an area that deals with the direct conversion of the energy of light radiation (solar radiation in the visible region) into electrical energy. Large-area semiconductor structures are used for this conversion. The photovoltaic phenomenon was observe first by Becquerel in 1839 on electrochemical cells. The development of semi-conductor technology in the 1950s, the preparation of the P-N junction and the knowledge of the physical processes in the P-N junction also created the preconditions for the production of a solar cell with reasonable efficiency [1][2].

A crucial reason for intensifying research into photovoltaics for terrestrial use was realizing the fact that humanity's energy resources on earth are delectable, with alarming estimates assuming that some resources may be depleted in a very short time. Although the forecasts have been gradually adjusted, current projects do not speak of periods longer than several centuries (for coal). The advantage of solar photovoltaic cells is that they work safely, quietly, they do not need any fuel, they do not produce waste, in most installations they have no moving parts and therefore they do not need maintenance [1][3].

Natural conditions in Slovakia are favorable for the permanent use of solar radiation as a source of energy for photovoltaic junction, similar to that in Austria, the Czech Republic or Germany. The development of PV in Slovakia supported by the binding EU requirement for wider involvement of RES in the overall energy balance. Although energy from photovoltaic modules cannot cover the total electricity demand, it will contribute to meeting the targets of covering the production of electricity from renewable sources, which means reducing dependence on energy supplies from abroad. In addition, PV represents a major manufacturing industry and has an impact on employment in the region [2][3][4].

II. Photovoltaic cell parameters

A. *Open-circuit voltage U_{OC}*

This is the maximum value of the voltage at the output terminals of the PV cell at zero current, without a connected load at a given temperature and light intensity. The U_{OC} value ranges from tens to hundreds of Volts.

B. *Short circuit current I_{SC}*

This is the maximum value of the current at a given lighting and at zero cell voltage. It is thus equal to the current generated by the light $I_{SC}=I_L$, assuming that the resistance R_S is zero. The magnitude of the short circuit current depends on the intensity of illumination, spectral sensitivity, irradiated area and temperature and varies in the order of units up to tens of A.

C. *Maximum power P_{MPP}*

The main unit is Wp (Watt Peak – top performance). This is the maximum power that the panel is able to deliver under STC conditions. The value of the maximum power is given by the relation: [6]

$$P_{MPP} = U_{MPP} * I_{MPP} \quad (1)$$

where the voltage U_{MPP} is at the point of maximum power of the panel and where the jet I_{MPP} is at the point of maximum power of the panel.

D. *Effectiveness η*

The efficiency of a photovoltaic panel is determined by the properties of the materials from which the panel and the photovoltaic cell are made. The material affects the spectral sensitivity of the cell to incident radiation, which means that the panel cell uses energy of different wavelengths with different efficiencies. The efficiency of a photovoltaic panel cell can be defined by the formula:

$$\eta_{ef} = \frac{P_{MPP}}{P_{rad}} = \frac{P_{MPP}}{E*A_c} \quad (2)$$

where P_{rad} is the power of the incident radiation, E is the light intensity under standardized test conditions; AC is the area of the photovoltaic cell.

E. Fill Factor FF

It depends on the quality of the contacts, the morphology of the material and the resistance of the active semiconductor layer. In translation, we know it as the load factor, and the higher this value, the more power the PV cell is able to deliver to the load. The average value for amorphous cells is around 0.6. Fill Factor is based on the relationship.

$$FF = \frac{P_{MPP}}{I_{SC}*U_{OC}} = \frac{U_{MPP}*I_{MPP}}{I_{SC}*U_{OC}} \quad (3)$$

F. Series resistance Rs

It is an indicator of the quality of a PV cell. It is a parasitic resistance which represents total resistance of the semiconductor material, the resistance of contacts and connections. A good cell should have the lowest R_S value, because its high value causes a voltage drop at the panel terminals.

G. Parallel resistance Rp

In most cases, it caused by extensive defects. Too low R_P value indicates a poor PV cell that behaves almost like a short circuit. The resistance value should be as high as possible [6] [7] [8].

III. Photovoltaic cell defects

When designing and operating a PV system, you need to strive for the lowest possible investment and operating costs. The key factors by which we can reduce these costs are to increase the efficiency, reliability and service life of PV panels. However, these panels are affected by various malfunctions and defects that reduce their effectiveness. According to the latest statistics, the performance of crystalline silicon panels decreases by about 0.8% per year. Therefore, in order to increase the reliability and service life of PV modules, it is necessary to understand why and how efficiency is reduced. This section deals with defects and failures of PV modules that occur during the production process, during installation or during operation, and the impact of defects on the efficiency, operation and life of the PV panel. Front / rear layer defects [6] [8] [9] [10].

As mentioned above, tempered glass is usually used for the front layer of the panel. When the glass is damaged, the intensity of sunlight can be reduced in the way that for-eign objects such as dust and other impurities can penetrate the cracks, obscuring the cell and thus reducing the performance of the PV panel – Fig. 1. If the glass is more damaged, water can penetrate the cells, which can cause the cell to corrode. Cell corrosion can lead to overall panel failure. If this layer is not significantly damaged during its entire life, this layer is constantly stressed by various thermal and mechanical stresses, which contribute to the slow degradation of the protective material, which causes a slight reduction in performance [11] [12] [13] [14].

If the backing layer is damaged due to various mechanical and thermal stresses, UV radiation can cause water to reach the cells, leading to damage to the cells.

Figure 1. Photovoltaic cell defects on the glass

A. Delamination of the PV panel

Delamination of the panel is a separation of the individual parts of the pan-el from each other and it is the separation of the EVA film from the protective glass or the PV cell itself. As result of this separation, various cavities and bubbles are form between the individual layers of the panel. This defect is very serious because water can enter these bubbles and cavities, which can cause corrosion of the PV cells, resulting in a malfunction of the PV panel and a decrease in the efficiency and performance of the PV panel due to decrease in the intensity of incident sunlight. In Fig. 2 we can see the delimitation of the panel [15].

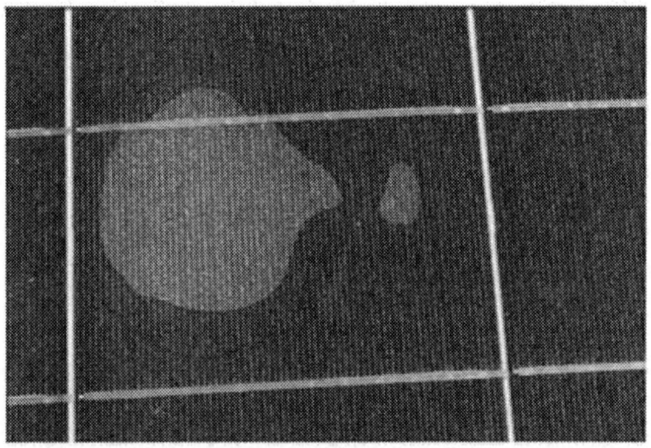

Figure 2. Delamination of the PV panel

B. Bypass diode defects

Bypass diodes are used to protect PV cells, even if one of the cells is shaded. In this case, the cell ceases to act as a source of electricity, and it begins to become an electrical appliance that consumes the energy produced by the other cells in the panel. The cell, which starts to behave like an appliance, reduces the output of the PV panel and, in addition, with this power dissipation, the cell begins to heat up, which can damage it. The amount of reduced power depends on how shaded the cells are. It is worse to shade one cell on a larger cell area than to shade several cells from a small cell area. In this case, the bypass diodes direct the current to flow around the cell and to prevent hot spots called HOT-SPOTS. Damage to the bypass diodes (Fig. 3) is rare and caused by insufficient current sizing, where in most cases they was damaged by excessive heating, which leads to the destruction of the entire diode. Ultimately, damage to the bypass diode is a major problem and can lead to a total failure of the panel [14].

Figure 3. Destroyed bypass diode

IV. Mathematical model of a photovoltaic cell

The P-N junction is a semiconductor diode. Depending on the polarity of the externally connected DC voltage, current (direct direction) or no current (closing direction) flows through the diode. The P-N junction current is given by the U-I characteristic given by the Shockley equation.

The replacement scheme of the photovoltaic cell is based on the principle of the Shockley equation and is shown in Fig. 4. The given scheme consists of a current source, an ideal diode, a parallel resistor R_{SH} expressing the stray current and a series resistor R_S, which represents the internal resistance [15] [16] [17] [18] [19].

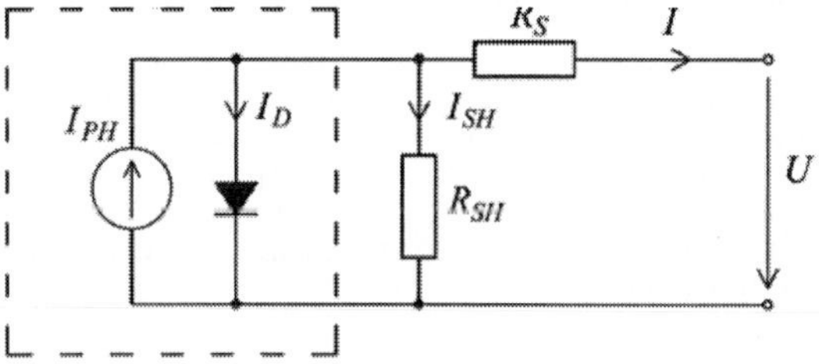

Figure 4. The replacement scheme of the photovoltaic cell [19]

A photovoltaic cell is a large-area semiconductor diode that operates in generator mode when light falls on it. In this mode, the diode operates in the 4th quadrant of the U-plane (Fig. 5). The limit values of the volt-ampere characteristic are the no-load voltage U_{OC} and the short-circuit current I_{SC}. It is interesting to note that the second expression in equation (10) is actually the equation of diode with a negative sign. That is, the graph of Equation (10) is actually the rotated volt-ampere characteristic (VACH) of the diode shifted upward by the I_{PH} current. In this way, the characteristic is moved to the 1st quadrant [15] [19] [120] [21].

External conditions such as the temperature of the solar cell or the intensity of the incident light have a significant effect on the shape of the VACH. Also, its shape is also affected by the internal properties of the cell (cell structure), such as the series resistance RS, which is required to have the lowest possible value, and the leakage resistance RSH, which is to reach a high value. The point on the VACH at which the solar cell reaches maximum power called the Maximum power point. The voltage and current values in this case will be indicate in fig. 6 as U_m or I_m. In the idle and short circuit state, no power is produce by the cell [15] [19] [22] [23].

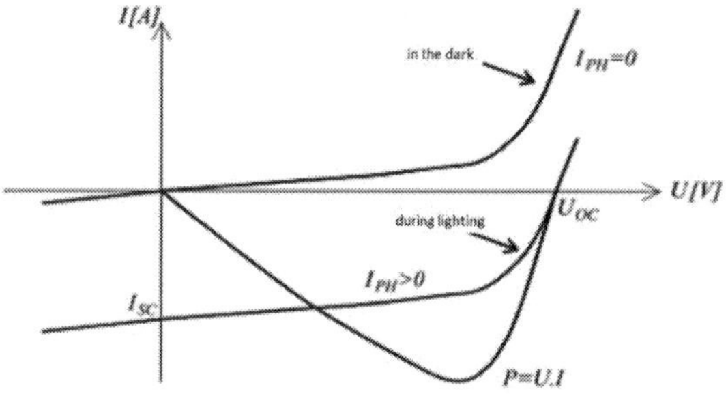

Figure 5. U – I and U – P characteristics of the photovoltaic cell - 4th quadrant [19]

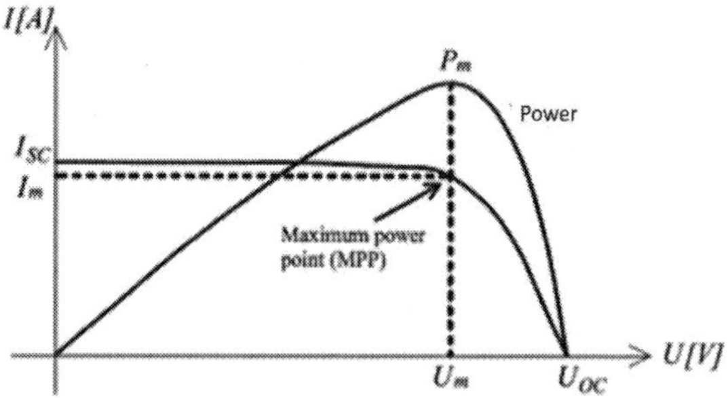

Figure 6. U – I and U – P characteristics of the photovoltaic cell - 1st quadrant [19]

V. Mathematical model of a photovoltaic cell

A. Change in solar radiation intensity

The intensity of sunlight changes constantly depending on the season and whether it is day or night. It also depends on the weather, geographical location, air pollution, etc. In this simulation, the value of the series resistor was set to 0.15 Ω, in the case of a parallel resistor it was the value of 150 Ω, also the value of the panel temperature was set to 25 °C, and all these values were constant during the simulation. Only the value of the battery voltage marked I has changed and this value represents the magnitude of the solar radiation intensity, while the value of 1 V represents the value of 1000 W/m^2 [17] [17] [24].

It is clear from the simulation results that the higher the solar intensity is, the more electricity the PV cell produces. Simulations were perform at five different intensities:

I=1000 (1 V), 800 (0,8 V), 600 (0,6V), 400 (0,4 V), 200 (0,2 V) W/m^2.

We have found that, for example, in a cloudy sky, there is a significant decrease in the output produced, which caused by the fact that the PV cell is exposed to much less solar radiation, i.e. there is significant decrease in the intensity of solar radiation. This decrease in the power produced is due to the fact that the curve itself changes so that by decreasing the intensity of the solar radiation the value of the short-circuit current I_{SC} decreases significantly, and therefore the no-load voltage U_{OC} also decreases, but not as intensely as in the case of current. From the simulation results, it is possible, for example, to determine the magnitude of voltages and currents in the MPP point at individual intensities, which shown in Fig. 7. It presented from the Fig. 7 that with decreasing radiation intensity, of course, the power of PMPP also decreases, and the current at the MPP point decreases, while the voltage at the MPP point increases. For example, at a radiation intensity of 1000 W/m2, the volt-age at the MPP point is 0.62 V, while at an intensity of 400 W/m2 the voltage increases by 0.05 V. Conversely, the current at the MPP point decreases sharply as the radiation intensity decreases.

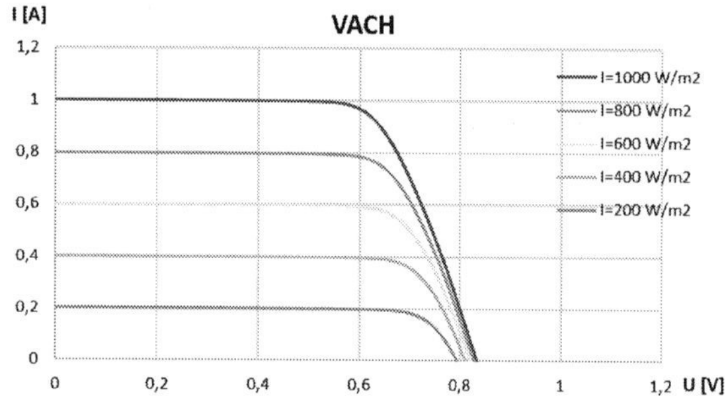

Figure 7. V-A characteristics of the photovoltaic cell - Change in solar radiation intensity

B. R_S series resistance change

This resistance represents losses on the panel and it is derived from the total resistance of the semiconductor material, the resistance of the contacts and connections, with the resistance value increasing at imperfect and damaged connections, which means higher transient resistance, leading to panel overheating. The magnitude of the resistance changes immediately after the panel put into operation, which caused by changes in the temperatures in which the panel is located.

It is clear from the simulation results that if the resistance value on the resistor R_S increases, the power of the PV cell decreases, which can be seen in the characteristics in Fig.8. In this figure, five different V-A characteristics are shown, each representing a different value of the resistors R_S, namely 0.1 values; 0.15; 0.3; 0.6 and 1 Ω. The values of the parallel resistor R_S, the intensity of solar radiation and the panel temperature were un-changed in the simulation. We have found that, ideally, the PV cell should have the resistance value of the resistor R_S as low as possible, then the cell has an almost ideal elbow shape respectively the V-A characteristic curve, which is shown in fig. 10 at a resistance value of 0.1 Ω. It can also be noticed that as the resistance increases, the shape of the curve changes in such a way that the current and voltage changes significantly with varying load, while the short-circuit current I_{SC} and no-load voltage U_{OC} remain the same, but this does not apply to the resistance value 1Ω, when the short-circuit current value ISC also changes. These curve shapes are undesirable because they cause a decrease in the power produced.

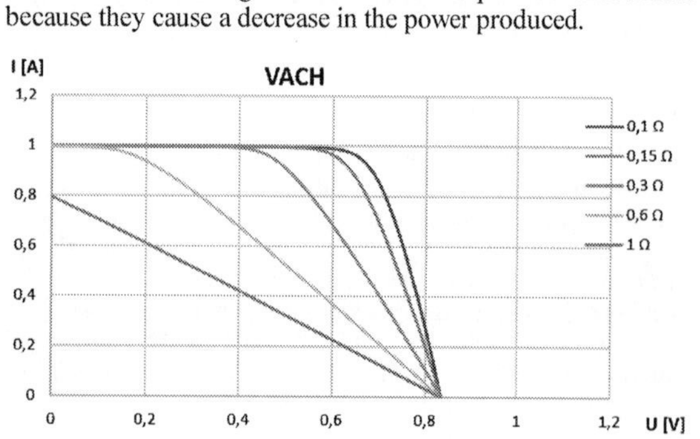

Figure 8. V-A characteristics of the photovoltaic cell - RS series resistance change

The results of the simulation show that as the series resistance of the PV cell increases, the U_{MPP} voltage, the I_{MPP} current and, of course, the P_{MPP} power decreases. For example, with a series resistance of a PV cell of 0.1 Ω, the voltage at the MPP point is 0.66 V, while if the resistance increases to 1 Ω, the voltage drops by about a third, i.e. to 0.43 V. The I_{MPP} current increases with resistance from 0.1 Ω to 1 Ω drops by more than half.

Based on the created simulations, we performed an experiment. We selected a photovoltaic cell that did not show signs of defects described in this paper. We measured the V-A characteristic on this photovoltaic cell. Measuring was performing by EKO MP-11 measuring instrument – see Fig.9. The V-A characteristic of the photovoltaic cell is shown in Fig. 13 - without defect. The resulting measured values were adjusted by the least squares method so that they could be compared with the values from the simulation.

Based on this measurement, we performed simulations on a mathematical model of a photovoltaic panel. The resistance values were set as follows - R_S = 0.15 Ω, R_P value 150 Ω. Subsequently, we changed the values of the resistors to achieve the same V-A characteristic as in the measurement. By searching for the correct R_S and R_P values, we achieved a mathematical model of a real photovoltaic panel. This panel may already have signs of defects.

Subsequently, we released the contacts of the photovoltaic cell, which increased the resistance of the contacts. In Fig. 10 it is possible to see the V-A characteristic after releasing the contacts (with defect). After this measurement, simulations were perform with a change in R_S resistance. The R_S resistance was changed with an interval of 0.01 Ω from 0.05 Ω to 2 Ω until a similar curve as the measurement was obtained. By simulation, it was found that the defect on the photovoltaic panel represents a resistance R_S = 0.35 Ω.

Based on the simulation and measurement, it is possible to observe the displacement of the MPP point. It is also possible to observe a decrease in the I_{MPP} current and the U_{MPP} voltage. This observation shows that the P_{MPP} will also change, which results in a decrease in the output of the photovoltaic cell and also a decrease in the production from the photovoltaic cell and the photovoltaic panel.

Based on the comparison of simulation and measurement on the PV panel, it is possible to identify the defect. According to the shape of the curve, it is possible to define the series and parallel resistance in the PV panel model. From these resistances we can determine what defect is located on the photovoltaic panel.

Figure 9. Measuring of V-A characteristic

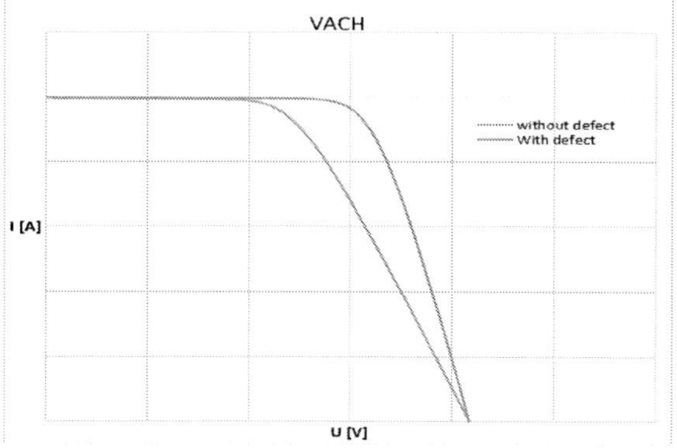

Figure 10. V-A characteristics real photovoltaic cell – with and without defect

VI. Conclusion and discussion

From the simulation results, we can observe the impact of photovoltaic cell / panel defects on the change of V-A characteristics and thus the impact on electricity production. It is necessary to realize that a defect on one panel will cause a decrease in electricity production in the entire serial string. Series and parallel resistances in the PF cell model rep-resent defects.

Measurements of volt-ampere characteristics were performing on the photovoltaic cell that has no defects. Based on the obtained V-A characteristic, simulations were created, where the values of R_S and R_P were adjusted so as to achieve a similar curve of V-A characteristic by measuring. In the next step, a defect created on the photovoltaic cell - the contacts of the photovoltaic cell released. In the mathematical model, this change represented by a change in the resistance of the R_S. In the mathematical model, the value of the resistance of the R_S was changed with the interval 0.01 Ω from the value 0.05 Ω to 2 Ω, until a similar course as the measurement was achieved. By simulation, it was found that the defect on the photovoltaic panel represents a resistance $R_S = 0.35$ Ω. The photovoltaic panel was new and therefore without internal defects. Therefore, a parallel resistance of 400 Ω was determined in the simulations. At higher values, there are no longer any changes in the V-A characteristic. If the panel shows signs of damage, it is possible to assume that the value of the parallel resistor would also change. If the panel is damaged, this value would decrease, which would also change the shape of the V-A characteristic curve. This would also change the power supplied from the photovoltaic panel and thus its production of electricity.

Acknowledgment

This work was supported by the Slovak Research and Development Agency under the contract No. APVV-19-0576 and by the Ministry of Education, Science, Research and Sport of the Slovak Republic and the Slovak Academy of Sciences under the contract VEGA 1/0757/21.

This work was supported by project SK-UA-21-0024 Energy management improvement of hybrid photovoltaic systems of local objects with storage batteries.

This work was supported by the Slovak Research and Development Agency under the contract No. APVV-19-0576 and by the Ministry of Education, Science, Research and Sport of the Slovak Republic No. APVV-21-0312.

References

[1] M. Pavlík, "Obnoviteľné zdroje energie vo všeobecnosti," 1.vyd. Technická univerzita v Košiciach, 2019. pp. 75. ISBN 978-80-553-3317-5.

[2] S.A. Kalogirou, "Solar Energy Engineering - Processes and Systems," lsevier Science & Technology. 2014. pp. 841. ISBN: 978-0-12-397270-5.

[3] M. Kolcun, D. Medveď, J. Petráš, R. Stolárik and Š. Vaško, "Výskum charakteristík fotovoltaických komponentov pre efektívne projektovanie solárnych systémov," 1.vyd. Technická univerzita v Košiciach. 2014. pp. 120. ISBN 978-80-553-1961-2.

[4] D. Medved, J. Zbojovsky, „Influence of the Position of the Electrical Contact on the Size of the Temperature Distribution." *Przegląd Elektrotechniczny*, vol. 97, pp. 230-233, 2021.

[5] M. Bereš, D. Kováč, T. Vince, I. Kováčová, J. Molnár, I. Tomčíková, J. Dziak, P. Jacko, B. Fecko and Š. Gans, "Efficiency Enhancement of Non-Isolated DC-DC Interleaved Buck Converter for Renewable Energy Sources." *Energies*, vol. 14, pp. 1-15, 2021.

[6] P. Jacko, M. Bereš, I. Kováčová and D. Kováč, "Remote IoT Education Laboratory for Microcontrollers Based on the STM32 Chips," *Sensors*, vol. 22, pp. 1-22, 2022.

[7] J. Dzmura, J. Petráš, J. Balogh, J. Kurimský, R. Cimbala, I. Kolcunová, B. Dolník and M. Kolcun, "Separation of solid particles from flowing gases by AC high voltage." *Journal of electrostatics*, vol. 88, pp. 158-166, 2017.

[8] Z. Conka, P. Hocko, M. Novák, M. Kolcun and G. Morva, "Impact of renewable energy sources on stability of EWIS transmision system," *14th International Conference on Environment and Electrical Engineering (EEEIC)*, 2014, pp. 1-5.

[9] I. Kolcunová, R. Cimbala, J. Petráš, J. Kurimský, B. Dolník, J. Džmura and J. Balogh, "Experimental observation of negative differential characteristic of corona discharge in ultraviolet spectrum." *Journal of electrostatics*, vol. 88, pp. 139-147, 2017.

[10] S. Said, A. Massoud, M. Benammar, S. Ahmed, "A Matlab/Simulink-Based Photovoltaic Array Model Employing SimPowerSystem Toolbox," *Journal of Energy and Power Engineering*, vol. 6, pp. 1965-1975, 2012.

[11] A. Zielińska, M. Skowron, A. and Bień, "Modelling of photovoltaic cells in variable conditions of temperature and intensity of solar insolation as a method of mapping the operation of the installation in real conditions," *2018 International Interdisciplinary PhD Workshop (IIPhDW)*, 2018, pp. 200-204.

[12] H.H. Azri, A.F. Siti, A.W.M. Zuhdi, M.S. Bahrudin, F. Zaabar and M.N. Harif, "Modelling and Simulation of Photovoltaic Solar Cell using Silvaco TCAD and Matlab Software," *2018 IEEE International Conference on Semiconductor Electronics (ICSE)*, 2018, pp. 214-217.

[13] J. Zhang, J. Liu, K. Ding, L. feng, F.U. Hamelmann and X. Chen, "Model Parameter Analysis of Cracked Photovoltaic Module under Outdoor Conditions," *2020 47th IEEE Photovoltaic Specialists Conference (PVSC)*, 2020, pp. 2509-2512.

[14] Z. Cui, "PROTEUS-based simulation platform to study the photovoltaic cell model under partially shaded conditions," *2011 International Conference on Electric Information and Control Engineering*, 2011, pp. 3446-3449.

[15] E. M. G. Rodrigues, R. Godina, E. Pouresmaeil and J.P.S. Catalao, "Simulation study of a photovoltaic cell with increasing levels of model complexity," *2017 IEEE International Conference on Environment and Electrical Engineering and 2017 IEEE Industrial and Commercial Power Systems Europe (EEEIC/I&CPS Europe)*, 2017, pp. 1-5.

[16] N. Das, A.A. Ghadeer and S. Islam, "Modelling and analysis of multi-junction solar cells to improve the conversion efficiency of photovoltaic systems," *2014 Australasian Universities Power Engineering Conference (AUPEC)*, 2014, pp. 1-5.

[17] S. Chtita, Y. Chaibi, A. Derouich, and J. Belkadid, "Modeling and Simulation of a Photovoltaic Panel Based on a Triple Junction Cells for a Nanosatellite," *2018 International Symposium on Advanced Electrical and Communication Technologies (ISAECT)*, 2018, pp. 1-6.

[18] T. Maniak,"Modelovanie parametrov fotovoltickýchh článkov," bakalárska práca, 2021, Technická univerzita v Košiciach, pp.77.

[19] A. Belan, "Model fotovoltického článku," Posterus, vol. 6, No.10, ISSN 1338-0087.

[20] J. Molnár, S. Kirešová, T. Vince, D. Kováč, P. Jacko, M. Bereš and P. Hrabovský, "Weather Station IoT Educational Model Using Cloud Services." *J. Univers. Comput. Sci.*, vol. 26, pp. 1495-1512, 2020.

[21] S. Z. Aljoaba, A.M. Cramer and B.L. Walcott, "Thermo-electrical modeling of light wavelength effects on photovoltaic cell performance," *2011 37th IEEE Photovoltaic Specialists Conference*, 2011, pp. 000192-000192.

[22] W. Hayder, B.H. Mouda and A. Aicha, "Modeling of a photovoltaic cell based on recurrent neural networks," *2017 International Conference on Green Energy Conversion Systems (GECS)*, 2017, pp. 1-5.

[23] J. Zbojovsky, A. Meszaros, P. Kurimsky, "Shielding of the electromagnetic field in the range from 0.9 to 6 GHz by experimental shielding coatings," *2018 IEEE International conference and workshop in Obuda on eletrical and power engineering*, 2018, pp. 273-276.

[24] D. Gupta, M. Kumari and A. Samadhiya, "Photovoltaic Modeling using Single Diode Model in MATLAB," *2020 IEEE International Conference on Computing, Power and Communication Technologies (GUCON)*, 2020, pp. 734-739.

Review of the Methods for Reducing Torque Ripple of Induction Motor Direct Torque Control

Nikolay Djagarov
Electrical Department
Nikola Vaptsarov
Naval Academy
Varna, Bulgaria
jagarov@ieee.bg

Hristo Milushev
Electrical Department
Nikola Vaptsarov
Naval Academy
Varna, Bulgaria
h.milushev@nvna.eu

Julia Djagarova
Nikola Vaptsarov
Naval Academy
Varna, Bulgaria
julija_d@abv.bg

Evgeniya Panova
Nosov Magnitogorsk State
Technical University,
Magnitogorsk, Russia
ea.panova@magtu.ru

***Abstract* — Modern electric drives have requirements for efficiency, a wide range of speed and torque control, high response, and reliability. Induction drives take the dominant part of drives due to their advantages. For more than thirty years, induction electric drives with direct torque control have been researched, designed and introduced. The article provides an overview of induction electric drives with direct torque control and considers the main methods and means that improve these electric drives and eliminate their main disadvantages - ripple of torque and currents.**

***Keywords* — electric drive, induction motor, direct torque control, torque ripple, torque ripple reduction.**

I. Introduction

Induction motors are most widely used for a variety of electric drives, due to their advantages - simplicity, reliability, low cost. Effective speed regulation of induction drives is a major factor in the design of electric drives.

For many years, the scalar V/f method was the main method for controlling the speed of an induction motor [1], as it provided optimal operation at all values of frequency and load and operation with an almost constant power factor, static stability margin and absolute slip.

However, the V/f method had insufficient control speed and in the 70s, using the idea of DC motor controlling the speed, a vector (field oriented) method for controlling the speed of an induction motor was proposed, in which, in addition to voltage and frequency, the angle of the stator voltage of the motor was also controlled in relation to the rotor field [2]. For this, independent control is used along two orthogonal axes: the field is controlled along the direct axis *d*, and the torque is controlled along the quadrature axis *q*. Thus, the control efficiency is increased due to the predictive calculation of the control signal, which is very important in transient operating modes.

In the 80s, a new method of direct torque control (DTC) of a motor was proposed by controlling the supply voltage in terms of amplitude, frequency and phase [3, 4, 5, 15-21]. In contrast to vector control, pulse-width control of the inverter gates is not used, but individual control of each inverter switch. Thus, the values are not averaged over a certain time interval and increases the control speed.

The study was funded by Bulgarian Research Fund, Competition for Financial Support for Basic Research Projects 2021, Project title: "Integrated wireless sensor networks for diagnostics of ship electrical equipment"

DTC overcomes the disadvantages of vector control: the need for current regulators; coordinate transformations; pulse width modulation. DTC also provides high efficiency, high power/torque ratio and high reliability. The ease of DTC implementation ensures good torque control.

However, DTC has some disadvantages: *1*. Difficulty in torque control at low speed; *2*. Strong ripple of current and torque; *3*. More mechanical vibrations.

The article provides an overview of the methods used to reduce the current and torque ripple of induction motors. In the first chapter, methods of induction motor speed controlling are considered, in the second chapter - the features of direct torque control, in the third chapter - methods for reducing the torque ripple and switching frequency.

II. Induction Motor Control

A. Classification of Induction Motor Control Methods

Accurate control of the instantaneous torque of an induction motor is essential in highly efficient drives. Motor torque is the product of motor current and flux. To ensure a high torque of the motor, it is necessary to maintain the flux at a sufficiently high optimal level that does not saturate the magnetic circuit [6]. In an induction motor, the field and current are caused by the stator current, which can be controlled. Optimal motor torque occurs with independent control of flux and torque, the vectors of which are orthogonal.

Induction motors operate optimally at the rated torque and speed. To maintain efficiency while reducing speed and load, it is necessary to adjust the field in accordance with the load and speed. Induction drive control is dynamic, recurrent and non-linear. Therefore, many control methods cannot be applied due to the following difficulties: *1*. Complex calculations are required; *2*. Nonlinearity of systems; *3*. Unpredictable (load and noise) disturbances; *4*. Changes in motor parameters.

Various principles and methods are used to control induction drives. The methods of controlling an induction motor can be divided into four main categories (Table 1) [7,8].

To identify the rotor field (flux) of the motor in vector control, Hall sensors were first used, which reduced the reliability of the induction motor. Therefore, an indirect vector method is used a motor mathematical model. Moreover, this method of control is called sensorless. The vector control method has two loops: internal, flow and torque control, with a different type of evaluation;

external, speed control and creating a standard of torque and magnetic flux. Various types of torque control methods are used: linear (PID, PI); nonlinear (direct self-control, on-line optimization with model forecasting, neural networks, fuzzy logic) [9,10].

TABLE I. INDUCTION MOTORS CONTROL METHODS

Control method	Controller	Advantages/ Disadvantages
Scalar control	1. Open-loop voltage/ frequency (V/Hz) 2. Slip frequency and voltage control 3. Slip frequency and current control	Stable/performance dependent on machine parameters
Vector control	4. Direct vector control 5. Indirect vector control 6. Direct self-control 7. Acceleration control	High performance, flux and torque are controlled, performance is dependent on machine parameters, integral error accumulation
Sensorless control	8. Model reference adaptive control 9. Luenberger observer	No speed sensor, performance is dependent on machine parameters
Intelligent control	10. Fuzzy control	Performance approximates that vector control, less dependent on machine parameters
	11. Neural-network control	High performance, robust structure
	12. Expert system control	High performance, performance is independent of machine parameter variations, no integral error accumulation, performance is dependent on high precision speed sensor

In vector control, various coordinate systems for obtaining the motor model are used, with the help of which sensorless control is carried out. Most often, a coordinate system is used that rotates synchronously with the rotor magnetic flux. At a constant rotor flux amplitude, there is a linear relationship between the control variables and the torque. Other coordinate systems are used to separate and linearize the motor equations. For example, non-linear methods, in which the speed and flux of the rotor are separated by means of feedback or input-output separation [11]. The variational method (passive control method) is also used.

Another problem is the time lag in discrete feedback of identification and evaluation of the stator and rotor flux due to the sampling and computation time of the control signal [12].

Sampling period Δt can be calculated from:

$$\Delta\psi_s^s = u_i^s.\Delta t \tag{1}$$

where: $\Delta\psi_s^s$ - stator flux change; u_i^s - stator voltage during sampling.

B. Mathematical Model of Induction Motor

Induction motor currents, currents, voltages can be represented as vectors in different coordinate systems. Depending on the coordinate system used, the mathematical models of the motor will have a different shape.

In d,q orthogonal rooting axes [13]:

Stator

$$\begin{aligned} u_d &= r_s \cdot i_d + \frac{d}{dt}\psi_d - \omega_k \cdot \psi_q \\ u_q &= r_s \cdot i_q + \frac{d}{dt}\psi_q + \omega_k \cdot \psi_d \end{aligned} \tag{2}$$

Rotor

$$\begin{aligned} 0 &= r_r \cdot i_{rd} + \frac{d}{dt}\psi_{rd} - (\omega_k - \omega_r).\psi_{rq} \\ 0 &= r_r \cdot i_{rq} + \frac{d}{dt}\psi_{rq} + (\omega_k - \omega_r).\psi_{rd} \end{aligned} \tag{3}$$

where: the stator and rotor parameters and variables are denoted by the indices s and r, respectively; ω_k, ω_r - respectively angular speed of rotation of the axes d,q, and the rotor.

The ratios for flux and currents:

$$\begin{aligned} \psi_d &= l_s \cdot i_d + m_{ad} \cdot i_{rd} \\ \psi_q &= l_s \cdot i_q + m_{ad} \cdot i_{rq} \\ \psi_{rd} &= m_{ad} \cdot i_d + l_r \cdot i_{rd} \\ \psi_{rq} &= m_{ad} \cdot i_q + l_r \cdot i_{rq} \end{aligned} \tag{4}$$

where: l_s, l_r - respectively self-inductance of the stator and rotor; m_{ad} - mutual inductance between the stator and rotor windings.

The projections of the variables in the axes d,q can also be obtained with the help of their projections in the axes α,β (Fig. 1):

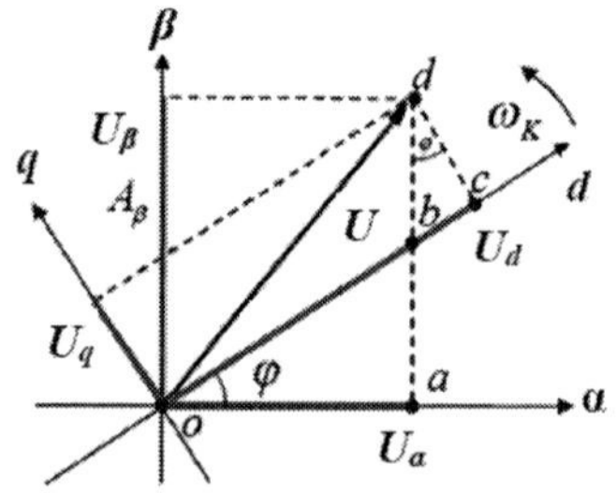

Fig. 1. Relationship between components of variables in the d,q and α,β axes

$$\begin{aligned} i_{s\alpha} &= i_{sd}\cos(\theta_k) - i_{sq}\sin(\theta_k) \\ i_{s\beta} &= i_{sd}\sin(\theta_k) + i_{sq}\cos(\theta_k) \end{aligned} \tag{5}$$

$$\begin{aligned} u_{s\alpha} &= u_{sd}\cos(\theta_k) - u_{sq}\sin(\theta_k) \\ u_{s\beta} &= u_{sd}\sin(\theta_k) + u_{sq}\cos(\theta_k) \end{aligned} \tag{6}$$

$$\theta_k = \int \omega_k dt \tag{7}$$

The torque of a three-phase induction motor depends on the rotor current, the magnetic flux interacting with the rotor and creating an electromotive force and power factor:

$$T_e = \psi.I_2 \cos\varphi_2 \tag{8}$$

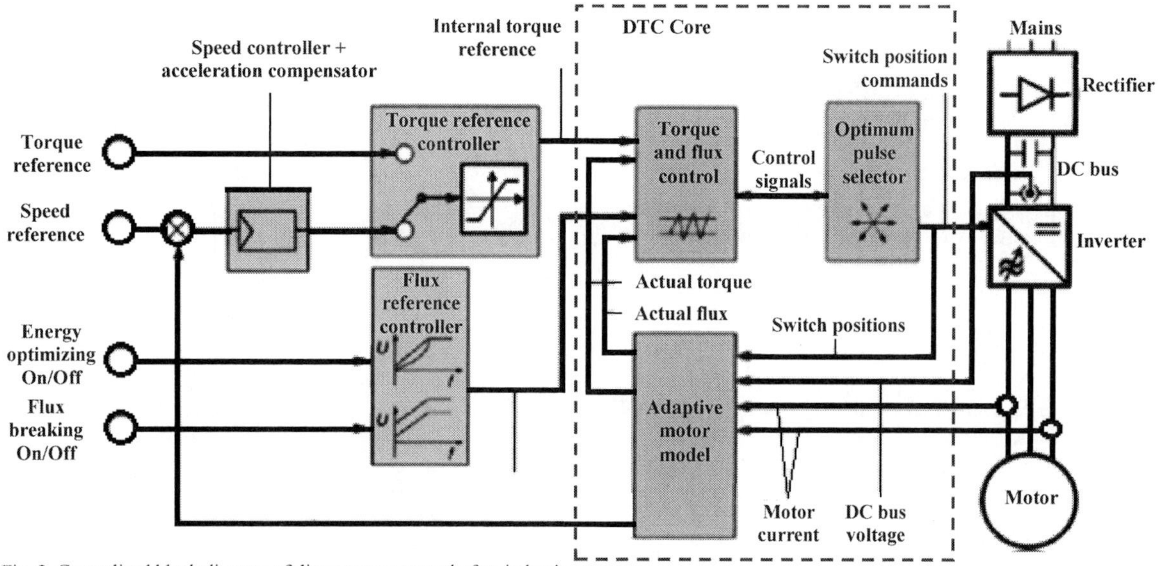

Fig. 2. Generalized block diagram of direct torque control of an induction motor

where: T_e - electromagnetic torque of the motor; ψ - flow; I_2 - rotor current; $\cos\varphi_2$ - rotor power factor.

The equation for the motor speed will take the following form [13]:

$$\frac{d}{dt}\omega_r = \frac{1}{\tau_m}\left(T_e - T_m\right) = \frac{1}{\tau_m}\left(\psi_{rd} \cdot i_q - \psi_{rq} \cdot i_d - M_m\right) = \frac{1}{\tau_m}\left[x_{ad}\left(i_{rd} \cdot i_q - i_{rq} \cdot i_d\right) - T_m\right] \quad (9)$$

where: $T_e = \psi_{rd} \cdot i_q - \psi_{rq} \cdot i_d = x_{ad}\left(i_{rd} \cdot i_q - i_{rq} \cdot i_d\right)$ motor electromagnetic torque; T_m – mechanical (brake) torque.

Transforming the equations of an induction motor in an orthogonal coordinate system allows you to independently control the flux and torque of the motor. However, you can choose other coordinate transformations to separate and linearize the motor equations. For example: non-linear transformation in which the speed and flux of the rotor are separated by feedback (linearization control with feedback or input-output separation). Or a passive control method based on the variational theory using the Euler - La Grange equations.

III. Classical Direct Torque Control

A. Main Characteristics

Due to its advantages, the method of direct torque control (DTC) of the induction motor is widely used in various electric drives. This method has high dynamic performance with very simple control circuits. Its main advantages: Simple direct control of stator torque and flux without speed feedback; Indirect regulation of stator currents and voltages; High static and dynamic speed control accuracy down to low speeds; Large starting torque; Torque linearity; Optimal control of inverter switch commutations depending on the load; Approximate sinusoidal form of stator currents and magnetic flux; No coordinate transformation.

Disadvantages: The need for a higher sampling rate (up to 40 kHz compared to 6-15 kHz with vector control), which increases the inverter switching losses; More complex motor model; Large ripple of torque and magnetic flux; Changing the switching frequency in accordance with the width of the hysteresis bands and the motor speed; Hysteresis method is used, the inverter has a continuously variable switching frequency what causes the acoustic noise spectrum; The choice of voltage vectors is not optimized within the magnetic flux hysteresis band; There are problems with motor starting; Torque and flux estimators are needed to identify the motor parameters.

Figure 2 shows the generalized block diagram of direct torque control of an induction motor [22]. The main blocks of the system, their functions, connections between them and signal transmission are shown. This scheme is universal, as it can be used to explain all existing DTC modifications.

The basic idea behind DTC is to select the best voltage vector that rotates the stator flux and produces the desired torque. During this rotation, the stator magnetic flux is maintained within the specified hysteresis band.

With vector control and with a constant rotor magnetic flux, torque control is performed by a transverse current:

$$T_e = \frac{l_m}{l_r}\psi^r . i_q^s = \frac{l_m}{l_r}\psi^r . i^s . \sin\delta \quad (10)$$

where: T_e - electromagnetic torque; ψ^r - rotor flux; i^s - stator current; δ - torque angle (Fig. 3); l_m - mutual inductance; l_r - rotor inductance.

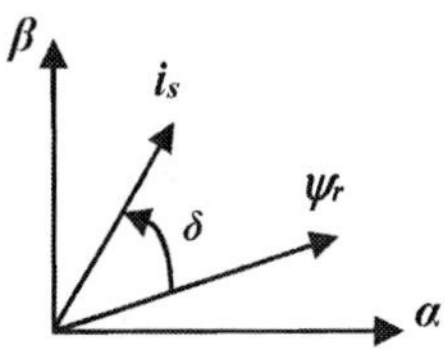

Fig. 3 FOC vector diagram

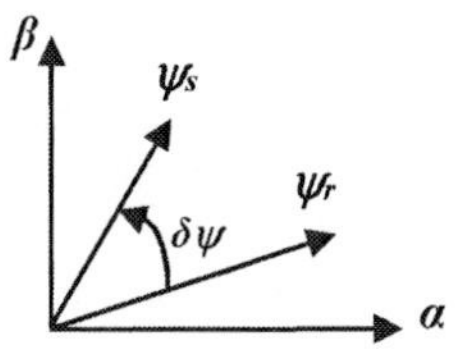

Fig. 4 DTC vector diagram

With direct torque control (DTC) and pulse width modulation (PWM), it is possible to control not only the stator current, but also the stator flux for torque control (Fig. 4):

$$T_e = \frac{l_m}{l_r}\psi^r.\frac{1}{\sigma.l_s}\psi^s.\sin\delta_\psi \qquad (11)$$

where: δ_ψ - torque angle (fig. 4).

Usually, a frequency converter uses a three-phase voltage inverter with two levels. Inverter output voltage vector:

$$\boldsymbol{U_s} = \sqrt{\frac{2}{3}}.U_{DC}\left[S_a + S_a^{\,j\frac{2\pi}{3}} S_a^{\,-j\frac{2\pi}{3}}\right] \qquad (12)$$

where S_a, S_b and S_c are switching functions that take a logical value of 0 or 1.

The DTC method independently controls the stator flux and the electromagnetic torque by supplying control signals to the inverter switches. The selection of a suitable voltage vector is done using a switching table, in which the selection depends on deviation in flux and torque. In this case, the magnetic flux and torque are limited by the boundaries of two hysteresis bands. The deviations of the magnetic flux and torque are fed to the inputs of the hysteresis controllers, and their outputs determine the corresponding voltage vector for each switching period.

The stator flux equation:

$$\begin{aligned}\psi_{sd} &= \int\left(u_{sd} - r_s.i_{sd}\right)dt\\ \psi_{sq} &= \int\left(u_{sq} - r_s.i_{sq}\right)dt\end{aligned} \qquad (13)$$

During sampling (torque control period) T_e, the switching functions S_a, S_b and S_c are not changed, that is, the switches are not switched. Then the expression for k+1 samples of the stator flux:

$$\begin{aligned}\psi_{s\alpha}(k+1) &= \psi_{s\alpha}(k) + \left[u_{s\alpha}(k) - r_s.i_{s\alpha}(k)\right].t_{sem}\\ \psi_{s\beta}(k+1) &= \psi_{s\beta}(k) + \left[u_{s\beta}(k) - r_s.i_{s\beta}(k)\right].t_{sem}\end{aligned} \qquad (14)$$

In vector form:

$$\boldsymbol{\Psi_s}(k+1) = \boldsymbol{\Psi_s}(k) + \left[\boldsymbol{U_s}(k) - r_s.\boldsymbol{I_s}(k)\right].t_{sem} \qquad (15)$$

At high speeds, the voltage drop across the stator resistance can be neglected and the equation (15) will take the form:

$$\boldsymbol{\Psi_s}(k+1) = \boldsymbol{\Psi_s}(k) + \boldsymbol{U_s}(k).t_{sem} \qquad (16)$$

Equation (13) shows that the stator flux can be changed by the stator voltage $\boldsymbol{U_s}$ during the sampling time t_{sam}. The end of the flux vector rotates in a circle in the direction defined by the voltage vector.

A two-level hysteresis comparator regulates the flux within a bandwidth determined by the width of the hysteresis. The hysteresis band depends on the switching frequency of the inverter (Fig. 5 and Fig. 6).

The logical outputs of the flux controller:

$$\begin{aligned}S_\psi &= 1 \ \text{ if } \ \Delta\psi_s > h_{\psi s}\\ S_\psi &= 0 \ \text{ if } \ \Delta\psi_s \le -h_{\psi s}\end{aligned} \qquad (17)$$

where $h_{\psi s}$ is hysteresis band of stator flux.

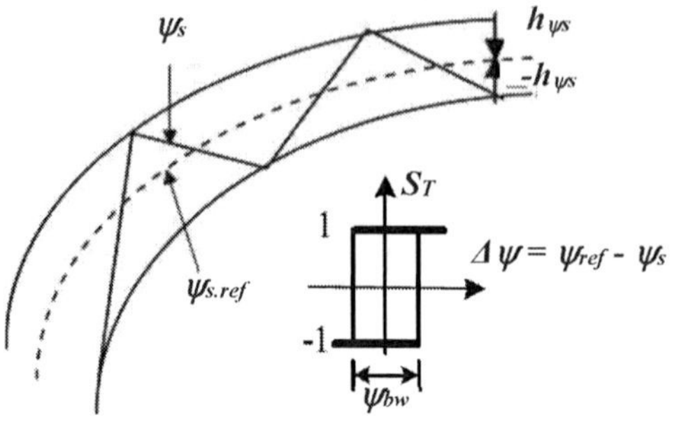

Fig. 5. Two-level hysteresis comparator for stator flux control

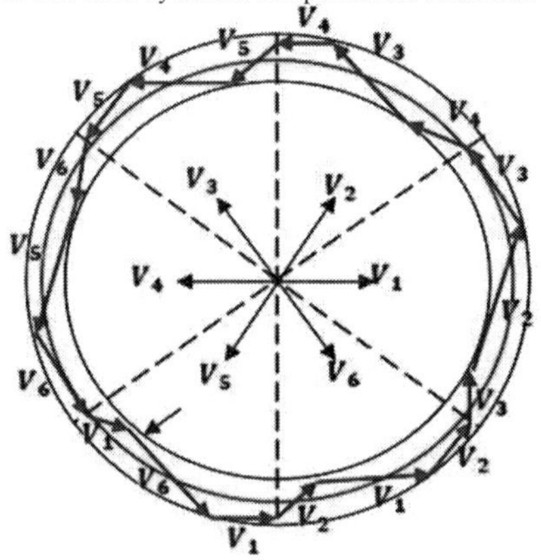

Fig. 6. Movement of the stator magnetic flux vector in the complex plane

The stator flux deviation depends on the difference between the reference value of flux and the actual estimated value:

$$\Delta\psi_s = \left|\psi_{s.ref}\right| - \left|\psi_s\right| \qquad (18)$$

During one sampling period, the rotor flux vector is constant. Electromagnetic torque T_e rotor and stator and rotor flux vectors $\boldsymbol{\Psi_s}$, $\boldsymbol{\Psi_r}$ are linked by the following relation:

$$T_e = k.\psi_s.\psi_r.\cos\delta \qquad (19)$$

Equation (8) also determines the electromagnetic torque as a function of the stator and rotor flux and torque angle δ_ψ .

The torque regulation performed by three-level hysteresis comparator (Fig. 7). The index bw stands for bandwidth.

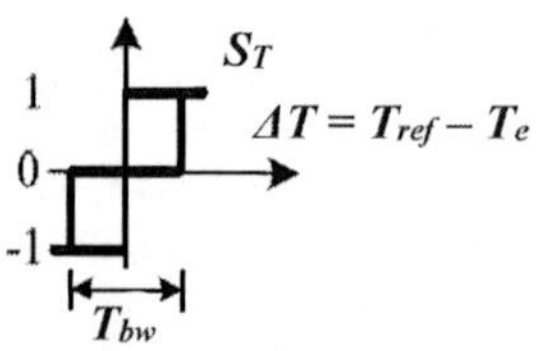

Fig. 7. Three-level hysteresis comparator for electromagnetic torque control

The logical outputs of the torque controller:

$$\begin{aligned} S_{Te} &= 1 \text{ if } \Delta T_e > h_{Te} \\ S_{Te} &= 0 \text{ if } -h_{Te} \le \Delta T_e \le -h_{Te} \\ S_{Te} &= -1 \text{ if } \Delta T_e < -h_{Te} \end{aligned} \tag{20}$$

where: h_{Te} - hysteresis band of torque.

The torque deviation depends on the difference between the reference's values of the torque and the actual estimated values:

$$\Delta T_e = T_{e.ref} - T_e \tag{21}$$

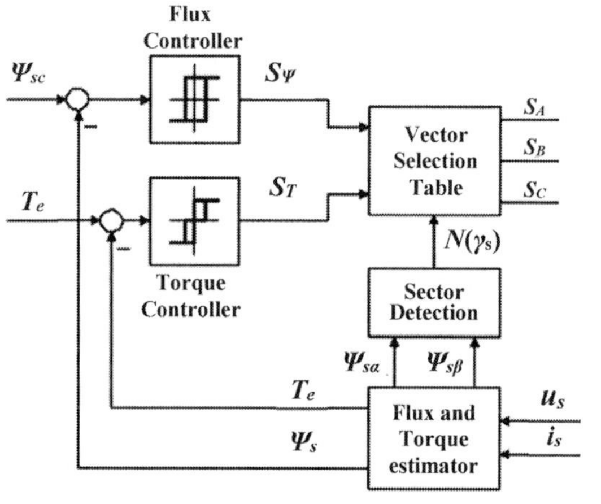

Fig. 8. Scheme of DTC with circular stator flux path

Independent torque and flux control is performed using two hysteresis comparators (Fig. 8). The choice of the voltage vector depends on the output signals of the comparators and the position of the magnetic flux. The circular path of the stator magnetic flux vector has six sectors (Fig. 6, Table 2).

TABLE 2. GENERALIZED SWITCHING TABLE

Parameter	Increases	Decreases
ψ_s	V_{i-1} and V_{i+1}	V_{i+2} and V_{i-2}
T_e	V_{i+1} and V_{i+2}	V_{i-1} and V_{i-2}

The vectors V_i and V_{i+3} are not considered, since they increase/decrease the torque in the sector in accordance with the position of the flux vector in the first or second sector. Zero vectors V_0 and V_7 stop the stator flux and its value does not change, but the torque decreases, not as much as when choosing active voltage vectors (Table 3).

TABLE 3. LOOK-UP TABLE FOR BASIC DIRECT TORQUE CONTROL

Flux	Torque	1	2	3	4	5	6	Comparator
$S_\Psi=1$	$S_T=1$	V_2	V_3	V_4	V_5	V_6	V_1	Two-level
	$S_T=0$	V_7	V_0	V_7	V_0	V_7	V_0	
	$S_T=-1$	V_7	V_1	V_2	V_3	V_4	V_0	Three-level
$S_\Psi=0$	$S_T=1$	V_3	V_4	V_5	V_6	V_1	V_2	Two-level
	$S_T=0$	V_0	V_7	V_0	V_7	V_0	V_7	
	$S_T=-1$	V_5	V_6	V_1	V_2	V_3	V_4	Three-level

Figure 9 shows the relationship between motor torque and stator voltage, where the thin lines are parallel to ψ_{rk}^s and the bold line intersects the q-axis with the ordinate $\omega_k.\psi_{sk}$ [24]. This graph, $\Delta T_k = \Delta T_{k1} + \Delta T_{k2}$, is used to visualize the effect of the voltage vector on the motor torque,

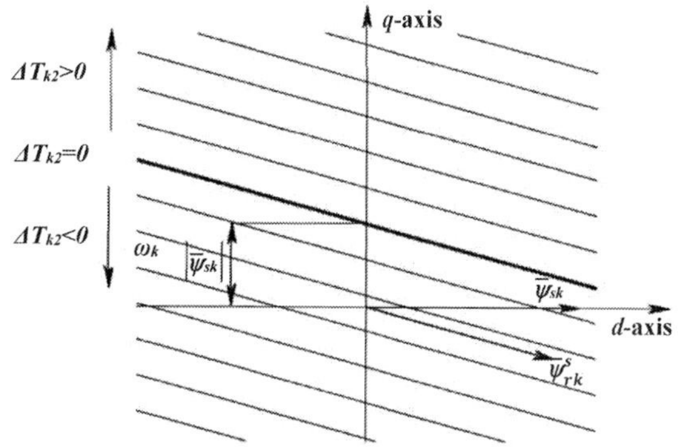

Fig. 9. Diagram of the torque variation $\Delta Tk2$

shown in Fig. 10 for zero motor speed and Fig. 11 for the motor rated speed.

B. Modifications of DTC

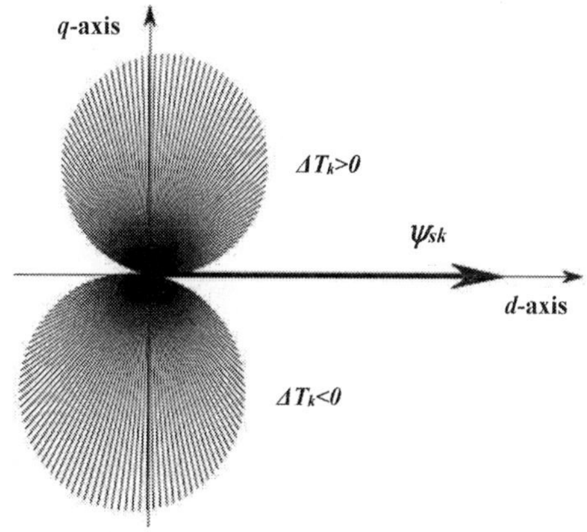

Fig. 10. Torque variations at zero speed and rated torque

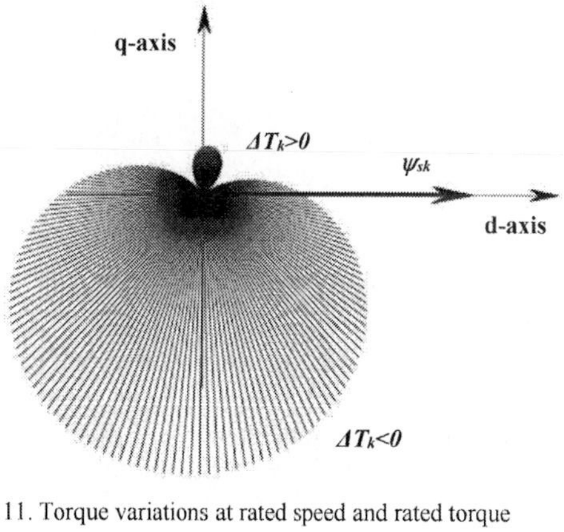

Fig. 11. Torque variations at rated speed and rated torque

In Fig. 12 is shown the classification of methods for improving direct torque control [23].

DTC can be divided into: 1. With space-vector modulation; 2. Circular flux trajectory; 3. Hexagon flux trajectory [11].

The size of the hysteresis band affects the performance of the drive. For example, harmonic current distortion, inverter switching frequency, torque ripple and

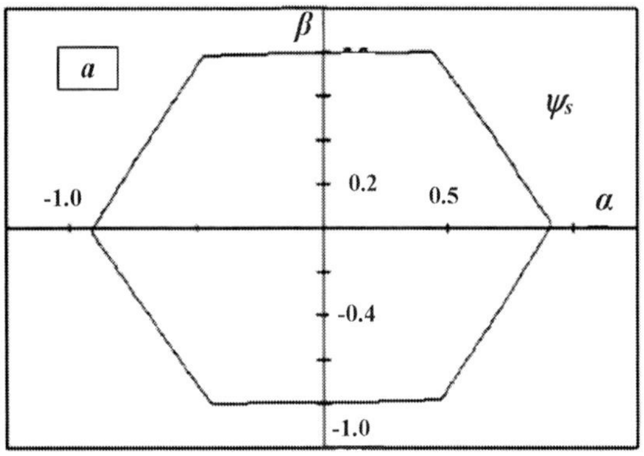

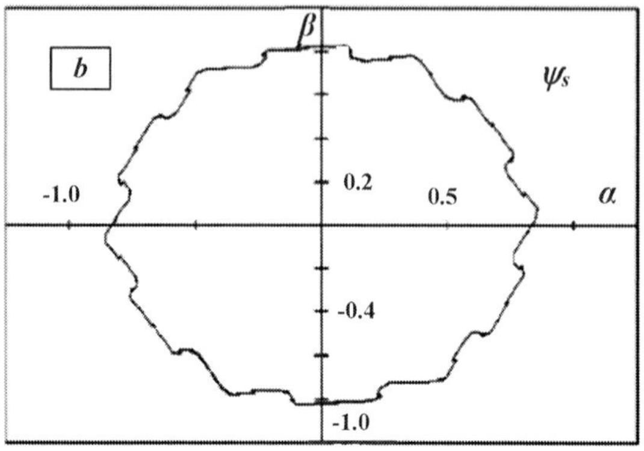

Fig. 12. Output vectors on stator flux under six-step operation and sinusoidal PWM

loss are highly dependent on the tape size [26]. With digital control, small tape values cause high switching frequencies without reducing ripple. The minimum ripple depends on the minimum change in torque and flux over the cycle period, which is dependent on the V_c and T_c values. Torque and current ripple depend on the switching frequency and their selection differs significantly at low and high speeds.

As can be seen from the above, the stator current and flux are controlled by the output voltage of the inverter, which can have six nonzero values (active vectors) and two zero values (zero vectors) (Fig. 9).

The output voltage of the inverter is a cyclic and symmetric sequence of active vectors, and the stator flux moves at a constant speed along a hexagonal trajectory (Fig. 10). Zero vectors stops the flux, but does not change its trajectory. This differs from the sinusoidal PWM mode of the active and zero vectors, and the stator flux moves along a trajectory that resembles a circle (Fig. 4). The rotor flow rotates continuously at synchronous speed along an almost circular path.

The relative motion of the two flux vectors affects the torque as they form a torque angle (Fig. 4), which determines the instantaneous torque of the motor. Switching the active voltage vector causes a rapid movement of ψ_s with respect to ψ_r, which increases the motor torque due to an increase in the δ_ψ angle. At a zero vector, the stator flux ψ_s begins to stop, and ψ_r continues to move forward, which reduces the angle δ_ψ, and the motor torque M_e. If the duration of the zero state t_0 is long enough, ψ_r will overtake s, as a result, the angle δ_ψ and the motor torque will change direction. Importantly, there is a direct relationship between torque fluctuations and the duration of zero states. By switching the cyclically active and zero vector, the motor torque is regulated. At very low speeds, the rotor current ψ_r moves too slowly and cannot quickly reduce the torque. In this case, the active vector moving backward ψ_s is selected, and not the zero vector. Zero vectors cannot be used in the field of field weakening, or by slowing down the δ_ψ phase.

The rate of change of the stator magnetic flux vector, expressed in a fixed stator frame of reference:

$$\frac{d}{dt}\boldsymbol{\Psi}_s = \boldsymbol{U}_s - r_s.\boldsymbol{I}_s \qquad (20)$$

where $\boldsymbol{U}_s$ and $\boldsymbol{I}_s$ are space vectors of stator voltage and stator current.

In DTC IM, flux control based on (20) generally neglects the drop-in stator resistance when selecting voltage vectors from a look-up table:

$$\Delta\boldsymbol{\Psi}_s = \boldsymbol{U}_s.\Delta t \qquad (21)$$

When the torque needs to be reduced, the zero voltage vectors are selected and the stator flux is stopped. In this case, the voltage drop across the stator resistance is considered, at which the change in the stator magnetic flux vector:

$$\Delta\boldsymbol{\Psi}_s = -r_s.\boldsymbol{I}_s.\Delta t \qquad (22)$$

As can be seen from the above, the stator current and flux are controlled by the output voltage of the inverter, which can have six nonzero values (active vectors) and two zero values (zero vectors).

The inverter output voltage is a cyclic and symmetrical sequence of active vectors, and the stator flux moves at a constant speed along a hexagonal path (Fig. 12.a). Zero vectors stop the flow, but does not change its direction. This differs from the sinusoidal PWM mode of the active and zero vectors, and the stator flux moves along a trajectory resembling a circle (Fig. 12.b). The rotor flow rotates continuously at synchronous speed along an almost circular path.

The relative motion of the two flux vectors affects the torque as they form a torque angle δ_ψ (Fig. 4), which determines the instantaneous torque of the motor. Switching the active voltage vector causes a rapid displacement of Ψ_s with respect to Ψ_r, which increases the motor torque due to an increase in the angle δ_ψ. At zero vector, the stator flux Ψ_s begins to stop, and Ψ_r continues to move forward, which reduces the angle δ_ψ,, and the motor torque me. If the duration of the zero state t_0 is long enough, Ψ_r will overtake Ψ_s, as a result, the angle δ_ψ and the motor torque will change direction. What is important is that there is a direct relationship between torque fluctuations and the duration of zero states. By switching the cyclically active and zero vector, the motor torque is regulated. At very low speeds, the rotor flux Ψ_r moves too slowly and cannot quickly reduce the torque. In this case, the active vector moving backward Ψ_s, is selected, and not the zero vector. Zero vectors cannot be used in the field of field weakening, or by slowing down the δ_ψ phase.

Figure 13 shows a diagram of DTC of IM with PWM inverter. Stator flux control adjusts the duration of active voltage vectors and torque control adjusts the duration of zero voltage vectors. During sampling, the voltage vector selection block defines the inverter switching state S_A, S_B, S_C.

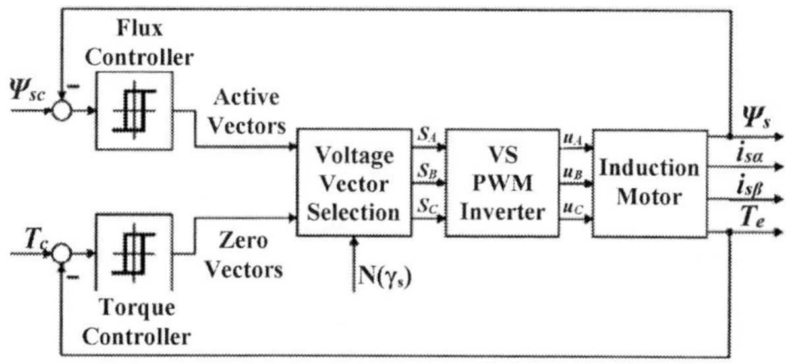

Fig. 13. Scheme of DTC of IM with PWM inverter

Features of power supply by inverter:

• The output voltage of the inverter can be either active or zero.

• Active forward vectors move Ψ_s at a constant speed, increasing the torque, and zero vectors stop Ψ_s, decreasing the torque.

• Active back vectors move the stator flux in the opposite direction.

• In 6-stage mode (active vectors only), the stator flux moves in a hexagonal path at a constant speed.

• With a sinusoidal PWM (active and zero vector) and a high switching frequency, Ψ_s follows an almost circular path at synchronous speed.

• The rotor flux always moves in a circular path at a synchronous speed

Figure 13 shows a voltage inverter PWM DTC scheme that contains two hysteresis controllers. Flux controller determines the duration of active voltage vectors, which move the stator flux along a reference trajectory, and torque controller which defines the duration of the zero vectors that hold the torque in a reference hysteresis band. At each sampling time, the voltage vector selection unit selects the inverter switching state (S_A, S_B, S_C), which reduces magnetic flux and torque deviations.

References for stator magnetic flux ψ_{sc} and torque T_c are compared with the current values and values in the corresponding controllers. Flux regulator outputs:

$d_\psi = 1$ for $\psi_s < \psi_{sc} - H_\psi$
$d_\psi = 0$ for $\psi_s > \psi_{sc} + H_\psi$

and a torque regulator as

$d_T = 1$ for $T_e < T_c - H_T$
$d_T = 0$ for $T_e = T_c$
$d_T = -1$ for $T_e < T_c + H_T$

where: $2H_\psi$ - flux tolerance band; $2H_T$ - torque tolerance band.

IV. Reducing torque ripple

There are various methods to increase the efficiency and improve the operation of the direct torque control of induction motors. These methods are based on the same principle of controlling the stator torque and flux and determining the inverter control signals from the switching table. Figure 14 shows a classification of methods for improving direct torque control [23]. Control methods are divided into two groups: typical; modern. The following abbreviations are used in the figure: SVM - space vector modulation; DSVM - discrete space vector modulation.

The use of artificial intelligence methods (neural network, fuzzy logic, genetic algorithm) improves the dynamic characteristics of DTC control, optimizes performance while reducing torque ripple and magnetic flux, reduces THD, and increases operating efficiency [27,28].

Space vector modulation (SVM) is used to control the voltage inverter, which improves the DTC [29,30,31]. This is more complex control, but fluctuations in flux and torque are reduced and the average switching frequency of the inverter becomes constant.

It uses pulse width modulation (PWM) with a constant switching frequency [32,33] for direct torque control of IM. This control method is developed in discrete time for digital implementation and DSP use. A conventional DTC uses a small number of voltage vectors, which causes unwanted fluctuations in torque, flux, and current [34,35]. The improvement in performance is achieved using the DTC algorithm using SVM for fixed time intervals. Thus, Discrete Space Vector Modulation (DSVM) is used using a five-level torque comparator to obtain more voltage vec-

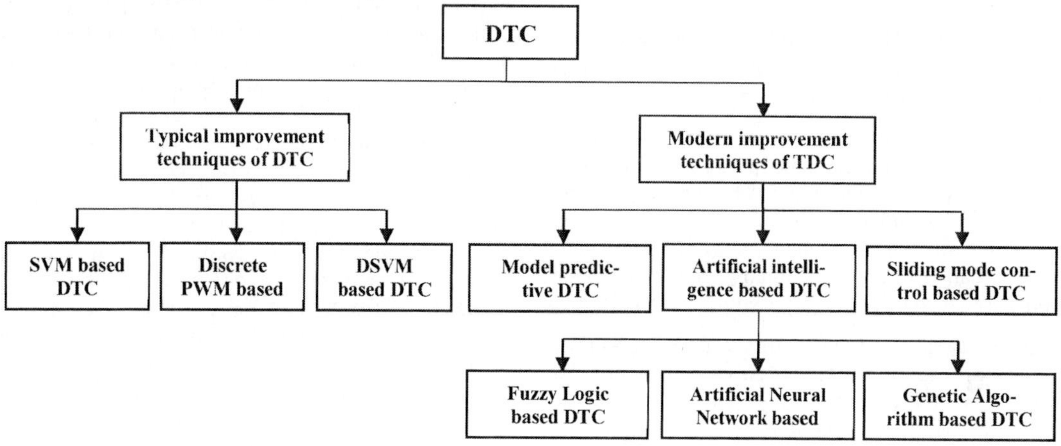

Fig. 14. Classification of methods for improving direct torque control

tor. A large number of control vectors minimizes the resulting pulsations in the steady state. They are very useful, especially in high power electric drives.

Sliding mode control is variable structure control [36], it is known for its stability to internal uncertainties (variations in machine parameters), external uncertainties (load disturbances) and approximations in modeling [37]. Its main characteristic is manifested in the modified control law in a discontinuous manner. This method has certain disadvantages: the appearance of the bounce phenomenon caused by the intermittent part of the control [38]. The system is under strong control all the time to ensure that it converges to what it wants.

With induction electric drives with time-varying dynamics, conventional PI controllers degrade the quality of control, therefore, controllers based on a neural network are introduced. To select the states of switches to a voltage inverter, use a neural network [39,40,41]. The idea is to replace the conventional switch table that defines the states of the inverter switches with a neural selector controlling switches. The structure is a multilayer neural network that allows both hysteresis comparators and a selection table to be replaced.

The application of neural networks to the DTC provides good response time for torque and flux control with a fixed switching frequency, which reduces the torque ripple and harmonic frequency of currents compared to other traditional methods. In addition, it is very robust against various uncertainties in motor parameters.

Fuzzy logic, or processing of uncertainties, is one of the classes of artificial intelligence [42,43,44] and is used in variable speed drives. This method improves the dynamic performance of the DTC. A new switch table can be used with fuzzy logic that optimally controls the flux and torque. Inference rules allow you to correct for differences between flux and torque reference and their estimated values.

An internal fuzzy speed controller is also used, which dynamically adjusts the integration coefficient and proportionality coefficient of the PI controller, depending on the deviation and changes in speed. In addition, a fuzzy controller to optimize the voltage vector selection replaces the flux and torque hysteresis.

In addition, fuzzy logic is used to control the electromagnetic torque hysteresis band, which minimizes torque ripples and improves dynamic performance.

The fuzzy logic method is used to study the effect of changing parameters on the DTC performance through switch tables. However, the main problems of this method are the complex tuning of the fuzzy logic parameters and the complexity of the implementation. Since a fuzzy controller uses many rules, a lot of tuning experiments need to be done.

Genetic algorithms (GA) represent a large group of stochastic optimization algorithms based on methods derived from natural evolution and genetics [45,46,47]. They provide one or more solutions for various problems, while they are robust to parameter changes and require low computing power. But it is difficult for them to choose the parameters, since they depend on the tasks and a priori information.

The use of GA for DTC induction motor improves their dynamic performance and optimizes the speed control gain values well. GA is also used to optimize PI controllers, which improves the static and dynamic characteristics and reduces the ripple of the torque and improves the speed control.

Modification of the band of the hysteresis of the magnetic flux and torque is applied, which reduces the harmonics of current, ripple and switching frequency of the converter switches [48,49]. With a constant band of the torque hysteresis, with an increase in the magnetic flux range, the flux hodograph approaches the hexagonal shape due to harmonic distortions. With a fixed band of the magnetic flux hysteresis and with an increase in the torque hysteresis band, the torque ripple increases.

The hysteresis band affects the drive characteristics of an induction motor with DTC - current harmonics, flux ripple and torque, as well as the switching frequency of the inverter switches.

(a) Influence of the flux hysteresis band: for a given torque band, the switching frequency of the inverter depends on the flux hysteresis width. A narrow bandwidth means a higher switching frequency. The hodograph of the stator magnetic flux vector approaches a circle, and the phase current shape is almost sinusoidal. With an increase in the band of the flux hysteresis, the switching frequency decreases, and the hodograph of the stator magnetic flux vector becomes a hexagon, and the shape of the phase current is similar to an induction machine.

(b) Influence of the torque hysteresis band in the range of 5% - 25% of the maximum torque: increasing the torque hysteresis width increases the torque ripple. If the band is too small, the torque will exceed the limits, resulting in a torque deviation outside the hysteresis band. Reverse voltage vector (instead of zero vector) will be selected to reduce torque. The reverse voltage vector will quickly decrease the torque and cause a torque surge and the torque ripple will decrease.

The PWM method is used with adaptive control of the hysteresis band [50]. The hysteresis bandwidth varies with load and power conditions to maintain a fixed modulation frequency. Thus, consistency of work is obtained at the expense of the complexity of the implementation.

PWM is used with synchronous sinusoidal triangular modulation, with the current deviation acting as a modulating curve [51,52]. The advantage is that the maximum switching frequency of the switches is limited to the frequency of the triangular waveform and well-defined harmonics are generated. The disadvantages are the multiple crossing of the ramp, which creates a problem with a higher rate of change of the deviation from the rate of change of the ramp and taking into account the deviation in magnitude and phase in line currents.

Figure 15 shows a control with sinusoidal hysteresis, in which the hysteresis bands change sinusoidal during the main period (Fig. 15 (b), and are not fixed (Fig. 15 (a)). In this case, the ripple changes depending on the magnitude of the current, and reduces ripple, which reduces harmonics while maintaining speed and ease of implementation.

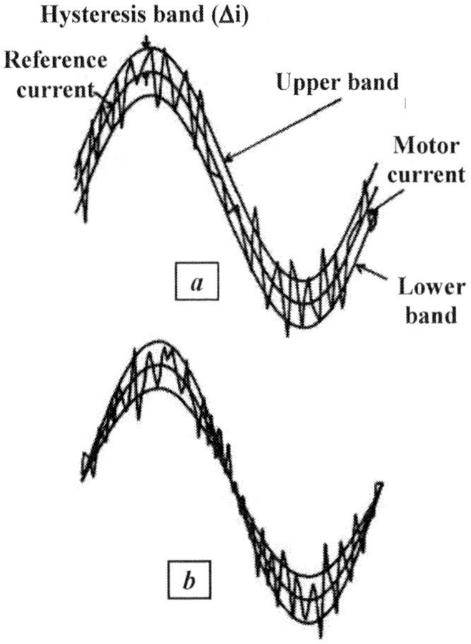

Fig.15. (*a*) Fixed band control. (*b*) Sinusoidal band control

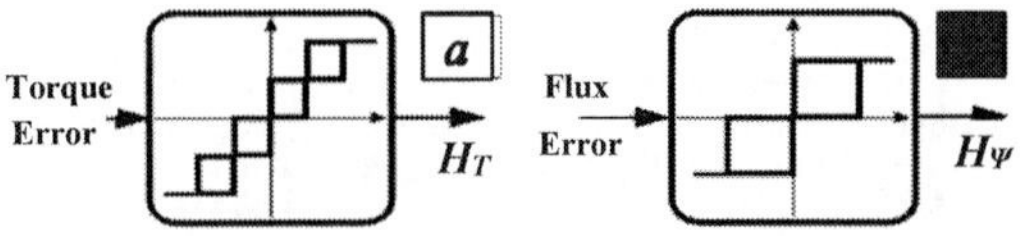

Fig.16. (*a*) 5-level torque regulator, (*b*) 3-level flux hysteresis regulator

An effective technique to improve the performance of the induction motor and reduce the ripple of torque and current during DTC control is to use inverters with many levels (Fig.16). This allows more voltage vectors with different amplitudes, less voltage distortion, less voltage in switches, less harmonic content, lower switching frequencies, reduced torque and flux ripple, and performance. improvement in low-speed operations, provided that the appropriate switching algorithm is performed [54,55,56,57].

V. CONCLUSION

The article provides an overview and analysis of methods for increasing the efficiency of direct torque control in an induction machine. The purpose of the analysis is to generalize the experience of controlling induction electric drives and identify further directions for improving the direct method of the motor torque control.

The analysis of the considered methods is made in terms of torque and magnetic flux ripple, switching frequency, sensitivity to parameters, steady state and dynamic response, and control complexity.

Classic DTC control represents the simplest control structure for electric drives with low switching frequency, which is the main problem. To overcome these problems and to reduce torque ripple, the sampling rate must be high. The use of artificial intelligence techniques for control reduces the ripple of torque and switching frequency at the same sampling rate, but the complexity of control increases.

It is very difficult to estimate which solution is the best for the DTC to work. The choice of control method depends on the application, cost, reliability and accuracy of the system.

REFERENCES

[1] M.P. Kostenko, "The work of a multi-phase asynchronous motor with a variable number of periods," *Electricity*, 1925, pp.87-95.

[2] F. Blaschke, "Apparatus for field-oriented control or regulation of asynchronous machines," United States Patent No.3,805,135, 1972.

[3] M. Depenbrock, "Direct self-control of the flux and rotary moment of a rotary field machine," United States Patent 4,678,248, 1985.

[4] M. Depenbrock, "Direct self-control of inverter-fed induction machine." *IEEE Transactions on Power Electronics*, vol. 3, pp. 420–429, 1988.

[5] I. Takahashi, T. Noguchi, "A new quick-response and high-efficient control strategy of an induction motor." *IEEE Transactions Industry Applications*, Vol. IA-22, pp.820–827, 1986.

[6] A.M. Trzynadlowski, *The Field Orientation Principle in Control of Induction Motors*, 1994 Springer, pp. 269.

[7] J. Yu, T. Zhang, J. Qian, "Electrical Motor Products." International Energy-Efficiency Standards and Testing Methods, 2011, 175 p.

[8] G.S. Buja, M. P. Kazmierkowski, "Direct torque control of PWM inverter-fed AC motors — A survey. " *IEEE Transactions on Industrial Electronics*, vol. 51, pp.744–757, 2004

[9] A. Chattopadhyay, "Advances in vector control of ac motor drives — A review." *Recent Advances*, vol. 22, pp. 797-820, 1997.

[10] S.M. Tripathi, R. Vaish, "Taxonomic research survey on vector-controlled induction motor drives," *IET Power Electronics*, vol. 12, pp. 1603-1615, 2019.

[11] G.S. Buja, M.P. Kazmierkowski, "Direct torque control of PWM inverter-fed AC motors — A survey." *IEEE Transactions on Industrial Electronics*, vol. 51, pp. 744–757, 2004.

[12] M. Bertoluzzo, G. Buja, R. Menis, "Analytical formulation of the direct control of induction motor drives," *IEEE International Symposium of Industrial Electronics*, 1999, pp. 14–20.

[13] N.F. Djagarov, "A method of transient electromechanical processes modeling in power systems," *IEEE Bucharest PowerTech Conference*, 2009, pp. 1-6.

[14] C. Lascu, I. Boldea, F. Blaabjerg, "Direct torque control of Sensorless induction motor drives: A sliding-mode approach," *IEEE Transactions on Industry Applications*, vol. 40, pp. 582–590, 2004.

[15] S.M. Gadoue, D. Giaouris, J.W. Finch, "Artificial intelligence-based speed control of DTC induction motor drives—A comparative study." *Electric Power Systems Research*, vol. 79, pp. 210–219, 2009.

[16] S. Gdaim, A. Mtibaa, M.F. Mimouni, "Direct torque control of induction machine based on intelligent techniques." *International Journal of Computer Applications*, vol. 10, pp. 29-35, 2010.

[17] C.M.F.S. Reza, M.D. Islam, S. Mekhilef, "A review of reliable and energy efficient direct torque controlled induction motor drives," *Renewable and Sustainable Energy Reviews*, vol. 37, pp. 919–932, 2014.

[18] T. Sutikno, N.R.N. Idris, A. Jidin, "A review of direct torque control of induction motors for sustainable reliability and energy efficient drives," *Renewable and Sustainable Energy Reviews*, vol. 23, pp. 548–558, 2014.

[19] A. Ammar, A. Benakch, A. Bourek, "Closed loop torque SVM-DTC based on robust super twisting speed controller for induction motor drive with efficiency optimization." *International Journal of Hydrogen Energy*, vol. 42, pp. 17940–17952, 2017.

[20] F. Niu, B. Wan B., A.S. Babel, K. Li, E.G. Strangas, "Comparative evaluation of direct torque control strategies for permanent magnet synchronous machines," *IEEE Transactions on Power Electronics*, vol. 31, pp. 1408–1424, 2016.

[21] R.H. Kumar, A. Iqbal, N.C. Lenin, "Review of recent advancements of direct torque control in induction motor drives–a decade of progress," *IET Power Electronics*, vol. 11, pp. 1–15, 2017.

[22] "DTC A motor control technique for all seasons,". White paper, ABB Finland, pp. 11.

[23] N. El Ouanjli, A. Derouich, A. El Ghzizal1, S. Motahhir1, A. Chebabhi, Y. El Mourabit1, M. Taoussi, "Modern improvement techniques of direct torque control for induction motor drives - a review," *Protection and Control of Modern Power Systems*, vol. 4, pp. 1-12, 2019.

[24] D. Casadei, G. Serra, A. Tani, L. Zarri, "Assessment of direct torque control for induction motor drives," Bulletin of the Polish Academy of Sciences Technical Sciences, vol. 54, pp. 237-254, 2006.

[25] A. Balashanmugham, M. Mockaisamy, S. Murugesan, *Torque ripple reduction in DTC induction motor drive*, Direct Torque Control Strategies of Electrical Machines, Edited by Fatma Ben Salem, Intech Open, 2020, pp. 24.

[26] N. Djagarov, H. Milushev, A. Varganova, J. Djagarova, "Investigation of the influence of hysteresis on the characteristics of direct torque control of induction motor," *IEEE 17-th International Conference on Electrical machines, Drives and Power Systems ELMA 2021*, 2021, pp. 1-4.

[27] C.M.F.S. Reza, M.D. Islam, S. Mekhilef, "A review of reliable and energy efficient direct torque controlled induction motor drives," *Renewable and Sustainable Energy Reviews*, vol. 37, pp.919–932, 2014.

[28] N. Goel, R.N. Patel, S. Chacko, "Torque ripple reduction of DTC IM drive using artificial intelligence. In International conference on electrical power and energy systems," *IEEE International Conference on Electrical Power and Energy Systems (ICEPES)*, 2016, pp. 1-5.

[29] J. Rodriguez, J. Pontt, C. Silva, S. Kouro, H. Miranda, "A novel direct torque control scheme for induction machines with space vector modulation," *IEEE 35th Annual Power Electronics Specialists Conference*, 2004, pp. 1392–1397.

[30] S. Belkacem, F. Naceri, R. Abdessemed, "Improvement in DTC-SVM of AC drives using a new robust adaptive control algorithm," *International Journal of Control, Automation, and Systems*," vol. 9, pp. 267–275, 2011.

[31] R. Abdelli, D. Rekioua, T. Rekioua, "Performances improvements and torque ripple minimization for VSI fed induction machine with direct control torque," ISA Transactions, vol. 50, pp.213–219, 2011.

[32] A.Del Pizzo, P. Marino, N. Visciano, "Harmonic and interharmonic impact of DTC-based induction motor drives on 3-wire network," *IEEE International Symposium on Industrial Electronics*, 2002, pp. 1201-1206.

[33] J. Maes, J. Melkebeek, "Discrete time direct torque control of induction motors using back-EMF measurement," *IEEE Industry Applications Conference*, 1998, pp. 407-414.

[34] D. Casadei, D., Serra, G., & Tani, A, "Implementation of a direct torque control algorithm for induction motors based on discrete space vector modulation." *IEEE Transactions on Power Electronics*, vol. 15, pp. 769–777, 2000.

[35] H.R. Keyhani, M.R. Zolghadri, A. Homaifar, "An extended and improved discrete space vector modulation direct torque control for induction motors," *IEEE 35th Annual Power Electronics Specialists Conference*, 2004, pp. 3414-3420.

[36] I.V. Utkin, "Sliding mode control design principles and applications to electric drives," *IEEE Transactions on Industrial Electronics*, vol. 40, pp. 23–36, 1993.

[37] A. Tanvir, A.R. Beig, K. Al-Hosani, K., "Sliding mode based DTC of three-level inverter fed induction motor using switching vector table," *IEEE 9th Asian Control Conference (ASCC)*, 2013, pp. 1-6.

[38] M. Rodic, K. Jezernik, K., "Speed-sensorless sliding-mode torque control of an induction motor." *IEEE Transactions on Industrial Electronics*, vol. 49, pp. 87–95, Feb. 2002.

[39] S. Jadhav, J. Kirankumar, B. Chaudhari, "ANN based intelligent control of induction motor drive with space vector modulated DTC," *IEEE International Conference on Power Electronics, Drives and Energy Systems (PEDES)*, 2012, pp. 1-6.

[40] L.M. Grzesiak, B. Ufnalski, B. "DTC drive with ANN-based stator flux estimator," *European conference on power electronics and applications*, 2005, pp. 1-10.

[41] R. Kumar, R. Gupta, S. Bhangale, H. Gothwal, "Artificial neural network based direct torque control of induction motor drives," *IET-UK International Conference on Information and Communication Technology in Electrical Sciences (ICTES 2007)*, 2007, pp. 361–367.

[42] C. Venugopal, "Fuzzy logic based DTC for speed control of matrix converter fed induction motor," *IEEE International Conference on Power and Energy*, 2010, pp. 753-758.

[43] S.A. Mir, D.S. Zinger, E. Elbuluk, E. "Fuzzy implementation of direct self-control of induction machines, " *IEEE Transactions on Industry Applications*, vol. 30, pp. 729-735, 1994.

[44] M. Hafeez, M.N. Uddin, M. N., R.S. Rebeiro, "FLC based hysteresis band adaptation to optimize torque and stator flux ripples of a DTC based IM drive," *IEEE Electrical Power & Energy Conference*, 2010, pp. 1-5.

[45] F.J. Lin, W.D. Chou, P.K. Huang, P. K. (2003) "Adaptive sliding-mode controller based on real-time genetic algorithm for induction motor servo drive." *IEE Proceedings on Electric Power Applications*, vol. 150, pp. 1-13, 2003.

[46] S.M. Gadoue, D. Giaouris, J.W. Finch, "Genetic algorithm optimized PI and fuzzy sliding mode speed control for DTC drives," *Proceedings of the World Congress on Engineering*, 2007, pp. 1-6.

[47] L. Hao, M. Qiuyun, Z. Zhilin, "Research on direct torque control of induction motor based on genetic algorithm and fuzzy adaptive PI controller," *International Conference on Measuring Technology and Mechatronics Automation*, 2010, pp. 1-6.

[48] J.-K. Kang, D.-W. Chung, S.-K. Sul, "Direct control of induction machine with variable amplitude control of flux and torque hysteresis bands," *IEEE International Electric Machines and Drives Conference IEMDC'99*, 1999, pp. 640-642.

[49] D. Casadei, G. Grandi, G. Serra, A. Tani, "Effect of flux an torque hysteresis band amplitude in direct torque control of induction motor," *IEEE 20th Annual Conference of IEEE Industrial Electronics IECON'94*, 1994, pp. 299-304.

[50] B.K. Bose, "An adaptive hysteresis-band current control technique of a voltage fed PWM inverter for machine drive system," *IEEE Transactions on Industrial Electronics*, vol. 37, pp. 402-408, 1990.

[51] D.M. Brod, D.W. Novotny, "Current Control of VSI-PWM Inverters," *IEEE Transactions on Industry Applications*, vol. IA-21, pp. 562-570, 1985.

[52] A. Tripathi, P.C. Sen, "Comparative analysis of fixed and sinusoidal band hysteresis current controllers for voltage source inverters," *IEEE Transactions on Industry Applications*, Vol. IA-39, pp. 63-73, 1992.

[53] Y. Farajpour, M. Alzayed, H. Chaoui, S. Kelouwani, "A novel switching table for a modified three-level inverter-fed DTC drive with torque and flux ripple minimization." *Energies*, vol. 13, pp. 1-18, 2020.

[54] J. Rodriguez, S. Bernet, B. Wu, J. Pontt, S. Kouro, "Multilevel voltage-source-converter topologies for industrial medium-voltage drives," *IEEE Transactions on Industrial Electronics*, vol. 54, pp. 2930–2945, 2007.

[55] W. Yao, H. Hu, Z. Lu, "Comparisons of space-vector modulation and carrier-based modulation of multilevel inverter." *IEEE Transactions on Power Electronics*, vol. 23, pp. 45–51, 2008.

[56] S. Busquets-Monge, S. Alepuz, J. Bordonau, J. Peracaula, "Voltage balancing control of diode-clamped multilevel converters with passive front-ends," *IEEE Transactions on Power Electrics*, vol. 23, pp. 1751–1758, 2008.

[57] L. Dalessandro, S. Round, J. Kolar, "Center-point voltage balancing of hysteresis current controlled three-level PWM rectifiers," *IEEE Transactions on Power Electronics*, vol. 23, pp. 2477–2488, 2008.

Operation of Micro Sources Connected to LV Distribution Grid

Martin Kaspirek
Grid management
EG.D (member of E.ON group)
Brno, Czech Republic
martin.kaspirek@egd.cz

Daniel Kouba
Grid management
EG.D (member of E.ON group)
Brno, Czech Republic
daniel.kouba@egd.cz

Abstract — **This paper describes the measurement campaign which is based on evaluation of 140 power quality measurements made in 70 different low voltage distribution grids with connected renewable energy sources (photovoltaic plants). The evaluation of these measurements focuses on parameters of power quality according to the EN 50 160 standard (voltage variations, flicker, harmonics) and power flows. Measured group of micro sources has a connection to the distribution grid in so-called simplified mode when power sources connected in this mode should not produce active power in the delivery point according to the grid code. The methodology for new grid connections of this type of power sources is proposed when this new methodology takes into consideration real operation conditions.**

Keywords: micro source; photovoltaic plant; power quality; distribution grid; renewable energy sources; EN 50 160 standard

I. Introduction

The supply territory of the company EG.D in the Czech Republic accounts for approximately 1.5 million customers when the most of them are connected to the low voltage distribution grid. More than 1000 MW of distributed energy sources (DES) were connected to this network mainly in photovoltaic (PV) sources. The mass operation of DES also results in worsening of voltage quality parameters in the distribution network. In case of accumulation of power sources in one part of distribution system overvoltage can be detected or else voltage variations will not comply with the requirements of the standard EN 50 160 [1]. Additionally various interest groups or entities recently suggest and enforce the increase of grid connectivity in delivery points of customers. According to the opinion of the EG.D company the steps directing to higher connectivity should not cause worse level of voltage quality in the distribution grid. We active support the factual discussion about the topic but we recommend determining of the impact of proposed measures in advance, including changes in the connectivity methodology. The large enough sample of grids should be analyzed because low voltage distribution grids are very diverse. If you evaluate only a few pilot grids then you can't conclude in the sense of correct verification of the methodology when this methodology is considered for general use (for all sorts of the grids).

II. Connectivity of Power Plants

A. Value of Power for Grid Calculations

The methodology was changed in 2020 when until then the installed power of photovoltaic panels was used. According to the current grid code [2] the nominal power of the inverter is used. The physical law is taken into consideration for calculation of impact of real active power flow. This step is positive perceived from the side of customer when the discussions were difficult before 2020. The new approach causes increasing of connectivity in the sense of nominal power of power plants (about approximately 10 percent on average). The real use-case is shown in Fig. 1 when limiting condition is the voltage change caused by the nominal power of the invertor never the nominal power of photovoltaic panels.

B. Value of Power Factor for Grid Calculations

According to the current grid code [2] the value 1 should be used for grid calculations. Consideration with other value is possible in individual cases or when distribution system operator (DSO) determined with the focus on local grid conditions. For example the value 0.95 (consumption of reactive power) can be used in cases when power plant connected to low voltage distribution grid has implemented regulation characteristics Q(U) and P(U) (Q – reactive power, P – active power, U – grid voltage in the delivery point).

C. Value of Real Active Power Flow for Grid Calculations

Some installations can guarantee lower value of power flows into the supply grid than the value of nominal power. The customer can guarantee zero power flow into the supply grid (overflow) in extreme case. This situation results in no impact of supply grid (from the point of view of voltage changes) but it is supposed the motivation to such a operation conditions. Yes, the customer is financial motivated to consume as much electricity as possible. Additionally, restrictions are possible from the side of DSO when overflow is higher than the value in contract between customer and DSO. It is only necessary to install convenient (measuring profiles) electric meter in delivery point of customer. The possible situation is described on the Fig.1. Nominal power of installed photovoltaic panels is 12 kW when the inverter decreases the active power to 10 kW because of its lower nominal power. The minimal consumption is 3 kW so that it is not possible to generate more than 7 kW from the view of delivery point (which consists of generation and consumption). The impact on supply grid (voltage changes) is influenced by the value 7 kW of active power and it is no reason from the side of DSO why do not calculate the connectivity with this value.

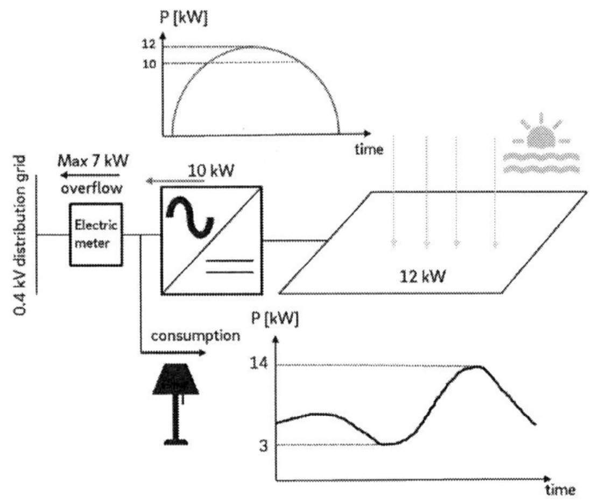

Figure 1. Usual principal scheme of the customer's electric installation

III. Description of Measured Grids

Seventy other representative low voltage (LV) distribution grids with connected PV plants (micro sources) were chosen and these grids were measured. Each measurement lasted at least one week according to the EN 50 160 standard. Two VQ measurements were made at the same time in each grid, first in the substation (on the LV level, on the output of the transformer 22kV/0.4kV – node U2 in the Fig. 1) and second in the point of connection of PV plant to LV distribution grid (supply terminal of PV plant – node U3 in the Fig. 1). The nominal active power (AP) of each PV plant is available. The measurements were made in February, in March and in April 2021 according to the free capacity of our technicians.

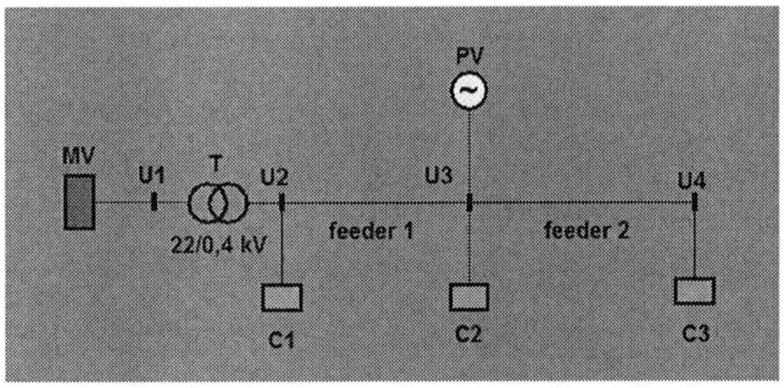

Figure 2. General block diagram of measured LV grids

Measured PV plants are connected in the node U3 when T is distribution MV/LV transformer and C1, C2 and C3 are general customers with the load profile – see general block diagram in the Fig. 2 (measured customer C2 operates the micro source and its delivery point consists of the PV plant and consumption). We wanted to discover the problem with active power overflow into the grid and if this problem occurs. We suggested to measure in the point of connection of PV plant to gird (node U3) for this reason. In case of accumulation of power sources in one part of distribution system, it is theoretically possible that overflow will be detected on the output of transformer MV/LV. We measured in the node U2 for verification of these assumptions.

IV. Requirements of the EN 50 160 Standard

A. Voltage Variations

All the LV measurements were evaluated according to the standard EN 50160 [1]. Under normal operating conditions excluding the periods with interruptions, supply voltage variations should not exceed ± 10% of the nominal voltage Un.

A test method for low voltage level (under normal operating conditions) follows these rules/requirements [1]:

- During each period of one week 95% of the 10 min rms-values of the supply voltage shall be within the range of Un ± 10%; and
- All 10 min rms-values of the supply voltage shall be within the range of Un + 10%/-15%

Also when one 10 min rms-value of the supply voltage exceeds during the week the limit Un + 10% (110% Un), it results in overvoltage and voltage quality or voltage variations will not comply with the requirements of the standard EN 50 160.

B. Flicker

Under normal operating conditions, during each period of one week the flicker Plt (long-term flicker severity) caused by voltage fluctuation should be less than or equal to 1 for 95 % of the time.

C. Harmonic Voltages

Under normal operating conditions, during each period of one week, 95 % of the 10 min rms-values of each individual harmonic voltage shall be less than or equal to the values given in [2] when the limit value for the 3rd voltage harmonic is 5%, for the 5th voltage harmonic is 6%, for the 7th voltage harmonic is 5% and for the 15th voltage harmonic is 0.5% of nominal voltage (only the most important harmonic voltages were selected). Moreover, the THD of the supply voltage (including all harmonics up to the order 40) shall be less than or equal to 8%.

D. Unbalance

Under normal operating conditions, during each period of one week, 95 % of the 10 rms-values of the negative phase sequence component (fundamental) of the supply voltage shall be within the range 0 % to 2 % of the positive phase sequence component (fundamental).

The power quality analyzer Fluke 1744 (class A according to the IEC 61000-4-30 standard [3]) was used for all the measurements. This analyzer enables to evaluate long-term values of active power (similar as electric meter works) and especially short-term values.

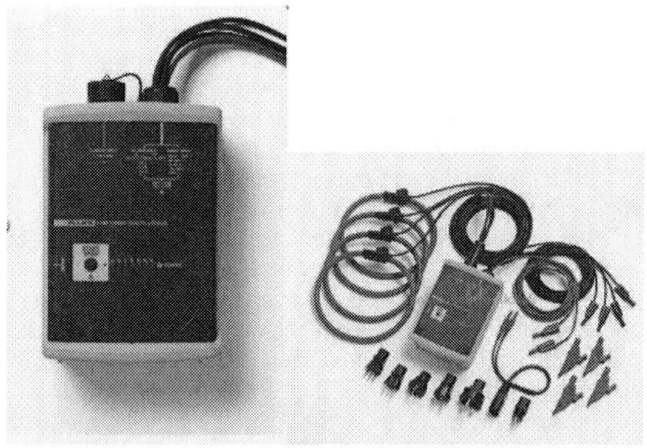

Figure 3. Power quality analyzer Fluke 1744

V. Reference Impedance

The standard IEC 60725 [4] sets the reference impedance value, or more precisely, a reference (relative) short circuit impedance (short circuit power) value for the LV distribution grids and electric equipment of nominal current less than 16/75 A. It is assumed that if the required impedance value is met at the point of network connection, an electric appliance will not produce any adverse retroactive impacts on the network therefore no interference in the network will occur.

Table 1. Reference impedance for equipment with current ratings ≤16A/75A

equipment with nominal current	Reference impedance Zphase-earth	Reference short circuit power Sk"
up to 16A	0.47 Ω	570 kVA
up to 75 A	0.35 Ω	760 kVA

The question is when the DSO and when the customer is responsible for poor voltage quality (VQ). It can be used the reference impedance as the guideline for decision about responsibility for poor VQ. The experience of DSO shows that keeping of the reference impedance will not guarantee full voltage quality for all LV grids. In this paper the short circuit impedance was evaluated and compared with the reference value. The real measured grid impedance was better (lower) than the reference value for all the seventy grids.

VI. Evaluation of Measurements

A. *Examples of Evaluation*

For illustration real representative examples of measurements are shown. 10 min values are average effective values and minimal values are 1s minimal values recorded during relevant 10 min interval. Appliance sign convention was selected when minus values of active power mean overflow to the supply grid.

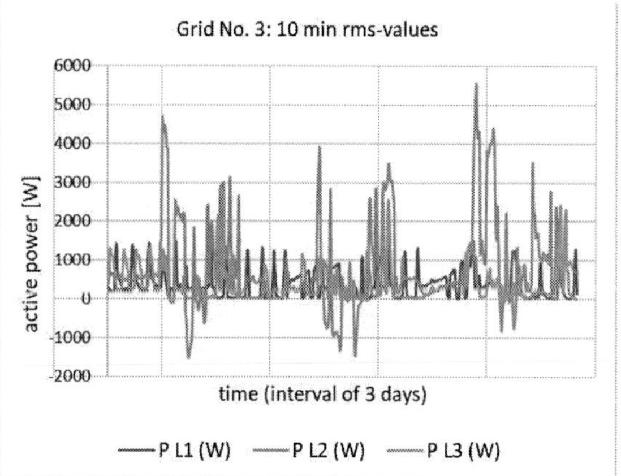

Figure 4. Values of AP (10 min) in the grid No. 3

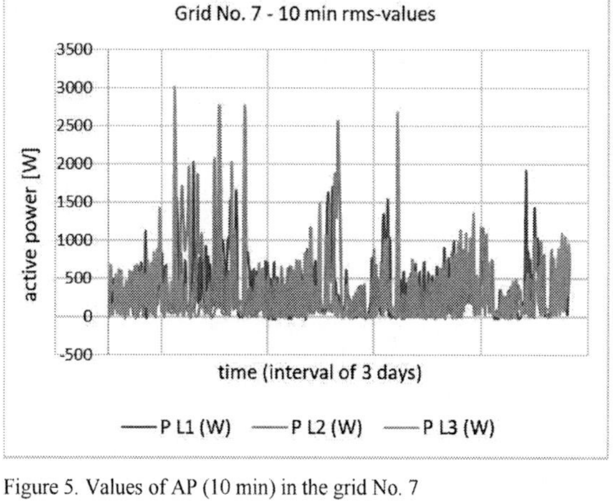

Figure 5. Values of AP (10 min) in the grid No. 7

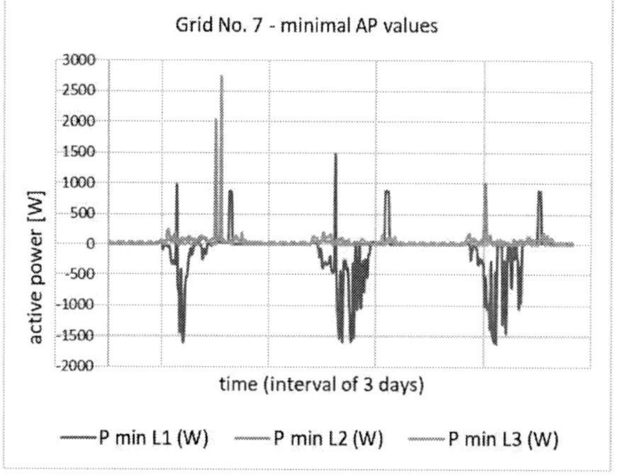

Figure 6. Values of AP (1s minimal values) in the grid No. 7

The Fig. 5 shows no overflow of active power from view of 10 min rms-values. But short-term overflow of active power (1 s values) was detected – see the Fig. 6. This means that it is necessary to measure and evaluate short-term values of active power (for example 1 s values or shorter) otherwise it is not possible to detect real power flow conditions in the delivery point of the customer. Usually used electric meters measure and evaluate 15 min values and are not able to detect short-term overflow of active power.

Several cases of overflow were detected on the output of the 22/0.4 kV transformer – see example on the Fig. 7. The overflow values were low (maximal 10 kW) but it was proved that generation of PV plants connected to LV level can influence the MV voltage level.

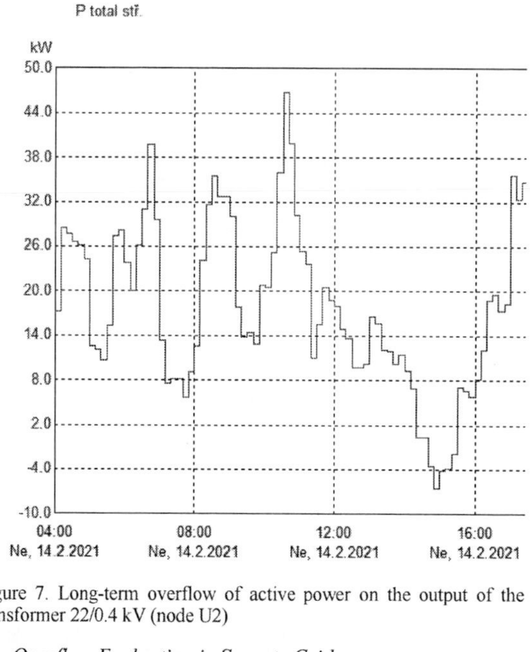

Figure 7. Long-term overflow of active power on the output of the transformer 22/0.4 kV (node U2)

B. *Overflow Evaluation in Seventy Grids*

The table 2 gives evaluation of active power overflow in all the seventy micro sources (node U3). No overflow was detected in 30 % cases. Only short-term (no long-term) active power overflow was detected in 37 % cases. Both short-term and long-term overflow was detected in 33% of cases and some overflow (either short-term or long-term) of active power was detected in

70 % of grids and it is alarming. The analysis shows that customers (micro sources) do not comply with the grid contract and should be restricted.

Table 2. Overflow Evaluation

Number of grids [pieces]	Number of grids [%]	Short-term (1 s) overflow	long-term (10 min) overflow
21	30	no	no
26	37	yes	no
23	33	yes	yes

Average value of short-term overflow of active power is 1.3 kW (90 % percentile is 2 kW) and average value of long-term overflow of active power is 1 kW (90 % percentile is 1 kW).

C. *Evaluation in Seventy Grids*

The table 3 gives evaluation of selected voltage quality parameters in all the seventy micro sources (node U3). No grids with inconvenient voltage variations were detected.

Table 3. Voltage Quality Evaluation – Compliance with the EN 50 160 Standard

Number of grids [pieces]	Number of grids [%]	Voltage variations	flicker	harmonics
56	80	yes	yes	Yes
10	14	yes	no	Yes
4	6	yes	yes	no

VII. Methodology for New Grid Connections

It is not possible to consider zero value of active power for connecting of micro sources because of overflows into the grid. We know that overflow of active power occurs in 70 % cases. We know the average value of overflow so we can estimate average operational conditions of the micro source. We suggest grid connectivity calculating with the value of 30 % of nominal power of the inverter. The situation can change after implementation of restrictions when penalty induce the customer to do measures. We consider implementations of penalties in September 2022. It will be necessary to evaluate the effectivity of its introduction.

VIII. Conclusion

It was demonstrated that operation of DES in the LV distribution grids causes poor voltage quality when problems with flicker and harmonics were detected. The scope of the paper was overflow analysis when overflow of active power was detected in 70 % of connected micro sources when the customers violate grid contract. Penalties are proposed as the solution for doing measures from side of the customer.

References

[1] EN 50160 Ed.3, Voltage characteristics of electricity supplied by public distribution systems. Brussels: European Committee for Electrotechnical Standardization, 2010. pp. 20.

[2] Pravidla provozování distribučních soustav – Příloha 4, Pravidla pro paralelní provoz zdrojů se sítí provozovatele distribuční soustavy (Czech grid code – part 4), Energy regulatory office, 2020.

[3] IEC 61000-4-30 Ed.3.0, Electromagnetic compatibility (EMC): Part 4-30: Testing and measurement techniques – Power quality measurement methods. Geneva: IEC.

[4] IEC/TR 60725:2012 Consideration of reference impedances and public supply network impedances for use in determining the distrurbance characteristics of electrical equipment having a rated current ≤75A per phase, IEC, ISBN 978-2-83220-161-9.

Internal Heating of Li-ion Battery Cell With Alternating Current

Samuel Bucko, Roman Cimbala, Marek Pavlík
Dept. of electric power engineering
Faculty of electrical engineering and informatics, Technical University of Košice
Košice, Slovakia
samuel.bucko@tuke.sk, roman.cimbala@tuke.sk, marek.pavlik@tuke.sk

Abstract—**The thermal management in lower temperatures is required to lifetime and usability battery cells which works on Li-ion technology. New experimental methods can improve lifespan of electrical devices like electric vehicles or battery farms which contain Li-ion battery cells, and which are used in lower temperatures. One of new experimental methods for preheating the battery cells before charging is internal AC heating, which can be possible replacement for convention external heating methods. This paper present the method of internal AC heating and thermal distribution on Li-ion battery cell. Experiment was focused on total time of heating to safe temperature level with different electrical parameters. Last part of experiment was focused on State of Health (SOH) of batteries, which were used in AC preheating cycle.**

Keywords: battery; internal heating; state of health; Li-ion

I. Introduction

During the last two decades chargeable electrochemical sources, mainly on the basis of Li-ion technology, have been a part of electricity production. Batteries with Li-ion cells are massively used with electric vehicles, for balance of consumption and production during a daily diagram or as an electricity storage from photovoltaic power plants.[1][2] Management of large battery storage is demanding thorough the year from the perspective of temperature variations or state of charge (SOC) and immediate use of storage. These and other factors adversely affect battery life and therefore the particular projects are more expensive as well as they have a negative impact on the environment.[3][4]

The critical aspect of battery storage is its use during winter season and mainly the charging process. At low temperatures, the resistance of Li-ion cells increases and that causes a decrease in the efficiency of electricity production during standard battery discharge. At low temperatures below zero, there is a risk of electrode surface crystallization with Li-ion technology and moreover, the risk of internal short circuit rises. Such a short circuit in large battery storages may result in an accident and subsequently significant economic and environmental damage. Therefore, Li-ion battery cells producers declare that the charging process shall run within a temperature range of 0° – 5°C.[5]

There are currently several scientific studies and methods which are used for battery heating in practice. The basic division from the point of view of heating is an internal and external heating. Frequently used external heating methods are contact heating method by means of heated air or liquid. These types of heating use an external heat source. Heat loss arises on the way from a heater to the batteries which are initially heated on their surface. These disadvantages of external heating were a motivation for the development of internal heating methods.

Intetrnal heating does not have the intermediate step of the heat propagation. The temperature distribution in battery pack between cells is uniform compared with external battery heating. The internal heating uses Joule heating and reaction heat generated by the alternate current flowing through battery cell.

The advantages of internal heating of a battery cell increase the heating efficiency and it does not require any other electrical and mechanical parts, which increase the weight and the price of storage. However, direct currents that the batteries operate with do not meet the requirements for such heating. For this method it is necessary to use properties of an alternating current which may generate losses equal to I^2R. In some cases studies [8][12], the internal heating works with voltage supply 50Hz (60Hz), when the requirements for additional devices are minimized. This method is economically very appealing, although the heating itself takes a relatively long time (more than 15 minutes). Another internal heating method is the use of higher/lower frequencies that voltage supplies. Yan ji, *et al.* in [10] used frequencies 0,1 – 10Hz with the variation of current. In [11], there were used higher frequencies, 300Hz and charging current lower than 2C.

It is necessary to observe parameters as SOC and SOH with AC battery heating because the use of alternating quantities which are used in DC may affect the degradation of Li-ion layer, electrode coating or a separator. Some studies monitor changes in battery parameters by means of electrochemical impedance spectroscopy. This method is based on the application of low voltage with a sinusoidal wave to the electrochemical system - the battery. The measured impedance is the function of voltage frequency, which generates alternating current with phase shift. The disadvantage of this method is a high price of a potentiostat which carries out the measurements.

Another method of monitoring voltages and currents during discharging/charging is the Coulomb Counting Method (CCM).

A. Coulomb counting method

The battery is total discharged when the voltage of battery fall under voltage limit called as „cut-off" voltage - V_{CO}. The releasable capacity, Q_{releas}, of an operating battery is the released capacity when voltage of battery cell is equivalent to V_{CO}. The SOC is defined as the percentage of releasable capacity relative to the battery rated capacity Q_{rated} :

$$SOC = \frac{Q_{releas}}{Q_{rated}} \cdot 100\% \tag{1}$$

When the battery is fully charged, the releasable capacity is on maximum, Q_{MAX}. This maximum capacity Q_{MAX} has different character as Q_{rated}, because the changeable with time of use the battery. The SOH is defined as the percentage of maximum capacity relative to the battery rated capacity:

$$SOH = \frac{Q_{MAX}}{Q_{rated}} \cdot 100\% \qquad (2)$$

Another parameter that point to actual state of discharge battery is DOD, depth-of-discharge. DOD is defined actually released capacity to the battery rated capacity:

$$DOD = \frac{Q_{released}}{Q_{rated}} \cdot 100\% \qquad (3)$$

With the measured charging or discharging current I_b the difference of the DOD in time period τ can be calculated as

$$\Delta DOD = \frac{-\int_{to}^{to+\tau} I_b(t)dt}{Q_{rated}} \cdot 100\% \qquad (4)$$

The sign before I_b is rated to charge (+) or discharge (-) mode of battery.[6]

In the Coulomb Counting method, the current leaving and entering the battery is calculated periodically, and the SOC of the battery is estimated by subtracting or adding the cumulative net charge. Usually, the initial SOC - at time 0s in discharge mode, can be estimated by the open-circuit voltage based approach or the extended Kalman filter based approach. The rated capacity, C, can be obtained from battery manufacturer or be measured in laboratory. The SOC, SOH, charge or discharge current, temperature the characteristics of the battery itself and other factors impose their influences on the coulombic efficiency in simulation process of battery cell.[7]

Fig. 1, which obtained from experimental measurements and the shape of curve is characteristic for Li-ion polymer batteries. This figure displays curves obtained under battery degradation by discharge/charge cycles. It can be seen that, as the curve reaches faster to the cut-off voltage, the battery charging is necessary much sooner and overall the capacity of battery is decreasing. In figure is graphically indicated, the curve changes in horizontal plane are associated with SOH level and the SOC can be determined by vertical advancement.

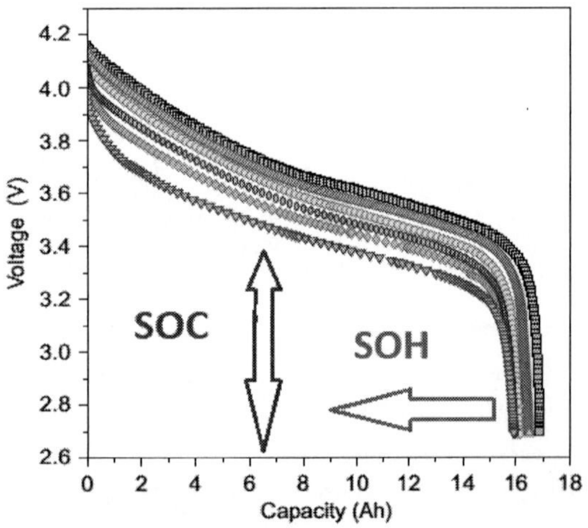

Figure 1 Discharge behavior of Li-ion battery cell with graphically indicated SOC and SOH

B. Degradation process in sub-zero temperature

The primary cause of the battery cell degradation at sub-zero temperatures is lithium plating. The temperature influences the resistivity of battery cell, which increases with lower temperature values. The mechanism of lithium plating is shown on Fig. 2. J. Zhu *et al.* describe the lithium plating depending on frequency and current of AC heating cycle.[8]

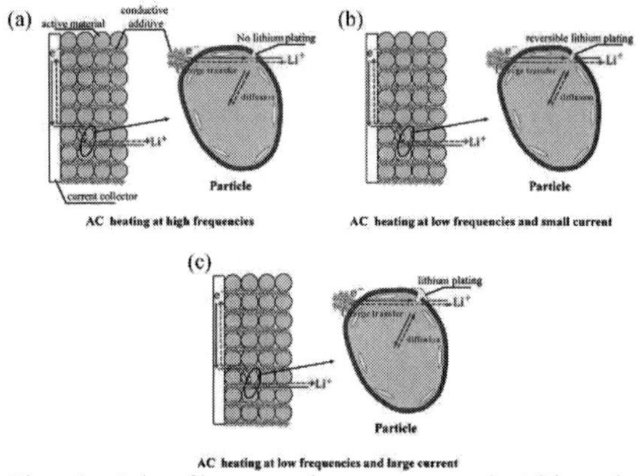

Figure 2 Aging of battery at sub-zero temperature by AC internal heating method [8]

From long term view, all parts of resistance have increasing character with more AC heating cycles. On the opposite side, the battery capacitance is decreasing but not rapidly. The lithium plating process increasing with decreasing of frequency used in AC heating cycle. In case when the low frequency and low currents are used for heating, the lithium plating has reversible character. These conditions don't have massive influence on battery cell degradation. In case when the low frequency and high current were used, the lithium plating is created, and it has irreversible character. Overall, the degradation of battery cell is accelerated. Losses of recyclable lithium in battery cell makes the cell capacity decreasing rapidly. The lithium plating causes the limitation of the SOC area of cell. The result of limitation od SOC is decreasing of battery usable time in discharging cycle.[9]

II. Experimental Setup

The commercial Lithium-Ion Polymer pouch battery cell for testing was selected. Manufacturer declared the capacity of new battery cell on 1100 mAh. The continuous discharge current at 2 C was set by resistor R_L and when V_{CO} was reached 2.7 V; the battery was switched into charge mode. Charging mode was done by Deltaelektronika SM-400 power supply in CC - CV modes. The charge/discharge process by relay and PC was controlled. Left side of Fig. 2 shown the charge/discharge process.

After 5 cycles charging/discharging at room temperature, the battery was soaked at -10,5 °C for more than 8 h. After 8h of freezing, the battery cell was inserted into thermal box with liquid nitrogen. The starting temperature was set at temperature under -25 °C. The first voltage level was set at manufacturer charging voltage 4.2 V with frequency 350 Hz. For AC heating process the AC supply GW Instek APS-7050E was used. For experiment was used arrangement of two batteries as shown at Fig. 1, at right side. The AC current of heating by resistor R_H of battery was set on 2 C. R_H to prevent temperature distortion was placed out of the box. Keysight DAQ 970A and four thermocouples collected temperatures on surface of battery. The heating process end when one of thermocouples reaches temperature above 0 °C. This criterion was set by manufacturer of battery, when they declare safety charge/discharge process. Second setup for AC heating was the same as above, but there was used higher voltage 8.0 V.

The charge/discharge process of battery cell and the AC heating was controlled by PC with Matlab software. During the

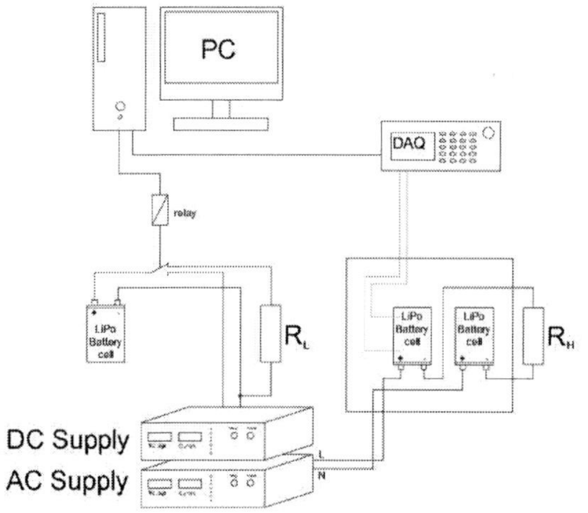

Figure 3 Measurement and acquisition system, left side – charge/discharge setup, right side – AC heating

experiment, the battery voltage and current are continuously sampled every 500 ms. Temperature was sampled every 2 s.

III. Results and discussion

The discharge characteristics of new battery cell and after freezing cycle are shown in Fig.3. Due to the small distances between curves, the detail of cut-off voltage was zoomed. The shape of curves has standard shape with relatively long band of time – from 700 s to 2000 s - with small voltage drop: 0.14 V. Typically for Lithium-ion battery is steep voltage decreasing, which end at cut-off voltage. Distance between current value of voltage of battery cell and cut-off value can be used as graphic representation of SOC. The changes in times of discharge of battery can represent the SOH of battery cell.

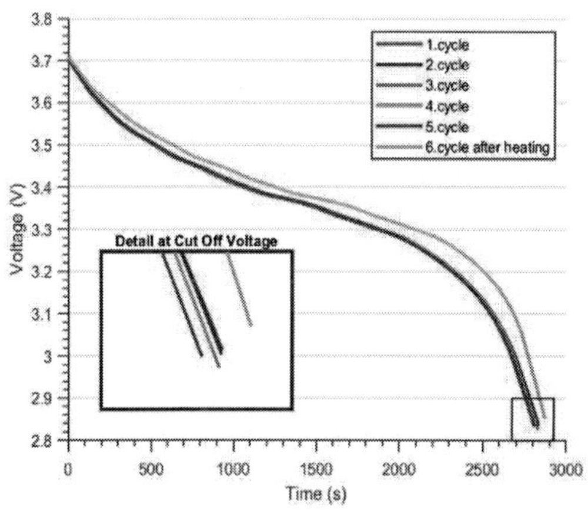

Figure 5 Discharge behaviors of battery cell after charge/discharge cycles

In case 5 discharge/charge cycles, the time of discharge has decreasing character with numbers of cycles. However, the freezing cycle not continued to reduction of battery cell capacity. It turned out that, the capacity of battery based on discharging time was improved by freezing and heating cycle. The possible reason for such increase of discharging time is not indicated clearly in the experimental studies. One possible reason could be that the battery discharge test was carried out after the battery is allowed to rest for freezing time – 8 h.

The surface temperatures of the battery cell during internal heating are shown in Fig. 4. It was observed that all of the temperatures increase consistently from temperature below -20 °C during internal heating. The maximum temperature difference measured by thermocouples was 4.14 °C for case 4.2 V. This difference was caused by non-uniform distribution of current in battery cell. Higher values were observed nearest at output electrodes of battery cell. Because of the limited amount of liquid nitrogen and imperfect insulation of thermal box, the ambient temperature of experiment was increased. Another source of undesirable heat in thermal box which had influence on thermocouple for monitoring of ambient temperature was battery AC heating itself. Fig. 4 exhibits that the battery cell is heated from - 20.5 °C to 0 °C within 738 s. The temperature of the cell as function of time and average temperature-rise rate is 1.67 °C / min for case 4.2 V AC.

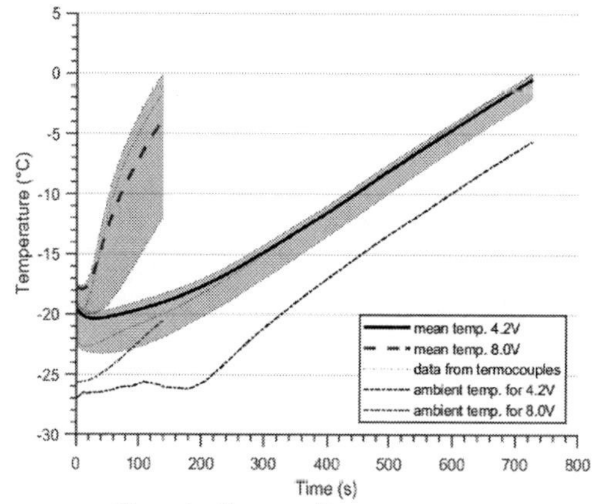

Figure 4 Battery cell temperature curves

Increasing voltage caused the short time of battery heating. The duration of heating process with 8.0 V was 155 s. The temperature-rise rate is 5.31 °C / min for case 8.0 V AC. Internal AC heating with double voltage 8.0 V for battery takes much less time like with nominal charging voltage 4.2 V. However, the battery cell after second heating with higher voltage was damaged and another load cycles could not be performed.

IV. Conclusion

The discharge characteristic of battery cell based on CCM has been presented in first part. There was an influence of different conditions discussed. From the comparison of measured values from charge/discharge cycles it can be stated that the SOH of battery is heading to shorter times of discharge. This phenomenon can be attributed to decreasing of capacity of battery cell.

On the other hand, after freezing cycle of battery cell was done, the discharge time of cell be higher as after first discharge/charge cycle of new battery cell.

For future research it necessary using the thermal chamber with low temperatures and better distribution of thermocouples for monitoring ambient temperature.

Acknowledgment

This work was supported by the Slovak Research and Development Agency under the contracts No. APVV-19-0576 and APVV-21-0312 and by the Ministry of Education, Science, Research and Sport of the Slovak Republic and the Slovak Academy of Sciences under the contract VEGA1/0757/21.

References

[1] N. Boyouk, N. Munzke and M. Hiller, "Peak Shaving of a Grid connected-Photovoltaic Battery System at Helmholtz Institute Ulm (HIU)," *IEEE PES Innovative Smart Grid Technologies Conference Europe (ISGT-Europe)*, 2018, pp. 1-5.

[2] Z. Conka, M. Kolcun, M. Kanalik, D. Medved and M. Vojtek. "Increasing Net Transfer Capacity (NTC) by UPFC in Central East Europe power system," *16th International Scientific Conference on Electric Power Engineering (EPE)*, 2015, pp. 132-136.

[3] M. Ivancak, M. Kolcun, D. Medved, and Z. Conka. "Modelling microgrid as the basis for creating a smart grid model," *Przegląd Elektrotechniczny = Electrotechnical Review*, vol. 95, pp. 41-43, 2019.

[4] Y. Wang, L. Li, Q. Ding, J. Liu and P. Chen, "Lithium-ion battery SOC estimation based on an improved adaptive extended Kalman filter," *IEEE 16th Conference on Industrial Electronics and Applications (ICIEA)*, 2021, pp. 417-421.

[5] Y. Xie et al. "A Comprehensive Study on Influence of Battery Thermal Behavior on Degradation and Consistency," *IEEE Transactions on Transportation Electrification*, 2022.

[6] Saeed Mian Qaisar. "Event-Driven Coulomb Counting for Effective Online Approximation of Li-Ion Battery State of Charge," *Energies*, vol.13, no. 21: 5600, 2020 doi: 10.3390/en13215600.

[7] Kong Soon Ng, et al. "Enhanced coulomb counting method for estimating state-of-charge and state-of-health of lithium-ion batteries." *Applied Energy*, vol. 86, pp. 1506-1511, 2009.

[8] J. Zhu, Z. Sun, X. Wei and H. Dai, "An alternating current heating method for lithium-ion batteries from subzero temperatures." *Int. J. Energy Res.*, vol. 40, pp. 1869–1883, 2016.

[9] G. Zhang, et al., "Investigation the Degradation Mechanisms of Lithium-Ion Batteries under Low-Temperature High-Rate Cycling." *ACS Appl. Energy Mater.*, 2022.

[10] Y. Ji, C.Y. Wang, "Heating strategies for Li-ion batteries operated from subzero temperatures." *Electrochimica Acta*, vol. 107, pp. 1-11, 2013.

[11] R. Howard, C. Vagg, "Effect of Internal AC Heating on the Temperature Homogeneity of Different Size Battery Cells." *Batteries*, vol. 8, pp. 1-11, 2022.

[12] Hu Xiaosong, et al., "Battery warm-up methodologies at subzero temperatures for automotive applications: Recent advances and perspectives." *Progress in Energy and Combustion Science*, vol. 77, pp. 1-28, 2020.

B: High Voltage Technique and Diagnostics in Power Engineering

Influence of Alternating Electric Field Intensity on the Dielectric Response of Hybrid Nanofluid-Paper Insulation System

Peter Havran, Roman Cimbala, Juraj Kurimský, Bystrík Dolník, Samuel Bucko
Department of Electrical Power Engineering
Technical University of Košice, FEE&I
Košice, Slovakia
peter.havran@tuke.sk, roman.cimbala@tuke.sk, juraj.kurimsky@tuke.sk, bystrik.dolnik@tuke.sk, samuel.bucko@tuke.sk

Abstract—**The issue of the application of magnetic nanofluids in high-voltage equipment is currently highly topical, as they have certain advantages and disadvantages. This study provides a dielectric relaxation response (DRR) of an oil-paper insulation system at different AC electric field intensities. As a liquid medium, a hybrid nanofluid with different content of magnetic nanoparticles Fe_3O_4 and fullerene nanoparticles C_{60} in their combination is investigated using six samples. DRR is analyzed by the dielectric spectroscopy method in the frequency domain, where the capacitance and loss ratios are described by complex permittivity and complex electric modulus in the frequency band. The results revealed the relaxation and conductivity processes of the examined samples. Gradually increasing the concentration of C_{60} in the samples with hybrid nanofluid leads to a gradual reduction of dielectric losses at operating frequencies and in the low-frequency band. This effect reduces the relaxation contributions of dielectric losses. Based on the Havriliak-Negami function, we confirm that the polarization losses dominate the conductivity losses in the entire frequency spectrum, regardless of the intensity of the alternating electric field.**

Keywords: complex electric modulus; complex permittivity; electric field; Havriliak-Negami function; hybrid nanofluid-paper

I. Introduction

To ensure the reliable operation of the power transformer with the best possible maintenance strategy, it is necessary to correctly evaluate the condition of the oil-paper insulation system [1][2]. There are many diagnostic methods for assessing the condition of an oil-paper insulation system. Recent ones include, for example, frequency-dependent dielectric spectroscopy (FDS), which is based on examining the dielectric properties of a material when applying relatively low electric field intensities. Based on this, it is considered a non-destructive diagnostic method. Quantities such as dielectric dissipation factor *tan δ*, complex permittivity ε^*, and complex electric modulus M^* can be used to accurately understand the state of the insulation system, as they can sensitively capture the response of the dielectric material to the applied electric field [1]. Dielectric spectroscopy also analyzes dielectric parameters, which are time-dependent and indicate the dielectric response of the material in the time band (TDS). By comparing the FDS and TDS methods, it is known that FDS has stronger resistance to interference, whether electromagnetic or mechanical, so it is a better alternative for studying the oil-paper insulation system in power transformers [3]. The dielectric response of the FDS method reflects the complex effects of dynamic electric charge movements at the microscopic level in interaction with an alternating electric field. These movements are related to polarization and conductivity processes. The study [4] pointed out that the FDS characteristics of dielectric losses of the oil-paper insulation system in the low-frequency band are related to the movement of free particles, thus the conductivity process of the material manifests itself. In the high-frequency band, dielectric losses are caused by polarization processes. However, the physical processes reflecting the characteristics of FDS are unpredictable in any case. Recent research points to the fact that in the low-frequency band, dielectric losses are not only caused by conductivity processes. Different dielectric materials exhibit several loss factors that are not only related to dielectric conductivity, as previously thought [2]. In a study [5] using complex electric modulus analysis M^*, they pointed out that the dielectric response of various polymeric materials (epoxy resin, polyethylene, etc.) in the low-frequency band is related to a polarization process that is covered by conductivity losses. It was also proved in the paper [6] that the conductivity losses overlap the polarization of the spatial charge in the material $CaCu_3Ti_4O_{12}$. This phenomenon has also been found in other studies, such as [7] and [8]. From these findings, it is clear that the analysis of the dielectric response in the low-frequency band using the complex permittivity ε^* is not sufficient. Several methods have been identified for eliminating the effect of conductivity on FDS characteristics, such as the Kramer-Kroning transformation. However, the results are still insufficient [8]. The analysis of the complex electric modulus M^* [8]-[11] can solve this problem. The complex electric modulus M^* shifts the relaxation peak to the higher frequency band (compared to *tan δ* and ε^* [12]) and eliminates the effect of conductivity, thus obtaining the FDS characteristic, which contains only the relaxation response of the material. This parameter also has its disadvantages. For example, it is not appropriate to describe the effect of moisture content in the insulating material using a complex electric modulus M^*. The best solution for the analysis of dielectric material from a complex point of view is to use a combination of dielectric parameters, complex permittivity ε^*, and complex electric modulus M^* [8].

The issue of the application of magnetic nanofluids in high-voltage devices is currently highly topical, as they have certain advantages and disadvantages [13][14]. The most important thing is to select the optimal carrier fluids with the lowest possible dielectric losses and then suspend them with a suitable concentration of magnetic nanoparticles [13]. The disadvantages of magnetic nanofluids are, for example, high dielectric losses and abrasive effects on cellulose insulation, which causes mechanical degradation, formation of polar substances, and further increase of dielectric losses [15][16]. This study aims to describe the dielectric response of paper impregnated with a hybrid nanofluid in the FDS frequency spectrum. This study provides a dielectric relaxation response (DRR) of an oil-paper insulation system at different AC electric field intensities. As a liquid medium, a hybrid nanofluid with different content of magnetic nanoparticles Fe_3O_4 and fullerene nanoparticles C_{60} in their combination is investigated using six samples. In studies [17]-[19], the concentration of fullerenes from 0 to 1.5 wt% reduces dielectric losses in the network frequency band 50 and 60 Hz. Therefore, the main motivation is to point out the influence of hybrid and non-hybrid impregnation fluid on dielectric losses of the oil-paper insulation system at different intensities of an alternating electric field.

II. Materials and Methods

The dielectric response study was performed on six different samples of the oil-paper insulation system. All samples were impregnated with Shell Diala S4 ZX-1 insulating base oil ("SD"). Each sample had a different concentration of nanoparticles suspended in SD. Magnetic Fe_3O_4 nanoparticles (10 nm) and C_{60} fullerene nanoparticles (1.1 nm) were used. The classification of the samples is given in Tab. 1.

Table 1. Overview of the measured samples of paper impregnated with hybrid nanofluid

Sample	Concentration of C_{60} (%w/V)	Concentration of magnetic nanoparticles (%w/V)
1.	0.01	0.01
2.	0.02	0.01
3.	0.03	0.01
4.	0.01	0
5.	0	0.01
6.	0	0

The first, second, and third samples represent an impregnating hybrid nanofluid (hereinafter referred to as "HIN-P") because they contain two different types of nanoparticles (Fe_3O_4 and C_{60}). The fourth, fifth, and sixth samples are non-hybrid impregnating nanofluids (NIN-P). The production of the hybrid nanofluid was carried out using a co-precipitation process, followed by stabilization of the magnetic nanoparticles by oleic acid in SD base oil. Subsequently, powdered fullerene with a density of 1600 kg/m^3 was dispersed in the magnetic nanofluid. The prepared liquid samples were applied to Petri dishes, impregnated with paper samples.

Experimental measurements were performed using the method of dielectric relaxation spectroscopy in the frequency domain. It is a non-destructive diagnostic method that works with low electric field intensities and points to the behavior of electrophysical parameters without any degradation of the investigated material. These requirements are met by a professional measuring instrument IDAX 300, which serves to diagnose the insulation system by changing the frequency of the applied AC voltage U and measuring the alternating current I. Using voltage and current, it calculates the impedance Z, from which dielectric parameters such as capacitance C and the dissipation factor $tan\ \delta$ are further expressed. Through two shielded wires, the IDAX 300 meter was connected to a high-voltage and low-voltage electrode, between which a sample of impregnated paper with a thickness of 0.42 mm was inserted. The electrode system consisted of two cylindrical Rogowski electrodes with a diameter of 54 mm. A decimal graduated voltage of 1 V, 10 V, and 100 V was generated from the measuring instrument. At the considered distance between the electrodes of 0.42 mm, these voltages generated alternating electric field strengths of 2.38 kV/m, 23.8 kV/m, and 238 kV/m. The circuit diagram of the measuring setup is shown in Fig. 1.

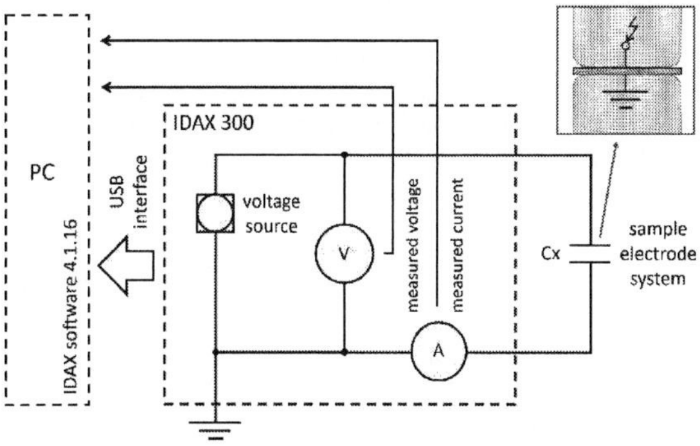

Figure 1. Wiring diagram of the experimental setup

The dielectric characteristics of capacitance C and dissipation factor $tan\ \delta$ in the frequency band 1 mHz – 10 kHz (2.38 kV/m and 23.8 kV/m) and 1 mHz – 3 kHz (238 kV/m) were captured using the IDAX 300 measuring instrument. At the highest electric field intensity, it was impossible to perform measurements in the frequency band 4 kHz – 10 kHz, because there was a decrease in capacitive reactance and consequent increase in electric current, which loaded the source of the IDAX 300 meter. The measured data of capacitance C and dissipation factor $tan\ \delta$ were obtained using IDAX 4.1.16 software on a computer. Subsequently, they were converted to the real part ε' and the imaginary part ε'' of the complex permittivity ε^* using equations (2) and (3):

$$\varepsilon^* = \varepsilon' - i\varepsilon'' \quad (1)$$

$$\varepsilon' = C/C_0 \quad (2)$$

$$\varepsilon'' = \varepsilon' \tan\delta \quad (3)$$

where C_0 is a geometric capacitance equal to 4.83×10^{-11} F. As already mentioned in the introduction, it is necessary to consider a complex electric modulus to analyze the dielectric response of the oil-paper insulation system M^* for a more detailed description of electrophysical processes. The complex electric modulus M^* was calculated from the complex permittivity ε^*, i.e., it's real M' and imaginary part M'' is equal to:

$$M^* = 1/\varepsilon^* = M' + iM'' \quad (4)$$

$$M' = \varepsilon'/(\varepsilon'^2 + \varepsilon''^2) \quad (5)$$

$$M'' = \varepsilon''/(\varepsilon'^2 + \varepsilon''^2) \quad (6)$$

The obtained data of complex permittivity ε^* and complex electric modulus M^* were plotted in the graphical dependencies, which are given below in part III. The experiment was performed under laboratory conditions with an air pressure of 1013 hPa, an air temperature of 294 K, and air humidity of 36 vol%.

III. Results of Dielectric Spectroscopy

A. Analysis of complex permittivity

Fig. 2 shows the spectra ε' in the frequency band 1 mHz – 10 kHz at an electric field strength of 2.38 kV/m. Samples HIN-

P and the 5th sample show an increase ε' in the frequency band 2 Hz – 1 mHz. We attribute this increase to dielectric relaxation. The increase in C_{60} concentration in HIN-P samples does not cause a direct dispersion ε' in the low-frequency band. Increasing the frequency to 10 kHz slightly decreases ε' for all samples. When comparing NIN-P samples, a C_{60} concentration of 0.01 %w/V (4th sample) causes an increase ε' of 0.05 in the measured frequency band compared to a sample impregnated with pure oil (6th sample). The 5th sample with a Fe_3O_4 concentration of 0.01 %w/V has the highest ε' compared to the 4th and 6th samples, due to the higher polarizability of the magnetic nanoparticles.

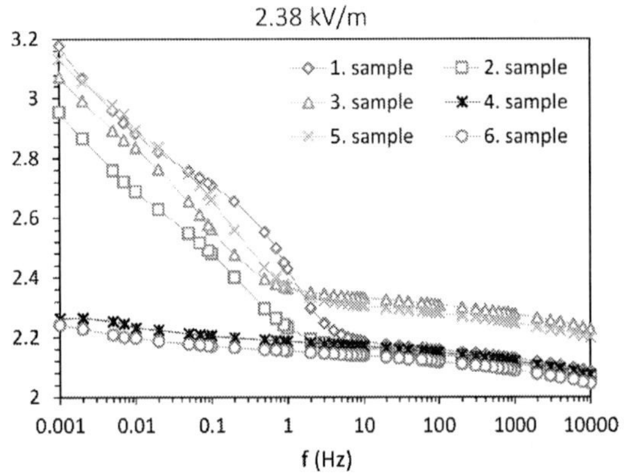

Figure 2. $\varepsilon'(f)$ of oil-paper samples at intensity 2.38 kV/m

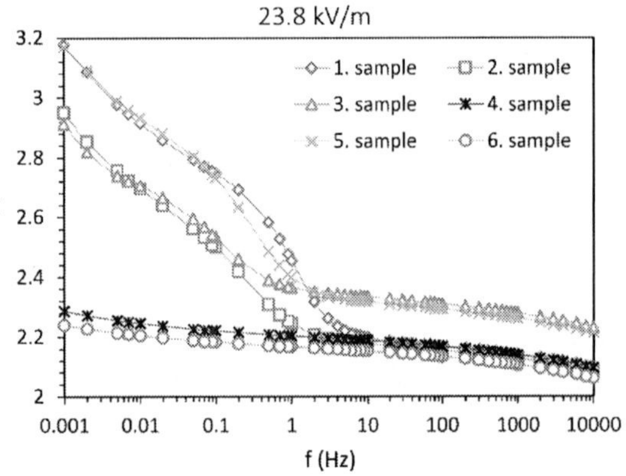

Figure 3. $\varepsilon'(f)$ of oil-paper samples at intensity 23.8 kV/m

From Fig. 2, Fig. 3, and Fig. 4, the decadic increase in electric field strength does not affect the ε' of the sample impregnated with pure oil (6th sample) and oil enriched with C_{60} (4th sample). The effect of a 10-fold increase in the electric field intensity to 23.8 kV/m (Fig. 3) does not cause significant changes in the dielectric behavior of the examined samples. An exception is the characteristic of the 3rd sample, which reduced its values in the low-frequency band. With a 100-fold increase in electric field strength to 238 kV/m in Fig. 4, there was a more significant change in the dielectric response of HIN-P samples and 5th sample, i.e., samples with a concentration of magnetic nanoparticles. The ε' values of these samples decreased in the low-frequency band 1 mHz – 0.2 Hz, with the consequence that the shape of the curves reveals another possible polarization process at a frequency of 1 mHz.

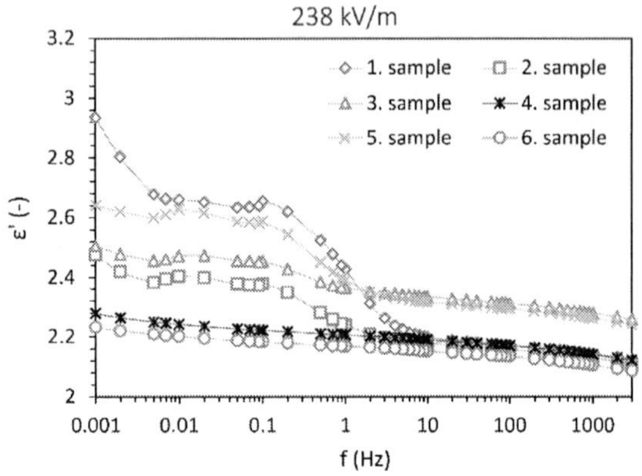

Figure 4. $\varepsilon'(f)$ of oil-paper samples at intensity 238 kV/m

Fig. 5 shows dielectric losses in the measured frequency spectrum by ε'' at an electric field strength of 2.38 kV/m. The values of ε'' are plotted on a logarithmic scale at base 2. The plot at base 10 was not used, due to the smaller number of numbers on the ε'' axis and the poorer orientation in the graph. A slight decrease in ε' in the higher frequency band correlates with an increase in ε''. This increase is attributed to interlayer polarization at the oil-paper interface, as it is detectable in all samples examined. By decreasing the frequency below 100 Hz, an increase ε'' is visible in the samples with the concentration of magnetic nanoparticles (HIN-P and 5th sample). These characteristics end with a peak signaling the relaxation process. Because samples with magnetic nanoparticles show the captured peak, we attribute this relaxation to electrical double-layer polarization. This polarization causes an increase in dielectric losses at 50 Hz and 60 Hz. It can be seen from the characteristics that the gradual increase of the C_{60} concentration leads to a gradual decrease in the dielectric losses at the operating frequencies. The influence of the increasing concentration of C_{60} nanoparticles on the relaxation peak of the electric double-layer polarization is chaotic and is not unambiguous at the applied electric field intensity of 2.38 kV/m. It is true for the size of the peak but not for its position in the frequency spectrum. Increasing the C_{60} concentration in HIN-P samples causes the characteristic to shift to lower frequencies. As the frequency of the applied electric field (1 mHz – 0.02 Hz) is further reduced, the dielectric losses of the samples with magnetic nanoparticles and similarly of the 4th and 6th samples increase again. From the given characteristics, it is impossible to accurately estimate whether the increase in dielectric losses is due to the conductivity or polarization process. In the low-frequency band, we also find that a higher concentration of C_{60} has a decreasing effect on HIN-P samples in terms of dielectric losses. By comparing the 4th and 6th samples, it can be seen that the concentration of C_{60} without the presence of magnetic nanoparticles has the opposite effect on the magnitude of dielectric losses. The proof is the higher dielectric losses of the sample enriched with C_{60}.

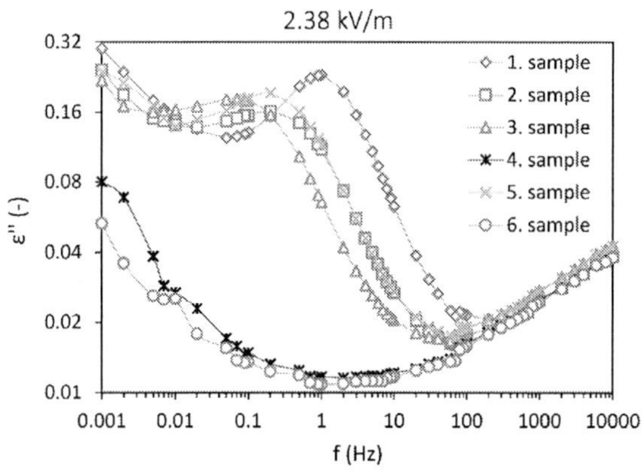

Figure 5. $\varepsilon''(f)$ of oil-paper samples at intensity 2.38 kV/m

In Fig. 6 we present the values of ε'' samples with a 10-fold increase in electric field strength to 23.8 kV/m. The change occurred in the unambiguous arrangement of the dependences of ε'' HIN-P samples around the frequency of 1 Hz. This means that a 10-fold increase in the electric field intensity only affects the electric double-layer polarization of the 3rd sample. This effect reduces the relaxation peak of the 3rd sample, respectively reducing its value. The same logarithmic scale of the ε'' axis as in Fig. 5 was also used in Fig. 6 and Fig. 7.

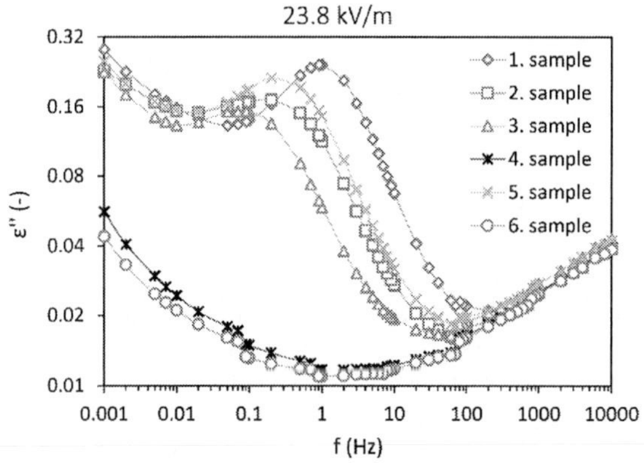

Figure 6. $\varepsilon''(f)$ of oil-paper samples at intensity 23.8 kV/m

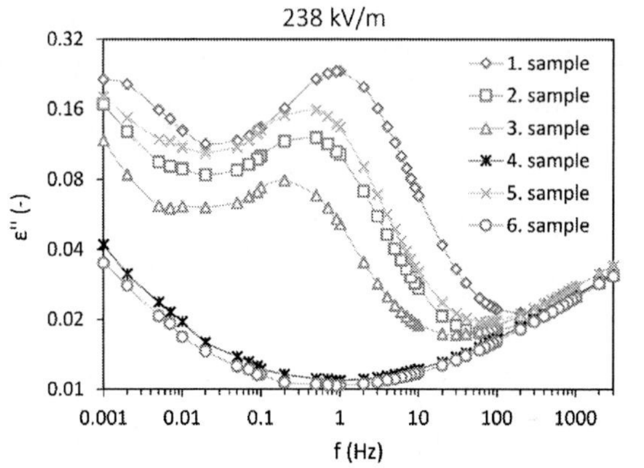

Figure 7. $\varepsilon''(f)$ of oil-paper samples at intensity 238 kV/m

With a further 10-fold increase in electric field strength to 238 kV/m in Fig. 7, the whole ε' frequency spectrum is already changing. Dielectric losses decrease across the frequency spectrum. As the intensity of the electric field is close to operating, the reduction of dielectric losses is a positive signal for applying these materials. The electric field intensity of 238 kV/m weakened the contribution of polarization and conductivity losses in the entire frequency spectrum. Increasing the concentration of C_{60} in HIN-P samples causes a reduction of dielectric losses in the band 1 mHz – 100 Hz. Thus, even at operating frequencies, which was reflected in all applied electric field intensities. It is true given the use of magnetic nanoparticles in HIN-P samples. Otherwise, there is a slight increase in dielectric losses. The proof is the comparison of the 4th and 6th samples. We register an increase in ε'' values in the low-frequency band even at an electric field intensity of 238 kV/m. All samples show an increasing tendency. The exception is the 1st HIN-P sample, where we capture a relaxation peak at a frequency of 1 mHz. This relaxation is related to the increase of the applied electric field to 238 kV/m because compared to the intensity of 2.38 kV/m and 23.8 kV/m there was no shift of the characteristic ε' due to higher frequencies.

B. Analysis of Complex Electric Modulus

The M^* analysis verifies the dielectric behavior of the investigated samples in terms of polarization processes. It is well known that M^* filters conductivity processes and provides only a relaxing response of the material (except for electrode polarization). Fig. 8 shows the characteristics of M' in the measured frequency spectrum at an electric field strength of 2.38 kV/m. A logarithmic scale was not used on the M' axis because otherwise, the measured spectra could not be plotted. By comparing the spectra of ε' and M'' when examining only liquid insulation (oils, nanofluids, etc.), the captured relaxation process shifts to higher frequencies. The relaxation characteristics do not shift to higher frequencies in the oil-paper samples we examined. This means that the combination of cellulose paper and liquid insulation does not change the position of the polarization peaks in the frequency spectrum. Thus, the loss characteristics of M'' do not reveal relaxation processes that are not captured by the characteristics ε''. In Fig. 8, Fig. 9, and Fig. 10 we find that the shapes of the characteristics of M'' are the same as the characteristics of ε''. Based on this, we can say that polarization processes dominate in the whole frequency spectrum and conductivity losses are suppressed by polarization losses. This applies to all applied electric field intensities of 2.38 kV/m, 23.8 kV/m, and 238 kV/m.

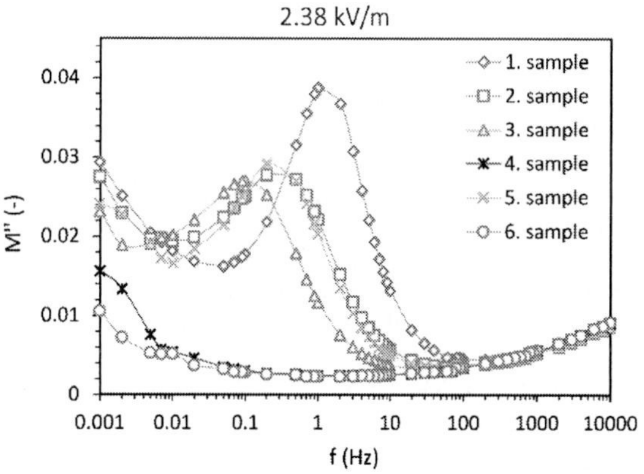

Figure 8. $M''(f)$ of oil-paper samples at intensity 2.38 kV/m

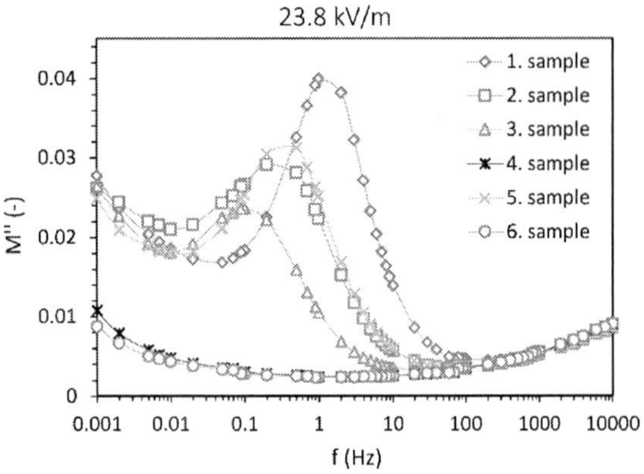

Figure 9. $M''(f)$ of oil-paper samples at intensity 23.8 kV/m

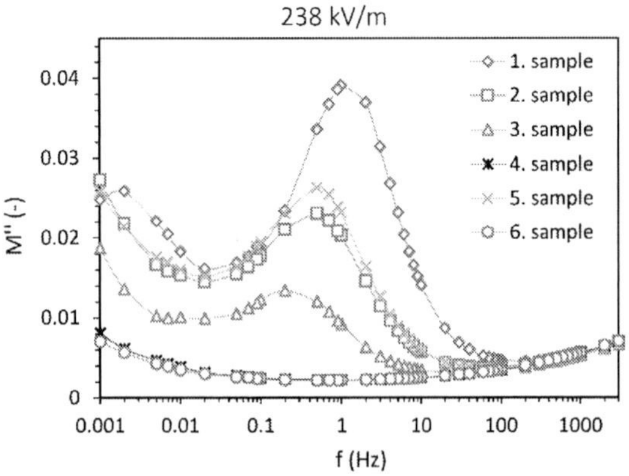

Figure 10. $M''(f)$ of oil-paper samples at intensity 238 kV/m

IV. Discussion

In part III. the results of dielectric relaxation spectroscopy of the examined samples were analyzed using ε^* and M''. The claim that polarization processes dominate the entire frequency spectrum and ε''_{DC} conductivity losses are suppressed by ε''_P polarization losses can be verified by modeling electrophysical processes using the Havriliak-Negami (H-N) function. The following relations apply to conductivity and polarization processes:

$$\varepsilon_{DC}''(\omega) = \sigma_{DC}/i\omega\varepsilon_0 \tag{7}$$

$$\varepsilon_P''(\omega) = +\mathrm{Im}\{\varepsilon_\infty+[(\varepsilon_s-\varepsilon_\infty)/(1+(i\omega\tau)^\alpha)^\beta]\} \tag{8}$$

where σ_{DC} is the unidirectional conductivity of the material, ω is the angular frequency, ε_0 is the permittivity of the vacuum (8.85×10^{-12} F/m), ε_s is the static permittivity, ε_∞ is the optical permittivity, τ is the relaxation time and α and β are the parameters of the distribution of relaxation times. Fig. 11 shows the modeling of electrophysical processes of the 1st sample (HIN-P) at an intensity of 238 kV/m using the H-N function. We confirm the statement that polarization losses dominate over conductivity losses in the measured frequency spectrum. Modeling dielectric losses ε''*model.* we find that HIN-P exhibits four relaxation processes and a conductivity process that is not dominant in the low-frequency band. The modeled dielectric loss characteristic is composed of the sum of the contributions of polarization (ε''_{P1}, ε''_{P2}, ε''_{P3}, ε''_{P4}) and conductivity losses (ε''_{DC}).

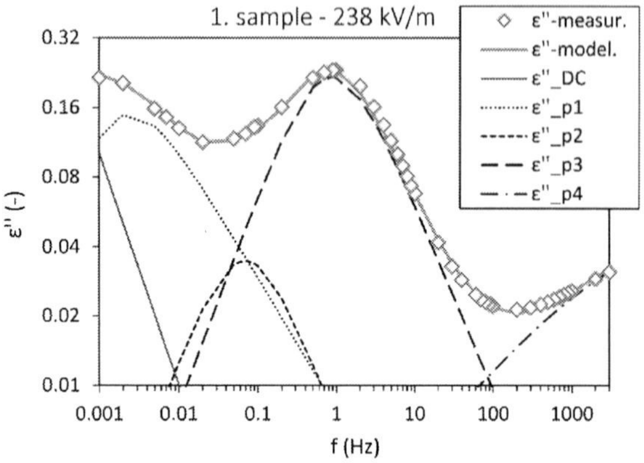

Figure 11. Processes modeling of the 1. sample using the H-N function

The captured relaxation at 2 mHz ε''_{P1} is caused by the spatial charge polarization with increasing electric field strength. The proof is the characteristic M'' 6th sample. In the vicinity of the frequency 0.07 Hz, we capture a relaxation peak with a relatively smaller contribution of dielectric losses ε''_{P2}. This polarization is related to magnetic nanoparticles, which tend to polarize with the liquid carrier medium. From Fig. 7, it can be seen that the magnetic nanoparticle samples (HIN-P and 5th sample) show the same trend behavior at a relaxation frequency of 0.07 Hz. However, dielectric losses are slightly reduced. It means that relaxation peak no. 2 is present exclusively due to magnetic nanoparticles. The proof is the characteristic of the ε'' 5th sample. C_{60} nanoparticles are not responsible for the presence of this relaxation, as the 4th sample has no contribution ε''_{P2} at this frequency (see Fig. 7). Interestingly, increasing the concentration of C_{60} in HIN-P reduces the contributions of ε''_{P1}, ε''_{P2}, ε''_{P3} and thus reduces dielectric losses. As mentioned in part III.A, the contribution ε''_{P3} is due to electrical double-layer polarization. Because all examined samples show an increasing trend in the frequency band 100 Hz – 10 kHz, the contribution ε''_{P4} is due to the oil-paper interfacial polarization. Tab. 2 shows the relaxation parameters of ε''*model.* modeling using the H-N function.

Table 2. Relaxation parameters of modeling the characteristics of the 1. sample using the H-N function

Parameter	p1	p2	p3	p4
ε_s (-)	3.06	2.65	2.69	2.16
ε_∞ (-)	2.65	2.57	2.17	1.51
α (-)	0.96	0.925	0.921	0.412
β (-)	0.6	0.95	0.881	0.201
τ (s)	100.14	2.45	0.205	2.2×10^{-5}

V. Conclusion

The evaluation of this study offers several interesting conclusions. Electrical double-layer polarization causes an increase in dielectric losses at operating frequencies of 50 Hz and 60 Hz. It can be seen from the characteristics of the HIN-P samples that the gradual increase of the C_{60} concentration leads to a gradual decrease of the dielectric losses at the operating frequencies and also in the low-frequency band. By comparing the 4th and 6th samples, it can be seen that the concentration of C_{60} without the presence of magnetic nanoparticles has the opposite effect on the magnitude of dielectric losses. Thus, increasing the concentration of C_{60} in HIN-P reduces the contributions of ε''_{P1}, ε''_{P2}, ε''_{P3}. This applies to all applied electric field intensities. Increasing the intensity of the electric field to 238 kV/m changes the whole frequency spectrum ε'' so with a reduction in dielectric losses, which is a positive signal for the application of these materials in practice. Comparing the ε'' and M'' spectra does not shift the relaxation characteristics to higher frequencies. Thus, the loss characteristics of M'' do not reveal relaxation processes that are not captured by the characteristics ε''. This means that the shapes of the M'' characteristics are the same as the ε'' characteristics. Based on the H-N function, we confirm that the polarization losses dominate over the conductivity losses in the measured frequency spectrum. By modeling dielectric losses, we find that the 1st sample of HIN-P exhibits four relaxation processes and a conductivity process that is not dominant in the low-frequency band. We present individual types of relaxation processes in the discussion. This study provides a useful insight into the analysis of changes in the dielectric response of modern insulating materials, which are investigated at different electric field strengths.

ACKNOWLEDGMENT

This research was funded by the Ministry of Education, Youth and Sports within the project VEGA 2/0011/20 and 1/0154/21 and the Slovak Agency for Research and Development based on contracts no. APVV-15-0438, APVV-17-0372, and APVV-18-0160.

REFERENCES

[1] S.D. Flora, J.S. Rajan, „Study of Frequency Domain Spectroscopy on Model Transformer Windings at Elevated Temperatures," 2015 IEEE 11th Inter. Conf. Prop. Appl. Diel. Mater. (ICPADM), IEEE press, July 2015, pp. 340-343.

[2] J. Xie, M. Dong, Y. Hu, T. Zhuang, R. Albarracín-Sánchez, J.M. Rodríguez-Serna, „Modeling Oil-paper Insulation Frequency Domain Spectroscopy Based on its Microscopic Dielectric Processes," IEEE Trans. Diel. Elect. Insul. Vol. 26, pp. 1788-1796, 2019.

[3] J. Liu, X. Fan, Y. Zhang, H. Zheng, J. Jiao, „Temperature Correction to Dielectric Modulus and Activation Energy Prediction of Oil-Immersed Cellulose Insulation," IEEE Trans. Diel. Elect. Insul. Vol. 27, pp. 956-963, 2020.

[4] M. Dong, Y. Liu, M. Ren, X.Z. Wu, H.C. Cheng, J.Z. Li, „Explanation Study of Frequency-domain Dielectric Spectroscopy for Oil-paper Insulation System," Proc. Csee. Vol. 35, pp. 1002-1008, 2015.

[5] F.Q. Tain, Y. Ohki, „Electric modulus powerful tool for analyzing dielectric behavior," IEEE Trans. Diel. Elect. Insul. Vol. 21, pp. 929-931, 2014.

[6] K.N. Wu, Y.W. Huang, J.Y. Li, S.T. Li, „Space charge polarization modulated instability of low frequency permittivity in $CaCu_3Ti_4O_{12}$ ceramics," Appl. Phys. Lett. Vol. 111, pp. 217-220, 2017.

[7] D. Linhjell, L. Lundgaard, U. Gäfvert, „Dielectric response of mineral oil impregnated cellulose and the impact of aging," IEEE Trans. Diel. Elect. Insul. Vol. 14, pp. 156-169, 2007.

[8] K. Yang, M. Dong, Y. Hu, J. Xie, M. Ren, „The Application of Dielectrc Modulus in the Oil-impregnated Paper Insulation," 2020 Elect. Insul. Conf. (EIC), IEEE press, June-July 2020, pp. 103-106.

[9] P. Havran, et al., „Dielectric Properties of Electrical Insulating Liquids for High Voltage Electric Devices in a Time-Varying Electric Field," Energies. Vol. 15, 391, 2022.

[10] I.M. Ward, „Anelastic and dielectric effects in polameric solids," Wiley: New York and London, 1967.

[11] L. Sheng-Tao, W. Hui, L. Chun-Jiang, L. Jian-Ying, „Dielectric modulus response of $CaCu_3Ti_4O_{12}$ ceramic," J. Phys. Vol. 62, pp. 474-479, 2013.

[12] M. Wübbenhorst, J. Turnhout, „Analysis of complex spectra. I. One-dimensional derivative techniques and three-dimensional modeling," J. Non-Cryst. Sol. Vol. 305, pp. 40-49, 2002.

[13] M. Šárpataky, J. Kurimský, M. Rajňák, „Dielectric Fluids for Power Transformers with Special Emphasis on Biodegradable Nanofluids," Nanomaterials. Vol. 11, pp. 1-28, 2021.

[14] J. Kudelcik, S. Hardon, L. Varacka, „Measurement of Complex Permittivity of Oil-Based Ferrofluid in Magnetic Field," Acta Phys. Pol. A. Vol 131, pp. 931-933, 2017.

[15] M. Kole, S. Khandekar, „Engineering applications of ferrofluids: A review," J. Magn. Magn. Mater. Vol. 537, 168222, 2021.

[16] S. Bucko, M. Krchňák, R. Cimbala, L. Kruželák, J. Zbojovský, „Abrasive properties of transformer oil-based magnetic nanofluid," 2018 19th Inter. Scie. Conf. Elect. Pow. Engin. (EPE), IEEE press, May 2018, pp. 1-4.

[17] P. Aksamit, D. Zmarzly, T. Boczar, M. Szmechta, „Aging properties of fullerene doped transformer oils," 2010 IEEE Inter. Sym. Elect. Insul., IEEE press, 2010, pp. 1-4.

[18] Z. Huang, et al., „Significantly Enhanced Electrical Performances of Eco-Friendly Dielectrc Liquids for Harsh Conditions with Fullerene," Nanomater. Vol. 9, 989, 2019.

[19] S. Bronnikov, A. Podshivalov, S. Kostromin, M. Asandulesa, V. Cozan, „Dielectric spectroscopy investigation on relaxation in polyazomethine/fullerene C_{60} nanocomposites," Euro. Pol. J. Vol. 78, pp. 213-225, 2016.

Use of Artificial Intelligence to Determine the Degree of Environmental Pollution

Ľuboš Pavlov, Ľuboš Skurčák
Department of Science, Research, Development and Analysis
VUJE, a.s.
Trnava, Slovakia
lubos.pavlov@vuje.sk, lubos.skurcak@vuje.sk

Ladislav Hluchy, Marcel Kvassay, Peter Krammer
Institute of Informatics
Slovak Academy of Sciences
Bratislava, Slovakia
hluchy.ui@savba.sk, kvassay.ui@savba.sk, peter.krammer@savba.sk

Martin Kremler
Air Quality Monitoring
Slovak Hydrometeorological Institute, Bratislava, Slovakia
Department of Astronomy, Physics of the Earth, and Meteorology
Faculty of Mathematics, Physics and Informatics Comenius University, Bratislava, Slovakia
martin.kremler@shmu.sk

Abstract—**The reliability, longevity and associated maintenance costs of the technological equipment of the electricity system depend on several factors. Throughout its operation, it is necessary to consider the impact of various meteorological conditions, fauna and flora, as well as human activity itself in the form of local environmental pollution. The degree of influence of these factors on the operation of technological equipment should be taken into account already in design and processing of project documentation. Selection of specific types of components of technological equipment such as insulators, masts, conductor types, etc. is implemented on the basis of proven technical standards using the results of long-term field measurements, which are economically demanding. At present, however, in developed economies, a lot of data is collected and archived in various digital databases. These data sources bring new possibilities of their use for the needs of electric power industry. However, new approaches to finding new unknown contexts need to be developed and tested. Artificial intelligence techniques are used for such tasks, which are characterized by the ability to understand and find connections between input and output data that are unknown, often non-linear, and it is currently not possible to describe them by other commonly used models. An important feature of artificial intelligence techniques, and especially neural networks, is their ability to recognize and classify situations characterized by the development of several parameters simultaneously.**

Keywords: pollution; artificial intelligence; classification; measurement, feedforward neural network

I. Introduction

In Slovakia, pollution measurements for the needs of the electric power industry were performed according to the STN 33 0405 "Design of external electrical insulation according to the degree of pollution". This standard was repealed in 2015 without compensation, which paved the way for the development of a new methodology for determining pollution based on artificial intelligence methods. The following text summarizes the experience and achieved results of the application of neural networks to determine the relationships between measurements performed in different regions of Slovakia according to STN 33 0405 and air pollution data monitored and collected by the Slovak Hydrometeorological Institute.

II. Pollutants Evaluated by SHMI

The Slovak Hydrometeorological Institute (SHMI), on the basis of Directive 2008/50 / EC of the European Parliament and of the Council [1] on ambient air quality and cleaner air in Europe, carries out continuous monitoring in order to best characterize air quality, taking into account public health protection. Air pollution can be caused by natural sources as well as human activity (anthropogenic sources) [2], for example:

- burning fossil fuels in electricity generation, transport, industry and households,
- industrial processes and the use of solvents, for example in the chemical and mineral processing industries,
- agriculture,
- waste treatment,
- volcanic eruptions, wind-blown dust, emissions of volatile organic compounds from plants.

Slovak Hydrometeorological Institute through a network of measuring stations called "National Air Quality Monitoring Network" (NAQMN) provides continuous measurements using automatic instruments and manual measurement based on sampling and chemical analysis in the SHMI Testing Laboratory. Manual monitoring covers the measurement of concentrations of heavy metals, volatile organic compounds and polycyclic aromatic hydrocarbons in the air, as well as air quality monitoring and precipitation quality analysis at regional background stations in the framework of the Monitoring Program for Monitoring and Evaluation of Long-range Transmission of Air Pollutants in Europe (EMEP) [3].

Air Act no. 137/2010 Coll. as amended sets out the procedure for air quality assessment and criteria in full compliance with EU directives and allows mathematical modelling for spatial air quality assessment (CEMOD and IDW-A models) to be used for air quality assessment in addition to monitoring station measurements [3] .

The model for spatial assessment of gaseous pollutant concentrations (CEMOD) uses meteorological and emission data in sequential form (time sequence in hours). The preparation of meteorological inputs for modelling includes the processing of data from meteorological stations as well as outputs from the ALADIN meteorological model, while in addition to information on the ground layer, vertical profiles of meteorological characteristics are also required. Emission input data are obtained from the database of the National Emission Information System (NEIS), from which the annual emission flows, positions and parameters of chimneys, temperature and flue gas velocity were processed.

The interpolation model for comprehensive air quality assessment (IDW-A) is used for spatial assessment of sites with frequent exceedances of limit values. This includes pollutants such as PM_{10}, $PM_{2.5}$, O_3 a NH_3. The calculation takes into account a large amount of relevant emission input data such as condensation of hot flue gases escaping from chimneys and the results of the calculation of resuspension of solid pollutants deposited on traffic routes. The IDW-A interpolation model makes it possible to determine the spatial distribution of pollutants in the territory of the Slovak Republic on the basis of measured data via the network of monitoring stations of the NAQMN network.

The most significant pollutants assessed by SHMI include:

- **Dust particles (PM_{10} and $PM_{2.5}$)** are small particles or droplets with an aerodynamic diameter of less than 10 µm (PM_{10}), resp. 2.5 µm ($PM_{2.5}$). The term "PM" comes from the English "particulate matter". PM freely dispersed in the air forms an atmospheric aerosol. Their origin is in various technological processes, they are released mainly during the combustion of solids, they are contained in the exhaust gases of motor vehicles. However, they also enter the air by agitating particles deposited on the earth's surface (secondary dust) [2].
- **Sulphur oxides (SO_X)** are compounds such as sulphur dioxide (SO_2), sulphur dioxide (SO_3) and aerosol H_2SO_4. Sulphur dioxide is a colourless reactive gas which, at higher concentrations, has a strong irritating odour. Sulphur oxide is normally a colourless liquid which, however, polymerises to shiny white needles when cooled [4].
- **Nitrogen oxides (NO_X)** are formed during combustion processes, their most important components are nitrogen dioxide (NO_2) and nitric oxide (NO). Nitrogen dioxide is a yellow-brown poisonous gas, nitric oxide is a reactive gas that oxidizes rapidly to NO_2. Nitrogen oxides, carbon monoxide and volatile organic compounds enter into reactions that affect ground-level ozone concentrations are so-called O_3 precursors [4].
- **Ozone (O_3)** is a triatomic oxygen molecule. It is a very soluble water in water, with strong oxidizing properties, which has acute and chronic health effects. While stratospheric ozone plays an important role in protecting against the sun's harmful ultraviolet radiation, tropospheric (ground-level) ozone has an adverse effect on human health, vegetation, architectural structures and is therefore classified as a pollutant [4].
- **Carbon monoxide (CO)** is a colourless, tasteless, odourless, poisonous gas that is highly toxic. It is formed during incomplete combustion and is part of the exhaust gases of motor vehicles [2].
- **Benzo (a) pyrene (BaP)** belongs to the group of polycyclic aromatic hydrocarbons (PAH), it forms yellow acicular water-insoluble crystals with a melting point of 179 °C. It is part of the fine fraction of atmospheric aerosol [4].

III. Determination of the Degree of Environmental Pollution

The company VUJE, a. s. in the past, it carried out measurements in order to determine the degree of pollution for the needs of the transmission system operator of the Slovak Republic. In the years 2008 to 2009, measurements were performed in power stations and in the period 2010 to 2013, measurements were performed in the corridors of high-voltage power lines. These measurements were processed in accordance with standard STN 33 0405. After 2015, when the standard in question was repealed, the determination of the degree of pollution in the territory of the Slovak Republic is evaluated in accordance with the standard ČSN 33 0405 "Design of outdoor electrical insulation according to the degree of pollution". These are identical standards; therefore the change of the standard does not change the evaluation procedure and the results obtained according to STN 33 0405 are fully comparable with the results according to ČSN 33 0405. For this reason, we present only the standard ČSN 33 0405.

The degree of environmental pollution was measured by means of distributed sedimentation vessels in selected localities and subsequent analysis of their content in accordance with the standard [6], [7] at six-week intervals. In addition to information on the direction of the prevailing winds, known sources of pollution according to SHMÚ data or other available information were also considered when designing the layout of sedimentation vessels.

The company VUJE, a. s. for the needs of evaluating the content of sedimentation vessels, it uses its own accredited testing laboratory of chemical regimes and physicochemical analysis.

The following values determined by the ČSN 33 0405 standard are determined for each measuring point and each collection:

a) the total amount of particulate matter converted per unit area per unit time S ($mg.cm^{-2}.day^{-1}$),

b) the amount of soluble matter in the sediment converted per unit area per unit time S_r ($mg.cm^{-2}.day^{-1}$),

c) specific electrical conductivity 0,2% solution of soluble substances $\gamma_{0,2}$ ($\mu S.cm^{-1}$).

Subsequently, the maximum values of the total amount of particulate matter, the amount of soluble substances and the conductivity for each measuring point are determined on the basis of the Weibull distribution. S_m, S_{rm}, $\gamma_{0,2m}$, that will not be exceeded with probability 0,5 %.

These values are used to calculate the parameter p, on the basis of which the appropriate degree of pollution is determined in accordance with the ČSN 33 0405 standard.

IV. Application of Neural Networks

Artificial neural networks (ANN) represent one example of machine learning that allows you to compute any computable function. In practice, neural networks are mostly used for classification and functional approximation or mapping of functions, i.e. assigning values from one domain to values in another domain, using training data. Neural networks are tolerant to uncertainties in the data intended for training [9].

The following approaches were analysed in the application of neural networks for the purpose of determining the degree of environmental pollution:

1. Determination of Weibull distribution parameters, on the basis of which the maximum value S_m, S_{rm}, $\gamma_{0,2m}$, was determined, which will not be exceeded with a probability of 0.5%, with subsequent determination of the degree of environmental pollution in accordance with the ČSN 33 0405 standard.
2. Determination of maximum values S_m, S_{rm}, $\gamma_{0,2m}$, which will not be exceeded with a probability of 0.5%. Subsequently, the degree of environmental pollution was determined in accordance with the ČSN 33 0405 standard.
3. Determination of the product of the maximum values S_m, S_{rm}, $\gamma_{0,2m}$, on the basis of which the degree of environmental pollution was defined in accordance with the ČSN 33 0405 standard.
4. Direct determination of the degree of environmental pollution in accordance with the ČSN 33 0405 standard.

The analysis of the data showed that the most promising approach is number 2, i.e. the determination of the maximum values of S_m, S_{rm}, $\gamma_{0,2m}$, which will not be exceeded with a probability of 99.5%.

Based on available foreign professional studies and publications in the field of artificial intelligence, an analysis was carried out, aimed at verifying the use of standard artificial neural networks applied to a similar type of task. The analysis resulted in several selected standard neural networks. The results show that none of the standard neural networks used achieved the required accuracy in determining the degree of pollution from SHMI data [14]. In the future, the possibilities of how to increase the accuracy of neural networks will be explored, while the expansion of relevant input data is expected, which can better describe the physical and chemical properties between measurements performed in accordance with ČSN 33 0405 and pollutant data evaluated by SHMÚ. These data can be e.g. data characterizing transport and agriculture, which are a relatively significant source of pollution, or satellite images.

A partial problem that needed to be solved was the reliability of determining the maximum values of S_m, S_{rm}, $\gamma_{0,2m}$, using the Weibull distribution. In the classical approach according to the method of moments and the method of maximum likelihood in accordance with the standard STN 01 0224, extreme deviations occurred in several cases of prediction of maximum values.

For this reason, several feedforward neural network configurations have been developed and tested. A schematic representation of an example of a feedforward neural network with three hidden layers is shown in Figure 1.

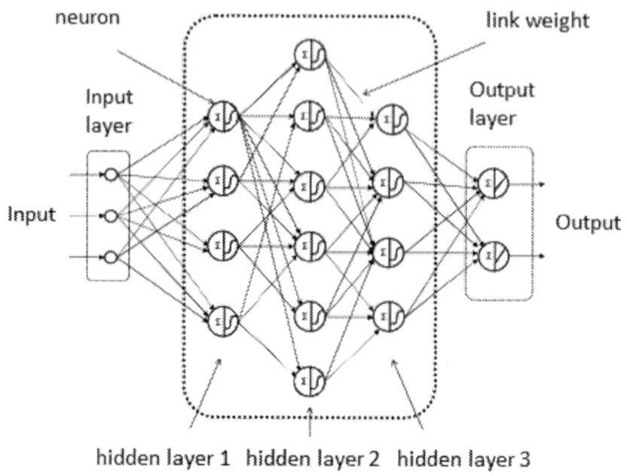

Figure 1. Example of a feedforward neural network with three hidden layers

For the needs of this task, the most suitable configuration appeared to be a feedforward neural network formed by three layers, i.e. input layer, hidden layer and output layer [9].

In this case, the input layer is represented by synthetic data (S, S_r a $\gamma_{0,2}$) which were obtained on the basis of measured values in accordance with ČSN 33 0405 (S, S_r a $\gamma_{0,2}$).

The output layer is represented by output parameters (S_m, S_{rm} a $\gamma_{0,2m}$). In established neural networks, each neuron receives at its input a weighted sum of signals from previous neurons, which it transforms via a transfer function into an output signal. It is sent to other neurons via connections characterized by weight coefficients and the process is repeated. The transmission functions of neurons in this case are nonlinear [9].

In the training process, the values of the input parameters are presented to the neural network, for which the output network calculates the required output values, while adjusting the weight coefficients of the connections between the neurons. This is repeated until the output values reach the required accuracy. The number of input neurons corresponds to the number of input parameters. The number of output neurons corresponds to the number of required output parameters. The most appropriate number of hidden layers and neurons in them is usually sought experimentally in the training process. As part of the solution, several specific feedforward neural networks with different numbers of hidden layers were tested. The best results were obtained using one hidden layer with 20 neurons. This network was used in further analyses.

Synthetic training data were prepared from available experimental fallout data for all pollution classes. The defined procedure, which is not the content of this paper and will still be subject to validation, obtained 250,000 training patterns (S, S_r, $\gamma_{0,2}$), which were standardly divided in the ratio of 70/15/15% for training, validation and testing set and used to train a set of 100 different ANN. The 20 best trained were then selected. The training quality criterion was the regression coefficient calculated from the test set of samples. The resulting values of the calculated maximum values S_m, S_{rm}, $\gamma_{0,2m}$, which will not be exceeded with a probability of 0.5%, were the average of these 20 ANN. This is a commonly used procedure to improve the predictive capabilities of the ANN.

Discussion

The maximum values (S_m, S_{rm} a $\gamma_{0,2m}$) were calculated using the classical approach according to the method of moments and the method of maximum likelihood and using trained feedforward neural networks with using synthetic data. The comparison of the results is graphically shown in the following figures (Figure 2 to Figure 4). The following values are color-coded in the figures:

- the results are shown in red (ANN),
- the values calculated by the classical algorithm according to the method of moments and the method of maximum likelihood in accordance with the standard STN 01 0224 are expressed in orange.

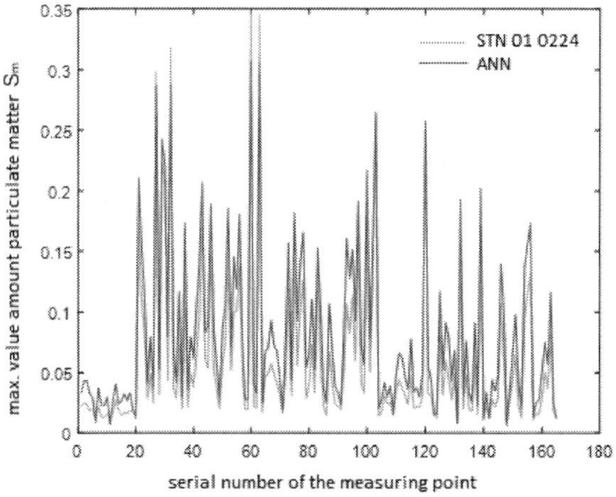

Figure 2. Maximum amount of particulate matter Sm in measured places

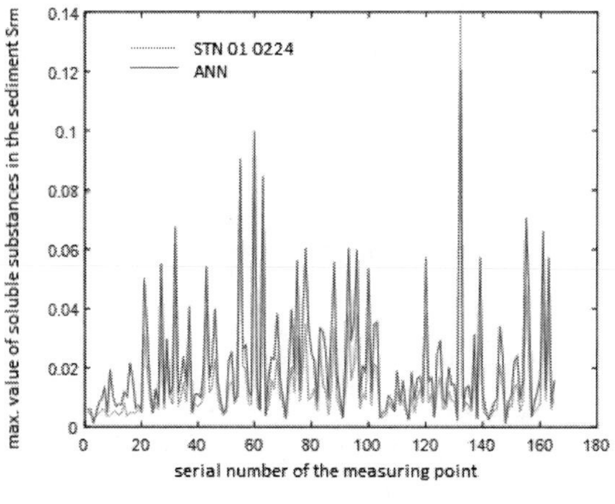

Figure 3. Maximum values of the amount of soluble substances at the sediment S_{rm} in measured places

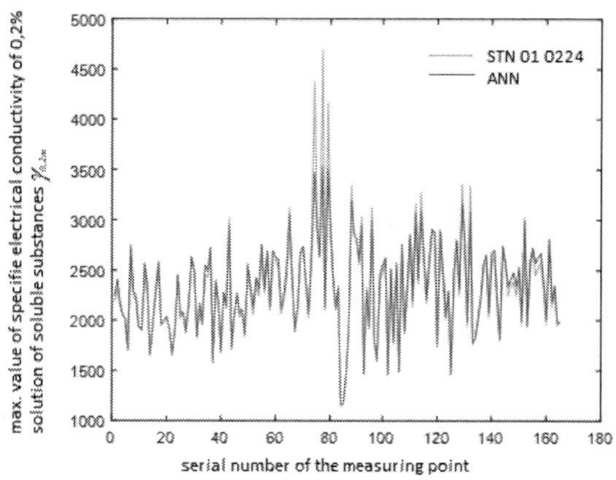

Figure 4. Maximum values of specific electrical conductivity of 0,2% solution of soluble substances $\gamma_{0,2m}$ in measured places

The results of the calculations of the magnitude of the degree of contamination by the above procedure, i.e. using artificial neural networks, were compared with the results of measurements of the degree of environmental pollution performed by VUJE, a. s. in the years 2008 to 2013 (Figure 5).

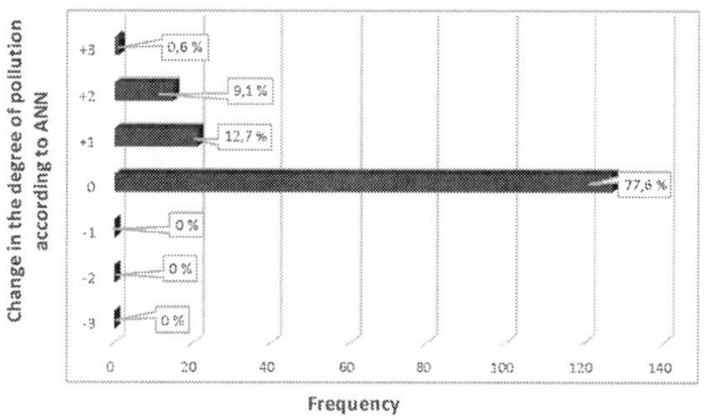

Figure 5. Comparison of pollution degree results

A comparison of the results shows that the success rate of this trained feedforward neural network procedure is at the level of 77.6%. A total of 165 sites were compared, where measurements were made in the past to determine the degree of pollution. In 128 cases, the same degree of pollution was determined as in the measurements according to ČSN 33 0405. With the help of ANN, the degree of pollution was not underestimated at any of the measured points. In only one case was a significant shift in the degree of environmental pollution identified. The calculation of the degree of pollution using ANN can be described as more conservative in comparison with the results of the calculation of the degree of pollution in the sense of ČSN 33 0405. The results will be validated in the next project solution.

Conclusion

The application of artificial intelligence techniques for the purpose of determining the degree of environmental pollution brings the solution of some specific tasks. The agreement of the ANN results with the classical procedure according to the maximum likelihood method, which we understand as a reference, is at the level of 77.6%. It should be emphasized that, assuming the use of a sufficiently wide range of Weibull distribution parameters, the synthetic training patterns covered all four pollution classes evenly, although the representation of higher classes in real measured data is sporadic. In the generated synthetic training patterns, the unlikely parameters S, S_r a $\gamma_{0,2}$, on the recognition of which ANN were trained, were also represented. However, when solving the system of equations in the method of maximum likelihood, such unlikely values could cause extreme deviations of the calculated maximum values of the parameters S, S_r a $\gamma_{0,2}$ from the expected values. Therefore, we consider the determination of maximum values of S_m, S_{rm} a $\gamma_{0,2m}$ using ANN as a better alternative.

Acknowledgment

This work was supported by the Slovak Research and Development Agency under the contract No. APVV-20-0548.

Appendix

For a more detailed demonstration of the trained classification model performance, we present the so-called Confusion Matrix, showing the distribution of records in the test set, from the point of view of the actual and predicted pollution class in Tab. 1.

The individual rows of the matrix represent the actual representation into classes, while the columns represent the predicted classes (values I. - IV. In both cases). The numbers in the matrix represent the frequencies of individual occurrences in the predicted data of the test set. Ideally, this matrix should consist of non-zero values exclusively on the main diagonal of the matrix. However, due to the high complexity of the task, we must expect a more realistic result.

In this case, some slight deviation to the higher pollution classes is evident. However, the trained model does not allow an error in which the degree of contamination would be underestimated (the values below the main diagonal of the matrix are zero in all cases).

The trained classification model can thus be considered significantly conservative, as the degree of pollution is not underestimated. Therefore, when using such a model, there is no risk associated with underestimating the degree of pollution. The disadvantage is in some cases an overestimation of the degree of pollution, which can result in higher costs.

Tab. 1: Confusion Matrix of created model for classification of pollution level.

		Predicted Class			
		I.	*II.*	*III.*	*IV.*
True Class	*I.*	109	13	8	1
	II.	0	0	1	7
	III.	0	0	0	7
	IV.	0	0	0	19

References

[1] Access to European Union law. "DIRECTIVE 2008/50 / EC OF THE EUROPEAN PARLIAMENT AND OF THE COUNCIL of 21 May 2008 on ambient air quality and cleaner air for Europe Official Journal of the European Union," OJ L 152, 11.6.2008, pp. 1-44.

[2] Slovak Hydrometeorological Institute. "Negative effects of monitored pollutants" Internet: www.shmu.sk/sk/?page=229, Jan. 2020 [Jan. 20, 2022].

[3] Report on air quality in the Slovak Republic, October 2021 Bratislava.

[4] Slovak Hydrometeorological Institute. "National Emissions Inventory System of the Slovak Republic" Internet: https://oeab.shmu.sk/, Dec. 2000 [Nov.11, 2021].

[5] SPIRIT. "Slovak National Emission Information System - software tools." Internet: http://www.air.sk/neis.php, Dec. 2000 [Nov.22, 2021].

[6] STN 33 0405 "Design of external electrical insulation according to the degree of pollution."

[7] CSN 33 0405 "Design of external electrical insulation according to the degree of pollution."

[8] STN 01 02 24 "Rules for setting estimates and confidence limits for Weibull distribution parameters."

[9] J. Kleine Deters, R. Zalakeviciute, M. Gonzalez, Y. Rybarczyk. "Modelling PM2. 5 urban pollution using machine learning and selected meteorological parameters," in *Journal of Electrical and Computer Engineering*, Hindawi, 2017. DOI: 0.1155/2017/5106045.

[10] Institute of Computer Science. "Forward neural networks", Internet: https://neuron.ics.upjs.sk/wp-content/uploads/2016/01/UNS15_03.pdf, Dec. 5, 2014 [Dec.12, 2021].

[11] L. Maraaba, Z. Alhamouz, H. Alduwaish. "A neural network-based estimation of the level of contamination on high-voltage porcelain and glass insulators," in *Electrical Engineering*, 2018. ISSN: 1545-1554, DOI: 10.1007/s00202-017-0634-z.

[12] C. C. Aggarwal. "Neural networks and deep learning a textbook," in *Springer*, 2018. pp. 7-52. ISBN: 978-3-030-06856-1, DOI: 10.1007/978-3-319-94463-0.

[13] Y.C. WU, A. Shiledar, YC. Li, J. Wong, S. Feng, X. Chen, C. Chen, K. Jin, S. Janamian, Z. Yang, Z. S. Ballard, Z. Gorocs, A. Feizi, A. Ozcan. "Air quality monitoring using mobile microscopy and machine learning," in *Light: Science &Applications*, 2017, 6(9). DOI:10.1038/lsa.2017.46.

[14] P. Krammer, M. Kvassay, R. Forgac, M. Ockay, L. Skovajsova, L. Hluchy, L. Skurcak, L. Pavlov. "Regression Analysis and Modelling of Local Environment Pollution Levels for Electric Power Industry," in *Computing and Informatics*, 2022. DOI: 10.2139/ssrn.4059473.

Aerial Diagnostics of High-Voltage Lines Using Modern Diagnostic Systems

Ľuboš Skurčák, Ľuboš Pavlov, Peter Bojda, Ján Gbelec
Department of Science, Research, Development and Analysis
VUJE, a.s.
Trnava, Slovakia
lubos.skurcak@vuje.sk, lubos.pavlov@vuje.sk, peter.bojda@vuje.sk, jan.gbelec@vuje.sk

Abstract—**Ensuring and maintaining reliable operation of the transmission or distribution system in large networks of overhead and high-voltage overhead lines is a challenging task. At the beginning of this process is a well-made design with project documentation, which is followed by the precise construction of external power lines. However, regular inspection and maintenance of lines is also important to maintain high operational reliability, the aim of which is to minimize the occurrence of emergency situations resulting in the interruption of electricity transmission or distribution. At the same time, it is important that maintaining the high reliability of the power lines is carried out in a cost-effective manner. Experience in the case of large energy networks confirms that the only effective and fast way to obtain information on power lines is to use aerial inspections. Other forms of inspections (in the case of power lines, such as walking and climbing inspections) also have their justification, and ideally their results are complementary. However, the importance of aerial revisions is irreplaceable in terms of the speed of assessing the real condition of large line structures. For this reason, electricity network operators are interested in an expert service in the form of aerial inspection of power lines, aimed at identifying defects and failures that may jeopardize the safety and reliability of operating equipment, analysis of their severity and processing of protocols. The presented paper is based on the practical experience of the authors and is focused on the presentation of the possibilities of using modern diagnostic systems for aeronautical diagnostics of external power lines.**

Keywords: reliability; aerial audits; diagnostics; external power lines

I. Introduction

Advanced multi-sensor diagnostic and measurement systems enabling automation and combination of several types of data are increasingly used in technologically developed countries as part of aerial revisions of linear energy structures. These trends are clearly identifiable in several professional articles and publications [1], [2], [3], [4]. The most well-known and currently already in common use is the LiDAR system. This system uses the method of remote distance measurement based on the calculation of the pulse propagation time of the laser beam reflected from the scanned object. It is thus possible to obtain a spatial image of the scanned objects and their surfaces. The current trend is also focused on digitizing outputs and providing data outside the visible spectrum. Such data sources include a thermal imaging camera scanning in the infrared spectrum, a corona camera operating in the ultraviolet spectrum, or hyperspectral cameras operating in the broad electromagnetic spectrum. By processing the outputs obtained through modern diagnostic systems, it is possible to assess in more detail the real state of line structures and capture the threats to their operation at an early stage.

II. Use of the LiDAR System for Linear Constructions

The output data of the LiDAR system represents a cloud of points in 3D space with a very precise position, which can be categorized into different classes after processing, create 3D models of scanned objects or georeferenced outputs in the form of a digital model of the surface or terrain.

The laser scanner used by VUJE, a. s. provides a maximum measuring range of up to 1350 m, an effective measuring speed of up to 820,000 measurements / sec. and an operating flight altitude of up to 530 m. Even in the most powerful mode, i.e. at the highest frequency of laser pulses, the system can reliably detect objects with a low reflectivity (20%) at a distance of 215 m at an operating flight altitude of up to 140 m above the ground. The minimum point density for one aerial laser scan declared by the manufacturer according to the scan head specification [5], for a scan speed of 820 kHz and a flight speed of 30 kn (approx. 55 km/h) reaches an average density of more than 50 points per m^2 at a distance of 150 m, for shorter distances always higher. The used technology for laser scanning in the sense of the manufacturer's data [5] allows to capture a practically unlimited number of reflections of one beam.

The recorded cloud points must be processed by specialized software enabling their georeferencing and classification into defined classes (conductor, mast, insulator, vegetation, ground surface without vegetation). After processing, the point clouds are saved in *.las, *.laz file formats. Such output data can then be used in a more detailed assessment of defects obtained from visual aerial control and for the analysis of spatial distances of classified point cloud classes. Within this analysis, it is possible to perform necessary measurements of the spatial distances of classified classes, namely conductor-ground, conductor-vegetation, conductor-foreign building and conductor-foreign infrastructure. The following text describes practical examples of using the LiDAR system.

A. Power Line Protection Zone Width Control

Pursuant to § 43 of the Act of the National Council of the Slovak Republic no. 251/2012 Coll. on energy and on the amendment of certain laws for the protection of system installations, protection zones are established [6]. The protection zone is an area in the immediate vicinity of the system equipment, which is designed to ensure reliable and smooth operation and to ensure the protection of life and health of persons and property. The figure (Figure 1) shows an example of a 220 kV overhead line control in the protection zone using data from the LiDAR system. Using the recorded spatial distances of the classified point cloud classes, it is possible to check the entire area of the protection zone by accurate measurements. In this particular case, a vegetation at a distance of approximately 14 m from the outermost conductor was identified in the protection zone, so the width of the protection zone was not sufficient.

Figure 1. Protection zone of the 220 kV overhead line

The spatial image of the scanned objects with a density of more than 50 points per m^2 enables detailed control and eventual identification of such facts as various illegal landfills, small buildings, gardens or undetected high vegetation. By measuring, it is possible to determine the exact height, width of objects, or the size of the area extending into the width of the protection zone of the overhead line. Figure 2 shows building in the protection zone of the 400 kV overhead line.

Figure 2. Building in the protection zone of the 400 kV overhead line

B. Control of Vegetation in the Protection Zone of the Power Line

The occurring vegetation in the protection zone of the power line must be at a sufficient distance from the live parts of the power line, especially with regard to possible deviations by the action of the wind, the fall of the vegetation and its branches. The minimum vegetation distances must comply with the requirements defined in the standard (STN 33 3300 [7] for lines built before 31 December 2006, STN EN 50341-1 [8] for lines built after 1 January 2007). The following figure (Figure 3) shows an example of measuring the distance of a vegetation below 400 kV of a power line. In this way, a medium-high vegetation was identified and documented (the distance of the vegetation from the phase conductors was less than 5 m).

Figure 3. Medium-high vegetation in the protection zone of the 400 kV overhead line

C. Controlling the Height of Conductors Over Crossed Objects

When checking the height of conductors over crossed objects, aerial scanning must be performed only when the lines are frost-free. The distances of conductors from crossing objects are determined by the standards STN 33 3300 [7] and STN EN 50341-1 [8]. The figures (Figure 4 and Figure 5) show examples of conductor height control over a motorway and an electrified railway line).

Figure 4. Crossing of a 400 kV power line with a highway

Figure 5. Crossing of a 400 kV power line with an electrified railway line

In both cases, the distances of the line conductors from the crossed objects according to the above-mentioned standards were observed.

D. Ground Wires Deflection Control

Using the LiDAR system data, it is possible to identify differences in the deflections of the phase conductors and ground wires of the overhead lines. The size of the deflections varies due to load and climatic conditions. The maximum deflections of the conductors within the design of external lines in accordance with the standard STN EN 50341-1 [8] are determined by calculation for the two most unfavourable states. The calculation must also take into account other contexts such as the location of the built external line, terrain profile, etc. The maximum deflections of the phase conductors and ground wires specified by the project must be observed in all circumstances. The figure (Figure 6) shows an example of the identified difference between two suspended ground wires on the right and left side of a DONAU mast. In this case, when checking the spatial distances, a difference in the deflection of these wires of up to 3 m was found.

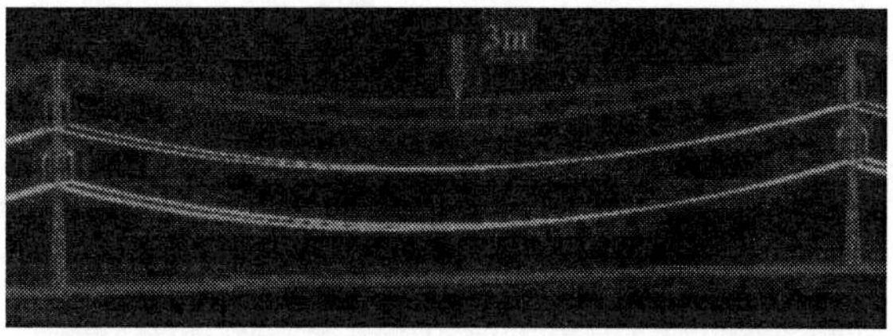

Figure 6. Ground wires deflection control on the 400 kV line

Through the processing of LiDAR data, with a sufficient number of points per unit area, it is also possible to implement:

- control of the shape and distance of the curtains,
- check the deflection of the insulator hinges,
- distance control of spacers and interfacial spacers,
- shape inspection of masts,
- checking the distances of mast anchorage and anchorage ropes.

III. Hyperspectral Imaging

Multispectral and hyperspectral imaging recordings allow data to be acquired outside the visible spectrum. The main difference between a multispectral and a hyperspectral image is the number and width of the recorded bands of electromagnetic radiation. Multispectral images generally record 3 to 10 bands. Hyperspectral images usually have 100-200 bands and the spectral width of these bands is very narrow (10-20 nm) and therefore provide sufficient information for detailed surface analyses [9], [10].

E.g. used HYSPEX system (Figure 7), consists of two cameras HySpex SWIR-384 and HySpex VNIR-1800, while it allows to record wavelengths from 400 to 2500 nm, with significantly higher spectral and spatial resolution compared to satellite images. The resulting hyperspectral records in the spectral range from 400 nm to 1000 nm are with a maximum sampling range of 3.26 nm per channel, with a dynamic range of at least 20 thousand shades of gray. Wavelengths in the range of 980 to 2500 nm are detected by a high-quality sensor cooled to 150 K, which achieves low background noise and a high dynamic range. Together with a maximum frame rate of 400 frames per second, aberration-corrected optics and high optical transmission of f/2.0, they ensure excellent levels even in darker areas of the image of highly dynamic scenes.

Figure 7. Hyperspectral system HYSPEX

Due to the development of digital technologies, the mentioned multispectral and hyperspectral images also find application in practice. At present, hyperspectral images are used mainly in archaeological prospecting. They are increasingly being used to ascertain and monitor the condition of forests. In the field of power energy, these systems have the potential to capture operational threats at an early stage. When analysing hyperspectral records capturing the surroundings of power lines, it is possible to identify the risk of a dangerous situation caused by the surrounding vegetation. Based on the spectral curves, it is possible to determine individual tree species, e.g. using spectral image classification by maximum likelihood methods and SAM (Spectral Angle Mapper) [11], [12], [13]. Various techniques and methods are used for image processing and analysis, e.g. Thresholding Method, Edge Based Segmentation method, Region Based, Clustering Based, Watershed, Partial Differential Equation Based, Artificial Neural Network) [14], [15], [16]. Reference data are also used in the analysis (such as height, thickness, type of trees, crown projections, etc.).

Hyperspectral image analyses make it possible to identify sites located along the protection zone of overhead lines in which potentially damaged or unhealthy trees can occur. Categorization of tree species and vegetation occurring in the protection zone of external lines is one of the other results of the analysis. The result of categorization are basic parameters such as the representation of trees (vegetation), height, thickness, age, etc. Based on these parameters, it is possible to determine the growth parameters of individual tree species [17]. By analysing the hyperspectral images, it is therefore possible to set the optimal excavation plan, resp. removal of vegetation occurring in the protection zone of overhead lines.

IV. Defect Identification Using a Thermovision Camera

Modern diagnostic systems enable the control of power lines even in the infrared spectrum. During the operation of overhead lines, it is possible to identify problematic current connections on the overhead line conductors. An important parameter is the resolution of the sensor used, which should be at least 640x512 px, ideally 1024x768 px.

The purpose of the measurement using a thermal imaging camera is to identify excessive heating of these connections and terminals. It is recommended practice to measure the heating of the terminals and connections at least 50% of the line load. When classifying the risk of failure due to heating of the terminals and connections compared to the connected conductor, it is necessary to convert the heating to 100% of line load. Based on the results, measures are subsequently recommended for the transmission or distribution line operator.

Figure 8 shows an example of a recorded thermal imaging fault at the ground wire terminal of a 110 kV line using a VarioCAM HD camera manufactured by InfraTec GmbH.

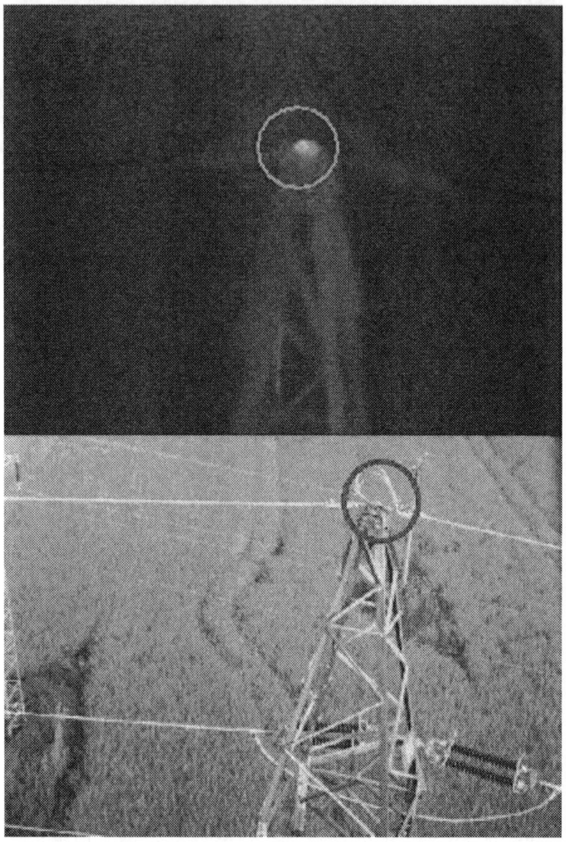

Figure 8. Identified warming at the ground wire clamp

Conclusion

As already mentioned in the text, the use of modern diagnostic systems for the purposes of aerial diagnostics of external power lines opens the way to obtain a lot of input data. The obtained data can be used not only to assess the real state of external lines, but also to predict various operating states using simulations and calculations. Based on data from the LiDAR system, it is possible to realize vectorization of phase conductors, ground wires, or masts with a sufficient point cloud density. Subsequently, it is possible to perform other calculations, e.g. for determining the ampacity of external lines, analysis of the influence of wind speed and direction as well as icing on the conductors and ground ropes of external lines. It is also possible to perform simulations of vegetation growth and identify the fall of trees.

Another way to use LiDAR data is to create a digital terrain model (DTM) and a digital surface model (DSM). The created DTM model in combination with cloud points in 3D space can also be used for simulations aimed at calculating the magnitude of the electromagnetic field and noise load, which are generated by the operated external high-voltage lines. The advantage of this procedure is high accuracy and detail that allows the assessment of even very complicated situations such as intersections of several lines, assessment of long concurrences of cycle routes with lines, etc. An example of calculating the intensity of the electromagnetic field below 400 kV power lines is shown in the following figure. Real data from the aerial scanning of the line and data of the new cycle route were used for the calculation.

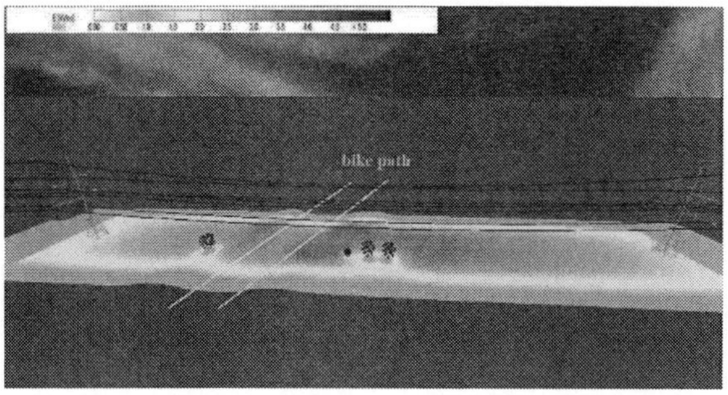

Figure 9. Example of calculating the intensity of the electromagnetic field below 2x400 kV power lines

ACKNOWLEDGMENT

This work was supported by the Slovak Research and Development Agency under the contract No. APVV-19-0214.

REFERENCES

[1] J. Kamaruzaman, Y. Norsuzila. "Mapping of Power Transmission Lines on Malaysian Highways Using UPM-APSB's AISA Airborne Hyperspectral Imaging System," in *Computer and Information Science*, 1(1), 2008, pp. 1-7. DOI: 10.5539/cis.v1n1p88.

[2] L. Matikainen, M. Lehtomäki, E. Ahkas, J. Hyyppä, M. Karjalainen, A. Jaakkola, A. Kukko, T. Heinonen. "Remote sensing methods for power line corridor surveys," in *ISPRS Journal of Photogrammetry and Remote Sensing*, vol. 119, September 2016, pp. 10-31. DOI: 10.1016/j.isprsjprs.2016.04.011.

[3] L. Matikainen, K. Karila, J. Hyyppä, P. Litkey, E. Puttonen, E. Ahokas. "Object-based analysis of multispectral airborne laser scanner data for land cover classification and map updating," in *ISPRS Journal of Photogrammetry and Remote Sensing*, vol. 128, June 2017, pp. 298-313. DOI: 10.1016/j.isprsjprs.2017.04.005.

[4] C. Coudrain, S. Bernhardt, M. Caes, R. Domel, Y. Ferrec, R. Gouyon, D. Henry, M. Jacquart, A. Kattnig, P. Perrault, L. Poutier, L. Rousset-Rouvière, M. Tauvy, S. Thétas, J. Primot. "Sieleters, an airborne infrared dual-band spectroimaging system for measurement of scene spectral signatures," in *Optics Express*, 23(12), 2015, pp. 16164-16176. DOI: 10.1364/OE.23.016164.

[5] RIEGL Laser Measurement Systems. "RiPRECISION UAV." Internet: http://www.riegl.com/products/software-packages/riprecision-uav/, January 2017 [Jan.20, 2022].

[6] National Council of the Slovak Republic. "Act 251/2012 Coll. of on energy and amending certain laws" Internet: https://www.slov-lex.sk/pravne-predpisy/SK/ZZ/2012/251/, 31 July 2012 [Jan.22, 2022].

[7] STN 33 3300 "Construction of external power lines."

[8] STN EN 50341-1 "External power lines with alternating voltage above 1 kV."

[9] P. Shippert. "Why use hyperspectral imagery?," in *Photogrammetric engineering & Remote Sensing*, vol. 70, 2004, pp. 377-380.

[10] E. Christophe, D. Léger, C. Mailhes. "Quality criteria benchmark for hyperspectral imagery," in *IEEE Trans. Geosci. Remote Sens*, 2005, 43(9), ISSN: 2103–2114.

[11] California Institute of Technology. "The ECOSTRESS spectral library." Internet: https://speclib.jpl.nasa.gov/library, January 2017 [Jan. 20, 2022].

[12] Decree no. 297/2011 Coll. "Decree of the Ministry of Agriculture and Rural Development of the Slovak Republic on forest management records".

[13] M. Swamer. "Extraction of tree crowns and heights using liadar." Internet: http://web.pdx.edu/~jduh/courses/geog493f12/Projects/Swamer Houser.pdf, May 2012 [Jan.20, 2022].

[14] Y. Tarabalka, J. Chanussot, J.A. Benediktsson. "Segmentation and classification of hyperspectral images using watershed transformation. Pattern Recognition," in *Elsevier*, 2010, 43(7), pp. 2367-2379.

[15] Er. Anjna, Er. Rajandeep Kaur. "Review of Image Segmentation Technique," in *International Journal of Advanced Research in Computer Science*. vol. 8, No. 4, May 2017 (Special Issue). ISSN: 0976 – 5697.

[16] Lesoprojekt Zvolen. "Tree growth tables." Internet: www2.slslhr.sk/ucivo/hul3_pdf/rasttab/1.pdf, October 2016 [Dec.16, 2021].

Anti-static Coating Electrical Parameters After Ageing Process

Jaroslav Džmura, Jaroslav Petráš
Department of Electric Power Engineering
Faculty of Electrical Engineering and Informatics, Technical University of Košice
Košice, Slovakia
jaroslav.dzmura@tuke.sk, jaroslav.petras@tuke.sk

Abstract — **The paper deals with one of the protection methods against the electro-static charge by conductive coating. We focus on methyl-silicone resin with the addition of conductive polymer. Individual electrical parameters are studied on these coatings because these parameters determine the ability of charge leakage. The purpose of the study was to find out if the coating changes its electrical parameters after some time and if the anti-static protection purpose of the coating is deteriorated. For the study, we have chosen coatings based on unary component paint, which are dissolved in xylene. Xylene is a methyl-silicone resin in which white titanium pigment is added. We have chosen 7 coating samples that were spread on both conductive and insulation material. Initial measurements were made on all samples in order to determine electrical properties of these materials before ageing. A part of the samples underwent the ageing process in dry and warm environment and the rest of the samples were placed and aged in a cold and wet environment. After a determined time, we have made electrical property measurements on the coating samples again. We used dielectric spectroscopy method for the measurements. The set of measurements was made by Keithley 6517B device and we measured the charge and discharge current. After ageing, another set of measurements was made and the results were compared. As the result, we can tell that by ageing process the electrical parameters of anti-static coatings considerably change, thus the anti-static protection capability of the coating is reduced. We can conclude that the anti-static coating in this case has to be checked and renewed in regular intervals.**

Keywords: antistatic coating; ageing; electrical property of paint; measurement of electrical properties

I. INTRODUCTION

Conductive polymers (CPs), sometimes referred to as organic semiconductors, have a wide range of potential applications due to their properties, therefore there is a high interest of scientific teams around the world [23]. They are recommended as alternative materials for a wide range of practical applications [1-7]. These materials have the possibility to change their specific electrical conductivity in the range beginning at values typical for insulators to values corresponding to metallic materials. Thus, a combination of the electrical properties typical for semiconductors with material properties that allow easy processing provides many advantages. Not only the structure of these materials can be changed but also their physical properties are modified according to the external environment. Therefore, they are sometimes referred to as smart materials [1, 8]. Unlike traditional polymers, CPs show so-called intrinsic electrical conductivity.

II. MATERIALS AND METHODS

Examples of such known CPs are polyacetylene (PA), polypyrrole (PPy), polythiophene (PTs), polyaniline (PANI), polyphenylene (PPE) or poly (p-phenylenevinylene) (PPV), poly (p-phenylenediamine) (PPDA), etc. [9-11]. The simplest CP in terms of structure is polyacetylene PA, whose conductivity by iodine vapor is increased very significantly (up to $5 \cdot 10^3$ $S.m^{-1}$). However, this material is not very suitable for technological applications due to its low stability in air [11]. Therefore, attention has been paid to more stable CPs since the late 1980s, especially polyaniline PANI and polypyrrole PPy [1]. PANI and PPy belong to p-type semiconducting polymers with relatively high conductivity, relatively high environmental and chemical stability in acidic environments. Its synthesis in the form of powder, film but also composite materials is easily reproducible [12-17]. If an inorganic or organic acid is used as the proton donor, the specific conductivity of these materials increases in the range of eight to ten orders of magnitude [9, 18].

A. Electric Conductivity of Conductive Polymer Materials

The conductivity of these polymers is based on the regular alternation of single and double bonds, i.e. by conjugation, and by the presence of charge carriers. Carriers are created by a process, which in analogy with conventional semiconductors, is called doping, while the increase or optimization of conductivity can be achieved by doping [1, 9, 18]. The nature of the dopant affects not only the electrical, mechanical and physical but also the morphological properties of the polymer. The conductivity of these substances varies within the range from $\times 10^{-4}$ to $\times 10$ $S.m^{-1}$ [1, 9]. Another common feature of CPs is their electroactivity and the possibility of their existence in several stable oxidation states. The transition between these oxidation states is generally possible, which results in a change in the electrical, mechanical, chemical and morphological properties of the material [19, 20].

B. Application Areas for Conductive Polymer Materials

The variability of CPs properties allows them to use in a wide range of applications. Conventional conductive composites consist of a non-conductive polymer matrix in which a conductive component - carbon black, graphite, or carbon fibers - is dispersed. This also includes the use in anti-corrosion coatings [21, 22, 25]. CPs, in which electrical conductivity is considered as the main property, are used to produce e.g.

electrostatic and conductive materials and wide variety of other applications.

C. Material Type for Measurement

The aim of the experimental part of our work was to test experimentally prepared paints based on solvent-type silicone acrylate binder containing $CaTiO_3$ and TiO_2 pigments coated with a layer of CPs based on PANI and PPY salt by chemical oxidative polymerization in terms of their electrical conductivity. The electrical properties of the formed protective coatings were compared with a protective coating without active pigments, which contained only an uncoated neutral limestone filler $CaCO_3$.

We have focused on how the antistatic paints were able to conduct the generated charge to earth potential. As mentioned above, tested antistatic materials/paints differ in their chemical composition and this influences the particular surface conductivity of these paints.

We have painted coatings based on single component paint that are soluble in Xylene onto laboratory samples. Xylene is a solvent for a methyl-silicone resin in which white titanium pigment is added. This pigment is titanium dioxide TiO_2, which is the most used compound of titanium. Typically, it has the highest covering capability among all other pigments. The filler of these paints is Omyacarb 2 VA. It is a carbonate in a form of white marble powder. Our coatings contained zinc; therefore, they have the paint form, as the weight of the zinc is considerably high. The paints have the same consistency as one-component paints.

We have used the paints on following samples:

- BISIL
- $CaTiO_3$
- PANI

The paints were applied on two different materials:

- a dielectric surface - ordinary sheet glass with 0.25 m × 0.10 m dimensions,
- a conductive surface - steel plate S 235 JRG2 with 0.10 m × 0.10 m dimensions.

For all of the paints 4 samples were prepared, 2 on dielectric surface and 2 on the conductive surface. The coatings we have used in our experiments are primarily designed for metallic constructions so we focused on volume resistance of that material with coating. All the samples with coating were prepared by same method and same process, thus the thickness of coating on all samples used in our experiments was equal to 0.18 mm and it had no effect during comparison validation of the samples.

However, when using dielectric material surface for coating application a few questions have to be discussed. The coating should be carefully grounded, to make it work effectively in particular application. Furthermore, if the conductivity of the coating is too high there is higher risk for spark discharge caused by electro static charge accumulated on dielectric pad. The appropriate coating material selection should take into account also the material type of the surface on which the coating is applied.

D. Measurement Conditions

We have made dielectric tests on antistatic paints at 3 different environmental conditions in order to observe the degradation degree of material conductivity:

- Initial measurement without any influence of outer environment – the samples were in measurement process 2 months after we have received them from the supplier so that the sample material had time to stabilize.
- the influence of higher temperature – the samples were exposed to higher temperature of 70°C during 2 months in a degradation chamber with stable air circulation.
- the influence of outer environment – the samples were exposed to unfavorable outer environmental conditions with combined physical parameter influence during 2 months.

The tested sample was placed on a pertinax pad with appropriate dielectric parameters and a conductive copper layer, thus we have eliminated the sample charging with charge from environment. The electrode system has the form of coaxial rings (Fig. 1) with moderate homogenous electric field. The outer radius of the inner electrode ring in our experiments was r = 24.5 mm; for outer electrode ring the inner radius was R = 25.5 mm.

E. Measurements Without Outer Environment Influence

In this measurement set-up, we have measured charging and discharging currents, capacitance values and loss factor after 2 months after the delivery from supplier. The samples were placed in a dark environment in laboratory with closed doors and windows at a stable room temperature. These conditions ensured the most real like electric parameters of the materials, which differ in chemical composition and production methods.

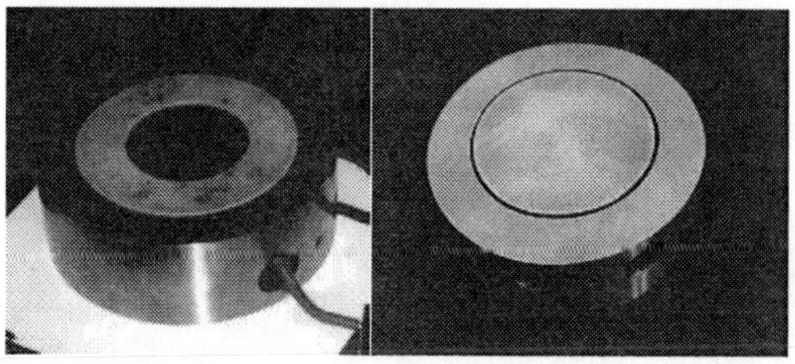

Figure 1: Measurement electrodes – coaxial rings

III. The Measurement of Charging and Discharging Currents

The first measurement was made with Keithley 6517B electrometer shown in Fig. 2. This method is a DC method. For current generation it is important to connect the electrodes correctly.

During 1000 s lasting measurement the sample is inserted between the pad and measurement electrode. Thus, the charging and discharging lasted both 500 s. The generator supplied current at constant voltage 10 V. We wanted to get the values of conductive currents I_G, which create a constant part of the depicted graph on Fig. 3. The I_G value was read from interpolations. The value of conductivity current directly determines the degree of conductive polymer representation in paint. All the read current values are compared with currents that flow through the dielectric material. For dielectric materials, this kind of current in unwanted as it represents the losses in insulation material, degrading thus the insulation capability.

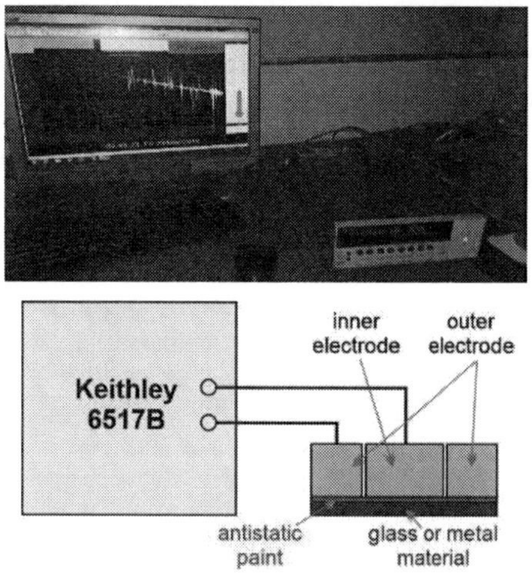

Figure 2: Measurement setup with Keithley 6517B electrometer

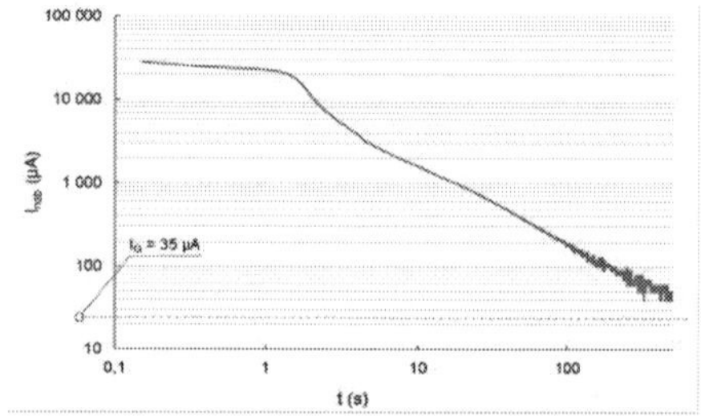

Figure 3: Charging current of non-conductive paint $CaTiO_3$ before the ageing process – glass sheet

We can say that the least conductive paints on a glass sheet were those containing $CaTiO_3$.

Following values for particular coating types are conductive currents with measured resistance and calculated surface resistivity (according to ANSI/ESD STM11.11-2006 Surface Resistance Measurement of Static Dissipative Planar Materials standard) (Tab. 1).

Table 1: Measured data on the Keithley 6517B electrometer

Coating material	*I_G (μA)*	*R (MΩ)*	*ρ_S (MΩ)*
PANI	*350*	*0,029*	*4,487*
BISIL	*110*	*0,091*	*14,278*
$CaTiO_3$	*210*	*0,048*	*7,479*

The exception is a PANI polymer. This sample did show conduction capability, as the I_G current value had 10 times higher value than the current of non-conductive paint: PANI: I_G = 350 μA (see Fig. 4)

We can also say that perovskite $CaTiO_3$ with combination of conductive polymers behave as absorber. A decrease of I_G value can be observed; thus clean perovskite has a higher value that the one with a conductive addition. The paints on glass containing TiO_2 without additions of calcium oxide have approximately same value of conductivity currents, as well as the commercially available paint BISIL (Tab. 1)

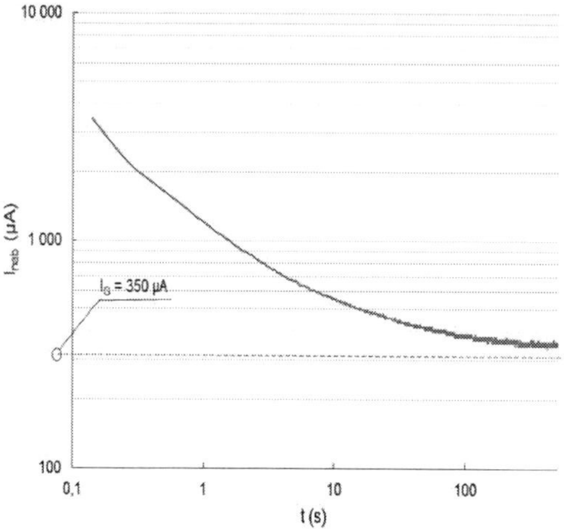

Figure 4: Charging current on conductive paint PANI before ageing process – glass sheet

For paints on steel pads we have measured higher value currents in general due to the pad material. There is also considerably higher effect of conductive polymers (Tab. 2).

Table 2: Measured data on the Keithley 6517B electrometer

Coating material	*I_G (mA)*	*R (kΩ)*	*ρ_S (kΩ)*
PANI	*0,55*	*18*	*2 856*
BISIL	*0,19*	*53*	*8 266*
$CaTiO_3$	*120*	*0,083*	*13,088*

We can also observe an opposite effect of $CaTiO_3$ in combination with metallic material comparing to expectations. The constant current flowing though the paints on glass sheet increased by 90 μA.

IV. Capacitance and Loss Factor Measurement

The second set of measurements focused on paint capacitance and loss factor measurements depending on frequency. The capacitance and loss factor measured before and after ageing process show electrical parameters of painting changed. The capacitance and loss factor influence the coating's capability to eliminate the charge because the resistance of the coating is inversely proportional to capacitance and loss factor measured. The measurement methods used in our experiments are usually applied in dielectric spectroscopy for electric parameter measurements in general and they can be applied also for the measurement of antistatic material electric parameter. They are AC methods and they were made by IDAX–300 insulation diagnostic analyzer (show on Fig. 5) in the frequency range:

- 10000 Hz, 9000 Hz, ... 1000 Hz
- 900 Hz, 800 Hz, ... 100 Hz
- 90 Hz, 80 Hz, ... 10 Hz

- 9 Hz, 8 Hz, ... 1 Hz
- $9\cdot10^{-1}$ Hz, $8\cdot10^{-1}$ Hz, ... $1\cdot10^{-1}$ Hz
- $9\cdot10^{-2}$ Hz, $8\cdot10^{-2}$ Hz, ... $1\cdot10^{-2}$ Hz
- $9\cdot10^{-3}$ Hz, $7\cdot10^{-3}$ Hz, $5\cdot10^{-3}$ Hz, $3\cdot10^{-3}$ Hz, $2\cdot10^{-3}$ Hz, $1\cdot10^{-3}$ Hz

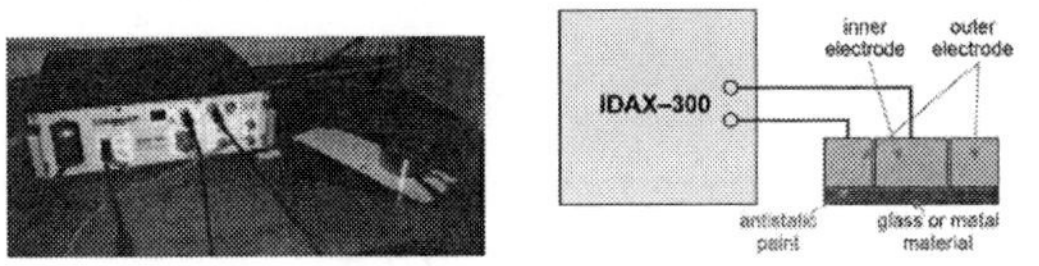

Figure 5: Measurement set-up by IDAX–300 insulation diagnostic analyzer

Measurements were made at constant voltage 10 V on a dielectric pad. These measurements are very time extensive, as at f = 1 mHz the measurement period lasts 1000 s. In our case the measurement lasted approximately 58 minutes.

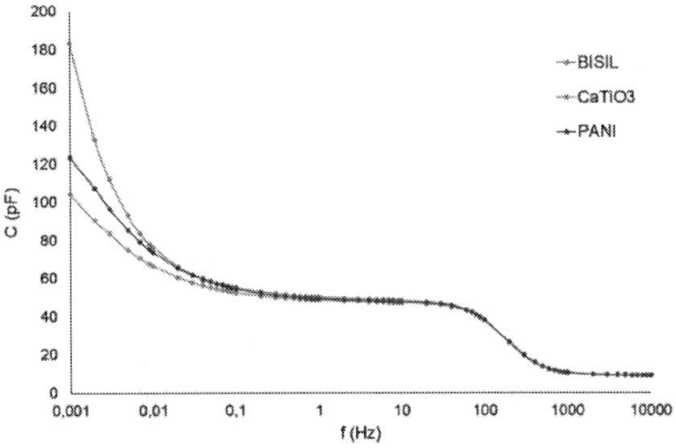

Figure 6: IDAX insulation diagnostic analyzer: Paint capacitance before ageing process – glass pad

The capacitance between electrodes (without sample) during AC measuring methods is compensated by the measuring device itself (it has a built-in compensation) so this capacitance was not measured in our experiments. The capacitance values of the samples on both pad materials were in the range of hundreds pF (Fig. 6 for glass pad and Fig. 7 for steel pad). The highest capacitance was observed on BISIL paint, on the other side $CaTiO_3$ had in both cases lowest capacity.

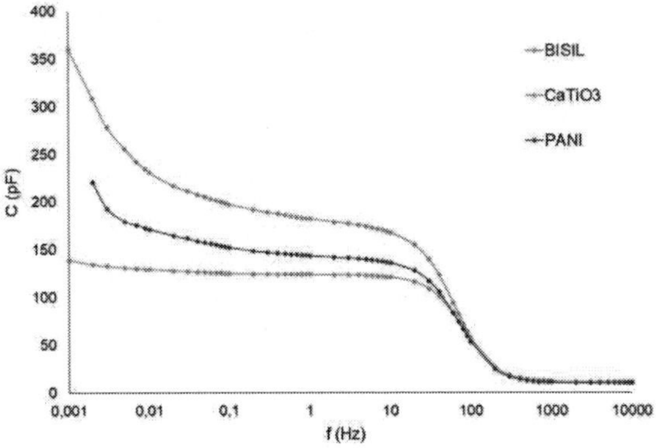

Figure 7: IDAX insulation diagnostic analyzer: Paint capacitance before ageing process – steel pad

The measurement of the loss factor quantifies the paint conductivity. We have found that the non-conductive paints on glass sheet have greater values of the loss factor (Fig. 8). The influence of conductive polymers can be observed in cases of steel pads (Fig. 9).

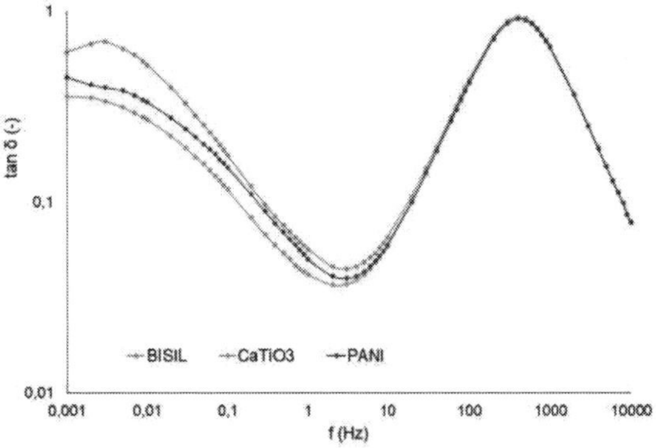

Figure 8: IDAX insulation diagnostic analyzer: The loss factor of paints before ageing process – glass sheet

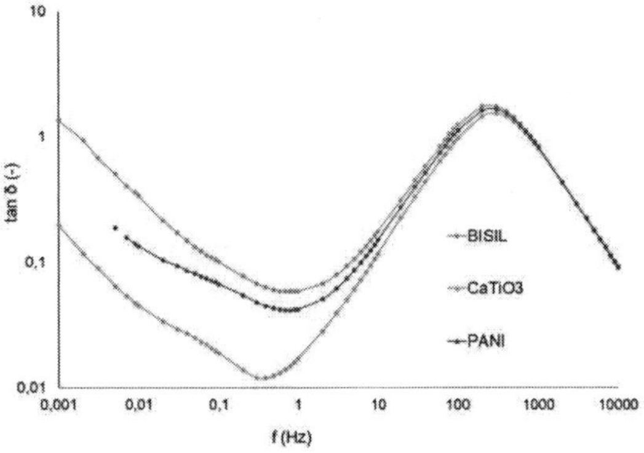

Figure 9: IDAX insulation diagnostic analyzer: The loss factor of paints before ageing process – steel pad

V. Samples After Ageing Process By Increased Temperature in Oven

Increased temperature was applied on the samples on the Memmert UF55plus oven. This oven is programmable and it has forced air circulation in the chamber. For this measurement set we wanted to compare the conductivity and capacitance changes after ageing process. The ageing process lasted 60 days at constant 70°C temperature and constant hot air circulation. The sample parameter change was supposed to be caused by chemical structure changes of the antistatic paints. Thus the paint degradation can cause the deterioration of the paint capability to conduct the charge to earth. The measurement sets previously described were repeated with the same samples after ageing process with same methods, devices and frequency ranges.

After ageing process, the conductive currents were lower, this is caused by the fact that the paint was dried, it had lower humidity. The samples with highest conductivity values before ageing process appeared to be the most influenced by ageing. E.g. PANI paint, which had $I_G = 350$ μA conductive current value for glass sheet case before ageing, the value decreased to $I_G = 20$ μA after ageing.

The ageing process also decreased the capacitance values of the samples on glass sheet. Before ageing these values were in the range 100 – 150 pF, after ageing the capacitance decreased to approximately 100 pF. BISIL paint on glass sheet had the highest capacitance at 1 mHz – $C_{BISIL} \approx 110$ pF. The same effect could be observed with the sample on steel plate: 350 pF before ageing and 200 pF after ageing for BISIL paint sample.

The loss factor tgδ also decreased after ageing process, especially in case of the samples with conductive polymer PANI. This is caused by sample drying.

VI. Samples After Ageing Process in Outer Environmetal Conditions

This ageing process took place in outer environmental conditions during winter months December and January (60 days). We can consider these months as the coldest in average and they have the highest volume of precipitation in the form of snow or rain. In addition, foggy weather with low air pressure appears often in these months.

The conductive currents were measured and they were significantly higher in comparison to the samples with ageing process in oven. This is caused by high humidity in the environment [24]. The typical conductive current value for PANI samples on steel plate was $I_G \approx 25$ μA. In cases with glass sheet, the current had values in the range from 2 to 10 μA with highest value for perovskite containing $CaTiO_3$.

The capacitance was similar to the samples with oven ageing; they had values between 150 to 200 pF for steel plate samples and 50 to 110 pF for glass sheet. BISIL coating had the highest capacitance value on glass sheet.

The difference between both ageing processes can be observed on loss factor parameter tgδ. This parameter was higher in comparison to the samples without ageing, especially for PANI paints. This set of measurements showed, that the most suitable paint type for electrostatic charge elimination is the PANI polymer.

VII. Results and Discussion

In following summary graphs on Fig. 10 and Fig. 11, the differences in capacitance changes and loss factor changes are depicted between two ageing types. The graphs are taken from measurement set with IDAX insulation diagnostic analyzer. The ageing process influence is shown on BISIL paint. The loss factor changes are depicted on PANI coating samples. In both cases, the samples were on steel plate.

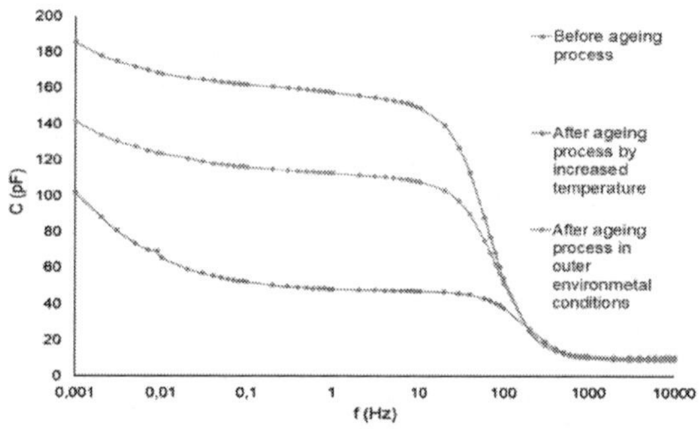

Figure 10: BISIL sample capacitance

We have measured current values for the samples according to the methods we used, so the resistivity can be calculated from Ohm's law $5\cdot10^{-6}$ Ohm surface resistivity of the glass plate.

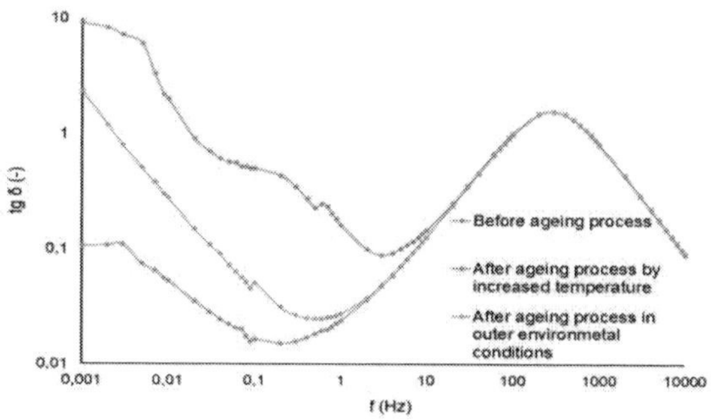

Figure 11: Loss factor of PANI paint

It is obvious that the capacitance increases during ageing process in oven. The loss factor tgδ had the highest value in initial state.

The reliability of antistatic paints was proven in these experiments. Presented paints have high value of conductivity in general. After experiments, the most reliable conductive paintings appeared to be the ones with PANI polymer. High conductivity of this polymer was proven by high conductive current values and higher loss factor values.

For reliable electrostatic charge elimination on non-conductive surfaces, the BISIL paint can be used. This paint had the highest capacitance value at various conditions. Our experiments also showed the influence of the underlying material on electrical parameters of the paints. The ageing process influenced the loss factor and at the same time the quality of the paint material. The paints in our experiments are suitable also for metallic underlying materials due to their anti-corrosion protection in extreme conditions, because they contain zinc.

Acknowledgment

This research was funded by the Slovak Academy of Sciences and Ministry of Education in the framework of project VEGA 2/0011/20, 1/0154/21, 1/0340/18, Slovak Research and Development Agency under the contract No. APVV-18-0160, APVV-15-0438.

This publication is the result of the project implementation: Innovative Testing Procedures for 21st Century Industry ITMS: 313011T565.

References

[1] J. Prokeš, J. Stejskal, M. Omastová. "Polyaniline and polypyrrole - two representatives of conductive polymers." *Chemical Letters* 95. 2001, pp. 484-492.

[2] J.W. Schultze, H. Karabulut. "Application potential of conducting polymers." *Electrochimica Acta* 50. 2005, pp. 1739-1745.

[3] B. Somboonsub, S. Srisuwan, M. A. Invernale, S. Thongyai, P. Praserthdam, D.A. Scola, G.A. Sotzing. "Comparison of the thermally stable conducting polymers PEDOT, PANi, and PPy using sulfonated poly(imide) templates." *Polymer* 51. 2010, pp. 4472-4476.

[4] J. Király, F. Ronkay. "Temperature dependence of electrical properties in conductive polymer composites." *Polymer Testing* 43. 2015, pp. 154-162.

[5] F.M. AL-Oqla, S.M. Sapuan, T. Anwer, M. Jawaid, M.E. Hoque. "Natural fiber reinforced conductive polymer composites as functional materials: A review. Synthetic Metals 206. 2015, pp. 42-54.

[6] R. Balinta, N.J. Cassidy, S.H. Cartmell. "Conductive polymers: Towards a smart biomaterial for tissue engineering." *Acta Biomaterialia* 10. 2014, pp. 2341-2353.

[7] N.K. Guimard, N. Gomez, Ch.E. Schmidt. "Conducting polymers in biomedical engineering." *Progress in Polymer Science* 32. 2007, pp. 876–921.

[8] N. Abu-Thabit, A. Makhlouf, "Handbook of Smart Coatings for Materials Protection" 17 – Recent advances in polyaniline (PANI)-based organic coatings for corrosion protection, pp. 608-659, 2014. ISBN 978-0-85709-680-7.

[9] J. Stejskal. "Polymers of phenylenediamines" in *Polymer Science* 41 pp. 1-31, 2015.

[10] U. Abaci, H. Guney, U. Kadiroglu, "Morphological and electrochemical properties of PPy, PAni bilayer films and enhanced stability of their electrochromic devices (PPy/PAni–PEDOT, PAni/PPy–PEDOT)." *Electrochimica Acta*, 95 214–224 2013.

[11] B. Somboonsub, S. Srisuwan, M. Invernale, S. Thongyai, P. Praserthdam, D. Scola, G. Sotzing, "Comparison of the thermally stable conducting polymers PEDOT, PANi, and PPy using sulfonated poly(imide) templates." *Polymer*, 51 4472–4476, 2010

[12] E. Sanches, J. Silva, J. Ferreira, J. Soares, A. Santos, G. Trovati, E. Fernandes, Y. Mascarenhas. "Nanostructured Polyaniline Emeraldine-base form (EB-PANI), a structural investigation for different neutralization times." *Journal of Molecular Structure*, 1074 732–737, 2014.

[13] T. V. Vernitskaya, O. N. Efimov. "Polypyrrole a conducting polymer; its synthesis, properties and Applications" *Ossian Chemici Reviews* 66 (5), pp. 446- 449, 1997.

[14] E. Kang, K. Neoh, K. Tan. "Polyaniline: A polymer with many interesting intrinsic redox states" *Prog Polym Sci*, 23, pp. 277-324, 1998.

[15] X. Li, M. Huang, W. Duan. "Novel Multifunctional Polymers from Aromatic Diamines by Oxidative Polymerizations" *Chem Rev*, 102, pp. 2925-3030, 2002.

[16] N. Blinova, J. Stejskal, M. Trchová, J. Prokeš. "Polyaniline prepared in solutions of phosphoric acid, Powders, thin films, and colloidal dispersions." *Polymer*, 47 42–43, 2006.

[17] J. Stejskal, P. Kratochvíl, M. Helmstedt. "Polyaniline Dispersions. 5. Poly(vinyl alcohol) and Poly(N-vinylpyrrolidone) as Steric Stabilizers." *Langmuir*, 12 3389–3392, 1996.

[18] T. Campos, D. Kersting, C. Ferreira. "Chemical synthesis of polyaniline using sulphanilic acid as dopant agent into the reactional medium." *Surface and Coatings Technology*, vol. 122, pp. 3-5, 1999.

[19] J. Stejskal, A. Riede, D. Hlavatá, J. Prokeš, M. Helmstedt, P. Holler. "The effect of polymerization temperature on molecular weight, crystallinity, and electrical conductivity of polyaniline." *Synthetic Metals*, 96 55–61, 1998.

[20] M. Trchová, P. Matějka, J. Brodinová, A. Kalendová, J. Prokeš, J. Stejskal. "Structural and conductivity changes during the pyrolysis of polyaniline base." *Polymer Degradation and Stability*, 91 114–121, 2006.

[21] V. Gedela, V. Srikanth. "Electrochemically active polyaniline nanofibers (PANi NFs) coated graphene nanosheets/PANi NFs composite coated on different flexible substrates." *Synthetic Metals*, 193 71–76, 2014.

[22] N. Y. Abu-Thabit, A. S. H. Makhlouf. "Handbook of Smart Coatings for Materials Protection", 17-Recent advances in polyaniline (PANI)-based organic coatings for corrosion protection pp. 459–486, 2014.

[23] M. Pavlik, I. Kolcunova, L. Lison. "Measuring the shielding effectiveness and reflection of electromagnetic field of building material", in: *proceedings of the 2015 16th international scientific conference on electric power engineering (epe), International Scientific Conference on Electric Power Engineering, 16th International Scientific Conference on Electric Power Engineering*, Kouty nad Desnou, Czech republic, ISSN2376-5623, pp. 56-59, May 2015.

[24] R. Cimbala, S. Bucko, L. Kruzelak, M. Kosterec. "Thermal Degradation of Transformer Pressboard Impregnated with Magnetic Nanofluid Based on Transformer Oil", In: *proceedings of the 2017 18th international scientific conference on electric power engineering (epe), International Scientific Conference on Electric Power Engineering, 18th International Scientific Conference on Electric Power Engineering,* Kouty nad Desnou, CZECH REPUBLIC, ISSN 2376-5623, pp. 715-719, May 2017.

[25] J. Zbojovsky, M. Pavlik, D. Medved, Z. Čonka, L. Kruzelak. "Shielding of the electromagnetic field in the range from 0.9 to 6 GHz by experimental shielding coatings", in: *ieee international conference and workshop in obuda on electrical and power engineering (cando-epe), IEEE International Conference and Workshop in Obuda on Electrical and Power Engineering (CANDO-EPE)*, Budapest, HUNGARY, ISBN 978-1-7281-1154-4, pp. 273-276, Nov. 2018

Additive Magnetic Coatings for the Purposes of Shielding the Electromagnetic Fields

Ján Zbojovský, Iraida Kolcunová, Juraj Kurimský, Roman Cimbala
Department of Electric Power Engineering
Technical University of Košice, FEI
Košice, Slovakia
jan.zbojovsky@tuke.sk, iraida.kolcunova@tuke.sk, juraj.kurimsky@tuke.sk, roman.cimbala@tuke.sk

Abstract — **The paper deals with the investigation of the shielding effectiveness of the electromagnetic field by means of experimentally created shielding magnetic fluids. It is focused on the frequency range 0.8-9 GHz. Mixtures of additive shielding coatings are applied on the MDF board, measurements of shielding effectiveness and reflection are performed in anechoic chamber. Subsequently, the attenuation of the material is calculated from the obtained data. The obtained results will provide an overview of the possibilities of using newly created coatings for the purposes of shielding the electromagnetic fields. (Abstract)**

Keywords: coatings; electromagnetic wave; frequency; absorption; reflection; shielding

I. Introduction

In the last 30 years, there has been a great development of the communication technologies in the world. This change has brought with it a number of modern technologies that are an integral part of today's life on earth. These modern technologies are often artificial sources of electromagnetic radiation, which is practically unavoidable today. We are most exposed to electromagnetic radiation in work areas or at home, so it is necessary to research materials designed for the possibilities how to shield electromagnetic fields. Although there are not definitively proven its negative effects on the population it is necessary to examine how to reduce its possible consequences by shielding materials. Building materials are one of the most important materials even in electromagnetic field shielding. Because people are the most exposed to electromagnetic field indoors, it is appropriate to examine how to prevent accumulation of this field. This is possible if we reduce its penetration between the rooms, from the building to the surroundings and also from the surroundings to the building. Therefore, it is necessary to explore these materials and options improving their shielding properties. The aim of this work is the investigation of the shielding effectiveness of the electromagnetic field by experimentally created shielding magnetic fluids. It is focused on the frequency range 0.8-9 GHz. Mixtures of additive shielding coatings were applied on the MDF board, measurements of shielding effectiveness and reflection were performed in anechoic chamber.[1][2][3]

II. Shielding Effectiveness of the Electromagnetic Field

Shielding effectiveness is most commonly used in practice and can be easily defined as the sum of electromagnetic field reflection and electromagnetic field absorption. The following figure (Fig.1) shows how the electromagnetic wave penetrate through the shielding material. It can be seen that part of the wave is reflected, part of the wave is absorbed by the shielding material and part of the wave passes through the material in the form of a transmitted wave. The basic unit of shielding effectiveness is decibel [dB]. [4][5]

Figure 1. Electromagnetic wave impigning on the shielding material

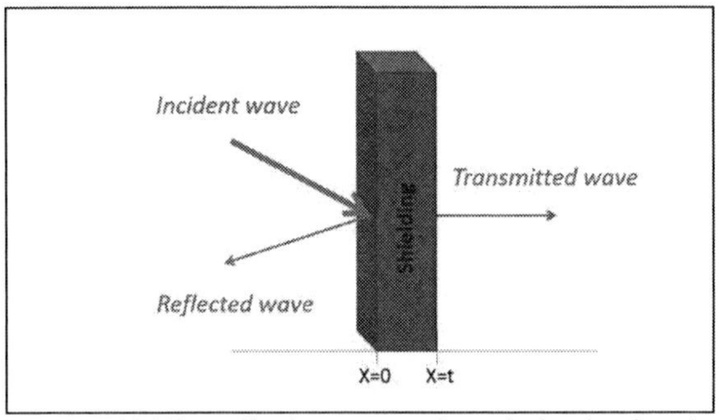

The shielding effectiveness can be calculated using the following formula, if the values of the transmitted signal are in logarithmic units.

$$SE = |E_1| - |E_2| \tag{1}$$

Where E_1 is the intensity of the electric field that impinges on the shielding material, E_2 is the intensity of the electric field, which is at a certain point behind the shielding material. The absorption of the electromagnetic field can be calculated on the basis of the relationship for shielding effectiveness:

$$SE = A + R \tag{2}$$

Where A is absorption and R is reflection of the wave.

There are a large number of shielding materials that have different shielding effectiveness. The table 1 describes the categories of shielding effectiveness.

Table 1. Categories of the shielding effectiveness

Category of shielding	SE [dB]
Insufficient shielding	0 - 10
Shielding form minimum requirements	10 – 30
Shielding sufficient for most common requirements	30 – 60
Very good shielding	60 – 90
Shielding of the very high quality	90 - 120

III. Measurement Setup

A. Measurement samples

Primalex Plus is a white interior paint suitable mainly for the application of mineral surfaces but also other absorbent surfaces. AF3 additive is a special magnetic additive for mixing with coatings containing long carbon fibers, which guarantees reliable grounding even without the use of grounding tape. HSF 54 magnetic shielding coating is a universal electrically conductive coating used for shielding of low and high frequency electromagnetic radiation. Composed of carbon and acrylic binder. It is frost-resistant and this makes it a widely usable coating, whether indoors or outdoors. The manufacturer states an attenuation of 39dB in one layer and 49dB in two layers at a frequency of 1GHz.

B. Workplace

The measurement were performed in an anechoic chamber. The anechoic chamber is used to measure the electromagnetic interference of electrical and electronic equipment. The measurements can be performed in the frequency range from 30Hz to 18GHz. It was necessary to use the following measuring devices, transmitting and receiving antennas and to connect them to the appropriate wires according to the wiring diagrams. The Agilent N5183A was used as a generator. Agilent N9038A MXE EMI was used as a spectrum analyzer. The receiving antenna was an R&S HF907. RF Spin DRH18-E was used as the transmitting antenna.[6][7][8]

On the following figurres are shown the wiring diagrams for the measuring the shielding and reflection.

Figure 2. Wiring diagram for measuring the shielding effectiveness

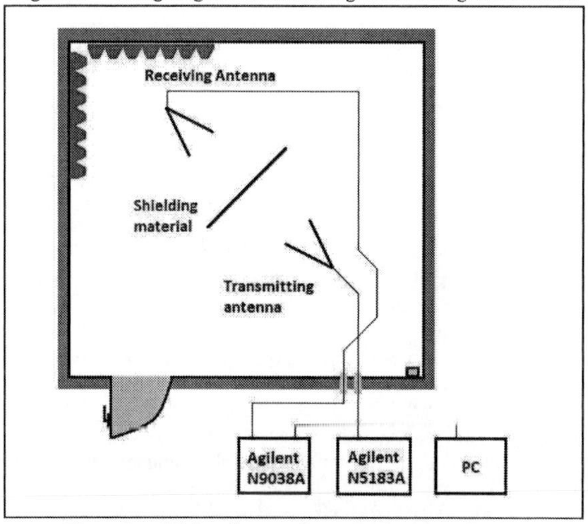

Figure 3. Wiring diagram for measuring the reflection

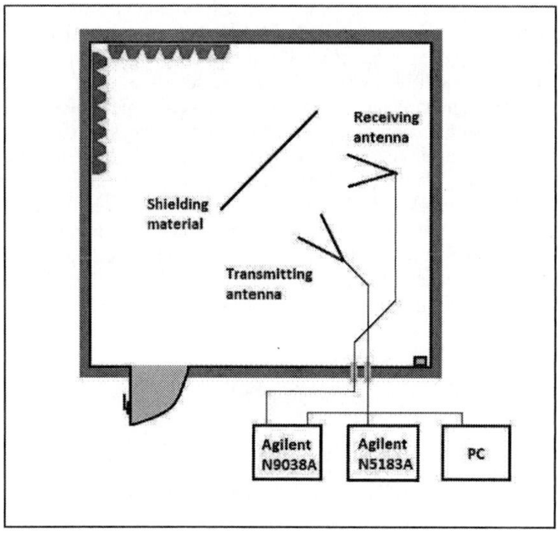

IV. Results from the Measurements

A. Measurement the shielding effectiveness

The following graph in the Fig. 4 shows a comparison of the shielding effectiveness of a clean MDF board and a board painted with non-conductive interior paint Primalex Plus. It can be seen that a pure board has a higher ability to shield the electromagnetic field at higher frequencies than a board painted with Primalex Plus. At a frequency of 3.5GHz, a clean board has a shielding effectiveness up to 3.163dB, which is a difference of more than 3dB compared to a board coated with Primalex. Furthermore, it can be seen that a board coated with Primalex Plus has a shielding effectiveness at a certain point at frequency 2 GHz only 1.120 dB, which is almost zero shielding effectiveness if we add inaccuracy or measurement errors to this value. At other frequencies, the shielding effectiveness values were only smaller. [9][10] [11]

Figure 4. Dependence the shielding effectiveness of MDF board and board painted with paint Primalex Plus on frequency

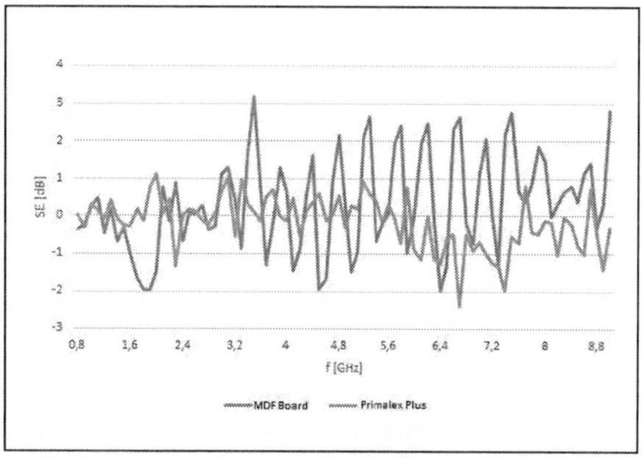

In the next graph in the Fig. 5 it is possible to see the shielding effectiveness of MDF board with Primalex Plus paint with was mixed with AF3 additives. This additive is suitable to improve the shielding effectiveness of the paint to which they are added. In our case, the first board were painted in a ratio of 10 g of AF3 additive with approximately 300 g of Primalex Plus. The second plate in the ratio of 30g AF3 to 300g Primalex Plus and the third plate in the ratio of 50g to 300g. It can be seen from the graph that the board with the highest ratio of additives has the best shielding properties, it means the board with a ratio of 50 g of AF3 additive. The maximum value of this shielding is 11.918dB, at a frequency of 7.2GHz, which represents the lower limit of shielding for minimum requirements. [12]

Figure 5. Dependence the shielding effectiveness of MDF board and board painted with paint Primalex Plus on frequency

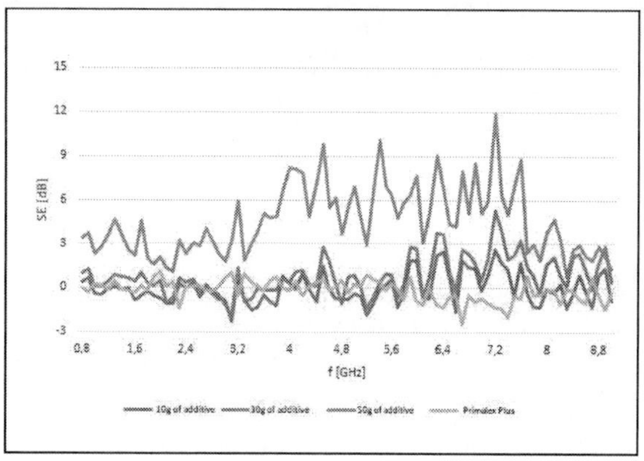

On the following graph are shown the comparison of all measured samples with professional paint HFS 54

Figure 6. Dependence the shielding effectiveness of all measured samples

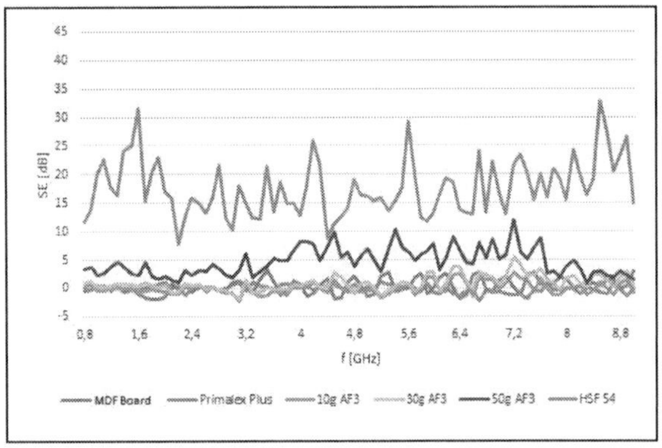

B. Measurement the reflection

From the graph in the Fig. 6 it can be seen that the professional coating HSF 54 achieved highest values of reflection. On the contrary, the sample coated with 10g AF3 additives reflects the electromagnetic waves at worst. Among the samples coated with additives, the sample with the highest concentration of additives achieved highest values of reflection. Samples coated with additives had approximately the same reflection course, but in the sample with 30g and 50g AF3 additives there were fluctuations at 2.4GHz, when the reflection value reached better values than with HSF 54 coating, namely 25.814dB in the sample with 30g AF3 and 26.238 dB for a sample with 50g AF3. The sample with professional coating HSF 54 has its maximum at the frequency of 2.2 GHz, where it reached the value of reflection 25.266dB and the minimum at the frequency of 1.2 GHz, when it reached the value of reflection only 3.528dB. Here it is possible to notice the anomaly, when in the range of 1GHz the given sample reached its maximum and minimum. As the frequency gradually increased, the reflection value slowly decreased. At the end of the measured frequency range, an anomaly occurred again, when the reflection value of the HSF 54 coated sample increased sharply.

Figure 7. Dependence the reflection of samples coated with AF3 additives mixed with Primalex Plus interior paint and dependence the reflection of sample with HSF 54 coating on frequency

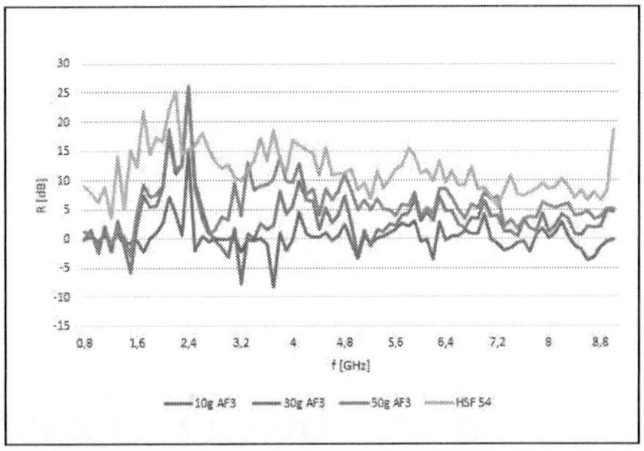

C. Calculation the absorption

It is not possible to measure the absorption of the electromagnetic field, because this measurement requires the placement of the antenna inside the measured object, in our case inside the MDF board. Absorption can be calculated by following formula:

$$A = SE - R \tag{3}$$

On the graph in the Fig. 8 is shown the calculation of the absorption of samples with the coating, either with AF3 additives mixed with Primalex Plus interior paint or with HSF 54 professional coating from the YShield. It can be seen that the HSF 54 coating has the best absorption properties in the whole frequency spectrum except for the frequency of 3.2GHz, when the 30g AF3 sample achieves better absorption properties than the just mentioned coating. All samples with AF3 additives have approximately the same absorption properties.

Figure 8. Dependence the absorption of samples coated with AF3 + Primalex Plus additives and dependence the absorption of sample with HSF 54 coating on frequency

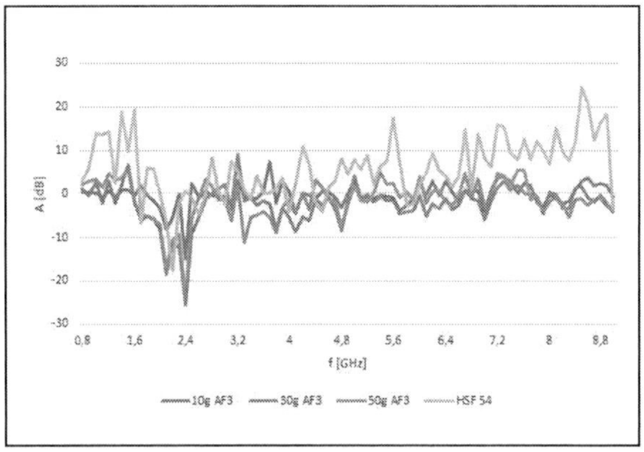

V. Conclusion

In this work, the issue of shielding, absorption and reflection of the electromagnetic field is discussed. Commonly available shielding coatings and additives were used for shielding the electromagnetic field. As expected, the best shielding effectiveness values were achieved by HSF 54 professional coating from YShield. AF3 YShield additives, added to Primalex Plus in low concentrations, did not affect shielding effectiveness much. To improve the shielding effectiveness it was necessary to add more additives to the paint. Larger amounts of AF3 additives (50g) provided a shielding for minimum requirements. Closely related to the measurement of effectiveness is the measurement of reflection, which was concluded that coating HSF 54 best reflect electromagnetic waves at higher frequencies. Coating with the addition of 10 g of AF3 additives to Primalex Plus does not reflect electromagnetic waves almost at all. Since the measurement of the absorption cannot be performed, as it is necessary to place the antenna directly in the structure of the material, the absorption calculations were performed after measuring the shielding and reflection effectiveness of the given samples. HSF 54 achieves the best values of absorption. The coating with the highest concentration of AF3 additives achieved the lowest absorption capacity, together with a sample containing 30 g of AF3 additives in Primalex Plus.

Acknowledgment

This work was supported by the Scientific Grant Agency of the Ministry of Education of Slovak Republic and the Slovak Academy of Sciences by the projects

APVV-17-0372, APVV-18-0160, VEGA 2/0011/20 and VEGA 1/0154/21.

REFERENCES

[1] World Health Organization, *Non-ionizing radiation, part 2: radiofrequency electromagnetic fields*, IARC monographs on the evaluation of carcinogenic risks to humans, International Agency for Research on Cancer, 2013, pp. 480, ISBN: 978-92-832-1325-3.

[2] I. Kenar, "Electromagnetic Pollution Awarness Scale", *Int. J. Edu. Sci.*, vol. 5, pp. 173-177, Sep. 2017.

[3] P. Vecchia, R. Matthes, G. Ziegelberger, J. Lin, R. Saunders, A. Swerdlow, *Exposure to high frequency elektromagnetic fields, biological effects and health consequences (100 kHz to 300 GHz)*, INCIRP Scientific Secretary, 2009, pp. 392.

[4] S. Bucko, M. Češkovič, J. Labun and P. Kurdel, "Mobile phone effect on quality of results of the sensitive measurements," *MOSATT 2017*, 2017, pp. 22-25.

[5] E. Lumnitzer, P. Liptai and R. Drahoš, "Measurement and Assessment of Pulsed Magnetic Fields in the Working Environment," *Elektroenergetika 2015*, 2015, pp. 331-333.

[6] IEEE Power Engineering Society, "Method for Measuring the Effectiveness of Electromagnetic Shielding Enclosures." U.S. IEEE Standard 299-2006, Feb. 28, 2007.

[7] M. Pavlik, D. Medved, "Measuring shielding effectiveness of electromagnetic field for degradation shielding paint," *Przeglad Elektrotechniczny*. vol. 97, pp. 226-229, Dec. 2021.

[8] N. Abbas, H.T. Kim, "Multi-Walled Carbon Nanotube/Polyethersulfone Nanocomposites for Enhanced Electrical Conductivity, Dielectric Properties and Efficient Electromagnetic Interference Shielding at Low Thickness," *Macromolecular Research*, vol. 24, pp. 1084-1090, Dec. 2016.

[9] A. Rusiecki, "Calculation and measurement of shielding effectiveness of slotted enclosure with built-in conductive stirrer," *Przegland Elektrotechniczny*, vol. 88, pp. 263-266, Oct. 2012.

[10] J.M. Kim, Y. Lee, M.G. Jang, C. Han, W.N. Kim, "Electrical conductivity and EMI shielding effectiveness of polyurethane foam–conductive filler composites," *Journal of Applied Polymer Science*, vol. 134, pp. 2017, Sep. 2016.

[11] M. Pavlík, "Compare of shielding effectiveness for building materials," *Przegląd Elektrotechniczny*, vol. 95, pp. 137-140, May 2019.

[12] P. Liptai, M. Moravec, M. Badida, "Analysis and comparison of electromagnetic fields radiated by the selected equipment," *13th International Multidisciplinary Scientific Geoconference: Ecology, Economics, Education and Legislation: conference proceedings*, 2013, pp. 17-20.

C: Power Systems Relaying, Faults Analysis

Shutdowns Assessment of Biomass Burned Unit

Aleš Hromádka
Department of Electric Power Engineering, UWB
Research and Innovation Centre for Electrical Engineering, UWB
Pilsen, Czech Republic
aleshrom@fel.zcu.cz

Zbyněk Martínek
Department of Electric Power Engineering, UWB
Pilsen, Czech Republic
martinek@fel.zcu.cz

Abstract—**Shutdown assessment has been performed at the biomass burned unit in the CHP plant in Pilsen. The assessment has shown a state of operated production technologies and its average availability in investigated years for a further assessment. The assessment has also identified the significant production technologies according to its availabilities. However, the paper has been mainly focused on typical shutdown causes of the biomass burned unit and how to solve it in future.**

Keywords: availability; biomass CHP unit; shutdown causes; renewable resorces; wood chip boiler

I. Introduction

The new biomass burned "green" unit has been implemented at the CHP plant in Pilsen in 2010. The unit has included fluidized bed combustion boiler B7 and condensing turbine TG3.

The boiler B7 with installed thermal power 37,5 MW has had unique construction with two thrusts. One of the thrusts is remoted due better wood chip or straw pellet burning. [1]

The turbine TG3 has been two-stage condensing turbine with installed electric power 10,5 MW (maximal electric power 13,5 MW). There has also been a possibility of using steam from the first stage of the turbine TG3 for a peak heater of the district heating. [1]

Both the production technologies have been interconnected by steam collector (SSC). The system of the unit could be simplified for the shutdowns assessment only to the boiler B7 as a heat source and the turbine TG3 as a heat engine and between them the collector SSC, which have been shown in Figure 1. Operability of the unit has been ensured in case of the B7, the TG3 and the SSC in operation simultaneously.

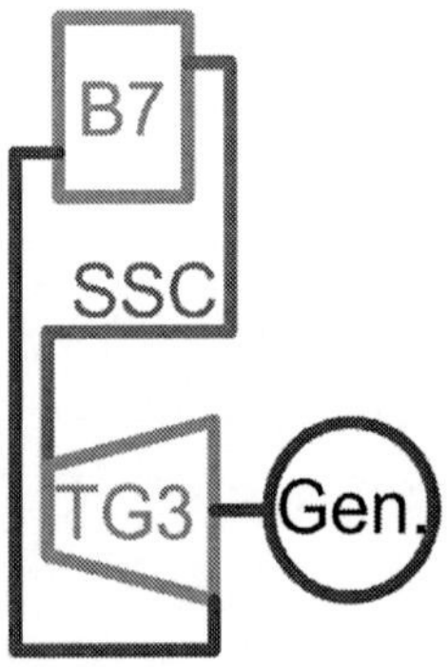

Figure 1. The biomass burned unit in the CHP plant in Pilsen [1]

The unit has been independent from the rest of the production technologies in the CHP plant. As it has mentioned, the unit has been operated since 2010. Therefore, the unit operational data has been available for years 2013 - 2017. There is a higher price of electric power due to renewable fuels, which could be burned in the boiler B7. Therefore, the unit has been sustained in operation continuously at the maximal possible electric power and each an electric production restriction would have caused an income loss. [1]

II. Assessment of the Unit

Now, the simplified shutdowns assessment of the unit will be investigated. There will be only respected both types of the shutdowns i.e., the unplanned shutdowns (US) and the planned shutdowns (PS) and their total shutdown times (UST and PST) without respect of the shutdowns cause. The following assessment will be divided into the assessment of the boiler B7 and the assessment of the turbine TG3. [2]

The occurrences of the US and the PS respectively the durations of the UST and the PST for the boiler B7 have been shown in table 1.

Table 1. Overview of both shutdown types of the boiler B7

Investigated year	**Overview of the US and the PS**			
	US B7 [occurrences]	***UST B7 [hours]***	***PS B7 [occurrences]***	***TPS B7 [hours]***
2013	22	218.25	3	218.37
2014	40	424.4	12	834.9
2015	13	493.48	4	174.7
2016	7	38.55	10	360.23
2017	10	150.98	4	218.58
Average	18.4	265.13	6.6	361.37

There have been enormous occurrence number of the US at the boiler B7. Therefore, the average occurrence number of the US has been relatively high as well (i.e., 18.4 occurrences per year). The average time of the US has been high as well (i.e., 265.13 hours per year). The state of the US during the investigated years has shown to significant availability problem related to the boiler B7.

Moreover, there have been relatively high occurrence number of the PS at the boiler B7 during the investigated years as well. It means that the average occurrence number of the PS has been relatively high as well (i.e., 6.6 occurrences per year). The average time of the PS has been enormously high as well (i.e., 361.37 hours per year). The state of the PS has also confirmed the availability problem of the boiler B7, which has been already mentioned in the paragraph with the US.

Now the following assessment has been focused on the specific causes of the US and the PS of the boiler B7.

Firstly, the specific US causes have been described more deeply in table 2.

Table 2. Overview of the specific US causes at the boiler B7

Causes of unplanned shutdowns	B7 overview	
	US B7 [occurrences]	*UST B7 [hours]*
To minimum temp. in CCh	33	281.97
Blockage of fuel paths (low quality of fuel)	8	161.21
To minimum temp. in CCh (fuel reservoir)	6	149.15
To maximum temp. in CCh	8	96.65
Cracked dilatation of the safety valve	1	90.68
Hydraulic circuit fault	1	77.07
Blockage of the exhaust paths	5	75.47
Leaks of the pressure unit in the boiler	2	70.41
Broken fuel feeder	2	62.08
To minimal water BD level (feed pump fault)	2	29.93
Blockage of the ash outlet in CCh	2	25.54
Position sensor fault on the left safety valve	1	3.73
Failure of electric line power EN32	4	3.48
Fault of FC at compensating valve	1	2.18
Fault in electric drive in the fuel feeder	4	1.5
Fault of the MFP due to minimum water level	1	0.75
To maximal steam temp. due to safety valve	1	0.45
To minimal temp, in CCh (fault of burner)	1	0.32
To minimal temp. in CCh (flow sensor)	1	0.25
Wrong function of filter in desulphurisation	1	0.07
Wrong reaction of DCP (fault of UPS)	1	0.03

The most significant US cause of the boiler B7 has been shutdown to minimum temperature in the combustion chamber (CCh), because of a bad fuel quality or a higher fuel moisture (i.e., 33 occurrences). These US causes have been blockage of the fuel paths to the boiler B7 (i.e., 8 occurrences) because of a bad fuel quality, the US to minimum temperature in the CCh (i.e., 6 occurrences) because of a fault in the fuel reservoir, shutdown to maximum temperature in the CCh (i.e., 8 occurrences) because of a rapid fuel quality change.

Obviously, the higher UST have mostly connected with the higher US. However, there have been some of the US causes, which have occurred only once and have been relatively time consuming. The typical failure of a component could be solved by a redundancy or by a better setup of repairs and an improving in a diagnostic control.

Next, the specific PS causes have been described deeply in table 3.

Table 3. Overview of the specific PS causes at the boiler B7

Causes of planned shutdowns	B7 overview	
	PS B7 [occurrences]	*PST B7 [hours]*
Big leakage from the drum of the B7	1	486.5
Problems with louver and electro separator	6	278.55
Reparation of fuel feeder and fuel paths cleaning	3	211.59
Reparation and cleaning of the B7	1	140.93
Blockage of the ash outlet in the CCh B7	2	80.83
Exhaust and ash outlet blockage	1	60.73
High vibration of safety valve	2	58.18
Leak of the pressure unit in the B7	2	52.68
Reparation and cleaning of fuel paths	1	46.7
Fault of fuel feed hydraulics	1	42.5
Reparation of exhaust paths	1	41.23
Problem with louver separator	1	40.38
Releasing of exhaust paths	3	32.48
Reparation of feeder at electro separator	2	25.65
Reparation of steam pump	1	30.85
Reparation and cleaning of electro separator	1	23.65
Bearings failure in fuel feeder	1	4.42

According to the PST, the most significant PS cause has been leakage of the pressure drum. The leakage of the pressure drum has occurred only once, but the leakage was enormous that has caused the longest single PS, which has been noticed in the data. The pressure drum had to be resealed, which has been very time consuming (i.e., 486.5 hours per year).

The second most significant PS cause according to the duration has been problems with the louver separator and the electric separator (i.e., 6 occurrences). The problems with the louver separator and the electric separator have been the most often occurred PS in the boiler B7. The total PST due to this PS cause has been relatively high as well (i.e., 278.55 hours per year).

The next significant PS causes according to the duration have been the fuel feeder reparation and the cleaning of the fuel paths, a general reparation, and the cleaning of the boiler B7, blockage of the ash outlet from the boiler B7, the unstable burning in the boiler B7, the blockage of the exhaust paths and the ash outlet under the electric separator. Probably the most of the previously mentioned PS causes have related to the inconsistent fuel quality as well.

The next production technology, which has been described by their US and PS, has been the turbine TG3. The main disadvantage of the turbine TG3 has been its dependence on the heating power from the boiler B7. If the boiler B7 has shutdown, the turbine TG3 cannot be in operation as well.

Table 4. Overview of both shutdown types of the turbine TG3

Investigated year	Overview of the US and the PS			
	US TG3 [occurrences]	*UST TG3 [hours]*	*PS TG3 [occurrences]*	*PST TG3 [hours]*
2013	7	26,87	1	1,18
2014	7	4,18	0	0,00
2015	5	1,72	0	0,00
2016	2	1,20	0	0,00
2017	0	0,00	0	0,00
Average	4,2	6,79	0,2	0,24

In table 4 is shown that the turbine TG3 has been neglectable source of the US and the PS in comparison with the boiler B7.

The average occurrence number of the US have been low (i.e., 4.2 occurrences). The same situation has been with the UST. The average duration in the US has been low as well (i.e., 6.79 hours per year).

The PS of the turbine TG3 has occurred only once according to the operational data. The rest of the PS have occurred mostly due to the PS of the boiler B7. Therefore, the PS of the turbine TG3 could be neglected completely, because the most PS causes have been caused by the boiler B7.

The specific US causes of the turbine TG3 have described in table 5.

Table 5 Overview of the specific US causes at the turbine TG3

Causes of unplanned shutdowns	US TG3 overview	
	US TG3 [occurrences]	*UST TG3 [hours]*
Leakage on drainage flange of HPRV	2	20.82
Low parameters of input steam	1	4.88
To minimal temperature of input steam	7	3.82
To maximal temp. of input steam	4	1.12
Reparation of output measurement	1	0.96
To condenser protection (pump fault)	2	0.48
Fault of cooling pumps for el. generator	1	0.32
To minimal pressure in regulation liquid	1	0.18
To minimal pressure in cooling oil	1	0.1

The most significant US cause has been the leakage on the flange of the heating drainage for the high-pressure regulation valve (HPRV). The leakage has been solved by resealing of the heating drainage.

The next US causes have been the low thermodynamic parameters at the turbine TG3 input, the US to minimal temperature at the turbine TG3 input and the US to maximal temperature at the turbine TG3 input. The mentioned US causes of the turbine TG3 have also related to the bad fuel quality.

III. Comparison of Investigated Shutdown Causes

All shutdown causes of the unit have been detailly described in the previous assessments. Pareto analysis of the unit could be done as well, which is shown in Figure 2.

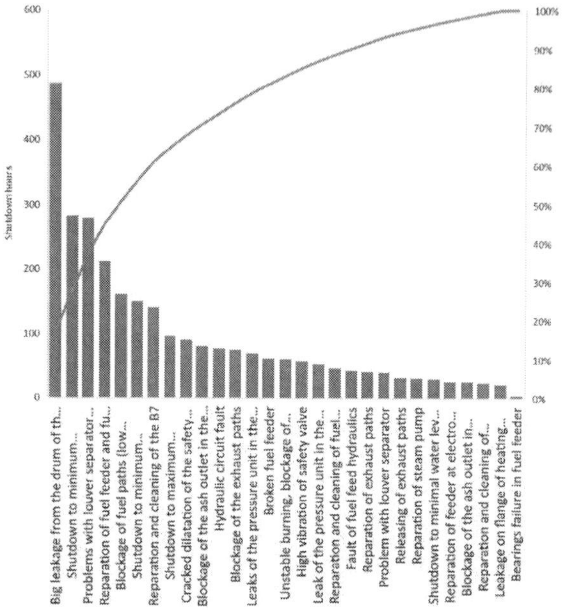

Figure 2. Pareto analysis of the investigated shutdown causes [3]

Thanks to the Pareto analysis, the key shutdown causes with the highest impact on the unit could be identified. On the other hand, neglectable shutdown causes, which took less 10 hours per year, have been neglected for the Pareto analysis.

Most the unit shutdowns have been related to the boiler B7. The boiler B7 has had the significant problem with the leakage of the pressure drum. The next significant the unit shutdown causes have also been related to the inconsistent fuel quality and with higher moisture in the fuel. Therefore, two drying lines for imported raw wood chips have been installed at the CHP plant in Pilsen. However, an improvement in a monitoring of the fuel quality and a fraction control to the boiler B7 could be a promising investment, which should minimize the shutdown causes related to the bad fuel quality.

A certain part of the boiler B7 shutdown causes have been already solved. However, there are still possibilities, which could improve the current state of the boiler B7.

IV. Solution of Availability of the Unit

The availability of the unit has been calculated and assessed in this chapter. The calculation has considered the average total shutdown duration per year for the boiler B7 and for the turbine TG3, i.e., that the boiler B7 shutdowns and the turbine TG3 shutdowns has been considered as the same for the following calculation. Each of the shutdowns have caused the shutdown of the whole unit. Another simplification is that the US and the PS have been also considered as the same. [4]

The average availability of the unit could be calculated by the average shutdown duration during the investigated operation years. The maximum operation time per year could be 8760 hours. If the average shutdown duration is divided by the maximum operation time, the average failure rate per year will be result. In table 6. have been shown the calculated availabilities of the boiler B7 and the turbine TG3 included in the unit.

Table 6. Availability comparison of the boiler B7 and the turbine TG3

Investigated technology	Availability comparison of investigated technologies per year	
	Average shutdown time [hours]	*Average availability [%]*
B7	626.49	92.85%
TG3	16.13	99.82%

The boiler B7 has had lower availability per year than the turbine TG3. On the other side, the turbine TG3 has had decent availability. The total availability of the unit (A_{UTot}) has been calculated quite simple as multiplication of both already mentioned availabilities in equation 1: [5], [6], [7]

$$A_{UTot.} = 0.929 * 0.998 = \mathbf{92.68}\ \%. \quad (1)$$

Obviously, the unit has not had such a good availability. As it has been mentioned, low availability of the unit has been mainly caused by several types of issues in the boiler B7, which have been often related to the bad fuel quality.

The average failure rate (FR) of the unit will be calculated by subtraction the average availability from one in equation 2.

$$FR = 1 - 0.9268 = \mathbf{7.32}\ \%. \quad (2)$$

Optimization of the total availability of the unit should be focused on certain weak segments, which could be easily improved, e.g., there have been human faults or problems with the leakage of the pressure drum have been occurred quite often. These causes could be removed by convenient precautions e.g., thoroughly resealing of the pressure drum and establishing the better operational regulations for employees.

V. Conclusion

The assessment of the US and the PS of the unit at the CHP plant in Pilsen has been described in this paper.

Firstly, the actual technological description of the unit at the CHP plant in Pilsen has been presented with its basic parameters.

Then the boiler B7 has been evaluated as much more problematic technology than the turbine TG3 according to their evaluations of availability. Therefore, the raw shutdown assessment of the boiler B7 has been done. Unfortunately, the assessment has provided no correlation between the US and PS, but there have been shown enormous number of shutdown occurrences and quite high shutdown duration, which must be solved in future. It has caused necessity of the next assessment creation, which would respect the specific shutdown causes of the boiler B7.

After that, the next assessment of the boiler B7 has been done according to the specific causes of the US and the PS as well. Majority of shutdown causes of the boiler B7 have been related to the bad fuel quality.

The next investigation has continued by the US and the PS at the turbine TG3. Only the US causes of the turbine TG3 have been considered in the investigation. Majority of the shutdown causes at the turbine TG3 have also related to the bad fuel quality, respectively with the boiler B7.

The most significant shutdown causes of the unit have been deeply assessed. The assessment has provided following information:

The boiler B7 has been according to its shutdowns more problematic production technology at the unit than the turbine TG3.

Majority of the shutdowns at the boiler B7 have related to the bad fuel quality or the high moisture in the fuel.

Another significant shutdown cause at the boiler B7 has also been the working medium leakage of the pressure drum.

Finally, the solution of the unit availability has been done. Thanks to simple topology of the unit (i.e., just one boiler and one turbine), the total availability could be calculated easily just for simple scenario (i.e., both production technologies at the unit had to be in operation simultaneously). The total availability of the unit has been 92.68 %, respectively the average failure rate has been 7.32 %. The boiler B7 has been determined as the more problematic production technology according to the calculated availability as well, which has related to the higher shutdown occurrences and higher shutdown duration as well.

The solution using the operational data allows to identify weak segments with the significant impact on the investigated unit. Then the weak segments can be improved by redundancy implementations, better repair intensity setup or improving in controlling and monitoring e.g., the fuel quality or the relative higher moisture in the fuel. [5]

The main contributions of the paper could be summarized as following:

The main contribution of the paper has been to show a motivation for the current state improving of the biomass burned unit at the CHP plant in Pilsen. The unit has been necessary to operate at nominal load because of an excellent economical possibility of the unit, which relates to the renewable source subsidies. That is the reason these aspects must be considered and solved as soon as possible.

The presented way of the assessment of specific shutdown causes at CHP plants based on the real operation data of the unit at the CHP plant in Pilsen.

The calculation of the unit availability, respectively the unit failure rate and the determination of the operational recommendations for the improvement of the current state, especially the pressure drum resealing at the boiler B7, the improvement of the fuel quality checking measurement at the boiler B7, a roofing of all fuel storages for improvement and sustaining fuel quality and using higher share of residual heat from the CHP process for the fuel drying.

The availability assessment is a valuable approach on how to reveal weak segments and to design a future improvement of already operated CHP plants. It relates to an increase of a system availability and better sustainability as well.

The similar assessment procedure, which is described in the sections 2,3 and 4, could be generalized and used in the most types of power plants if the deep operational data about the shutdowns are available.

The methodology and the conclusions can be also generalized and applied for an availability improvement at a lot of CHP plants in Czechia.

This paper has been based on available data about the US and the PS of the unit at the CHP plant in Pilsen. However, the newer operational data has not been available. The complete dataset with all operation years would provide better knowledge about the unit availability progress during the operation years. This could be done in the future investigation of the unit.

ACKNOWLEDGMENT

This paper has been supported by the student research project SGS-2018-023.

REFERENCES

[1] A. Hromádka, M. Sirový, and Z. Martínek. "Innovation in an Existing Backpressure Turbine for Ensure Better Sustainability and Flexible Operation," in *Energies*. 2019, 12(14), iss. 2652. ISSN: 1996-1073, DOI: 10.3390/en12142652.

[2] Z. Martínek, A. Hromádka, and J. Hammerbauer. "Reliability Characteristics of Power Plants," in *Advances in Electrical and Electronic Engineering*. 2017, 15(1), pp. 37-45. ISSN: 1804-3119, DOI: 10.15598/aeee.v15i1.2043.

[3] A. Hromádka, J. Jiřičkova, and Z. Martínek. "Analysis of the unplanned shutdowns and estimation of reliability in Temelin," in *19th International Scientific Conference on Electric Power Engineering (EPE)*, 2018, pp. 1-6. ISSN: 2376-5631, doi:10.1109/epe.2018.8395953.

[4] M. Valdma, , M. Keel, H. Tammoha, and K. Kilk. "Reliability of electric power generation in power systems with thermal and wind plants," in *Oil Shale*. 2007, 24(2), pp. 197–208. ISSN: 0208-189X.

[5] V. Mentlík, P. Trnka, M. Trnková, and L. Šasek. " Spolehlivostni aspekty elektrotechnolgie," in *BEN – technická literatura*, vol. 1, 2011. Praha. ISBN: 978-80-7300-412-5.

[6] J. Tuma, Z. Martínek, M. Tesařová, and I. Chemisinec. "Security, quality and reliability of electrical energy," in *Conte*, 2007, Praha. ISBN 978-80-239-9056-0.

[7] P. Skvarka. "Spolahlivost v Jadrovej Energetike," in *Alfa*, 1990, Bratislava. ISBN: 80-05-00095-2.

Model of Unsymmetrical Three-Phase Power System

Karel Máslo
Transmission System Analysis Dept.
ČEPS, a.s.
Prague, Czech Republic
maslo@ceps.cz

Jan Koudelka, Václav Vyčítal
Dept. of Electrical Power Engineering
Faculty of Electrical Engineering and Communication
Brno University of Technology, Brno, Czech Republic
xkoude20@vut.cz, vycital@vut.cz

Abstract — **This paper presents extension of the symmetrical components network model to the three-phase in a simulation engine, which is used in the MODES program and in the dispatch training simulator DTS. This extension allows to model network asymmetries in both the fault-free state and faulted operation, even to simulate contemporary multiple asymmetric short-circuit faults. The extended model can be used both for analytical calculations (such as setting protections) and for training dispatchers in solving faults in the distribution. The reliability of the created model is verified using a test system by comparing simulation results with another program that allow three-phase calculation. The test system is topological simple, but it contains several components important for model testing – e.g. arc suppression coil automatic tracking and directional earth fault relay.**

Keywords: arc suppression coil automatic tracking, dynamic models, directional earth fault relay, faulted phase earthing, network asymmetries, three phase modelling

I. Introduction

Since the beginning of development, MODES network simulator has undergone a vast development and several improvements making it a multi-purpose tool for analysis of power system dynamics. Various ways of its usage has been already presented in previous years. Some of them were black start capability and system restoration [1], calculations of short circuit currents [2] and [3], gas and combined gas cycle dynamics modelling [4] and [5], modelling of fast valving, power system stabilizers and double-fed induction generator [6] , [7] and [8], last but not least, dynamic models of various system components and stability assessment [9] - [13]. The number of presented problems demonstrates the robustness and applicability of MODES for analysis of various problems. In addition, MODES simulation engine is applied for on-line transient stability assessment [14] and in dispatch training simulator [15].

Previously mentioned applications have in common that analyses are done for RMS values (phasors) and investigated issues are mostly from balanced transmission system operation. It is worth mentioning that the simulation engine uses the symmetrical components method for network modelling allowing to specify except positive sequence parameters also zero sequence parameters. Thus, it is suitable for analyses of simple unbalance network operation (e.g. single unbalance short circuit calculation). The sequence representation of the power system speeds up the simulation, but it has limitations originating from its basis. It is known that the symmetrical components representation is not very applicable in case of simultaneous faults.

MODES development until now has been focused on transmission system analysis, which is, in general, considered to be symmetrical (as lines are transposed and loads are approximately balanced). Current development of simulation engine is oriented to distribution system analysis, which is quite different. Distribution system is basically unsymmetrical as loads are not evenly distributed through phases and lines are not transposed, thus the system is unsymmetrical even in fault-free state. In addition, unbalanced faults are predominant in distribution system and even simultaneous unbalanced faults can occur. This implies the need of a significant simulation engine extension and development, including computation in phase values (as was mentioned, symmetrical components are not sufficient). The extended simulation engine can be used for distribution system analysis in MODES and also for dispatch training simulator DTS. Modelling approach is applicable in general; thus its usage is possible also for simultaneous unbalanced faults in transmission system, which are, however, rare and not the primary scope.

This paper describes the fundamentals of power system elements modelling in phase values in Section II. A simple test system used for simulations is described in Section III.; it is a MV distribution system with 2 branches. Application of extended simulation engine for analyses of the test system under faulted operation – a typical earth fault in reactance grounded MV distribution system is considered – is demonstrated in Section IV. The final Section V introduce the future development and application of the grid model.

II. Three-Phase Models of Network Components

As mentioned in the introduction, the method of symmetrical components is commonly used for analytical calculations (load flow, short circuits and dynamics). Transition between both systems – symmetrical components (0, 1, 2) and phase values (A, B, C) – for matrices of impedances $\underline{\mathbf{Z}}$, admittances $\underline{\mathbf{Y}}$ and vectors of voltage $\underline{\mathbf{U}}$ and current $\underline{\mathbf{I}}$ phasors are given by (1).

$$\underline{\mathbf{Z}}_{012}=\underline{\mathbf{T}}\,\underline{\mathbf{Z}}_{ABC}\,\underline{\mathbf{T}}^{-1} \quad \underline{\mathbf{Y}}_{012}=\underline{\mathbf{T}}\,\underline{\mathbf{Y}}_{ABC}\,\underline{\mathbf{T}}^{-1}$$
$$\underline{\mathbf{U}}_{012}=\underline{\mathbf{T}}\,\underline{\mathbf{U}}_{ABC} \quad \underline{\mathbf{I}}_{012}=\underline{\mathbf{T}}\,\underline{\mathbf{I}}_{ABC} \qquad (1)$$

$\underline{\mathbf{T}}$ stands for the transformation matrix, $\underline{\mathbf{T}}^{-1}$ for its inverse respectively, which are given by (2).

$$\underline{\mathbf{T}}=\frac{1}{3}\begin{bmatrix}1 & 1 & 1\\ 1 & \underline{\alpha} & \underline{\alpha}^2\\ 1 & \underline{\alpha}^2 & \underline{\alpha}\end{bmatrix} \quad \underline{\mathbf{T}}^{-1}=\begin{bmatrix}1 & 1 & 1\\ 1 & \underline{\alpha}^2 & \underline{\alpha}\\ 1 & \underline{\alpha} & \underline{\alpha}^2\end{bmatrix} \quad \underline{\alpha}=e^{j120^o} \qquad (2)$$

For a limited scope of the paper, we will present only the phase matrices for overhead line, two-winding transformer and generator.

A. Overhand line (in fault-free and faulted operation)

A three-phase power line with 3 phase conductors (A, B, C), neutral N, alternatively with ground wires, can be described with self-impedances, mutual impedances, self and mutual capacitances. In case of power system of multiple wires (e.g. LV line – A, B, C, N), the technique of Kron reduction [16] is applied to obtain a 3-wires equivalent as shown in Figure 1. Shunt conductance (representing leakage and corona losses) is not depicted in Figure 1, but it is considered in further calculations.

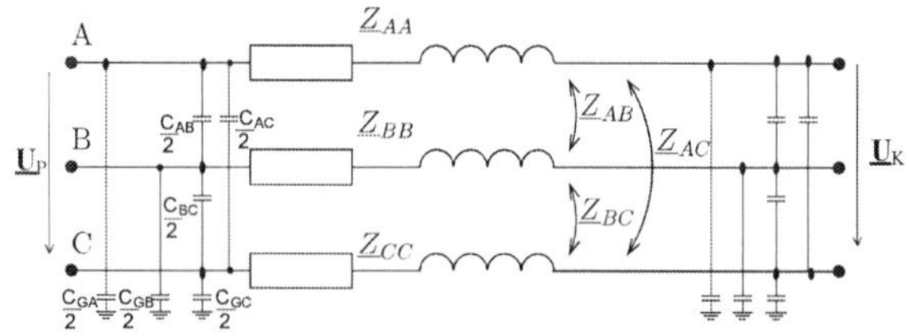

Figure 1. Line segment with impedances and capacitances

The nodal admittance matrix (3) gives the relation between current injections to the beginning and terminal bus $\underline{\mathbf{I}}_P$, $\underline{\mathbf{I}}_K$ and nodal voltages $\underline{\mathbf{U}}_P$, $\underline{\mathbf{U}}_K$.

$$\begin{bmatrix}\underline{\mathbf{I}}_P\\ \underline{\mathbf{I}}_K\end{bmatrix} = \begin{bmatrix}\underline{\mathbf{Z}}^{-1}+\frac{1}{2}\underline{\mathbf{Y}} & -\underline{\mathbf{Z}}^{-1}\\ -\underline{\mathbf{Z}}^{-1} & \underline{\mathbf{Z}}^{-1}+\frac{1}{2}\underline{\mathbf{Y}}\end{bmatrix}\begin{bmatrix}\underline{\mathbf{U}}_P\\ \underline{\mathbf{U}}_K\end{bmatrix} \qquad \underline{\mathbf{Y}} = \mathbf{G} + j\omega\mathbf{C}$$

$$\underline{\mathbf{Z}} = \begin{bmatrix}\underline{Z}_{AA} & \underline{Z}_{AB} & \underline{Z}_{AC}\\ \underline{Z}_{AB} & \underline{Z}_{BB} & \underline{Z}_{BC}\\ \underline{Z}_{AC} & \underline{Z}_{BC} & \underline{Z}_{CC}\end{bmatrix} \tag{3}$$

$$\mathbf{C} = \begin{bmatrix}C_{AA} & -C_{AB} & -C_{AC}\\ -C_{AB} & C_{BB} & -C_{BC}\\ -C_{AC} & -C_{BC} & C_{CC}\end{bmatrix} \qquad \mathbf{G} = \begin{bmatrix}G & 0 & 0\\ 0 & G & 0\\ 0 & 0 & G\end{bmatrix}$$

Short circuits and phase interruptions occurring on the line can be easily simulated by modification of admittance matrix. Short circuit or earth fault means the addition of an infinite value admittance to the corresponding bus (assuming the fault occurred close to the bus), thus the corresponding element of admittance matrix has an infinite value. Phase interruption means self-admittance of the faulted phase in admittance matrix is equal to zero.

B. Two-winding transformer

Let's assume the transformer connection wye-wye with neutral points grounded on both sides through non-specified impedances $\underline{Y}_P$ on primary side and $\underline{Y}_S$ on secondary side. Neglecting magnetization current and considering resistance R and leakage reactance X only, we can write a similar matrix (4) for the transformer as for the line.

$$\begin{bmatrix}\underline{\mathbf{I}}_P\\ \underline{\mathbf{I}}_K\end{bmatrix} = \begin{bmatrix}\underline{\mathbf{Y}}' & -\underline{\mathbf{Y}}'\\ -\underline{\mathbf{Y}}' & \underline{\mathbf{Y}}'\end{bmatrix}\begin{bmatrix}\underline{\mathbf{U}}_P\\ \underline{\mathbf{U}}_K\end{bmatrix} \qquad \underline{\mathbf{Y}}' = \underline{Y}\begin{bmatrix}1-\underline{K} & -\underline{K} & -\underline{K}\\ -\underline{K} & 1-\underline{K} & -\underline{K}\\ -\underline{K} & -\underline{K} & 1-\underline{K}\end{bmatrix}$$

$$\underline{K} = \frac{\underline{k}_P + \underline{k}_S}{\underline{k}_P\underline{k}_S + 3(\underline{k}_P + \underline{k}_S)} \qquad \underline{k}_P = \frac{\underline{Y}_P}{\underline{Y}} \quad \underline{k}_S = \frac{\underline{Y}_S}{\underline{Y}} \tag{4}$$

$$\underline{Y} = \frac{1}{R + jX}$$

$\underline{K}$ factor in (4) varies from 0 for both neutral points solidly grounded to 1/3 for both neutral points insulated.

C. Generator

Generator can be modelled as an induced voltage $\underline{E}$ behind the impedance – reactance X. Generator is considered to be symmetrical, thus voltage phasors $\underline{\mathbf{E}}$ are simply expressed as (5).

$$\underline{\mathbf{E}} = \begin{bmatrix}1\\ \underline{a}\\ \underline{a}^2\end{bmatrix}\underline{E} \tag{5}$$

For simplified calculations, induced voltage $\underline{E}$ is considered to have constant value.

III. Test System Description

To verify the proper functioning of the created three-phase network model, simple network from the textbook [17] was used. It is a simplified 22 kV reactance-grounded distribution network, fed from 110 kV system equivalent and comprising two branches as shown in Figure 2, where grid parameters are also specified.

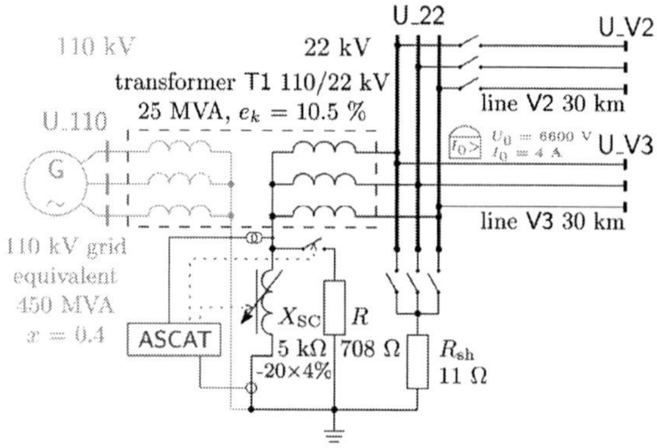

22 kV line parameters:

$$\underline{\mathbf{Z}} = \begin{vmatrix}1.1+j1.4 & 0.6+j0.7 & 0.6+j0.7\\ 0.6+j0.7 & 1.1+j1.4 & 0.6+j0.7\\ 0.6+j0.7 & 0.6+j0.7 & 1.1+j1.4\end{vmatrix}\ \Omega/\mathrm{km}$$

$$\underline{\mathbf{Y}} = \begin{vmatrix}0.342+j3.14 & 0 & 0\\ 0 & 0.342+j3.3 & 0\\ 0 & 0 & 0.342+j2.98\end{vmatrix}\ \mu\mathrm{S/km}$$

Figure 2. Single-line diagram of the test system

The arc suppression coil (ASC) with reactance X_{SC} = 5 kΩ is connected to the neutral point of the transformer secondary wye winding. ASC tuning is modelled discretely in 20 steps, 4 % each, i.e. within range 1÷5 kΩ. In reality, the tuning process of ASC is continuous, which is unfortunately unfeasible in described model. Each change of ASC reactance means change of K factor in (4), thus the change of admittance matrix (4) and finally, the whole nodal matrix needs to be recalculated making the simulation time-consuming.

ASC is equipped with a parallel resistance R = 708 Ω (value corresponding to 1 Ω connected to the auxiliary 500 V winding), which is connected for a short time to increase the active part of earth fault current, thus for proper operation of protection devices (see e.g. [18] for more info about it). Both ASC and resistor are controlled by an automatic tracking system ASCAT (see e.g. [19]). The ASCAT model measures the voltage U_0 on ASC (which is basically a zero sequence voltage component) and if the voltage change is more than 20 %, it starts to switch the ASC taps until the peak of the resonance curve is reached.

For fault identification, the test system contains the directional earth fault protection DEF (ANSI 67N). Models of DEF and ASCAT have been standard part of the MODES and they have been only adapted for use in a three-phase model.

The test system was also equipped by shunting system, composed of the shunting resistor $R_{sh.}$ = 11 Ω, practically used in case of an earth fault with high resistance to reduce the residual current in fault location. The shunting method (earthing the affected phase in the transformer substation) is described in [20] and [21].

IV. Dynamic simulation on a test system

Test system was used for performance of various scenarios corresponding to situations occurring in the real system. Scenarios including an earth fault were validated by simulation in OpenDSS. Results of validation simulation are plotted in dashed lines, together with results obtained using MODES.

A. Tuning of ASC in fault-free operation

First simulation demonstrates ASC operation in fault-free state in the distribution system comprising the line asymmetry, which was modelled by asymmetry of capacities to ground (see Figure 1). For a discrete model of ASC, this process can be called the automatic tap tracking. Usually, ASC is tuned into resonance to compensate the inductive part of earth fault current. In case of a dispatch control in the distribution network (switching of some branch(es)), ASCAT is initialized and launch the tuning of ASC to obtain the resonance mode. This scenario was considered for the first simulation case, which starts with ASC tuning into resonance with capacitance for only one branch (V3) being in operation. Then another branch (V2) is switched at t = 1 s, followed by tuning of ASC to resonance with the new value of grid capacitance. Finally, at t = 2.5 s, second branch (V2) is switched off. ASCAT detects change and activated the tuning of ASC again. Simulation results (ASC voltage) are shown in Figure 3.

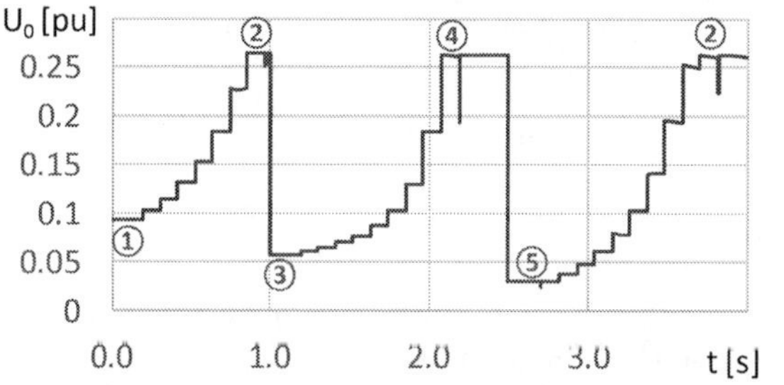

Figure 3. ASC voltage during the automatic tap tracking

Tuning of ASC is simulated as discrete change of its reactance, which is evident from Figure 3 (in reality, reactance change is continuous). After reaching the peak at time t = 0.98 s, the tuning ends; ASC reactance is in resonance with grid capacity to ground. When the second line is switched on at t = 1 s, ASC is no more in resonance. This detuning is detected by ASCAT according to the voltage drop and a new tuning process starts to reach the peak (of the new resonant curve). After switching off the line at t = 2.5 s, the tuning is repeated and the ASCAT achieves the peak of the original resonance curve. The tunning process is shown in Figure 4.

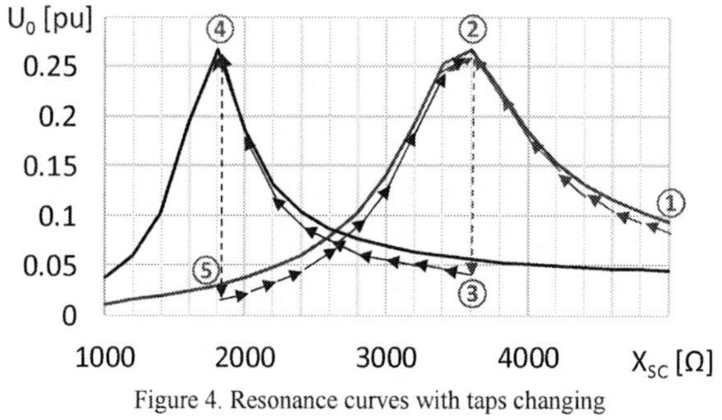

Figure 4. Resonance curves with taps changing

The initial resonance curve is drawn in blue in the right part of the graph and the arrows indicate the switching of the taps. Reduction of ASC reactance is marked, when the line is switched off and the resonance curve moves to the left (drawn in black). After switching off the line again, the resonance curve moves to starting position. The ASCAT reaches its peak, this time by increasing the ASC reactance.

B. Earth fault with auxiliary resistor connection

Another ASCAT function is to connect the shunt resistance to the ASC in case of an earth fault for proper functioning of protection devices. The small magnitude of the fault current makes the fault location difficult. Connecting the shunt resistance eliminates the compensating effect of ASC, thus causes an increase of fault current to the value which can be detected by earth fault protection.

Threshold value of ASC voltage U_0 for earth fault indication is set to 30 %. If ASCAT detects the earth fault, the auxiliary resistor is connected with a certain delay for a specified short period of time needed for fault detection (typically for 1 second).

The second simulation was a demonstration of the described process. After reaching the steady state with only operation of only one branch (V3) of the test system and ASC tuned to resonance (X_{SC}= 3600 Ω), the earth fault at the end of the line V3 is applied at time t = 0.1 s. ASCAT system automatically detects the earth fault and after the set delay of 0.1 s connects the auxiliary resistor for 0.1 s (for simulation purposes processes were fastened). Simulation results are plotted in Figure 5.

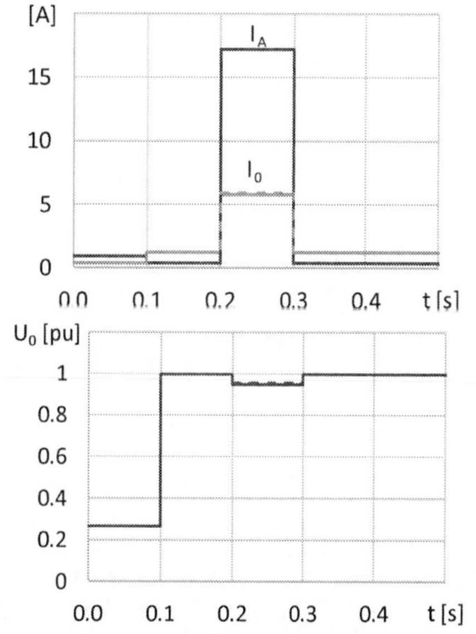

Figure 5. Waveforms of currents and zero sequence voltage during ground fault and auxiliary resistor connection

In the moment of the earth fault occurrence (0.1 s), the voltage U_0 increases to the nominal phase voltage (1 p.u.). Auxiliary shunt resistance connection at t = 0.2 s leads to a drop in voltage - a certain ASC detuning. This is much more evident in the phase current I_A of the line, which is significantly increased as well as its zero component I_0, which exceeds the value of 5 A. The directional earth fault protection (ANSI 67N) with threshold of 4 A detects the fault and issues a message. This message will help with fault localization (faulted branch is identified).

The figure shows that the waveforms obtained by both tools (MODES solid line and OpenDSS dashed) are practically identical.

C. *Shunting system*

The last simulation is a demonstration of the grounding of affected phase in case of resistive (arc) earth fault. In this case, the test system has both lines V2 and V3 switched on, the ASC is tuned to resonance (X_{SC} = 1800 Ω). A resistance earth fault at the end of line V3 occurs at time t = 0.1 s (fault resistance is set to 500 Ω). The ASC voltage and V3 line currents waveforms are shown in Figure 6.

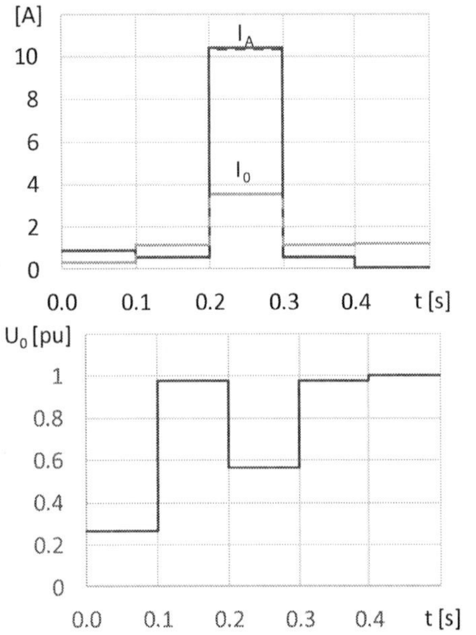

Figure 6. Waveforms during ground fault and shunting

The initial simulation results correspond to previous case. Earth fault occurrence at t = 0.1 s leads to increase of voltage U_0 close to the nominal value (1 p.u.), connection of auxiliary shunt resistor at t = 0.2 s for a limited period (0.1 s) means a significant increase of the phase current I_A of the line, which is slightly smaller than in the previous case. The zero component of the current is lower than 5 A, so the directional earth fault relay does not detect the fault. At time t = 0.4 s, the affected phase is earthed in the substation through a resistance of 11 Ω. The fault is thus transmitted to the substation and the fault current I_A drops to zero.

Shunting system brings the desired result – the fault is transmitted from the unknown location, where it can cause risks (due to possible arc burn), to the substation with a good grounding system. Technically, an artificial earth fault is done in the substation with a small resistance, which means that the current in the location of original fault is very small. The consequent problem is the original fault location.

Again, an excellent agreement of the results of both tools can be seen and the differences would only be seen in high resolution.

V. Conclusion

For analysis of distribution power systems are required simulation tools dealing with power system asymmetry. Unlike transmission networks, distribution networks contain significant asymmetry in fault-free operation mainly due to asymmetry of power lines not perfectly transposed and load unbalance. In addition, more common faults are the unbalanced ones – mainly earth faults, simultaneous unbalanced faults are also of importance. Therefore, a three-phase calculation is preferred in such cases.

The paper presents the approach to build a three-phase model for MODES network simulator, which was used for transmission system analysis so far. Computations in both symmetrical components and phase values were implemented in order the tool is suitable for distribution system analysis. Its functionality is demonstrated on simple cases imitating real faults in a simple distribution test system. Processes and protections in a distribution network were simulated – ASC tuning, earth fault, directional earth fault relay or faulted phase earthing. Simulation results were validated using a different simulation tool – OpenDSS.

The presented model is suitable for analysis of the distribution system operation and planning, which includes also protection settings. It will be also used in dispatch training simulator. The model will be improved for simulation of non-standard operation states and faults (e.g. double earth fault) and extended by models of other grid elements (e.g. three winding transformer model).

Acknowledgement

This research work has been carried out in the Centre for Research and Utilization of Renewable Energy (CVVOZE) and supported by the Ministry of Education, Youth and Sports of the Czech Republic under BUT specific research programme [grant number FEKT-S-20-6449].

References

[1] K. Máslo, J. Fantík, "Electromechanical and electromagnetical phenomena during power system restoration," IX. Mezinárodní vědecká konference ELEKTROENERGETIKA, 1998

[2] K. Máslo, J. Novák, J. Doležal, "Analýza výpočtu zkratových proudů," 1. Mezinárodní vědecké sympozium ELEKTROENERGETIKA 2001

[3] K. Kósa, K. Máslo, "Simulácia skratovej skúšky v sústave 400 kV ES SR v rozvodni Veľký Ďur," 2. Mezinárodní vědecké sympozium ELEKTROENERGETIKA 2003

[4] J. Anděl, K. Máslo, "Využití modelu plynové turbíny při návrhu elektráren a tepláren," 1. Mezinárodní vědecké sympozium ELEKTROENERGETIKA 2001

[5] K. Máslo, "Model a testování ostrovního provozu paroplynového cyklu," 3. mezinárodní vědecké sympozium ELEKTROENERGETIKA 2005

[6] K. Máslo, M. Bíca, I. Petružela, "Zvýšení stability jaderné elektrárny Temelín pomocí rychlého řízení ventilů," 2. Mezinárodní vědecké sympozium ELEKTROENERGETIKA 2003,

[7] K. Máslo, K. Witner, A. Kasembe, M. Škach, Z. Brettschneider, "Systémové stabilizátory – požadavky a jejich prověření," 4. Mezinárodní vědecké sympozium ELEKTROENERGETIKA 2007

[8] J. Rusnák, K. Máslo, P. Trubač, "Modelování větrné elektrárny s dvojitě napájeným asynchronním generátorem," 5. Mezinárodní vědecké sympozium ELEKTROENERGETIKA 2009

[9] K. Máslo, "R&D projects and solution of power flows problems in the Central European region," Proc. 7th International Scientific Symposium on Electric Power Engineering, Sept. 2013, pp. 23-27

[10] K. Máslo, "Improved methods for future power system operation and development assesment," Proc. 8th International Scientific Symposium on Electric Power Engineering, Oct. 2015

[11] K. Máslo, "Frequency stability in the context of the European network codes," Proc. 9th International Scientific Symposium on Electric Power Engineering, Sept. 2017, pp. 436-442

[12] K. Máslo, A. Kasembe, "Frequency Stability of Continetal Europe Synchronnous Area", Proc. 10th International Scientific Symposium on Electric Power Engineering, Sept. 2019, pp. 342-345

[13] K. Máslo, "Dynamic Models in the Context of the EU Commission Regulations," Proc. 10th International Scientific Symposium on Electric Power Engineering, Sept. 2019, pp. 26-30

[14] M. Chladová, K. Máslo, V. Krištof, "Short circuit calculations and Dynamic stability assessment within dispatch control environment, operational requirements and SW reality," Proc. 10th International Scientific Symposium on Electric Power Engineering, Sept. 2019,

[15] K. Máslo, M. Kolcun, "Simulation engine for dispatcher training and engineering network simulators," Proc IFAC CIGRE/CIRED Workshop on Control of Transmission and Distribution Smart Grids (CTDSG'16), Oct. 2016, pp. 395-399

[16] W. H. Kersting, Distribution System Modelling and Analysis, Third Edition, ISBN 978-1-4398-5622-2, CRC Press, 2012

[17] Z. Trojánek, M. Chladová: Přechodné jevy v ES, University textbook ČVUT FEL, 1988, p.145

[18] D. Topolanek, M. Lehtonen, P. Toman, J. Orsagova, J. Drapela, "An earth fault location method based on negative sequence voltage changes at low voltage side of distribution transformers," Int J Electr Power Energy Syst 2020;118: 105768. https://doi.org/10.1016/j.ijepes.2019.105768.

[19] X.B. Li, L.S. Wei,C.L. Wang, R. Liang, "Review on the Arc Suppression Coil with Automatic Tracking and Compensation in China," Australian Journal of Electrical and Electronics Engineering 7 (3), Jan. 2010, pp. 289-296, DOI: 10.1080/1448837X.2010.11464280

[20] D. Kouba, K. Procházka, "The Analysis of Efficiency of Shunt Resistor During a Single-phase Earth Fault Using the Two-port Network Theory," Proc.of Conference CIRED, 2013. pp. 1-4, DOI:10.1049/CP.2013.1066

[21] D. Topolánek, P. Toman, J. Orságová, J. Dvořák, "The Method of the Additional Earthing of the Affected Phase During an Earth Fault and Its Influence on MV Network Safety", Proc. of the IEEE PowerTech 2011, p.1- 8

Reliability Estimation of High Voltage SF_6 Circuit Breakers, Considering Uneven Wear

Illia Diahovchenko
Department of Electric Power Engineering
Sumy State University
Sumy, Ukraine
i.diahovchenko@etech.sumdu.edu.ua

Michal Kolcun, Zsolt Čonka, Samuel Bucko,
Dušan Medveď
Department of Electric Power Engineering
Technical university of Košice
Košice, Slovakia
michal.kolcun@tuke.sk, zsolt.conka@tuke.sk,
samuel.bucko@tuke.sk, dusan.medved@tuke.sk

Abstract — **A high-voltage circuit breaker is among the most critical elements of the switching nodes of the power system. The localization of failures and the restoration of the normal operational mode of the power system are largely determined by the reliable operation of circuit breakers when switching off short-circuit currents. This paper presents a fuzzy logic-based method to estimate the reliability of high-voltage SF_6 circuit breakers and to set priorities for the maintenance schedule. In this method a realistic mathematical model is combined with a fuzzy inference system, and its effectiveness was tested with a MATLAB-Simulink model. Based on the simulation results, a preventive maintenance strategy for high-voltage SF_6 circuit breakers can be proposed for particular operational state and conditions.**

Keywords: circuit breaker; reliability; maintenance; fuzzy logic; failure-free operation

I. Introduction

High voltage circuit breakers are among the crucial elements of power systems, responsible for defense against failures of transmission lines, transformers and equipment of electric substations. At the end of the last century, a breakthrough in the technology of high voltage switching devices took place; oil and air circuit breakers began to gradually give way to vacuum and gas-insulated SF_6 analogues, which better meet the requirements of compactness, reduced maintenance time, and service life [1]. SF_6 gas has an excellent property of arc extinguishing: the arc's extinction time is nearly 100 times faster compared to air [2]. Additionally, replacement of high-voltage oil and air circuit breakers with SF_6 ones allows to increase reliability, reduce operational costs and receive an economically positive effect [3].

Estimation and improvement of a circuit breaker's reliability requires a large amount of data to predict multiple aspects of the equipment. In order to create a general reliability model, it would be necessary to carry out extensive laboratory experiments for various types of equipment, its modifications and generations, considering different manufacturers. It is very difficult to conduct such a wide range of experiments. Additionally, it is barely possible to directly estimate the probability of a failure-free operation of a circuit breaker with standard bench tests, as only the main sample is tested, and test centers cannot test the circuit breaker for long periods of time. Therefore, various statistical and mathematical models are used to predict the reliability performance of the circuit breakers through their service life.

In the article [4] a statistical approach was used to examine the remaining lifetime and the failure rate of high voltage SF_6 circuit breakers. Analytical models for reliability assessment of high-voltage air circuit breakers, taking into account influence of different degrees of overload with various durations, are introduced in [5], [6]. The work [7] presents a methodology to identify when a circuit breaker is in most need of maintenance, considering its physical and functional characteristics that can and contribute to the occurrence of failures. An analytical approach to identify breaker's active failure events involving different order of contingencies is given in [8]. The authors of the paper [9] developed a hybrid prognostics approach for circuit breakers, combining uncertainty in the failure threshold and dynamic reliability techniques. In [10] an improved kernel extreme learning machine was employed for fault diagnosis of high-voltage circuit breakers. An artificial intelligence-based methodology, which combines fuzzy logic and machine learning, was introduced in [11] for assessment of the SF_6 circuit breakers condition.

In general, statistical and purely mathematical approaches [4]-[6], [8] are more time consuming and less flexible to variations in operational conditions. The methodology from [7] uses the history of past incidents and interventions, and the accuracy of the suggestion for maintenance is very dependent on the opinion of professionals. The Markov processes used in [9] and machine learning techniques employed in [10], [11] are computationally expensive and require expert tuning.

Unlikely, this study combines an analytical approach with a fuzzy logic system and allows creation of a computationally effective method, which is supported by both the mathematical formulations and knowledge of experts in drawing up rules for predictive maintenance planning. The main contributions of this paper are as follows.

II. Mathematical Model of SF_6 Circuit Breaker Reliability, Taking Into Account Uneven Wear

A. Formulation of the Mathematical Model

Let us consider a model that describes the operation of circuit breakers at different degrees of overload with various durations. Mathematical evaluation of the reliable operation of the circuit breaker is based on the failure model, which takes into account the initial failures and failures due to wear of the arc suppression device [5], [6]. The probability of absence of initial and sudden failures is determined according to long-term statistics for protective devices of similar type. The switching resource and the number of operations with various currents on the connection where the circuit breaker is installed, can be defined at bench tests. The failure of a circuit breaker depends on the expected intensity of switching and the available resource before the period of operation, as well as the degree of the resource degradation at each switching. The probability of a failure-free operation of the high-voltage circuit breaker for time t can be determined as [5]

$$P(t) = P_0 P_1(t) P_2(t), \quad (1)$$

where P_0, $P_1(t)$, and $P_2(t)$ are the probabilities of no infant failures, no sudden failures, and no failures due to wear of the arc suppression device, respectively. The probability of no-failure operation of the arcing chamber $P_2(t)$ is determinant for assessment of the probability of a failure-free operation of the circuit breaker with uneven wear.

The probability of a circuit breaker's failure at the very beginning of operation due to manufacturing defects or mistakes during installation is in the range from 0.001 to 0.006 [12], and is quite stable. Therefore, when predicting the reliability of SF_6 circuit breakers, the probability of no initial failure can be considered as $P_0 = 0.994 \div 0.999$. Some other reasons of the infant failure may include operation not according to the standards, installation of the components which do not meet the requirements, lack of manufacturing quality [4].

The probability of no sudden failures changes in time by the exponential law (i.e., absolutely continuous distribution, which simulates the time between two consecutive endings of the same event) with the rate parameter λ_0:

$$P_1(t) = \exp(-\lambda_0 t) \quad (2)$$

The value of λ_0 can be determined with sufficient accuracy only basing on the experience of similar equipment's operation. The intensity of sudden failures depends on a large number of different reasons. As equipment and technology improve, some of the causes are eliminated. However, it is impossible to completely prevent these failures, as each new sudden failure is caused by a new set of reasons that almost never recurs. Failures of SF_6 circuit breakers can be caused by failures of their components and mechanisms, such as hydraulic valves, pressure control switchgear, air compressors, SF_6 density relays, switching counters, control and protection systems etc. [4]. Therefore, sensing of the condition of these "bottleneck" functional nodes is seen as an important measure to decrease the amount of no sudden failures [13], [14].

The life cycle of a circuit breaker depends on its age, the number of trips and severity of tripping it has performed [7]. For the circuit breakers protecting overhead lines, the degradation of the arc suppression device occurs unevenly. Short circuits are possible at different distances from the place of installation, as well as sharp increase in the intensity of short circuit events is often supervised during the periods of thunderstorms and ice deposits on overhead transmission lines [12]. The resource of the arc suppression device depends on the distance to the short circuit point, the value of a short circuit current being disconnected, and the speed of voltage recovery. It is important to take the uneven wear into account, because neglecting it can lead to unreasonably higher reliability assessment results.

The probability of the no-failure operation with uneven wear of the arcing chamber depends on the short-circuit current (i.e., distance of the short-circuit point from the source) and the intensity of the failure flow:

$$P_2(t) = f\left(t, I^{(3)}_{s.c.Ki}, \lambda_i\right), \quad (3)$$

where $I_{s.c.Ki}{}^{(3)}$ is the three-phase short-circuit current at the point K_i, λ_i is the intensity of the failure flow.

A diagram of a transmission line protected with a SF_6 circuit breaker is shown in Fig. 1. A short circuit is possible at each of the three points K_i, which are at different distances from the power substation. The value of the short circuit current depends on the distance of the K_i point from the power substation and determines the wear rate of the circuit breaker's arcing chamber and contacts, which is referred as the resource usage per one switching off [6].

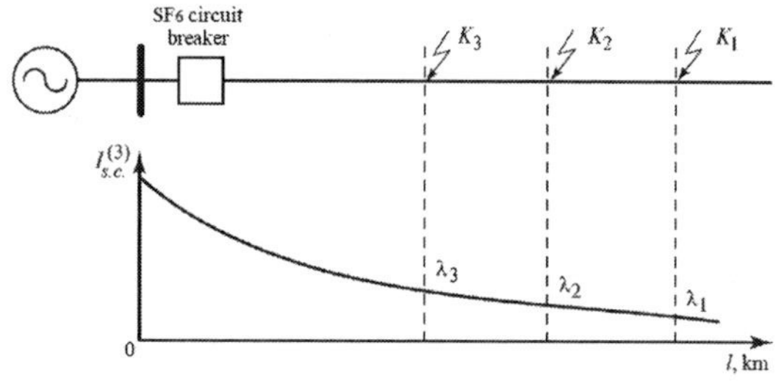

Figure 1. Diagram of a transmission line protected with a SF_6 circuit breaker with different points of short circuit

Let us consider an object that at any given time can be in any of a finite number of states. In this case, E_1, E_2... E_N are the states of a failure-free operation, and E_0 is the state of failure. Transitions can occur as follows: to the state E_{N-k} from the states E_{N-k+1}, E_{N-k+2}..., E_{N-1}, E_N, to the state E_0 – from any state. E_N would be the initial state of the object with the whole resource, and E_1 would be the last state of the object prior to a failure. The constants λ_1, λ_2, ..., λ_N are the parameters of the flows of possible transitions. The graph of this random process is shown in Fig. 2. Arrows indicate the direction of transitions.

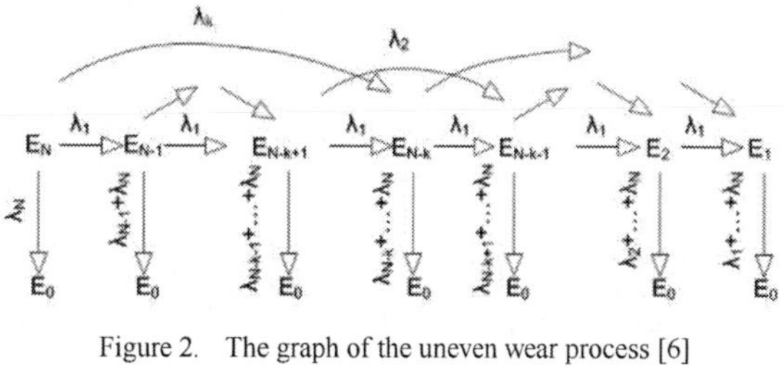

Figure 2. The graph of the uneven wear process [6]

Transitions with the parameters λ_1, λ_2, ..., λ_N can be considered as results of influences of various overloads. The number N indicates the available margin of safety, or resource of the object. Different overloads reduce the available resource to varying degrees. E.g., an overload with the parameter λ_1 reduces the resource by $1/N$, an overload with λ_2 – by $2/N$, and an overload with λ_N immediately leads to a failure. The probability of a failure-free operation will be defined as the sum of the probabilities of states from N to 1 [6]:

$$P(t) = \sum_{i=1}^{N} P_i(t) \quad (4)$$

If N is the initial resource, then each i-th state is characterized by the remainder of the resource i and the used resource $N - i$. The differential equation for the probability of this state is:

$$P_i'(t) = -\sum_{j=1}^{N} \lambda_j P_j(t) + \sum_{j=1}^{N-i} \lambda_j P_{i+j}(t), \tag{5}$$

where j is the usage index of the j/N-th part of the initial resource.

This model can be adapted for the assessment of $P_2(t)$ of a high-voltage SF_6 circuit breaker of the BГТ-110 series [15], [16]. According to the manufacturer's data [15], these circuit breakers can withstand and switch off:

- 50 short circuit currents $I_{s.c.K1}^{(3)} \approx (0.3 \div 0.6) I_{trip}^{max}$;
- 28 short circuit currents $I_{s.c.K2}^{(3)} \approx (0.6 \div 0.8) I_{trip}^{max}$;
- 20 short circuit currents $I_{s.c.K3}^{(3)} \approx I_{trip}^{max}$.

The maximum short circuit tripping current I_{trip}^{max} is determined during bench tests. For the SF_6 circuit breaker of the BГТ-110 series it is assumed $I_{trip}^{max} = 40$ kA [16].

Let us consider the case when the initial resource of the circuit breaker $N = 50$. The following notation is taken: λ_1 is the failure rate (i.e., the number of short circuit events) at the point K_1, with each trip of a short circuit event one unit of resource is used, the circuit breaker can withstand 50 trips; λ_2 is the failure rate at the point K_2, 1.4 resource units are used, the circuit breaker withstands 28 trips; λ_3 is the failure rate at the point K_3, 2.5 resource units are consumed, the circuit breaker withstands 20 trips.

Denote by m the number of disconnections with λ_1, by p — with λ_2; by k — with λ_3. The remaining resource n is determined by the formula

$$n = N - (m + 1.4p + 2.5k) \tag{6}$$

Solving of the formulas (4), (5) gives the formula for $P_2(t)$ for any $n \leq N$:

$$P_2(t) = \exp(-a_1 - a_2 - a_3) \sum_{k=0}^{A_k} \sum_{p=0}^{A_p} \sum_{m=0}^{A_m} \frac{a_3^k a_2^p a_1^m}{k!\,p!\,m!}, \tag{7}$$

where $a_1 = \lambda_1 t$; $a_2 = \lambda_2 t$; $a_3 = \lambda_3 t$;

$$A_k \leq \frac{n-1}{2.5};\ A_p \leq \frac{n-1-2.5k}{1.4};\ A_m \leq n - 1 - 2.5k - 1.4p$$

The mathematical model (7) allows [6]:

- to determine $P_2(t)$ at any arbitrary time t for known resource N and given values of the short circuit flow parameters λ_1, λ_2, λ_3;
- to determine the time for carrying out a preventive repair with a known resource N and values λ_1, λ_2, λ_3 to ensure the specified value of $P_2(t)$.

B. *Probability of a Failure-free Operation Under Different Conditions*

The available resource and the protection level of a transmission line are two parameters crucial for high reliability satisfaction. Let us consider the dependences of the probability of a failure-free operation of the arcing chamber P2(t) for various values of the initial resource in the range $N = (2 \div 50)$ for a transmission line with a BГТ-110 circuit breaker.

The following values of the parameter λ_i are considered for a transmission line:

- $\lambda_1 = 15$ year^{-1}, $\lambda_2 = 0.15$ year^{-1}, $\lambda_3 = 1.5$ year^{-1} for a faulty line;
- $\lambda_1 = 10$ year^{-1}, $\lambda_2 = 0.1$ year^{-1}, $\lambda_3 = 1$ year^{-1} for a troubled line;
- $\lambda_1 = 5$ year^{-1}, $\lambda_2 = 0.05$ year^{-1}, $\lambda_3 = 0.5$ year^{-1} for a vulnerable line;
- $\lambda_1 = 2$ year^{-1}, $\lambda_2 = 0.02$ year^{-1}, $\lambda_3 = 0.2$ year^{-1} for a well-protected line.

The "faulty" and the "troubled" lines are considered to be poorly protected, while the "vulnerable" and the "well-protected" lines are rather properly protected. The corresponding dependence curves $P(t)$ built according to the formula (1) are shown in Fig. 3 (only the values of λ_1 are denoted in the legend for the corresponding curves). For calculation of the probability of no sudden failures $P_1(t)$, it is assumed that $\lambda_0 \approx 0.01$ year^{-1} [6], and the probability of no initial failure is assumed to be $P_0 = 0.999$.

As it can be seen, the probability of a failure-free operation of the high-voltage circuit breaker for time t is drastically decreases with the increase of the parameter λ_i, i.e., the worse is the transmission line's protection, the lover is the probability of a failure-free operation, and vice versa. For instance, if the desired value of a no-failure operation probability for a year is $P(t) = 0.9$, then the time of no-failure operation of the circuit breaker with the resource value $N = 30$ (see Fig. 3b) will be 1.55 years in case of a faulty line, 2.25 years in case of a troubled line, 4.35 years in case of a vulnerable line, and 9.25 years in case of a well-protected line. It should be noted, that the λ_1 parameter has dominant influence on the probability of no-failure operation of the arcing chamber $P_2(t)$, while the impact of the parameters λ_2 and λ_3 is rather small, and in some cases it can be neglected without a risk to lose accuracy in calculations.

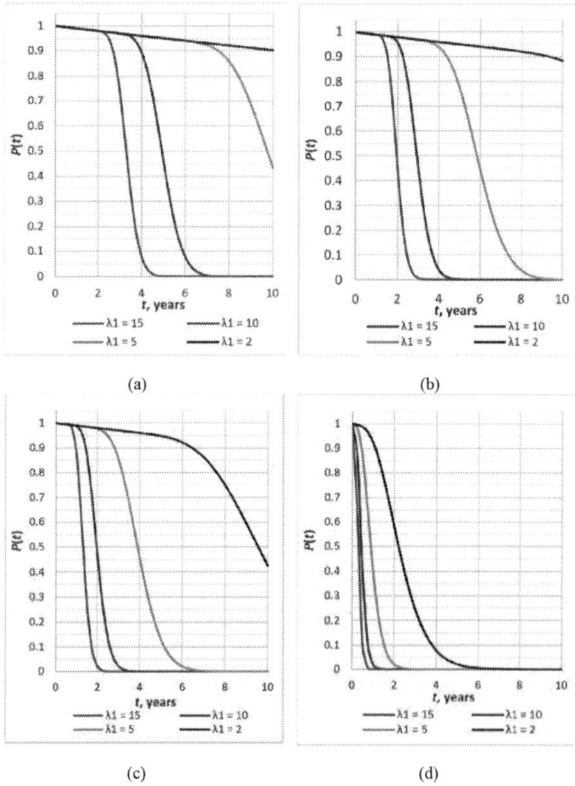

Figure 3. Dependencies of $P(t)$ for faulty, troubled, vulnerable, and well-protected lines: (a) for $N = 50$; (b) for $N = 30$; (c) for $N = 20$; (d) for $N = 5$

III. Algorithm and Model Description for Fuzzy Logic-Based Diagnostics of the SF_6 Circuit Breakers

To simulate the operation of a 110 kV transmission line, protected with a SF_6 circuit breaker, the corresponding model was created in the MATLAB-Simulink graphical programming environment, as shown in Fig. 5. The load is placed in the end of the transmission line, and the line itself is split into four segments of different lengths. Three points of the three-phase faults (i.e., short circuit events) were determined between the corresponding line segments. These points correspond to the points K_1, K_2, and K_3 in Fig. 1, and the failure rates are λ_1, λ_2, and λ_3, respectively. The further is the short circuit point from the source, the smaller is the amplitude of the short circuit current. As it was specified before, the circuit breaker can withstand 50 trips at the point K_1, 28 trips at the point K_2, and 20 trips at the point K_3.

Next, a diagnostic tool is to be created for strategic maintenance of circuit breakers, basing on the mathematical model for reliability assessment from Section 2.

Designing of a diagnostic tool for power application can be challenging, since it needs to be tuned to a specific system, and usually several measurements need to be taken and compared to obtain reasonable performance [17]. In this paper a mathematical tool of fuzzy logic is used to assess the reliability state of the SF_6 circuit breaker. Such an instrument is appropriate for system engineers and transmission system operators (TSOs), since it helps to ease decision-making about the maintenance or replacement of the circuit breakers, basing on their resource and taking into account proximity of a short circuit point and protection rate of a transmission line.

Fuzzy logic is well suited for this type of analysis, because it allows to conduct qualitative assessments of both input data and output results and gives a possibility of rapid modeling of dynamic systems and their comparative analysis with a given degree of accuracy [18]. Using signals from the fuzzy logic controller, a TSO will be able to make timely decisions regarding the circuit breaker's maintenance.

The flowchart in Fig. 5 demonstrates the proposed fuzzy-logic-based algorithm. After a short circuit event has happened and the protected line has tripped, the remained resource of the circuit breaker is to be re-estimated with (7).

After the fault has been isolated and the actual resource of the circuit breaker is calculated, then a fuzzy logic control system for diagnostic of the circuit breaker is to be employed. It consists of a set of input membership functions (MFs), a rule-based controller and a defuzzification process. The membership functions are curves that define how each point in the input space is mapped to a membership value between 0 and 1 [19]. The diagnostics part of the algorithm was implemented in the MATLAB-Simulink, and it is shown in Fig. 6.

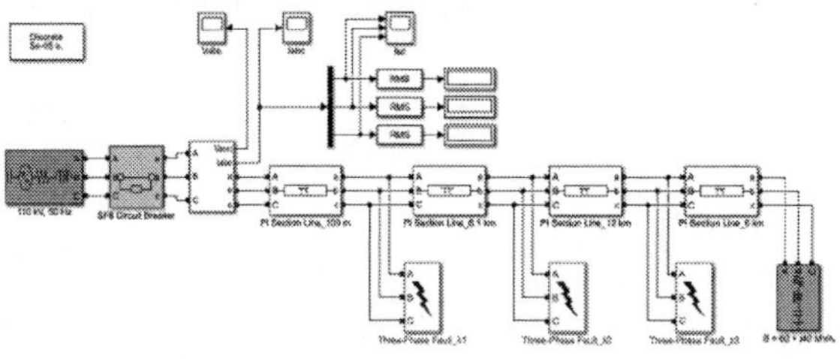

Figure 4. MATLAB-Simulink model of a transmission line protected with a SF_6 circuit breaker with different points of short circuit

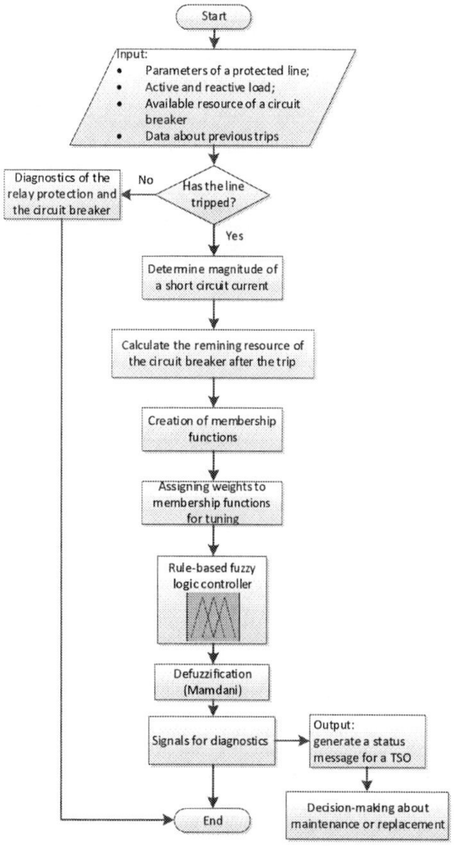

Figure 5. The flowchart of the proposed fuzzy-logic-based approach

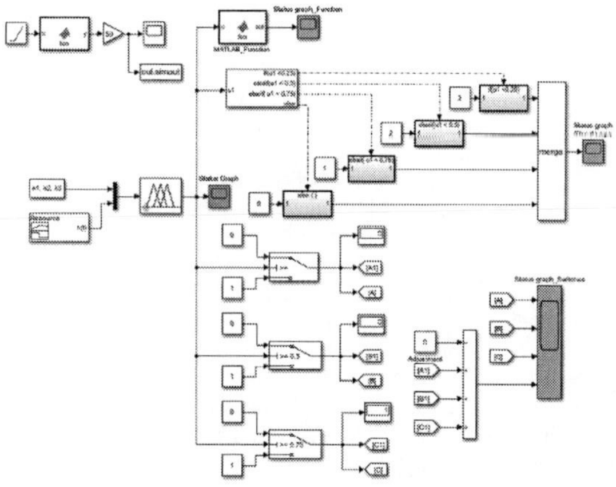

Figure 6. The diagnostics model in MATLAB-Simulink

The fuzzy logic input uses MFs to determine the fuzzy value of the input [19]. The protection level of a transmission line and the available resource of the circuit breaker are the two inputs to the fuzzy logic system for reliability assessment and maintenance scheduling, and each one of these inputs has several membership functions. The attributes of MFs for each input are then manipulated to add weight to different inputs. The Mamdani fuzzy system with the centroid method for defuzzification was selected [18]. The developed system for reliability assessment uses pre-determined values for the protection level of a transmission line, which are to be input by a user.

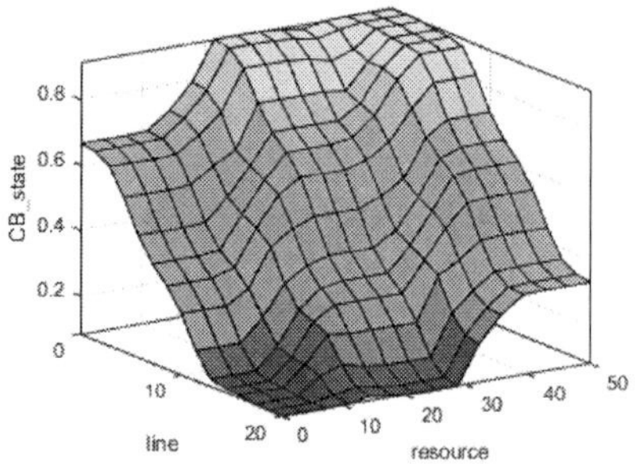

Figure 7. Fuzzy control rule surface for the reliability assessment system

The output also has a set of membership functions. These MFs define the possible responses and outputs of the studied system [17]. In the created model the output MFs are Gaussians and define the reliability results with the mean 0 as a "replacement", with the mean 0.333 as a "warning", with the mean 0.667 as an "observation" and with the mean 1 as a "healthy" circuit breaker.

Next, all of the output MFs are combined into one aggregated fuzzy set. A crisp value for representing uncertain data from this aggregated topology is obtained through a defuzzification process. The corresponding rule surface for the proposed fuzzy system is demonstrated in Fig. 7.

When the fuzzy control system is developed, it can be exported into the Simulink model. After the inputs are processed by the fuzzy logic controller, the fuzzy output is decoded and split into four ranges, and then an integer from 0 to 3 is sent to scopes and also to workspace variables. These integers reflect the circuit breaker's state and can be decoded into a degree of caution (i.e., a message) for a TSO, which is of the following four types: "healthy", "observation", "warning", "replacement". In case of a "replacement", an alarm signal can be sent to a transmission system operation center.

The proposed fuzzy logic-based method estimates the reliability of high-voltage SF_6 circuit breakers and advices on the strategic maintenance decisions. When a SF_6 circuit breaker is "healthy", no additional actions are required, and it can continue to operate with minimal risks of a failure. The "observation" status suggests that a SF_6 circuit breaker can be kept in operation, but some testing and a visual revision are suggested. Additionally, such a circuit breaker should be seen as a candidate for the upcoming scheduled maintenance and repairs. A circuit breaker with a "warning" status should be prepared for unplanned maintenance and repairs or to an overhaul. And in case of a "replacement" degree of caution, a SF_6 circuit breaker should go through a major overhaul or replacement without delay.

IV. Simulation Results and Discussion

During estimated life cycle of the switching devices, they must be subjected to periodic preventative maintenance, at predetermined schedule, following the manufacturer's recommendations or prescriptions of specific standards. Instead of following a fixed maintenance timetable similar for all devices, a decision about advancing or postponing maintenance can be made based on the actual condition of a circuit breaker, taking into account reliability performance of a protected transmission line and resource of the circuit breaker.

Let us demonstrate the work of the proposed method for a high voltage SF_6 circuit breaker reliability assessment, when it is installed at the head of a line with different values of the parameter λ_i. In particular, we consider a "well-protected" line, a "vulnerable" line, a "troubled" line and a "faulty line with the line protection characteristics as per Section 2. The depiction of the circuit breaker's switching resource n was modelled with the exponential function $n(t) = \exp(-0.25n)$ for a 10 years period, as shown in Fig. 8. The available resource starts from $N = 50$, as it is assured by the manufacturer [15], and by the end of the tenth year it exponentially drops down to 4. Note, that the synthesized depletion curve Fig. 8 is for demonstration of the method only, and in real life the depletion processes can go smoother, so that the SF_6 circuit breakers' service life can up to 30-40 years with rear maintenance [15], [16].

The outputs of the fuzzy logic-based diagnostics of the 110 kV SF_6 circuit breaker are sent to the scopes (see Fig. 6). The integer values from the scopes are to be decoded into messages for a TSO. The outputs can be summarized as follows.

- For a well-protected line with $\lambda_1 = 2$ year^{-1}, $\lambda_2 = 0.02$ year^{-1}, $\lambda_3 = 0.2$ year^{-1} the status of the circuit breaker during the first 3.8 years is "healthy", and then it changes to "observation". The latter requires to "keep an eye" on the circuit breaker and prepare it for the upcoming scheduled maintenance and repairs.
- For a vulnerable line with $\lambda_1 = 5$ year^{-1}, $\lambda_2 = 0.05$ year^{-1}, $\lambda_3 = 0.5$ year^{-1} the status of the circuit breaker during the first 3.8 years is "observation". The circuit breaker should be tested and supervised during its operation, and it should be prepared for the upcoming scheduled maintenance and repairs. If no maintenance has been performed, the circuit breaker's status will change to "warning", and unplanned maintenance and an overhaul will be expected.
- For a troubled line with $\lambda_1 = 10$ year^{-1}, $\lambda_2 = 0.1$ year^{-1}, $\lambda_3 = 1$ year^{-1} the "healthy" operation with a minor risk of a fault can be supervised during the first 1.1 year, and further the status changes to "observation", and after 6.05 years – to "warning". The amendments related to repairs and maintenance are similar to the previous cases.
- For a faulty line with $\lambda_1 = 15$ year^{-1}, $\lambda_2 = 0.15$ year^{-1}, $\lambda_3 = 1.5$ year^{-1} the "observation" status can be observed during the first year. Further the circuit breaker's status flips to "warning", and if no unplanned maintenance or an overhaul are performed, it deteriorates to the "replacement" status. The latter requires a major overhaul or replacement without delay.

Basing on the simulation results, a preventive maintenance strategy for high-voltage SF_6 circuit breakers can be generalized. For high values of λ_i (i.e., $\lambda_1 > 10$) maintenance and repairs are necessary after a few years of exploitation, followed by overhauls due to severe wear and replacement. For medium values of λ_i (i.e., $2 \leq \lambda_1 < 10$) scheduled maintenance and repairs take place after 3-5 years of exploitation. For low values of λ_i (i.e., $\lambda_1 < 2$) mostly observation is required; maintenance is less frequent, and it can be scheduled for several years ahead.

More attention should be paid to the circuit breakers' reliability state before and after the thunderstorm and ice periods, since unplanned maintenance might be required after periods with more frequent tripping of the transmission lines. Periodical testing and maintenance of circuit breakers is crucial even for "low-risky" locations (i.e., for well-protected lines), since 60% of failures result from ageing, with a steep slope in the wear-out region of a bathtub curve [4].

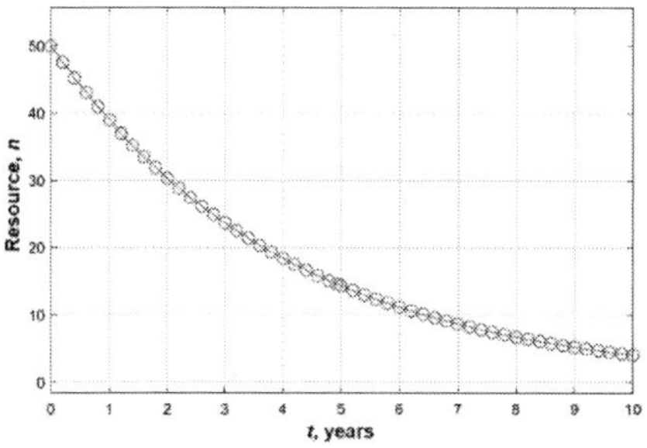

Figure 8. Circuit breaker's resource depletion curve

V. Conclusions

This study presents a fuzzy-logic-based method supported by mathematical formulation to establish a maintenance priority schedule for the SF_6 circuit breakers at the transmission level, with respect to reliability. Instead of following a fixed maintenance timetable equal for all devices of the same type, a decision about advancing or postponing maintenance is made based on the actual condition of a circuit breaker, taking into account reliability performance of a protected transmission line and resource of the circuit breaker. The diagnostics of the circuit breaker's state is based on a set of evaluation criteria to derive all the fuzzy logic parameters, namely membership functions and rule-base. The results obtained from the fuzzy logic inference system range from 0 to 1 p. u. To translate them into a term recognizable to operators, a crisp partition of the interval from 0 to 1 is adopted with the following terms: 0 – a "healthy" circuit breaker, 1 – "observation", 2 – "warning", 3 – "replacement". These terms provide an intuitive indication for the need and the priority of a maintenance or an overhaul action to be launched. In case of a "replacement" message, a TSO needs to take a prompt action to avoid a failure.

The mathematical formulation for the proposed method is based on the probabilistic reliability assessment of a high-voltage SF_6 circuit breaker. Additionally, deterioration of switching devices depends not only on their physical condition, but also on climatic (e.g., thunderstorm and ice seasons) and operational conditions to which they are subjected. Therefore, the outputs of the fuzzy diagnostic system cannot guarantee that a circuit breaker with the lowest performance score would fail first, if it remains without attention. However, a TSO will receive insights about the equipment's state, and it is possible to expect that more deteriorated equipment is more prone to failure.

The fuzzy logic system allows the application of knowledge of experts in drawing up rules, which consider the effect of aging or wear and tear in the evolution of failure rates. This design allows to automatically detect threatening regimes of the circuit breaker's operation, while a human (e.g., a TSO) is involved in the loop for decision-making to avoid any potential machine mistake.

Acknowledgment

This work was supported by the Ministry of Education, Science, Research and Sport of the Slovak Republic and the Slovak Academy of Sciences under the contract VEGA 1/0757/21. This work was supported by the Slovak Research and Development Agency under the contract No. APVV-19-0576.

References

[1] Ruben D. Garzon (2002). "High Voltage Circuit Breakers: Design and Applications," in *CRC Press.*, 2nd edi., 2002, pp. 425. DOI: 10.1201/9780203910634.

[2] International Standard IEC 60376:2018. "Specification of technical grade sulphur hexafluoride (SF6) and complementary gases to be used in its mixtures for use in electrical equipment," 3rd edition, 2005–06.

[3] V. A. Nepomnyashchiy, P. V. Ilyushin. "Evaluation of the effectiveness of the use of SF6 circuit breakers in the reconstruction and new construction of electric grid facilities," in *Reliability and Safety of Energy*, 4(19), pp.13-21, 2012.

[4] X. Zhang, E. Gockenbach, Z. Liu, H. Chen, L. Yang. "Reliability estimation of high voltage SF6 circuit breakers by statistical analysis on the basis of the field data," in *Electric Power Systems Research*, vol. 103, pp. 105-113, 2013.

[5] U. B. Guk. "Theory of reliability in the electrical engineering," in *Energoatomizdat*, 1990.

[6] I. E. Vasiliev. "Reliability of power supply," in *MPEI Publishing House*, 2014.

[7] E. A. Vianna, A. R. Abaide, L. N. Canha, V. Miranda "Substations SF6 circuit breakers: Reliability evaluation based on equipment condition," in *Electric Power Systems Research*, vol. 142, pp. 36-46, 2017.

[8] K. R. Timalsena, P. Piya, R. Karki "A Novel Methodology to Incorporate Circuit Breaker Active Failure in Reliability Evaluation of Electrical Distribution Networks," in *IEEE Transactions on Power Systems*, 36(2), pp. 1013-1022, 2021. DOI: 10.1109/TPWRS.2020.3010529.

[9] J. I. Aizpurua, V. M. Catterson, I. F. Abdulhadi, M. S. Garcia. "A Model-Based Hybrid Approach for Circuit Breaker Prognostics Encompassing Dynamic Reliability and Uncertainty," in *IEEE Transactions on Systems, Man, and Cybernetics: Systems*, 48(9), pp. 1637-1648, 2018. DOI: 10.1109/TSMC.2017.2685346.

[10] L. Chen, W. Shuting. "Intelligent fault diagnosis of high-voltage circuit breakers using triangular global alignment kernel extreme learning machine," in *ISA transactions*, vol. 109, pp. 368-379, 2021. DOI: 10.1016/j.isatra.2020.10.018.

[11] M. Žarković, Z. Stojković "Artificial intelligence SF6 circuit breaker health assessment," in *Electric Power Systems Research*, vol. 175, pp. 105912, 2019. DOI: 10.1016/j.epsr.2019.105912.

[12] U. B. Guk. "Analysis of the reliability of electric power installations," in *Energoatomizdat*, 1988.

[13] V. V. Volokhin, I. M. Diahovchenko, B. V. Derevyanko, "Prospects of nanomaterials use in current and voltage hall sensors to improve the measurements accuracy and reduce the external impacts," in *2017 IEEE 7th International Conference Nanomaterials: Application & Properties (NAP)*, 2017, pp. 03NE08-1-03NE08-5. DOI: 10.1109/NAP.2017.8190239.

[14] I. Diahovchenko, R. Mykhailyshyn, D. Danylchenko, S. Shevchenko. "Rogowsky coil applications for power measurement under non-sinusoidal field conditions," in *Energetika*, 65(1), pp. 14-20, 2019. DOI: 10.6001/energetika.v65i1.3972.

[15] SF6 circuit breakers of ВГТ-110 series. User Manual., IVEZH.674122.002 RE. [Online]. Available: https://www.elec.ru/files/2020/11/18/ВГТ-110.pdf. [Accessed Mar. 31, 2022].

[16] SF6 circuit breakers of ВГТ-110 series. Catalogue, 2018. [Online]. Available: https://www.uetm.ru/fs/sources/d5/0d/52/48/e556efe739df0247f9316b8e.pdf. [Accessed Mar. 31, 2022].

[17] B. R. Klingenberg and P. F. Ribeiro. "Fuzzy Logic Application for Time-Varying Harmonics," in *Time-Varying Waveform Distortions in Power Systems*, 2009, pp. 372. ISBN: 978-0-470-71402-7.

[18] Bart Kosko. "Fuzzy Thinking: The New Science of Fuzzy Logic," in *Q. Rev. Biol.*, 1995.

[19] "Foundations of Fuzzy Logic," MathWorks, Inc., 2022. [Online]. Available: https://www.mathworks.com/help/fuzzy/foundations-of-fuzzy-logic.html. [Accessed Mar. 31, 2022].

Data driven approaches in fault detection of wind turbines

Dušan Krokavec, Anna Filasová
Department of Cybernetics and Artificial Intelligence
Technical University of Košice
Košice, Slovakia
dusan.krokavec@tuke.sk, anna.filasova@tuke.sk

***Abstract*—The paper gives an approach to fault residuals construction performing the data driven technique. The method comes out from the system input and output representations and is constructed on a collection of measured data of normal and faulty system behavior. The paper evaluates the performance of an FDI structure based on subspace identification technique to highlight the some usefulness of the data driven approaches. Proceeding along the same lines it is outlined the connection to input/output data and given an approach to identification of parameterized matrices and the fault detection logic. The principle is applied directly in analysis of the data generated by monitored system wind turbine model parameters.**

Keywords: fault detection; fault residuals; data driven techniques; wind turbine models.

I. Introduction

Beginning in the 70's of the last century, the analytical redundancy approach, instead of hardware redundancy, forces the concept of fault detection filters, constructed on the observations derived from the system measurements using Kalman filters [1] and system state observers [2]. The principal book on the fault detection and diagnosis was oriented on chemical and petrochemical processes [3], the first presentation of the data-driven method is also linked to this application area [4]. Thus, in the presented context, fault detection and isolation (FDI) part of such diagnostic system has to give an efficient solution to allow small fault detection time delay.

Many model based FDI techniques have been proposed including the above mentioned Kalman filters [5], unknown input observers [6] and H_∞/H_2 observer-based residual filter schemes [7]. The underlying idea behind the model based FDI is to use a mathematical model of the system as the source of redundant information, and to produce the state estimate based fault residuals by using the systems measured outputs. For all representations based on system models the fundamental question is the system observability. The disadvantage of this approach is the need for an accurate model of the process.

An alternative approach known as data driven FDI (DD FDI) uses a collection of measured data to unearth (discover) patterns of normal and faulty system behavior. Data driven FDIs prevailingly reflect results in data mining, machine learning and computational intelligence. The most popular DD FDI techniques are data feature-based prediction techniques using the nonlinear time series based prediction principle [8], the correlation analysis and Bayesian inference principle [9], the fault feature extraction and classification using computational intelligence principle [10], and projection-based techniques [11]. The additive fault estimators can also be considered as data driven procedures, usually applicable in the case of poor analytical knowledge of the system dynamics and external disturbances [12], [13].

The wind turbine technology is one of the appropriate ways for increased use of renewable energy. Thus, in order to improve a wind turbine behavior and reliability, faults prevention in the parts of the wind turbine and fault tolerant control stay relevant objectives. A way to ensure these tasks consists in introducing advanced fault detection, isolation, and accommodation schemes [14], [15], where FDIs are designed to estimate the generator speed by using a bank of state observers and bank of unknown input observers, respectively. To improve diagnosis of the particular wind turbine faults the wind farm level can be applied, when a wind turbine is considered in comparison to another turbine of the wind farm [16].

The first data-driven solution in wind turbine diagnosis relies on Takagi–Sugeno (T-S) fuzzy models that are derived from a clustering c-means algorithm, followed by an identification procedure related to the noise rejection [17]. To achieve maximization of the wind power extraction, T-S fuzzy models of reduced dimension are adapted in the given context [18]. However, the expert's knowledge is needed to be employed to define the proper fuzzy rules. The power of the approaches can be amplified when combining the wind power extraction T-S models by DD FDI and the weighting two or more metrics related with analytical FDI [19], [20]. In all schemes, a properly large data set should be available as a-priori knowledge to train the structure in the fault-free case.

In this paper a middle ground is taken and focused on data driven approaches, based on subspace identification methods, for fault detection of wind turbines. As the system state-space description also maps the system state into output variables, the natural way for constructive data driven approaches using input and output data sequences is to suppress the system model description and to identify only the parameterized matrices of a system. This leads to explicit formulas for the pseudo inverse of the predetermined data map matrices that seems to be basic. Of course, this use of the orthogonal complement of non-square data matrices for identification of parameterized matrices of the system is not new and the presented way seems to be a new adaptation.

The considered wind turbine model description, focused on the compromise between the power and rotor speed relation, with projection in the turbine operating range, seems to be subordinated to the data driven algorithm for fault residuals. This makes hard relation with the input/output data representation of the wind turbine behavior to generate the redundant outputs.

The paper is organized as follows. Ensuing introduction in Sec. I, and continuing by the system description using lagged variables in Sec. II, the residuals generation is analyzed in Sec. III and identification of parameterized matrices is outlined in Sec. IV. In the task relation, Sec. V describes substantial details in the parameterized matrices construction of wind turbines and applicability of the proposed method using benchmark model parameters. Finally, Sec. VI gives some prioritized concluding remarks.

For sake of convenience, throughout this paper used notations reflect usual conventionality so that $\boldsymbol{x}^T$, $\boldsymbol{X}^T$ mean transpose of the vector $\boldsymbol{x}$ and matrix $\boldsymbol{X}$, respectively, the notation $\boldsymbol{X}^{\ominus 1}$ denotes the Moore-Penrose pseudo inverse of a non-square matrix, $\boldsymbol{X}^\perp$ reflects the orthogonal complement of a non-square matrix $\boldsymbol{X}$, $\boldsymbol{I}_n$ is the n-th order identity matrix, $\mathbb{R}^{n \times r}$ signifies the set of $n \times r$ real matrices, $\mathbb{R}^n$ is the set of n-dimensional real vectors and $\mathcal{Z}_+$ is the set of positive integers.

II. System Description by Lagged Variables

Considering that the task is focuses on developing the FDI method in the data-driven fashion, the linear discrete-time model of the square system is used in the standard state space representation given by

$$\boldsymbol{q}(i+1) = \boldsymbol{F}\boldsymbol{q}(i) + \boldsymbol{G}\boldsymbol{u}(i) + \boldsymbol{v}(i) \tag{1}$$

$$\boldsymbol{y}(i) = \boldsymbol{C}\boldsymbol{q}(i) + \boldsymbol{w}(i) \tag{2}$$

where $\boldsymbol{q}(i) \in \mathbb{R}^n$ is the state vector, $\boldsymbol{u}(t) \in \mathbb{R}^m$, $\boldsymbol{y}(t) \in \mathbb{R}^m$ are input and output vectors, respectively, while $\boldsymbol{F} \in \mathbb{R}^{n \times n}$, $\boldsymbol{G} \in \mathbb{R}^{n \times m}$, $\boldsymbol{C} \in \mathbb{R}^{m \times n}$ and the system is corrupted by the stochastic disturbances $\boldsymbol{v}(i) \in \mathbb{R}^n$, $\boldsymbol{w}(i) \in \mathbb{R}^m$, which are zero means and normally distributed white noise.

Taking (1), (2) the following calculations can be performed for the square system ($r = m$)

$$\begin{aligned}\boldsymbol{y}(i) &= \boldsymbol{C}\boldsymbol{q}(i) + \boldsymbol{w}(i)\\ \boldsymbol{y}(i+1) &= \boldsymbol{CF}\boldsymbol{q}(i) + \boldsymbol{CG}\boldsymbol{u}(i) + \boldsymbol{C}\boldsymbol{v}(i) + \boldsymbol{w}(i+1)\\ &\;\;\vdots\\ \boldsymbol{y}(i+s-1) &= \boldsymbol{CF}^{s-1}\boldsymbol{q}(i) + \left[\boldsymbol{CF}^{s-2}\boldsymbol{G} \cdots \boldsymbol{CG}\,\boldsymbol{0}\right]\boldsymbol{u}_o(i) +\\ &\quad + \left[\boldsymbol{CF}^{s-2} \cdots \boldsymbol{C}\,\boldsymbol{0}\right]\boldsymbol{v}_o(i) + \boldsymbol{w}(i+o-1)\end{aligned} \tag{3}$$

where

$$\boldsymbol{u}_o(i) = \begin{bmatrix}\boldsymbol{u}(i)\\ \vdots\\ \boldsymbol{u}(i+s-1)\end{bmatrix}, \quad \boldsymbol{v}_o(i) = \begin{bmatrix}\boldsymbol{v}(i)\\ \vdots\\ \boldsymbol{v}(i+s-1)\end{bmatrix} \tag{4}$$

which can be written as

$$\boldsymbol{y}_o(i) = \boldsymbol{P}\boldsymbol{q}(i) + \boldsymbol{R}\boldsymbol{u}_o(i) + \boldsymbol{S}\boldsymbol{v}_o(i) + \boldsymbol{w}_o(i) \tag{5}$$

where, for $s \geq n$,

$$\boldsymbol{y}_o(i) = \begin{bmatrix}\boldsymbol{y}(i)\\ \vdots\\ \boldsymbol{y}(i+s-1)\end{bmatrix}, \quad \boldsymbol{w}_o(i) = \begin{bmatrix}\boldsymbol{w}(i)\\ \vdots\\ \boldsymbol{w}(i+s-1)\end{bmatrix} \tag{6}$$

$$\boldsymbol{P} = \begin{bmatrix}\boldsymbol{C}\\ \boldsymbol{CF}\\ \vdots\\ \boldsymbol{CF}^{s-1}\end{bmatrix}, \quad \boldsymbol{S} = \begin{bmatrix}\boldsymbol{0} & \boldsymbol{0} & \cdots & \boldsymbol{0} & \boldsymbol{0}\\ \boldsymbol{C} & \boldsymbol{0} & \cdots & \boldsymbol{0} & \boldsymbol{0}\\ \vdots & \vdots & \ddots & \vdots & \vdots\\ \boldsymbol{CF}^{s-2} & \boldsymbol{CF}^{s-3} & \cdots & \boldsymbol{C} & \boldsymbol{0}\end{bmatrix} \tag{7}$$

$$\boldsymbol{R} = \begin{bmatrix}\boldsymbol{0} & \boldsymbol{0} & \cdots & \boldsymbol{0} & \boldsymbol{0}\\ \boldsymbol{CG} & \boldsymbol{0} & \cdots & \boldsymbol{0} & \boldsymbol{0}\\ \vdots & \vdots & \ddots & \vdots & \vdots\\ \boldsymbol{CF}^{s-2}\boldsymbol{G} & \boldsymbol{CF}^{s-3}\boldsymbol{G} & \cdots & \boldsymbol{CG} & \boldsymbol{0}\end{bmatrix} \tag{8}$$

$\boldsymbol{y}_o(i), \boldsymbol{u}_o(i) \in \mathbb{R}^{sm \times m}$, $\boldsymbol{w}_o(i) \in \mathbb{R}^{sm}$ and $\boldsymbol{P} \in \mathbb{R}^{sm \times n}$, $\boldsymbol{R} \in \mathbb{R}^{sm \times sm}$, $\boldsymbol{S} \in \mathbb{R}^{sm \times sn}$.

Thus, applying of input and output data sets, the related data structures can be investigated, representing the dependence of future outputs $\boldsymbol{y}_o(i)$ on the past inputs and outputs $\boldsymbol{z}_p(i)$

$$\boldsymbol{z}_p(i) = \begin{bmatrix}\boldsymbol{y}(i-r)\\ \vdots\\ \boldsymbol{y}(i-1)\\ \boldsymbol{u}(i-r)\\ \vdots\\ \boldsymbol{u}(i-1)\end{bmatrix} = \begin{bmatrix}\boldsymbol{y}_p(i)\\ \boldsymbol{u}_p(i)\end{bmatrix}, \quad \boldsymbol{z}_o(i) = \begin{bmatrix}\boldsymbol{y}_o(i)\\ \boldsymbol{u}_o(i)\end{bmatrix} \tag{9}$$

and the future inputs $\boldsymbol{u}_o(i)$, where $\boldsymbol{z}_p(i) \in \mathbb{R}^{2r \times m}$, $\boldsymbol{z}_o(i) \in \mathbb{R}^{2s \times m}$ and it is used for $s \geq r \geq n$.

Constructing from a limited number of structural relations (5) it can be parameterized in such a way that

$$\begin{aligned}\left[\boldsymbol{y}_o(i) \cdots \boldsymbol{y}_o(i+p-1)\right] &= \boldsymbol{P}\left[\boldsymbol{q}(i) \cdots \boldsymbol{q}(i+p-1)\right] +\\ &+ \boldsymbol{R}\left[\boldsymbol{u}_o(i) \cdots \boldsymbol{u}_o(i+p-1)\right] +\\ &+ \boldsymbol{S}\left[\boldsymbol{v}_o(i) \cdots \boldsymbol{v}_o(i+p-1)\right] +\\ &+ \left[\boldsymbol{w}_o(i) \cdots \boldsymbol{w}_o(i+p-1)\right]\end{aligned} \tag{10}$$

and using the lagged variables and the corresponding data matrices defined in such a way that

$$\boldsymbol{Q}_i = \left[\boldsymbol{q}(i) \cdots \boldsymbol{q}(i+p-1)\right] \tag{11}$$

$$\boldsymbol{Y}_o = \left[\boldsymbol{y}_o(i) \cdots \boldsymbol{y}_o(i+p-1)\right] \tag{12}$$

$$\boldsymbol{U}_o = \left[\boldsymbol{u}_o(i) \cdots \boldsymbol{u}_o(i+p-1)\right] \tag{13}$$

$$\begin{aligned}\boldsymbol{D}_o &= \left[\boldsymbol{S}\boldsymbol{v}_o(i)+\boldsymbol{w}_o(i) \cdots \boldsymbol{S}\boldsymbol{v}_o(i+p-1)+\boldsymbol{w}_o(i+p-1)\right]\\ &= \boldsymbol{S}\boldsymbol{V}_o + \boldsymbol{W}_o\end{aligned} \tag{14}$$

$$\boldsymbol{V}_o = \left[\boldsymbol{v}_o(i) \cdots \boldsymbol{v}_o(i+p-1)\right] \tag{15}$$

$$\boldsymbol{W}_o = \left[\boldsymbol{w}_o(i) \cdots \boldsymbol{w}_o(i+p-1)\right] \tag{16}$$

then (10) can be governed by the relation

$$\boldsymbol{Y}_o = \boldsymbol{P}\boldsymbol{Q}_i + \boldsymbol{R}\boldsymbol{U}_o + \boldsymbol{D}_o \tag{17}$$

where $\boldsymbol{Y}_o, \boldsymbol{U}_o \in \mathbb{R}^{sm \times pm}$, $\boldsymbol{D}_o \in \mathbb{R}^{sm \times p}$, $\boldsymbol{V}_o, \boldsymbol{W}_o \in \mathbb{R}^{sm}$, $\boldsymbol{Q}_i \in \mathbb{R}^{sm \times p}$.

When new samples $\boldsymbol{y}_o(i)$ and $\boldsymbol{u}_o(i)$ are available, updated matrices are formulated by appending the new columns and deleting the first columns of $\boldsymbol{Y}_o$, $\boldsymbol{U}_o$, respectively. before them modifying appropriately $\boldsymbol{y}_o(i)$ and $\boldsymbol{u}_o(i)$.

Remark 1: Noting that for a non-square matrix it yields also that $\boldsymbol{P} = \boldsymbol{P}$, then multiplying this equality from the right-hand side by the identity matrix $\boldsymbol{I}_n$ it yields

$$\boldsymbol{P} = \boldsymbol{I}_n\boldsymbol{P} = \boldsymbol{P}\boldsymbol{P}^{\mathrm{T}}(\boldsymbol{P}\boldsymbol{P}^{\mathrm{T}})^{-1}\boldsymbol{P} \tag{18}$$

which implies

$$\boldsymbol{I}_n = \boldsymbol{P}^{\mathrm{T}}(\boldsymbol{P}\boldsymbol{P}^{\mathrm{T}})^{-1}\boldsymbol{P} = \boldsymbol{P}^{\ominus 1}\boldsymbol{P} \tag{19}$$

where

$$\boldsymbol{P}^{\ominus 1} = \boldsymbol{P}^{\mathrm{T}}(\boldsymbol{P}\boldsymbol{P}^{\mathrm{T}})^{-1} \tag{20}$$

is the Moore-Penrose pseudo inverse of $\boldsymbol{P}$ and $\boldsymbol{P}^{\ominus 1} \in \mathbb{R}^{n \times sm}$.

Writing formally the above equality as

$$\boldsymbol{P} = \boldsymbol{P}\boldsymbol{P}^{\ominus 1}\boldsymbol{P} \tag{21}$$

it can be derived that

$$\begin{aligned}\boldsymbol{P}(\boldsymbol{I}_n - \boldsymbol{P}^{\mathrm{T}}(\boldsymbol{P}\boldsymbol{P}^{\mathrm{T}})^{-1}\boldsymbol{P}) &= \boldsymbol{P}(\boldsymbol{I}_n - \boldsymbol{P}^{\ominus 1}\boldsymbol{P})\\ &= \boldsymbol{P}\boldsymbol{P}^{\perp}\\ &= \boldsymbol{0}\end{aligned} \tag{22}$$

where

$$\boldsymbol{P}^{\perp} = \boldsymbol{I}_n - \boldsymbol{P}^{\ominus 1}\boldsymbol{P} \simeq \boldsymbol{0} \tag{23}$$

is the orthogonal projector onto the kernel of $\boldsymbol{P}$ (orthogonal complement of $\boldsymbol{P}$).

Remark 2: The subspace method aided data-driven fault detection can be detailed for systems with the additive faults, including the actuator faults. This application request to alter the state equation (1) as

$$\boldsymbol{q}(i+1) = \boldsymbol{F}\boldsymbol{q}(i) + \boldsymbol{G}\boldsymbol{u}(i) + \boldsymbol{H}\boldsymbol{f}(i) + \boldsymbol{v}(i) \tag{24}$$

where $\boldsymbol{H} \in \mathbb{R}^{n \times r_f}$, $\boldsymbol{f}(i) \in \mathbb{R}^{r_f}$, and leads to the relation formally analogous with (17)

$$\boldsymbol{Y}_o = \boldsymbol{P}\boldsymbol{Q}(i) + \boldsymbol{R}\boldsymbol{U}_o + \boldsymbol{D}_o \tag{25}$$

but with the additive elements

$$\boldsymbol{f}_o(i) = \begin{bmatrix}\boldsymbol{f}(i)\\ \vdots\\ \boldsymbol{f}(i+s-1)\end{bmatrix}, \quad \boldsymbol{R}_f = \begin{bmatrix}\boldsymbol{0} & \boldsymbol{0} & \cdots & \boldsymbol{0} & \boldsymbol{0}\\ \boldsymbol{CH} & \boldsymbol{0} & \cdots & \boldsymbol{0} & \boldsymbol{0}\\ \vdots & \vdots & \ddots & \vdots & \vdots\\ \boldsymbol{CF}^{s-2}\boldsymbol{H} & \boldsymbol{CF}^{s-3}\boldsymbol{H} & \cdots & \boldsymbol{CH} & \boldsymbol{0}\end{bmatrix} \tag{26}$$

and the modified data matrix constructed for $l = i, \ldots, i+p-1$ as

$$\boldsymbol{D}_o = \left[\boldsymbol{D}_o(i) \cdots \boldsymbol{D}_o(i+p-1)\right] \tag{27}$$

$$\boldsymbol{D}_o(l) = \boldsymbol{R}_f\boldsymbol{f}_o(l) + \boldsymbol{S}\boldsymbol{v}_o(l) + \boldsymbol{w}_o(l) \tag{28}$$

which the same construction of a residual generator.

III. Residuals Generation Using Lagged Variables

The main idea of data-driven residuals generation from the system input and output data is expanded in this section by applying the lagged variables.

Reordering $\boldsymbol{RU}_o$ (17) can be rewritten as

$$\boldsymbol{Y}_o - \boldsymbol{RU}_o = \boldsymbol{PQ}_i + \boldsymbol{D}_o \tag{29}$$

Since $\boldsymbol{Q}_i$ is unknown, the relation (29) can be pre-multiplied from the left-hand side by $\boldsymbol{P}^{\perp}$ to eliminate the effect of the unknown state matrix $\boldsymbol{Q}_i$ to get

$$\begin{aligned}\boldsymbol{P}^{\perp}(\boldsymbol{Y}_o - \boldsymbol{RU}_o) &= \boldsymbol{P}^{\perp}\boldsymbol{PQ}_i + \boldsymbol{P}^{\perp}\boldsymbol{D}_o\\ &= \boldsymbol{P}^{\perp}\boldsymbol{D}_o\end{aligned} \tag{30}$$

utilizing the property (22) to construct the residual as

$$\boldsymbol{\Omega} = \boldsymbol{P}^{\perp}\boldsymbol{Y}_o - \boldsymbol{P}^{\perp}\boldsymbol{RU}_o \tag{31}$$

$$\boldsymbol{\omega}(i) = \boldsymbol{P}^{\perp}\boldsymbol{y}_o(i) - \boldsymbol{P}^{\perp}\boldsymbol{Ru}_o(i) \tag{32}$$

Thus, the data-driven residual generation can be achieved if $\boldsymbol{P}^{\perp}$, $\boldsymbol{P}^{\perp}\boldsymbol{R}$ can acquired purely from the system input and ouput data.

Note, the residual is indirectly related to the extended observability matrix $\boldsymbol{P}$ as well as to the block triangular Toeplitz matrix $\boldsymbol{R}$ and, to obtain the residual, it is necessary to identify $\boldsymbol{P}^{\perp}$ as well as $-\boldsymbol{P}^{\perp}\boldsymbol{R}$.

When $\boldsymbol{\omega}(i)$ is the residual vector at the time instant i, the statistic test can by applied using the covariance matrix of the residual matrix in the fault-free case, computed as

$$\boldsymbol{\Xi}_{\omega} = \frac{1}{p-1}\boldsymbol{\Omega}^{\mathrm{T}}\boldsymbol{\Omega} \tag{33}$$

to generate the fault occurrence threshold with on-line data as

$$t_{\omega} = \boldsymbol{\omega}^{\mathrm{T}}(i)\boldsymbol{\Xi}_{\omega}^{-1}\boldsymbol{\omega}(i) \tag{34}$$

and to introduce the detection logic

$$\begin{cases} t_{\omega} < t_h, & \text{no fault}\\ t_{\omega} \geq t_h, & \text{a fault is detected}\end{cases} \tag{35}$$

where $t_h > 0$, $t_h \in \mathbb{R}$ is a defined threshold.

Note, when there is no faults, then $t_{\omega} \sim \chi^2(k)$, where $\chi^2(k)$ is the chi-squared distribution with k degrees of freedom [11].

Assuming that the measurements are with followed with a Gaussian distribution, then the kernel density estimation can be utilized to determine the threshold [21].

IV. Identification of Parameterized Matrices and Fault Detection

When the system is corrupted by the white system and measurement noise, the instrumental variables $\boldsymbol{z}_p(i)$ can be used to suppress their effects, exploiting the property that the past data $\boldsymbol{z}_p(i)$ are uncorrelated with the future noise.

In the sense of the given assumptions, the general condition can be written as follows

$$\lim_{p\to\infty}\frac{1}{p}\boldsymbol{D}_o\boldsymbol{Z}_p^{\mathrm{T}} = \lim_{p\to\infty}\frac{1}{p}\left(\boldsymbol{SV}_o\boldsymbol{Z}_p^{\mathrm{T}} + \boldsymbol{W}_o\boldsymbol{Z}_p^{\mathrm{T}}\right) = 0 \tag{36}$$

where the associated forms are obtained by constructing $\boldsymbol{Z}_p$, $\boldsymbol{Z}_o$ from (9) as follows

$$\boldsymbol{Z}_p = \begin{bmatrix}\boldsymbol{Y}_p\\ \boldsymbol{U}_p\end{bmatrix} = \begin{bmatrix}\boldsymbol{y}_p(i) & \cdots & \boldsymbol{y}_p(i+p-1)\\ \boldsymbol{u}_p(i) & \cdots & \boldsymbol{u}_p(i+p-1)\end{bmatrix} \tag{37}$$

$$\boldsymbol{Z}_o = \begin{bmatrix}\boldsymbol{Y}_o\\ \boldsymbol{U}_o\end{bmatrix} = \begin{bmatrix}\boldsymbol{y}_o(i) & \cdots & \boldsymbol{y}_o(i+p-1)\\ \boldsymbol{u}_o(i) & \cdots & \boldsymbol{u}_o(i+p-1)\end{bmatrix} \tag{38}$$

Remark 3: Using the orthogonal complement of $\boldsymbol{U}_o$, defined analogously as

$$\boldsymbol{U}_o^{\perp} = \boldsymbol{I}_n - \boldsymbol{U}_o^{\ominus 1}\boldsymbol{U}_o = \boldsymbol{I}_n - \boldsymbol{U}_o^{\mathrm{T}}(\boldsymbol{U}_o\boldsymbol{U}_o^{\mathrm{T}})^{-1}\boldsymbol{U}_o \tag{39}$$

and the instrumental variable $\boldsymbol{Z}_p$, then (29) implies

$$\begin{aligned}\boldsymbol{\Phi} &= \lim_{p\to\infty}\frac{1}{p}(\boldsymbol{Y}_o - \boldsymbol{RU}_o)\boldsymbol{U}_o^{\perp}\boldsymbol{Z}_p^{\mathrm{T}}\\ &= \lim_{p\to\infty}\frac{1}{p}\boldsymbol{Y}_o\boldsymbol{U}_o^{\perp}\boldsymbol{Z}_p^{\mathrm{T}}\\ &= \lim_{p\to\infty}\frac{1}{p}\boldsymbol{PQ}_i\boldsymbol{U}_o^{\perp}\boldsymbol{Z}_p^{\mathrm{T}}\end{aligned} \tag{40}$$

then, realizing the singular value decomposition (SVD) of the left side of $\boldsymbol{\Phi}^{\mathrm{T}}$, it yields

$$\boldsymbol{\Phi}^{\mathrm{T}} = \begin{bmatrix}\boldsymbol{\Psi_P} & \boldsymbol{\Psi_{P^{\perp}}}\end{bmatrix}\begin{bmatrix}\boldsymbol{\Delta} & \boldsymbol{0}\\ \boldsymbol{0} & \boldsymbol{0}\end{bmatrix}\begin{bmatrix}\boldsymbol{\Upsilon}_{\boldsymbol{P}}^{\mathrm{T}}\\ \boldsymbol{\Upsilon}_{\boldsymbol{P}^{\perp}}^{\mathrm{T}}\end{bmatrix} \tag{41}$$

and it can be chosen [22]

$$\boldsymbol{\Psi_{P^{\perp}}} = \boldsymbol{P}^{\perp\mathrm{T}} \tag{42}$$

Knowing $\boldsymbol{P}^{\perp}$, it can be written

$$\begin{aligned}&\boldsymbol{P}^{\perp}\lim_{p\to\infty}\frac{1}{p}(\boldsymbol{Y}_o - \boldsymbol{RU}_o)\boldsymbol{Z}_p^{\mathrm{T}}\\ &= \boldsymbol{P}^{\perp}\left(\lim_{p\to\infty}\frac{1}{p}\boldsymbol{Y}_o\boldsymbol{Z}_p^{\mathrm{T}} - \boldsymbol{R}\lim_{p\to\infty}\frac{1}{p}\boldsymbol{U}_o\boldsymbol{Z}_p^{\mathrm{T}}\right)\\ &= \boldsymbol{P}^{\perp}\begin{bmatrix}\boldsymbol{I} & -\boldsymbol{R}\end{bmatrix}\lim_{p\to\infty}\frac{1}{p}\boldsymbol{Z}_o\boldsymbol{Z}_p^{\mathrm{T}}\\ &= \boldsymbol{0}\end{aligned} \tag{43}$$

which implies

$$\begin{aligned}\boldsymbol{P}^{\perp}\boldsymbol{\Phi}_{YZ} &= \boldsymbol{P}^{\perp}\lim_{p\to\infty}\frac{1}{p}\boldsymbol{Y}_o\boldsymbol{Z}_p^{\mathrm{T}}\\ &= \boldsymbol{P}^{\perp}\boldsymbol{R}\lim_{p\to\infty}\frac{1}{p}\boldsymbol{U}_o\boldsymbol{Z}_p^{\mathrm{T}}\\ &= \boldsymbol{P}^{\perp}\boldsymbol{R}\boldsymbol{\Phi}_{UZ}\end{aligned} \tag{44}$$

where

$$\boldsymbol{\Phi}_{YZ} = \lim_{p\to\infty}\frac{1}{p}\boldsymbol{Y}_o\boldsymbol{Z}_p^{\mathrm{T}}, \quad \boldsymbol{\Phi}_{UZ} = \lim_{p\to\infty}\frac{1}{p}\boldsymbol{U}_o\boldsymbol{Z}_p^{\mathrm{T}} \tag{45}$$

Rewriting (45) as

$$\boldsymbol{P}^{\perp}\boldsymbol{\Phi}_{YZ}(\boldsymbol{\Phi}_{UZ}^{\mathrm{T}}\boldsymbol{\Phi}_{UZ})^{-1}\boldsymbol{\Phi}_{UZ}^{\mathrm{T}}\boldsymbol{\Phi}_{UZ} = \boldsymbol{P}^{\perp}\boldsymbol{R}\boldsymbol{\Phi}_{UZ} \tag{46}$$

then

$$\boldsymbol{P}^{\perp}\boldsymbol{R} = \boldsymbol{P}^{\perp}\boldsymbol{\Phi}_{YZ}(\boldsymbol{\Phi}_{UZ}^{\mathrm{T}}\boldsymbol{\Phi}_{UZ})^{-1}\boldsymbol{\Phi}_{UZ}^{\mathrm{T}} = \boldsymbol{P}^{\perp}\boldsymbol{\Phi}_{YZ}\boldsymbol{\Phi}_{UZ}^{\ominus 1} \tag{47}$$

where

$$\boldsymbol{\Phi}_{UZ}^{\ominus 1} = (\boldsymbol{\Phi}_{UZ}^{\mathrm{T}}\boldsymbol{\Phi}_{UZ})^{-1}\boldsymbol{\Phi}_{UZ}^{\mathrm{T}} \tag{48}$$

Note, a principal component analysis (PCA) formulation, suitable for $\boldsymbol{P}^{\perp}$ and $\boldsymbol{P}^{\perp}\boldsymbol{R}$ computation, can be found in [23].

The idea of the algorithm for the residual construction and the fault evaluation can be presented as follows

(1) generating instrumental variables $\boldsymbol{Z}_o$ and $\boldsymbol{Z}_p$ with fault-free data,
(2) performing $\boldsymbol{U}_o^{\perp}$ and constructing $\boldsymbol{\Phi}$ by (39), (40), respectively,
(3) performing SVD of $\boldsymbol{\Phi}^{\mathrm{T}}$ and separating $\boldsymbol{P}^{\perp\mathrm{T}}$ by (41), (42),
(4) constructing $\boldsymbol{\Phi}_{YZ}$, $\boldsymbol{\Phi}_{UZ}$ by (45), performing $\boldsymbol{\Phi}_{UZ}^{\ominus 1}$ by (48) and computing $\boldsymbol{P}^{\perp}\boldsymbol{R}$ according to (47),
(5) constructing residual $\boldsymbol{\Omega}$ according to (31) from fault-free data,
(6) calculating the covariance matrix $\boldsymbol{\Xi}_{\omega}$ in the fault-free case according to (33),
(7) generating $\boldsymbol{Y}_o$ and $\boldsymbol{U}_o$ with on-line operating data using (12), (13) to construct operating residual according to (31),
(8) determining the fault occurrence threshold by (34),
(8) evaluating the residual according to the detection logic (35).

V. Model of the Wind Turbine

The overall model is given by interconnecting the models of the individual subsystems, considering a generic three blade horizontal wind turbine, driven by variable speeds, with a full converter coupling [24].

The aerodynamic power extracted from the wind is [25], [26]

$$\tau_r(t) = \frac{1}{2}\rho\pi R^3 C_q(\lambda(t),\beta(t))v^2(t) \tag{49}$$

where $v(t)$ is actual speed of wind [m/s], R is rotor radius [m], $C_q(*)$ is torque coefficient expressing what fraction of the power in the wind is extracted by wind turbine and ρ is air density [kg/m^3]. The torque coefficient is depending on the tip speed ratio λ [rad] between the tangential speed of the tip of a blade and the actual speed of wind, parameterized by the blades pitch angle β [rad], where

$$\lambda(t) = \frac{\omega_r(t)R}{v(t)}, \quad C_q(\lambda(t),\beta(t)) = \frac{C_p(\lambda(t),\beta(t))}{\lambda(t)} \tag{50}$$

ω_r is rotor angular speed [rad/s], $C_q(\lambda(t),\beta(t))$ is the rotor torque coefficient. Consequently, the turbine speed is controlled to track the reference $\omega_{rt}(t)$, to maintain the tip-speed ratio at its optimal value λ_o, by relation implying from (50)

$$\omega_{rt}(t) = \lambda_m \frac{v(t)}{R} \tag{51}$$

and, substituting $\lambda(t)$ into (49), it can be obtained

$$\tau_r(t) = \frac{1}{2\lambda^2(t)}\rho\pi R^5 C_q(\lambda(t),\beta(t))\omega_r^2(t) \tag{52}$$

Then the proposed two-mass drive train model is

$$\begin{aligned} J_r\dot{\omega}_r(t) &= \tau_r(t) - (B_{dt}+B_r)\,\omega_r(t) - \\ &- K_{dt}\theta_\Delta(t) + \frac{B_{dt}}{N_g}\omega_g(t) \end{aligned} \tag{53}$$

$$\begin{aligned} J_g\dot{\omega}_g(t) &= \frac{\mu_{dt}K_{dt}}{N_g}\theta_\Delta(t) + \frac{\mu_{dt}B_{dt}}{N_g}\omega_r(t) - \\ &- \big(\frac{\mu_{dt}B_{dt}}{N_g^2} + B_g\big)\omega_g(t) - \tau_g(t) \end{aligned} \tag{54}$$

$$\dot{\theta}_\Delta(t) = \omega_r(t) - \frac{1}{N_g}\omega_g(t) \tag{55}$$

where $\omega_r(t)$ is rotor angular speed [rad/s], $\omega_g(t)$ is the generator rotating speed [rad/s], $\theta_\Delta(t)$ is torsion angle [rad], $\tau_g(t)$ is aerodynamic torque [kg m^2/s^2], J_r is rotor inertia moment [kg m^2], N_g is gear ration, B_{dt} is torsion damping coefficient [Nms/rad], B_r is rotor external damping [Nms/rad], K_{dt} is torsion stiffness [Nm/rad], B_g is generator external damping [Nms/rad] and μ_{dt} is efficiency od drive train.

The hydraulic pitch system is modeled using second-order differential equation [27]

$$\ddot{\beta}(t) + 2\zeta\omega_n\dot{\beta}(t) + \omega_n^2\beta(t) = \omega_n^2\beta_r(t) \tag{56}$$

since, in order to limit the control effort, the reference speed signal is passed through the second order linear filter, while $\beta(t)$ is measured pitch angle [rad], $\beta_r(t)$ is reference pitch angle [rad], ω_n is pitch system filter natural frequency [rad/s], and ζ is damping factor.

Finally, the generator and converter dynamics are modeled by the first-order differential equation [17]

$$\dot{\tau}_g(t) + \alpha\tau_g(t) = \alpha\tau_{gr}(t) \tag{57}$$

where $\tau_g(t)$ is the generator torque [Nm], $\tau_{gr}(t)$ is the generator torque reference [Nm] and α is the generator and converter model parameter [1/s].

Defining the state model variables

$$\begin{aligned} &q_1(t) = \omega_r(t), \quad q_2(t) = \omega_g(t), \quad q_3(t) = \theta_\Delta(t) \\ &q_4(t) = \dot{\beta}(t) \quad\quad q_5(t) = \beta(t) \quad\quad q_6(t) = \tau_g(t) \\ &u_1(t) = \tau_{gr}(t) \quad u_2(t) = \beta_r(t) \end{aligned} \tag{58}$$

then

$$\boldsymbol{A} = \begin{bmatrix} a_{11} & a_{12} & -\frac{K_{dt}}{J_r} & 0 & 0 & 0 \\ a_{21} & a_{22} & a_{23} & 0 & 0 & 0 \\ 1 & -\frac{1}{N_g} & 0 & 0 & 0 & 0 \\ 0 & 0 & 0 & -2\zeta\omega_n & -\omega_n^2 & 0 \\ 0 & 0 & 0 & 1 & 0 & 0 \\ 0 & 0 & 0 & 0 & 0 & -\alpha \end{bmatrix} \tag{59}$$

$$\boldsymbol{B}^{\mathrm{T}} = \begin{bmatrix} 0 & 0 & 0 & 0 & 0 & \alpha \\ 0 & 0 & 0 & \omega_n & 0 & 0 \end{bmatrix}, \quad \boldsymbol{C} = \begin{bmatrix} 1 & 0 & 0 & 0 & 0 & 0 \\ 0 & 1 & 0 & 0 & 0 & 0 \end{bmatrix} \tag{60}$$

where

$$\boldsymbol{q}^{\mathrm{T}}(t) = \begin{bmatrix} q_1(t) & q_2(t) & q_3(t) & q_4(t) & q_5(t) & q_6(t) \end{bmatrix} \tag{61}$$

$$\boldsymbol{u}^{\mathrm{T}}(t) = \begin{bmatrix} u_1(t) & u_2(t) \end{bmatrix}, \quad \boldsymbol{y}^{\mathrm{T}}(t) = \begin{bmatrix} y_1(t) & y_2(t) \end{bmatrix} \tag{62}$$

and

$$\begin{aligned} a_{12} &= \tfrac{B_{dt}}{J_r N_g}, & a_{21} &= \mu_{dt}\tfrac{B_{dt}}{J_g N_g} \\ a_{23} &= \mu_{dt}\tfrac{K_{dt}}{J_g N_g}, & a_{22} &= -\tfrac{\mu_{dt}B_{dt}}{J_g N_g^2} - \tfrac{B_g}{J_g} \end{aligned} \tag{63}$$

with any fixed value with relation to

$$a_{11} = \tfrac{\rho\pi R^5 C_q(\lambda(t),\beta(t))\omega_r(t)}{2J_r\lambda^2(t)} - \tfrac{B_{dt}+B_r}{J_r} \tag{64}$$

The parameters used in the model are [20]

$$\begin{array}{lll} \rho = 1.225\,\frac{\mathrm{kg}}{\mathrm{m}^3} & R = 57.5\,\mathrm{m} & N_g = 95 \\ B_{dt} = 775.49\,\frac{\mathrm{Nms}}{\mathrm{rad}} & B_r = 7.11\,\frac{\mathrm{Nms}}{\mathrm{rad}} & B_g = 45.6\,\frac{\mathrm{Nms}}{\mathrm{rad}} \\ K_{dt} = 2.7\times10^6\,\frac{\mathrm{Nm}}{\mathrm{rad}} & J_g = 390\,\mathrm{kg\,m^2} & J_r = 55\times10^6\,\mathrm{kg\,m^2} \\ \mu_{dt} = 0.97 & \omega_n = 11.11\,\frac{\mathrm{rad}}{\mathrm{s}} & \zeta = 0.6 \\ \alpha = 50\,\frac{1}{\mathrm{s}} & & \end{array}$$

which implies

$$\begin{array}{lll} a_{12} = 1.484\times10^{-2} & a_{13} = -49.0909 & a_{21} = 0.0203 \\ a_{23} = 7.0688\times10^{4} & a_{22} = -0.1171 & a_{32} = -0.0105 \\ a_{44} = -13.332 & a_{45} = -123.4321 & a_{66} = -50 \end{array}$$

and using the constant speed wind profile implying the following parameters

$$\beta = 0.05 \quad \lambda_s = 2.4 \quad \omega_r = 1.25 \quad C_q = 0.25 \quad a_{11} = 1.0499$$

Converting to the discrete-time system model (1), (2) using the Matlab function "$c2d(*)$" with the sampling time $t_s = 0.02s$ the system matrix difference equations were used to generate the input and output data to construct the instrumental variable in the simulation to generate residuals and to evaluate the residual according to the detection logic.

As the results Fig. 1 presents the residual response, reflecting a short impulse-like fault on the pitch sensor measurement, corrupted at the time step $i = 2600$ (time instant $i_t = 52$ s), as well as the residual evaluation. The FDI was trained according to the algorithms discussed in the paper and the results are analyzed in terms of the detection times. In order to generate a suitable training data set interval with the proportion balance of the faulty-free and the faulty events there were set both interval parameters as $s = r = 10$.

This examples illustrate the power that can be invoked through the prescribed method properties.

VI. Concluding Remarks

In the paper, a benchmark model is used for building and simulation testing the fault detection scheme based on subspace method data-driven principle, applicable to wind turbine benchmark. To construct the reference residual from fault-free data, the discrete-time system parameters are used for the prediction data set generation in detection of pitch sensor faults. Compared with the model-based fault detection method, prior knowledge of complex system model is not needed and only the parameterized matrices need to be identified.

Exploited data-driven methodology has sufficient adaptability in correspondence to different operating points at which faults can

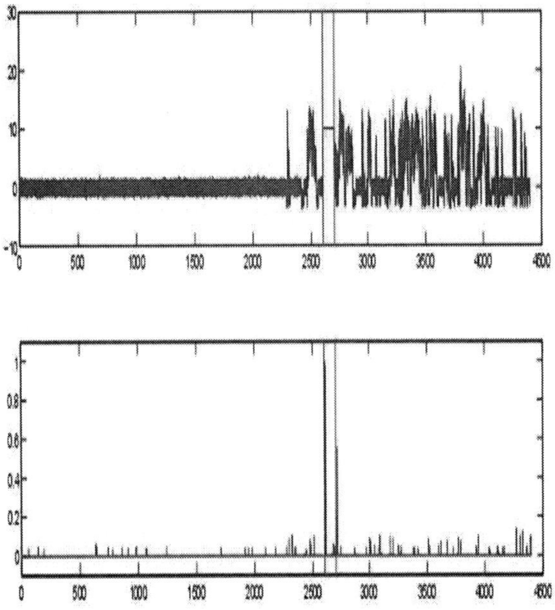

Figure 1. The residual time response

occurred, while primary quality of fault detection is covered by the non-trivial fault occurrence threshold generation and iterative detection logic. Simulation studies indicate that the residual performance is satisfactory.

Modifications of this unifying framework to propose hybrid solutions, combining neural networks and fuzzy inference system approaches, is the subject of future research in this algorithmic field. From fault diagnosis point, the hybrid input-output representations and fault feature generators are challenging subsequent research works in the wind turbine diagnosis. It is worth to point out that the proposed method can be simply adjusted to disturbances and noise, which would find applications in wide industrial systems.

ACKNOWLEDGEMENT

The work presented in this paper was supported by VEGA, the Grant Agency of Ministry of Education and Academy of Science of Slovak Republic, under Grant No. 1/0483/21. This support is very gratefully acknowledged.

REFERENCES

[1] R.K. Mehra and J. Peschon, "An innovations approach to fault detection and diagnosis in dynamic systems," Automatiea, vol. 7, pp. 637-640, 1971.

[2] A.S. Willsky, "A survey of design methods for failure detection in dynamic systems," Automatica, vol. 12, pp. 601-611, 1976.

[3] D.M. Himmelblau, Fault Detection and Diagnosis in Chemical and Petrochemical Processes, Amsterdam: Elsevier, 1978.

[4] V. Venkatasubramanian, R. Vaidyanathan, and Y. Yamamoto, "Process fault detection and diagnosis using neural networks - I. Steady-state processes," Computers & Chemical Engineering, vol. 14, no. 7. pp. 699-712, 1990.

[5] Z. Wang, Y. Wang, and Z. Ji, Advances in Fault Detection and Diagnosis Using Filtering Analysis, Singapore: Springer Nature, 2022.

[6] D. Du, S. Xu, and V. Cocquempot, Observer-Based Fault Diagnosis and Fault-Tolerant Control for Switched Systems, Singapore: Springer Nature, 2021.

[7] S.X. Ding, Advanced Methods for Fault Diagnosis and Fault-tolerant Control, Berlin: Springer Nature, 2021.

[8] J. Zhao, W. Wang, and C. Sheng, Data-Driven Prediction for Industrial Processes and Their Applications, Cham: Springer Nature, 2018.

[9] J. Wang, J. Zhou, and X. Chen, Data-Driven Fault Detection and Reasoning for Industrial Monitoring, Singapore: Springer Nature, 2022.

[10] G. Niu, Data-Driven Technology for Engineering Systems Health Management. Design Approach, Feature Construction, Fault Diagnosis, Prognosis, Fusion and Decision, Singapore: Springer Nature, 2017.

[11] Z. Chen, Data-Driven Fault Detection for Industrial Processes. Canonical Correlation Analysis and Projection Based Methods, Wiesbaden: Springer Nature, 2017.

[12] Q. Jia, W. Chen, Y. Zhang,and Y. Jiang, "Robust fault reconstruction in discrete-time Lipschitz nonlinear systems via Euler-approximate proportional integral observers," Mathematical Problems in Engineering, vol. 2015, pp. 1-14, 2015.

[13] D. Krokavec and A. Filasov'a, "Descriptor approach in design of discrete-time fault estimators," WSEAS Transactions on Systems and Control, vol. 13, pp. 481-490, 2018.

[14] W. Chen, S.X. Ding, A. Haghani, A. Naik, A.Q. Khan, and S. Yin, "Observer-based FDI schemes for wind turbine benchmark," IFAC Proceedings Volumes, vol. 44, no. 1, 2011, pp. 7073-7078, 2011.

[15] P.F. Odgaard and J. Stoustrup, "Fault tolerant control of wind turbines using unknown input observers," IFAC Proceedings Volumes, vol. 45, no. 20, pp. 313-318, 2012.

[16] P.F. Odgaard and J. Stoustrup, "Fault tolerant wind farm control - a benchmark model," Proc. IEEE International Conference on Control Applications CCA 2013, Hyderabad, India, pp. 412-417, 2013.

[17] S. Simani, S. Farsoni, and P. Castaldi, "Wind turbine simulator fault diagnosis via fuzzy modelling and identification techniques," Sustainable Energy, Grids and Networks, vol. 1, no. 1, pp. 45-52, 2015.

[18] X. Liu, Z. Gao, and M.Z.Q. Chen, "Takagi-Sugeno fuzzy model based fault estimation and signal compensation with application to wind turbines," IEEE Transactions on Industrial Electronics, vol. 64, no. 7, pp. 5678-5689, 2017.

[19] D. Krokavec and A. Filasová, "One approach to design the fuzzy fault detection filters for Takagi-Sugeno models," Advanced and Intelligent Computations in Diagnosis and Control, Cham: Springer, pp. 19-33, 2016.

[20] D. Krokavec and A. Filasová, "Model-based fault diagnosis of wind turbines built on Takagi–Sugeno fuzzy observers," Proc. 10th International Scientific Symposium on Electrical Power Engineering Elektroenergetika 2019, Stara Lesna, Slovakia, pp. 377-382, 2019.

[21] B.W. Silverman, Density Estimation for Statistics and Data Analysis, Boca Raton: CRC Pres, 1998.

[22] Z. Chen, H. Fang, and Y. Chang, "Weighted data-driven fault detection and isolation. A subspace-based approach and algorithms," newblock IEEE Transactions on Industrial Electronics, vol. 64, no. 5, pp. 3290-3298, 2016.

[23] J. Wang and S.J. Qin, "A new subspace identification approach based on principal component analysis," Journal of Process Control, vol. 12, pp. 841-855, 2002.

[24] P.F. Odgaard, J. Stoustrup, and M. Kinnaert, "Fault-tolerant control of wind turbines. A benchmark model," IEEE Transactions on Control Systems Technology, vol. 21, no. 4, pp. 1168-1182, 2013.

[25] K.E. Johnson, M.J. Pao, L.Y. Balas, and L.J. Fingersh, "Control of variable-speed wind turbines. Standard and adaptive techniques for maximizing energy capture," IEEE Control Systems Magazine, vol. 26, no. 3, pp. 71-81, 2006.

[26] J.F. Manwell, J.G. McGowan, and A.L. Rogers, Wind Energy Explained. Theory, Design and Application, New York: John Wiley & Sons, 2009.

[27] H. Sanchez, T. Escobet, V. Puig, and P.F. Odgaard, "Fault diagnosis of an advanced wind turbine benchmark using interval-based ARRs and observers," IEEE Transactions on Industrial Electronics, vol. 62, no. 6, pp. 3783-3793, 2015.

[28] S. Simani, S. Farsoni, and P. Castaldi, "Data–driven techniques for the fault diagnosis of a wind turbine benchmark," International Journal of Applied Mathematics and Computer Science, vol. 28, no. 2, pp. 247-268, 2018.

D: Power System Control, Liberalisation of Electricity Market

Various Approaches for a Design of the Hydraulic Turbine Model

Daniela Perduková, Marek Fedor
Dept. of Electrical Engineering and Mechatronics
Faculty of Electrical Eng. and Informatics TU Kosice
042 00 Kosice, Slovakia
daniela.perdukova@tuke.sk,
marek.fedor.4@student.tuke.sk

Pavol Fedor
Dept. of Electrical Engineering and Mechatronics
Faculty of Electrical Eng. and Informatics TU Kosice
042 00 Kosice, Slovakia
pavol.fedor@tuke.sk

Abstract—**The article describes a brief overview of the current knowledge and ways in the field of a design of the hydraulic turbine model as the most important subsystem of the hydropower plant. The hydraulic turbine is considered as a heart of the hydropower plant because of its important function in the whole process of electrical energy production. Turbine´s efficiency calculation is analytically demanding and dependent on parameters that are often obtained only by a theoretical estimation. Therefore, using its accurate model, which takes into account its efficiency, is necessary to achieve realistic and relevant results. In this article, a fuzzy system was used to create turbine´s efficiency model based on a measured operating data of the turbine flow rate and its mechanical speed without any need of knowing exactly the turbine parameters. It is shown, that this fuzzy model has practically reached the same accuracy as the analytical models. Such a model can be used in the design and development process of an optimal energy control of hydraulic turbine, as well as in the effective determination process of an actual turbine's wear rate. The achieved results were obtained using the computational tools of the Fuzzy Toolbox in the MATLAB program.**

Keywords: hydraulic turbine, small hydropower plant, fuzzy system, efficiency

I. Introduction

Hydro-electricity plays an important role in the safe, stable and efficient operation of the electric power system. In recent years design of small hydro power plants has been examined by various groups throughout the world due to its merits of offering better performance than the conventional fossil fuels to meet the energy need.

Hydroelectric power generation complies with ecological criteria aimed at environment protection. There is a lot of renewable energy accumulated in the surrounding water resources (lakes, rivers, artificial water reservoirs, etc.). This hydropower can first be converted into rotational mechanical power using hydraulic turbines, then, using synchronous and asynchronous power generators, it can be converted into electrical energy, and finally this whole energy conversion process can be controlled taking into account the requirements of the electrical energy end user.

In the energy conversion process, an essential role is played by the efficiency of the hydraulic turbine, both in terms of the overall efficiency of the conversion, and in terms of wear and possible maintenance or replacement.

The article presents an overview of the most common simulation model schemes of the hydraulic turbine model as the most important subsystem of the hydropower plant. The hydraulic turbine is considered as a core of the hydropower plant whose efficiency calculation is analytically demanding and dependent on parameters that are often obtained only by theoretical estimation. In the article a fuzzy system was used to create the turbine´s efficiency model based on measured operating data on the turbine flow rate and the mechanical speed without the need to know the turbine parameters. Such a model is applicable in practice for the design of an energy-optimized hydro-turbine control as well as to the effective determination of the deterioration rate. The correctness of the results was verified by simulation measurements in the MATLAB program.

II. Simulation Model of a Small Hydropower Plant

In general, the simulation model of a small hydropower plant (SHP) consists of five main sub-systems, i.e. the Governor – representing a turbine control system; the Servodrive, that serves as an actuator to regulate the flow of the water throughout the hydraulic turbine via controlling the valves, or guide vanes according to the Governor output signal; the simulation model of a Hydraulic Turbine – representing the process of conversion of the energy; as well as a block for the Electric Generator and another one for the Grid. In Figure 1, there are all the previous-stated main sub-systems depicted, creating the basic simulation model of SHP.

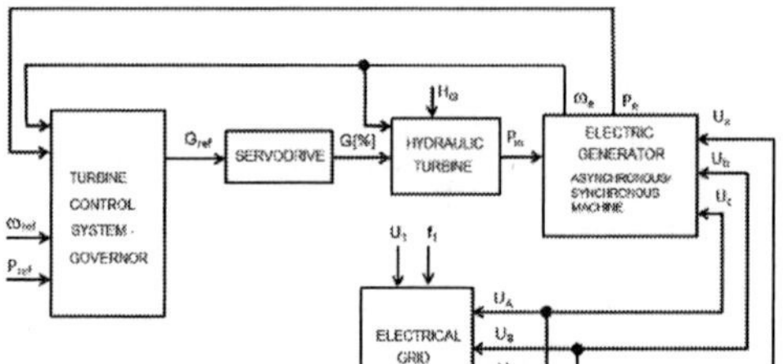

Figure 1 Block diagram of a small hydropower plant

The vast majority of the scientific papers and literature use a model, wherein all the parameters are being normalized, i.e. their value is between 0 and 1 or between 0 and 100% [1-3].

The phenomena of the water hammer, cavitation and traveling waves are deeply explored in literature [1-4], while including even the way how simulating and modelling their impact on the functionality of hydropower stations and how to reduce it. On the other hand, in case of the SHP, where the length of the penstock and the height of the water head available are small, the impact of these phenomenon can be neglected.

III. HYDRAULIC TURBINE MODEL

The hydraulic turbine is considered as a heart of the hydropower plant because of its important function in the whole process of electrical energy production. Therefore, using its accurate model is necessary to achieve realistic and relevant results. The turbine converts the energy of the falling water into a rotational mechanical energy and consequently to an electric energy through the use of the generators that are connected to the turbines. The model of a hydraulic turbine is a relatively complex non-linear dynamic system, some of the parameters of which are often obtained only by a theoretical estimation. However, correct and accurate modelling of the turbine is crucial in terms of accurate diagnosis and specification of the response of the entire hydropower plant to external influences, as well as in terms of correct tuning of its control.

Several approaches to hydraulic turbine modelling have been listed in the literature [1-3], differing in the manner they take into account the efficiency of the turbine or its impact on the final mechanical output of the turbine *P*m.

A. Model Taking into Account Turbine Efficiency Through Turbine Flow Value Q_{nl} at Zero Load

The development of a hydraulic turbine model is based mainly on theory and on the mathematical description given in expert publications [5] and [6]. The resulting block diagram of the hydraulic turbine structure obtained from program MATLAB is shown in Fig.2. The variable *u* here represents the input to the given block.

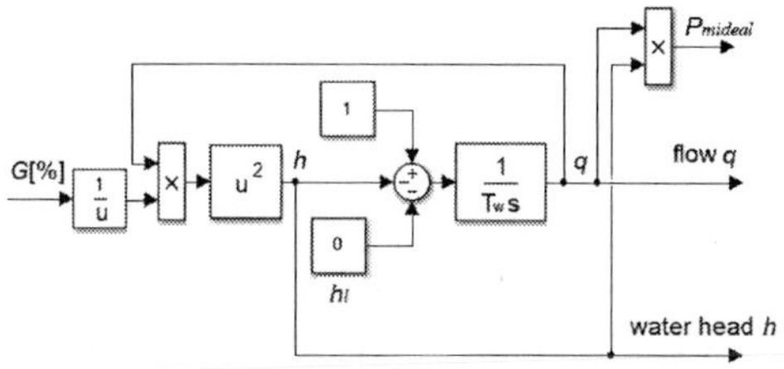

Figure 2 Hydraulic turbine model

Modeling the penstock, as a part of the depicted model's structure, was done assuming the incompressibility of the water and under the assumption that the penstock is strong enough with a length *L* and a cross-sectional area *A*. The loss of the water column caused by the penstock characteristic is so far neglected. Variable *G*[%] represents guide vane opening position.

Equation 1 represents the basic formula used in the hydraulic turbine's simulation model development.

$$\frac{dq}{dt} = (h_0 - h - h_l) g \frac{A}{L} \tag{1}$$

The parameters used in the Equation 1are listed in Table 1.

Table 1. List of parameters in Equation1

Symbol	Quantity	Unit
q	Turbine's flow	[m³/s]
A	Cross-sectional area of the penstock	[m²]
L	Length of the penstock	[m]
g	Gravitational acceleration	[m/s²]
h_0	Static water head (water column)	[m]
h	Water head (at the turbine's inlet area)	[m]
h_l	Loss of the water head caused by the friction in the penstock	[m]

Using the per-unit values of the parameters in the Equation (1), the Equation (2) is:

$$\frac{dq}{dt} = \frac{(1-h-h_l)}{T_w} \tag{2}$$

where the parameter of h_l, representing the loss of the water head, is being neglected. Plus, the coefficient of T_W used in the equation (2) is a time constant. It is usually called in the literature as a water time constant, or a water starting time. To determine the value of the T_W, the equation (3) is used [2]

$$T_w = \left(\frac{L}{A}\right) \frac{q_{base}}{g h_{base}} \tag{3}$$

where: h_{base} is the total static head available and q_{base} is the water flow in case of guide vane opening position.

By introducing values h_{base} and q_{base} we are able to express, by means of the turbine characteristic, the value of water flow at G=1 as

$$q = G\sqrt{h}\text{, and thus } q_{base} = 1\sqrt{h_{base}} \tag{4}$$

From Eq.4 follows that the turbine's water flow q is a function of guide vane position G and water head h. Thus ideally, the mechanical power of the hydraulic turbine would be characterized as a turbine's water flow multiplied by the water head available (in case of 100% turbine's efficiency)

$$P_{mideal} = qh \tag{5}$$

B. Hydraulic Turbine Model Taking into Account Penstock Losses

In real operation, a turbine does not operate at 100% efficiency and hence the requirement to consider this fact in the process of hydraulic turbine modelling. There are several ways of incorporating the impact of turbine efficiency into a simulation model [3, 7-9].

One of the simplest ways is to use the so-called zero-load turbine flow. This flow, referred to as q_{nl}, is a constant that characterizes the value of the turbine's constant power losses, and in the formula this constant is subtracted from the actual turbine flow q. It is then this final flow value that is multiplied by the level h as is given in equation for mechanical power P_m calculation:

$$P_m = A_t h(q - q_{nl}) - DG\Delta\omega_m \tag{6}$$

Another factor present during the operation of the hydro power plant is the turbine damping effect, which is a function of the differences in speeds, servo position and water supply valve position. The $\Delta\omega_m$ variable represents the change in mechanical shaft speed, D is the penstock diameter (also as the turbine damping coefficient), and G is the above-mentioned Guide Vane Opening Position [%].

The A_t parameter is considered to be the hydro turbine's gain, which calculation differs in various scientific publications, e.g. according to [3] it can be calculated as follows:

$$A_t = \frac{\text{Turbine Output [MW]}}{(\text{Generator Output [MVA]} h_r (q_r - q_{nl}))} \tag{7}$$

where the quantities with the subscript r can be considered as rated. This means that q_r is the turbine flow at rated load, and h_r is the level required for achieving the mentioned rated flow of q_r. These parameters are usually part of the technical documentation of the particular systems in a hydro power plant. The model for computing hydraulic turbine losses is illustrated in Fig. 3.

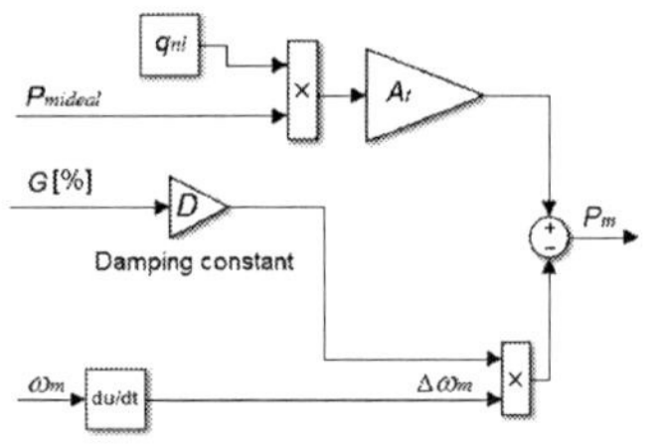

Figure 3 Model for computing hydraulic turbine losses

The block scheme of hydraulic turbine model taking into account losses is shown in Fig.4.

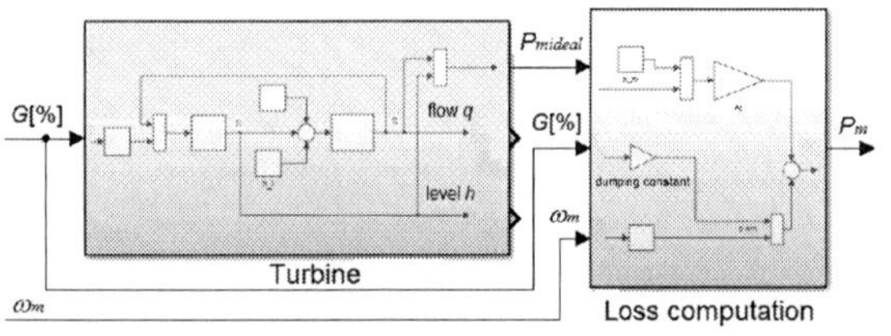

Figure 4 Hydraulic turbine model taking account losses

C. Model Taking into Account Turbine Efficiency - Direct efficiency calculation

Another way how to take into account the turbine's efficiency is to calculate its concrete value. This way of modeling a hydraulic turbine's sub-systems is based on the very basic formula for mechanical power calculation P_m:

$$P_m = \eta \rho g q h = \eta \rho g P_{mideal} \tag{8}$$

where η represents a hydraulic turbine's efficiency; ρ is a water density (usually used value of 1000kg/m^3); q represents an actual turbine's flow; g is a gravitational acceleration (9.81m/s); and h that is an actual water head available.

As already mentioned, there are several ways of expressing hydraulic turbine efficiency. In this case we present the computation quoted in [3]. This efficiency calculation method requires knowing exactly some of the turbine's characteristic parameters, e.i. radius of the turbine's blades R_{blades} [m] and the size of the blades area A_{blades} [m^2]. The actual values of its mechanical angular velocity ω_m, as well as the actual turbine's water flow q, are as well necessary for the correct calculation. Then turbine's efficiency can be calculated:

$$\eta(\lambda, q) = \left[\frac{1}{2}\left(\frac{90}{\lambda_i} + q + 0.78\right) \exp\left(\frac{-50}{\lambda_i}\right)\right] (3{,}33q) \tag{9}$$

where λ_i is

$$\lambda_i = \left[\frac{1}{(\lambda + 0{,}089)} - 0{,}0035\right]^{-1} \tag{10}$$

and λ is defined as

$$\lambda = \frac{R_{blades} A_{blades} \omega_m}{q} \tag{11}$$

In this case the input into the computation of hydraulic turbine efficiency would be the actual turbine flow q and also the mechanical angular velocity of the turbine ω_m. The mathematical description is represented in the block diagram in Fig. 5.

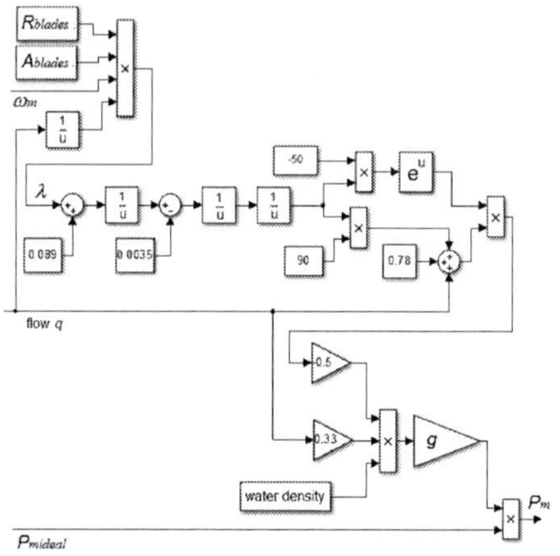

Figure 5 Hydraulic turbine efficiency computation model

The block scheme of hydraulic turbine model taking into account efficiency is shown in Fig.6.

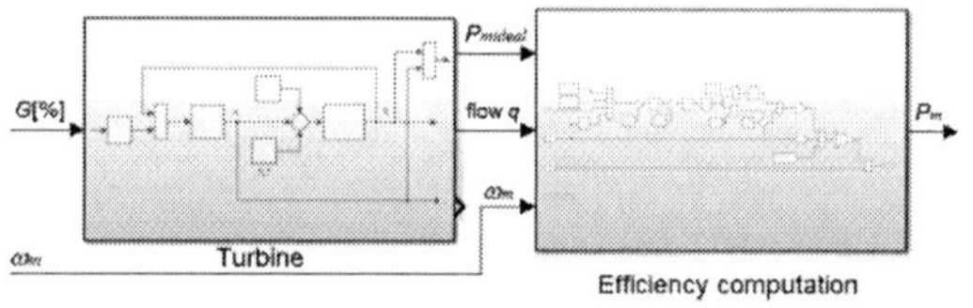

Figure 6 Hydraulic turbine model taking into account efficiency

IV. Hydraulic Turbine Efficiency Fuzzy Model

All the previous-mentioned ways of calculating the turbine's efficiency are based on quite complicated and not always accurate calculations, mainly because of the need of detailed parameters knowledge, or only estimated parameter's values. For example, according to equation (11), this efficiency depends on the parameters of the blades of a particular turbine, which can change significantly throughout operation. Furthermore, this dependence is strongly non-linear (see equations (9-11) and is dependent on several only approximately determined parameters. For example, the efficiencies at various turbine flows and mechanical angular velocities for the radius of the turbine's blades R_{blades} = 1.9 m and the blades area A_{blades} = 9 m^2, are illustrated in Fig. 7.

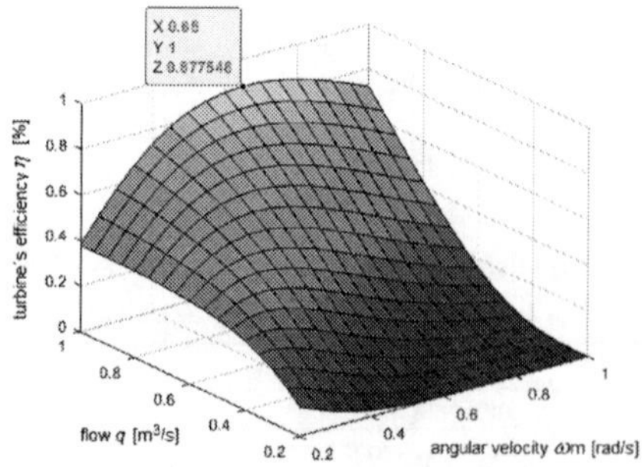

Figure 7 The efficiency of the hydraulic turbine (R_{blades}=1.9m and A_{blades}=9m^2) – Maximum turbine's efficiency is 87,75%

When the circumfluence area of the blades changes (e.g. as a result of wear) to $A_{blades} = 8\ m^2$, the turbine efficiency will change, as shown in Fig. 8. The maximum efficiency of the turbine in this case is 76,67%, while in the previous case it was approx. 87,75%.

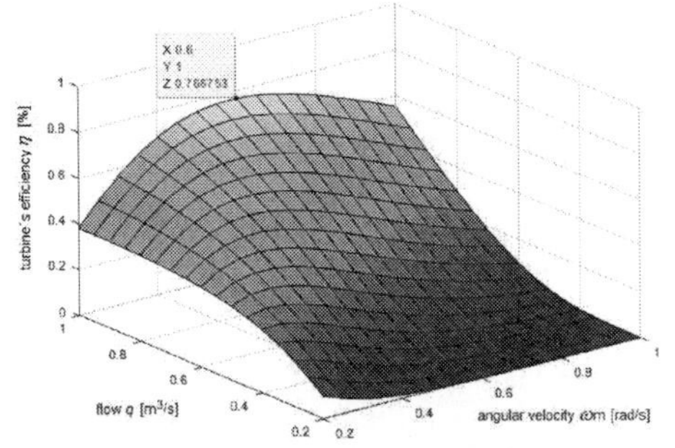

Figure 8 The efficiency of the hydraulic turbine (R_{blades}=1.9m and A_{blades}=8m²) – Maximum turbine's efficiency is 76,67%

In Fig. 7 and Fig.8 respectively, there is an obvious difference in the maximum possible efficiency of the turbine, caused only by the wearing of the turbine's blades, or its runner, and thus by the change in the blades inner area A_{blades}, which is one of the key values in the calculation defined by Eq. (11). This relation between the turbine's wear rate and its efficiency is strongly nonlinear and therefore, it is often difficult to implement this phenomenon into the simulation model.

Therefore, the efficiency of the hydraulic turbine is possible to reach via measuring the water flow q; turbine's mechanical speed ω_m and the turbine's efficiency at the steady state operation points for R_{blades}=1.9m and A_{blades}=9m², e.g. according to Table 2.

Table 2. Measured efficiency at different steady state operation points

Efficiency η [%]		Speed ω_m				
		0.2	0.4	0.6	0.8	1
Flow q	0.2	12.77	30.36	36.23	37.83	37.97
	0.4	2.74	26.40	49.60	64.41	73.24
	0.6	0.35	13.64	40.90	66.92	87.22
	0.8	0.04	5.68	27.20	56.26	84.31
	1	0.00	2.11	16.07	42.03	72.49

The data from Table 2 can be used for searching fuzzy inference structure (FIS) of the modelled efficiency describing the measured relation between $[\omega_m, q] \rightarrow \eta$.

The fuzzy model can be designed based on this database using standard cluster analysis tools with adaptive approaches resulting in improved quality of modelling and reduced development time. In terms of the adaptive approach, a hybrid arrangement based on a fuzzy inference tool in connection with a neural network was used [10-12]. From the large number of methods for adaptive fuzzy system development [12-13] we chose the adaptive neuro-fuzzy inference system ANFIS in MATLAB [14]. The basic characteristics of cluster analysis are the reduction of the fuzzy rules number and good initial rule parameters setting. We used subtractive clustering as a fast and robust data analysis method, with the following parameters: Range of influence = 0.5, Squash factor = 1.1, Accept ratio = 0.45, Reject ratio = 0.005.

As a result, we obtained a static Sugeno type fuzzy system with twelve rules, as shown in Fig. 9.

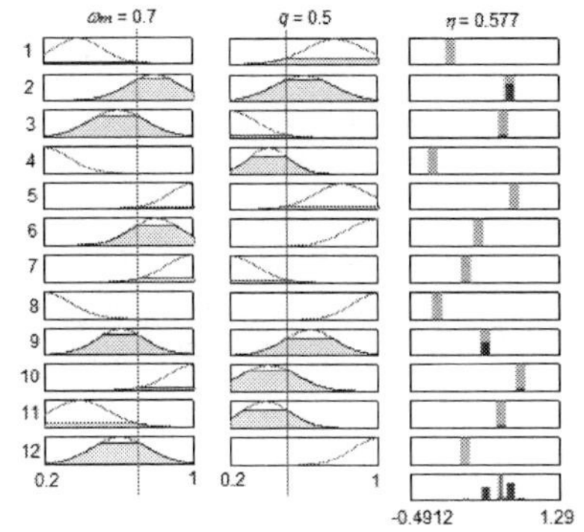

Figure 9 Efficiency fuzzy model – Sugeno type with 12 rules

The proposed fuzzy model replaces the numerical calculation of the turbine efficiency based on equations (9)-(11). These equations, as already mentioned, contain several parameters that we cannot determine quite accurately. In addition, the efficiency calculation in the simulation model remains constant when the parameters R_{blades} and A_{blades} change. This shortcoming is eliminated by the proposed fuzzy model, which changes the value of the turbine efficiency when the inputs, which significantly affect the size of the parameters R_{blades} and A_{blades}, to the proposed fuzzy system change (the water flow q and turbine's mechanical speed ω_m).

From Fig. 10 we can see that the efficiency based on the fuzzy model for ω_m = 0.65 and q = 1 is η=87,62 %. Based on the comparison of the efficiency obtained from the fuzzy model and the efficiency calculated analytically (see Fig. 7) we can conclude that the proposed fuzzy model achieves almost the same accuracy as the analytical model. This is also shown by a graphical comparison of the efficiencies obtained analytically and based on the fuzzy model shown in Fig. 11.

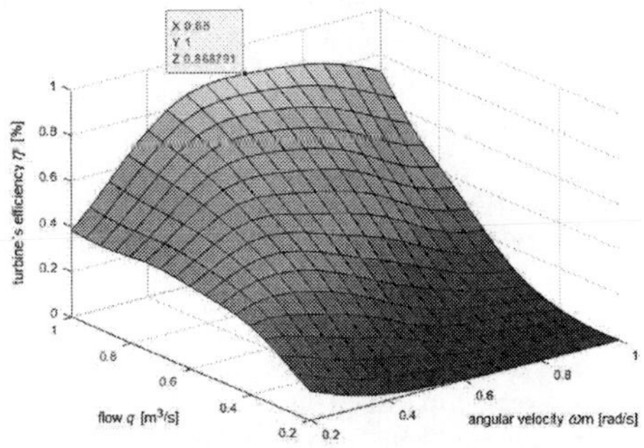

Figure 10 Fuzzy model of turbine efficiency based on measured data the water flow q and turbine's mechanical speed ω_m

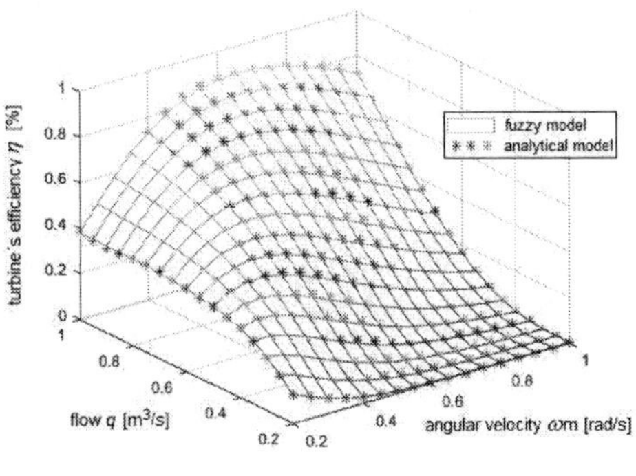

Figure 11 Comparison of analytical and fuzzy model of turbine efficiency

The block scheme of hydraulic turbine model with hydraulic turbine efficiency fuzzy model is shown in Fig.12.

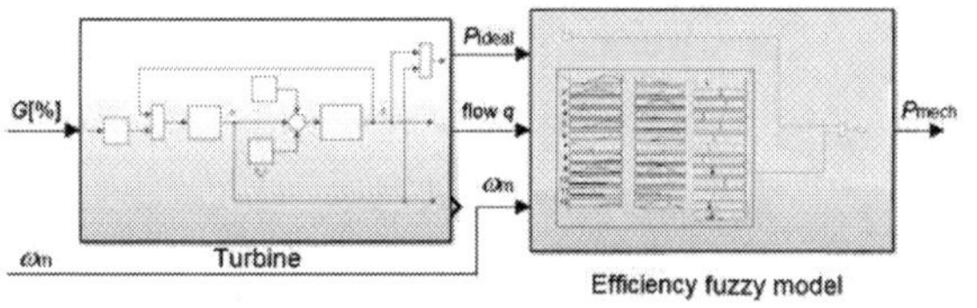

Figure 12 Hydraulic turbine model based on the efficiency fuzzy model

V. Conclusion

Computer models of energy systems make it possible to significantly simplify and speed up their whole design and development process, as well as to save financial costs in their implementation. The calculation of a hydraulic turbine's efficiency is analytically demanding and often depends on the parameters, which are determined only by a theoretical estimation.

The presented article describes a brief overview of the current knowledge and ways in the field of modelling hydropower turbine, as well as it describes hydraulic turbine model based on its efficiency fuzzy model. The mentioned fuzzy model can be obtained from a measured data of turbine's water flow, and its mechanical speed. It is shown that the model obtained in this way practically equals the analytical models obtained by analytical methods. The advantage is that the determination of the efficiency value based on the fuzzy model takes into account changes in important parameters caused by the operation of the turbine. This cannot be achieved by using an analytical model (eq. 9-11), where the values of the parameters remain constant throughout the simulation.

To compile the fuzzy model, it is only necessary to collect the database of the input data of the turbine only once. The design can be performed and solved on a computer. It follows up that the proposed methodology can be considered as automated fuzzy modelling technique for obtaining not only accurate, but also low complexity fuzzy model from system measurements. Thus the presented automated fuzzy approach presents a very suitable tool for application in the industrial practice.

With this method no principal limitations for the investigated system´s nonlinearities are defined, and therefore there is good reason to assume that the presented fuzzy model can be used in the design and development process of an optimal energy control of hydraulic turbines, as well as in the effective determination process of an actual turbine's wear rate.

Acknowledgment

The article was published with the support of the project No APVV-16-0206 and APVV-19-0210.

References

[1] A.M-H. German, P.M. Sa´ad, I.J. Dewi. "Modeling and Controling Hydropower Plants," in *Springer*, 2013, London. ISBN: 978-1-4471-2290-6, DOI: 10.1007/978-1-4471-2291-3.

[2] M. Sattouf. "Simulation Model of Hydro Power Plant Using Matlab/Simulink," in *International Journal of Engineering Research and Applications*, 4(1), pp. 295-301. 2014. ISSN: 2248-9622.

[3] A. Acakpovi, E.B. Hagan, F.X. Fiftatin. "Review of Hydropower Plant Models," in *International Journal of Computer Applications*, 108(18), pp. 33-38. 2014. DOI: 10.5120/19014-0541.

[4] P. Fedor, D. Perduková. "Power System Emulator Based on DCS," in *Innovations International Scientific Journal*, No.2, pp. 77-79. 2019.

[5] Working Group on Prime Mover and Energy Supply Models for System Dynamic Performance Studies. "Hydraulic Turbine and Turbine Control Models for System Dynamic Studies," in *IEEE Transactions on Power Systems*. 1(7), pp. 167-179. 1992.

[6] D.J. Littler. "Turbines, Generators and Associated Plant," in *Elsevier*, 3rd edition, 1991. ISBN: 9781483287485.

[7] I. Neborak, V. Sladecek, M. Kuchar. "Modelling and simulation of induction machine and frequency converter considering power losses," in *16th International Scientific Conference on Electric Power Engineering (EPE)*, 2015, pp. 251-255. DOI: 10.1109/EPE.2015.7161100.

[8] C.D. Tran, P. Palacky, M. Kuchar, P. Brandstetter, B.H. Dinh. "Current and Speed Sensor Fault Diagnosis Method Applied to Induction Motor Drive," in *IEEE Access*, 2021, vol. 9, pp. 38660-38672. DOI: 10.1109/ACCESS.2021.3064016.

[9] M. Kuchar, P. Palacky, P. Simonik, J. Strossa. "Self-Tuning Observer for Sensor Fault-Tolerant Control of Induction Motor Drive," in *Energies*, 2021, 14(9), article no. 2564. DOI: 10.3390/en14092564.

[10] D. Perduková, P. Fedor. "Design rules of a fuzzy controller using genetic algorithms," in: *Proc. of the 7th International Scientific Symposium on Electrical Power Engineering ELEKTROENERGETIKA 2013*, 2013, pp. 513-516. ISBN: 978-80-553-1441-9.

[11] M. Kuchar, P. Brandstetter, M. Kaduch. "Sensorless induction motor drive with neural network," in *IEEE 35th Annual Power Electronics Specialists Conference*, 2004, Vol. 5, pp. 3301-3305. DOI: 10.1109/PESC.2004.1355058.

[12] P. Girovsky, J. Timko, J. Zilkova, V. Fedák. "Neural Estimators for Shaft Sensorless FOC Control of Induction Motor," Proc. of 14th International Power Electronics and Motion Control Conference (EPE-PEMC), Ohrid, Macedonia, 2010. DOI: 10.1109/EPEPEMC.2010.5606907.

[13] R. D. Gaceanu, F. Poph. "A fuzzy incremental clustering approach to hybrid data discovery," in *Acta electrotechnica et informatica*, 12(2), pp. 16-23. 2012

[14] P. Fedor, D. Perduková. "Model Based Fuzzy Control Applied to a Real Nonlinear Mechanical System," in *Iranian Journal of Science and Technology, Transactions of Mechanical Engineering*, 40(2), pp. 113-124. 2016.

[15] J. S. R. Jang. "ANFIS: Adaptive-Network-Based Fuzzy Inference System," *in IEEE Transactions on Systems, Man, and Cyber*. 23(3), pp. 665-685. 1993.

Inverter-Based Resources Modelling and Simulation

Karel Máslo
Transmission System Analysis Department.
ČEPS, a.s.
Prague, Czech Republic
maslo@ceps.cz

Abstract—**The paper presents inverter-based resource models implemented in the MODES network simulator. These models are carried out as current and voltage sources by the Norton and Thevenin equivalents. Dynamic behavior of different model types is presented by simulations of power balance coverage on a simple network model.**

Keywords: dynamic model, inverter-based resources, renewable energy sources

I. Introduction

The number of sources connected to the network via power electronics is constantly growing. These sources include wind turbines, photovoltaic panels and battery systems are also referred to inverter-based resources (IBRs). Most of them are so-called grid following inverters, but the second type grid-forming inverters are being developed and becoming commercially available - see e.g. [1]. This article also mentions a need for advanced tools to study new phenomena and controls and requirements for high-fidelity models.

An example of such a tool is the MODES network simulator used in the Czech Republic and Slovakia for research and teaching purposes. Paper [2] reported a model of wind turbine with the permanent magnet generator and full power converter. This IBR was modelled using a Thevenin equivalent connected to the grid. The paper [3] dealt with a photovoltaic power plant model connected to the grid using the Norton equivalent (as a current source). The paper [4] again dealt with the wind farm connected to the grid also using the Norton equivalent to inject a pre-defined active and reactive powers into the grid. amount for power output. The paper [5] showed an improved photovoltaic power plant model with added a limited over frequency sensitive mode capability.

However, the development of renewable energy sources (RES) models continues. A new concept of the so-called virtual synchronous generator (VSC) has appeared in the literature [6] - [9]. The VSC mimics the dynamics of the synchronous generators in this concept. So, VSC can offer similar robustness as the conventional sources with synchronous generators. For example, there are high-levels control diagrams for VSM convertor controller in the paper [6].

It is not the purpose of this article to analyse in detail the capabilities of these VSC models. We would rather focus on the features of the existing IBR models and show the differences in their dynamics compared to conventional synchronous machines.

This paper presents different IBR dynamic models implemented in the MODES network simulator. These models are carried out as current and voltage sources by the Norton and Thevenin equivalents - see Figure 1. They can thus be used for modelling the so-called grid following and grid forming IBR in dynamic stability calculations.

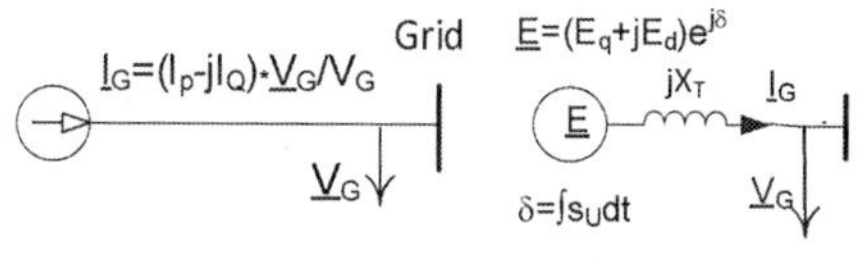

Figure 1. Model as a current source (left) and a voltage source (right)

The paper is organized as follows. A simple test system used for simulations is described in Section II. It is well known WSCC nine node test system modified by adding another IBR unit. IBR models are mentioned in Section III - mostly only by reference to already published articles, only the grid forming wind turbine generator model is described in more detail. Section IV shows the results of simulation calculations covering the power deficit in the test network. The final Section V summarizes important findings from simulation calculations.

II. Test System

To test the functionality of the newly created models, it is appropriate to use known system models whose dynamic behaviour is transparent and the results have already been published. One of the most common is the already mentioned IEEE system, also known as the WSCC system, which is widely used for transient stability studies. Therefore, this system was used as the default network model in this paper. The model was extended by a IBR connected to node 8 (like the reference [9]), as shown in the following figure.

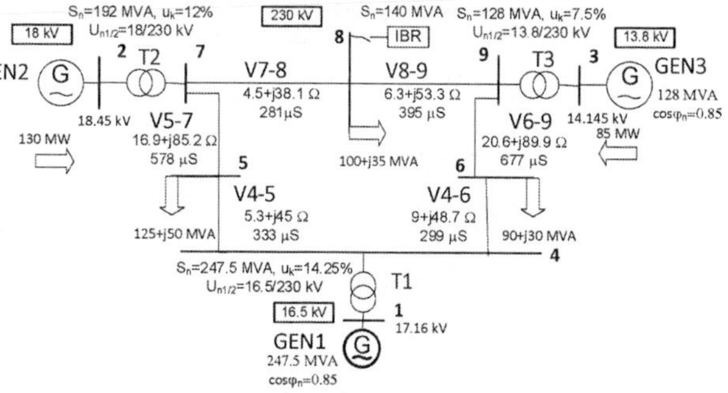

Figure 2. One line scheme of the modified WSCC system

The load was modelled without the frequency self-regulating effect, simply by a constant admittance.

The data of the generators were taken from the original book [10] and are listed in Table 1 after conversion to the nominal values of S_n.

Table 1. Parameters of generators

Gen	X_d (-)	X_q (-)	X_d' (-)	X_d'' (-)	T_{d0}' (s)	T_{d0}'' (s)	T_{q0}'' (s)	T_m (s)	X_q' (-)	T_{q0}' (s)
1	0.36	0.239	0.15	0.1	8.96	0.06	0.4	19.14	0.239	0.6
2	1.72	1.66	0.23	0.18	6	0.06	0.4	6.66	0.3	0.535
3	1.68	1.61	0.232	0.18	5.89	0.06	0.4	4.7	0.32	0.6

As in the reference [11], the generator model was supplemented by an excitation system IEEE type 1 model. This model combines voltage regulator with an exciter, and it corresponds to the model ExcAVR1 from IEC standard [12] Parameters from this standard were used for calculations are in Table 2:

Table 2. Typical parameters for ExcAVR1 model (without saturation)

K_A (-)	T_A (s)	T_B (s)	T_E (s)	T_F (s)	K_F (-)	V_{Rmin} (-)	V_{Rmax} (-)
500	0	0.2	1	1	0.12	-6	7

The IEC turbine-governor model GovHydro1 for GEN1 and GovSteam1 was used for GEN2 and GEN3. GovHydro1 model for the hydro turbine is depicted in Figure 3:

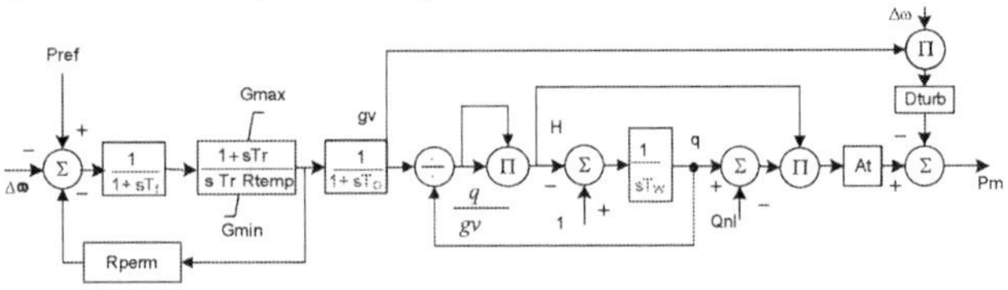

Figure 3. Block diagram of GovHydro1 model

Parameters for the GovHydro1 model are in Table 3:

Table 3. Typical parameters for GovHydro1 model

T_f (s)	T_r (s)	T_G (s)	R_{perm} (-)	R_{temp} (-)	G_{min} (-)	G_{max} (-)	T_W (s)	Q_{nl} (-)	A_t (-)	D_{turb} (-)
0.05	5	0.5	0.04	0.3	0	0	1	0.08	1.2	0.5

GovSteam1 model was described in detail in [4], so only their parameters according to the IEC standard are collected in Table 4:

Table 4. Typical parameters for GovSteam1 model

d_{b1} (-)	T_1 (s)	T_2 (s)	K (-)	T_3 (s)	uc (1/s)	uo (1/s)	P_{min} (-)	P_{max} (-)
0.0	0	0	20	0.1	-10	1	0	1

T_4 (s)	T_5 (s)	T_6 (s)	K_1 (-)	K_3 (-)	K_5 (-)
0.3	5	0.5	0.2	0.3	0.5

III. IBR Models

IBR use wind and solar energy and also power from energy storage devices. The latter type is used in some countries (which suppress conventional sources) for primary frequency control (now renamed 'frequency containment process' control thanks to EU regulation [14]).

In this article, we focus on the first two types - photovoltaics and wind turbines, although models of battery systems already exist (see e.g. [15] or [16]). A photovoltaics dynamic model was published in [5] and it is depicted in Figure 4:

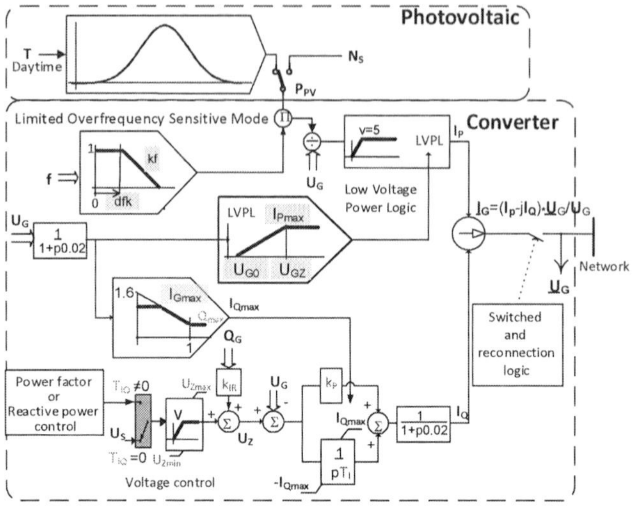

Figure 4. Simplified block diagram of photovoltaics model from MODES

The model enables so called limited over frequency sensitivity mode – to decrease power output when the frequency falls below a limit. A wind turbine generator (WTG) Type 4 (synchronous generator with full power converter interface - see [17] for more information) was published in [4]. This model enables not only a limited over frequency sensitivity mode, but as well limited under frequency sensitivity mode (implemented in the wind turbine model) and a fast primary frequency and inertial controls (implemented in the converter model). Both converter models (for photovoltaics and wind turbines) use the Norton equivalent - they operate in grid following mode and inject power into the grid.

A grid forming WTG Type 4 model was already published in [2] and it uses the Thevenin equivalent. This model was tested in separated operation with isolated load only. A wind farm consisting of 21 WTGs was even able to start an asynchronous drive in a thermal power plant's self-consumption (see for [2] more details).

The transmission of the source power P_S (WTG, photovoltaics or even the energy storage) for the power transmitted to the AC grid P_G is ensured by the inverter, the scheme of which is shown in Figure 5.

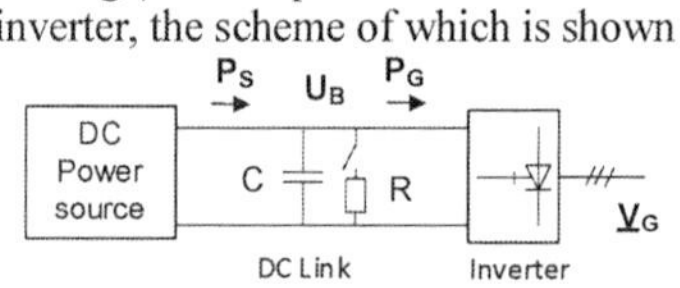

Figure 5. Basic scheme of inverter

DC link comprises a capacitor C and a discharge sunt resistor R. The capacitor allows some power accumulation within the allowable U_B voltage changes. If the voltage exceeds the upper limit, the shunt is switched and the excess power (given by the difference between P_S and P_G) changes to heat until the voltage drops below the permissible value. The DC link voltage U_B can be calculated according to the relation:

$$U_B(t+\Delta t) = \sqrt{U_B(t)^2 + \frac{2}{C}(\mathbf{P_S} - \mathbf{P_G})\Delta t} \qquad (1)$$

The equation is adapted for the integration method, where the new voltage value at time t + Δt is calculated from the old voltage value and the mean difference between the power supplied P_S and the power transmitted to the P_G network during the integration step Δt. This equation plays a similar role as the swing equation for synchronous generators – to maintain power balance between source and load.

Simplified block diagram of the grid forming WTG Type 4 model is in Figure 6.

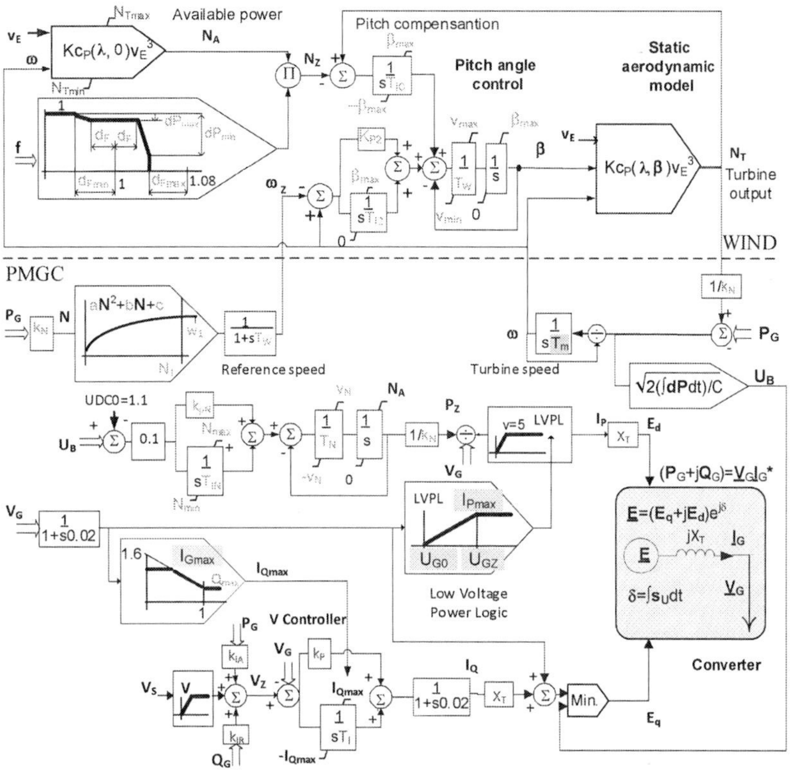

Figure 6. Simplified block diagram of grid forming WTG model from MODES

The model consists of two parts: the permanent magnet generator with converter (PMGC) and the detailed wind turbine model (WIND) with the pitch angle control a dependency on wind velocity v_E. The block diagram of the WIND model is simplified to speed control mode.

Following tables show parameters for these models.

Table 5. Parameters for PMGC model

T_m (s)	I_{Gmax} (-)	I_{Pmax} (-)	U_{g0} (-)	U_{gZ} (-)	v (%/s)	U_{Zmin} (-)	U_{Zmax} (-)	K_P (-)	T_I (s)	K_{IA} (-)	K_{IR} (-)
5	1.1	1.11	0.4	0.9	10	0.8	1.2	2	0.04	0	0

N_1 (-)	w_1 (-)	T_{IN} (s)	T_W (s)	T_N (s)	k_{pN} (-)	a (-)	b (-)	c (-)	v_N (-/s)
1	1.47	0.6	50	0.05	6	-0.75	1.59	0.63	0.45

Table 6. Parameters for WIND model

B_{max} (°)	K_{p2} (-)	T_{IC} (-)	T_{I2} (s)	T_W (s)	dP_{min} (%)	dF (%)	dF_{min} (%)	dF_{max} (%)	dP_{max} (%)
27	150	0.2	0	0.3	95	0	0.4	4	15

The model consists of two parts: the permanent magnet generator with converter (PMGC) and the detailed wind turbine model (WIND) with the pitch angle control a power dependency on wind velocity v_E. The block diagram of the WIND model is simplified to speed control mode.

The PMGC model consist of the loops for active and reactive currents I_P and I_Q. The active current loop controls DC link voltage U_D, which is calculated according to equation (1). This part of the model is unique and does not occur in any model according to the IEC or WECC standard [18].

The reactive current loop controls terminal voltage V_G to the reference value V_S. The complete model also allows to regulate to the reactive power Q_G. or to the specified power factor (this is not drawn in the diagram for simplicity).

The model enables to increase the reactive current limit I_{Qmax} during a voltage drop below the nominal value and vice versa, it reduces the limit for active current when the voltage drops below the value U_{GZ} (Low Voltage Power Logic).

The converter model simply transforms internal voltages E_q and E_d from one coordinate system to another with help of the angle δ, which is calculated integrating of the voltage frequency deviation s_U.

IV. Simulations and Discussion

67 MW step change of the load in the node 5 (same change as in reference [11]) was simulated on the modified WSCC system for these cases:

1. without IBR,
2. with photovoltaics IBR model from [5],
3. with grid following WTG model from [4],
4. with grid forming WTG model from .

The following figures show results of these simulations. Generator active powers (full lines) and turbine outputs (dashed) are plotted for each unit.

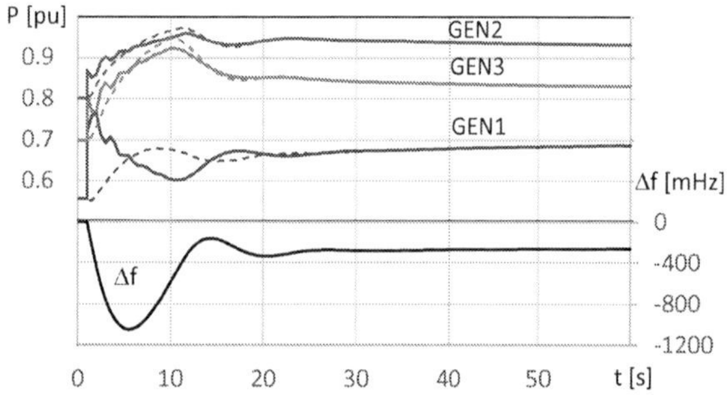

Figure 7. Powers and frequency deviation for system without IBR

Power and frequency deviation waveforms are very illustrative to explain the process after the power imbalance appears. When a power imbalance occurs, it is instantly (when neglecting very fast electromagnetic phenomena) by covered by generator active powers as can be seen from the step change of the power, the magnitude of which depends on the electrical distance of the generator from the fault location. The imbalance is then compensated by the kinetic energy of the inertial masses of the rotating machines connected to the network. this leads to a decrease in the speed and thus the frequency of the network, to which the turbine speed governor responds, which is historically known as primary frequency control. While the first two events are spontaneous responses of an AC power system - secured by physical laws (Kirchhoff's first law and energy conservation law), the third action is already artificial - provided by a speed governor (historically developed from the Watt's flyball governor).

The IBR unit is in operation in the second case. Photovoltaics works at half power due to the partly cloudy sky around noon. This allows the steam unit GEN3 to be shut down.

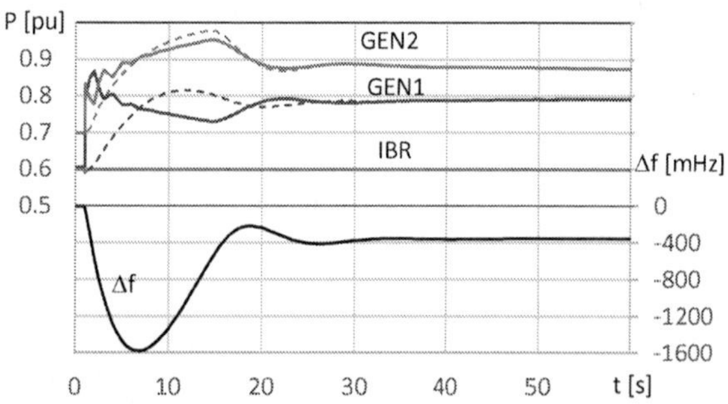

Figure 8. Powers and frequency deviation - system with photovoltaic

IBR power is constant - IBR does not react to power imbalance or frequency deviation. Due to this indolence, the frequency drop is much larger than in the previous case and in the real system would cause under frequency load shedding (now called under EU regulation [19] low frequency demand disconnection). The larger frequency drop was caused by the reduced kinetic energy from 3.3 to 3 GJ and the regulation energy (also called the network power frequency characteristic of primary control in [20]) from 252 MW/Hz to 197 MW/Hz compared to the previous case.

In the next case, we examine the dynamics of the grid following WTG model with the parameters published in the reference [21] or [22]. The exception is the parameter dPmax, which determines the size of the reserve for primary frequency control, i.e. the turbine power achievable at a given wind speed and the power at zero frequency deviation. This reserve is maintained by the pitch angle control. This parameter has been increased from 5 to 10%.

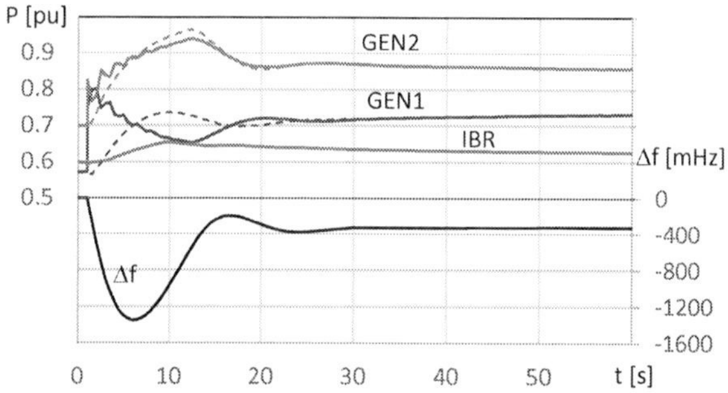

Figure 9. Powers and frequency for system with grid following WTG

By maintaining a power reserve (and thus supplying less renewable energy to the grid) and having the frequency containment function activated, the turbine responds to the frequency deviation and contributes to its reduction. However, the unit still does not contribute to the inertia of the network or take over the initial power imbalance. However, the IBR unit still does not contribute to the inertia of the system or take over the initial power imbalance.

However, the situation will change if the IBR has the grid forming capability, which will be reflected in the model by replacing the Norton equivalent with Thevenin (see Figure 6).

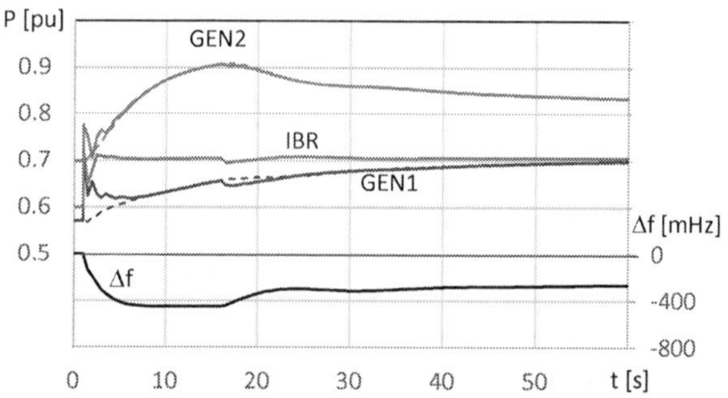

Figure 10. Powers and frequency for system with grid forming WTG

The IBR is now able to take over much of the load change immediately, with much of the energy being covered by the charged capacitor in the inverter's DC link. Therefore, not so much kinetic energy from rotating machines is needed and the grid frequency does not decrease so much. It even declines less than in the first case of three synchronous machines. Moreover, the IBR will be able to switch to island operation and cover the load in its microgrid if the power balance is approximately balanced.

We will be convinced of this by another simulation, when at time t = 1s we switch off both lines from node 8 and the IBR starts to supply only its local load, which we reduced from 100 to 95 MW before outages (5 MW was taken over by node 6).

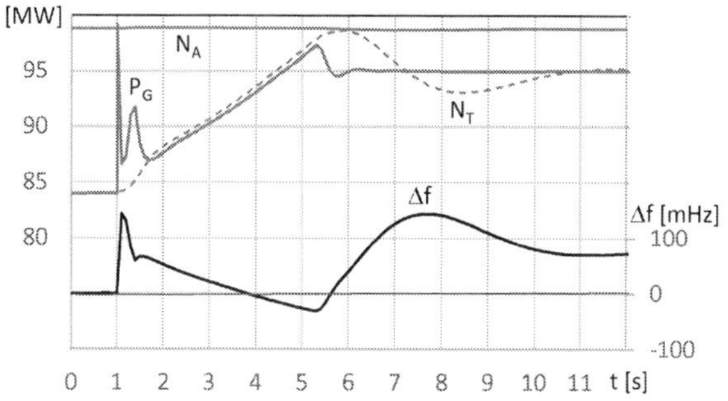

Figure 11. Island operation with grid forming WTG

The regulation process after the transition to the island consists of three parts. In the first part, the power imbalance of 11 MW is covered by the energy accumulated in the capacitor. Then the speed controller starts to increase the wind turbine output power N_T by means of the pitch angle control. After the fifth second, the turbine power restores the voltage on the capacitor. The recharging process is completed in the eleventh second.

The success of the transition to the island is conditioned by four factors:

1. the turbine power reserve (the initial turbine power N_T must be smaller the available power N_A),
2. the energy supply in the capacitor (acts as an energy storage),
3. the regulation of the turbine power to cover the load power and
4. to maintain the voltage on the capacitor and finally the ability to regulate network voltage.

Of course, a wind turbine must have a sufficient source of primary energy - that is, wind. In the event of its variability, it is appropriate to supplement the microgrid with backup sources and/or energy storage devices.

V. Conclusions

The article shows the dynamic behavior of different types of resources when covering power balance. The firs case analyzes the process of power balance covering in a conventional system with only synchronous machines. This process is desribed in detail, including the various stages of deficit coverage involving electromagnetic energy accumulated in the circuits of synchronous machines, kinetic energy accumulated in the rotating masses of rotating machines, accumulated energy in primary drives and finally energy in fuel or water. Importantly, the first two phases using the accumulated electromagnetic and kinetic energy take place completely automatically and spontaneously only on the basis of the laws of physics.

Other cases analyze the dynamic behavior of inverter-based resources (IBRs) in different configurations - grid following and grid forming. It turns out that theoretically the grid forming IBRs can have better dynamic properties than conventional synchronous machines. However, their management and control must be more complex than with existing IBRs. This is the only way to artificially secure the capabilities that take place spontaneously on conventional synchronous machines.

If IBRs dominated networks emerge, this development will be necessary. In that case, we are likely to encounter stability issues that we do not yet know, but that will be a different story.

References

[1] J. Matevosyan, J. MacDowell, N. Miller, B. Badrzadeh, D. Ramasubramanian, A. Isaacs, R. Quint, E. Quitmann, R. Pfeiffer, H. Urdal, T. Prevost, V. Vittal, D. Woodford, S. H. Huang, J. O'Sullivan. "A Future With Inverter-Based Resources," in *IEEE Power and Energy Magazine*, 19(6), 2021, pp. 18-28. DOI: 10.1109/MPE.2021.3104075.

[2] K. Máslo, M. Pistora. "Modelování separátního provozu větrné turbíny," in *Proc. 12th International Conference Electric Power Engineering*. 2011. ISBN: 978-1-62993-917-9.

[3] M Pistora., K. Máslo. "Impact of embedded renewables on transmission and distribution network," in *Proc 10th International Scientific Conference Control of Power Systems*, 2012, pp. 1-11.

[4] K. Máslo. "Frequency stability in the context of the European network codes," in *Proc 9th International Scientific Symposium on Electric Power Engineering*, 2017, pp. 436-442.

[5] K. Máslo, "Dynamic Models in the Context of the EU Commission Regulations," in *Proc. 10th International Scientific Symposium on Electric Power Engineering*. 2019, pp. 26-30.

[6] A. J. Roscoe, M. Yu, A. Dysko, C. Booth, R. Ierna, J. Zhu, H. Urdal. (2016). "A VSM (Virtual Synchronous Machine) convertor control model suitable for RMS studies for resolving system operator / owner challenges," in *Proc. 15th Wind Integration Workshop*. [On-line]. pp. 1-8. Available: http://strathprints.strath.ac.uk/58053 [May 11., 2022].

[7] H. A. Alsiraji, J. M. Guerrero. "A new hybrid virtual synchronous machine control structure combined with voltage source converters in islanded AC microgrids," in *Electric Power Systems Research*. No. 193. 2021.

[8] V. Mallemaci, F. Mandrile, S. Rubino, A. Mazza, E. Carpaneto, R. Bojoi. "A comprehensive comparison of Virtual Synchronous Generators with focus on virtual inertia and frequency regulation," in *Electric Power Systems Research*. No. 201 .2021.

[9] B. Muftau, M. Fazeli. "The Role of Virtual Synchronous Machines in Future Power Systems: A Review and Future Trends," in *Electric Power Systems Research*. No. 206. 2022.

[10] P.M. Anderson, A.A. Fouad. "Power system control and stability," in *IEEE Press Power systems engineering*. 1993, pp. 464. ISBN: 978-0780310292.

[11] A. Delavari, I. Kamwa, P. Brunelle. "Simscape Power Systems Benchmarks for Education and Research in Power Grid Dynamics and Control," in *Proc. 2018 IEEE Canadian Conference on Electrical & Computer Engineering (CCECE)*, 2018, pp. 1-5, DOI: 10.1109/CCECE.2018.8447645.

[12] International Standard. IEC 61970-302 Common Information Model (CIM) dynamics. June 2018.

[13] K. Máslo, A. Kasembe, M. Kolcun. "Simplification and unification of IEEE standard models for excitation systems," in *Electric Power Systems Research*, vol. 140, 2016, pp. 132–138.

[14] European commission. "COMMISSION REGULATION (EU) 2017/1485 establishing a guideline on electricity transmission system operation," in *Official Journal of the Eorupean Union*, pp. 10. 2017.

[15] K. Máslo, A. Kasembe. "Extended long term dynamic simulation of power system," in *Proc. 52nd International Universities' Power Engineering Conference (UPEC)*, 2017, pp. 1-5. ISBN: 978-1-5386-2344-2, DOI: 10.1109/UPEC.2017.8232006.

[16] K. Máslo, A. Kasembe. "Long Term Dynamic of Power System with renewable energy sources and energy storage systems," in *Proc. International Conference on Renewable Energies and Power Quality (ICREPQ)*, vol. 17, 2019, pp. 1-4. ISSN 2172-038 X, DOI: 10.24084/repqj17.253.

[17] A. Ellis, Y. Kazachkov, E. Muljadi, P. Pourbeik, J. J. Sanchez-Gasca. "Description and technical specifications for generic WTG models — A status report," in *Proc. IEEE/PES Power Systems Conference and Exposition*, 2011, pp. 1-8, ISBN: 978-1-61284-788-7.

[18] P. Pourbeik, "WECC Second Generation Wind Turbine Models," in *Final Report*, Jan. 2014.

[19] European commission. "COMMISSION REGULATION (EU) 2017/2196 of 24 November 2017 establishing a network code on electricity emergency and restoration ," in *Official Journal of the Eorupean Union*, page 64. 2017.

[20] ENTSO-E. "UCTE OH – Policy 1: Load-Frequency Control," March 2009.

[21] R. Jankowski, M. Wilk, M. Haller, P. Centeno, D. Petrescu, P. Ziółek, M. Małecki, J. Setreus, K. Elkington, K. Máslo, J. Warichet, C. Strotmann, R. Pestana, N. Machado, E. Carlini, S. Moroni, T. K. Vrana, N. Grisey. "e-HIGHWAY 2050 Modular Development Plan of the Pan-European Transmission System 2050," [Online]. pp. 118, Available: https://core.ac.uk/reader/154669779?msclkid=a313b1f4af2b11ecb2abba255718a30a.

[22] K. Maslo, A. Kasembe, S. Moroni, R. Pestana. "Frequency stability modelling of the future continental Europe power system," in *Proc. 52nd International Universities' Power Engineering Conference* , 2017, pp. 1-4, DOI: 10.1109/UPEC.2017.8232003.

Possible Role of Biomass in Energy Transformation – Economic Competitiveness and Policy Implications on Case Example of the Czech Republic

Jaroslav Knápek, Martin Horák, Tomáš Králík
Dept. of Economics, Management and Humanities
CTU in Prague, Fac. of Electrical Engineering
Prague, Czech Republic
knapek@fel.cvut.cz, horakm11@fel.cvut.cz,
tomas.kralik@fel.cvut.cz

Kamila Vávrová, David Outrata
Department of Phytoenergy and Biodiversity
VÚKOZ, Publ. Res. Inst.
Průhonice, Czech Republic
vavrova@vukoz.cz, outrata@vukoz.cz

Abstract—**The paper focuses on the determination of the potential of biomass from energy crops purposefully grown on agricultural land in relation to the biomass production price and on the development of a cost curve for purposefully grown biomass. To model the potential of energy crops, an approach based on deriving crop yields from soil and climatic parameters on individual plots of the analyzed area is used. For the construction of the cost curve of purpose-grown biomass, a model based on the derivation of the biomass production price based on the opportunity cost principle - the forgone economic effect from growing conventional crops - is developed. The methodology presents 5 basic approaches to the allocation of agricultural land for energy crops in terms of priority of food production, maximization of biomass production, minimization of total production costs, even geographical distribution of energy crops over the analyzed area and at the same time in terms of maximization of non-productive effects provided by perennial energy crops. The developed methodology is applied to the example of the Czech Republic with a maximum allocation of 15% of agricultural land for energy crops. The results of the analyses show that assumptions about the land allocation method significantly influence both the shape of the biomass cost curve and the biomass potential for a given area of land for energy crops. The biomass potential for the different land allocation assumptions for the conditions of the Czech Republic ranges from 37.5 to 76 PJ.**

Keywords: biomass potential, biomass potential cost curve, GIS modelling, perennial energy crop

I. Introduction

Biomass plays an important role in the overall renewable energy portfolio. Within the EU, biomass contributes to about 60% of all renewable energy sources [1], with solid biomass (for combustion) playing a dominant role (75%) [2]. In the Czech Republic, the total contribution of biomass to renewable energy sources (RES) is even higher, reaching (including liquid biofuels) about 86%, while solid biomass accounts for about 76% of the total biomass contribution. Current strategic documents foresee a further development of biomass use, e.g. the National Energy and Climate Plan foresees an increase in bioenergy use between 2016 and 2030 by about 30% or about 47 PJ in absolute terms. This represents more than 60% of the total expected increase in renewable energy by 2030. Approximately 70% of this increase, or about 33 PJ in absolute terms, is expected to come from increased use of combustible forms of biomass for heat and power generation. The main increase is to be provided by dendro-mass, whose current consumption is expected to increase by more than 20% (+ approx. 23 PJ) [3].

Solid biomass is already significantly used in the Czech Republic for electricity and heat production. The combustion of solid biomass produces about 2.5 TWh (2020), which is about 24% of the total gross electricity production from RES. The largest share is contributed by the combustion of wood chips (e.g. from forest residues) and other waste biomass - about 1.25 TWh. Solid biofuels (briquettes and pellets) contribute about 0.236 TWh [4]. At the same time, solid biomass contributes about 17 PJ to the useful heat supply from large district heating systems systems (about 17%), with a dominant share of coal (61%) [5]. The use of solid biomass as a substitute for coal in district heating systems is expected to increase significantly in the horizon until 2030. According to the original Ministry of Trade and Industry assumptions from autumn 2020, (solid) biomass was expected to reach up to 23% share in gross heat production in large heat sources in 2030 [6].

The potential of readily available and relatively cheap residual and waste biomass, especially from the wood processing industry, has already been largely exhausted. At the end of this decade, the availability of so-called forest residues will be significantly reduced due to the bark beetle calamity [7]. Thus, purpose-grown biomass on agricultural land will play an important role in the future. In order to objectively estimate the possible contribution to the total biomass potential from this source, it is necessary to respect the context in which this potential is determined so as not to increase it inadequately

The main objective of the paper is to highlight the importance of different assumptions about land allocation for determining the biomass potential of energy crops grown on agricultural land. At the same time, the paper presents the results of modelling on the example of the Czech Republic, where the biomass potential of perennial energy crops is determined for different land allocation assumptions and a biomass potential cost curve is constructed.

II. Methodology

A number of approaches to determining biomass potential are discussed in the literature. Some of them are based on the principle of so-called technical potential, where ecological, legislative, economic constraints are not considered and the potential is derived only from the rate of land allocation for energy crops. Other approaches are based on respecting different combinations of constraints resulting from the agro-technologies used, ecological or technological constraints. For example, the procedure for estimating biomass potential described in [17] is based on basic physical and biological factors at a given location. This then leads to the determination of the technical biomass potential, where no additional constraints (agrotechnical, ecological, economic, etc.) are considered for the cultivation or extraction of biomass. On the contrary, [18] emphasizes the long-term sustainability of biomass and the projection of this requirement in the determination of biomass potential (biodiversity protection, socio-ecological aspects, climate protection, etc.). Another aspect that significantly influences biomass potential is the priority of land use for conventional agriculture and food production - see e.g. [19]. Geographical information systems (GIS) are often used to model biomass potential, both at a global scale (see e.g. [20]) and at regional and local scales (e.g. [25]). The biomass potential is then derived by default from the soil and climate conditions on a given plot, respecting the resolution of the analysis (see e.g. [21]). At the same time, a number of authors, especially recently, emphasize the aspect of ecological benefits of perennial energy crops, with the understanding that this aspect should be considered when determining biomass potential - see e.g. [22], [23]. Other authors (e.g. [24]) also emphasise the competitiveness aspect of biomass when determining biomass potential. An overview of methods for determining biomass potential can be found, for example, in [8].

The biomass cost curve is an important factor in assessing the potential of biomass for energy use and in designing and evaluating bioenergy development strategies while respecting the economic context. For example, [26] presents the results of the S2Biom project mapping biomass resources including biomass cost curves at local, regional and European level.

[27] presents a comprehensive analysis of biomass potential from both purpose grown energy crops and residual biomass from conventional crops. It incorporates a number of constraining conditions into the modelling as well as the economic aspect of biomass competitiveness, which it then uses to generate biomass potential cost curves.

The methodology used here for modelling biomass potential is based on estimating yields of conventional and energy crops according to the conditions of the site (land), especially soil and climatic characteristics. The model used uses the system of soil ecological rating units (BPEJ), and the part of the BPEJ containing information on climatic and soil conditions on the land (HPKJ) is used for the modelling. For more information on the BPEJ soil classification system and the estimation of crop yields according to the BPEJ value, see e.g. [9].

From long-term crop yield observations, each combination of soil and climate characteristics is assigned an expected biomass yield for each conventional and energy crop. The decision to establish plantations of perennial energy crops for biomass production for energy purposes means that the farmer loses the economic effect of conventional crop production on the land. Assuming economically rational decision-making, he will only decide to produce energy biomass if this production gives him at least the same economic effect as the production of conventional crops over the lifetime of the perennial energy crop plantation (opportunity cost principle). The production price of biomass from energy crops is thus determined for a given parcel of land from the balance of the net economic effects from the production of conventional and energy crops over the lifetime of the plantation of energy crops. A cash flow simulation for both types of production is used for the calculation. Knowing the expected yields of both conventional and energy crops, the level of subsidies, the costs of growing both types of crops and the prices of conventional crops, the variable to be calculated is the biomass production price. The calculation has to consider agronomic constraints such as the rotation of conventional crops on a given plot. The principle of calculating the biomass production price is given in the following formula (1):

$$c_{alt}: \sum_{t=1}^{T_{j,lf}} \left(c_{alt,j,t} \times Q_{j,t} - SUB_{j,t} - EXP_{j,t}\right) \times \left(1 + r_{n,ec}\right)^{-t} = \sum_{t=1}^{T_{j,lf}} \left(c_{i,t} \times Y_{i,t} - SUB_{i,t} - EXP_{i,t}\right) \times \left(1 + r_{n,cp}\right)^{-t} \quad (1)$$

$$c_{alt,j,t} = c_{alt,j,1} \times (1 + i)^{t-1} \quad (2)$$

Where:

$c_{alt,j,t}$ ...the production price of biomass from the jth energy crop on a given plot in year t [EUR/GJ]

$Q_{j,t}$ ...energy yield of the j-th energy crop on the plot in year t [GJ/ha]

$SUB_{j,t}$...the subsidy for the j-th energy crop in year t [EUR/ha]

$EXP_{j,t}$...the cost of growing the j-th energy crop in year t [EUR/ha]

$r_{n,ec}$ ...nominal discount for perennial energy crops [-]

$T_{j,lf}$ ...plantation life of the jth energy crop [years]

$c_{i,t}$ ...market price of the i-th conventional crop [EUR/tonne]

$Y_{i,t}$ ...hectare yield of the i-th conventional crop [EUR/tonne]

$SUB_{i,t}$...subsidy for the i-th conventional crop in year t [EUR/ha]

$EXP_{i,t}$...cost of growing the i-th conventional crop in year t [EUR/ha]

$r_{n,cp}$ ...nominal discount for conventional crops [-]

i ...expected average long-term inflation

All calculations are in current year prices. Base year prices (commodity prices, costs, subsidies, etc.) are used for the calculations, which are then escalated by appropriate escalation factors. For the basic calculations, both price and cost escalations can be applied using expected long-term average inflation.

Growing perennial energy crops is different from conventional crops in terms of business risk. The cultivation of conventional crops is defacto a one-year decision-making cycle, which is associated with relatively low risk and low commitment of the invested funds. In contrast, the establishment of a multiannual plantation of energy crops represents a significant expenditure (in terms of total present value over the lifetime of the plantation) and is usually associated with higher risks, e.g. in terms of the risk of bad weather after the plantation is established, which can significantly affect the quality of the plantation. This higher risk can then be reflected in a higher discount value. Details on the methodology for estimating the production price of biomass from perennial stands of energy crops can be found in [10].

The methodology described above is based on determining the biomass production price of each energy crop for each plot described by a unique combination of soil and climate conditions. Different energy crops differ in their requirements - e.g. fast-growing tree plantations tolerate wetland conditions well, whereas miscanthus plantations, for example, are significantly more drought tolerant.

When constructing a cost curve for the biomass potential of energy crops, it is thus necessary to decide according to which criteria a given energy crop is allocated to a given plot. The following five basic approaches can be used here:

1. Biomass yield in tonnes per hectare or converted energy yield in GJ/ha without further restrictions on the distribution of energy crops across the country, on the production price, on food security and on environmental aspects.

2. Biomass production price derived according to formula (1) without restrictions on the distribution of energy crops across the territory of the country up to the limit of land allocation for energy crops

3. Biomass production price derived according to formula (1), limitation on the allocation of energy crops is implemented on a regional basis

4. Biomass production price (formula (1)), respecting the preference of land use for food production and the limitation on the allocation of energy crops per region

5. Criteria based on maximising the environmental benefit (the so-called non-productive function) of energy crops on a given plot of land.

While approach 1 maximises the energy yield from growing energy crops, it is based on the unrealistic assumption that the most fertile land will be used. At the same time, this approach does not consider the production cost of biomass, which is high due to the valuation procedure used. This is due to the high income from the sale of conventional crops, which has to be compensated by the high production price of biomass.

Approach 2 results in energy crops being heavily unevenly allocated across the country, leading to a situation where in some regions the share of energy crops is significantly higher than the nationally set land allocation value. Approach 2 thus again leads to an unrealistic result.

Approach 3 is closer to reality in that it respects the requirement for an even distribution of energy crops across the national territory. This is important not only in terms of ensuring local biomass production but also in terms of the environmental benefits associated with perennial energy crops.

Approach 4 combines the considerations of even distribution of energy crops across the national territory with a preference for food production. The most fertile land is prioritised for conventional agriculture. The food production aspect is considered through wheat yields, and energy crops are allocated to land with the lowest wheat yields.

Approach 5 is based on allocating energy crops to land where they can contribute most to long-term sustainable land use. For this purpose, a 'land risk index' is constructed for each plot, combining 5 basic ecological factors such as landscape connectivity and heterogeneity, threat of drought, threat of water and wind erosion. Energy crops are then allocated to the plots with the highest index value [11].

III. Case Study for the Czech Republic – Input Data

The actual calculations and modelling were carried out in a GIS environment. For modelling purposes, a primary database was created by intersecting the BPEJ and LPIS databases. LPIS is a register of agricultural land and its basic unit is a farm block (a continuous area of one owner with one crop). The intersection of these databases creates polygons that are characterised by their geographical location, BPEJ value and area. The primary database has about 1.5 million of these polygons. A more detailed description can be found e.g. in [8].

For modelling the production price of biomass from energy crops, UZEI data on the cost of growing individual conventional crops [12] at the 2019 price level are used. Prices of conventional crops were taken from [13]. Data on the costs of growing perennial energy crops were taken from [8] and [9].

The biomass calorific value was considered to be 6.92 GJ/t(FM) (SRC chips) and 13.6 GJ/t(FM) (Miscanthus), respectively. The lifetime of the SRC plantation was considered to be 24 years, while the lifetime of the Miscanthus plantation was considered to be 10 years.

The same nominal discount rate for both energy and conventional crops was used for the calculations, namely 10%, while a long-term average inflation of 2% was assumed. The assumption on the inflation rate is based on the long-term inflation target applied by the Czech National Bank [14]. Despite rapid inflation growth in the last quarter of 2021 and especially in the first half of 2022, the CNB expects a return to the long-term inflation target in Q3 of 2023 year. The value of the nominal discount is derived from the value of the discount (return on total capital employed) for RES-based electricity generation plants (including biomass) used to set electricity feed-in tariffs (6.3-7%) [15]. This value is increased by an estimate of the risk premium for growing biomass for energy purposes. The impact of changes in long-term average inflation and the level of the discount on biomass production prices and on the shape of the cost curve is discussed in Section 4 "Results and Discussion".

The "land risk index" values were calculated for two model regions - Central Bohemia region (STC) and South Moravia region (JMK) using the methodology described in [11].

The upper limit of possible allocation of (arable) land for energy crops was set at 15%. This means 369 thousand hectares of the total land area included in the analysis (2.460 million hectares).

IV. Results and Discussion

The result of modelling the biomass potential according to Approach 1 is shown in Fig. 1.

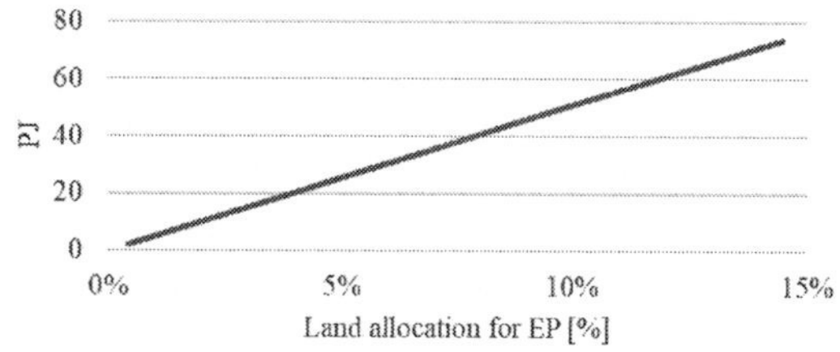

Figure 1. Dependence of solid biomass potential from energy crops by % of land allocation, priority of energy gain from the land

According to this approach, where only the area of land allocated to energy crops is limited, the biomass potential from energy crops is about 76 PJ. The average (weighted) biomass production price is then 4.55 EUR/GJ.

Approach 2 leads to a similar value of biomass potential (71 PJ), with a decrease in the average biomass production price to 4.25 €/GJ. The biomass potential cost curve for approach 2 is shown in Fig. 2.

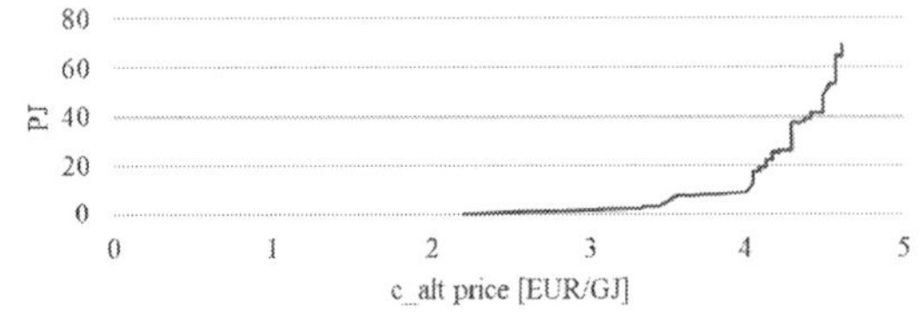

Figure 2. Cost curve of biomass potential - application of approach 2 (only limitation to 15% of the area for the whole Czech Republic)

The other approaches mentioned above lead to significantly lower values of biomass potential. If we consider the limit for energy crops in each region separately, then the biomass potential is significantly reduced to about 55 PJ. Along with this, the cost curve of biomass potential changes significantly - see Fig. 3. The average biomass production price then rises to 4.42 EUR/GJ.

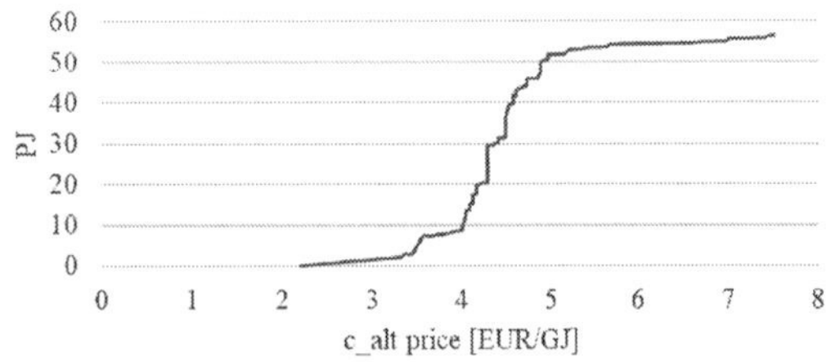

Figure 3. Cost curve of biomass potential - application of Approach 3 (limitation to 15% of the area for each region of the Czech Republic)

Approach 4, where the most productive land is prioritised for conventional agriculture and food production, leads to a further significant reduction of the biomass potential to 37.5 PJ. At the same time, there is a significant change in the cost curve and a large increase in the average biomass production price to 8.45 EUR/GJ - see Fig. 4.

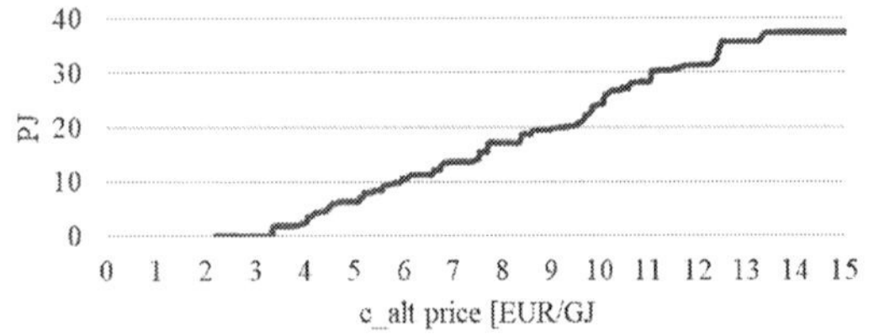

Figure 4. Cost curve of biomass potential - application of Approach 4 (limitation to 15% of the area for each region of the Czech Republic and preference for conventional production)

The impact of the fifth possible approach to land allocation of energy crops, where ecological benefits are preferred (land is allocated for energy crops according to the "land risk index" - LRI) is summarised in Tab. 1.

The first row in Tab. 1 for the given region always presents the results of the calculation of the potential and average production price when allocating land according to the Land risk index. The second row then presents the results for the allocation according to the production price up to the limit of 15% for the region, and the third row presents the results of the modelling for the preference of land use for conventional crops.

The results of the modelling of the biomass potential of perennial energy crops do not consider the fact that growing energy crops means that the relevant part of the agricultural land is not used for conventional production, thus eliminating the contribution of straw that is not used for livestock production (bedding, feed preparation). Under realistic conditions, the net contribution of energy crops to the biomass potential would be lower than the values reported here.

Table 1. Results of biomass potential modelling for the South Moravian Region (JMK) and the Central Bohemian Region (STC)

Region/alloocation approach	**Potential [PJ]**	**C_alt [EUR/GJ]**
STC, 5th approach	7,8	10,6
STC, 2nd approach	12,2	4,4
STC, 4th approach	7,5	9,3
JMK, 5th approach	5,1	10,6
JMK, 2rd approach	7,6	4,8
JML,4th approach	5,4	9,0

At the same time, the results of modelling the biomass potential from energy crop plantations demonstrate the crucial influence of the assumptions used for the calculation on the estimation of the biomass potential. If only the energy yield of the land is considered as a basic criterion for land allocation to energy crops, the biomass potential from energy crops is in principle directly proportional to the area of land allocated to energy crops (in the observed range of 0-15% of land allocation) and the biomass potential from energy crops reaches a relatively high value of 76 PJ for a 15% land allocation (approach 1). In this case, however, the allocation of energy crops to the most suitable land for them occurs regardless of the biomass production price and at the same time regardless of the geographical distribution of the biomass potential. This results in a concentration of energy crops on geographically limited sites. If we select land not according to energy yield but according to production price (again, up to a limit of 15% of land, regardless of regional distribution), there is a slight decrease in biomass potential from 76 to 71 PJ and a similar decrease in average biomass production price (in both cases by about 7%). This approach, similar to approach 1, leads to a concentration of energy crops only in certain regions. The substantially lower biomass potential (55 PJ, a decrease of about 22.5%) leads to the (closer to practice) assumption of an even allocation of land for energy crops across the country, with land in a given region being allocated according to the biomass production price. A further reduction in biomass potential occurs if we assume that the most profitable land would be used for conventional agricultural production. This then leads to a drastic reduction in biomass potential (subject to a land allocation limit of 15% in each region) to only 37.5 PJ (i.e. 47.2% compared to Approach 2). At the same time, this leads to a substantial increase in the biomass production price, to an average of 8.45 EUR/GJ. The last fifth approach, where land is allocated for energy crops according to their ecological contribution, leads to a slight increase in biomass potential compared to approach 4, but this is accompanied by an increase in the biomass production price. This is due to the fact that energy crops are also allocated to productive land that would have been used for conventional crops. Due to the high yield of conventional crops on these plots, the production price of biomass also increases (due to the opportunity cost principle).

Approach 4 is the closest to reality, with the expectation that it will be combined with Approach 5 in the near future (due to the desire to take sustainable land management considerations much more into account). For this reason, a conservative approach should be taken in estimating the potential contribution of perennial energy crops to the biomass potential.

In addition to land allocation assumptions, the biomass potential and cost curve of biomass potential is also influenced by the biomass production price. On many land parcels, the biomass production price is too high, which in practice would make biomass uncompetitive on the energy commodity market and the real contribution of such biomass to the overall biomass potential would be limited or zero. The production price of biomass from energy crops is mainly influenced by the yield of conventional crops on the land in question and the market prices of conventional crops, in addition to the actual costs of growing conventional and energy crops. The increase in conventional crop prices (which can be observed in the first quarter of 2022) leads (due to the opportunity cost principle) to a significant increase in the production price of biomass from energy crops. This in turn may limit their real contribution to the biomass potential, or significantly increase the overall price level of biomass, or even of solid biofuels produced from it.

Other factors that affect the cost price of biomass and the shape of the biomass potential cost curve are the level of expected long-term inflation and discount value. This can be documented in the sensitivity analysis for Approach 2, where the calculation of biomass potential is constrained by only a 15% land allocation for energy crops. With an increase in long-term expected inflation from 2% to 4% and an associated increase in the discount rate from 10% to 12%, the marginal biomass price (i.e., the biomass price of the land parcels ranked in ascending order of biomass production price that close the balance of the land allocation for energy crops) increases from about 4.5 EUR/GJ to 5.3 EUR/GJ. If long-term inflation were to increase to 6% and the discount to 14%, this marginal biomass price would increase to 6.4 EUR/GJ. This shifts the cost curve to the right. This shift is mainly due to the fact that energy crops have high initial costs associated with establishing plantations. The higher discount value then reduces the weight of future cash flows associated with biomass production. Conventional crops do not have these high upfront costs and the effect of increasing discounting is thus less pronounced.

In modelling the biomass production price, a maximum transport distance of 10 km from the plantation to the logistics point is assumed [16]. Transport costs account for 7-13% of total biomass production costs (in present value over the lifetime of the plantation). Other things being equal, transport costs increase proportionally with increasing transport distance. Transporting biomass over significantly greater distances would thus significantly increase the production costs of biomass.

The fifth approach to land allocation for energy crops is based on the preference (maximization) of the ecological benefits from plantations of perennial energy crops. The priority consideration is therefore the provision of non-productive functions that are not included in the (market) price of biomass. A fundamental barrier to the implementation of this approach is the current set-up of support in agriculture or biomass use as the energy source. Current support schemes do not reflect these positive benefits (positive externalities) in the price of production and therefore these benefits are not currently taken into account in farmers' land use decisions.

Given the increasing demand for sustainable land management and addressing the environmental problems associated with intensive agriculture, we can expect an increasing preference for Approach 5 in the near future, i.e. the allocation of perennial energy crops to land where they provide the highest environmental benefits. Often this is on land with high yields of conventional crops, where the biomass production price is then significantly higher. In practice, without specifically targeted support, these plots would often not be allocated to perennial energy crops. The above methodology for calculating the biomass production price for the climatic and soil conditions of individual plots, combined with an assessment of the ecological risk of the land (land risk index), makes it possible to set a differentiated level of support that would be sufficiently motivating for farmers and at the same time limit inefficient use of subsidies.

Linking agrotechnical, ecological and economic constraints allows modelling a realistic level of biomass potential from agricultural land given different land use strategies. As can be seen from the methodology used, the economic potential depends on a number of factors - the input parameters for determining its level. The biomass potential is influenced by factors linked to land and to per hectare crop yields - the area of agricultural land, the percentage of agricultural land allocated to energy crops, advances in breeding and agro-technologies (affecting per hectare crop yields), the impact of climate change (affecting both crop yields and the geographical distribution of biomass potential), as well as factors linked to agricultural commodity and energy markets. Land use strategies with respect to environmental or food production considerations also play an important role in determining biomass potential. Determining the biomass potential of agricultural land is thus a dynamic task where biomass potential can change significantly over time. Similarly, the production price of biomass is a dynamic variable. Therefore, the biomass potential needs to be continuously updated in the light of changing parameters that significantly affect its level.

The above methodology can be used for a number of tasks, such as objectifying the level of support for biomass-fired power/heat plants. The basic philosophy of the methodology can also be used to determine the production price (and thus the biomass potential) of annual crops used as inputs to biogas or biomethane plants. This allows to objectify the inputs for the analysis of the setting of possible support for biogas and biomethane plants, or to assess their potential with respect to the economics of electricity production in biogas plants and biomethane in biomethane plants.

V. Conclusions

The contribution of energy crops grown on agricultural land to the total biomass potential depends on the assumptions used for its calculation. If the limitation is applied only to the area of land (maximising the energy yield), the contribution to the biomass potential is relatively high, reaching about 76 PJ when 15% of arable land is allocated to perennial energy crops. However, this is a so-called technical potential, which does not consider other priorities (food production, even distribution across the territory of the Czech Republic) or the economic aspect of biomass production. If these aspects are considered, the contribution of perennial energy crops to the biomass potential is substantially reduced to only about half (37.5 PJ). Respecting these realistic constraints then leads to the conclusion that the increase in biomass availability (e.g. for replacing fossil fuels, in the case of the Czech Republic mainly coal) due to its production from agricultural land should be conservatively assessed while respecting all economic and ecological contexts.

Acknowledgment

This work was supported by the Technology Agency of the Czech Republic [TK04010166].

References

[1] N. Scarlat, J. Dallemand, N. Taylor, M. Banja, J. Sanchez, M. Avraamides. "Brief on biomass for energy in the European Union," in *Publications Office of the European Union*, 2019.

[2] Statista, "Bioenergy in Europe - Statistics & Facts," Internet: https://www.statista.com/topics/5219/bioenergy-industry-in-europe, 2021 [Apr. 4, 2022].

[3] MPO. "Vnitrostátní plán České republiky v oblasti energetiky a klimatu." Internet: https://mpo.cz/cz/energetika/strategicke-a-koncepcni-dokumenty/vnitrostatni-plan-ceske-republiky-v-oblasti-energetiky-a-klimatu--252016/, 2020 [Apr. 4, 2022].

[4] MPO. "Obnovitelné zdroje energie v roce 2020." Internet: https://www.mpo.cz/assets/cz/energetika/statistika/obnovitelne-zdroje-energie/2021/9/Obnovitelne-zdroje-energie-2020.pdf, 2021 [Apr. 4, 2022].

[5] ERU. "Roční zpráva o provozu ES ČR pro rok 2020." Internet: https://www.eru.cz/zpravy-o-provozu-elektrizacni-soustavy#2020, 2021 [Apr. 4, 2022].

[6] MPO. "Strategie stabilizace a rozvoje soustav zásobování tepelnou energií." 2020. Pracovní materiál MPO.

[7] J. Knápek, K. Vávrová, T. Králik, D. Outrata. "Biomass potential - Theory and practice: Case example of the Czech Republic region," in *Energy Reports*. 2020, vol .6, pp. 292-297. ISSN 2352-4847, DOI 10.1016/j.egyr.2020.11.146.dd.

[8] J. Knápek, T. Králík, K. Vávrová, J. Weger. "Dynamic biomass potential from agricultural land," in *Renew. Sustain. Energy Rev.*, vol. 134, Dec. 2020. DOI: 10.1016/j.rser.2020.110319.

[9] K. Vávrová, J. Knápek, J. Weger. "Short-term boosting of biomass energy sources Determination of biomass potential for prevention of regional crisis situations," in *Renewable and Sustainable Energy Reviews*, vol. 67. Elsevier Ltd, pp. 426–436, Jan. 01, 2017. DOI: 10.1016/j.rser.2016.09.015.

[10] J. Knápek, T. Králík, K. Vávrová, M. Valentová, M. Horák, D. Outrata. "Policy implications of competition between conventional and energy crops," in *Renewable and Sustainable Energy Reviews*. 2021, vol. 151, pp. 1-11. ISSN 1364-0321. DOI: 10.1016/j.rser.2021.111618.

[11] VÚKOZ, v.v.i., "Metodika alokace porostů energetických plodin na zemědělské půdě zohledňující mimoprodukční funkce v krajině." Certifikovaná metodika, VÚKOZ, v.v.i. č. 3/2021-054.

[12] UZEI, "Costs and revenues of selected agricultural products.," Internet: www.uzei.cz/data/usr_001_cz_soubory/210105_naklady2019.pdf, 2019 [Apr. 4, 2022].

[13] SZIF, "Market (monthly) reports for cereals, oilseeds and feed," Internet: www.szif.cz/cs/zpravy-o-trhu?year=2020&cdr=05&ino=0, 2021 [Apr. 4, 2022].

[14] ČNB, "Monetary policy forecast – spring 2022." Internet: www.cnb.cz/en/monetary-policy/inflation-target/, 2022 [May 25, 2022].

[15] T. Králík. "Story of the Czech RES Support from the beginning to the present process of over-compensation," in *ENERGY ECOLOGY ECONOMY 2018. Energetika - Ekológia - Ekonomika 2018*, Vysoké Tatry - Tatranské Matliare. 2018. pp. 62-65. ISSN 2585-9684.dd.

[16] K. Vávrová, J. Knápek, J. Weger, T. Králík, J. Beranovský. "Model for evaluation of locally available biomass competitiveness for decentralized space heating in villages and small towns," in *Renewable Energy*. 2018, vol. 129, pp. 853-865. ISSN 0960-1481.

[17] R. Slade, R. Saunders, R. Gross, A. Bauen. "Energy from Biomass: The Size of the Global Resource," in *Imperial College Centre for Energy Policy and Technology and UK Energy Research Centre*, London, Nov. 2011. https://ukerc.ac.uk/project/energy-from-biomass/, [May 25, 2025].

[18] D. Lauka, A. Barisa, D. Blumberga. "Assessment of the availability and utilization potential of low-quality biomass in Latvia," in *Energy Procedia*, vol. 147, pp. 518-524, Aug. 2018. DOI: 10.1016/j.egypro.2018.07.065.

[19] H. Long, X. Li, H. Wang, J. Jia. "Biomass resources and their bioenergy potential estimation: A review," in *Renewable and Sustainable Energy Reviews*, vol. 26, pp. 344-352, Oct. 2013. DOI: 10.1016/j.rser.2013.05.035.

[20] M. Hoogwijk, A. Faaij, B. Eickhout, B. de Vries, W. Turkenburg. "Potential of biomass energy out to 2100, for four IPCC SRES land-use scenarios," in *Biomass and Bioenergy*, 29(4), pp. 225-257, Oct. 2005. DOI: 10.1016/j.biombioe.2005.05.002.

[21] M. Beccali, P. Columba, V. D'Alberti, V. Franzitta. "Assessment of bioenergy potential in Sicily: A GIS-based support methodology," in *Biomass and Bioenergy*, 33(1), pp. 79-87, Jan. 2009. DOI: 10.1016/j.biombioe.2008.04.019.

[22] I. Vera, B. Wicke, P. Lamers, A. Cowie, A. Repo, B. Heukels, C. Zumpf, D. Styles, E. Parish, F. Cherubini, G. Berndes, H. Jager, L. Schiesari, M. Junginger, M. Brandao, N. S. Bentsen, V. Daioglou, Z. Harris, F. van der Hilst. "Land use for bioenergy: Synergies and trade-offs between sustainable development goals," in *Renew. Sustain. Energy Rev.*, vol. 161, pp. 112409, Jun. 2022. DOI: 10.1016/J.RSER.2022.112409.

[23] D. Scordia, S. Cosentino. "Perennial Energy Grasses: Resilient Crops in a Changing European Agriculture," in *Agriculture*, 9(8), pp. 169, Aug. 2019. DOI: 10.3390/agriculture9080169.

[24] G. M. Souza, M. Victoria, R. Ballester, C. H. de Brito Cruz, H. Chum, B. Dale, V. H. Dale, E. C.M. Fernandes, T. Foust, A. Karp, L. Lynd, R. M. Filho, A. Milanez, F. Nigro, P. Osseweijer, L. M. Verdade, R. L. Victoria, L. Van der Wielen. "The role of bioenergy in a climate-changing world," in *Environmental Development*, vol. 23. Elsevier B.V., pp. 57-64, Sep. 2017. DOI: 10.1016/j.envdev.2017.02.008.

[25] M. Haase, C. Rösch, D. Ketzer. "GIS-based assessment of sustainable crop residue potentials in European regions," in *Biomass and Bioenergy*, vol. 86, pp. 156-171, Mar. 2016. DOI: 10.1016/j.biombioe.2016.01.020.

[26] J. van Stralen, C. Kraan, A. Uslu, M. Londo. "Integrated assessment of biomass supply chains and conversion routes under different scenarios," in *Technical report*, 2016, pp. 72. DOI: 10.13140/RG.2.2.26538.21441.

[27] P. Alexander, D. Moran, P. Smith, A. Hastings, S. Wang, G. Sünnenberg, A. Lovett, M. J. Tallis, E. Casella, G. Taylor, J. Finch, I. Cisowska. "Estimating UK perennial energy crop supply using farm-scale models with spatially disaggregated data," in *GCB Bioenergy*, 2014, vol. 6, pp. 142-155. DOI: 10.1111/gcbb.12121.

Equivalent Dynamic Model of the Continental Europe Synchronous Area for Frequency Stability Assessment

Tomáš Hába, Oldřich Rychlý, Karel Máslo
Transmission System Analysis Department
ČEPS, a.s.
Prague, Czech Republic
haba@ceps.cz, rychly@ceps.cz, maslo@ceps.cz

Abstract — **The paper deals with frequency stability simulations for power system disturbances. It presents the equivalent dynamic model of the Continental Europe Synchronous Area which is suitable for assessment of the power system frequency response to a disturbance such as abrupt power deficit in the system. Presented model allows to simulate individual implemented measures with an impact on the dynamic behavior. The paper describes load flow and dynamic input data and shows the process of verifying the model on real events that have occurred in the system. The tuned model can then be used to assess the effectiveness of the implemented measures both in the normal operation state and in the system defense plan. The network simulator MODES is used as a simulation tool.**

Keywords: dynamic model, frequency stability, Continental Europe, power system disturbance, MODES

I. Introduction

The dynamic stability assessment (frequency, rotor angle and voltage) is an important part of ensuring safe and reliable power system operation. As the European electricity sector is undergoing a significant transformation these days including unbundling, market development and generation mix changes, the dynamic stability assessment becomes more and more important. The power systems are operated closer to their stability limits, which is caused by an increase in power exchanges and power production in intermittent and difficult dispatchable power-generating modules.

System frequency is a sensitive indicator of active power balance in the interconnected grid. Frequency stability is related to the ability of the power system to maintain steady frequency if power imbalance occurs in the system. It also means whether the system is able to maintain/restore equilibrium between load and generation [1]. Stability of the system frequency can be well assessed particularly when the active power balance is abruptly violated. Frequency stability assessment is usually carried out for large system disturbances while the power system condition is analyzed and evaluated.

The ongoing transition from conventional generation of electricity, which is based on large turbo or hydro generator units (typically synchronous power-generating modules), to intermittent inverter-based generation (typically dominant types of renewable energy sources) leads to a decrease of natural power system inertia. Synchronous power-generating modules have an inherent capability to resist or slow down frequency deviations, a characteristic which many renewable energy sources do not have. Therefore, the investigation of the system frequency behavior is getting more interest. Except for evaluating the frequency deviation after power imbalance, another important indicator of the instantaneous system state related to frequency stability is often assessed. This indicator is so-called Rate of Change of Frequency (RoCoF), which is the time derivative of the power system frequency. RoCoF must not exceed a certain limit for ensuring safe power system operation, particularly proper functionality of protection system, control loops and other equipment.

Special computational tools – so-called network simulators are used for the dynamic stability assessment of the power system. This paper focuses specifically on the simulation in the network simulator MODES.

The paper describes possible approach to build an equivalent model for frequency stability assessment of the Continental Europe Synchronous Area power system. First, a brief description of the investigated incident is presented in Section II. The equivalent single busbar model with modelling approach is described in Section III. Section IV shows the simulation results. Finally, Section V summarizes important findings, and the conclusion is drawn.

II. Continental Europe Synchronous Area Incidents

The Continental Europe Synchronous Area (CE) is one of the largest interconnected synchronous grid (by system load) in the world. Transmission system operators (TSOs) in CE have the responsibility to maintain steady frequency at nominal 50 Hz as an important value for proper operation of all power system components. Although CE is a very well interconnected system, thus highly reliable operation is secured, there have been several incidents affecting the whole system in the past.

The well-known incident is split of European power system UCTE (former CE) occurred on November 4, 2006. The system was separated into three asynchronous islands [2]. Other examples of system split may be mentioned: CE splitting into two asynchronous islands on January 8, 2021 [3] or the Iberian Peninsula separation from CE on July 24, 2021 [4].

Moreover, disturbances caused by a large active power imbalance have been observed in CE. As an example, local incident in Poland´s Rogowiec substation occurred on May 17, 2021. A manipulation incident in the substation led to a short circuit, which further caused the tripping of ten generation units in Bełchatów power plant with total loss of 3 322 MW of generated active power [5]. The total generation loss during the incident was higher than the Reference Incident of CE, which is based on the N-2 criterion, thus tripping of two largest nuclear power units, with the total capacity of 3 000 MW. Frequency Containment Process (FCP), formerly known as primary frequency control, is dimensioned to cover the power imbalance of the Reference Incident.

These incidents provide valuable data for further (post-mortem) analysis and the functionality of equivalent models and computational tools themselves can be verified. The use of these tools helps to verify the application of countermeasures or suggest a new countermeasure.

This Rogowiec incident was selected to verify dynamic model for CE frequency stability assessment in the MODES network simulator and will be further elaborated.

III. Dynamic Model for Frequency Stability Assessment

Time range of transients in the power system varies with respect to the studied dynamics of the event. Therefore, the type of transients should be specified first. The investigated transient of frequency stability is within seconds, thus electromechanical phenomena predominate. Turbine-governor parameters and synchronous machine inertia constant are main inputs that represent dynamic behavior of the power system in case of frequency stability studies in general. Power sources can be modeled with constant power supply for this type of transients.

Since frequency is a global parameter of the power system, detailed monitoring of power flows among nodes of the system are secondary for this type of investigated phenomena. Therefore, it is advantageous to utilize simplified equivalent model with aggregated generation and load as shown in Fig. 1. The drawback of such simplification is that the model cannot replicate phenomena such as interarea oscillations. However, the equivalent single busbar model is suitable for simulating the general frequency response of the power system. This approach is also used at ENTSO-E level for evaluation of the system frequency stability [3].

Single-line diagram of the power system equivalent model for Rogowiec incident from May 17, 2021 at 16:34:11 is depicted in Fig. 1. The total load and generation of CE has been estimated based on historical data (with the use of ENTSO-E Transparency Platform) and prediction models for 2022. The generation from renewables (marked as RES in the diagram) was quite low at the time of the incident – estimated at 25 % for photovoltaics and 20 % for the wind.

Load L1 is equipped by low frequency demand disconnection (LFDD) scheme with load shedding values according to Table 1. Interruptible load L2 is disconnected with steps shown in Table 2. The interruptible load response can be found in [3]. During the system split on January 8, 2021, the interruptible load was disconnected in France and Italy in a response to a frequency drop. This interruptible load is included in individual automatic special protection schemes as a reduction of industrial loads with dedicated interruptible contracts. Self-regulating effect of the load L1 and L2 is set to $K_L = 2\%/Hz$.

Table 1. LFDD scheme applied for load L1

f_{LS} (Hz)	49.00	48.87	48.74	48.61	48.49
Step (%)	2.5	2.5	2.5	2.5	2.5

Table 2. Interruptible load L2 disconnection scheme

f_{IL} (Hz)	49.82	49.75	49.70	49.65	49.20
Step (MW)	1 500	150	1 435	150	700
Step (%)	38	4	36	4	18

Total infeed from renewables (photovoltaics and wind) is divided into several units. The part marked as RES in Fig. 1 is modeled as constant delivery. No frequency response is expected for these inverted-based sources, although the setting of their frequency protections corresponds to the frequency range requirement arising from Commission Regulation (EU) 2016/631 (RfG) [6]. Therefore, these sources aggregated in RES are assumed to be capable of operating in the frequency range 47.5 – 51.5 Hz. The parts with risk setting of the frequency protections are modeled as separate deliveries RES1 – RES6. These units are switched-off by frequency relays that are defined as automatics. The frequency relays disconnect the unit when the frequency limit (according to Fig. 1) is reached. The volume of each unit was simply estimated from the report [7] with the fact that the ongoing retrofit process in CE has been approximated. The units RES1 – RES6 are disconnected at set frequency limits with time delay of 50 ms.

The units SYNC and GAS represent aggregated sources with synchronous generators, thus they provide inertia to the power system with T_m = 8.324 s (T_m is based on nominal apparent power S_n, $T_m = 2H$). Frequency Containment Reserves (FCR) are allocated on SYNC unit with the total capacity of 3 000 MW as primary control reserves, while GAS unit decreases power if frequency falls below 49.5 Hz with a gain of 0.1/ Hz up to 3 %. The overall acceleration time constant of the power system for the investigated operating condition according to the definition in [8] is T_N = 10.2 s. Common 6th order Park synchronous machine model is used for both SYNC and GAS units.

All components in Fig. 1 are connected to BUS through the branches with very low reactance X = 0.001 Ω (except for the units SYNC and GAS which are connected to BUS directly). The small reactance value guarantees that there is no voltage change during power change which is additionally maintained by a constant automatic voltage regulator (AVR) on SYNC and GAS unit. The block diagram of the AVR model is shown in Fig. 2 together with Table 3 containing typical parameters of AVR. The paper [9] describes the AVR modeling in more detail. The parameter $K_{stat} = 0$ guarantees that the AVR maintains voltage directly at the generator terminals, i.e. directly at BUS.

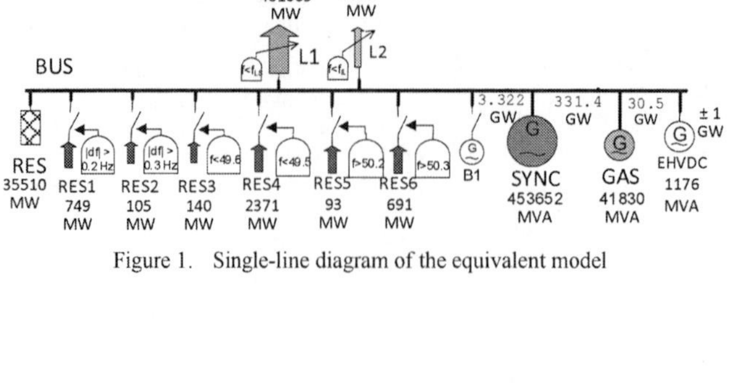

Figure 1. Single-line diagram of the equivalent model

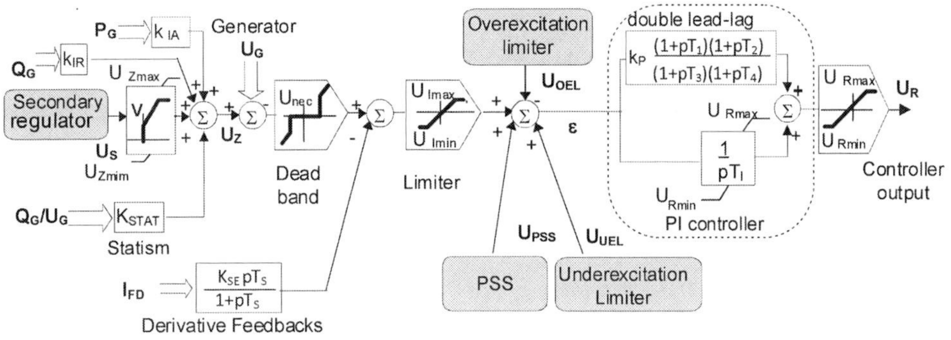

Figure 2. Block scheme of the AVR model

Table 3. Parameters of the AVR model

AVR	U_{zmin}	U_{zmax}	U_{nec}	K_{stat}	K_p	K_{se}	T_I (s)	T_s (s)	T_1 (s)	T_3 (s)	v (%/s)	U_{rmax}	U_{rmin}	U_{lmin}	U_{lmax}	T_2 (s)	T_4 (s)	K_{IA}	K_{IR}
RBA	0	1.2	0	0	50	0	0.14	1.96	0	0	10	9.9	-9.9	-9.9	9.9	0	0	0	0

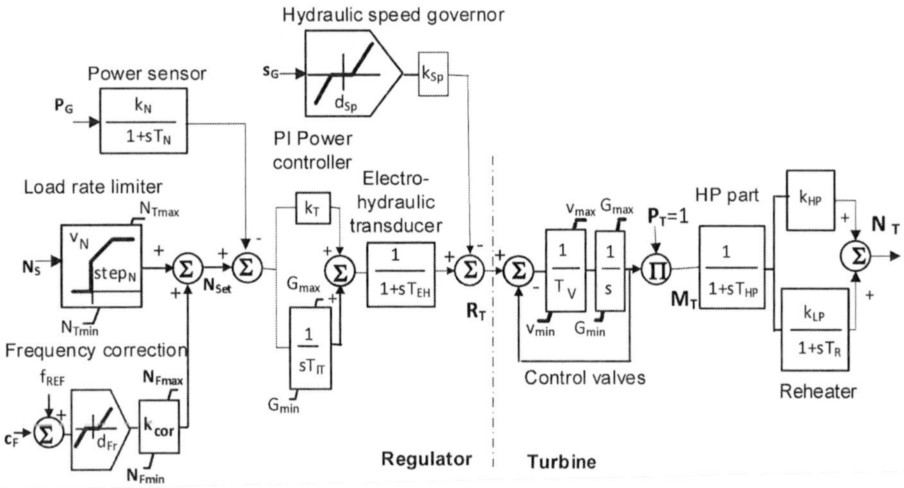

Figure 3. Block scheme of turbine-governor model

Table 4. Parameters of the regulator model

Regulator	T_{IT} (s)	T_{IB} (s)	T_N (s)	T_{EHP} (s)	k_T	k_B	k_{Sp}	k_{Fr}	K_{COR}	k_{Pres}	k_{For}	GEN	v_N (%/min)	stepN (%)	d_{Fr} (%)	d_{Sp} (%)	d_{Pres} (%)	d_P (%)	N_{Fmax} (%)	N_{Fmin} (%)
UCTEP	10	100	1	0.1	1	1	3	0	2.5	0	0.5	1	2.5	0	0	0.05	0	0	1.0	-1.0

Table 5. Parameters of the turbine model

Turbine	k_N	T_V (s)	T_{HP} (s)	T_R (s)	V_{min}	V_{max}	G_{min}	G_{max}	K_{LP}	K_{HP}
UCTEP	1.176	0.8	0.4	5	-1	0.1	0	1.1	0.25	0.75

Block scheme of the turbine-governor model for SYNC unit is shown in Fig. 3 including the parameters used (Table 4 and 5). Since the controller also contains speed governor, the controller must operate in the closed control loop. Therefore, a parallel arrangement of the power controller and the speed governor is used. The speed governor is applied in a transient dynamic process (accelerates the response) and the resulting steady state is determined by the power controller which has a proportionally integrative (PI) character.

The dynamic response of the turbine can also significantly affect the course of the transient process. The parameter K_{HP} determining the share of production in the high-pressure part and parameter K_{LP} determining the share of production in the rest of the turbine with a delayed response caused the passage of steam flow through the reheater. The constant value of the pressure of admission steam is considered in the model. Detailed description of the turbine-governor models is available in [9].

HVDC connection with CE neighboring synchronous areas is modeled as synchronous unit EHVDC with zero active power in the initial state. Proportional response to frequency deviation is emulated with dead-band ± 0.1 Hz, frequency droop 5 %, ramp 10 MW/s and reserve ± 1 000 MW. Since HVDC interconnection represents inverter-based source that does not contribute to power system inertia, the acceleration time constant for the EHVDC unit has been reduced to a minimum (T_m = 0.4 s) so as not to distort the simulation results.

IV. Simulation Results

Rogowiec incident from May 17, 2021, with the tripping of ten generation units in Bełchatów power plant has been simulated in the network simulator MODES using the dynamic models presented above. Bełchatów generation units are aggregated in unit B1 (see Fig. 1) with total capacity of 3 322 MW. Thus, incident scenario has been simplified so that all units are expected to trip at the same moment, which is, however, in good accordance with real situation. All ten units tripped within one second [5].

Fig. 4 shows frequency waveforms obtained from the simulation result in MODES (blue line) with a comparison of the real incident waveform (black line) obtained by Wide Area Monitoring System (WAMS) measurement with a high recording period (up to 50 samples per second). After the outage of Bełchatów generation units the frequency begins to decrease. The power imbalance is compensated by the kinetic energy of the inertial masses naturally provided by the rotating machines connected to the power system at the first moments. After a few seconds, primary control reserves (today known as FCR) begin to release as a reaction of the turbine governors to the power system frequency deviation. This leads to stabilization of the frequency with a certain deviation. The limit for normal operation state (50.0 ± 0.2 Hz) was not exceeded during the fault.

A comparison of both curves shows that the simulation results from MODES correspond well to the real measurement. The dynamic response of real system FCR seems to be slightly faster than in the model. The maximum frequency deviation (from nominal frequency 50 Hz) has reached the value of 171 mHz for the simulation, which is slightly higher than the measurement with the value of 161 mHz.

Contributions of individual model components to cover power imbalance are depicted in Fig. 5 (minus means a reduction in the imbalance). The unit SYNC has the main contribution by providing FCR, as expected, followed by self-regulating effect of load L1. The power response of the unit SYNC has dropped at first due to the effect of the power controller. However, then the unit has increased the power quickly because both a speed governor and a frequency corrector in the power controller have acted. HVDC interconnection in the model has minor contribution (maximum approx. 20 MW) because the set dead-band threshold has been exceeded.

Also, the self-regulating effect of load L2 has participated in covering the power imbalance to a lesser extent whereas GAS and all RES units have no frequency response as their frequency thresholds have not been reached.

Additional simulation was performed on the same equivalent model to verify the settings and functionality of individual model components, this time with 5% power deficit (approx. 20 GW). The frequency waveform for this simulation is depicted in Fig. 6. Due to the large power deficit the frequency has dropped below 49 Hz. Then, first step of LFDD scheme applied to the load L1 has been activated, which resulted in stabilization of the situation and a new equilibrium of the power system has been found. Fig. 7 shows the contributions of individual model components. Disconnection of RES units with

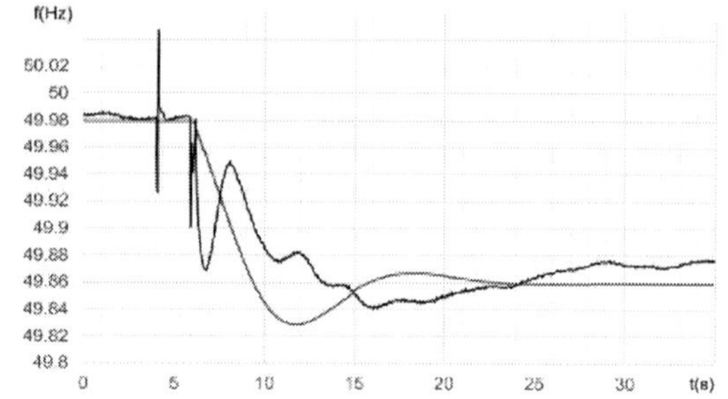

Figure 4. Comparison of simulated (blue line) and measured (black line) frequency of the power system after 3 322 MW outage

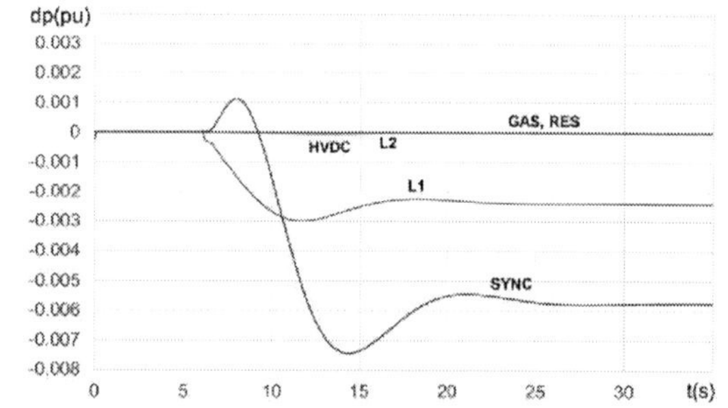

Figure 5. Power responses dp (related to PL = 405 GW) with contribution of individual components

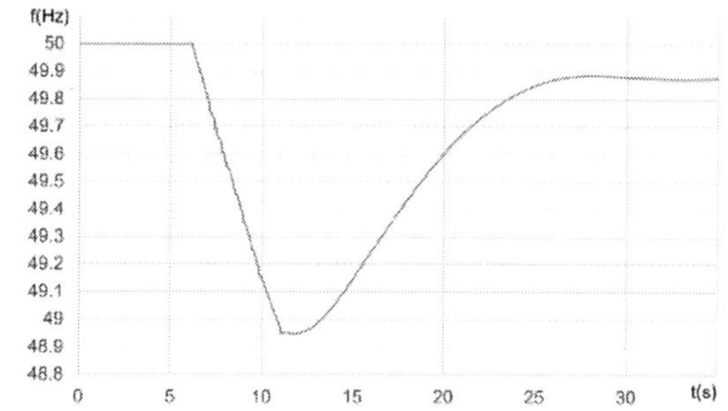

Figure 6. Time course of frequency for 5% power deficit simulation

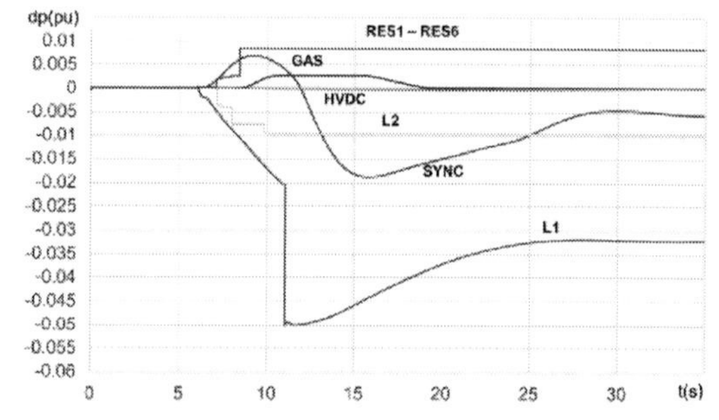

Figure 7. Power responses dp (related to PL = 405 GW) for 5% power deficit simulation

risk settings of frequency relays is already noticeable here as well as the reaction of interruptible load L2.

Regarding the simulation results, it can be stated that the equivalent single busbar model provides credible and reliable results. Thus, the model can be used for simplified simulations of CE frequency response during significant disturbances.

V. Conclusion

This paper presents simplified equivalent model for frequency stability assessment of CE, thus for very complex and interconnected power system. The equivalent model was validated for a relatively serious disturbance. The Rogowiec incident exceeded the limit value of the Reference Incident for CE (generation outage higher than 3 000 MW). Therefore, it can serve as a suitable event for model validation.

The paper described the approach of building equivalent single busbar model for frequency stability assessment and settings of individual components. Results of the simulation were compared with the real measurement and waveforms were described in detail.

The results revealed that the simplified model cannot replicate all transients and model limitations should be always kept in mind. However, global behavior of the power system and steady-state frequency deviation is credibly reproduced. The global already existing measures or the measures under consideration may be validated with such a tuned model. This is important, for example, when designing the defense plan measures. The model can thus be expanded with new components.

References

[1] P. Kundur, J. Paserba, V. Ajjarapu, G. Andersson, A. Bose, C. Canizares, N. Hatziargyriou, D. Hill, A. Stankovic, C. Taylor, T. Van Cutsem, V. Vittal. "Definition and classification of power system stability IEEE/CIGRE joint task force on stability terms and definitions," in *IEEE Transactions on Power Systems*, vol. 19, no. 3, pp. 1387-1401, Aug. 2004, doi: 10.1109/TPWRS.2004.825981.

[2] "Final report system disturbance on 4 November 2006", [Online]. Available: https://eepublicdownloads.entsoe.eu/clean-documents/pre2015/publications/ce/otherreports/Final-Report-20070130.pdf, Jan. 2007.

[3] "Continental Europe synchronous area separation on 08 January 2021 – Final Report", [Online]. Available: https://eepublicdownloads.azureedge.net/clean-documents/SOC%20documents/SOC%20Reports/entso-e_CESysSep_Final_Report_210715.pdf, Jul. 2021.

[4] "Continental Europe synchronous area separation on 24 July 2021 technical report", [Online]. Available: https://eepublicdownloads.azureedge.net/clean-documents/SOC%20documents/SOC%20Reports/entso-e_CESysSep_210724_211112.pdf, Nov. 2021.

[5] "Local grid incident in Rogowiec (Poland) substation on 17 May 2021 – final report", [Online]. Availabe: https://eepublicdownloads.entsoe.eu/clean-documents/news/2022/220318_Final_report_Rogowiec_incident.pdf, Mar. 2022.

[6] "Commision Regulation (EU) 2016/631 of 14 April 2016 establishing a network code on requirements for grid connection of generators", April 2016.

[7] "Dispersed generation impact on CE region security", ENTSO-E SG SPD Dynamic Study, [Online]. Available: https://eepublicdownloads.entsoe.eu/clean-documents/Publications/SOC/Continental_Europe/141113_Dispersed_Generation_Impact_on_Continental_Europe_Region_Security.pdf, Dec. 2014.

[8] "Frequency stability evaluation criteria for the synchronous zone of Continental Europe", ENTSO-E RG-CE System Protection & Dynamics Sub Group, [Online]. Available: https://eepublicdownloads.entsoe.eu/clean-documents/SOC%20documents/RGCE_SPD_frequency_stability_criteria_v10.pdf, Mar. 2016,

[9] K. Maslo and T. Hába, "Testing of dynamic models for stability studies," *TRANSACTIONS ON ELECTRICAL ENGINEERING*, vol. 7, no. 3, pp. 64–71, 2020, doi: 10.14311/TEE.2018.3.064.

The role of flexibility resources in the energy transition

Nikolaos Koltsaklis
Dept. of Economics, Management and Humanities
CTU in Prague, Fac. of Electrical Engineering
Prague, Czech Republic
nikkoltsak@gmail.com

Jaroslav Knápek
Dept. of Economics, Management and Humanities
CTU in Prague, Fac. of Electrical Engineering
Prague, Czech Republic
knapek@fel.cvut.cz

Abstract —**Increasing shares of variable renewable energy sources call for highly flexible power systems. Limited flexible capacity is often mentioned as a constraint on the amount of variable renewable energy sources that can be added to the grid. This work presents a generic optimization framework for the optimal scheduling of an islanded power system characterized by various flexibility resources. In particular, it includes a significant amount of intermittent RES and penetration of electric vehicles with both charging and discharging options. The model also considers the installation of energy storage capacities and the active market participation of demand response programs. The scheduling framework is daily with an hourly time step, and an illustrative case study of an islanded power system interconnected with a mainland power system has been selected to test the applicability of the proposed approach. The results highlight the importance of all those flexibility resources in the optimal energy management and the supply of the ancillary services. The proposed approach can be utilized by market operators, policymakers, and regulatory authorities for the determination of the appropriate system development, market design, and portfolio formation.**

Keywords: optimization, islanded system, flexibility, electric vehicles, energy storage, demand response

I. Introduction

For any energy system, the composition of its energy supply mix depends on three main factors: (i) the energy resources' domestic availability and/or the options for importing them via available transportation means, (ii) the system-wide energy requirements that must be satisfied, and (iii) applied energy policy decision-making determined by historical, economic, social, demographic, environmental, geopolitical, and/or institutional factors.

A series of works have been presented in the literature investigating the combined optimization of energy and reserves. A review of the modeling practices and trends in the co-optimization of the energy and reserves market is provided in [1]. The authors in [2] developed two energy and spinning reserve market clearing methods, investigating the impacts of demand participation in the reserve offers. The impact of demand flexibility on the energy and reserves market-clearing was also examined in [3]. The authors in [4] presented a mixed-integer linear programming (MILP) model for the thermal unit commitment problem. A similar unit commitment formulation has also been developed in [5]. The authors in [6] developed an optimization approach for the co-optimization of energy-reserve power markets. In the same context, the authors in [7] provided a MILP model for the joint clearing of energy and reserve power exchanges, combining the EUPHEMIA (pan-EUroPean Hybrid Electricity Market Integration Algorithm) hourly offers module with specific unit commitment constraints. Moreover, the same authors in [8] investigated EUPHEMIA's market products for the interconnected Greek power system. Furthermore, the authors in [9] developed an optimization model to determine energy and reserve dispatch in electricity markets. Focusing on the bidding strategies, the authors in [10] developed a methodological framework for the integration of bidding schemes in day-ahead energy and spinning reserve markets. In the same frame of reference, the authors in [11] proposed an optimization model for the optimal bidding strategy problem for a generation company to maximize its economic profit in day-ahead energy and spinning reserve markets. Finally, focusing on islanded power systems, the authors in [12] highlighted the contribution of electricity interconnection to the coverage of reserve requirements.

This work presents a methodological framework for the optimal energy and reserves scheduling of an islanded power system, considering the participation of thermal power units, renewable energy sources (RES), an energy storage system (ESS), electric vehicles (EVs), and demand response programs (DRPs). A key contribution of the submitted work is that it considers the activation of electricity interconnections as flexibility services providers. The key decisions to be determined by the mathematical model include (i) energy generation mix of the studied power system, (ii) reserve provision mix of the studied power system, and (iii) quantification of the flexibility provision from ESS, EVs, DRPs, and electricity interconnection.

The rest of the paper is structured as follows: Section 2 defines the problem to be solved and details the formulation of the mathematical model. In addition, a description of the case study with its relevant input data is given in Section 3, while Section 4 provides a detailed discussion of the model outputs, underscoring its key findings. Last but not least, Section 5 sums up the key concluding remarks.

II. Methodology

A. Objective function

The optimization model co-optimizes the energy and reserves scheduling of an island that is interconnected with the mainland power system from the system operator's perspective. The mathematical model's objective function concerns the net daily cost minimization (1), including the following components: (i) energy production cost of installed thermal power units, (ii) start-up costs of thermal power units, (iii) RES curtailment cost, and (iv) operating reserve supply cost in both upward and downward directions.

$$
\begin{aligned}
&MinC^{oper} = \\
&\overbrace{\sum_i \sum_t C_{i,t}^{prod}}^{\substack{\text{Energy supply cost}\\ \text{from thermal power units}}} + \overbrace{\sum_i \sum_t C_i^{stup} \cdot y_{i,t}}^{\substack{\text{Thermal power units}\\ \text{start-up cost}}} \\
&+ \overbrace{\sum_r \sum_t C_{r,t}^{curt} \cdot p_{r,t}^{curt}}^{\text{RES curtailment cost}} + \overbrace{\sum_n \sum_t C_{n,t}^{up} \cdot r_{n,t}^{up}}^{\text{Reserve up cost}} \\
&+ \overbrace{\sum_m \sum_t C_{m,t}^{dn} \cdot r_{m,t}^{dn}}^{\text{Reserve down cost}}
\end{aligned} \quad (1)
$$

B. Energy demand balance

Equation 2 defines the energy demand balance. In particular, the energy supply from thermal power units ($\sum_i p_{i,t}^{conv}$), RES ($\sum_r p_{r,t}^{res}$), power discharge from ESSs ($\sum_w p_{w,t}^{dis}$) and EVs ($\sum_v p_{v,t}^{dis} \cdot n_v^{tot}$), and the electricity imports from the mainland power system (imp_t) must meet the cleared energy demand ($\sum_y d_{y,t}^{cl}$), the electricity exports to the mainland power system (exp_t) and the charging load from both ESSs ($\sum_w p_{w,t}^{ch}$) and EVs ($\sum_v p_{v,t}^{ch} \cdot n_v^{tot}$) in each time period.

$$
\begin{aligned}
&\sum_i p_{i,t}^{conv} + \sum_r p_{r,t}^{res} + \sum_v p_{v,t}^{dis} \cdot n_v^{tot} + \sum_w p_{w,t}^{dis} \\
&+imp_t = \sum_y d_{y,t}^{cl} + \sum_v p_{v,t}^{ch} \cdot n_v^{tot} + \sum_w p_{w,t}^{ch} + exp_t
\end{aligned} \quad \forall t \; (2)
$$

C. Operational constraints

Equation (3) calculates the production cost of each thermal power unit ($C_{i,t}^{prod}$), dependent on its power output ($p_{i,t}^{conv}$) and its cost coefficients (a_i, b_i, and c_i). Constraints (4) and (5) describe the maximum (P_i^{max}) and minimum (P_i^{min}) operating limits of each thermal power units, considering its participation in energy ($p_{i,t}^{conv}$) and reserves-up ($r_{i,t}^{up}$) and down ($r_{i,t}^{dn}$) scheduling, subject to the decision of its operation or not ($x_{i,t}^{oper}$). In addition, constraints (6) and (7) determine the ramp-up (RU_i) and down (RD_i) limits of each thermal power unit. Moreover, constraint (8) defines the minimum uptime (UT_i) of each thermal power unit after its start-up decision ($y_{i,t}$), while constraint (9) imposes the minimum downtime (UT_i) of each thermal power unit after its shut-down decision ($z_{i,t}$). Finally, equation (10) formulates a logical relation among operation, start-up, and shut-down decisions.

$$C_{i,t}^{prod} = a_i \cdot (p_{i,t}^{conv})^2 + b_i \cdot p_{i,t}^{conv} + c_i \quad \forall i,t \; (3)$$

$$p_{i,t}^{conv} + r_{i,t}^{up} \leq P_i^{max} \cdot x_{i,t}^{oper} \quad \forall i,t \; (4)$$

$$p_{i,t}^{conv} - r_{i,t}^{dn} \geq P_i^{min} \cdot x_{i,t}^{oper} \quad \forall i,t \; (5)$$

$$(p_{i,t}^{conv} + r_{i,t}^{up}) - (p_{i,t-1}^{conv} - r_{i,t-1}^{dn}) \leq RU_i \quad \forall i,t \; (6)$$

$$(p_{i,t-1}^{conv} + r_{i,t-1}^{up}) - (p_{i,t}^{conv} - r_{i,t}^{dn}) \leq RD_i \quad \forall i,t \; (7)$$

$$\sum_{t'=t-UT_i+1}^{t} y_{i,t'} \leq x_{i,t}^{oper} \quad \forall i,t \; (8)$$

$$\sum_{t'=t-DT_i+1}^{t} z_{i,t'} \leq 1 - x_{i,t}^{oper} \quad \forall i,t \; (9)$$

$$y_{i,t} - z_{i,t} = x_{i,t}^{oper} - x_{i,t-1}^{oper} \quad \forall i,t \; (10)$$

D. Renewable energy generation

Constraint (11) describes the maximum generation potential of each RES in each time period ($F_{r,t}$), being equal to the amount that is directly utilized ($p_{r,t}^{res}$) and the other one that is curtailed ($p_{r,t}^{curt}$). Also, constraint (12) imposes an upper bound on the reserve supply potential in the downward direction of each RES in each period ($r_{r,t}^{dn}$).

$$p_{r,t}^{res} + p_{r,t}^{curt} \leq F_{r,t} \quad \forall r,t \; (11)$$

$$p_{r,t}^{res} - r_{r,t}^{dn} \geq 0 \quad \forall r,t \; (12)$$

E. Electricity trading modeling

Constraints (13)-(16) describe the energy and reserve exchanges with the mainland power system. In particular, constraint (13) expresses the upper bound (IMP_t^{cap}) of energy imports (imp_t) and operating upward reserve (r_t^{up}) from the mainland power system, as well as constraint (14) sets the corresponding maximum potential of downward reserve supply (r_t^{dn}). Constraints (15) and (16) provide the same information for the case of energy exports and reserve exchanges to the mainland power system.

$$imp_t + r_t^{up} \leq IMP_t^{cap} \quad \forall t \; (13)$$

$$imp_t - r_t^{dn} \geq 0 \quad \forall t \; (14)$$

$$exp_t + r_t^{dn} \leq EXP_t^{cap} \quad \forall t \; (15)$$

$$exp_t - r_t^{up} \geq 0 \quad \forall t \; (16)$$

F. Energy storage systems modeling

Constraint (17) describes the ESS modeling to calculate the state-of-charge level in each period. ($soc_{w,t}^{st}$), taking into account both charging (SF_w^{ch}) and discharging efficiencies (SF_w^{dis}) for charging ($p_{w,t}^{ch}$) and discharging ($p_{w,t}^{dis}$) decision-making, respectively. Constraints (18) and (19) impose the maximum charging (G_w^{ch-max}) and discharging limits ($G_w^{dis-max}$), respectively, taking into account participation in both energy ($p_{w,t}^{ch}$ and $p_{w,t}^{dis}$) and reserve ($r_{w,t}^{dn}$ and $r_{w,t}^{up}$) scheduling. On top of that, constraints (20) and (21) ensure that ESS energy level and reserve schedules are within the allowable energy limits, both minimum (SOC_w^{min}) and maximum (SOC_w^{max}), correspondingly.

$$soc_{w,t}^{st} = soc_{w,t-1}^{st} + p_{w,t}^{ch} \cdot SF_w^{ch} - \frac{p_{w,t}^{dis}}{SF_w^{dis}} \quad \forall w,t \; (17)$$

$$p_{w,t}^{ch} + r_{w,t}^{dn} \leq G_w^{ch-max} \quad \forall w,t \; (18)$$

$$p_{w,t}^{dis} + r_{w,t}^{up} \leq G_w^{dis-max} \quad \forall w,t \; (19)$$

$$soc_{w,t}^{st} - \frac{r_{w,t}^{up}}{SF_w^{dis}} \geq SOC_w^{min} \quad \forall w,t \; (20)$$

$$soc_{w,t}^{st} + r_{w,t}^{dn} \cdot SF_w^{ch} \leq SOC_w^{max} \quad \forall w,t \; (21)$$

G. Electric vehicles modeling

Constraints (22-26) describe the corresponding EV modeling in a similar way to ESS modeling. In particular, equation (22) determines the state-of-charge level in each period ($soc_{v,t}^{ev}$), constraints (23) and (24) set the upper bounds on charging (G_v^{ch-max}) and discharging decisions (G_v^{ch-max}) considering their participation in both energy and reserve scheduling. Finally, constraints (25) and (26) ensure that EV energy levels and reserve schedules are within the allowable energy limits.

$$soc_{v,t}^{ev} = soc_{v,t-1}^{ev} + p_{v,t}^{ch} \cdot SF_v^{ch} \cdot n_v^{tot} - \frac{\left(p_{v,t}^{dis} + E_{v,t}^{cons}\right)}{SF_v^{dis}} \quad \forall v,t \ (22)$$

$$p_{v,t}^{ch} + r_{v,t}^{dn} \leq G_v^{ch-max} \quad \forall v,t \ (23)$$

$$p_{v,t}^{dis} + r_{v,t}^{up} \leq G_v^{dis-max} \quad \forall v,t \ (24)$$

$$soc_{v,t}^{ev} - \frac{r_{v,t}^{up}}{SF_v^{dis}} \geq SOC_v^{min} \quad \forall v,t \ (25)$$

$$soc_{v,t}^{ev} + r_{v,t}^{dn} \cdot SF_v^{ch} \leq SOC_w^{max} \quad \forall v,t \ (26)$$

H. Demand response modeling

Constraints (27-30) describe the demand response modeling. More specifically, equation (27) determines the amount of the modified energy demand ($d_{y,t}^{cl}$), considering the demand response measures ($d_{y,t}^{up}$ and $d_{y,t}^{dn}$) upon the reference energy demand ($D_{y,t}^{ref}$). Equation (28) guarantees that the total energy demand remains unchanged over the scheduling time horizon. Finally, constraints (29-30) describe the demand variation ranges, both minimum ($D_{y,t}^{min}$) and maximum ($D_{y,t}^{max}$), correspondingly.

$$d_{y,t}^{cl} = D_{y,t}^{ref} + d_{y,t}^{up} - d_{y,t}^{dn} \quad \forall y,t \ (27)$$

$$\sum_t d_{y,t}^{up} = \sum_t d_{y,t}^{dn} \quad \forall y \ (28)$$

$$d_{y,t}^{cl} - r_{y,t}^{up} \geq D_{y,t}^{min} \quad \forall y,t \ (29)$$

$$d_{y,t}^{cl} + r_{y,t}^{dn} \leq D_{y,t}^{max} \quad \forall y,t \ (30)$$

I. System reserve requirements

Constraints (31) and (32) set the system operating reserve requirements in both upward (R_t^{up}) and downward (R_t^{dn}) directions, correspondingly.

$$\sum_i r_{i,t}^{up} + \sum_w r_{w,t}^{up} + \sum_v r_{v,t}^{up} + \sum_y r_{y,t}^{up} + r_t^{up} \geq R_t^{up} \quad \forall t \ (31)$$

$$\sum_i r_{i,t}^{dn} + \sum_w r_{w,t}^{dn} + \sum_v r_{v,t}^{dn} + \sum_y r_{y,t}^{dn} + \sum_r r_{r,t}^{dn} + r_t^{dn} \geq R_t^{dn} \quad \forall t \ (32)$$

The overall problem is formulated as a mixed-integer quadratically constrained programming problem involving the cost minimization objective function (1) and is subject to constraints and equations (2)-(32).

III. CASE STUDY

An illustrative case study of an islanded power system has been selected to test the applicability of the proposed approach. In particular, the selected islanded power system consists of 4 thermal power units in the form of diesel generators, the detailed techno-economic characteristics of which are presented in Tab. 1 and 2. These contain the cost coefficients a,b, and c, the start-up cost, the technical minimums and maximums, the ramp limits, as well as the minimum uptimes and downtimes.

Table 1: Economic data of the studied power system

Thermal unit i	a (€/MW²)	b (€/MW)	c (€)	Start-up cost (€)
G1	3	20	100	50000
G2	4.05	18.07	98.87	50000
G3	3.99	19.21	107.2	50000
G4	3.88	26.18	95.31	50000

Table 2: Technical data of the studied power system

Thermal unit i	Pmin (MW)	Pmax (MW)	Ramp-up and down (MW/min)	Min.uptime (h)	Min.down time (h)
G1	28	150	4	16	1
G2	20	120	3	16	1
G3	30	100	3	16	1
G4	20	200	5	16	1

In addition to the thermal power units, there is wind and solar power installed capacity, amounting to 280 MW, of which wind turbines account for 200 MW and solar photovoltaics the other 80 MW. Fig. 1 shows the daily reference electricity demand and the RES potential.

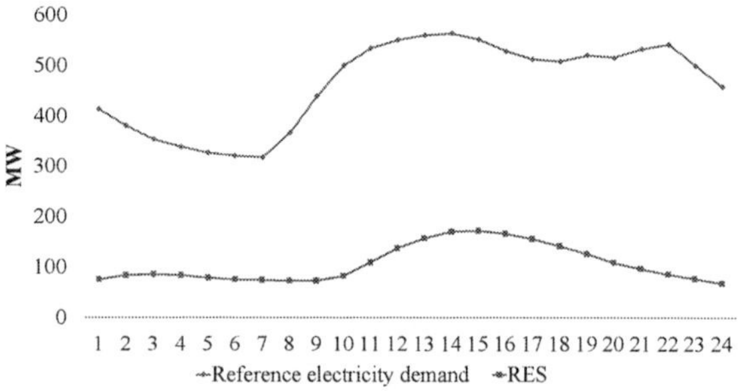

Figure 1. Reference electricity demand and RES potential in each time period

Table 3: ESS operational data

ESS data	Value
Battery capacity (MWh)	50
Initial storage level (MWh)	25
Charging efficiency (%)	0.95
Discharging efficiency (%)	0.9
Minimum battery energy storage level (MWh)	5
Charging rate (MW)	50
Discharging rate (MW)	50

Tab. 3 presents the operational data of the considered installed ESS, including the battery capacity, the initial energy storage level, the charging and discharging efficiencies, the minimum allowable battery energy storage level, as well as the charging and discharging rates.

Tab. 4 presents the operational data of the considered EVs, including the number of EVs, the battery capacity, the charging and discharging efficiencies, as well as the charging and discharging rates. The minimum battery energy storage level of each EV is assumed to be 10% of its battery capacity, as well as the initial energy storage level at hour "0" is considered 50% of its battery capacity.

Table 4: EVs technoeconomic data

Electric vehicle type	Number of electric vehicles	Charging, discharging efficiency (%)	Charging, discharging power (MW)	Battery capacity (MWh)
EV1	300	95	0.0072	0.0173
EV2	50	95	0.011	0.1
EV3	250	95	0.0037	0.0076
EV4	150	95	0.0046	0.0358
EV5	250	95	0.0037	0.023

The reserve requirements are assumed to be 20% of the reference electricity demand, in both upward and downward directions. The reserve provision price is assumed to be 10 €/MW for thermal power units, 12 €/MW for ESS, 15 €/MW for EVs, 20 €/MW for the grid, and 25 €/MW for the demand response programs.

Finally, the capacity of the electricity interconnection with the mainland power system is considered to be 250 MW for both imports and exports.

IV. RESULTS AND DISCUSSION

The problem has been globally optimized using the CONOPT solver within the General Algebraic Modeling System (GAMS) Studio 37 [13]. An optimality gap of 0% has been achieved.

The total daily operating cost of the islanded power system amount to around 1.77 million €, of which the electricity purchases net cost accounts for 71% of the total, and the rest is allocated between diesel oil units production cost (26%) and reserve provision cost (3%).

Fig. 2 depicts the examined islanded power system's energy supply and demand mix on an hourly basis. ESSs and EVs facilitate energy allocation for several hours of the day with their charging and discharging cycles. The results highlight that the contribution from the mainland power system's net imports is mainly responsible for almost 49% of the total electricity supply, followed by diesel-fired power generation with 28% of the total and RES with the remaining 23%, as depicted in Figure 3 presenting the energy supply allocation on a daily basis.

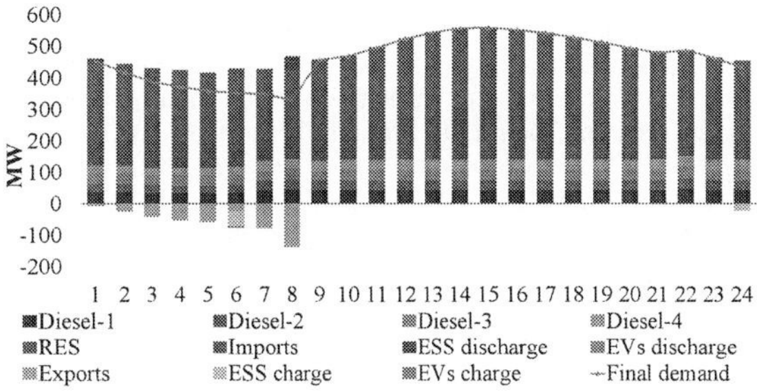

Figure 2. Energy supply and demand mix of the examined islanded power system on an hourly basis

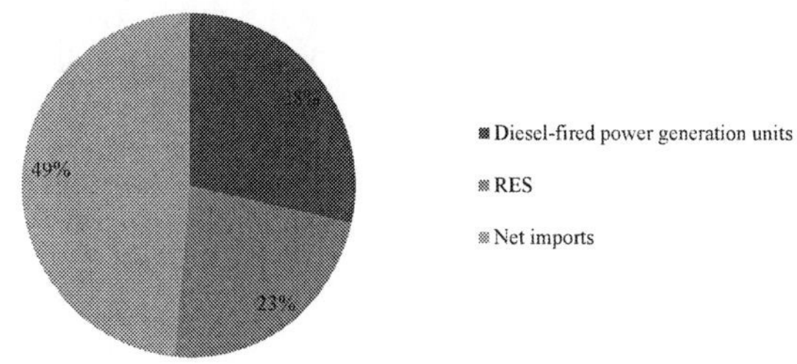

Figure 3. Energy supply allocation of the examined islanded power system on a daily basis

Fig. 4 portrays the operating-up reserve provision mix of the examined islanded power system. This service is exclusively covered by diesel-fired units, which are considered the most economical option for the coverage of that service. They all remain operational during the whole day, and they operate close to their technical minimums, as depicted in Fig. 5; thus, they have the potential to increase their power output, providing this upward service.

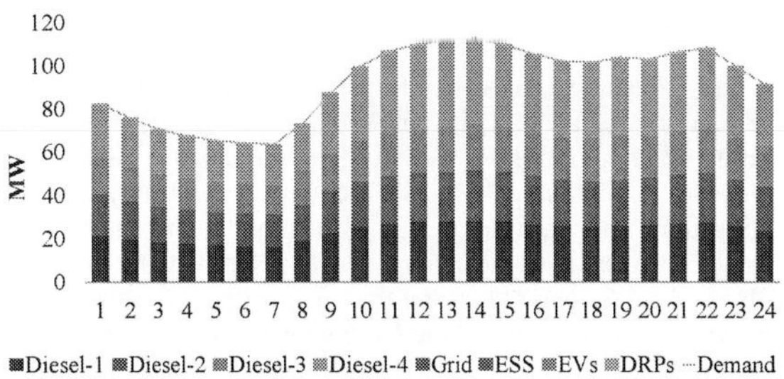

Figure 4. Operating-up reserve provision mix of the examined islanded power system

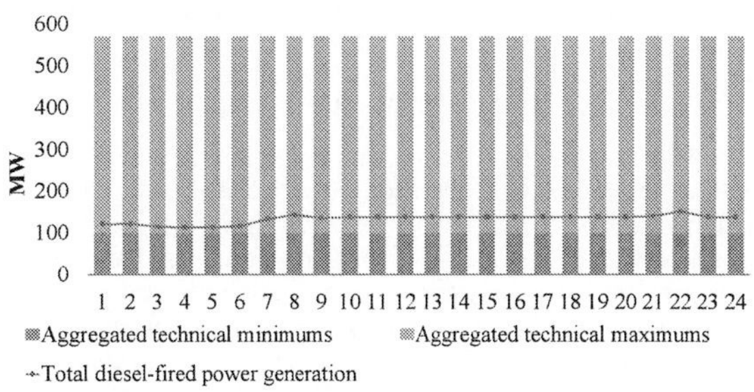

Figure 5. Hourly total diesel-fired power generation of the examined islanded power system on an hourly basis

Fig. 6 shows the operating-down reserve provision mix of the examined islanded power system. The ESS provides 45% of the total requirements of that service on a daily basis. The diesel-fired units contribute around 37% of the whole needs, as well as EVs, stand for almost 5% of the daily requirements. It is worth mentioning that the electricity interconnection with the mainland power system takes part in the coverage of that service, providing around 13% of the total daily needs.

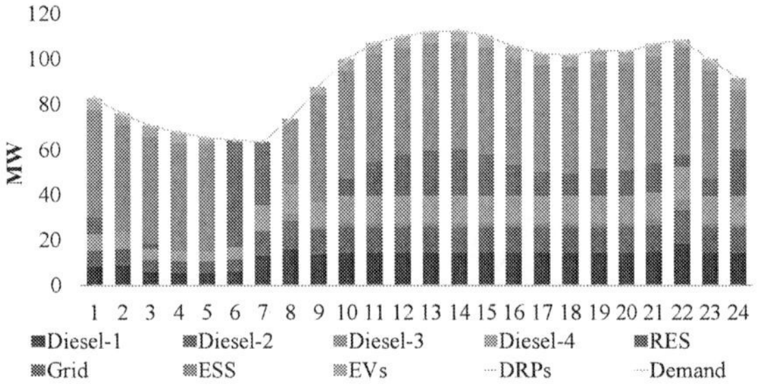

Figure 6. Operating-down reserve provision mix of the examined islanded power system

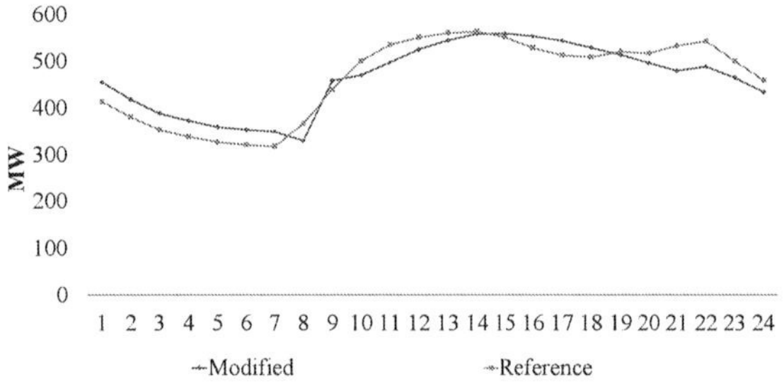

Figure 7. Modified and reference electricity demand of the examined islanded power system

Fig. 7 presents the modified and the reference electricity demand of the examined islanded power system. As it can be observed, although the total daily load remains the same, there are some different patterns in the hourly electricity demand allocation. The modified electricity demand is characterized by some increases in comparison with the reference one during 1-7h, 9h, 15-18h, namely during the hours when the net reference electricity demand is characterized by its lowest levels. Not surprisingly, there are some decreases in the amount of modified electricity demand compared to the reference one during 8h, 10-14h, and 19-24h. These hours comprise the peak periods of the net reference electricity demand, and thus the model determines as an optimal scheduling strategy to mitigate these peaks in order to cover the net load in a more economical way.

V. CONCLUSIONS

Islanded power systems face multiple challenges in their effort to cover their electricity requirements in a reliable, economical, and sustainable way. Within that context, this work proposes an optimization framework based on mixed-integer quadratic programming to optimally implement an islanded power system's energy and reserves scheduling through the consideration of various flexibility providers (ESS, EVs, DRPs), and in addition, it considers the participation of electricity interconnections as both energy and reserve providers. The model outputs highlight the importance of considering these resources in the energy and reserves mix, which can be of further significance in the case of very high RES penetration. Future challenges include the design of all those flexibility providers and the detailed testing of their performance on mid- and long-term planning frameworks.

ACKNOWLEDGMENT

THIS RESEARCH RECEIVED NO EXTERNAL FUNDING.REFERENCES

[1] P. González, J. Villar, C.A. Díaz, and F.A. Campos, "Joint energy and reserve markets: Current implementations and modeling trends". Electric Power Systems Research, vol. 109, 2014, pp. 101-111, doi: 10.1016/j.epsr.2013.12.013.

[2] S.S. Reddy, P.R. Bijwe,, and A.R. Abhyankar, "Optimum day-ahead clearing of energy and reserve markets with wind power generation using anticipated real-time adjustment costs". International Journal of Electrical Power & Energy Systems, , October 2015, vol. 71, pp. 242-253, doi: 10.1016/j.ijepes.2015.03.002.

[3] G. Liu, and K. Tomsovic, "A full demand response model in co-optimized energy and reserve market". Electric Power Systems Research, June 2014, vol. 111, pp. 62-70, doi: 10.1016/j.epsr.2014.02.006.

[4] M. Carrion, and J. M. Arroyo, "A computationally efficient mixed-integer linear formulation for the thermal unit commitment problem". IEEE Transactions on Power Systems, Aug. 2006, vol. 21, no. 3, pp. 1371-1378, doi: 10.1109/TPWRS.2006.876672.

[5] A. Viana, and J.P. Pedroso, "A new MILP-based approach for unit commitment in power production planning". International Journal of Electrical Power & Energy Systems, January 2013, vol. 44, no. 1, pp. 997-1005, doi: 10.1016/j.ijepes.2012.08.046.

[6] D. Divényi, B. Polgári, Á. Sleisz, P. Sőrés, and D. Raisz, "Algorithm design for European electricity market clearing with joint allocation of energy and control reserves". International Journal of Electrical Power & Energy Systems, October 2019, vol. 111, pp. 269-285, doi: 10.1016/j.ijepes.2019.04.006.

[7] N.E. Koltsaklis, A.S. Dagoumas, "Incorporating unit commitment aspects to the European electricity markets algorithm: An optimization model for the joint clearing of energy and reserve markets". Applied Energy, December 2018, vol. 231, pp. 235-258, doi: 10.1016/j.apenergy.2018.09.098.

[8] N. Koltsaklis, A. Dagoumas, "Policy Implications of Power Exchanges on Operational Scheduling: Evaluating EUPHEMIA's Market Products in Case of Greece". Energies, October 2018, vol. 11, 2715, doi: 10.3390/en11102715.

[9] M. Zugno, and A.J. Conejo, "A robust optimization approach to energy and reserve dispatch in electricity markets". European Journal of Operational Research, December 2015, vol. 247, no. 2, pp. 659-671, doi: 10.1016/j.ejor.2015.05.081.

[10] F. Wen, and A.K. David, "Coordination of bidding strategies in day-ahead energy and spinning reserve markets". International Journal of Electrical Power & Energy Systems, May 2002, vol. 24, no. 4, pp. 251-261, doi: 10.1016/S0142-0615(01)00038-2.

[11] M.E. Nazari, and M.M. Ardehali, "Optimal bidding strategy for a GENCO in day-ahead energy and spinning reserve markets with considerations for coordinated wind-pumped storage-thermal system and CO_2 emission". Energy Strategy Reviews, November 2019, vol. 26, pp. 100405, doi: 10.1016/j.esr.2019.100405.

[12] E. Lobato, L. Sigrist, and L. Rouco. "Value of electric interconnection links in remote island power systems: The Spanish Canary and Balearic archipelago cases". International Journal of Electrical Power & Energy Systems, 2017, vol. 91, pp. 192-200, 10.1016/j.ijepes.2017.03.014.

[13] A.Brooke, D. Kendrick, A. Meeraus, and R. Raman, "GAMS: A User's Guide", GAMS Development Corporation, Washington D.C., 1998.

Implementation of the Requirements Guidelines for New Electricity Generation Units in the EU to Operation Electricity System and Electricity Balancing

Ján Poničan, Matej Sadloň, František Janíček
Institute of Power and Applied Electrical Engineering
FEI STU
Bratislava, Slovakia
jan.ponican@stuba.sk, matej.sadlon@stuba.sk, frantisek.janicek@stuba.sk

Štefan Čerba, Branislav Vrban, Jakub Lüley, Vendula Filová
Institute of Nuclear and Physical Engineering
FEI STU
Bratislava, Slovakia
stefan.cerba@stuba.sk, branislav.vrban@stuba.sk, jakub.luley@stuba.sk, vendula.filova@stuba.sk

Abstract— **One of the major objectives of the EU energy policy is to create a non-discriminatory condition for the internal electricity market. The transposition of the European package into national law is intended to lay down uniform and non-discriminatory conditions for the connection of new power-generating modules to the electricity system and uniform conditions for market participants. Until 2019 there has been no harmonisation in the field of network connection requirements for power-generating modules across the European Union (EU). This has been changed by the so-called "RfG" network code on requirements for grid connection of generators regulation adopted by the EU Commission. The implementation procedure required EU member states to amend their network codes and to restructure them to make them compatible with the regulation. This Article therefore analyses the implementation procedure, explains the structure of this regulation, its main features, assesses its impacts and discusses the future of the respective regulation. This directive of RfG regulation is important from the point of view of the technical conditions for the operation of electricity systems. The article also analyses other legislation related to the implementation of the EU package, known as the "winter energy package". It contains aspects of regulation focused on guidelines so-called („SOGL") establishing a guideline on electricity transmission system operation.**

Keywords — electricity grid requirements, generators, network code, EÚ harmonization

I. Introduction

One of the major objectives of the EU energy policy is to create an internal electricity market. This process has been in place already since 1999. Upon its completion it should provide all consumers with a real choice in the supply, fair business opportunities and increased cross-border trade to reap the benefits of greater efficiency, competitive prices, higher service standards and to contribute to security of supply and sustainability. This is however a long and difficult process requiring a restructuring of the whole system.

The backbone of an internal market should be a well-functioning interconnected electricity grid. An interconnected system however requires a common understanding of the grid connection requirements for power generators. A system with significantly different grid connection requirements in different EU member states (MSs) represents a major barrier of the internal market. Moreover, system security is partly dependant on the technical capabilities of the power generators. Within a synchronous area, frequency change in one MS would immediately impact frequency and could damage equipment in all the others. All transmission (TSO) and distribution (DSO) system operators should be thus eager to unify the grid connection rules. Secure system operation is only possible if there is close cooperation between power-generating facility owners and system operators. Here sufficient robustness to cope with disturbances and an ability to help to prevent a disruption or to facilitate restoration of the system after a collapse are fundamental prerequisites [1].

The Commission Regulation (EU) 2016/631 establishing a network code on requirements for grid connection of generators (RfG Regulation) was an important milestone taking the sector closer to a level-playing field in terms of grid connection rules. Its importance lies in the unification of the grid connection requirements for power generators throughout the whole EU. It is stated in the recitals of the Act that harmonised rules for grid connection should be set out to provide a clear legal framework for grid connections, facilitate Union-wide trade in electricity, ensure system security, facilitate the integration of renewable electricity sources, increase competition, and allow more efficient use of the network and resources, for the benefit of consumers [1].

Hence it is clear that the EU sees transparent and harmonised grid connection rules as an important precondition in the internal market creation. Unlike most of the national rules, EU rules are usually subject to a long process of creation and adoption. RfG regulation was recommended to the European Commission already in 2013. In 2015 it was unanimously approved by the representatives of all the MSs. As it is a technical Act with significant implications a transition period of 3 years was adopted. TSOs and national authorities needed time for implementation of the rules into the national grid codes and both system operators and power generators needed time to adapt to the new requirements. The entire harmonised code entered into force in April 2019. The RfG regulation is not solely made up of exhaustive requirements strictly binding the MSs. It has also introduced non-exhaustive requirements that provided a framework and the details had to be shaped by the MS. Typical example being a rule defining upper or lower bounds and the MS must choose a level within the respective range. Thus, it might be assumed that there are completely new, slightly different grid codes in all over the EU and market participants are getting used to the new conditions. Commission Regulation (EU) 2017/1485 establishing a guideline on electricity transmission system operation (SOGL) has been so implemented in the technical conditions of the Slovak TSO. Technical conditions have been created for TSO, which must comply with EU law. These technical conditions are divided into several parts.

II. CATEGORIES OF POWER-GENERATING MODULES

The main novelty of the requirements is the distinction of the power-generating modules based on 2 factors. The first one being the size and effect on the overall system. Categories A, B, C and D are distinguished. The second being the generator type, i.e., whether the generator is synchronously connected to the grid, or the connection is made via a power converter. This is justified by the importance of the synchronous generators as they have an inherent capability to resist or limit frequency deviations. This is in contrast with technologies connected to the grid via power converters, usually renewable sources (RES), that do not have this characteristic. Grid stability requires countermeasures to be adopted and thus the RfG regulation adopts specific rules for devices connected via a power converter. Synthetic inertia being the most relevant one. Based on that, the RfG regulation sets out general requirements applicable to all generators, specific requirements for synchronously connected devices, specific requirements for generators connected via a power converter, i.e., Power Park Modules and specific requirements applicable to AC connected offshore generators.

Type A requirements represent minimal conditions that need to be met by the generator to be connected to the grid. A type of generator is only required to provide a limited automated response and minimal system operator control. Simply explained, the source must be able to stay in operation at small frequency deviations. The aim is to minimize the risk of simultaneous outage of small generating units even at small frequency disturbances. B type requirements provide for a wider range of automated response and higher resilience to operational events. This category must meet all the conditions related to category A plus conditions specific for type B. These generators are required to provide a significantly higher level of system operator control and information. They are also required to provide for an automated response to mitigate the impact of system events. Information exchange being the key to enable TSOs and DSOs to maintain the system stability. Operators need to have both an online overview of the state of the system and a possibility to give the power-generating modules direct operational instructions. For example, to adjust the output of the generator to ensure a system security. A substantial tightening of the requirements comes with category C. These generators are required to provide for a stable and highly controllable real-time dynamic response and should be able to provide ancillary services to ensure the security of supply. These sources should be able to respond to both intact and system disturbed situations and should provide the information and control necessary to utilise generation in different situations [1]. Specificity of the category D is a higher voltage level. These generating units should meet the highest level of requirements as they are crucial in the operation of the whole interconnected system.

As it is mentioned above, categories A, B, C, D are distinguished based on the voltage level and rated power of the generating unit. The main reason why network connection requirements need to be regulated is security of supply and stability of the network as such. From the frequency stability perspective, there is only a minor difference between the outage of one 1,000 MW thermal unit and of 100,000 small 10 kW household plants [3]. As the number of distributed generators increases and this goes hand in hand with their impacts on the grid stability a threshold discussion should not focus only at large generating units. The A category threshold is therefore set to 0.8 KW. This means that once the generating unit has such an installed capacity, it needs to meet the grid connection requirements, otherwise it may not be connected. Thresholds for categories B, C and D are on the other hand defined only with the highest possible value. MSs are therefore free to lower them within the implementation procedure.

Fig. 1 shows that many MSs used this option and lowered the threshold significantly. Good example is Slovakia that has decided to lower the B threshold from 1 MW to 100 kW, C threshold from 50 MW to 5 MW and the D threshold from 75 MW to 20 MW [4]. An official guidance on the RfG regulation [5] explains that the thresholds shall be determined based on the national generation structure, both current and future. Shares of each module type and a share of the modules connected via a power converter should be taken into consideration.

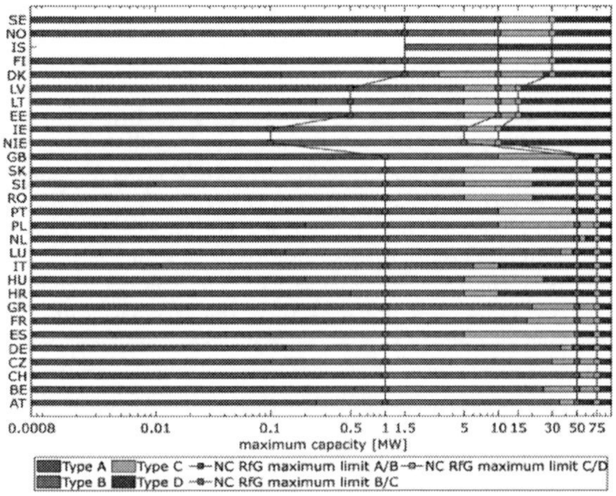

Figure. 1 Thresholds adopted in MSs for A,B,C and D categories [4]

III. SLOVAK IMPLEMENTATION OF RFG REGULATION

As in all the other MSs, the procedure was in the hands of the TSO and the national regulatory authority. The process resulted in a major amendment of network codes covering the technical operation of the transmission and regional distribution grids. However, even the EU harmonisation has not managed to change the fact that the grid connection of significant sources in Slovakia was basically prohibited due to technical constraints connected with the impact of renewable power generation on the system security. Slovak TSO has explained that new connections will only be allowed after an increase in interconnection capacity on the border with Hungary. This condition has been fulfilled only recently. Two 400 kV power lines connecting Veľký Ďur, Gabčíkovo and Gönyű and one 400 kV power line connecting Rimavská Sobota and Sajóivánka came into an operation on April 6, 2021. This has unlocked further grid connection capacity and gave an end to the so-called stop-state for new generating modules connection that has been in place since 2012.

First amendments of the distribution network codes were already visible in 2017. This were although only negligible and covered the new EU-wide distinction of power-generating modules into 4 categories based on their size and voltage level. More detailed implementation took place only in 2019.

Although it is important to highlight the fact that only the recent capacity opening gave a real practical power to the novel regulation.

An important observation made by a comparison of novel distribution network codes is that individual codes implemented the regulation almost identically. However, some non-exhaustive requirements allowing for a choice in details were set differently. Examples are differences in the setting of protections, slightly different voltage time courses during a short circuit operation and reactive power requirements for synchronous units of type C, D.

IV. Slovak Implementation of SOGL Regulation

The SOGL Regulation was implemented only within the technical conditions of the Slovak TSO. In the technical conditions, the regulation was implemented mainly in document "D" (dispatching control of the electricity system), which specifies the scope of data exchange of the TSOs, DSOs and major network users. The document specifies requirements for the safety and quality of transmission system operation in accordance with the Regulation. The scope of the data exchange specifies the requirements for the scope of data that are needed in the documents for the annual preparation of the operation of power facilities for the year "n+2" and "n+3". As part of the annual preparation of the transmission system operation, scenarios are analysed by calculations, which assess the impact of the transmission system on the safety of operation. Scenarios must analyse the real resource base, cross-border market mechanisms, production and consumption patterns, variations in production and much more [2].

Regulation requires analysing the dynamic stability of the power system. The aim of the analysis is to define stability limits and possible stability problems in the electricity system. If a potential impact on neighbouring TSOs is identified, possible corrective actions shall be defined. The analysis of monitoring and control of the dynamic stability of the power system is performed at least on an annual basis, where analyses of angular stability, frequency stability and voltage stability are performed. The TSO must monitor dynamic stability through field studies. The assessment of dynamic stability has to be coordinated with the other transmission system operators in each synchronous area. The assessment of dynamic stability is also focused on the possibilities of damping interregional oscillations that may affect other transmission system operators. In the event of a breach of the stability limits in each area, corrective measures shall be proposed and activated by the transmission system operator. The troubleshooting times shall be less than the critical fault clearing time calculated by the transmission system operator in the dynamic stability analysis. Regulation requires all market participants to cooperate in managing this analysis and to provide the required input data [2].

The regulation stipulates that the requirements for the safety and quality of the TSO are applied to the technical conditions of the transmission system operator in accordance with the regulation. The responsibilities of the TSO are also defined for voltage and reactive power control. The voltage limits for the continental European area must be during normal operation 0.9 – 1.118 p.u. for 110 – 300 kV level and 0.9 – 1.05 p.u. for 300 – 400 kV level. Operating safety limits are also defined for short-circuit current values in accordance with the regulation. The calculation of maximum and minimum short-circuit currents is each carried out by the transmission system operator in accordance with the Regulation and the currently valid standard STN EN 60909-0: 2016. The maximum short-circuit current at which the rated resistance of the elements in the network is exceeded must be determined by the transmission system operator. The maximum short-circuit current at which the rated resilience of the elements in the network is exceeded must be determined by the TSO.

V. Content of the Amendments

The central change is the introduction of 4 categories of power-generating modules based on which the complexity of requirements is defined. As already mentioned, thresholds distinguishing the categories were set uniformly for whole Slovakia. Sources connected to the distribution grid below 110 kV level may be covered by any of the A, B, C categories depending on the size of their installed capacity. Units connected to the 110 kV distribution grid, or the transmission system fall exclusively into category D. Each individual category is linked with a list of requirements that need to be meet for the generator to be connected to the system.

The requirements are mainly focused on the security of the system and the impacts a power-generating module may have on the grid. The requirements can be divided into several categories, such as Frequency and voltage; Requirements for reactive power; Load-frequency control related issues; Short-circuit current; Requirements for protection devices and settings; Fault-ride-through capability and Provision of ancillary services.

The first set of requirements is connected to frequency. Primary requirement being generating unit's capability to remain in operation for a defined period in case of a frequency deviation within a range of 47.5 to 51.5 Hz. Specific rates of change of frequency and specific timeframes are set for individual A, B, C and D categories. C and D type generating units also need to fulfil various conditions related to primary and secondary frequency regulation. These may be described as their active power frequency response capabilities enabling them to react to overfrequency and underfrequency. Their control systems must be able to adjust the set value of active power in accordance with the instructions of the relevant DSO.

Voltage requirements are built on unit's fault-ride-through and automatic connection after faults or planned disconnections capabilities. The new regulation defines the exact voltage and frequency ranges for individual categories of generating units, under which they can be reconnected back into the system by means of a phasing element. Proper behaviour during short circuits is crucial for maintaining the system safety in terms of their participation in short circuit detection, maintaining voltage during a short circuit and its post-short circuit restoration and ensuring frequency stability by quick post-short circuit active power recovery, is particularly important in small systems [3]. Fault-ride-through requirements are again distinguished based on the category of the generator and defined by a time course of the voltage under fault conditions at the connection point, in which the generating unit needs to remain connected and to continue in stable operation after the fault clearing. Different time courses are defined for synchronous and asynchronous sources. Asynchronous modules of type C and D with a fault-ride-through capability are for example required to be able to supply active power no later than 150 ms from the fault inception and to be capable of contributing to damping power oscillations. Figure 2 compares the fault-ride-through time course of the voltage for type B and type C synchronous and asynchronous generating modules with type D synchronous and asynchronous modules.

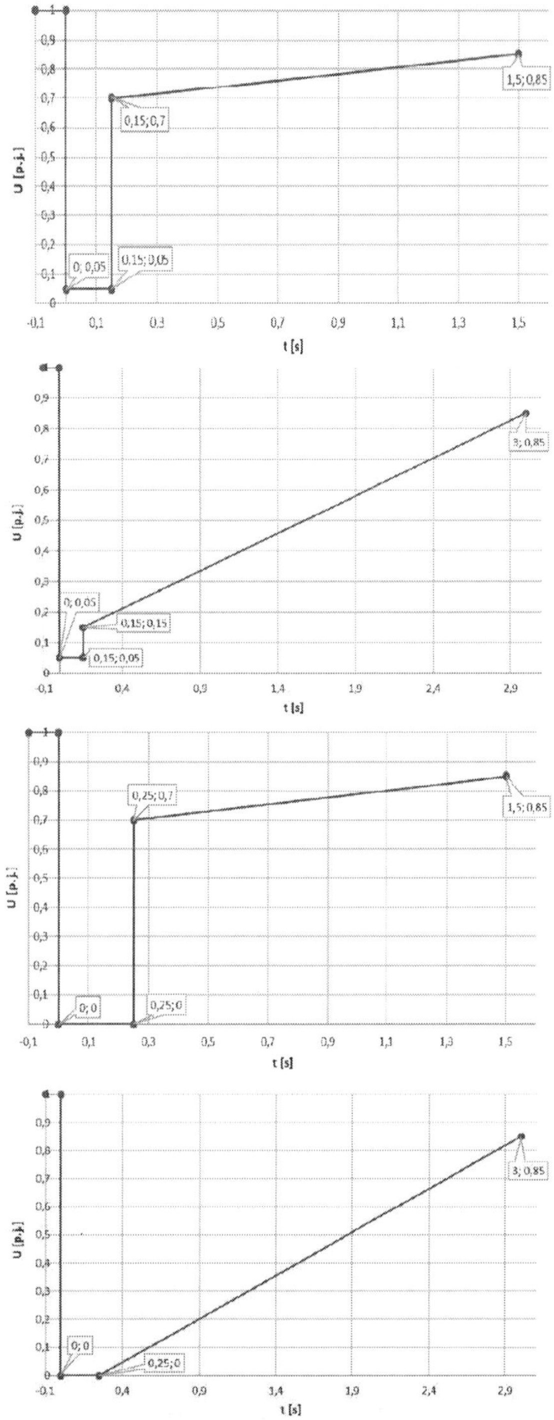

Figure 2 Comparison of fault-ride-through time courses of the voltage for various types of modules [6]

Third set of requirements is connected to the reactive power. Slovak network codes only apply reactive power related conditions to C and D sources even though the DSO could define relevant reactive power conditions also for category B. Although the conditions are set separately for synchronous and asynchronous modules, overall rules are very similar for both types. The main requirement is that in case of a maximum active power output, a synchronous power-generating module must be able to operate within the inner envelope of the diagram shown in figure 3. In case of a lower output, it is required to operate within its PQ diagram. Similar applies to asynchronous modules. They should be able to operate within the inner envelope of the figure 3 in case of a maximum active power output and in case of a lower output, it is required to operate within the limits of the figure 4. Individual DSOs adopted slightly different limits of the inner envelope, but in principle still very similar regulations.

An interesting finding based on a comprehensive comparison of Slovak network codes and the original EU regulation may be seen upon an analysis of the figure 4 and 5. The profile in question forms a binding part of the network code adopted by the western distribution system operator. However, other distribution network codes adopted similar diagrams. The figure says that axis y represents voltage. However, the original profile enacted in the EU regulation specifies that the y axis in question represents active power expressed by the ratio of its actual value and the maximum capacity. We might suppose that this is an implementation error. Interesting fact is that it has been adopted by all three DSOs.

Figure 6 shows a diagram of the operation of non-synchronous power modules for type D. It represents the ability of a non-synchronous production facility to provide reactive power when operating at a power other than its maximum power. The above graphs in terms of voltage levels of the electricity system in Slovakia are for sources connected to the transmission system in accordance with the regulation. For the purposes of the RfG, the y-axis for non-synchronous production facilities is represented by the active power relative to the nominal values of the production unit.

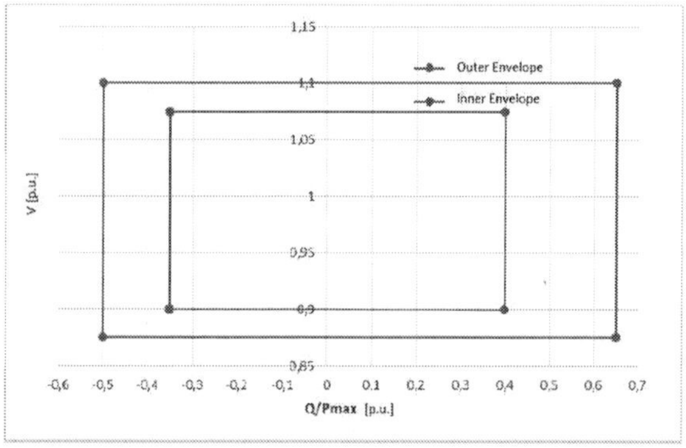

Figure 3 U-Q/Pmax-profile of a synchronous power-generating module [7]

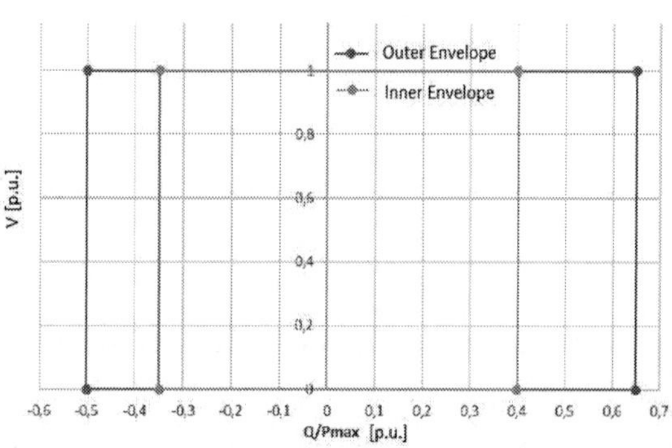

Figure. 4 U-Q/Pmax-profile of an asynchronous power-generating module [6]

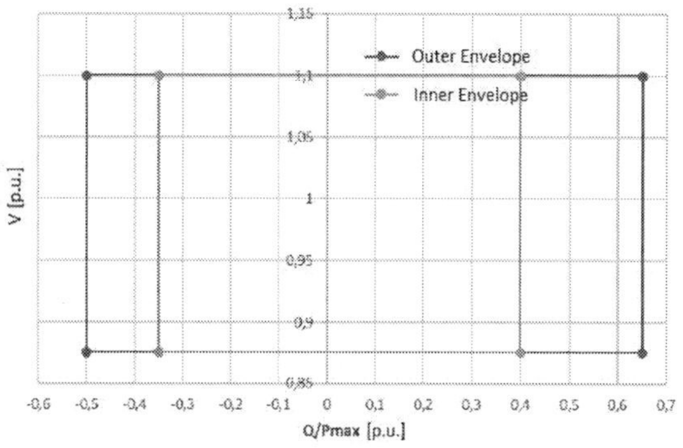

Figure 5 U-Q/Pmax-profile of an asynchronous power-generating module [6]

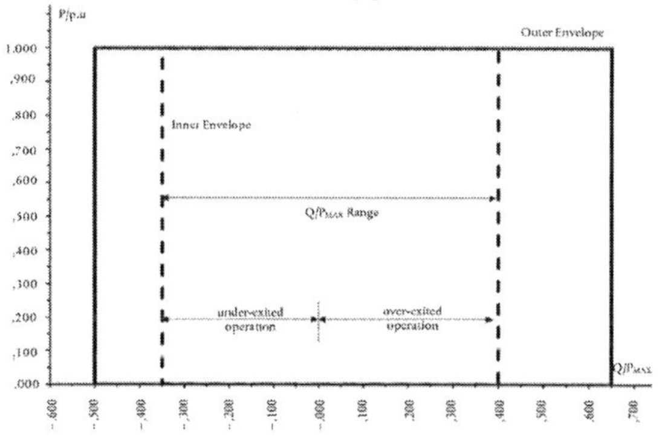

Figure 6 U-Q/Pmax-profile of an asynchronous power-generating module [8]

It is important to explain that requirements for A and B modules are not that extensive. A significant increase in prerequisites comes with the C category, i.e., in Slovak conditions in case the capacity of the generating module exceeds 5 MW. Type C and D generators must for example meet various requirements concerning the black start capability, island operation, houseload operation, the loss of angular stability and the instrumentation of the generating facility. The instrumentation needs to cover requirements related to facility to provide fault recording and monitoring of both the operation and of the dynamic system behaviour. Voltage, active power, reactive power and frequency parameters have to be recorded in defined time periods. C and D sources shall also provide monitoring the behaviour of the system by measuring frequency oscillations and regarding loss of angular stability or loss of control, these power-generating modules shall be capable of disconnecting automatically from the network to help preserve system security or to prevent damage. As mentioned above, C and D units are required to be capable of island operation and to be galvanically isolated at the main disconnection point. Another important requirement is related to the obligation to provide the system operator with simulating modules reflecting the behaviour of the unit in both steady-state and dynamic simulations or in electromagnetic transient simulations.

Authors need to express their opinion that the individual requirements in all new Slovak network codes are not organized in a transparent manner. If we compare the structure of the original regulation and the structure of the network code, we must conclude that the structure of the network code is very chaotic. Slovak codes miss the systematic order of the requirements based on the category of the generating unit. Thus, it will be very difficult for a laic investor to understand the code and to find out which are the requirements that must be met. Guidelines provided by the TSOs and DSOs would be very relevant.

VI. Conclusion

RfG regulation needs to be considered as a significant step towards a continental grid unification, a big step towards a level-playing field, transparency, and an EU wide internal electricity market. Up until its adoption, network connection requirements have only been regulated at the level of the continental Europe's synchronous system by provisions of an Operational Handbook, Policy 1 and Policy 5. As opposed to national network codes, these standards haven't been binding. TSOs' adherence to it was voluntary and they only became binding pursuant to provisions of a multilateral interconnection agreement [3]. Moreover, Handbook has never laid down concrete requirements for generating modules. It has rather focused on requirements that had to be met by the regulatory area as such. In practice it was still in the hands of the national authorities do adopt their own rules reflecting the needs to follow the Handbook. These were however often significantly different.

RfG regulation may be seen as a platform for further legislative processes and closer harmonisation. This has also been acknowledged by the EU commission. Its comprehensive comparison of implementation results in the individual MSs has shown that non-exhaustive rules gave the MSs a significant room, that in case of many requirements ended up in substantial differences across the whole EU [4]. Differences may have various negative impact and are a barrier in the internal market. One of the negative impacts being the fact that different thresholds introduce a challenge for generator manufacturers who plan to sell their products outside their own country. This is because a module of a particular type in one MS may be regarded as another type in another MS and therefore the product will have to comply with more requirements [4].

The findings from this article are important for the possibilities of higher integration of renewable energy sources into the electricity grid. Generating units connected to the distribution system should be able to provide regulatory services. Reactive power regulation is key to maintaining the safety and reliability of the system, and currently this option is provided by a minimum amount of generating units in the distribution grid. This feature is key to the higher development of decentralized energy and smart grids.

Acknowledgement

This publication was created thanks to support under the Operational Program Integrated Infrastructure for the project: International Center of Excellence for Research on Intelligent and Secure Information and Communication Technologies and Systems – 2. stage, ITMS code: 313021W404, co-financed by the European Regional Development Fund, by the Scientific Grant Agency of the Ministry of Education of Slovak Republic No. VEGA 1/0615/21 and partially by the ESFR-SIMPLE project funded by the European Union's Horizon Europe research and innovation program.

References

[1] Regulation (EU) 2016/631. "Establishing a network code on requirements for grid connection of generators." Internet: https://eur-lex.europa.eu/legal-content/EN/TXT/PDF/?uri=CELEX:32016R0631&from=EN [Apr. 15, 2021].

[2] Regulation (EU) 2017/1485. "Establishing a guideline on electricity transmission system operation." Internet: https://eur-lex.europa.eu/legal-

content/EN/TXT/PDF/?uri=CELEX:32017R1485&from=SK [Jan. 1, 2022].

[3] J. Rychlak. "National-level Implementation of the Commission Regulation (EU) 2016/631 of 14/04/2016 Establishing a Network Code on Requirements for Grid Connection of Generators." Internet: http://yadda.icm.edu.pl/baztech/element/bwmeta1.element.baztech-c7a1f6e5-ea84-4d5e-9773-48755c323faa/c/Rychlak_National_acta_energetica_2_2018.pdf [Apr. 20, 2021].

[4] Publications Office of the EU. "Implementation of the network code on requirements for grid connection of generators." Internet: https://op.europa.eu/en/publication-detail/-/publication/ee9ecda7-6788-11eb-aeb5-01aa75ed71a1/language-en?WT.mc_id=Searchresult&WT.ria_c=37085&WT.ria_f=3608&WT.ria_ev=search [Apr. 15, 2021].

[5] Entsoe. "Implementation Guidance Documents." Internet: https://www.entsoe.eu/network_codes/cnc/cnc-igds/ [Apr. 15, 2021].

[6] Západoslovenská distribučná, a. s. "Technické podmienky prevádzkovateľa distribučnej sústavy spoločnosti Západoslovenská distribučná, a.s. účinné od 1.5.2019." Internet: https://www.zsdis.sk/Uvod/Spolocnost/Dokumenty/Predpisy-prevadzkovatela [May. 5, 2021].

[7] Stredoslovenská distribučná, a. s. "Technické podmienky prevádzkovateľa distribučnej sústavy Stredoslovenská distribučná, a. s., platné od 1.4.2019 do 14.02.2021." Internet: https://www.ssd.sk/dokumenty/technicke-podmienky/archiv?page_id=6295 [May. 5, 2021].

[8] Slovenská elektrizačná prenosová sústava, a. s. "Účinné Technické podmienky prístupu a pripojenia, pravidlá prevádzkovania prenosovej sústavy." Internet: https://www.sepsas.sk/sk/dokumenty/technicke-podmienky/ [February. 5, 2022].

Smart Battery as Elementary Asset of Virtual Power Plant – Data from Pilot Operation

Michal Chudý[1], Peter Cvacho[2], János Kurcz[1], Peter Laurinec[2], Mikuláš Praščák[2], Juraj Štrkolec[3]

[1]Institute of power and applied electrical engineering, Slovak university of Technology in Bratislava, Ilkovičova 3, 841 04 Bratislava, Slovak republic
[2]PowereX j.s.a., Nam. 1. Maja 18, 811 06 Bratislava, Slovak republic
[3]AEN Batteries s.r.o., Nám. 1. Mája 18 811 06 Bratislava, Slovak republic

Email addresses: michal.chudy@stuba.sk, peter.cvacho@powerex.io, janos.kurz@stuba.sk, peter.laurinec@powerex.io, miki.prascak@powerex.io, juraj.strkolec@a-en.com

Correspondence: michal.chudy@stuba.sk

Abstract—**Virtual power plants (VPP) are a new and innovative concept of smart grid aggregation. As this concept is a strong candidate to be applied on a wide scale in the near future, this paper presents real data and functionalities of an operational smart battery, which is ready to be aggregated to VPP. This battery is installed at a single node of a power grid, and it is coupled together with solar photovoltaic (PV) production and real industrial customer - a factory. This kind of set-up is considered to be a basic behind the meter element aggregated in a VPP used for flexibility. This kind of smart system is legally in full control of the power consumer; however, this control can be delegated to an aggregator. The system can work as part of VPP, but also as a standalone system, which is the case presented in this paper. Nevertheless, in both cases they have capabilities of improving resiliency and stability of the power grid and contribute to energy sector decarbonization by providing carbon free flexibility. This paper brings data from a real operation of such a system and shows advantages of intelligent control of energy assets.**

Keywords—virtual power plant, aggregation, smart battery, BESS

I. INTRODUCTION

Climate change is recognized as a serious threat for our planet. Electricity production is a major greenhouse gas emissions contributor taking in the fact that 25 % of US greenhouse gas emissions are coming from the electricity production sector [1]. Renewable energy production can be part of the solution. The Integrated national energy and climate plan for years 2021 - 2030 shows that the aim is to increase the share of renewable energy sources on electric energy production by 2030 to 27,3% [2-4]. Although, the stochastic character of renewables such as wind and solar makes such a high contribution very challenging. Their usage requires more flexibility of the electric network system, i.e., extended capacity of spinning reserve and tougher requirements on ancillary services providers. The problem of reserves and imbalance shaving of the grid is the main obstacle for applying more renewable generators to the existing power grid. Efficient solution of this problem can be demand&response management of distributed energy sources which also includes a concept of Virtual Power Plant (VPP).

VPPs are usually presented as a new innovative concept of aggregation of distributed energy production. They usually contain flexible generators, energy storage interconnected with data communication in such a way that can provide demand&response management of distributed generators or energy storage units. These storages of energy could be beneficial for the power system as stated in [5]. Smart Grid is a necessary prerequisite for VPPs. The principle of VPPs is data interconnection between distributed power generators and storage units such as solar panels, wind turbines, diesel generators, batteries, pumped hydro plants and small hydro plants. A lot of theoretical work about VPPs introduced the concept from a technical perspective [6-7]. Usually, optimization of VPPs is the main theoretical problem of the concept. [8–10]. The market-based research of VPP is much less discussed, though this is often one of the major problems which is slowing down the real-life usage of VPPs. Nevertheless, e.g., Shi You et al. [11] are dealing with this problem.

VPP can contain assets in front-of-the-meter and act as a typical power producer, but it can aggregate also behind-the-meter energy assets in order to supply power grid with flexibility. This is a case, where even smaller distributed solar PV with Battery Energy Storage Systems (BESS) can be useful assets of VPPs. As each power market has some special attributes and each customer has different expectations of his energy asset, it can be very useful to operate such an asset with a local AI based energy management system, which is fully integrable to the VPPs. This paper presents such a system with a battery controlled by an AI driven energy management system – smart battery. Operation and benefits of such a system are compared with a system, where the battery is controlled by a simpler algorithm integrated in a PLC unit.

II. System Description

Smart battery operates in the set-up described by Fig. 1. The microgrid system contains an industrial consumer with real daily production. Industrial consumer has installed 225 kWp of rooftop PV solar. This PV production can be used "in the house", but it can be also exported to the power grid, and it is sold according to the hourly based market price. Installed battery has a storage capacity of 120 kWh. It is an indoor Battery Storage System using 55kW Siemens SINAMICS S120 inverter. Elementary battery functionalities are controlled by a PLC unit Simatic S7-1200. Intelligent Energy Management System (EMS) is built above this PLC unit in such a way that this PLC unit is used as a primary data gateway for a cloud based Intelligent EMS, which was designed and is being operated by PowereX [12].

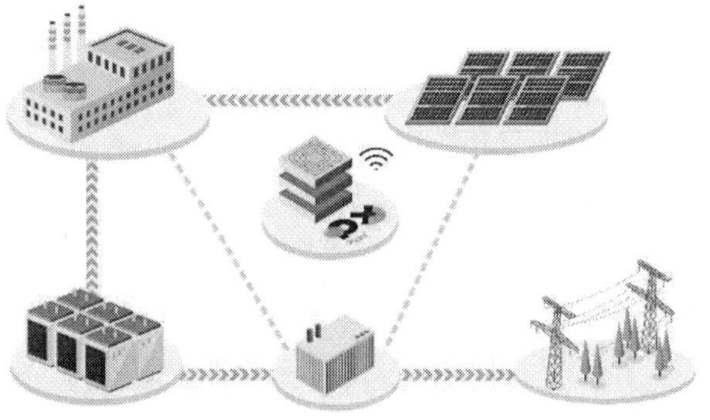

Fig. 1. Simplified sketch of a system with industrial customer, solar PV and smart battery. VPP, active connection with balancing authority is crucial for support of ancillary services.

More detailed description of the BESS system is shown in the block diagram in Fig.2, which also describes important electrometer readings of the whole system. Crucial parts of the BESS system can be listed as follows:

- 55kW inverter ALM
- Inverter controller CU320-2 DP
- Active Interface Module 55kW (for precharging of internal circuits)
- Battery rack with batteries 120kWh
- 2 winding dry transformer, IT (LV side) - avoiding oil change management
- AC filter
- Control system
- Battery management system (BMS)
- External cabling
- Energy Management System – PowereX

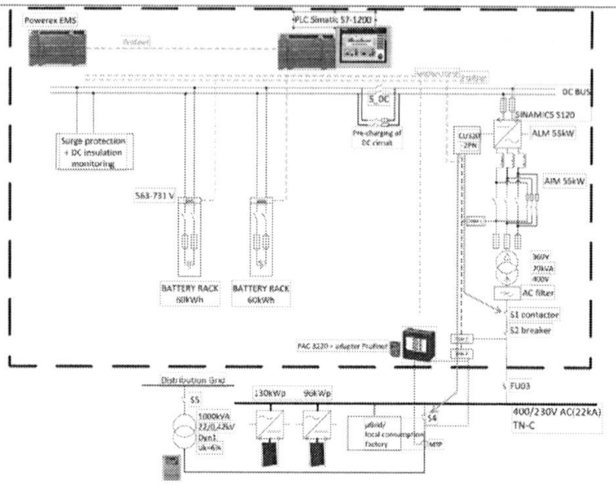

Fig. 2. Block diagram of BESS.

It is important to note that Solar PV is rather passive installation and behavior of the whole microgrid is driven by the battery.

III. AI Driven and PLC Energy Management Systems

The crucial part of the BESS is an intelligent EMS driven by AI (AI EMS in the following text), which is also the key distinguishing factor of the system from other standard systems. The AI EMS is cloud based, but there would not be much difference in functionality if the AI EMS is running locally, with internet connection to the required web services. Although the cloud-based solution offers more reliability, redundancy and together with lower price, it makes the more favored option. The crucial attribute of the AI EMS is secured connection done by VPN tunnel. Redundancy of the system is performed by two independent backup internet connection options. Redundancy of the smart EMS system is covered by the cloud system operator who guarantees 99.9 % constant availability of its services (tier 3+ data center). The AI EMS processes data from local solar PV and local industrial consumption; however, it also uses data from local energy market, energy imbalance responsible party, distribution system operator (DSO), and transmission system operator (TSO). More specifically, the AI part of the EMS is responsible for providing accurate forecasts data for the short-term electricity market, local solar PV production, factory consumption as well as system imbalance forecasts of the whole area of the TSO. It means that presented AI EMS goes far beyond smart EMS presented in [13] AI EMS smart algorithms evaluate all this data and calculate the best possible move of the battery for actual time in order to make the highest economic benefit. In addition, the AI EMS is capable of making a plan for the battery up to 36h in advance with high accuracy of forecasts at the beginning and slowly decreasing accuracy for further timesteps as within natural market behavior. Although this plan is constantly modified with upcoming actual data from forecasts and from the market. The AI EMS has a function of guarding contracted reserved capacity (prepaid demand charges). The other vital property is the possibility to be aggregated with other similar systems into a VPP, where optimal battery commands are being calculated considering also other assets present in the power grid.

On the other hand, the PLC based EMS (PLC EMS in the following text) is a standard device of most of the battery deliveries. It is supposed to be able to control the battery by simple pre-programmed algorithms. Nowadays, most comparable installations still work on similar simple energy management systems. In this case, the local EMS control is supplied by Simatic S7-1200. In this installation, local EMS is capable of full BESS control, however; this is activated only in the case of connection loss, malfunction of AI EMS or deliberate switching off the AI EMS. The PLC EMS algorithm works as follows: Battery is being discharged as fast as possible in order to minimize power consumption of the microgrid. It tries to reduce power consumption to zero if possible. Although this is done only until discharging threshold of the BESS is reached. In this case, this value is set to 30 % SOC (State of Charge). If SOC is higher or equal to 30% SOC, the battery is charged only if solar PV produces more energy than the factory is consuming. This usually happens during sunny days. If SOC falls under 30%, BESS is charged also from the grid to reach this level. This is due to the importance of having the BESS ready in the case of exceeding contracted reserved capacity of the client. This is also a reason why this threshold value is set.

IV. Results and Discussion

To compare functionalities and differences between operation of both EMS systems, an experiment where one day the system was operated by PLC EMS and the following by AI EMS was performed. Both days were chosen during the working week to keep the consumption of the factory on appx. same levels. The days were chosen also according to the weather forecast to keep solar PV production similar. As both days shown in Figs 3 and 4 are relatively similar (with standard deviation of consumption lower than 0.15 and standard deviation of solar PV production lower than 0.2) and going one after the other, it is correct to compare the operation of both EMS systems. It is important to mention that PLC EMS systems with presented functionality have better use cases in the microgrids where solar PV cannot feed the power grid with surplus energy from the microgrid. In that case, PLC EMS created value during the sampled day could be higher. If it comes to the numbers, PLC EMS did 146.39 kWh of discharging and 84.55 kWh of charging. Discharging value is higher as the BESS was fully charged at midnight when the operation of the PLC EMS started. This wouldn't be the case if during the second day of PLC EMS operation as it is obvious from Fig. 5 showing SOC. Therefore, it is correct to calculate saving only 89.55 kWh of energy from solar. Taking average price on the Slovak short term market during days 7-8.4 2022 which was appx. 150 EUR per MWh. Additional distribution and transmission fees make additional appx. 40 EUR per MWh, it is possible to calculate economic benefit of PLC EMS operation to 17.01 EUR and 3.582 EUR for cases with possibility of feeding the grid and without this possibility respectively. Contracted reserved capacity was not reduced as it was a decision of the customer, who was not ready to take the risk of reserved capacity decrease, even though the opportunity of financial savings was real. Although, this benefit would be approximately similar using both EMS systems as also the PLC EMS unit was perfectly ready for this function.

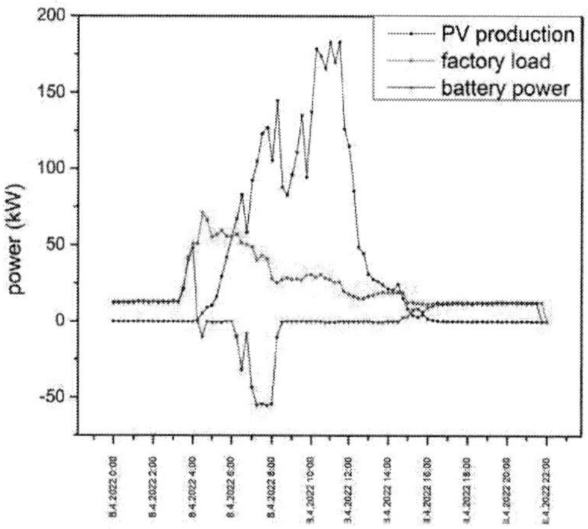

Fig. 3. Battery in microgrid controlled by PLC EMS (time is in UTC).

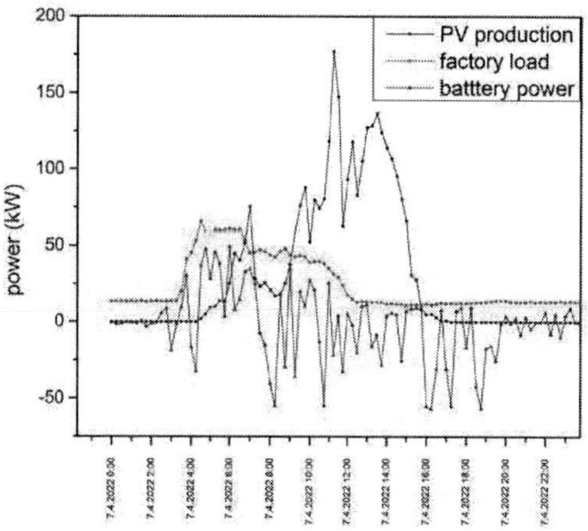

Fig. 4. Battery in microgrid controlled by AI based smart EMS (time is in UTC).

The AI EMS did 226.46 kWh of discharging and 176.37 kWh of charging. This is equivalent to (176.37+226.46)/240 = 1.67 of battery cycle. As battery lifetime is 10 years or 6000 cycles, it is optimal to make 1.64 of the battery cycle per day, which would fulfill 6000 cycles in 10 years. From this value, it is obvious that AI EMS considers this issue. On the contrary, PLC EMS does not use the full battery potential as it ran (146.39 + 84.55)/240 = 0.96 of the battery cycle per day. This means that by the end of its lifetime, the number of allowed cycles will not be fully used, and the battery will be degraded by age. The financial benefit of AI EMS is a bit more difficult to evaluate as in this case the aggregator takes responsibility for the energy imbalance of this particular system. Therefore, correct day-ahead nominations have much higher importance as in the case above where all imbalance responsibility is delegated to the electricity supplier. As each supplier may give different rules for nomination, that makes the final electricity bill savings very different, it will be useful to ignore the role of the nomination and estimate the pure battery impact regardless of the nomination. This could be calculated as follows:

$$Income = P\_arb + RE\_plus + RE\_minus - RE\_P\_plus - RE\_P\_minus$$

$$RE\_plus = C\_RE\_plus * E\_dis * succ\_rate$$

$$RE\_P\_plus = C\_RE\_plus * E\_dis * (1 - succ\_rate)$$

$$RE\_minus = C\_RE\_minus * E\_char * succ\_rate$$

$$RE\_P\_minus = C\_RE\_minus * E\_char * (1 - succ\_rate)$$

where,

P_arb - price arbitrage income

RE_plus - income for positive regulatory energy

RE_minus - income for negative regulatory energy

RE_P_plus - penalty for unsuccessful activation positive regulatory energy

RE_P_minus - penalty for unsuccessful activation of negative regulatory energy

C_RE_plus - cost of positive regulatory energy (current value 200 EUR/MWh)

C_RE_minus - cost of negative regulatory energy (current value 27EUR/MWh)

E_dis - energy discharging (226.46 kWh)

E-char - energy charging (176.37 kWh)

succ_rate - regulatory energy activation success rate

Taking a very conservative case with zero income for price arbitrage and relatively low success of imbalance prediction of 80%, resulting income would be 30.13 EUR. Optimistic case with 90% success of the imbalance prediction and 15 EUR income on price arbitrage (that is perfectly possible as daily low on the market was 14.33 EUR/MWh and max 280.47 EUR/MWh), the income can be up to 54.93 EUR.

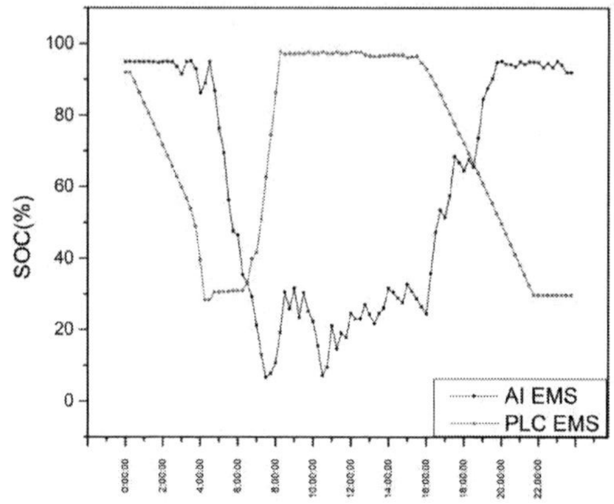

Fig. 5. Battery SOC during both experimental days of PLC EMS and AI EMS operation.

V. CONCLUSIONS

This paper described a system with a smart battery and compared operation of different energy management systems controlling this battery. In the first case, the battery was controlled by a simple EMS based on a Simatic S7-1200 PLC unit with a basic energy management algorithm. In the second case, AI based EMS provided by PowereX was employed. This AI EMS is capable of aggregating assets into a VPP, it uses real time electricity market data, power grid data, AI based forecast and advanced optimization algorithms for optimal battery management. Each EMS got an opportunity to run for 24h and the final benefits of both EMS systems were analyzed and compared. While calculation of the created value for PLC based EMS was rather straightforward, estimation of benefit for smart AI based EMS was rather tricky due to energy imbalance responsibility of the aggregator of the microgrid and the role of correct nomination according to conditions of the energy supplier. Therefore, a more correct methodology of pure battery impact calculation was proposed. Created revenue during the experimental day of AI EMS operation was calculated for optimistic and pessimistic scenarios. AI EMS created from 1.76 – 15.33 times higher value than PLC EMS. It was also shown that PLC based EMS creates only a very low income if the system is allowed to feed energy back to the power grid. In addition, flexibility supplied to the grid in this way contributes to carbon emission mitigation as this flexibility is supplied by a zero-carbon system. To conclude, experiment presented in this paper have shown strong financial and environmental benefits of intelligent control of distributed energy assets by AI driven EMS.

ACKNOWLEDGMENT

This work was supported by the Slovak Research and Development Agency under the contract No. APVV-20-0157.

REFERENCES

[1] U.S. Energy Information Administration (2019). "Electricity Explained - Basics." Internet: https://www.eia.gov/energyexplained/electricity/electricity-in-the-us.php [Apr. 4, 2022].

[2] Ministry of economic of Slovak republic. "Integrated national energy and climate plan for years 2021 – 2030." Internet: www.mhsr.sk/uploads/files/zsrwR58V.pdf, Oct., 2019 [Apr. 5, 2022].

[3] European Commission., "Eurostat DATA browser." Internet: www.ec.europa.eu/eurostat/databrowser/view/TEN00123__custom_2438627/default/table?lang=en, Mar. 18, 2022 [Apr. 5, 2022].

[4] Regulatory Office for Network Industries, "Podiel výroby elektrickej energie z obnoviteľných zdrojov energie." Internet: https://www.urso.gov.sk/podiel-vyroby-elektrickej-energie-z-obnovitelnych-zdrojov-energie/, [Apr. 5, 2022].

[5] K. Worthmann, C.M. Kellett, L. Grüne, and S.R. Weller. "Distributed and Decentralized Control of Residential Energy Systems Incorporating of Battery Storage." *IEEE Transaction on Smart Grid*, vol. 6, pp. 1913-1923, 2015.

[6] P. Asmus, "Microgrids, Virtual Power Plants and Our Distributed Energy Future," *Electr. J.*, vol. 23, pp. 72–82, 2010.

[7] H. Saboori, M. Mohammadi and R. Taghe, "Virtual Power Plant (VPP), Definition, Concept, Components and Types," *2011 Asia-Pacific Power and Energy Engineering Conference*, 2011, pp. 1-4.

[8] R. Caldon, A. R. Patria and R. Turri, "Optimisation algorithm for a virtual power plant operation," *39th International Universities Power Engineering Conference, 2004. UPEC 2004.*, 2004, pp. 1058–1062.

[9] P. Lombardi, M. Powalko and K. Rudion, "Optimal operation of a virtual power plant," *2009 IEEE Power & Energy Society General Meeting.*, 2009, pp. 1–6.

[10] H. Pandžić, I. Kuzle, and T. Capuder, "Virtual power plant mid-term dispatch optimization." *Appl. Energy*, vol. 101, pp. 134–141, 2013.

[11] S. You, C. Traeholt, and B. Poulsen, "A market-based Virtual Power Plant," *2009 International Conference on Clean Electrical Power.*, 2009, pp. 460–465.

[12] PowereX j.s.a, Internet: www.powerex.io, [Apr. 06, 2022].

[13] M.A. Hossain, R.P. Hemanshu, C.M. Moreno, "Real-time Battery Energy Management for Residential Solar Power System." *IFAC-PapersOnLine*, vol. 52, pp. 407-412, 2019.

E: Smart Grids and Electromobility

Modelling of Sulphur-Iodine Water Splitting Cycles for Hydrogen Production

Miroslava Farkaš Smitková
Department of Electrical Power Engineering and Applied Electrical Engineering
Faculty of Electrical Engineering and Information Technology of SUT
Bratislava, Slovakia
miroslava.smitkova@stuba.sk

František Janíček
Department of Electrical Power Engineering and Applied Electrical Engineering
Faculty of Electrical Engineering and Information Technology of SUT
Bratislava, Slovakia
frantisek.janicek@stuba.sk

Abstract — **Hydrogen could be an ideal energy carrier in the foreseeable future, and it can play an important role in future energy systems. Variety of methods and sources such as renewables as well as non-renewables can be used for hydrogen production. On the other hand, renewable energy sources and nuclear energy are more confident for the future large-scale hydrogen production. One of the most promising methods for large-scale hydrogen production is the thermo-chemical water decomposition using heat energy from nuclear, solar, and other sources. The water splitting thermo-chemicals cycles are processes where water is decomposed into hydrogen and oxygen via chemical reactions using intermediate substances which are recycled. There are over one hundred possible WSTCs that use different chemicals, some of them are pure thermochemical, and others use also electrolytic steps. One of the most promising thermochemical cycles for hydrogen production, selected for our study is the Sulphur-Iodine cycle.**

Detailed simulation of the Sulphur-Iodine was performed, and its thermodynamic analysis was calculated to evaluate the cycle´s efficiency, with respect to its thermodynamic theoretical values. The main results from the simulation and thermodynamic analysis are presented in the paper with respect to the advantages and disadvantages of the SI cycles and its possibilities for the future usage in the large-scale hydrogen production.

Keywords: hydrogen, hydrogen production; water spliting thermochemical cycyles; Sulphure-Iodine cycle; energy; energy efficiency

I. INTRODUCTION

In the last decade, the power industry notices a wide spectrum of technological progress which allowed an increase in the use of renewable energy sources (RES). Besides the renewable, it is very important to have an overall view of the energy sector - including fuels, power grid systems, accumulation of energy, cooling, and heating, economic factor, etc. Moreover, the current energy sector is facing many challenges - including safe energy supplies, an increase in energy demands, orientation on environmentally friendly energy production, problems with grid systems due to the rise of renewable energy sources, etc. Resent process in the world has shown us how easy it could become unsafe.

The main problem of the energy sector in the European Union (EU) is the high energy dependency. Energy supplies in the EU are strongly dependent on Russia. EU´s energy dependence was 57.497% in 2020, for comparison it was 44.4% in 1990 [2]. Germany was the largest consumer of natural gas in Europe and the seventh-largest energy consumer in the world in 2020 [1,4]. Natural gas represented 25% of Germany's total primary energy consumption, and consumption of natural gas increased slightly in 2019 [1,2,3,4]. Moreover, German authorities consider natural gas as ´a green source´ - the German government respects natural gas as it is an important technology on the way to greenhouse gas neutrality against nuclear energy and coal-fired power generation [4]. Since Russia's invasion of Ukraine, Germany and other countries rethink their position towards nuclear energy. Each country must revalue its energy strategy and try to become less dependent on suppliers. Renewable energy sources are not able to resolve the situation in energy supplies, especially not in a short time, and due to the problems in the grid system a new energy transition will be needed. In this light a new energy strategy is welcome. One of the possible ways could be large-scale hydrogen production where hydrogen is an ideal energy carrier and storage medium.

A variety of methods and sources such as renewables as well as non-renewables could be used for hydrogen production. On the other hand, renewable energy sources and nuclear energy are more confident for the future large-scale hydrogen production. One of the most promising methods for large-scale hydrogen production is thermo-chemical water decomposition using heat energy from nuclear, solar, and other sources.

The outline of the paper is as follows: the overall view of possibilities of the hydrogen production is briefly described, the focus is on water splitting thermochemical cycles and one selected WSTC – the Sulphur-Iodine cycle. Follows its thermodynamical analyses and finally, advantages and disadvantages of the SI cycle are discussed.

II. HYDROGEN PRODUCTION

A. Hydrogen Production – Overall View

The first step towards the hydrogen economy is a cheap and effective source for hydrogen production. Nowadays, hydrogen is produced mainly from fossil fuels (over 95%, [7]) because hydrogen combined with carbon makes compounds such as methane, coal, or petroleum. A variety of process technologies could be used for hydrogen production including chemical, biological, electrolytic, photolytic, and thermo-chemical. On the one hand, they have different impacts on the environment and on the other hand, their investment and operating costs are also different. All these facts must be considered.

In recent years, the research and development in hydrogen production are focused on green technologies and processes, mainly renewables and nuclear energy.

Depending on the materials and production method, hydrogen can be classified as grey, blue, and green [16]. Green hydrogen is produced from renewable energy sources, and it is considered the cleanest energy resource. Currently, this production represents only a small amount of the overall production, mainly due to its expensiveness. In the future, it will have the dominant role and green hydrogen production is expected to replace fossil fuels in hydrogen production that have negative environmental effects and nowadays play the dominant role in the hydrogen production [17]. Grey hydrogen is made from fossil fuels such as coal and natural gas. This production significantly results in the global warming because of CO_2 which is produced as a by-product [16,17]. Hydrogen produced from natural gas via steam reforming is marked as blue. It is obtained from grey hydrogen with a carbon capture and storage (CCS) system to avoid most of the greenhouse gas emissions. Hydrogen made through solar energy and nuclear energy is sometimes indicated as yellow and pink. Sources and processes for hydrogen production are schematically shown in Fig. 1.

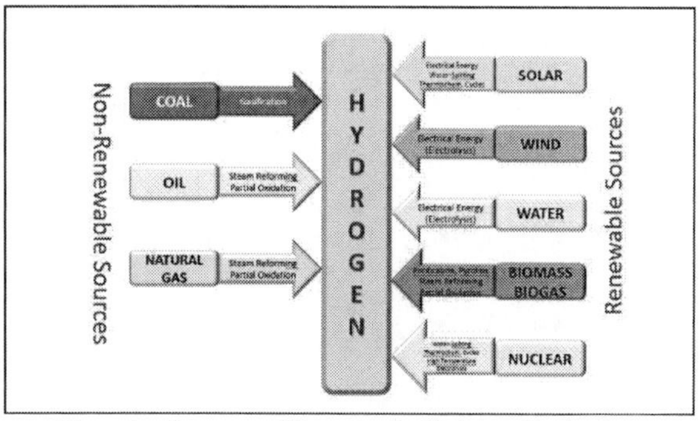

Figure 1. Sources and Processes for Hydrogen Production

Direct water decomposition is considered an ecologically friendly way of hydrogen production. Especially, when solar energy is used, the process is very clean. There are two possibilities - photovoltaic water electrolysis and direct thermal decomposition of water. Nowadays, photovoltaic water electrolysis is a leading technology [7]. It is limited by photovoltaic conversion efficiency (considering laboratory efficiency lower than 30%) and moreover, photovoltaic hydrogen production is still one of the most expensive technologies among the new alternatives using solar energy [7]. The main problem why a direct water dissociation is a non-practical way to produce hydrogen is in relatively high temperatures (and coupled material problems) and a small fraction of hydrogen at the thermodynamic equilibrium. For example, a thermal dissociation of water is needed temperature in the range 2 000 – 2 500K [7].

It is possible to overcome temperature problems and furthermore produce hydrogen in a larger scale by using water splitting thermo-chemicals cycles (WSTC). These are processes where water is decomposed into hydrogen and oxygen via chemical reactions using intermediate substances which are recycled [11,12,13,16,17]. According to [11,16,17], WSTC are one of the ecological ways using heat to decompose the water into hydrogen and oxygen where most of chemicals are recycled in the system. First studies about WSTCs were published in 1964, thereafter over one hundred WSTC were researched.

Water splitting thermo-chemicals cycles are very promising with the cooperation of renewables and nuclear energy. As was already aforementioned, there are over one hundred possible WSTCs, which use different chemicals, some of them are pure thermochemical, and others use also electrolytic steps. One of the most promising thermochemical cycles for hydrogen production was chosen for our study: Sulphur-Iodine cycle (SI cycle).

B. Sulphure-Iodine Cycle

The Sulphur-Iodine cycle (SI cycle) consists of three pure thermo-chemical steps. The SI cycle can be divided into three major subsystems, based on three main reactions of the cycle [13,14,15]: *Gibbs reactor, Bunsen reactor, and Equilibrium reactor.* The sub-systems are schematically shown in Fig. 2.

- ***Gibbs reactor***: Sulfuric acid H_2SO_4 is concentrated and decomposed here. The small amount of Sulfur trioxide SO_3 is founded in the outlet of the Gibbs reaction. Later SO_3 reacts with water and produces diluted H_2SO_4 which is recycled in the Bunsen reactor. [9]
- ***Bunsen reactor:*** two immiscible aqueous acids are produced and separated here - a light liquid (H_2SO_4/H_2O) which is a lower density phase and a heavy liquid ($HI/I_2/H_2O$). Hydrogen is lately generated from the heavy phase. The reaction proceeds exothermically and iodine I_2, and water are later recycled in the cycle. [10, 11]
- ***Equilibrium reactor:*** Hydrogen iodine *HI* is concentrated and thermally decomposed at a moderate temperature 450°C. The result from the equilibrium reactor is split in a liquid-gas separator. The hydrogen product and some *HI* are separated from most of iodine, which is returned to the main solution reaction in the Bunsen reactor. The gaseous hydrogen is then separated from *HI*, which is sent back to the equilibrium reactor, using a membrane. Pure hydrogen is the final product.

In our work, a complete simulation model of the SI cycle was performed in Aspen Plus program, which has the capability of modelling electrolytes via several different modelling techniques including an electrolytic version of the non-random two liquid (NRTL) technique. An electrolytic NRTL model (ELECNRTL) should be able to handle the chemistry of the Sulphur-Iodine cycle. A model has been developed based on the dates and recommendations from [12] and it is described in detail e.g., in [13]. Data for thermodynamic analysis were obtained from Aspen Plus simulation.

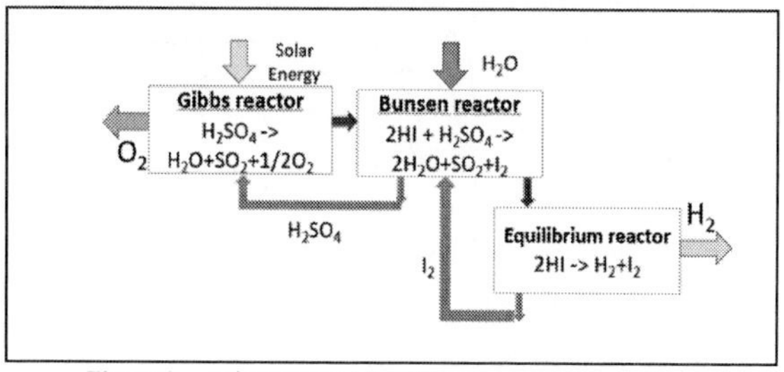

Figure 2. Schematic model of Sulphure-Iodine cycle

III. Thermodynamic Analysis

The water splitting thermo-chemicals cycles offer a possibility to transform solar energy into storable and transportable fuel – hydrogen. The main challenges of the SI process are improving efficiency and reducing the production costs.

Thermodynamics is a powerful tool used in the field of solar thermochemistry. The complex interactions between solar flux, reactant feed conditions, and chemical kinetics are important for designing reactors that convert solar energy efficiently into chemical fuels. It is important to note that the thermal cycle receiving thermal energy from a high-temperature reservoir and rejecting it to a low temperature reservoir. If we consider an ideal system, the maximum efficiency is associated with the Carnot efficiency. In our study, a process runs between upper temperatures T_{SR} - temperature of solar reactor, products leave solar reactor at this temperature and lower temperature of surroundings T_0, therefore it could be calculated follows [11, 12, 13]:

$$\eta_{Carnot} = 1 - \frac{T_0}{T_{SR}} \qquad (1)$$

The efficiency of the closed-cycle η_{cycle} is calculated as a ratio between net useful work W_{net} (kJ/mol) obtained from the products leaving the reactor at temperature T_{SR} and total solar heat input Q_{solar} (J/mol).

$$\eta_{cycle} = \frac{W_{net}}{Q_{solar}} \qquad (2)$$

Net useful work W_{net} (kJ/mol) is the maximum amount of work that the products leaving the reactor could produce when combined at temperature T_0. The amount is calculated directly in the simulation.

Finally, the SI cycle effective efficiency η_{eff} was calculated, it was considered a mass flow rate of hydrogen - the amount of hydrogen produced in the SI cycle M_{H2} (kg/h) and low calorific value of hydrogen LCV_{H2} (121.48 MJ/kg). The work needs in pumps W_P (kJ/mol) and compressor W_{Comp} (kJ/mol) were also taken into account. Both amounts are calculated directly in simulation. Therefore, the SI cycle efficiency η_{eff} is calculated following:

$$\eta_{eff} = \frac{(M_{H2}.LCV_{H2}) - (W_P + W_{Comp})}{Q_{Solar}} \qquad (3)$$

The Aspen Plus was chosen as the process simulator for SI cycle. It has the capability of modelling electrolytes via several different modelling techniques including an electrolytic version of the non-random two liquid (NRTL) technique. An electrolytic NRTL model (ELECNRTL) can handle the chemistry of the Sulphur-Iodine cycle. The Aspen Plus enables to obtain an output information about mass flows, pressure, temperature, heat, or work needed during the process. The model has been developed based on the data and recommendations from [12]. Selected inputs and outputs of the SI cycle are schematically shown in fig. 2 and summarized in table 1. Selected variables of SI cycle and calculated thermodynamical values are shown in table 2.

Table 1. Selected inputs and outputs of the SI cycle

	Input	Unit
H_2O	4313.28	kg
I_2	25.46	kg
Electric power	4769.55	kW
HI	1442.7	kg
H_3PO_4	1763.91	kg
	Output	**Unit**
H_2	372.73	kg
O_2	2898.12	kg
HI	1442.7	kg
I_2	25.46	kg
H_2O	42.61	kg

The main impact of the low effective efficiency of the SI cycles has the HI decomposition process which must be improved. The work needs in the distillation process were not considered in the SI cycle calculation. The effective efficiency will be lower when this contribution is considered.

Table 2. Selected variables of the SI cycle

		Value	Unit
Temperature of solar reactor	T_{SR}	1103	K
Temperature of surroundings	T_0	363	K
Mass flow rate of hydrogen	M_{H2}	372.7	kg/hr
Work needs in compressor and pumps	$W_{P,Comp}$	3.14	kJ/mol
Net useful work	W_{net}	168.34	kJ/mol
Solar energy input in the solar reactor	Q_{solar}	473.89	kJ/mol
Effective efficiency	η_{eff}	0.19	-
Carnot efficiency	η_{Carnot}	0.67	-
Efficiency of the closed-cycle	η_{cycle}	0.36	-

The separation of HI from the heavy liquid ($HI/I_2/H_2O$) is the most critical and extremely complex phase of the SI cycle. H_2O/I_2 is a highly immiscible, H_2O/HI is an azeotropic and strong electrolytic system and I_2 has a high triple point (387 K). At ambient temperature, the azeotrope of HI (correspondent to a molar fraction) is 13 - 15%. Because of azeotrope, the mixture of compounds (HI/H_2O) cannot be changed by simple distillation. When an azeotrope is boiled, the resulting vapor has the same ratio of constituents as the original mixture of liquids. [12, 13, 14, 15]

After optimization of the HI decomposition process, which is the most problematic part of the SI cycle, effective efficiency will be recalculated.

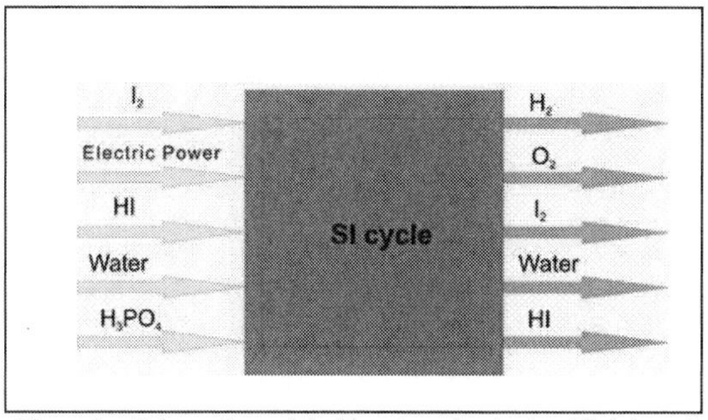

Figure 3. Schematic model of inputs and outputs of Sulphure-Iodine cycle

IV. SWAT Analyzis of the SI Cycle

Evaluation of the Strengths, Weaknesses, Opportunities, and Threats (SWOT analysis) of the Sulphur-Iodine cycle was performed. It is summarized as follows:

Strengths

- Pure thermo-chemical cycle.
- Use of available and non-toxic chemicals.
- Hydrogen is produced at a lower temperature in comparison with direct water decomposition.
- Overcome conventional condensation cycle in power plants to obtain electric energy, because the main source is heat obtained from the solar reactor.

Weaknesses

- Separation of heavy and light phase increases energy demands, and usage of other chemicals (H_3PO_4).
- Reactants loses, reduce backward reactions.
- Low efficiency.

Opportunities

- Decrease of dependence on fossil fuels import.
- Decrease of emissions and negative impact on the environment.
- Recycling of used chemicals.
- Hydrogen production at a massive production rate.

Threats

- Corrosion problems (use of aggressive materials, e.g., sulphur compound).
- High hydrogen production costs.
- High investment costs.

V. Conclusion

Hydrogen is an environmentally attractive alternative to displace fossil fuels, but current grey hydrogen production is orientated on natural gas steam reforming or coal gasification. The use of hydrogen reduces greenhouse gases only if green hydrogen is produced using non-fossil energy sources such as renewables.

A large-scale hydrogen economy will need a new source of hydrogen. Hydrogen production through thermochemical water splitting cycles represents an environmentally attractive way to produce hydrogen without using fossil fuels. There is still space for improvement of the WSTC cycles and their usage needs more investigation in the field of materials, reduction of temperature, using chemicals, etc. As s future work, our simulation could be optimized e.g., in the separation of H_2SO_4/HI mixture which is the most critical part of the SI cycle. It helps also to improve the cycle's efficiency and maximize the hydrogen production.

The main impact to lower effective efficiency of the SI cycles has HI decomposition process which must be improved. The work needs in the distillation process were not considered in the SI cycle calculation. Effective efficiency will be much lower when also this contribution is considered. After optimization HI decomposition process, which is the most problematic part of the SI cycle, effective efficiency will be recalculated.

Acknowledgment

This work was supported by the Slovak Research and Development Agency under the contract No. APVV-20-0157.

References

[1] "Germany". Independent statistics and analysis. EIA – U.S. Energy Information Administartion. Internet: <www.eia.gov/international/analysis/country/DEU> [May 02, 2022]

[2] Statistical Review of World Energy 2020 | 69th edition. BP p.l.n. 2020.

[3] *"Gas"*. IEA – Interantional Energy Agency. Internet: <www.iea.org/fuels-and-technologies/gas> [May 05, 2022]

[4] "Germany welcomes EU 'green' energy plan on gas, still opposes nuclear". Reuters. Internet: <www.reuters.com/markets/europe/germany-welcomes-eu-green-energy-plan-gas-still-opposes-nuclear-2022-01-02> [Apr. 09, 2022]

[5] "Energy 2010. A strategy for competitive, sustainable and secure energy". Luxembourg: Publications Office of the European Union, 2011. ISBN 978-92-79-18869-5.

[6] S.Z. Baykara. "Hydrogen production by direct solar thermal decomposition of water, possibilities for improvement of process efficiency," in International Journal of Hydrogen Energy, 29(14), 2004, Pages 1451-1458, ISSN 0360-3199.

[7] A. LE Duigou, J.M. Borgard, B. Larousse, D. Doizi, C. Eysseric, R. Allen, B.C. Ewan, G.H. Priestman, R. Elder, G. Cerri, C. Salvini, A. Giovannelli, G. de Maria, M. Roeb, N. Monnerie, M. Schmitz, C. Sattler, A. Buenaventura, D. de Lorenzo, S. Dechelotte, O. Baudouin. "Hythec : a search for a long term massive hydrogen production route," in *Proceedings International Hydrogen Energy Congress and Exhibition, IHEC 2005* Istanbul, Turkey, 13-15 July 2005.

[8] L.C. Brown , R.D. Lentsch, G.E. Besenbruch, K.R. Schultz, J.E. Funk. "Alternative Flowsheets for the Sulphur-iodine Thermochemical Hydrogen Cycle," in General Atomics. Report Num. GA–A24266, 2004..

[9] S. Goldstein , X. Vitart, J.M. Bordgard. "General comments about efficiency of the Iodine-Sulphur cycle couplet to the high temperature gas cooled reactor," in *International Scientific Journal for Alternative Energy and Ecology ISJAEE*, 3(11), 2004.

[10] A. T-Raissi. "Analysis of Solar Thermochemical Water-Splitting Cycles for Hydrogen Production". In: *Hydrogen, Fuel Cells, and Infrastructure Technologies*, FY 2003 Progress Report. [On-line]. Available: www.fsec.ucf.edu/en/research/hydrogen/analysis/documents/FY03_ProgressReport.pdf, [May. 10, 2022].

[11] M. Smitková, F. Janíček, J. Riccardi. "Life Cycle Analysis of Processes for Hydrogen Production," in International Journal of Hydrogen Energy, 36(13), 2011, 7844-7851, ISSN: 0360-3199.

[12] J. Riccardi, M. Schiavetti, I. Fastelli, M. Smitková. "Modelling of Westinghouse and Sulphur-Iodine Water Splitting Cycles for Hydrogen Production", *International Journal of Energy and Environmental Engineering* (IJEEE), 2(3) 2011, pp. 49-62, ISSN: 2008-9163.

[13] M. Smitková, J. Riccardi, F. Janíček. "Integrated Analysis of Hydrogen Produstion Focused on Water Spliting Thermochemical Cycles," in: *Optimization Using Exergy-Based Methods and Computational Fluid Dynamics : International Conference*. Berlin, Germany, 2009. ISBN: 978-3-86948-050-3.

[14] B. A. Casper, I. T. B. Kenneth, M. R. Jr. Edgar." "Prospects and challenges for green hydrogen production and utilization in the Philippines," *International Journal of Hydrogen Energy*, 47(41), 2022, pp. 17859-17870, ISSN: 0360-3199.

[15] J. Yu, Y. Dai, X. Wu, Z. Zhang, Q. He, C. Cheng, Z. Wu, Z. Shao, M. Ni. "Ultrafine ruthenium-iridium alloy nanoparticles well-dispersed on N-rich carbon frameworks as efficient hydrogen generation electrocatalysts," *Chemical Engineering Journal*, vol. 417, 2021, pp. 128105, ISSN: 1385-8947.

[16] M. Orfila, M. Linares, A. Pérez, I. Barras-García, R. Molina, J. Marugán, J.A. Botas, R. Sanz: "Experimental evaluation and energy analysis of a two-step water splitting thermochemical cycle for solar hydrogen production based on La0.8Sr0.2CoO3-δ perovskite," *International Journal of Hydrogen Energy*, 2022, ISSN: 0360-3199.

[17] M. Tolga Balta, "Thermodynamic performance assessment of boron based thermochemical water splitting cycle for renewable hydrogen production", *International Journal of Hydrogen Energy*, Volume 45, Issue 60, 2020, Pages 34579-34586, ISSN 0360-3199.

Influence of EV Charging Stations to Grid Operation

Jozef Király, Michal Kolcun, Dušan Medveď
Department of Electric Power Engineering
Faculty of Electrical Engineering and Informatics
Technical University of Košice
jozef.kiraly@tuke.sk, dusan.medved@tuke.sk, michal.kolcun@tuke.sk

Abstract — **The article deals with influence of operation of conventional electric vehicle chargers to modelled grid (including AC and DC charging) and also of influence of usage "vehicle to grid" charging stations and his contribution to network stability. Simulation of distribution grid behaviour due to mentioned influences was made in simulink environment and focused to available power in grid and his redistribution. The used power of EV chargers was from 11kW to 22kW AC and also up to 50kW in DC. In whole clusters was also considered influence of 125kW DC charging hub.**

Keywords: EV charging; influence of AC and DC charging; Vehicle to grid; Charging hub influence, V2V; V2G; V2X

I. Introduction

Operation of electric vehicles (EV) and with this inseparable charging of these vehicles can strongly affect operation of distribution grid to a greater extent. One of the significant factors is the type of charging such vehicles are charged. There are currently more than 3 million electric vehicles in operation in Europe, in general we can talk about battery electric vehicles and plug-in hybrid electric vehicles (just hybrid vehicles doesn't affect grid operation in any way). Interannual increase of BEV and PHEV vehicles number has over whole Europe almost exponential course - for example in Norway was increase approx. 35% [1] [2]. It is necessary to evaluate operation of these vehicles in many ways due to various way of use.

For example, charging of commercial fleet vehicles where almost every vehicle has defined certain range of use and these vehicles are charged after workday by AC chargers with controlled output power in cluster is quite different from view of influence to power grid as fast public DC charging by frequently used motorway [5]. The different influence shouldn't be considered just from view of power peak, but also from view of possibility of forecast these types of peaks in distribution grids. And as was mentioned at beginning of this paragraph, controlled AC charging of huge amount of EV in same time in scale of maximum usage of reserved grid capacity for certain point of grid can significantly affect grid load forecast, because this type of behavior is not usual from view of daily load curve [3]. Also, the possibility of usage BEV as battery energy storage system using EV charger that allows to operate vehicle in vehicle 2 grid mode (V2G) represents with presence of renewable energy sources (photovoltaic) completely different type of load as usual household [6]. These factors must be taken into account, in particular, when connecting these loads or sources to the distribution system, and these sources must also be managed in way of efficient use of electric power [4].

II. Types of EV Chargers and The Charging Process

A. AC Charging Stations

One of most frequent way of charging is using built in board charger of EV and charging by AC chargers. These chargers can be the Wallbox or is possible to use standard AC socket with EV charging cable. Usual limiting factor of these charging type is power of EV on board charger and not power of external charger. Power of onboard charger is from 3,6kW to 22kW and depends on car type. From view of load asymmetry is this value very important, because for example onboard charger with power 7,2kW can burden power grid on one phase by 32A. Especially by small household electric installation and using of renewables with inverters this type of asymmetry can be problem. benefit of AC charging is especially in possibility of charger output power management in easy way. This control is often provided by sub measuring system which is installed on side of main circuit breaker of household and provides information of available power to EV charging without affect to household. As was mentioned before, in this way is used full capacity defined by main circuit breaker. therefore, when designing especially larger residential sites, it is necessary from the point of view of the distribution system to take into account this feature of charge management and to consider when dimensioning power lines and sources.

B. DC Charging Stations

Unlike of ac charging, DC charging is characterized by rectifying of AC voltage directly in charger usually in power interval from 50kW to 360kW. So, in this way is power supply from side of charger directly connected to battery of EV, and main part of charging power regulation is caused by battery thermal management. Main goal of battery thermal management is to prevent battery damage and ensure long lifetime of battery. It is the use of these chargers that represents the most complex type of load from the point of view of operation, as without the possibility of prediction, power peaks represent a significant load, especially at charging hubs, due to the installation location (most often transit routes, highways, etc.). For this reason, an important part of the integration into the distribution system is the use of communication of these charging hubs with battery storages, renewable sources or. other elements of smart grids that can affect local conditions in the network.

III. Vehicle 2 Grid, Vehicle 2 Vehicle and Vehicle 2 Everything Operation

One of the positive effects of the use of EV in combination with smart grids is the use of these vehicles to compensate for the load in the microgrid networks by the local contribution of power from the electric vehicle battery at the time of shortage in the network, taking into account the required residual, resp. maintained EV battery charge capacity. In this way, it is possible to use the capacity of these batteries as a battery storage, resp. implements these batteries into the microgrid for optimal use of the available capacity of renewable resources.

The second possible mode of operation for homogeneous use of available network capacity is the use of vehicle 2 vehicle systems, which can achieve the sharing of residual battery capacity in fleet vehicles, where less used vehicles can supply energy to more exposed vehicles and thus eliminate the requirement for higher charging power. for these vehicles, resp. performance requirement at a time when the smart grid does not allow it, resp. it would be economically disadvantageous, e.g. due to the current availability of energy from renewable sources.

The main challenge of integrating vehicles into the smart grid is the future deployment of two-way communication between the vehicle and the individual components of the smart grid (vehicle 2 device), vehicle and pedestrian safety (vehicle 2 pedestrian), vehicle and infrastructure (vehicle 2 infrastructure) and vehicle and network (vehicle 2 network). In summary, we can call this communication model vehicle 2 everything (V2X).

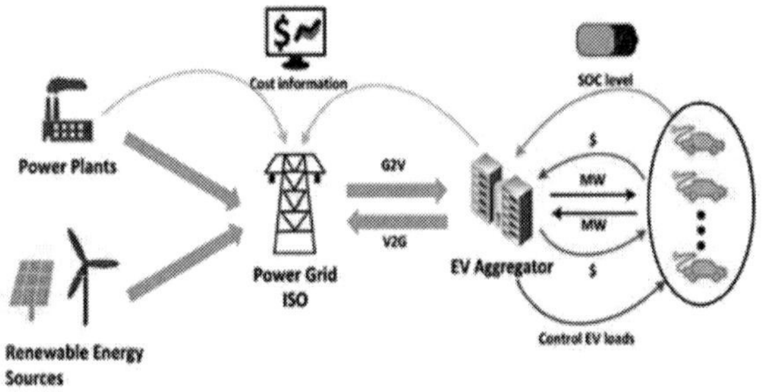

Figure 1. V2X topology [7]

In general, from view of electric power engineering and transportation, V2X is using any communication technology and tools, which can provide enhanced charging services to BEV and also bring another benefit for distribution grid operators (cost saving, effectively usage of grid etc.).[8]

IV. Battery Energy Storage Systems

One of devices that are related to smart grid and microgrid operation are battery energy storage systems (BESS). These BESS often provides load balance in certain spot of distribution grid for DC chargers, where isn`t sufficient power capacity for operation of these chargers during highest load in power grid. According to fact, that BESS can be considered as a next phase of BEV batteries lifecycle.

V. Modelled Network and Network Stability

Network stability affected by the installation of charging stations in residential areas (AC), resp. charging hubs (DC) can be described in particular in two ways. The first is the voltage asymmetry in phases (AC chargers), the second is the frequency changes due to the current peaks at the charging hubs (DC). The focus of this article is to look at the microgrid in terms of 24-hour operation and the impact of the presence of BEV and charging of these vehicles by different power and type of charging.

The modeled network for the purpose of demonstrating the impact of charging stations consists of the components listed in next table:

Table 1. Components of modelled grid

Component	Type	Power [kW]	Description
Distribution grid	22kV/0,4kV grid	6300	standard grid for cca 1200 households
Photo voltaic decentralised renewable energy source (PV DRES)	5kW PV	2500	500 households with PV
Wind farm	250kW	250	one farm
Electric vehicle charge (EVCH and vehicle 2 grid (V2G)	AC/DC/V2G	var.	Battery electric vehicle (BEV)

The investigated influence of chargers in this case is the operation of BEV in this closed microgrid, while for the purposes of simulation we considered the following model cases:

1. Charging by AC chargers 11kW with V2G (250 BEV)
2. Charging by AC chargers 22kW with V2G (250 BEV)
3. Charging by DC chargers 50kW with V2G (50 BEV)
4. Charging by DC chargers 125kW with V2G (50 BEV)

Vehicles were connected to grid in three different way. At first case were vehicles connected to home charger until 6AM and then charged at different place of microgrid (also AC charging) from 9AM to 4PM. The next stage of charging was from 7PM until morning. In the second case were vehicles connected in similiar way, but afternoon was charging procedure shifted by an hour. In this case was vehicles always charged at home. The third group of vehicles were vehicles which stay all day at home connected to a charger. All these groups were examined for chargers power as was mentioned at chapter beginning.

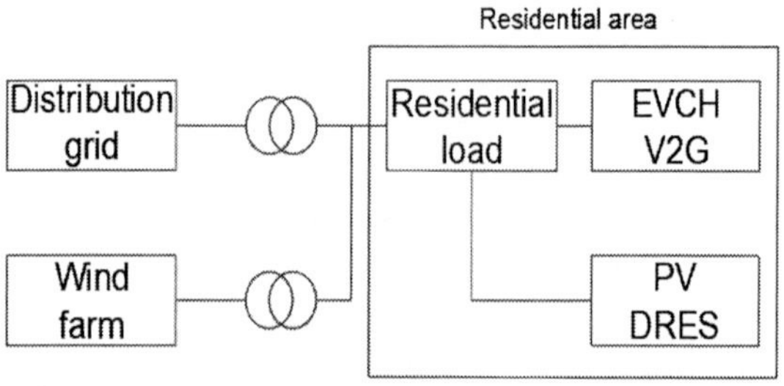

Figure 2. Topology of modelled microgrid

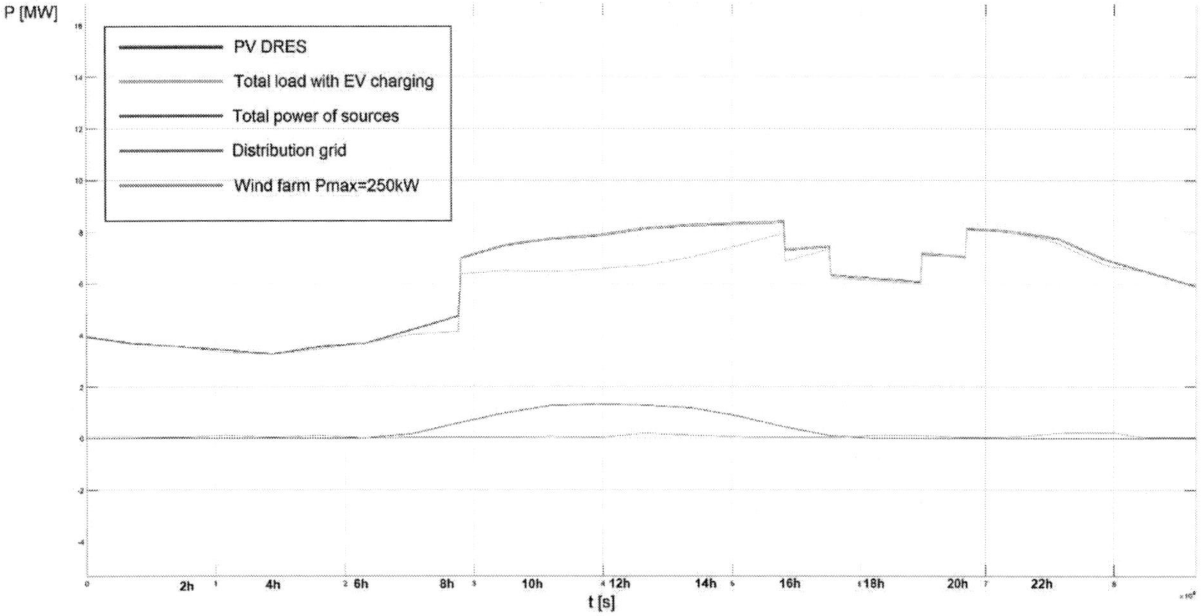

Figure 3. Charging of 250 vehicles in microgrid with AC 11kW and V2G enabled

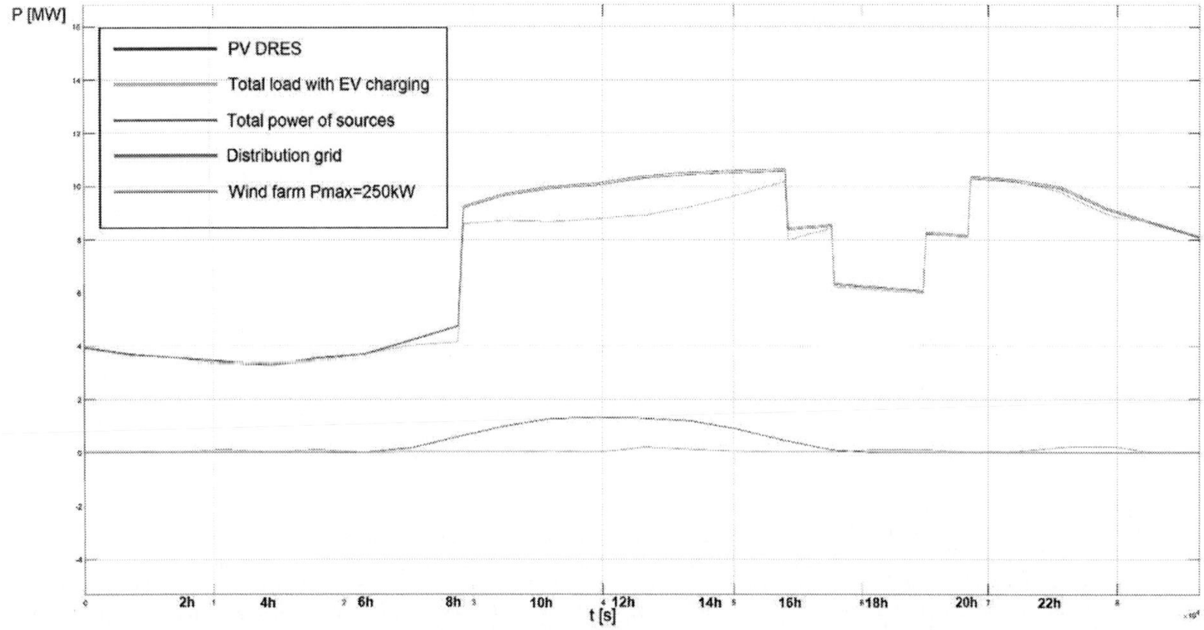

Figure 4. Charging of 250 vehicles in microgrid with AC 22kW and V2G enabled

VI. Simulated Power in Microgrid

The time interval examined for each experiment was 24 hours. This time was chosen due to the fact that just such a daily routine is the most likely scenario of the load on the distribution network. As can be seen in all four courses, a contribution from PV and an equally wind farm was considered on the resources side of the network. The course of the contribution from the PV side was a standard sunny day and it is therefore necessary to note that this is a more or less ideal situation. The use of these resources was chosen precisely because of the increasing expansion of this type of decentralized resource in the network. The wind farm was chosen with an output of 250 kW, which does not represent a significant contribution in terms of the overall size of the surveyed grid. In any case, it was appropriate to consider the contribution, as it manifests itself in the same way after sunset.

The most important part of the simulation was to study the effect of charging. As shown in Figure 3, charging with AC chargers starting at 8:00 will cause a significant increase in load from about 4MW to 6MW. The increase in the contribution from PV, which began to offset this increase to a small extent, was already a positive effect at this time. Figure 4 shows the case if the vehicles were charged with a power of 22kW. This example is rather theoretical since charging with such power would be

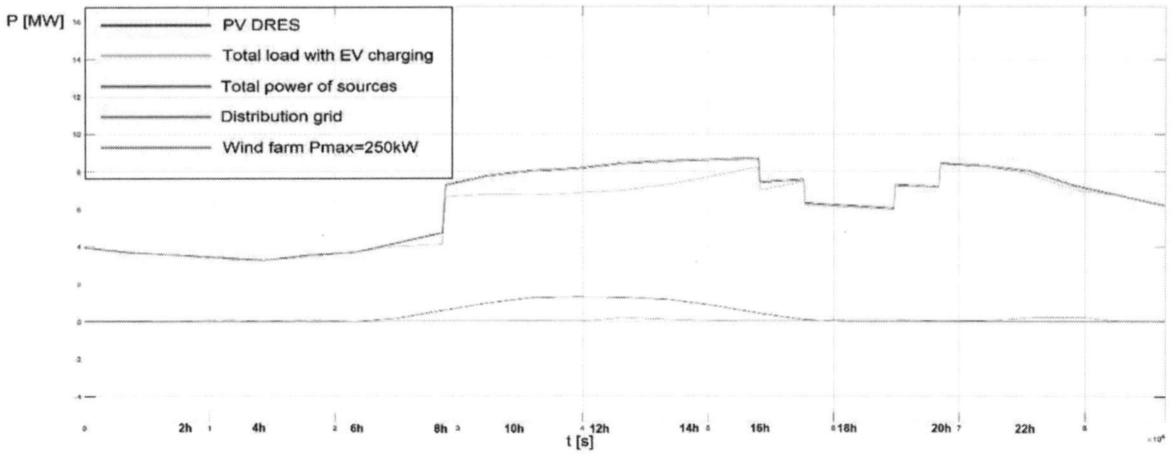

Figure 5. Charging of 50 vehicles in microgrid with DC 50kW and V2G enabled

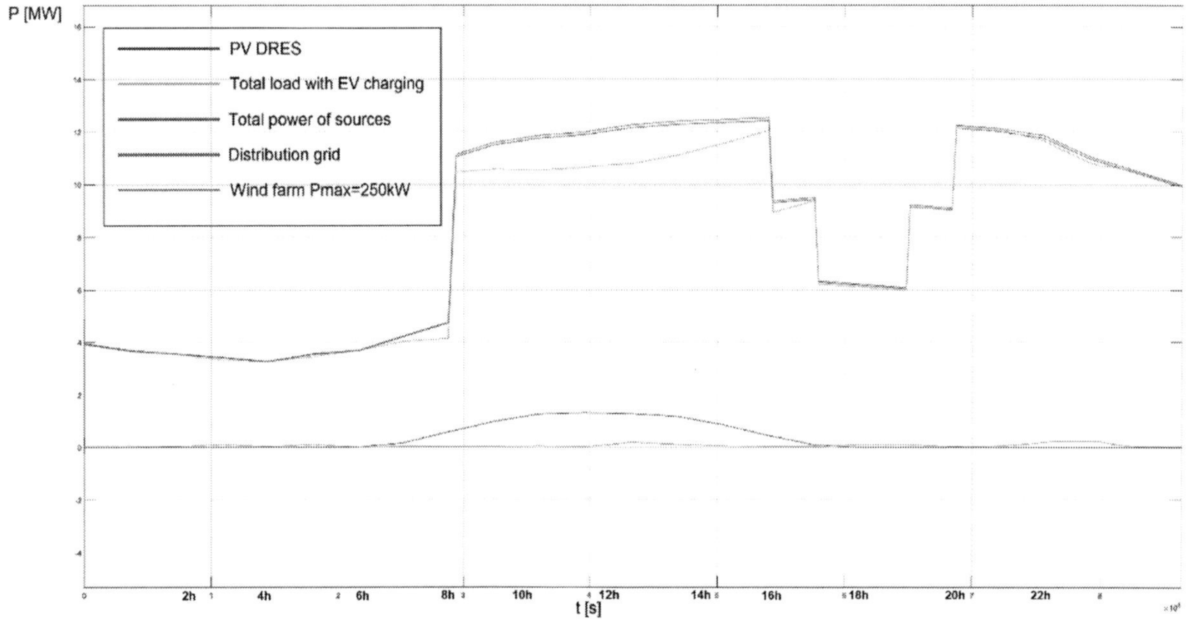

Figure 6. Charging of 50 vehicles in microgrid with DC 125kW and V2G enabled

unrealistic for most EVs given the limits given in the previous chapters. It is important to note that in this case there was an increase in the required power. In both cases, a symmetrical charging load was considered, so there is no need to consider the effect of asymmetry in the system. In the real network, however, we can consider an approximately even distribution of these chargers between the individual phases. At the time of charging the vehicles in the morning, the load is even, the last peak being the state of disconnection of vehicles at the time of moving the vehicles and their subsequent connection to the network. As it was important to shift the charging times in the afternoon to emphasize the different behavior of the users, in this part of the characteristic it is possible to observe sharp decreases in load and subsequent peaks after reconnection to the network. It is important to note that a group of vehicles (50) were still considered, which were constantly connected to the network in V2G mode and were considered as a source.

In the case of DC charging, lower numbers of vehicles were chosen, as this charging would take place through the charging HUBs. As can be seen, although a smaller number of vehicles were charged, their effect on charging power at 50kW was almost equivalent to 11kW of AC charging. Example of using 125kW DC charger was just for comparison of power peaks which can occurs in grid by using this type of chargers. Operation of a V2G system with such power is still unlikely due to the thermal management of the batteries, but it is possible to consider it in theory. In general, whole course of available power and load acts in predictable way, but this is caused only by fact, that model considers homogenous behavior of EV users. In general, whole course of available power and load acts in predictable way, but this is caused only by fact, that model considers homogenous behavior of EV users.

VII. Conclusion

The operation of distribution networks has been changing in recent years due to the impact of decentralized renewable energy sources and, more recently, an increase in the numbe of BEV and PHEV vehicles. As the simulation found, the number of vehicles, together with the power used for charging, represent a significant factor both in terms of sizing local resources, capacity of distribution lines and also in terms of future network development. In this regard, we can assume that the deployment of smart grids in the sense of using the vehicle 2 grid mode, respectively. vehicle 2 everything will be a necessity for efficient and stable management of the distribution system. With a significant expansion of charging stations and especially high-performance hubs, we can assume that the role of the distribution system operator will not only monitor the stability of network operation, but also actively intervene and control equipment connected to this network to achieve stability and efficiency with local production and consumption. The future recycling of electric vehicle batteries in BESS and the deployment of these devices will also have a positive effect on the local compensation of power fluctuations in the distribution network and may also contribute to the installation of DC charging hubs in locations with insufficient network capacity. The future focus of research should be on managing the performance of charging stations based on the available power in the network and the overall design of algorithms and rules that would contribute to the stability and security of the network.

Acknowledgment

This paper was supported by the Ministry of Education, Science, Research and Sport of the Slovak Republic and the Slovak Academy of Sciences under the contract VEGA 1/0757/21 and by the Slovak Research and Development Agency under the contract No. APVV-19-0576.

This work was supported by the Slovak Research and Development Agency under the contracts No. APVV-19-0576 and APVV-21-0312 and by the Ministry of Education, Science, Research and Sport of the Slovak Republic and the Slovak Academy of Sciences under the contract VEGA 1/0757/21.

References

[1] Registered vehicles. (2022) "Statistics Norway" [online]. Norway: *National Statistics Institute of Norway.* Available: https://www.ssb.no/en/transport-og-reiseliv/landtransport/statistikk/bilparken 2021 [Apr. 29 2022].

[2] H. Saele and I. Petersen. "Electric vehicles in Norway and the potential for demand response." In: *53rd International Universities Power Engineering Conference (UPEC),* pp. 1-6, 2018. doi: 10.1109/UPEC.2018.8541926.

[3] R. Cimbala, S. Bucko, L. Kruželák, M. Kosterec. "Thermal degradation of transformer pressboard impregnated with magnetic nanofluid based on transformer oil," in: *EPE 2017*, Ostrava : VŠB-TU, pp. 715-719, 2017. ISBN 978-1-5090-6405-2.

[4] J. Zbojovský, A. Mészároš, P. Kurimský. "Modelling the high frequency electromagnetic field propagation through the polystyrene," In: *Elektroenergetika* 2015, Košice: TU, pp. 556-559, 2015. ISBN 978-80-553-2187-5.

[5] J. Zhang, Y. Wang, J. Li and Y. Xu. "Day-Ahead Operation of EV Charging Station in a Logistics Center Considering Charging Behaviors of Different Types of EVs," in: *2021 IEEE 11th Annual International Conference on CYBER Technology in Automation, Control, and Intelligent Systems (CYBER),* pp. 626-631, 2021, doi: 10.1109/CYBER53097.2021.9588267.

[6] I. Sami, Z. Ullah, K. Salman, I. Hussain, S. M. Ali, B. Khan, C. A. Mehmood, U. Farid. "A Bidirectional Interactive Electric Vehicles Operation Modes: Vehicle-to-Grid (V2G) and Grid-to-Vehicle (G2V) Variations Within Smart Grid," in: *2019 International Conference on Engineering and Emerging Technologies (ICEET),* 2019, pp. 1-6, doi: 10.1109/CEET1.2019.8711822.

[7] C. Corchero and M. Sanmarti, "Vehicle- to- Everything (V2X): Benefits and Barriers," in: *15th International Conference on the European Energy Market (EEM),* pp. 1-4, 2018, doi: 10.1109/EEM.2018.8469875.

[8] M. Roccotelli, A. M. Mangini and M. P. Fanti, "Smart District Energy Management With Cooperative Microgrids," in *IEEE Access,* vol. 10, pp. 36311-36326, 2022, doi: 10.1109/ACCESS.2022.3163724.

Modelling of SLBESS Application in the Distribution Grid with the Help of a Benchmark Model

Marek Rybárik, Peter Braciník, Marek Roch
Department of Power Systems and Electric Drives
Faculty of Electrical Engineering and Information Technology, University of Žilina
Žilina, Slovakia
marek.rybarik@feit.uniza.sk, peter.bracinik@uniza.sk, marek.roch@uniza.sk

Abstract — **In recent years, we can see an enormous rise of distributed energy resources (DER) of which the biggest interest is in renewable energy sources (RES). To support their smooth integration, battery storage systems (BESS) are very often recommended and used to balance the RESs' output power fluctuations. Considering the proposed rise of electric mobility, the used batteries of electric vehicles can be applied for second life batteries energy storage systems (SLBESS) to support the operation of the grid with renewables. To do it efficiently, the operation of SLBESS has to be simulated in advance. As a support of these activities, the CIGRE Task Force has developed a few European network benchmarks for real-world simulations. The aim of the paper will be the presentation of the use of a such benchmark network, created in pandapower, for the simulation of SLBESS integration and application.**

Keywords: renewables, second life batteries, modelling, Python, benchamrk network

I. Introduction

Nowadays, we can observe a general trend in the decarbonization of energy, industrial as well as automotive sectors. In the energy sector, there is an ongoing pressure to use as much renewable energy sources as possible, with the focus on photovoltaic and wind generation technology. It results in the integration of distributed generation connected to the network through inverters what leads to the decrease of an inertia in the power system caused by the decline of rotating generators.

On the other hand, in the automotive sector, the fleet of electric vehicles is rapidly rising. EVs include a battery system, which serve as a source of energy for the electric drive. In United States as well as in Europe the battery life is defined by law in terms of its capacity. The battery is no longer suitable as a source of energy in the automotive and traction systems once its capacity drops below 80 % of its nominal capacity. After reaching this point, the first life cycle of the battery end and it need to be used in other applications. Most of the discarded batteries end up in recycling process that is economically demanding or in landfill as an environmental waste. Due to this phenomenon, the idea of secondary life usage for batteries in the energy sector came to life [1].

Application of second life battery storage systems (SLBESS) can be divided into two separate categories. The first category covers power system applications, such as support of renewable energy sources integration, participation in ancillary services for the transmission and distribution systems or the regulation of voltage and frequency. The other category covers application for modern commercial or residential buildings with smart metering and photovoltaic power generation as well as energy support of small industry facilities, commercial customers, and off grid microgrids [2], [3].

To ensure an efficiency and applicability of intended applications, the operation of SLBESS need to be simulated in advance. As a support of these activities, the CIGRE Task Force has developed a few European network benchmarks for real-world simulations. The aim of this paper is the presentation of the use of one such a benchmark network [4], created in pandapower [5], for the simulation of SLBESS integration and applications.

Section 2 describes shortly the application needs for SLBESS technology, Section 3 introduces the benchmark network model of the local distribution network, which was created by CIGRE Task Force C6.06.20. Section 4 is focused on modelling of a local distribution network with renewable energy sources and battery systems and the presentation of obtained simulation results. The last part is devoted to the conclusion.

II. Second Life Battery Energy Storage Systems

Today, the battery energy storage systems are designed in a wide range of size in the capacity depending on the specific application. For EVs, HEVs, PHEVs and traction purposes are used battery systems from 20 up to 630 kWh. In energy areas such as emergency power, local energy storage, communication base stations, UPS are stationary battery systems with sizes from 250 Wh up to 5 MWh. For ancillary services in the transmission system for the purpose of peak shaving, load leveling and spinning reserve, are used systems with a capacity from 5 MWh apt u 100 MWh [6].

The biggest problem of SLBESS technology is the lack of standards, practical tests, experiments, and technological research which will provide information and rules for their use in different applications areas. Moreover, their usage must be economical and easy to implement for integration into energy sector to be become a viable choice. Most batteries are designed in certain standards to suit specific attributes of electric vehicles, but disassembly process and subsequent assembly into a different shape and size is far from easy and completed batteries must pass various certifications and test to be installed in the grid [7].

Nowadays, there is huge research going on in the Europe that is focused on batteries. Different research groups and application teams test various types of battery technologies with the aim to find the best technologies to support the use of batteries in the applications for electricity sector. With the increased number of electric vehicles (EVs) running throughout the Europe, our attention must be focused to the use of used EVS batteries in SLBESS applications.

As already having reached the end of their first life cycle, the SLBESS have different "starting" parameters that need to be tested with respect to expectations required from the technologies supporting the operation of power system before their real application. Not only we need to check if the reassembled SLBESS systems can meet Network Codes requirements [8], but the modeling of their operation can also show us target values and limits of SLBESS operation parameters that should be meet during the reassembling process. Such an approach can significantly reduce development costs.

To provide results acceptable by researchers as well as by industry community a suitable benchmark network model ought to be used for the modelling. If done properly, it doesn't matter if the benchmark model is implemented in a tool that is not yet so popular or taken as a standard one.

III. Benchmark Network Model

The issue of distributed generation (DG) integration was researched by the CIGRE Task Force in section C6.06.20, that evaluated benchmark network models for the study of DG integration [3]. Methodology for benchmark network was created based on specific configurations and compromises. The methodology consists of a hierarchical structure with 4 levels. Starting from the highest level represented by the electric power system, through the second level represented by resources and the network, down to the third level representing the power distribution and high-voltage transmission, and finally to the lowest level represented by low-voltage (LV) urban network and medium voltage (MV) rural network. From there it expands according to individual specifications into levels that hold importance for DG benchmark models [9].

In case of benchmark models, the source node is becoming more and more important because many DG integration principles and techniques are dependent on control on the source side and power converters. In electrical gird the transmission and distribution networks must differ from each other, with characteristics of the distribution network varying significantly depending on whether it is categorized as urban or rural. Together, LV urban networks, MV rural networks and MV transmission make basic points for benchmark modelling [10].

A. Characteristic of the Benchmark Model

In general, benchmark networks are created to address the study of specific network and the effects of new techniques or technology. This specific model is supposed to help in the study of DG integration from perspective of electrical grid. The distribution benchmark network used in in this paper is taken from the German MV distribution network (Fig. 1). Network has a rural character and supplies small town and its surroundings, including residential an industrial customer. The nominal voltage is of 20 kV level supplied from a 110/20 kV transformer. For transport of electrical energy are mostly used cables but few areas of network use overhead lines. Overall length of lines in network with renewables and battery storage systems (Sub-network 1) is 15 km [9].

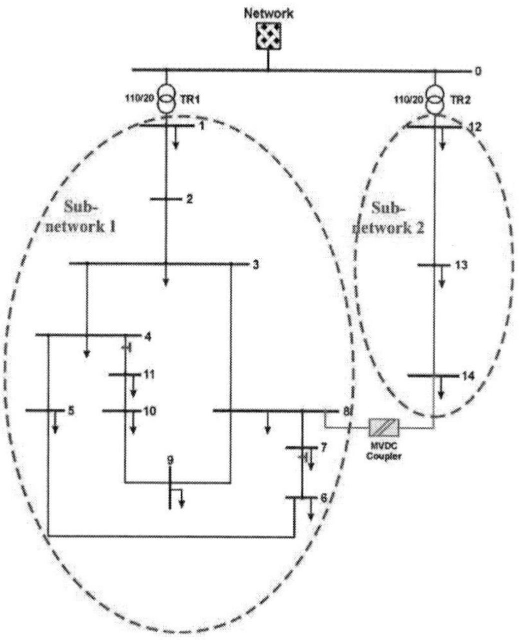

Figure 1. Benchmark model

B. Data for Demand Simulations

The demand of loads in the used benchmark network model (Fig. 1) was set according to normalized type demand diagrams used by one of the Slovak distribution system operators [11]. The normalized type demand diagram is the sequence of values of average hourly electricity consumption per year. It serves as a good replacement for a continuous measurement [11]. We used demand values typical for April (Fig. 2), when one can observe the highest generation in renewables, especially in photovoltaic power plants (PV).

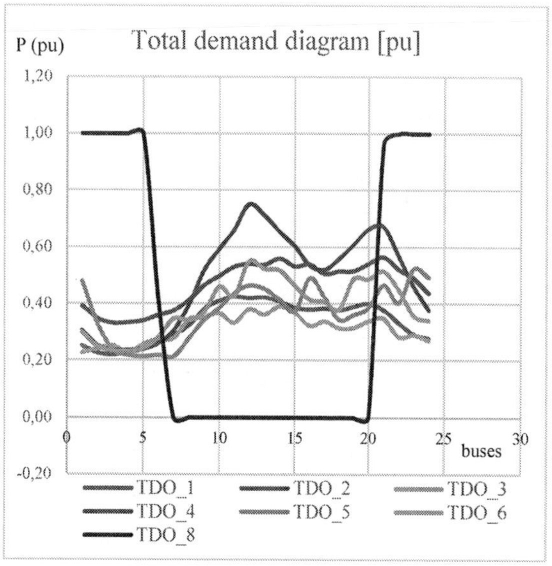

Figure 2. The values of used normalized type demand diagrams for April

C. *Benchmark Model Implmentation*

While looking for a simulation tool suitable for a present and future SLBESS operation modelling and analyses, we decided to use pandapower [5]. It's a simulation tool based on a Python language. It is maybe not so user friendly at the first look as e.g. Matlab, but the Python background provides a very easy cooperation with other Python modules and packages for optimization, data processing and visualization, machine learning, etc. So, the more complex calculations would be possible in the future work. The other important reason was the fact that the selected CIGRE benchmark model had been implemented and verified by pandapower team and was available for free [12].

IV. Simulations Results

A. *Simulation Description*

1. Basic model

The basic model represents exactly the benchmark model designed by the CIGRE Task Force. It consists only of typical power systems devices – power transformers, switching elements, lines, busbars, and loads representing the electric demand. There are two types of loads in the model. First one is for residential load and the other for industrial one. Basic model serves as a starting point to which all other simulation results (with renewables and SLBESS) were compared.

2. Model with renewables

In this case, the renewables (RES) were added to the basic model. The total output of RES is 1.71 MW. It's provided by 8 PV power plants with total installation capacity 0.21 MW and one wind turbine generator with installed capacity 1.5 MW. The connection of RES is shown in Table 1.

3. Model with RES and battery system

In the last case, the focus was put on the cooperation of RES and battery systems. There were 3 battery systems, with the total capacity 2.3 MW, added to the model (for connection information check Tab. 1). Following situation were analyzed:

- connection of a 500 kW battery,
- connection of 500 kW and 800 kW batteries,
- connection of all three batteries but they are all in charging mode,
- two batteries are discharging, while 1 MW battery is charged.

Table 1. Connection of RES and SLBESS

Node	P [MW]	Battery [MW]
1	-	0.5
3	0.02	-
4	0.02	-
5	0.03	-
6	0.03	-
8	0.03	0.8
9	0.03	1.0
10	0.04	-
11	0.01	-
7	1.50	-

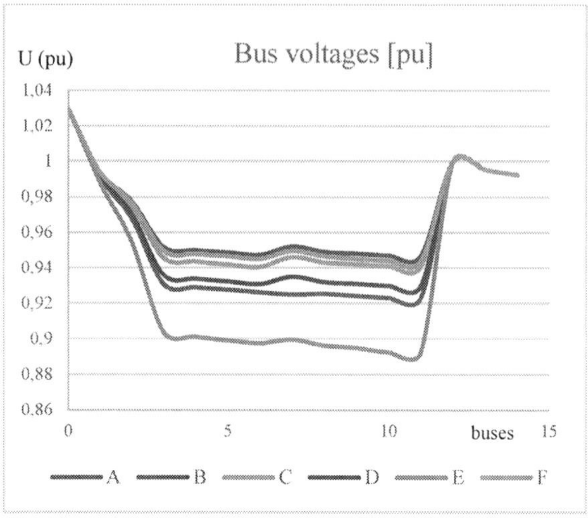

Figure 3. Bus voltages

B. *Simulation Scenarios*

To test the performance of the simulation model as well as its ability to correctly represent basic network quantities (bus voltages, power flow and lines' loading) the following simulation scenarios were done:

A) basic model,
B) basic model with RES,
C) scenario B with 1 battery (500 kW),
D) scenario B with 2 batteries (500 kW, 800 kW),
E) scenario B with 3 batteries (only charging mode),
F) model with RES and battery system case – case 3.

Bus voltages in sub-network 1 (Fig. 1) are presented in Fig. 3, where x-axis represents specific buses. The biggest voltage drop occurred in scenario E due to the charging of all battery systems. The voltage drop is around 0.9 kV compared to scenario A. Scenario F, when two batteries with total output power 1.3 MW are discharging, shows the best result from the bus voltage drops point of view. Bus voltage is stable around 0.95 pu, which is 20.9 kV.

The loading of network's power lines is shown in Fig. 4. The highest loading is on lines connected to buses that are supplied by transformers. In scenario A, the loading of these lines is 96.5 %, what is close to the upper limit. After RES implementation in scenario B, this loading decreased to 60.6 %, but the loading of the line between nodes 5 and 6 raised from 2 % up to 27.8 %.

The worst case was represented by scenario E, with three batteries being charged at the same time, causing lines 1 and 2 to exceed their ampacity limits. Such an operation state is technically as well as according to standards not acceptable because it can lead either to the damage of power line's conductor [13] or some other technical failure or even worse it could lead to the injury.

Scenario F represents a situation when the demand is partially covered by discharging battery systems connected to the nodes 1 (500 kW) and 8 (800 kW). Their output helped to cover the energy needed to charge the battery system connected to node 9 and it also helped to reduce the power fed by transformer TR1. It resulted in the significant decrease of the load of power lines 1 and 2 to the value 65 %. The buses' voltage levels and short circuit current Ik were not exceeded in the system as well.

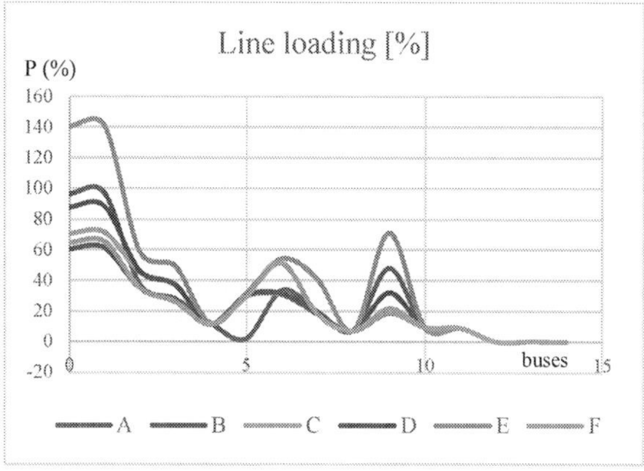

Figure 4. Loading of powerlines

We analyzed the effect of RES and SLBESS operation on the power transformers TR1 and TR2 (Tab. 2) as they are the most important elements in the modelled power system. Transformer TR2, responsible for supplying sub-network 2, didn't change its values because this area was permanently separated by MV coupler.

In the basic model (scenario A) TR1 was slightly overloaded to 101.4 %. After the installation of RES into network (scenario b), the load of the transformer reduced to 93.8 %. During the worst case (scenario F) TR1 is overloaded to 110.6 %, what can result in faster aging of power system's components, such as isolation, what can in turn lead to other faults.

V. Conclusions

The growing interest in possible application of used electric vehicles' batteries inside the future SLBESS systems supporting the operation of power system raises the need to model their operational behavior with the goal to identify the best application areas, services, and operational limits. Because there is a lack of data acquired from real SLBESSs, the only way how to make the research in this field now is through the simulations [14].

To make the results acceptable for other researchers as well as for the industry, the acceptable benchmark network model needs to be used. That's why the CIGRE benchmark network model designed for the distributed generation modelling and its implementation was used.

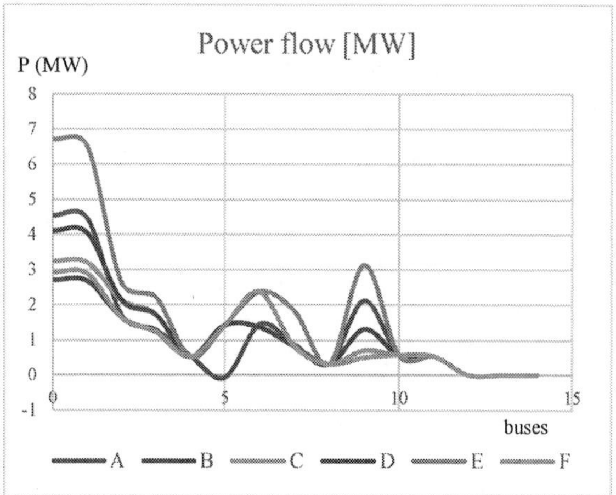

Figure 5. Active power flow in power lines

Table 2. Transformers data for simulated scenarios

	P [MW]		Ik [kA]		Load [%]	
	TR1	*TR2*	*TR1*	*TR2*	*TR1*	*TR2*
A	24.43	20.62	0.73	0.61	101.4	84.7
B	22.58	20.62	0.68	0.61	93.81	84.7
C	23.11	20.62	0.69	0.61	96.00	84.7
D	23.98	20.62	0.72	0.61	99.56	84.7
E	26.60	20.62	0.80	0.61	110.63	84.7
F	22.29	20.62	0.67	0.61	92.71	84.7

The simulation results confirmed expected impacts of RESs and battery systems on network operation. When the parameters of the real second life batteries will be acquired through the future testing or their ordinary usage in EVs, these parameters will be added to the presented CIGRE benchmark model and the effect of SLBESSs on the power system will be analyzed more deeply in the future.

References

[1] C. Koch-Ciobotaru, A. Saez-de-Ibarra, E. Martinez-Laserna, D.-I. Stroe, M. Swierczynski, and P. Rodriguez."Second life battery energy storage system for enhancing renewable energy grid integration." in *Energy Conversion Congress and Exposition (ECCE)*, 2015 IEEE, 2015, pp. 78-84: IEEE.

[2] P. B. L. Neto, O. R. Saavedra, and L. A. d. S. Ribeiro, "A DualBattery Storage Bank Configuration for Isolated Microgrids Based on Renewable Sources." *IEEE Transactions on Sustainable Energy*, vol. 9, no. 4, pp. 1618-1626, 2018.

[3] L. Prokop, S. Mišák, "Off - Grid energy concept of family house.", *Proceedings of the 13th International Scientific Conference Electric Power Engineering 2012, EPE* 2012, 2, pp. 753-758.

[4] Report of Task Force C6.04.02: "Benchmark Systems for Network Integration of Renewable and Distributed Energy Resources", 2014.

[5] L. Thurner, A. Scheidler, F. Schäfer, J. H. Menke, J. Dollichon, F. Meier, S. Meinecke, M. Braun. "pandapower - an Open Source Python Tool for Convenient Modeling, Analysis and Optimization of Electric Power Systems", *in IEEE Transactions on Power Systems*, vol. 33, no. 6, pp. 6510-6521, Nov. 2018.

[6] Electropaedia. "Battery Applications" [Online]. Available: https://www.mpoweruk.com/applications.htm, [Mar. 14, 2022].

[7] Batteries [Online]. Available: www.batteries2020.eu, [Mar. 14, 2022].

[8] Entsoe. "What are Network Codes?" [Online]. Available: https://www.entsoe.eu/network_codes/, [Mar. 14, 2022].

[9] K. Rudion, A. Orths, Z. A. Styczynski and K. Strunz, "Design of benchmark of medium voltage distribution network for investigation of DG integration." *2006 IEEE Power Engineering Society General Meeting*, 2006, pp. 6 pp.-, doi: 10.1109/PES.2006.1709447.

[10] K. Strunz, S. Barsali, Z. Styczynski, "CIGRE Task Force C6.04.02: Developing Benchmark Models for Integrating Distributed Energy Resources." *CIGRE Study Committee C6 Colloquium*, Cape Town, South Africa, 24 Oct. 2005.

[11] Stredoslovenská distribučná. [Online]. Available: https://www.ssd.sk/dodavatelia-elektriny-1/tdo-typovy-diagram-odberu?page_id=4331, [Mar. 15, 2022].

[12] Fraunhofer IEE and University of Kassel. "Pandapower documentation." Internet: pandapower — pandapower 2.9.0 documentation [Mar. 20, 2022].

[13] M. Krátký, S. Mišák, P. Gajdoš, P. Lukáš, R. Bača, P. Chovanec, "A Novel Method for Detection of Covered Conductor Faults in Medium Voltage Overhead Line Systems." *IEEE Transactions on Industrial Electronics,* 65 (1), art. no. 7962192, pp. 543-552., 2018, DOI: 10.1109/TIE.2017.2716861.

[14] M. Cenký, J. Bendík, Ž. Eleschová, A. Beláň, B. Cintula, P. Janiga, "Probabilistic Model of Electric Vehicle Charging Demand to Distribution Network Impact Analyses", in *2019 20th International Scientific Conference on Electric Power Engineering (EPE)*, pp. 1-6, IEEE, 2019.

Operation Control of Photovoltaic Power Plants in Smart Grids

Emil Dvorsky, Lenka Rakova, Pavla Hejtmankova
Department of Electrical Power Engineering
University of West Bohemia
Pilsen, Czech Republic
dvorsky@fel.zcu.cz, lencar@fel.zcu.cz, hejtman@fel.zcu.cz

Vladislav Sitar
Faculty of Mechanical Engineering
J. E. Purkyne University in Usti nad Labem
Usti nad Labem, Czech Republic
Vladislav.Sitar@ujep.cz

Abstract — **The future planned development of decentralized electricity generation may have a negative impact on the operation and management of the electricity system. Above all, problems can be expected with maintaining voltage and frequency stability in island microgrids, where the only sources of electricity will be photovoltaic power plants. With the growing consumption of electricity and the growing demand for the quality of the supplied electricity, there is a need to regulate these renewable energy sources more appropriately. One of the possibilities could be the primary and secondary regulation of voltage and frequency using the so-called virtual frequency and voltage method. The main aim of this paper is to analyze and compare the current regulators of the voltage inverters of the photovoltaic power plants with the proposed inverter regulator enabling primary and secondary regulation. The functionality of the solved control circuits of inverters is assessed on the basis of the results of the simulations of the small island microgrid with a symmetrical and asymmetrical load in the MATLAB Simulink software.**

Keywords: microgrid; virtual frequency and voltage method; standard controller; photovoltaic power plant; regulation

I. Introduction

In accordance with the State Energy Policy [1], the National Action Plan for Smart Grids 2019 - 2030 [2] and the requirements of the ENTSO-E, in the future it is expected in the Czech Republic to increase the amount of decentralized electricity sources, mainly of the renewable energy sources so as the wind power plants and the photovoltaic (PV) power plants. Another of the main aims of the energy sector in the Czech Republic are the requirements to increase the reliability, quality and security of electricity supply and at the same time to reduce the interruption of the electricity supply. The operation of the PV power plants should contribute to the accomplishment of these requirements too. This results in a new requirement for the inverters controllers of the PV power plants and for the photovoltaic systems themselves. If PV power plant operation control is required, it is necessary to have a battery storage device connected to the PV power plant and to modify the inverter control circuits used in current PV systems. Maintaining frequency and voltage stability in the electricity system will be always by means of the large conventional power stations than by means of the PV plants, which will be only the supporting local sources. The main aim of this paper is to compare of the control circuit, which is currently used by inverters of the photovoltaic power plants, with the control circuit of the inverter, which was created by the authors of this paper. This control circuit was created on the basis of the virtual frequency and voltage control method that was presented in detail in [3] and [4]. Both variants of the inverter controllers are compared and analyzed on the basis of the results of simulations of the operation of a small island microgrid in the MATLAB Simulink software.

II. Model of the PV Inverter Controller in MATLAB Simulink

One author of this paper in [3] proposed a model of the inverter controller using the virtual frequency and voltage method. This model was created only on the basis of expert knowledge and expert articles mostly published in IEEE (from [5] to [11]). However, this model has never been compared with the inverter controller model that is currently used in PV power plants. Therefore, one of the main aims of this paper is to compare these two different models of the inverters regulators and evaluate their advantages and disadvantages in their application in the model of the solved island microgrid. In Fig. 1 is a brief explanation of the general inverter control circuit model of current PV power plants.

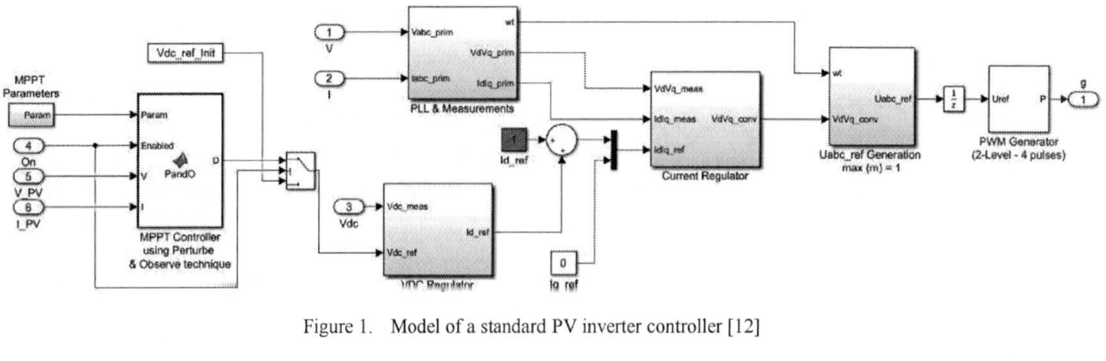

Figure 1. Model of a standard PV inverter controller [12]

This model is created in the MATLAB Simulink program and is taken from freely available models from the MathWorks website [12]. From the analysis of the available models, it was found that all already created models of the control circuits of three-phase inverters have an almost identical configuration and they use the Park's and Clark's transformation. As can be seen in Fig. 1, at the input to this regulator circuit is an MPP tracker. At the output of the MPP tracker is the voltage reference value that is required on the DC side of the inverter (Vdc_ref). However, if the output power of the inverter is to be changed, the MPP tracker must be disconnected. Therefore, the MPP tracker is not further used here and the reference value of the DC voltage (Vdc_ref) is entered as a constant value according to the nominal parameters of the PV system.

The Vdc_ref parameter enters a block named VDC Regulator. Here, the reference value of the current (Id_ref) is calculated from the deviation of the measured and reference value of the DC voltage and by means of the PI controller. Since it is assumed that the PV power plant has only active power, the reference value of the current in q coordinate, i. e. Iq_ref, is zero. At the input of the inverter controller, there are also the instantaneous values of voltage and current, which are measured at the connection point (Point of Common Coupling - PCC) of the PV power plant to the electricity system, marked V and I in this model. The measured values of voltages and currents are converted into the d-q coordinate system in the PLL & Measurements block using the Park's and Clark's transformation. From these measured quantities, the frequency (or angular velocity) in the electrical system is determined by means of the phase-locked loop PLL.

The main part of the controller circuit is the Current Regulator block, its scheme can be seen in Fig. 2. This block allows to calculate the required voltage value on the AC side of the inverter based on Kirchhoff's law. This is the sum of the voltage at the point of connection of the PV system to the electrical system (PCC) and the voltage drop on the inverter LC filter.

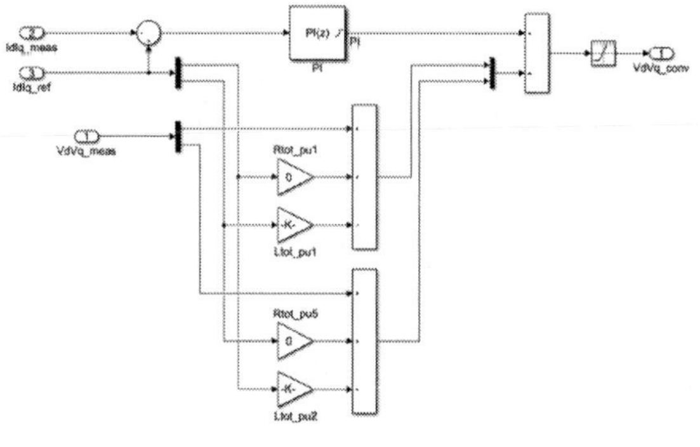

Figure 2. Block diagram of the Current Regulator [12]

In the following block, the output required voltage in PCC is transformed back into a three-phase system using the inverse Park's and Clark's transformation. Through the PWM Generator block, pulses are generated and transmitted to the IGBT voltage inverter based on the required voltage value in the PCC using the pulse width modulation.

This control circuit of the PV system inverter completely characterizes the inverters used today. Their main advantages are maintaining the frequency value at its nominal value and maximizing the use of solar radiation when connecting the MPP tracker.

If part of the electricity system is to be operated as an island micro-grid and only PV plants are decentralized sources in this island micro-grid, then the PV systems should be able to continuously change the output power to maintain the nominal values of frequency (f_n) and voltage (U_n) in this island microgrid. However, with this inverter control circuit, it is not possible to continuously regulate the voltage value on the AC side of the inverter. If the load increases, there is a risk that the voltage value may drop below the allowable limit, for example ± 10% U_n defined for low voltage level. Voltage drop can be prevented by modifying the control circuit of the inverter, for example the application of primary and secondary regulation of frequency and voltage using so-called the virtual methods.

III. Theory of Virtual Control Methods

The control circuit enabling primary and secondary regulation of frequency and voltage using virtual methods has a very similar scheme as the standard controller presented in Fig. 1. The difference is only in the input part of the circuit. This controller does not have an MPP tracker and instead of the VDC Regulator block there is a primary and secondary control block. At the output from the block of the primary and secondary control are the voltage and frequency reference values, which are required at the PCC point. From these quantities, the reference current is then calculated using the part of the control circuit respecting the voltage drop on the LC filter. The rest of the control circuit is almost identical so as the control circuit in Fig. 1.

Due to the fact, that the solved island microgrid is at the low voltage level, the frequency value is more influenced by the power balance of reactive powers and the voltage magnitude is mainly influenced by the flow of active powers. This statement is explained in [3] and [4]. The classical static frequency and voltage characteristics cannot be used here due to different impedance ratios (R/X) in this microgrid compared to the transmission system, but virtual methods can be used here. In the solved microgrid model, the method of virtual frequency and voltage was used for primary and secondary regulation, which is based on classical frequency and voltage characteristics. The principle of this method is briefly introduced in the left side of Fig. 3.

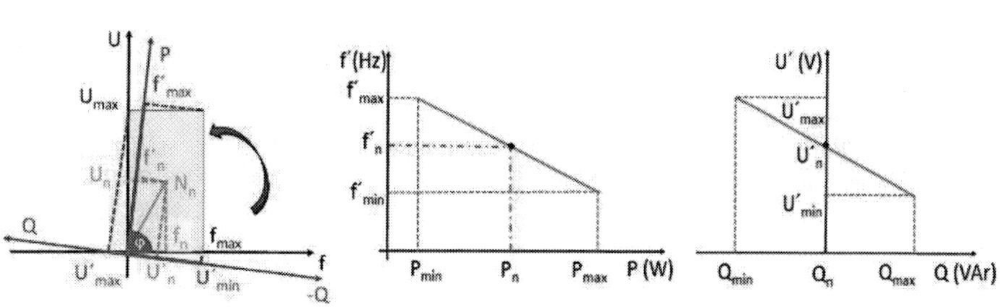

Figure 3. Principle of virtual frequency and voltage method [3]

This is a planar rotation of the coordinate axes of voltage U and frequency f by an angle φ to a position that corresponds to their location in the high-voltage transmission networks, where X >> R. The axis of the virtual voltage U′ is then parallel to the axis of the reactive power Q and the axis of the virtual frequency f′ is, on the contrary, parallel to the axis of the active power P. The resulting static characteristics are shown on the right side of Fig. 3. The defining equations of virtual frequency and voltage static characteristics are given by equations (1) and (2).

$$f - f_n = k_{pf}\frac{X}{Z}(P_n - P) - k_{qf}\frac{R}{Z}(Q_n - Q) \quad (1)$$

$$U - U_n = k_{pf}\frac{R}{Z}(P_n - P) + k_{qf}\frac{X}{Z}(Q_n - Q) \quad (2)$$

Where f (f_n) is frequency value (nominal frequency value) in [Hz], U (U_n) is voltage value (nominal voltage value) in [V], Z is line impedance in [Ω], X [Ω] is line reactance, R [Ω] is line resistance, and k_{pf} [Hz/VA] k_{qf} [V/VA] are coefficients of static characteristics.

From the above mentioned definitional relations, a block of the primary regulation is created in the MATLAB Simulink. On its input the reference frequency, voltage and power are compared with their measured values at points of PCC. The deviations of these quantities are further adjusted according to the given algorithm and the required reference voltage and frequency values are then at the output of this block. The reference values of active and reactive power at the input to the primary control block are given from the control circuit of the secondary control, see Fig. 4 where it is presented scheme of the primary and secondary control circuits.

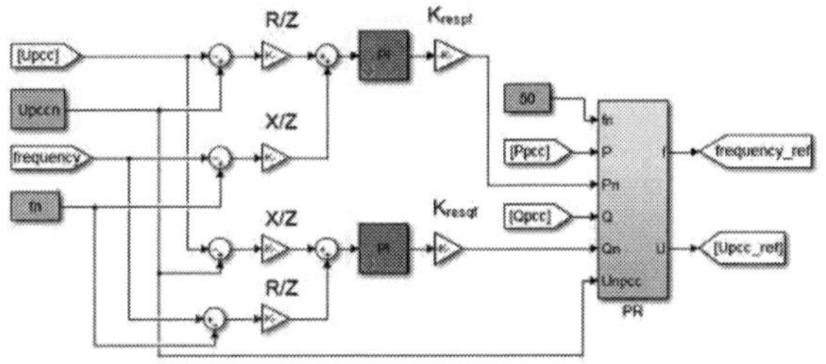

Figure 4. Model of primary and secondary control [3]

The recovery speed of nominal values of frequency and voltage in the microgrid by means of secondary control is given by the size of K_{respf} and K_{resqf}, i. e. coefficients of secondary control, especially time constant τ. The method is described and explained in more detail in [3].

IV. Model of the Solved Microgrid

A simple model of an island microgrid was chosen to compare the examined control circuits. This model is composed of a PV plant with a rated output of 9.2 kW_p, a three-phase IGBT voltage inverter, an inverter control circuit, an LC filter and of the variable loads marked in this model Z1 and Z2. The model of the solved island microgrid was created in the MATLAB Simulink software. Its scheme is presented in Fig. 5. The PV power plant with a nominal power 9.2 kW_p is divided into two parts, they are marked FVE_1 and FVE_2 in this model. A total was used of 40 panels of the ET Solar Industry ET-P660230 with a nominal power of one 230 W_p panel. The connection point of the PV power plant to the grid is marked the PCC. The capacitance of the LC filter is 15 μF and the inductance is 17 mH.

The controller scheme is always selected according to the specific analyzed variant of the inverter controller. In the basic microgrid model, the parameters of the IGBT voltage inverter are taken from the manufacturer's data sheet for the PowerOne / ABB 10.0 inverter. If the PV power plant is in a hybrid connection, the inverter with the higher maximum output power is selected. The control range of a controller, that use the virtual method, is assumed to be 0 kVA to 18.4 kVA or 0 kW to 18.4 kW, i. e. the PV power plant has half of the control range and the battery has half of ones too. The rated voltage is 230 V (assessed effective value) with a tolerance ± 10 % U_n. The frequency is still maintained at a constant nominal value of 50 Hz. The parameters of the PV panels and of the inverter IGBT are taken from the MATLAB Simulink library.

V. Simulation Results of Symmetrical Load Change

Both models of the voltage inverters controls were first tested using simulations of the symmetrical load changes. Throughout these simulations, the solar radiation intensity was 1,000 W/m² and the temperature was 25 °C, i. e. the nominal operating state corresponding to the maximum possible output of the PV power plant was modeled. In the first three simulations, a standard inverter controller of photovoltaic power plants was used, and in the fourth simulation, an inverter controller using a virtual method was applied.

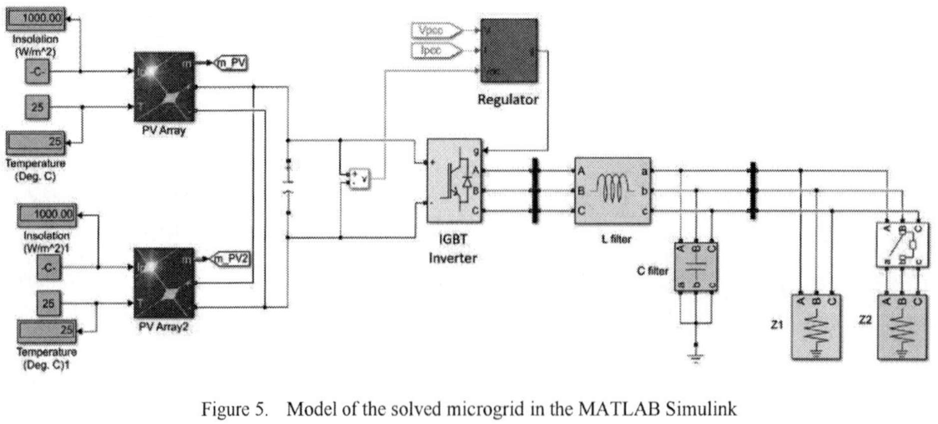

Figure 5. Model of the solved microgrid in the MATLAB Simulink

A. First Simulation - Standard Controller

In the first simulation, an island microgrid model with PV power plant without a storage device was used, the resulting graphs are shown in Fig. 6 and Fig. 7. At the beginning of the simulation, only a Z1 load with a power input of 9 kW was connected. The load Z2 is connected for one second at time 3 s, whereby the total load is increased by 4.2 kW. From the resulting waveforms it is clear, that in the microgrid was the nominal operation state to the time 3 s. When the Z2 load was connected, the power balance between the production and consumption of the electricity was disturbed, thus reducing the voltage value both at the PV plant and at the PCC point. This voltage drop was lower than the allowable ± 10 % U_n at the PCC point. The courses of the active and reactive powers measured at the PCC point were constant except for the small fluctuations when connecting and disconnecting the load.

B. Second Simulation - Standard Controller

In the second simulation, an attempt was made to eliminate the voltage drop below the permissible limit by connecting an accumulation device. At the beginning of the simulation, the Z1 load with a power input of only 7 kW was connected, which increased the control range of the source. In the time interval from 2 s to 3 s, the load Z2 with a power input of 7 kW was connected. From the resulting waveforms shown in Fig. 8 and in Fig. 9, it can be seen that the PV power plant was able to supply almost all required power Z1 and Z2, i. e. 14 kW, thanks to the battery. Since the inverter controller cannot continuously change the voltage on the AC side of the inverter, the voltage value at the PCC point has decreased with increasing load. Compared to the first variant of the simulation, the voltage drop is not outside the permitted limit rate. However, for a quality power supply, it is necessary to keep the voltage at the nominal value in the microgrid. Especially if a load sensitive to voltage changes is connected.

C. Third Simulation - Standard Controller

In the third simulation, a possible extreme case was simulated, when the load Z2 would increase to 10 kW. The other parameters of the model remained the same as in the second simulation.

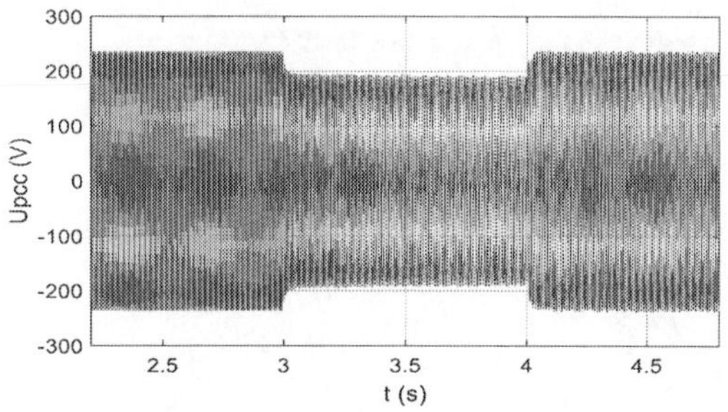

Figure 6. Voltage waveform at the PCC point - 1st simulation

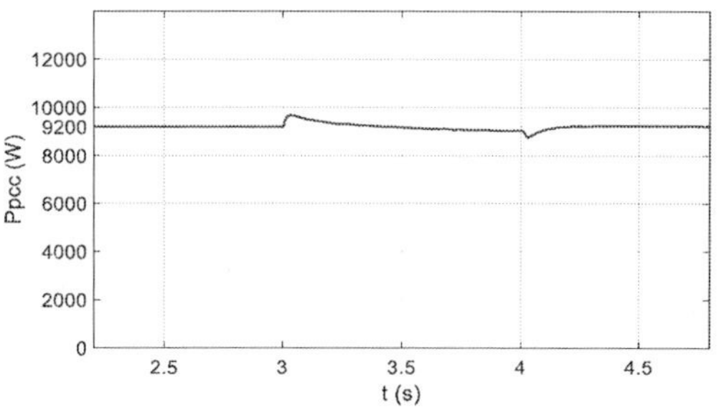

Figure 7. Active power waveform at the PCC point - 1st simulation

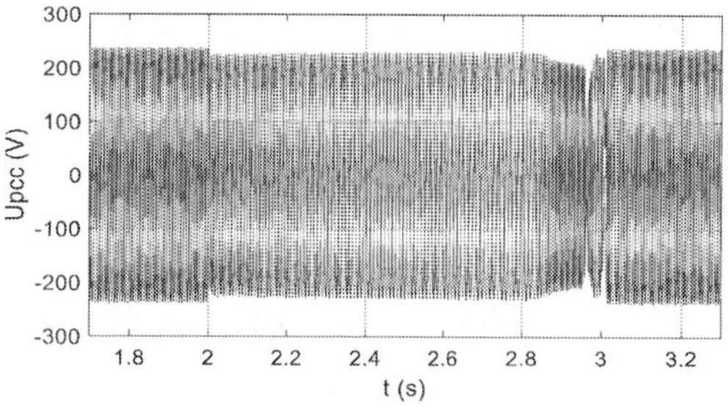

Figure 8. Voltage waveform at the PCC point – 2nd simulation

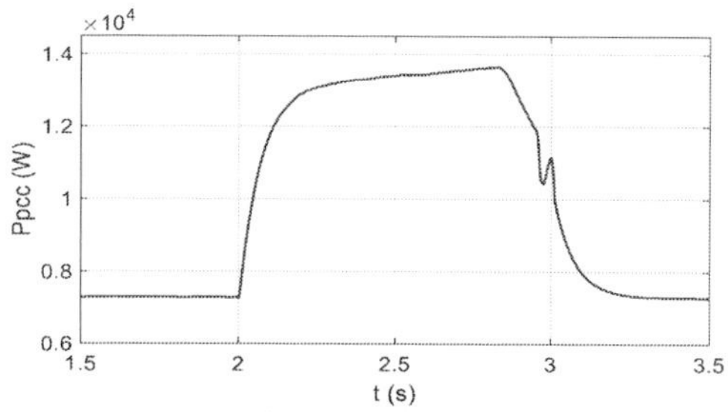

Figure 9. Active power waveform at the PCC point – 2nd simulation

The measured power and voltage waveforms at the PCC point are presented in Fig. 10 and Fig. 11. From the resulting waveforms it can be said that until the time of 2 s, when only the load Z1 (7 kW) was connected, the magnitude of the voltage at the PCC point was to the nominal value. In comparison with the waveforms of the quantities from the second simulation, it can be said that from 2 s to about 2.5 s, the trends of the waveforms of the quantities are very similar, i. e. the supplied power increases and the voltage value decreases. The electrical energy supplied to the load within 2.5 s is equal to the electrical energy supplied by discharging the battery. In a time period from 2.5 s to 3 s, the inverter tries to maintain the set nominal power value of 9.2 kW at the output of the PV power plant

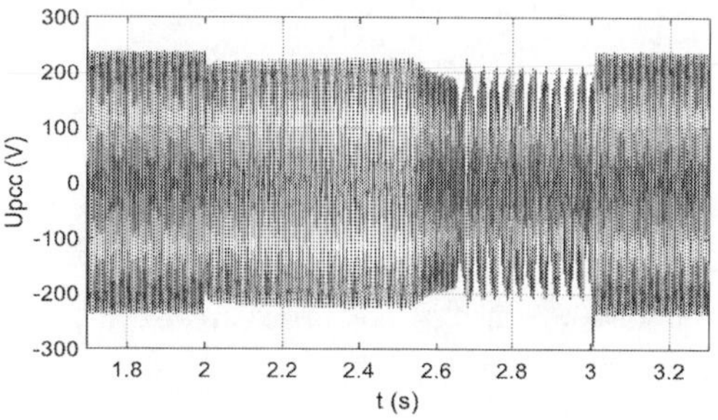

Figure 10. Voltage waveform at the PCC point – 3rd simulation

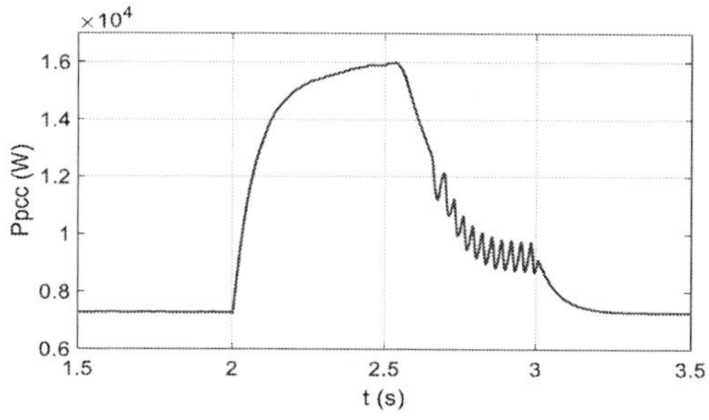

Figure 11. Active power waveform at the PCC point – 3rd simulation

However, the power consumption is much higher, and therefore the power balance is disturbed. This leads to an even greater voltage drop and oscillations of all quantities. This is until the Z2 load is disconnected. Subsequently, the nominal operating state in the microgrid occurs again. In the real situation, the consequences of the voltage drops and the oscillations would not be so great. The protections would evaluate the decrease of the values of the quantities in time and the load Z2 would be disconnected. If there were more sources in the microgrid, then in the event of the failure of the PV power plant with higher power, this extreme situation could occur. This situation can be eliminated using the virtual methods and a suitable storage device.

D. Fourth Simulation - Controller With a Virtual Method

The control circuit of the inverter, which uses the virtual frequency and voltage method, is analyzed using the fourth simulation of the solved microgrid model. The microgrid model is the same as the model that was used in the third simulation in the analysis of the standard controller. The loads change times are the same and the loads sizes Z1 = 7 kW and Z2 = 10 kW as well as. Replacement of the inverter controller is therefore the only change in the solved model of the island microgrid. Thanks to the primary and secondary frequency and voltage control, the inverter controller makes it possible to continuously change the rated output power of the PV power plant together with the output power of the storage device. This keeps the frequency and voltage values at their nominal values. This statement can be confirmed by analyzing the simulation results, which are shown in Fig. 12 and in Fig. 13. When increasing load is decreased voltage. The regulator of the primary control responded to the voltage drop by increasing the output power from the PV system. And before the voltage value could be stabilized at the new stable value, the secondary control regulator reacted and thus the reference voltage value stabilized at its nominal value of 230 V. The higher the value of the time constant, the slower the regulation to the nominal voltage value, see more [3].

Here the time constant was chosen $\tau = 60$. Due to the primary and secondary regulation, the nominal power is changing continuously, and thus the power balance and voltage and frequency stability are completely maintains.

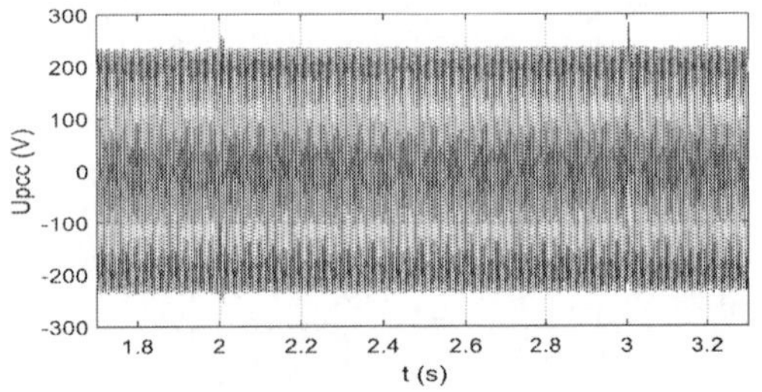

Figure 12. Voltage waveform at the PCC point – 4th simulation

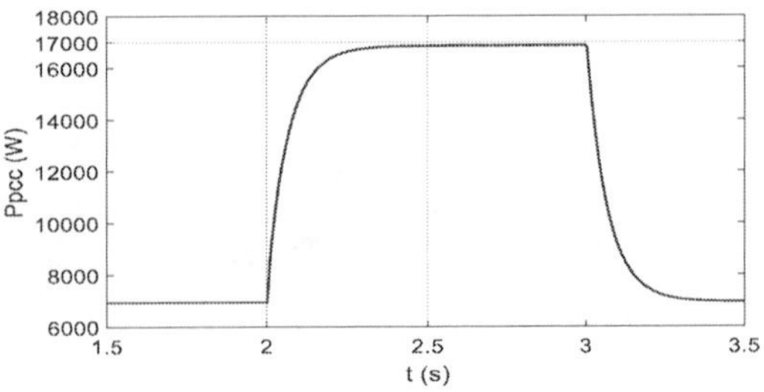

Figure 13. Active power waveform at the PCC point – 4th simulation

The voltage at the PCC point is still constant except for small fluctuations at the time of connection and disconnection of the load Z2. From the waveform of the measured active power at the PCC point, it is evident that the required power is supplied to the load. This is despite the fact that the size of the storage device has not changed compared to the simulation of the same operating state of the microgrid, where was a standard PV inverter controller. This confirms that the model of the controller using virtual methods is more advantageous compared to the standard controller, because it can continuously change the nominal power of the source and thus maintain the frequency and voltage stability in the given microgrid.

VI. Simulations Results of Asymmetric Load Change

The following simulations concern the analysis of the operation of an island microgrid with an asymmetric load. The microgrid model with all parameters is the same as the model used in the third and fourth simulation of the symmetric load. The only change is in the load values. The three-phase load Z1 = 8 kW and the single-phase load Z2 = 3 kW, that is connected in time from 2 s to 3 s. First, the functionality of a classical inverter controller was evaluated and then the inverter controller was analyzed, that uses the virtual frequency and voltage method.

A. Standard Inverter Controller

When using a standard inverter regulator, it is not possible to continuously change the voltage on the AC side of the inverter. At the beginning and at the end of the simulation, the load Z1 is connected, that is about 1 kW smaller than the nominal load. Therefore, the voltage value is higher than the nominal voltage. On the contrary, in the time from 2 s to 3 s because the Z2 load is connected, the magnitude of the voltage decreases in the PCC. This again confirmed that when using a conventional inverter controller, the PV power plant with battery supply almost all the required power into the loads, but the PV system cannot change the voltage continuously, see Fig. 14. In addition, due to the unbalanced load, a current and voltage asymmetry occurred in the island microgrid, see the detailed courses of voltage shown in Fig. 15. The three-phase waveforms of this voltage show the voltage asymmetry that arises due to the single-phase load Z2.

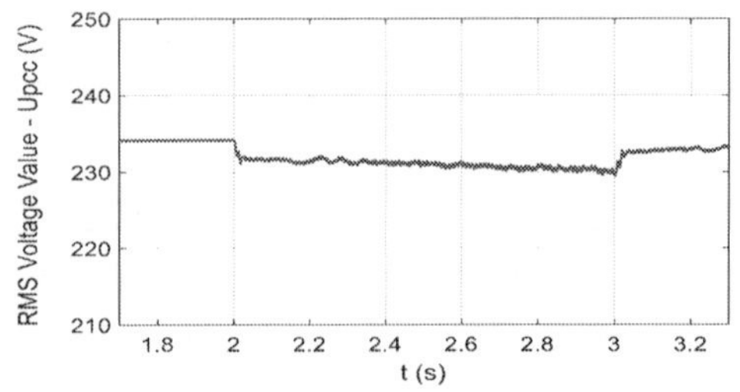

Figure 14. RMS voltage value at the PCC point – standard controller

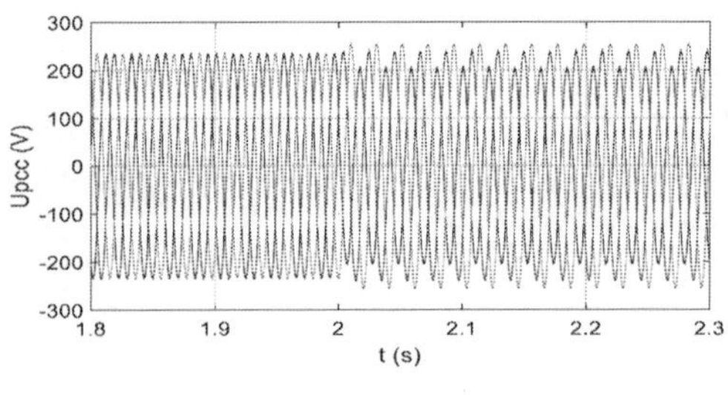

Figure 15. Detail of the voltage waveforms at the PCC point – standard controller

B. *Inverter Controller With a Virtual Method*

If an inverter controller is used, which uses primary and secondary control by means of the virtual method, the voltage value and the reference power of the photovoltaic system (PV power plant with battery) change according to the current load. The small deviation in the balance of electricity supply from its required consumption is due to the asymmetry of current and voltage, which were created by connecting a single-phase load Z2 (3 kW). However, the RMS voltage value is equal to the nominal value (230 V) at the PCC point, see Fig. 16. In contrast to the standard inverter controller, here the voltage is still maintained at the nominal value.

From a comparison of the details of the voltage waveforms from Fig. 15 and Fig. 17 shows that when using an inverter with primary and secondary voltage control, the asymmetry is lower than when using a conventional inverter controller. The voltage decreases most in the phase to which the unbalanced load Z2 is connected. The voltage waveforms in the other two phases are symmetrical. From the point of view of the asymmetrical electrical load in the island microgrid, it is more advantageous to apply an inverter controller using the virtual frequency and voltage method than a standard inverter controller. However, even this condition is not ideal. It would be best if the asymmetry of voltages or currents did not occur.

VII. Conclusion

Finally, a summary of the information obtained from the analysis of both PV inverter controllers can be made. The standard controller is undoubtedly suitable in that it can tracking the MPP, which the controller using the virtual methods does not allow. If decentralized production really increases in the future and one of the main decentralized sources will be PV power plants, then the current control circuit of PV inverters needs to be modified. One of the possibilities is the use of primary and secondary regulation of frequency and voltage based on the virtual methods together with the connection of storage equipment with sufficient capacity.

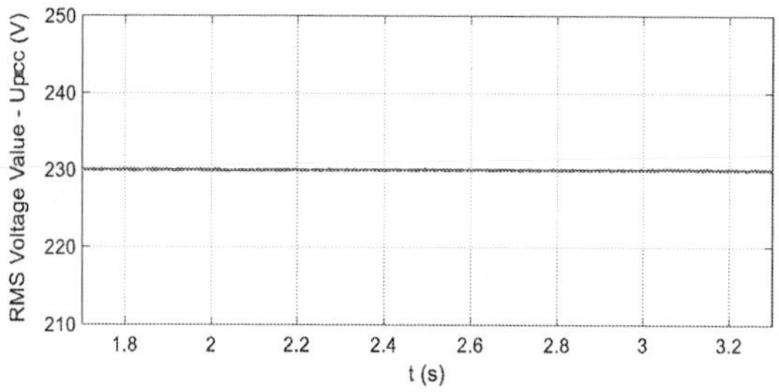

Figure 16. RMS voltage value at the PCC point - controller with a virtual method

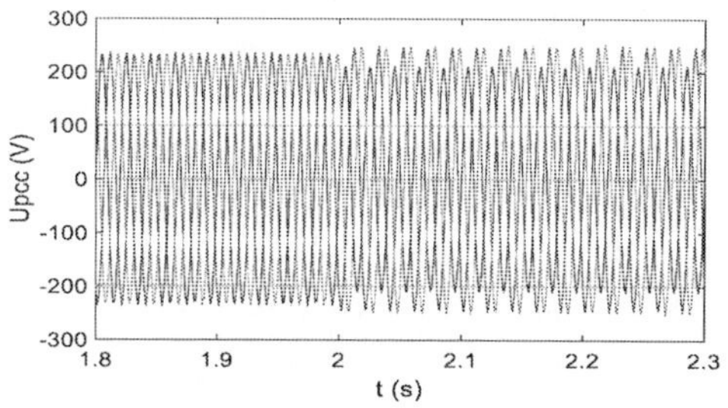

Figure 17. Detail of the voltage waveforms at the PCC point – controller with a virtual method

In the future, hybrid PV plants could also partially contribute to maintaining frequency and voltage stability in the electricity system, or at least in island microgrids. Ideally, it would be appropriate to combine the advantages of both controllers into one and create a controller that can tracking the MPP and at the same time can continuously regulate the output power of the PV plant and the storage device, including its charging and discharging. At the same time, it would be appropriate to ensure the elimination of voltage and current asymmetries in the island microgrid.

Acknowledgment

This paper was supported by the student research project SGS 2021-018.

References

[1] MPO. "State Energy Policy." MPO, Prague, Internet: https://www.mpo.cz/dokument158059.html, [Feb. 1, 2022].

[2] MPO, "National Action Plan for Smart Grids 2019 – 2030" Prague, Internet: https://www.mpo.cz/cz/energetika/strategicke-a-koncepcni-dokumenty/narodni-akcni-plan-pro-chytre-site/narodni-akcni-plan-pro-chytre-site-2019---2030---aktualizace-nap-sg--248894/ [Feb. 1, 2022].

[3] L. Rakova. "Frequency and Voltage Control in the Islanded Microgrids with Photovoltaic Power Systems." *Dissertation (Ph.D.), University of West Bohemia in Pilsen,* Faculty of Electrical Engineering, Pilsen.

[4] L. Rakova, E. Dvorsky, P. Hejtmankova, "Voltage control and load redistribution between decentralized power sources in microgrids in island mode." In *Proceedings of the 9th International Scientific Symposium on Electrical Power Engineering (ELEKTROENERGETIKA 2017).* Košice: Technical University of Košice, 2017. s. 165-169. ISBN: 978-80-553-3195-9.

[5] Y. LI, Y. W. LI, "Virtual frequency-voltage frame control of inverter based low voltage microgrid." In *Electrical Power & Energy Conference (EPEC),* 2009 IEEE, pp. 1-6, 22-23 Oct. 2009.

[6] Y. LI, Y. W. LI, "Power Management of Inverter Interfaced Autonomous Microgrid Based on Virtual Frequency-Voltage Frame." In: *2011 IEEE Trans. on Smart Grid,* Vol. 2, No. 1, pp. 30 – 40, March 2011. ISSN: 1949-3053.

[7] P. Kanakasabapathy, I. V. V. Rao, "Control strategy for inverter based micro-grid." In *Power and Energy Systems Conference: Towards Sustainable Energy,* 2014, Bangalore, 2014, pp. 1-6. ISBN: 978-1-4799-3421-8.

[8] A. Raghami, M.T. Ameli, M. Hamzel, "Primary and secondary frequency control in an autonomous microgrid supported by a load-shedding strategy." In: *Power Electronics, Drive Systems and Technologies Conference (PEDSTC),* 2013 4th, pp. 282-287.

[9] J. M. Guerrero; M. Chandorkar, T. Lee; P. C. Loh, "Advanced Control Architectures for Intelligent Microgrids—Part I: Decentralized and Hierarchical Control." In *Industrial Electronics, IEEE Transactions* on, vol.60, no.4, pp. 1254-1262, April 2013.

[10] M. Baharizadeh, M. S. Golsorkhi, M. Shahparasti and M. Savaghebi, "A Two-Layer Control Scheme Based on P – V Droop Characteristic for Accurate Power Sharing and Voltage Regulation in DC Mi-crogrids." *in IEEE Transactions on Smart Grid,* vol. 12, no. 4, pp. 2776-2787, July 2021, doi: 10.1109/TSG.2021.3060074.

[11] M. S. Golsorkhi, D. J. Hill and M. Baharizadeh, "A Secondary Control Method for Voltage Unbalance Compensation and Accurate Load Sharing in Networked Microgrids." *in IEEE Transactions on Smart Grid,* vol. 12, no. 4, pp. 2822-2833, July 2021, doi: 10.1109/TSG.2021.3062404.

[12] MathWorks. Internet: https://www.mathworks.com.

Concept for the Use of the Existing Trolleybus Power Supply Networks as Infrastructure for the Electric Vehicle Charging Stations

Valentin Gyurov, Vladimir Chikov, Nikola Makedonski, Yuliyan Yordanov
Department of Electric Power Supply and Equipment
Technical University of Varna
Varna, Bulgaria
valentin.giurov@tu-varna.bg, vladimir.chikov@tu-varna.bg, n.makedonski@tu-varna.bg, ypyordanov4@gmail.com

Abstract — **The present study introduces a concept and provides practical technical solutions to the use of the existing expensive trolleybus transport infrastructure as a power supply system for electric vehicle charging stations, which will be particularly effective at night. The proposed solution is subject to the general availability of trolleybus' poles and overhead contact lines, and considers the possibility of using the existing or constructing new parking lots, which will be equipped with a new specific type of DC/DC and DC/DC/AC charging station architecture.**

Keywords: charging stations for electric vehicles; eletric cars; trolleybus transport; power supply systems for electric transport; electromobiles; DC-DC converets.

I. Introduction

Technical solutions to electric power supply systems for trolleybus transport are commonly known and extremely widespread among the cities with urban electric transport. Distinctive features of their topology include medium voltage (MV) power lines, MV/LV (DC) rectifiers and catenary. The requirements for ensuring optimal voltage levels of the catenary, low voltage and power losses and attainment of optimal power load, determine the presence of a uniform distribution of rectifier stations as graphically represented in the cities' cadastral maps. In view of the fact that the trolleybus transport runs through major transport arteries, the catenary lines cover principal city roads, which belong to the highest category of roads according to the classification of national roads (main streets, boulevards, etc.). The traction infrastructure inclusive of its power lines, rectifier stations, poles and catenary, is a technical system with a high investment value of millions of euros and has a relatively high transmission capacity of 1.5-5MW [3]. Trolleybus transport is defined by a specific load profile and characteristic load schedules determined by the different intensity of the traffic load during different hours of the day. Thus, peak values are normally observed in the hours between 8:00-11:00 a.m. and 15:00-18:00 p.m., as against the 23:00-7:00 a.m. period when the power consumption is near zero and the system is in idle. The development of EV- charging infrastructure is a topical issue arising from the vital necessity for the provision of vast sources of power. [1], [2]. This, in turn, corresponds closely with the need for maximized nighttime power generation that would be used for a more cost-effective EV charging. Even more, the proposed concept has two further extremely beneficial effects. The first addresses the provision of another efficient way to use the regenerative braking energy that is fed back to the catenary systems for the operation of the new generations of trolleybuses during daytime braking and power the electric vehicles directly if there is no other trolleybus getting catenary supply at that time, and only if the rectifier station is that of uncontrolled rectifier type that blocks energy transfer to medium voltage level. The second effect is related to the integration of photovoltaic systems into the trolleybus transportation sector. It is generally acknowledged that the most effective place for their connection is the low voltage side of the power transformers in rectifier stations. Subsequently, the generated renewable energy can thereby be used by the trolleybuses, and in the absence of sufficient load, it can be directly consumed by the EV- charging stations. The presented concept is an innovative solution with three interrelated aspects of successful application in the electric power supply systems for urban electric transport.

II. Concept for Technical Solutions of Novel Charging Stations

A. Basic Idea of the Solution.

The proposed technical solution supports the development of a specific DC/DC and/or DC/DC&AC converter powered by the trolleybus overhead catenary system. Explored, further, is the possibility for such a specialized converter to be mounted internally, inside the utility pole, or externally, in a special electric box and at a height out of the zone of unintentional human activity of H = 2.5 m. The converter, thus projected, will supply EV-charging stations that are in close proximity to the trolleybus line, located in the vehicle parking lane, parking zones, etc.

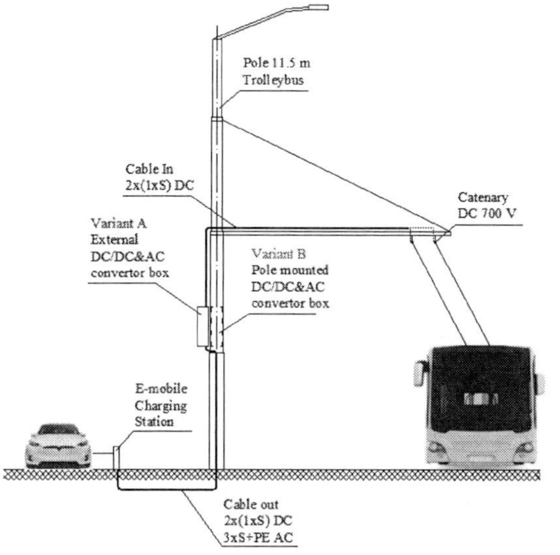

Figure 1. Representation of the tehcnical solution for specilized DC/DC and/or DC/AC EV-charging stations.

Depicted in Figure 1 is an illustrative example solution to the above formulated concept as to the construction of charging stations powered by the trolleybus catenary system.

B. Scematic Representation and Electrical Design Methodology.

Simulation-based studies were conducted into the specifically created for the purposes of the present paper DC-DC converter model. The functional pattern follows that of a Phase Shift Full-Bridge DC-DC converter. The respective block diagram is shown in Figure 2. The technical solution is in accordance with the most common standards for EV DC charging stations – particularly North American SAE J1772 and Japanese JEVS G106 with rated power of the charging station P_{RATED} being 40 kW and rated voltage V_{RATED} of 500V.

The elements L and C are calculated by formulas (1) and (5).

$$L = V_L \cdot \frac{D \cdot T_{SW}}{I_{RIPPLE}\%} \cdot 100 \tag{1}$$

where: V_L - voltage drop on L, [V]; I_{RIPPLE} – deviation setting of charging current [%],

$$V_L = V_{IN} \cdot \frac{V_1}{V_2} - V_{BAT} \tag{2}$$

where: V_{IN} - input voltage in the converter [V]; V_1 - nominal voltage on the primary side of the high frequency transformer [V]; V_2 - nominal voltage on the secondary side of the high frequency transformer [V]; V_{BAT} - rated battery voltage at 100% charge [V].

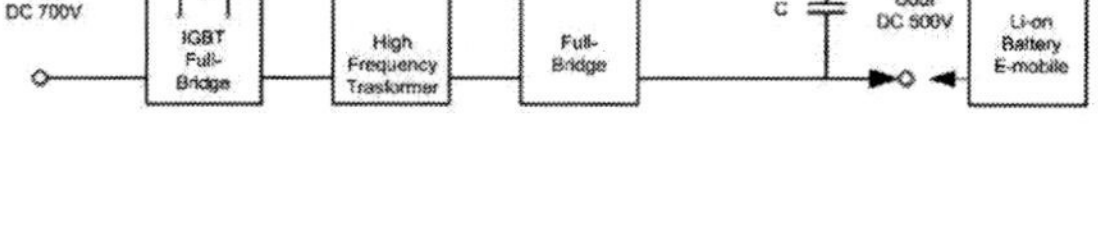

Figure 2. Block diagram of specilized model for EV DC/DC charging stations.

$$D = 0.5 \cdot \frac{V_1}{V_2} \tag{3}$$

$$T_{SW} = \frac{1}{f_{SW}} \tag{4}$$

$$C = I_{RIPPLE} \cdot \frac{D \cdot T_{SW}}{V_{RIPPLE}} \tag{5}$$

where: V_{RIPPLE} - deviation setting of charging voltage [%].

The proposed solution aims to achieve the prescriptive values stated by the North American SAE J1772 and Japanese JEVS G106 standards for Fast-Charging as shown in Table 1. [4], [5].

Table 1. Typical parameters for DC Fast-charging Stations [1]

Typical parameters for charging up to 80% of the battery capacity			
Distance traveled	***Estimated Energy Consumption [kWh]***	***Power [kW]***	***Aprox. charging time [min]***
25	5.2	40	8
50	10.4		16
75	15.6		25

The developed technical solution for DC/DC converter is founded on a technical solution for recharging the most common EV Li-ion battery types with a capacity within the range of 5 to 100 kWh and V_{BAT} = 350V. Considered carefully is a prime case of a 20 kWh (48Ah) battery at V_{BAT} = 350V (408V at full charge). The relevant set and calculated block parameters are recorded in Table 2.

Table 2. Calculated parameters of DC Fast-charging Stations.

Typical charging parameters for 20 kWh battery			
V_{IN} [V]	700	I_{RIPPLE} [%]	10
V_1 [V]	735	V_{RIPLLE} [%]	1
V_2 [V]	700	D	0.483
V_{BAT} [V]	350	f_{SW} [kHz]	25
$V_{BAT, FULL}$ [V]	408	T_{SW} [s]	40.10^{-6}
C [F]	$5.48.10^{-6}$	L[H]	278.10^{-6}

III. Simulation Study of Specilized DC/DC Converter Powered by Trollebus Electric Power Supply System

The simulation model reproduces the block diagram depicted in Figure 2. The parameters of the model are listed in Table 2. The physical realization of the model requires special design of the individual modules and a certificate attesting their relevance for a particular application - power supply systems for urban electric transport. This pertains mostly to the passive elements - high-frequency impulse transformer and current-limiting choke, filtering capacitor, which are thought appropriate for interior installation inside the trolleybus pole. Depicted in Figure 3 is the block diagram of such a simulation model. The impact of the switching and protection equipment (circuit breakers, DC contactors, etc.) is considered negligible for the purposes of the present simulation study due to their low resistance in normal operation. Their sizing is entirely subject to the design decision as to the construction of the prototype samples of the developed technical system. [4], [5], [6].

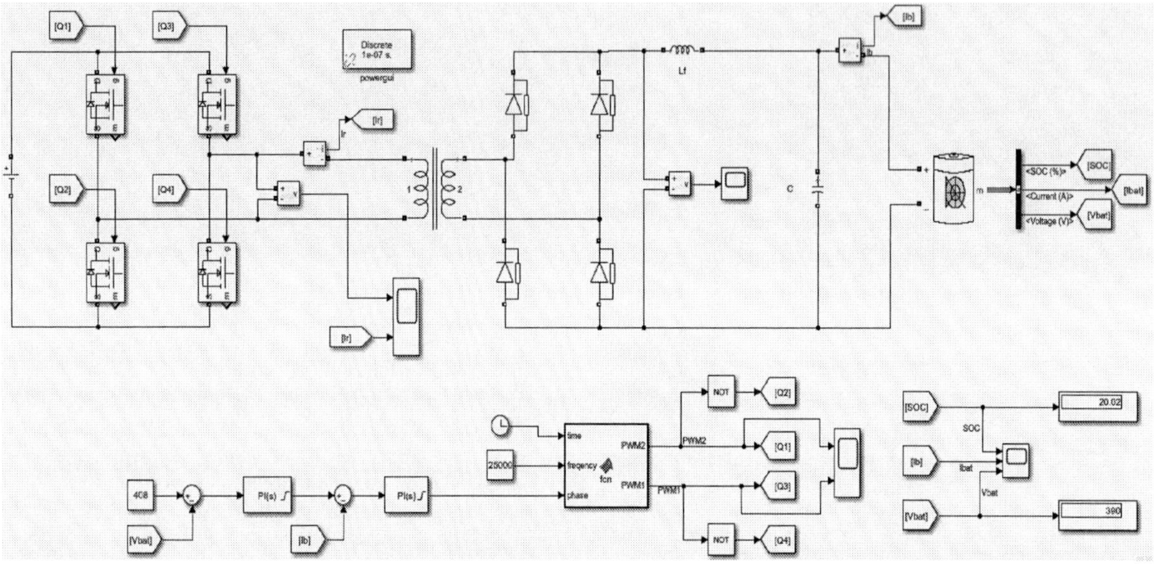

Figure 3. Simulation model of specilized DC/DC EV-charging stations

A distinct advantage is taken in the present study of the predetermined values of the target charging current I_{CHARGE} = 100A and I_{RIPPLE} = ± 10%. Outlined in Figure 4 and Figure 5 is the change in voltage V_1 and current I_1 on the primary side of the high-frequency transformer, i.e. those at the input of the circuit. The graphs show the expected shape of the pulse voltage and current for a Full Bridge-Phase Shift converter. As observed, -the current I_1 is steadily controlled within the set limits.

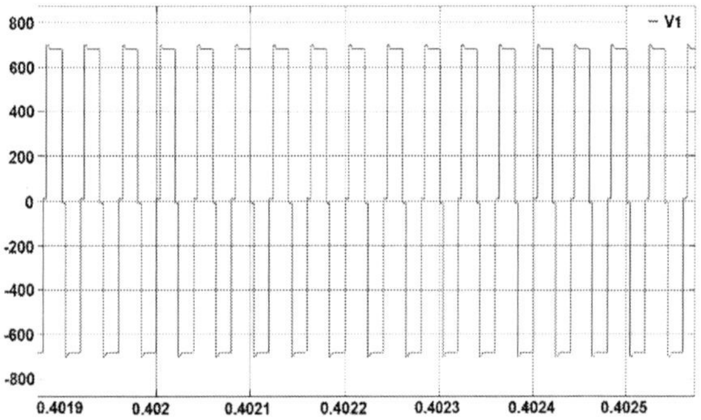

Figure 4. Voltage V_1 on the primary side of the high frequency transformer of the DC / DC converter.

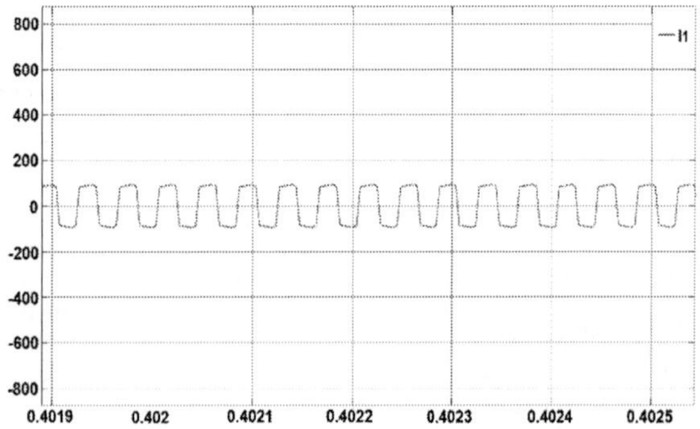

Figure 5. Current I_1 on the primary side of the high frequency transformer of the DC / DC converter.

The simulation results with the subsequent change in the V_{BAT} battery voltage and the charging current I_{CHARGE} can be clearly perceived in Figure 6 and Figure 7. The maximum allowable charge voltage is set by the formula:

$$V_{BAT,FULL} = V_{BAT} \cdot \frac{V_{CELL,MAX}}{V_{CELL}} \quad (6)$$

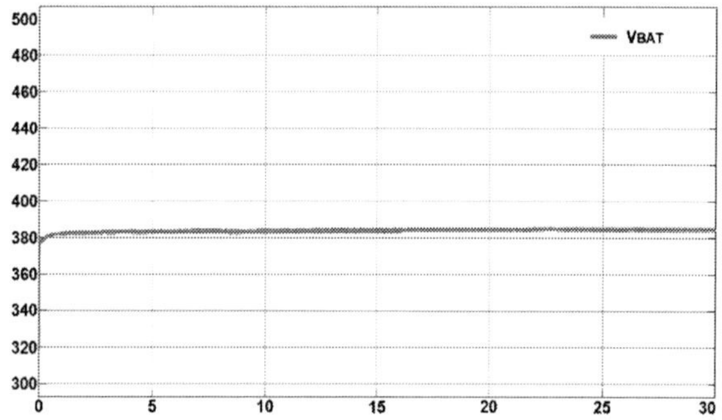

Figure 6. Changing the V_{BAT} battery charging voltage for a period of 30 seconds.

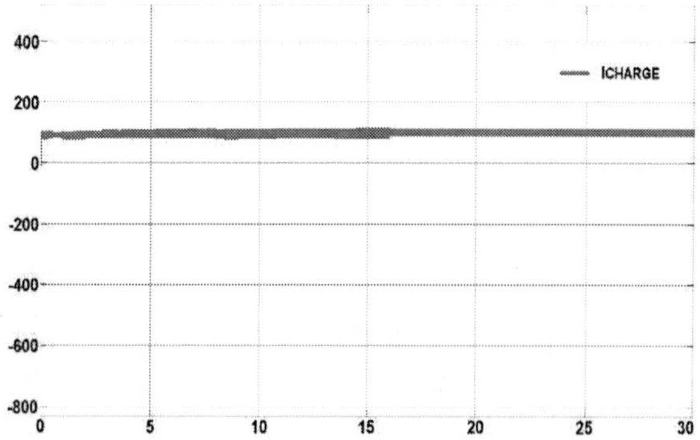

Figure 7. Changing the I_{CHARGE} battery charging voltage for a period of 30 seconds.

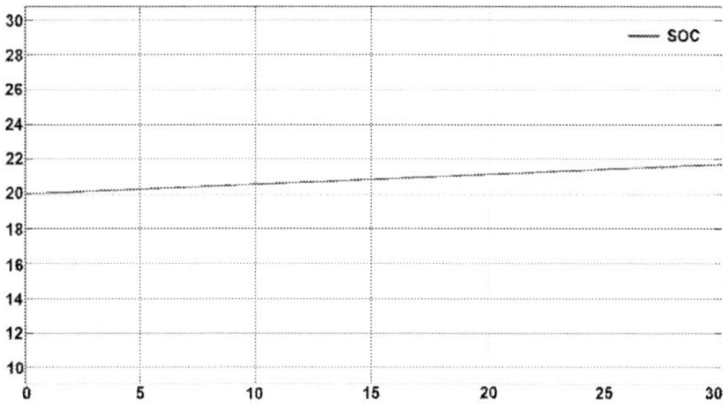

Figure 8. Changing the battery state of charge or its capacity - *SOC* for a period of 30 seconds.

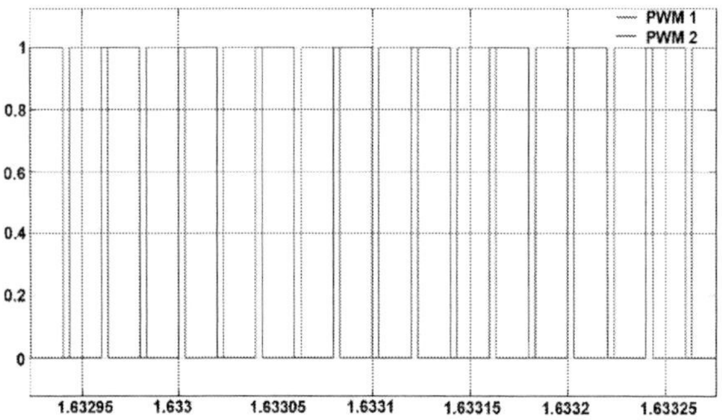

Figure 9. IGBT Bridge control pulses.

The equation (6) assumes values characteristic of Li-ion rechargeable batteries - V_{CEL} = 3.6V and $V_{CEL, MAX}$= 4.2 V.

The results of the current research study were obtained for a simulation with a specified time for 30 seconds. The technical solution maintains I_{CHARGE} = const. within the approved change of variable limits. The *SOC* change is of linear character and increases by 1.8% or 3.6%/min for the duration of 30 sec. This means that the technical solution will charge a completely discharged battery up to 80% within 22 minutes, which is fully consistent with the standard values for Fast-Charging Stations.

A distinctive feature that characterizes the solution under consideration is the IGBT Bridge principle of the control, the block diagram of which is depicted in Figure 3. The introduced Phase-Shift Full-Bridge (PSFB) DC/DC converter is used in systems with galvanically isolated DC sources and voltage leveling requiring the application of high-frequency transformer. The transistors of the input bridge operate with a duty cycle of 50% and the regulation of the energy transferred through the transformer is done by changing the phase of one of the half-bridges (Half-Bridges). The transmitted power varies from 0 at phase shift 0° to maximum value at phase shift 180°. The voltage on the secondary side is supplied by the Full Bridge Rectifier (FBR). The use of LC filter is associated with reduced output current and voltage ripples, improved (Electromagnetic Compatibility EMC) and enhanced performance of the feedback system. The negative feedback control system consists of two consecutive units. The first unit determines the voltage to which the battery is charged via a PI hysteresis regulator. The preset hysteresis, in turn, determines the charging current, provided that the nominal voltage of the battery is not reached, i.e. constant current mode (CC). The DC voltage mode is set by the constant block (before the adder) and it is the leading one, i.e. controls the charging current provided that the maximum battery voltage (CV mode) is reached. The second unit, comprising, again, an adder and a hysteresis comparator, limits the phase depletion of the control signals on one of the half-bridges to 180° by comparing the charge current limit against the level of the charging current. In real systems, the charging process is terminated by the Battery Management System (BMS) as conditioned by the prescribed SoC (State of Charge).

IV. Conclusion

The presented concept for the development of a specialized class of EV-charging stations, powered by the trolleybus contact network (catenary), claims a technical solution with significant technical novelty and has the character of an invention. The case study for a DC/DC charging station fully complies with the North American SAE J1772 and Japanese JEVS G106 standards, covering a wide range of electric vehicle models on the American and Asian markets, commonly used in the European Union. The mere fact that most modern cars are manufactured to charge on both standards, with the Fast-Charge being strongly preferred by consumers due to the shorter charging time, clearly sets out the reason why it acts as the primary focus of the current study and as a first step towards the implementation of this innovative solution. A distinct advantage of the proposed solution is the real possibility of using the entire existing trolleybus traction infrastructure as a state of capacity that will be particularly effective during trolleybus low load times (at night), which actually coincides with the most popular and highly convenient charging time for electric vehicles. This is precisely the part of the day that is viewed as the most expedient one for EV charging not only because of the normal operational needs but also for the sake of the opportunity to take maximum advantage of the much cheaper electricity rate during those off-peak hours at night when demand is low.

Acknowledgment

This work was supported by the Bulgarian Scientific Fund, Operational Programme "Fundamental Research", Session 2019 under contract №06-37/28 Date: 16.12.2019. „Research on Electricity Consumption Regimes in Electric Power Supply Systems for Urban Electric Transport with Two-way Power Flow".

The project discussed herein is supported by the European Regional Development Fund through the Operational Programme "Science and Education for Smart Growth" under contract UNITe № BG05M2OP001-1.001-0008. Project name: "National center of mechatronics and clean technologies".

References

[1] R. Wolbertus, R. Hoed, M. Kroesen, C. Chorus, "Charging infrastructure roll-out strategies for large scale introduction of electric vehicles in urban areas: An agent-based simulation study", *Journal of Transportation Research Part A* 148, pp. 262–285, 2021.

[2] M. Falvo, D. Sbordone, I. S. Bayram and M. Devetsikiotis, "EV charging stations and modes: International standards," *2014 International Symposium on Power Electronics, Electrical Drives, Automation and Motion*, 2014, pp. 1134-1139.

[3] G. Pavlov, L. Sekulov, "Study and analysis of efficiency of recuperative energy utilization in ground urban electric transport", *Conference on Electrical Machines, Drives and Power Systems, ELMA 2021 – Proceedings*, 2021, pp. 1-4.

[4] Japan Electric Vehicle Association Standards JEVS, G106. EV inductive charging system: General requirements.

[5] SAE J1772 (IEC 62196 Type 1) Plugs, socket-outlets, vehicle connectors and vehicle inlets - Conductive charging of electric vehicles - Part 1: General requirements.

[6] IEC 61851-23: DC electric vehicle charging station.

Analysis of the Two-way Power Transmission and Recuperation Efficiency of Solaris Trollino 18 Trolleybuses

Ilko Tarpov, Martina Tomcheva
Department of Power Supply and Electrical Equipment of Transport
Todor Kableshkov University of Transport
Sofia, Bulgaria
itarpov@vtu.bg, mtomcheva@vtu.bg

Abstract — **The paper explores the overall performance test results evaluating the operation of Solaris Trollino 18 trolleybus equipped with regenerative braking system supplemented with a traction battery energy storage system. The trolleybus has been put in operation for provision of proper public transportation services in the city of Sofia (Bulgaria). The research study, therefore, was conducted under real operating conditions distinctive of this particular type of trolleybus and an accurate energy efficiency assessment of the vehicles under discussion was subsequently carried out. The research is based on computer simulation models in addition to the real observations of the main modes of operation. Meticulous attention was paid to the energy recovery mechanism of the braking modes in which, electricity is fed back to the grid. The graphical representation of the results allows for a set of conclusions to be drawn about the trolleybus modes of power consumption and regenerative braking.**

Thereupon, on the analysis of the measurement results obtained, recommendations for improved energy efficiency of the vehicle operation are made.

Keywords: trolleybuses, urban electric transport, recuperation, energy efficiency

I. Introduction

Modern vehicles are highly regarded for their improved energy efficient movement in mode of traction and braking. The acquired reversibility and the more effective traction control make it possible for the energy to proceed in two-way power flows, i.e. from and to the power supply network. Enhanced, thereby, is the overall energy efficiency of the entire transport supply system.

Reported, currently, under real operating conditions, are no problems with respect to shortages of the generated air contact network capacities for trolleybus transport in the city of Sofia. This is primarily due to the separation of the trolleybus lines into small feeder zones, which enables the possibility for coping with large traction loads after steep and prolonged slopes, increased trolleybus traffic, inclusion of new feeder sections, and others.

When braking, the compositions under consideration are likely to perform conventional friction, rheostatic, and recuperative braking. If during rheostatic braking the generated energy is extinguished in rheostatic resistors, then, during regenerative braking this energy, through traction motors, is converted into a type of supply and when the voltage potential reaches values higher than those in the network it flows in the opposite direction. If the recovered energy cannot be fed into power supply network, it is most certainly extinguished in the rheostatic resistors. Such processes, therefore, should be constantly monitored and all the necessary measures should be adopted to guarantee full utilization of the generated energy, which is of system-type recuperative nature.

The primary focus of the present paper are Solaris Trollino 18 trolleybuses, which are equipped with a regenerative braking system and furnished with a traction battery for energy storage to be later used for movement in the absence of power, breakdowns or emergencies, traffic accidents or temporary changes of the route.

In accordance with [1], numerous studies should be performed under different conditions during initial operation, with the main objective of determining the algorithms of control in modes of operation and recuperative or electrodynamic braking.

The study was undertaken in real operating conditions specific to this type of trolleybuses. Offered is an assessment of the energy efficiency of the surveyed vehicles with particular consideration on the modes of braking, during which the energy is fed back to the power supply network.

II. Research and Analysis of the Two-Way Process of Energy Transfer

A. Theoretical Part

The surveyed Solaris Trollino 18 trolleybuses are articulated vehicles operated by the trolleybus depot „Nadezhda". 30 trolleybuses are currently used for public transportation services in the city of Sofia (Bulgaria). The compositions under consideration have asynchronous traction electric drive supporting a mode of selective regenerative-rheostatic electric braking.

Solaris Trollino 18 trolleybuses are equipped with a rechargeable traction battery to supply power to the traction motor control systems in the absence of a contact network. The traction accumulator battery is electrically connected to the intermediate circuit next to the recuperative AC/DC converter. In those cases where the trolleybus is connected to the contact network, it recharges the battery automatically via a built-in charger mounted on the roof. The maximum charge the battery can store during operation is likely to reach up to a maximum of 100 % of its capacity.

The Solaris Trollino 18 trolleybuses operated by the Sofia transportation system are provided with an on-board microprocessor system for meticulous measurement and detailed recording not only of the electric power consumption of the separate systems, but also of the energy generated and fed back in the traction network during regenerative braking. The resultant data about the energy consumption and recuperated to traction power in regenerative braking do not allow for an immediate analysis of the trolleybuses' total energy efficiency. This, in turn, necessitates the additional processing of the collected data in order to meet the purposes of the present research study.

The main parameters determining the mode of traction and ensuring minimal energy consumption are conditioned by the effective increase in speed. This parameter is computed by means of some traction calculations and by the formula:

$$V_r = -38{,}3.l_n^2 + 44{,}1.l_n^2 + 1{,}76, km/h \quad (1)$$

where l_n^2 – length of movement, km; V_r – rational acceleration speed, m/s.

If $l_n^2 > 0{,}5\ km$, then the assumed $V_r \approx 55\ km/h$.

Further optimization estimates were made in order to select rational speeds of acceleration for a definite length of the section between those two stops where the lowest electricity consumption is obtained. Recorded, accordingly, in Table 1 are some statistical acceleration speeds, determined experimentally, on the basis of the length of the route. Obtained from them, by the method of least squares, is the dependency 2.

$$P(A) = \sum_{i=1}^{n} P(H_i).P\left(A/_{H_i}\right) \quad (2)$$

Where A – an event triggered by other H_i events, known as hypotheses;

n – the number of hypotheses.

Table 1. Statistical acceleration speeds

V_r, m/h	5,55	7,64	9,17	10,4	11,5	12,5	13,2	13,5	14,2
l_n, km	0,1	0,15	0,2	0,25	0,3	0,35	0,4	0,45	0,5

Depicted in Fig. 1. is the V_r dependency on the length of the route (movement) l_n computed via traction calculations and through the use of Formula 1.

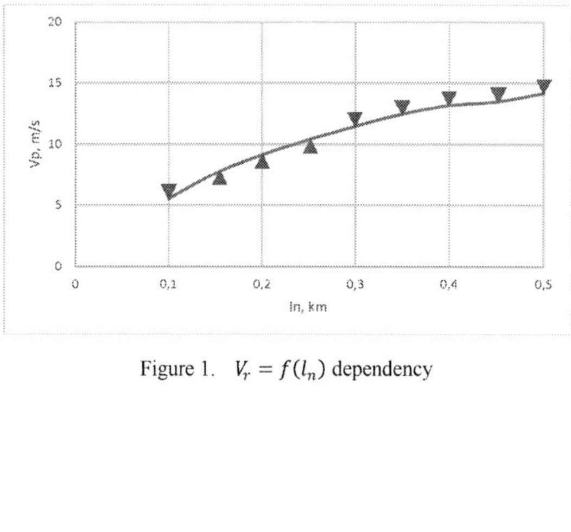

Figure 1. $V_r = f(l_n)$ dependency

Developed, further, are dependencies regarding the time of movement.

The rational speeds of movement calculated according to the methodology described above can be further incorporated into a dependency determining, with sufficient accuracy, the trolleybus's time of movement:

$$t_{ttm} = 5{,}67.l_n + 17{,}3; \quad (3)$$

where t_{ttm} is the trolleybus time of movement in s;

l_n – the length of movement in km.

Here, t_{ttm} should be adjusted to account for the effects of the inclinations and regulated intersections. Hence, formed is the dependency:

$$t_{tm0} = \frac{t_{tm}}{1+K.i_E} + t_{tsri} \quad (4)$$

where i_E is the equivalent inclination, ‰;

K – correction factor, K=0,0045 for trolleybuses;

t_{tsri} – time spent at regulated intersections.

The impact of the random factors upon the time of trolleybus movement is statistically estimated and, in most cases, such factors tend to exert undue influence of delay with a time-lag effect.

Hence, the total time of movement along the route is regulated by the dependency.

$$t_{tm} = t_{tm0} + t_{ttm} = \frac{t_{tm}}{1+K.i_E} + t_{tsri} + t_{ttm} \quad (5)$$

The average speed of the vehicle movement in that section - V_{sm} and the time of movement during traction - t_T are determined by:

$$V_{sm} = \frac{l_n}{t_{asm}}, km/h; \quad t_T = \frac{l_T}{V_{asm}} \approx \frac{2.l_T}{V_r}, s \quad (6)$$

where V_{asm} – the average speed of movement in the section during traction;

l_T – the distance travelled in the mode of traction.

There is a full range of additional parameters affecting the process of movement, one of which is the time spent at the stops - t_{ts}. This particular time consists of fixed and random components. The fixed components shall be determined by the dependency:

$$t_{ts} = \frac{1}{\eta}.(n_1.t_1 + n_2.t_2) + 6, s \quad (7)$$

where n_1, n_2 is the number of passengers getting on and off;

t_1, t_2 – the amount of time it takes for passengers to get on and off;

η – coefficient that measures the coincidence rate when getting on and off.

The mathematical model, thus presented, allows for an adequate assessment of the trolleybus level of recuperation implementation.

In order for all the braking processes to be recuperative for the specific conditions, there should be at least 22pcs of trolleybuses in actual operation within a section.

B. Research Part

Tracking down the trolleybuses two-way flow of energy was performed under normal operating conditions in their mode of movement along the lines № 1 and № 7. The technical solutions for the consumption of the recuperated energy should be further supplemented by an analysis of the process of recuperation within the trolleybus itself and in the traction power supply system. In view of this, the research studies should be conducted for real processes making allowances for the inclusion of fixed and random factors [2, 3].

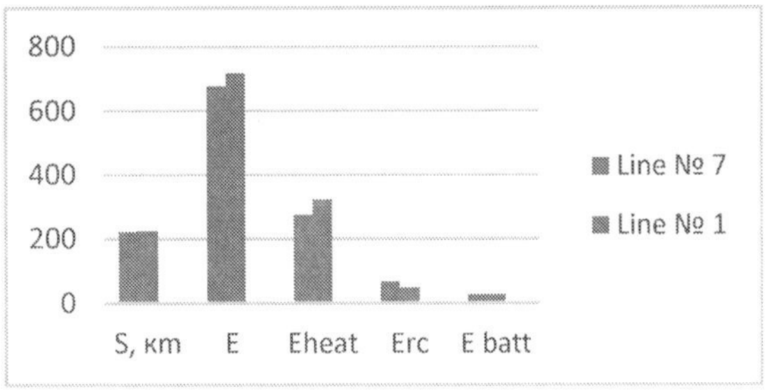

Figure 2. The average daily values of the surveyed energies along the Lines 1 and 7

The length of movement for one day is approximately the same for all the lines under consideration and amounts to 224,5 km. The average daily results for the values of the examined consumed electricity - E, energy for heating– E_{heat}, energy regained by recuperation - E_{rc} and the energy for recharging the accumulator batteries – E_{batt}/ of the surveyed trolleybuses are graphically represented in Figure 1.

The daily relative energy consumption – e* for each individual line by itself and in total is calculated via the dependency (1):

$$e^* = \frac{E}{S}, kWh/km \tag{8}$$

where E is the registered energy consumption over a given period of time, kWh;

S –the length of movement for that period of time, km.

In compliance with [4,5], using the indicator defined above, determined is the energy consumption per one kilometer length of movement. In practice, this is a fundamental indicator and provides for an energy analysis of a particular type of electric vehicle and technical assessment of its energy intensity and energy efficiency. As for the lines under discussion, that indicator is respectively 3,11 kWh/km aggregated for the two lines. The difference in the relative energy consumption for the individual lines approximates 4,3 %. The average energy consumption for heating is 300 kWh per day, with the respective hourly electricity consumption for heating being 19,33 kWh on average. The energy recovered by recuperation reaches 57,5 kWh with the mean time needed for the traction batteries to recharge being 28 kWh.

To calculate the regenerative braking efficiency consistent with (9), the coefficient K_{rc} should be expressed as a percentage.

$$K_{rc} = \frac{E_{rc}}{E} \tag{9}$$

The coefficient Крк is likely to reach up to 10-20 % along steep sections if the consumption of the recovered energy is high and provided that the descent load flow rate is greater than that of the ascent. Accordingly, for the vehicles under study, this coefficient adds up to 8,28 % on average.

To increase the efficiency of trolleybuses equipped with regenerative braking, it is of utmost significance for the braking process to commence at higher speed rates and then to set braking force smoothly and gradually. Failure to do so, will result either in automatic rheostatic or pneumatic braking, deriving no benefit from the advantage of regenerative braking.

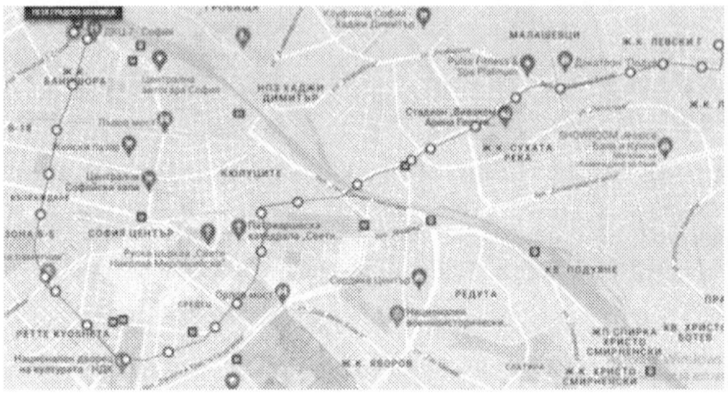

Figure 3. Topographic representation of Line 1 route

Trolleybus drivers should strive to ensure that the regenerative braking constituent part prevail over the aggregate amount of the other parts when applying other types of braking throughout the entire braking. The low operating speed in public transport predetermines the low energy recovery values.

Figure 4. Topographic map coverage of Line 7 route

Recuperation is generally more effective when braking rather than moving downhill or maintaining the maximum permissible speed of movement.

Clearly, for the present, the recuperation efficiency in Sofia public transport is low. Normally, the underlying reasons for that are prescribed to the poor unfavorable conditions for recuperation in the capital city, associated with the busy traffic, the service times, the configuration and the section arrangements along the traction power supply network, as well as the short length of the separate feeder zones.

The denivelation in the different directions is within the range of 50÷60 m. The number of the stops along the route of one complete round of Line 1 is 51 and for Line 7 - 66. The relevant routes can be easily traced in Fig. 3 and 4. The route profile of the two lines is depicted in Fig. 5.

The number of unexpectedly occurred stoppages in the surveyed urban transport environment is between 5 and 15. The average distance between the stops is 440 m. Furthermore, the trolleybus driver's style of the vehicle control has an extra concrete and discernible impact upon the efficiency of the regenerative braking. Outlined in Fig. 6 is a diagram of the ascending curve lines of the speed excluding the time spent at stops from the entire journey time.

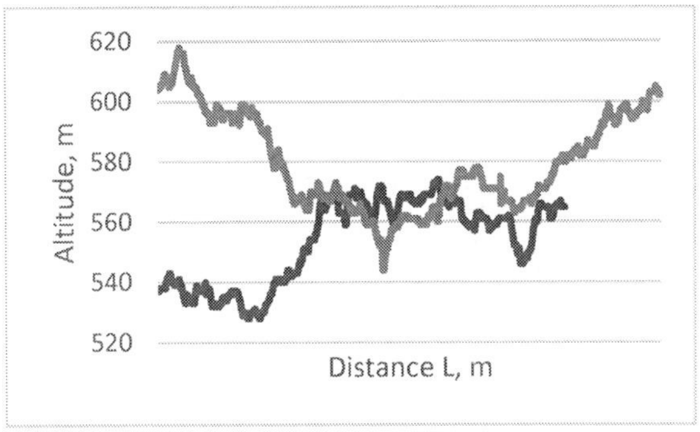

Figure 5. Elevation changes along lines 1 and 7

Is easily perceived that for the longest time the vehicles move at a speed of 10÷30 km/h, while movement at maximum speed occurs for the shortest period of time. This, eventually, presupposes the necessity for searching possible opportunities for increase in the number of sections allowing for the trolleybuses to run at maximum permissible speed, in addition to a subsequent increase in the distances between stops.

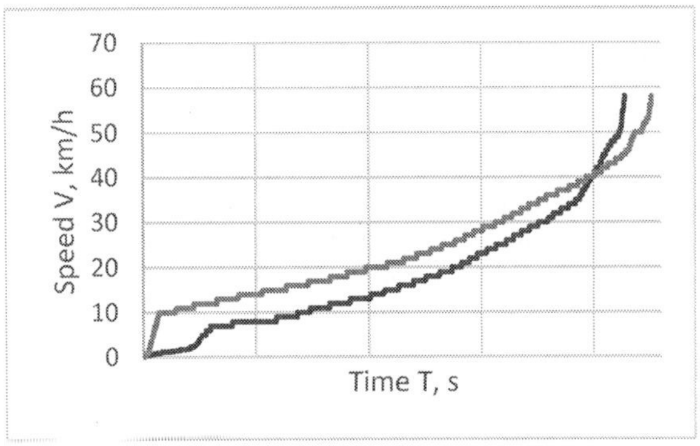

Figure 6. Ordered diagram of the speed as a function of time alomg the lines 1 and 7

It can be assumed form the data recorded in Tables 2 and 3 that the share of recuperated energy is subject not only to the number of stops but also to the distance travelled and the number of electric power substations.

Table 2. . Line 1 and 7 statistics

Line №	Stops, pcs.	*S*, km	*T*, min	V_{avg}, km/h	EPS, pcs.
1	51	22	67	20	8
7	66	30	119	15	7

Table 3. Energy data for lines 1 and 7

Line №	*E*, kWh	E_{rc}, kWh	*e**, kWh/km	K_{rc}, %
1	71,8	4,8	3,17	6,68
7	96,58	9,57	3,04	9,9

The increased number of power substations limits the possibility for the presence of another vehicle in the same feeder zone to consume the energy recovered. Technical prerequisites are, thus, created for a shift from regenerative to rheostatic braking which deprives the process of recuperation from its efficiency.

III. CONCLUSION

The conclusions reached from the research results are as follows:

- The share of the recuperated energy is up to 10 % of the energy consumed and depends exclusively on the speed of movement, the number of stoppages and, to the greatest extent, on the vehicle's driver;
- Maintaining temperature comfort in the surveyed trolleybuses accounts for more than 35 % of the total energy consumption;
- It is necessary to look for an opportunity to extend the segments with driving at maximum speed, as well as to increase the distances between the stations;
- Improved regenerative braking efficiency is related to a more spacious feeder zones;
- Further studies should be carried out to determine the distribution of the air contact network sections in view of improving the efficacy of the energy recovered during regenerative braking.

ACKNOWLEDGMENT

This work was supported by by the Bulgarian Scientific Fund, Operational Programme "Fundamental Research", Session 2019 under contract №06-37/28 Date: 16.12.2019. „Research on Electricity Consumption Regimes in Electric Power Supply Systems for Urban Electric Transport with Two-way Power Flow".

REFERENCES

[1] L. Sekulov., G. Pavlov, J. Isaev, R. Stoitsev. "Experimental study of the main modes of operation of the MAN trolleybus" (Експериментално изследване на основните режими на работа на тролейбус MAN), *Mechanics, Transport, Communications Magazine*, Volume 15, Issue 3, p. X-18, article № 1509, 2017.

[2] G. Dimitrov, G. Pavlov. "Analysis of the efficiency of the regenerative braking systems of Škoda Solaris trolleybuses in real operating conditions" (Анализ на ефективността на рекуперативните спирачни системи на тролейбуси Škoda Solaris в реални експлоатационни условия), *Journal of Mechanics, Transport, Communications*, Volume 14, Issue 3/2, p. X-24, Article № 1380, 2016.

[3] I. Petrov, G. Dimitrov, T. Lalev, "Experimental study of the energy efficiency of the metro trains of the capital metro in real operating conditions" (Експериментално изследване на енергийната ефективност на метровлаковете на столичния метрополитен в реални експлоатационни условия), *Yearbook of the Technical University* - Sofia, volume 67, book 1, 2017.

[4] V. Gyurov, N. Bezhanov, "Study on electric consumption regimes in rectifier stations of trolleybus transport", *Proceedings of the 11th Electrical Engineering Faculty Conference (BulEF)*, 2019, Varna, pp.1-4, ISBN: 978-1-7281-2698-2.

[5] V. Gyurov, N. Bezhanov, "Possibilities for energy planning in electric power supply systems of urban electric transport", *Proceedings of the 11th Electrical engineering faculty conference (BulEF)*, 2019, Varna, pp.1-6, ISBN: 978-1-7281-269.

Simulation Study of the Electric Consumption Regimes in Trolleybus Power Supply Networks

Valentin Gyurov, Plamen Parushev, Nikola Makedonski, Milen Duganov
Department of Electric Power Supply and Equipment
Technical University of Varna
Varna, Bulgaria
valentin.giurov@tu-varna.bg, pvparushev@abv.bg, n.makedonski@tu-varna.bg, m.duganov@tu-varna.bg

Abstract — **Brought forward in the paper is a simulation study into a section of the power supply system for the trolleybus transport in the city of Varna, Republic of Bulgaria. The simulation reproduces the conditions of the city's current medium voltage power supply system (MV), that of the rectifier stations, the catenary and the pertaining trolleybuses. The simulation models allow for a more accurate estimation of the respective regimes and voltage levels in the catenary lines and rectifiers under different modes of movement.**

Keywords: regenerative brakings; trolleybus transport; energy efficiency in trolleybus transport; urban electric transport.

I. Introduction

The computer simulation modelling reproduces the current medium voltage power supply system (MV) in the city of Varna, Republic of Bulgaria, that of the city's rectifier stations, the catenary and the pertaining trolleybuses. The simulation is developed through the use of those component parameters that correspond closely to the actual existing ones, which, basically, presupposes a subsequent parameterization of the specific power cables for medium voltage, type SAHEmT, power transformers 20/0.6 kV, rectifiers with power rating 1-3 MW, catenaries, inclusive of their actual lengths and cross sections. The immediate task of the research is to establish the energy parameters, the regimes of electric consumption and recuperative power generation, and the energy regenerative efficiency. The study recreates the main type of trolleybuses that provide public transportation services in the city-i.e. Skoda Solaris 26Tr, and focuses entirely on their individual operation in one section of the catenary system, together with the sequence in which two trolleybuses in a section would switch or transfer from one mode of operation to another. Examined further in the study are the possible modes of accelerating, decelerating (braking) and inertia. The research facilitates dynamic analyses of the modes and processes occurring in the operation of modern trolleybuses with frequency vector control in different modes of movement - motor, braking and inertia. Considered, additionally, is the mutual influence of two concurrent trolleybuses in one section of the catenary system and their major effect on the voltage regime.

The simulation model can be used for other similar types of electrically driven public-transport vehicles, subsequent to successfully accomplished parameterization. The results of the study might be most helpful for researchers taking particular interest in the field of trolleybus system sustainability and development, and for the field-specific practitioners as well.

II. Object of the Study

The object of the study is one rectifier station (RS) of the trolleybus traction network, namely, RS "South", Varna, Asparuhovo Quarter. The schematic diagram of the substation comprises power transformers 2x1000 kVA, 20/0.515 kV, DYn5 with rated current I_1=20*A* (on 21 kV), 21A (on 20 kV) or 22A (on 19 kV); I_2 = 22A and short-circuit voltage $u_{SC\%}$ = 7.26%. The power supply of RS "South" operates on medium voltage in closed-loop control scheme with Substation "Lazur", as depicted in Figure 1. The main power line is MV Line "Plana" with cable aluminum SAHEmT 3x1x185 mm², *L*=1073m. The RS "South" encompasses power rectifiers 6x500 kVA 515AC/630V DC which supply the Asparuhovo - Varna Center catenary section. The catenary is furnished with R100 copper conductor with a cross-sectional area of 100 mm². Calculated, as part of the current research study, were the parameters of the power transformers, the medium voltage cable lines and the contact network of the trolleybus transport–. The power transformer parameters were computed in relative and absolute units, which allows for their adequate adjustment in the simulation models of Matlab Sim Power Systems.

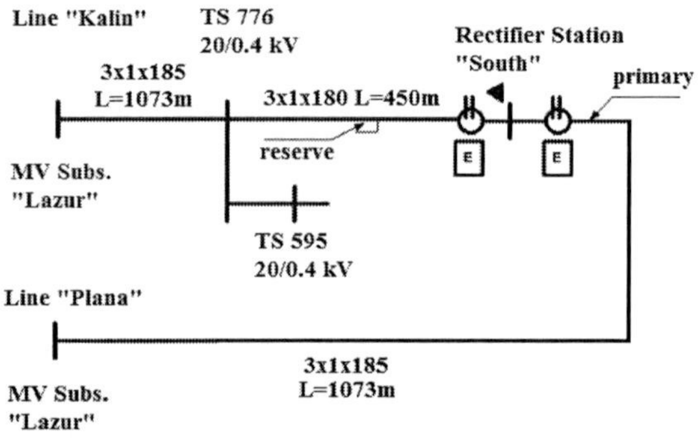

Figure 1. 20 kV Medium voltage network of the rectifier station "South", Varna, Bulgaria

Table 1. Calculated parameters of cable line and catenary

Parameters of MV AC cable and DC catenary			
AC Cable Line SAHEmT 3x1x185 mm² – RS "South"			
C [μF/km]	0.2717	I_C [A/km]	3.072
X_0 [ohm/km]	0.099	*L* [mH/km]	0.36
R [ohm/km]	0.215	*L* [m]	1073
DC Conductor R100 Catenary			
R [ohm/km]	0.183	*S* [mm²]	100

Table 2. Calculated parameters of transformers

Parameters of TM 1000 20/0.515V			
R_m [ohm]	$2.01.10^4$	R^*_m [p.u.]	461.538
X_m [ohm]	$4.039.10^3$	X^*_m [p.u.]	92.726
R_1 [ohm]	0.254	R^*_1 [p.u.]	$5.833.10^{-3}$
X_1 [ohm]	1.282	X^*_1 [p.u.]	0.029
R_2 [ohm]	$2.808.10^{-4}$	R^*_1 [p.u.]	$5.833.10^{-3}$
X_2 [ohm]	$1.416.10^{-3}$	X^*_1 [p.u.]	0.029
L_m [H]	12.857	L^*_m [p.u.]	not used in simula-tion
L_1 [H]	4.08.10-3	L^*_1 [p.u.]	
L_2 [H]	4.509.10-6	L^*_2 [p.u.]	

The main parameters of the cable lines, catenary and transformers are reported in Table 1 and Table 2.

III. Simulation Study in MatLab Sim Power System

The simulation study reproduces an actual power supply system and the developed model is parameterized in accordance with the individual components of the network. The study aims to recreate the possible modes of movement of two trolleybuses operating concurrently in one and the same section. The simulation model consists of power supply network, medium voltage cable lines, power transformers, catenary with distributed parameters. Considered is a common classic case - Trolleybus №1 is located at the beginning of the line, while Trolleybus №2 - in the middle of the route. The section of the catenary covers line №88, which is operated by a modern generation of Skoda Solaris 26tr trolleybuses.

Table 3. Parameters of Skoda Solaris 26tr

Parameter	Description
Type of motor	Asynchronous, 4 poles
Quianties of motors [pcs]	1
Nominal power of motor [kW]	160
Rated power of inverter [kVA]	205
Weight of trolleybus [kg]	10900
Total weight [kg]	18000
Passengers	86

The trolleybuses are equipped with vector control-based IGBT inverters to ensure efficient use of the regenerative braking energy to help 'recharge' another trolleybus in the network. The values of the returned energy are generally acknowledged to vary from 14 to 20% of the total energy consumption [1], [2], [3], [4], [5]. The key parameters of Skoda Solaris 26tr trolleybuses are listed in Table 3.

The simulation model is based on a pilot example developed by MathWorks „Direct Torque Control with Space Vector Modulation of an Induction Motor Drive". The simulation was subsequently extended by adding parameterized models of two motors, the power supply system, power transformers and redesigned for a 20/0.515 kV network at a frequency of 50Hz. The simulation model specifically developed for the present study is visually represented in Figure 2.

The trolleybus model parameters were described as per the data for the Skoda ML3450 K4 engine, 160kW, 1477 rpm. For mechanical simulation of the braking modes, a secondary moment of inertia is added to the main rotor inertia of the motor. The calculated value of the added moment of inertia is $J = 7$ $kg/m^2.rad^{-2}$ when the trolleybus mass is 18 t., the initial speed 50 km/h and the slowing down with acceleration - 2 m/s^2 including the reductor ratio. The mechanical moment of the motor loading is set within the range of 50-75% of the motor rated torque, i.e. 500 Nm. The simulation model defines different time intervals of event occurrence (starting-up, steady movement and stopping), making them shorter than the actual ones, but computed to approximate the determined values of the quantities, which proves to be quite effective in reducing the computational time.

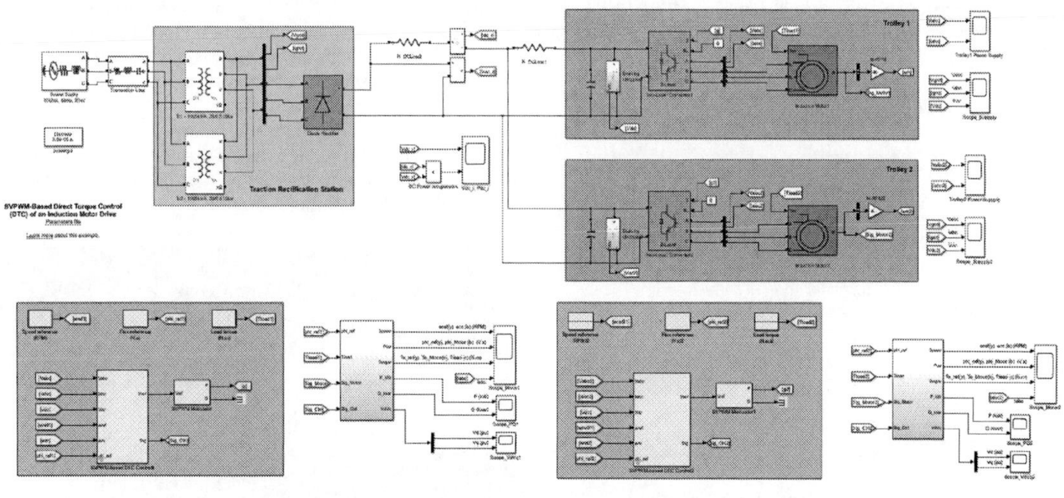

Figure 2. Simulation model of DTC SVC in MatLab Sim Power Systems

The IGBT voltage control is regulated by theory of Direct Torque Control with Space Vector Modulation (DTCSVM) which is schematically represented by a block diagram in Figure 3 [6], [7], [8], [9], [10].

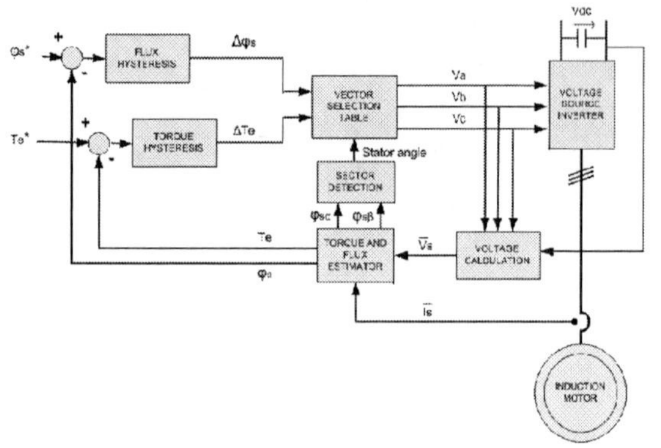

Figure 3. DTCSVM block diagram

The motor stator voltage V_S can be computed from Formula 1 [11].

$$\overline{V}_S = \overline{R}_S \times \overline{I}_S + \frac{d\overline{\varphi}_S}{dt} \qquad (1)$$

The DTC-SVM control uses the orientation of stator flux, which is clarified by Formula 2.

$$\overline{\varphi}_S = \overline{\varphi}_{So} + \int_0^T \left(\overline{V}_S - \overline{R}_S \times \overline{I}_S \right) dt \qquad (2)$$

The electromagnetic torque can be identified by Formula 3.

$$T_e = p \cdot \frac{3}{2} \left(\varphi_{S\alpha} \cdot I_\beta - \varphi_{S\beta} \times I_\alpha \right) \qquad (3)$$

IV. Simulation Results

The simulation strives to explore the individual modes of operation, as well as the concurrent operation of two trolleybuses. The individual modes of starting, running (steady state) and stopping of the two trolleybuses are shown in Figure 4. Reproduced, over the period of 0-5sec., is the individual operation of Trolleybus №1 with a load of 75%. As can be observed, at the moment when t=5sec. Trolleybus №2 with a load of 50% goes into motion, while Trolleybus №1 switches to regenerative braking mode at t=7sec. Figure 5 reconstructs the changes in the driving torque of the trolleybuses throughout the period under survey.

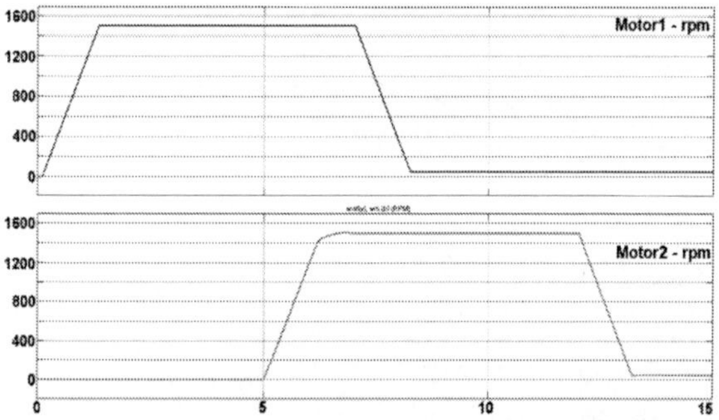

Figure 4. Operating modes of Trolleybus №1 (Motor 1) and Trolleybus №2 (Motor 2).

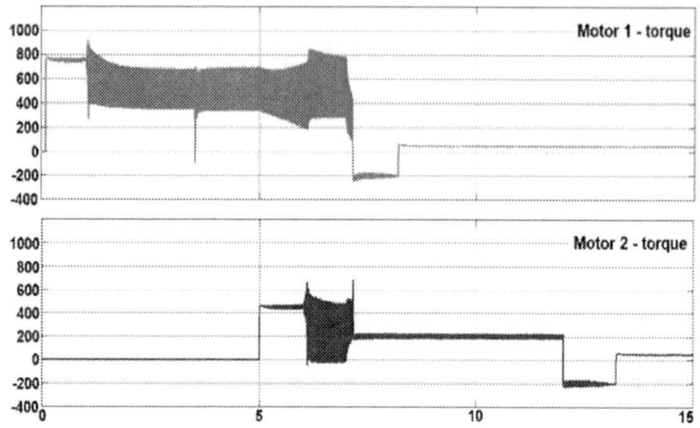

Figure 5. Torque control of Trolleybus №1 and Trolleybus №2

A notable result from the research is the dynamics of the voltage change in the catenary (DC side), as outlined in Figure 6. The test was performed at the maximum permissible voltage of + 15% of AC side or beyond 592V of the power transformer (EN 50160).

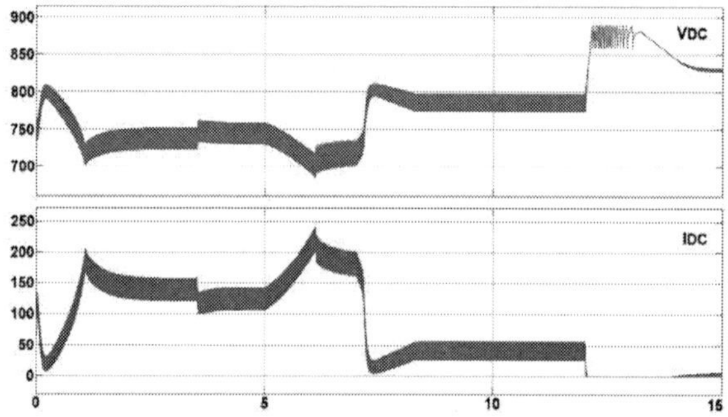

Figure 6. Voltage and current at DC catenary

The simulation given more advanced features for better visualization offers protection against a maximum voltage of 900V. In practice, with the actual trolleybuses such protection is provided with 800V. The characteristic changes in the voltage become visible during the process of recuperative braking of the first trolleybus when t=5sec, power consumption from the second trolleybus in t=7sec, which is designed to limit the rise in voltage. Another distincive feature is made prominent in the case of regenerative braking of only the second trolleybus when t=12sec., which induces an increase in the voltage to levels close to the maximum permissible ones. The low power consumption in regenerative braking modes can be further analyzed from the graphs showing the changes in the voltage and the current of the AC side in the rectifier stations (Figure 7).

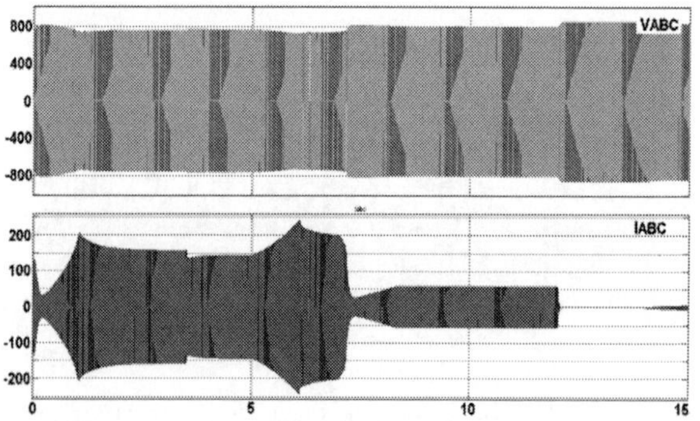

Figure 7. Voltage and current at AC bus in rectifier station

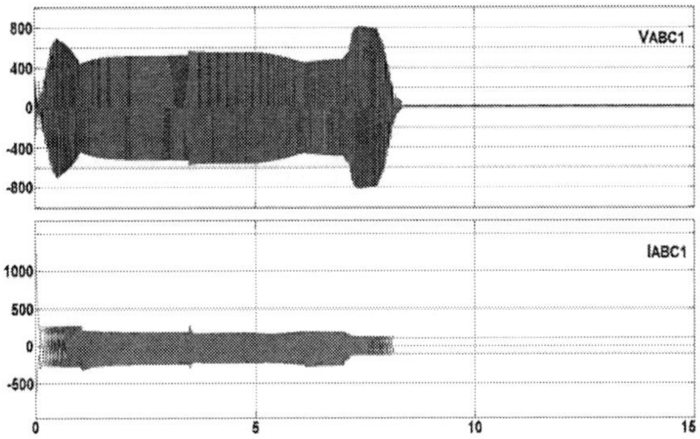

Figure 8. Voltage and current at AC side on inverter of Trolleybus №1

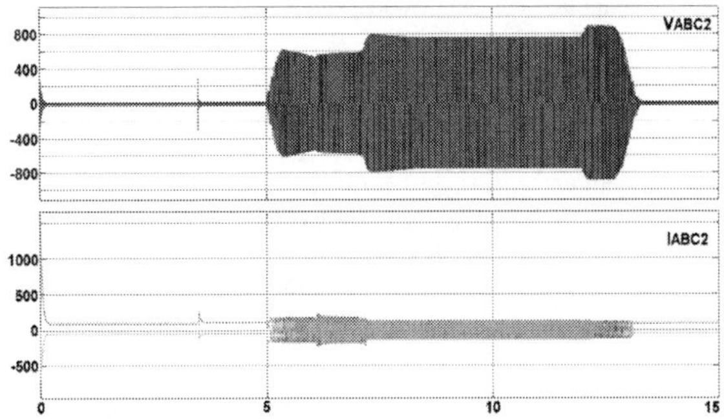

Figure 9. Voltage and current at AC side on inverter of Trolleybus №2

The changes in the voltage and current of the AC side of the inverters of Trolleybus №1 and Trolleybus №2 are captured in Figure 8 and Figure 9. The dynamics of the power consumption and voltage supply across the different specific modes is closely reflected in the change of the voltage and current. A comparative analysis of Figure 6 and Figure 9 shows that trolleybus №2 at its point of departure starts, in fact, consuming power from trolleybus №1 in the process of braking. 12 seconds later, during the mode corresponding to the of Trolleybus №1 stopping, the regenerative braking leads to an increase in the catenary voltage, as there is no other trolleybus to consume the generated energy. This particular increase in the catenary voltage does not, in reality, affect the voltage regime of the AC part in the rectifier station, which is reasonable to expect, given that rectifiers are diode three-phase bridge circuits. It should be borne in mind, however, that if the voltage rises to dangerous levels, this can lead to a break in the diodes and an accident at the rectifier station. And this is the particular reason why trolleybuses are equipped with overvoltage protection circuits.

V. Conclusion

The proposed simulation study is a close analysis of an actually existing technical system of the trolleybus transport in Varna, Republic of Bulgaria. Reconstructed in a simulation environment is an ordinary section - a rectifier station "South" positioned in Asparuhovo Quarter, comprising a medium voltage cable line, power transformers, rectifiers, catenary and trolleybuses. The simulation model consists of vector control inverters representative of traction vehicles – algorithm Direct Torque Control with Space Vector Modulation (DTCSVM) – an algorithm that is likely to respond more swiftly to the torque when contrasted with other methods for vector control. The time intervals used in the study do not correspond to the actual ones, as this would take a lot of computational resources, but they are suitable for the attainment of the mode-based steady- state approach, which makes the simulation physically accurate. The obtained results can be used for quantitative and qualitative assessment of the regimes of power consumption and recuperative generation in the electric power supply system for the trolleybus transport with its fleet of modern vehicles with vector control variable frequency drive (VFD) and a wider framework for further research and development.

Acknowledgment

This work was supported by the Bulgarian Scientific Fund, Operational Programme "Fundamental Research", Session 2019 under contract №06-37/28 Date: 16.12.2019. „Research on Electricity Consumption Regimes in Electric Power Supply Systems for Urban Electric Transport with Two-way Power Flow".

The project discussed herein is supported by the European Regional Development Fund through the Operational Programme "Science and Education for Smart Growth" under contract UNITe № BG05M2OP001-1.001-0008. Project name: "National center of mechatronics and clean technologies".

References

[1] M. Tomcheva, I. Tarpov, " Study of the power modes of skoda solaris 26Tr and 27Tr trolleybuses", In: *Proceedigs of 17th Conference on Electrical Machines, Drives and Power Systems, ELMA*, 2021, pp. 1-4.

[2] G. Dimitrov, G. Pavlov, "Analisys of the efficiency of regenerative braking systems of trollebys Skoda Solaris under real operation condition", *Journal of Mechanic, Transport, Communication*, Vol 14, No 3/2, 2016, pp. 24-31.

[3] F. Berrabah, A. Chebabhi, S. Zeghlache, S. Saad, "Direct torque control of induction motor fed by three-level inverter using fuzzy logic",In: *Advances in Modelling and Analysis* C 72(4), 2017, pp. 248-265.

[4] E. Simonec, "Analysis of trolleybus energy consumption", In: *Advance in Science and Tehcnology Research Journal*, Vol.7, Issue 18, pp.81-84, June 2013.

[5] J. Mwambeleko, T. Kulworawanichpong and K. Greyson, "Tram and trolleybus net traction energy consumption comparison", In: *18th International Conference on Electrical Machines and Systems (ICEMS)*, 2015.

[6] S. Kouro, R. Bernal, H. Miranda, J. Rodrıguez, J. Pontt. "Direct torque control with reduced switching losses for asymmetric multilevel inverter fed induction motor drives" In: *Conference Record of the 2006 IEEE Industry Applications Conference Forty-First IAS Annual Meeting* 35(4): 2441-2446.

[7] F. Wang, Z. Zhang, X. Mei, J. Rodríguez, R. Kennel. "Advanced control ctrategies of induction machine: field oriented control, direct torque control and model predictive control", In: *Journal "Energies"*, vol. 11, no. 120, 2018, pp. 1-13.

[8] D. Casadei, F. Profumo, G. Serra, A. Tani. "FOC and DTC: Two viable schemes for induction motors torque control", In: *IEEE Ttransactions on power electronics*, vol. 17, 2002, pp. 779-787.

[9] H. Hussein, A. Jaber. "Open loop & closed loop vector control of three phase version induction motor drives", *Diyala Journal of Engineering Sciences*, vol. 10, no. 2, 2017, pp. 27-38.

[10] B. Ganthia, P. Rana, S. Pattanaik, K. Rout, S. Mohanty. "Space vector pulse width modulation fed direct torque control of induction motor drive using Matlab-Simulink", *Proceedings of 3rd International Conference on Electrical, Electronics*, 2016, pp. 1-6.

[11] A. Bendaikha, S. Saad, A. Abdou, N. Defdaf, Y. Laamari. "A study of SVM-DTC and conventional DTC for Induction Motors drive fed by five-level inverter", *European Journal of Electrical Engineering*, vol. 21, no. 1, 2019, pp. 85-91.

Verification of Photovoltaic Inverters Properties and Their Compliance With Grid Code Requirements in the Czech Republic

Martin Vojtek, Petr Mastný, Jan Morávek, Jiří Drápela, Michal Vrána
Department of Electrical Power Engineering
Brno University of Technology, FEEC
Brno, Czech Republic
vojtek@vut.cz, mastny@vut.cz, moravek@vut.cz, drapela@vut.cz, xvrana10@vut.cz

Abstract —**Photovoltaic inverters represent the interface between the photovoltaic source and the distribution grid which can be significantly affected in a positive as well as a negative way. With the growing number of photovoltaic sources in the electric power system, the reliability of operating characteristics of inverters is becoming increasingly important. Therefore, this paper deals with testing of selected functions of photovoltaic inverters in order to verify compliance with operational requirements in the Czech Republic. Simple test procedures designed for this purpose were used. Tests were conducted at BUT in the laboratory called ResLab and a total of three types of inverters were tested. Test results are presented, compared and discussed. Overall, it can be stated that none of the tested devices meets the requirements to the full extent. The final part of the paper contains a summarization of the findings and provides recommendations of the authors towards distribution system operators in order to maintain the current level of quality and reliability of electricity supply.**

Keywords: generation module, photovoltaic, inverter, testing, RfG, grid code

I. Introduction

In recent years, the use of renewable resources and their share in the energy mix of almost all countries around the world has been steadily increasing. Photovoltaic power plants are among the largest representatives in the category of renewable sources [1]. They belong to the non-synchronous generation technologies because they are connected to the grid via power electronics - photovoltaic inverters. Unlike large conventional power plants, there are many sources, which often have small power output and are connected to medium or low voltage level. Therefore, they are often referred to as distributed generation (DG).

This way of electricity generation represents a change in the approaches used so far and brings many challenges from the electric power system operation control point of view. Therefore, the requirements for the connection and operation of the DG began to be formulated gradually. Due to the fact that the distribution systems of different countries have different specifics, these requirements were mostly formulated at the national level, or even at the level of the individual distribution systems operators (DSO), commonly referred to as network or grid codes [2]-[5]. In 2016, also higher level of network code was introduced in form of European commission regulation, which is applicable throughout European Network of Transmission System Operators (TSO) for Electricity (ENTSO-E) members′ operated networks. Regulation related to synchronous and non-synchronous generators is commission regulation (EU) 2016/631 of April 14th, 2016, establishing a network code on requirements for grid connection of generators (RfG) [6]. It contains several exhaustive and mandatory requirements which are integral part of regional grid codes, because ENTSO-E members (hereinafter as members) have been required (as stated in [6]) to introduce requirements from RfG into their respective network rules by April 27th, 2019. Nearly all members reported successful implementation (apart from several non-EU members and Cyprus, which is excluded) as stated in [7]. However, RfG is not meant to replace regional specific network codes and it is therefore not within its scope to address regional specific needs, because members are responsible to define ranges of non-exhaustive requirements and it is at their discretion to make non-mandatory requirements mandatory. RfG assumes close cooperation on requirements implementation between TSOs and DSOs and involves the same general requirements in both transmission and distribution systems. There are multiple approaches that the ENTSO-E members chose to introduce RfG into their regional rules due to various government-backed bodies and other interested bodies as stated in [7], but connecting installations or facilities must always comply with electrical and control requirements presented in network code applicable to its type based on its technology, power etc.

The network code currently in force in distribution system of the Czech Republic (called "The rules for distribution system operation" [3]) is established by individual DSOs and approved by Energy Regulatory Office. It follows "The rules for transmission system operation". Both respecting RfG requirements and also taking various European technical standards into account, [8], [9] in particular.

This contribution is focused on verification of selected requirements imposed on micro-sources, sources connected to low voltage distribution grid with rated current up to 16 A per phase and with total output up to 10 kW, as defined in network code [3]. Selected requirements that have been subject of conducted tests are described in the following section.

II. Verified Requirements for Generators

Overview of the verified requirements is in the Table 1. All of them are further described together with specific settings according to Czech grid code in following subsections.

Table 1. Overview of verified requirements

Requirement	RfG (article)	EN 50549- 1 (clause/subclause)
Rate of change of frequency (ROCOF) immunity	13.1(b)	4.5.2
Power response to overfrequency	13.2	4.6.1
Automatic connection and starting to generate electrical power	13.7	4.10
Voltage support by reactive power	17.2, 20.2(a)*	4.7.2
Voltage related active power reduction	-	4.7.3

*defined for type B, C and D power generation modules

A. Rate of Change of Frequency (ROCOF) Immunity.

This requirement defines the withstand capability during rapid frequency changes. It means that the generation module must remain connected and is able to operate during frequency changes with a specified rate. It is stated both in RfG[6] and EN 50549-1 [8] (hereinafter as EN), that relevant system operator (responsible party) defines this value. ±2Hz/s is required by Czech network code [3].

B. Power Response to Overfrequency

A power-generating module must be able to activate the active power-frequency response when the frequency is higher than threshold for activation. This function is referred as limited frequency sensitive mode – over-frequency (LFSM-O) in RfG [6] and power response to overfrequency in EN [8]. Frequency threshold and frequency droop settings are specified by relevant transmission system operator (TSO) for its control area in coordination with the TSOs of the same synchronously interconnected area to ensure minimal impact on neighboring areas. The default frequency threshold Δf_1 is 50.2 Hz and the frequency droop $s_2 = 5$ % according to Czech network code [6].

EN [8], which is essentially a product standard, requires that inverters must also allow to set a threshold for deactivating this function. It ensures that the power value to which an inverter has been limited must remain constant until the frequency decreases below the threshold value. Subsequently, it is possible to increase the power output gradually, with a certain gradient defined in the requirements for automatic connection and starting to generate electrical power (since these requirements were also subject of verification, they are described in the following section). This feature is required by Czech grid code since 2021 and the active power can be increased again only after the frequency decreased back to the value 50.05 Hz or lower. Principal of power response to overfrequency with deactivation threshold is shown in Fig. 1. P_{ref} is the reference active power to which the change in active power at the output of the production module is related. For non-synchronous generation technologies like photovoltaic inverters, $P_{ref} = P_M$ – the actual AC output power when the frequency threshold for activation is reached.

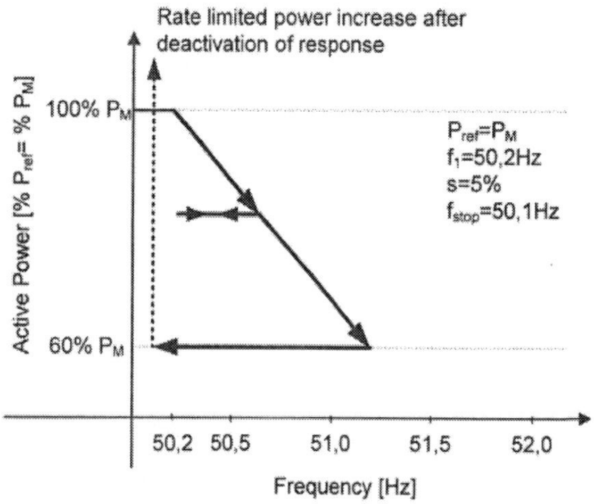

Figure 1. Example of active power frequency response to overfrequency with configured deactivation threshold

C. Automatic Connection and StartingtTo Generate Electrical Power

According to RfG [8], the relevant TSO shall specify the conditions under which a power-generating module is capable of connecting automatically to the network. Those conditions include frequency ranges within which an automatic connection is admissible, corresponding delay time and maximum admissible gradient of increase in active power output. EN [8] also requires option to determine voltage range within which an automatic connection is admissible, providing ranges both for voltage and frequency together with default settings, which should be used when no settings are specified by the DSO.

Czech grid code [3] allows automatic connection and starting to generate electrical power after tripping when the following criteria are meet:

- Voltage and frequency are within the following ranges for 300s (delay time):
  - voltage – 85% - 110% of nominal value,
  - frequency – 47,5 – 50,05 Hz
- Gradual rise to available power from zero power with a maximum gradient of 10% P_n/min

D. Voltage Support by Reactive Power

Voltage fluctuations must remain within the permitted limits defined in [10]. Generation modules must be able to contribute to this requirement during normal network operation.

RfG [6] requires voltage support by reactive power from generation modules of category B and higher. Essential requirements for generation modules with lower power output (category A) are defined in EN [8], which is focused on low voltage grids. There are requirements for reactive power capability and various control modes. The Czech network code[3] requires that the generation module must participate in the voltage support by reactive power in the range of the power factor between 0,9 capacitive and 0,9 inductive throughout the voltage and frequency operating range. It allows DSOs to individually choose and request different ways of reactive power control for each installation depending on the specific conditions at the point of common coupling, but it assumes the

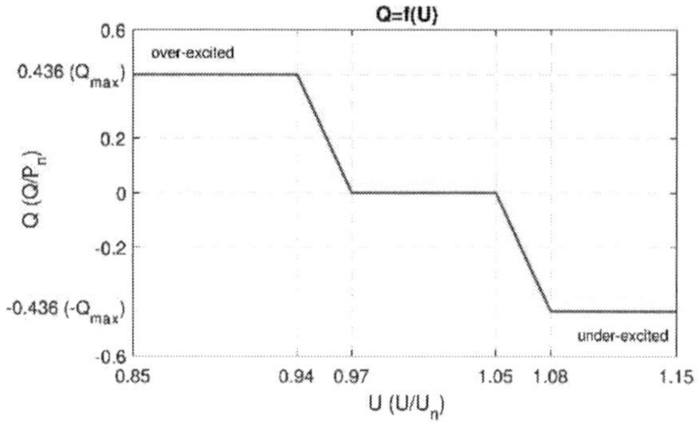

Figure 2. Q(U) control mode characteristic

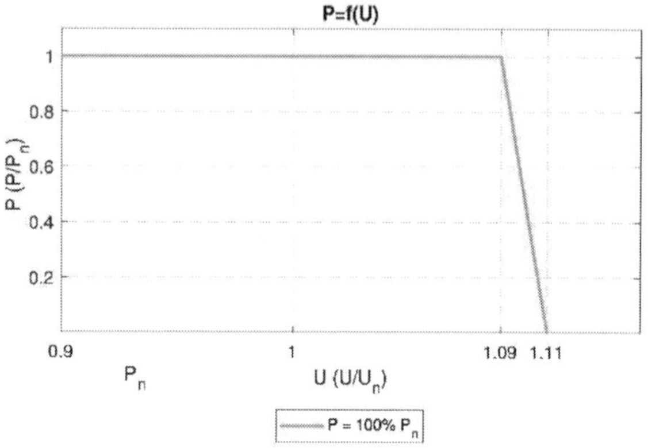

Figure 3. P(U) control mode characteristics

widest deployment of the control mode, in which the value of reactive power is dependent on voltage Q(U). Currently in practice, in fact every DSO requires specifically this function for generation modules connected to low voltage grid.

Q(U) control mode is autonomous and controls the reactive power based on voltage at point of common coupling according to predefined characteristic, which is defined by 4 points (voltage values). Characteristic, which is required in the Czech Republic by DSOs is shown in Fig. 2.

E. Voltage Related Active Power Reduction

In order to avoid disconnection due to overvoltage protection, generating modules are allowed to reduce active power as a function of this rising voltage according to EN [8]. In the Czech grid code is stated, that every power generating facility connected to the low voltage grid using inverter must be equipped with this logic. Specific points (voltage values) will be determined according to network conditions at the point of common coupling or connectivity study eventually. In practice, there is only one setting which is commonly used in the Czech Republic (Fig. 3).

III. Test Platform

An advanced test platform was assembled in the Renewable Energy Sources Laboratory (RESLAB) at the Brno University of Technology (BUT). The purpose of the test system is to create simulated alternating low voltage system, enabling the simulation of selected phenomena at the output voltage, separately from the public distribution grid. It is used for verification of settings and various functions of different types of power inverters. The use of selected laboratory equipment allows to perform repeatable tests always with the same parameters according to predefined test procedures. The arrangement of the laboratory equipment is shown in the following figure (Fig. 4) and wiring diagram in Fig. 5.

The test platform consists of the programmable 4-quadrant power supply with 50 KVA power output powered from the DS. Device under test (DUT), in this case PV invertor, is connected to the power supply and powered by PV string simulator (programmable DC power supply). There is also necessary measurement equipment, which ensures the monitoring of selected quantities.

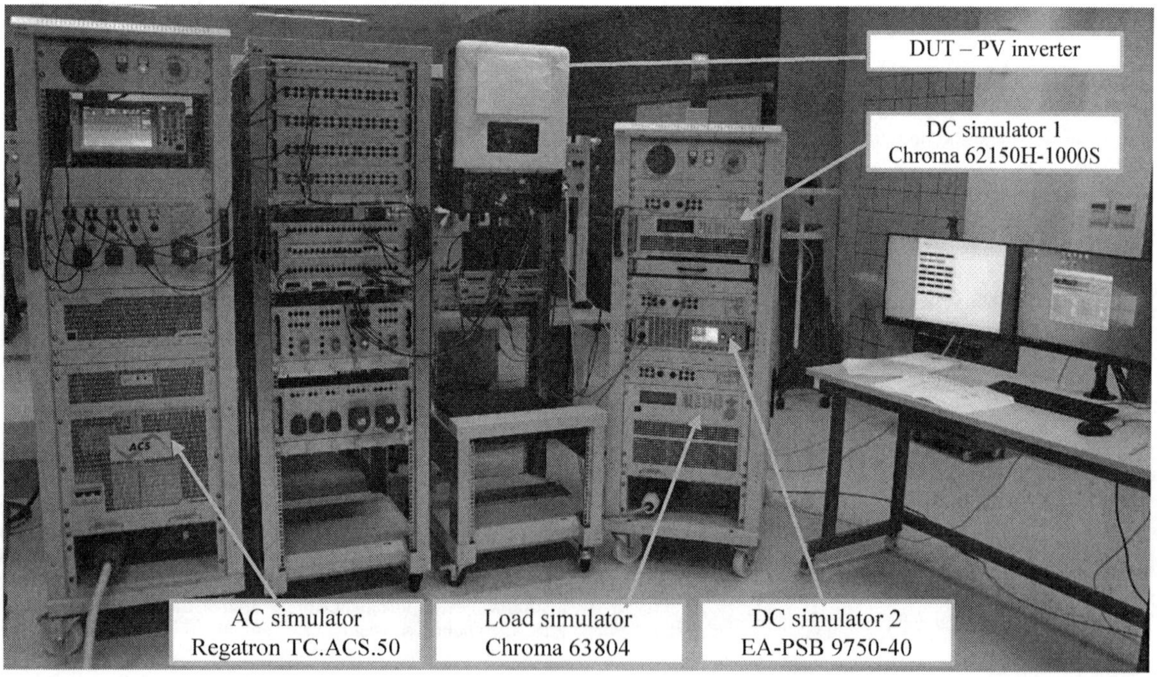

Figure 4. Equipment and arrangement of a test platform with individual simulators

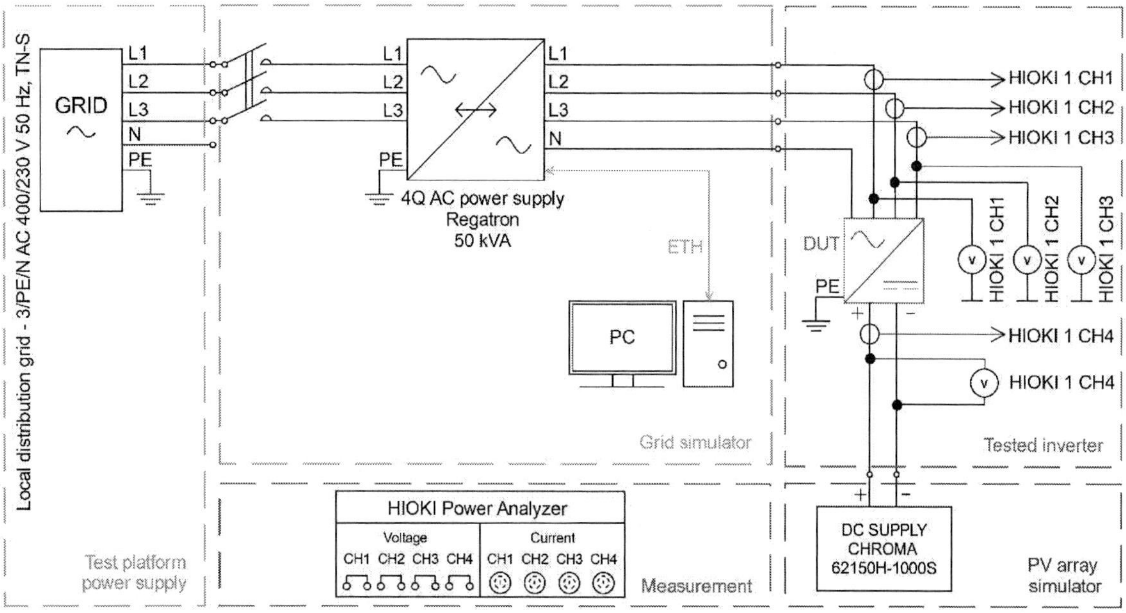

Figure 5. Wiring diagram of the test platform during the verification measurements

Programmable power supply is controlled using a personal computer via the Ethernet. The computer is equipped with the appropriate control and monitoring SW provided by the manufacturer. It enables programming of various testing sequences using several features within amplitude and frequency modulations.

This configuration makes it possible to achieve the network parameters necessary to verify the compliance of static (Q(U), P(U), P(f), etc.) and dynamic response (ROCOF immunity, fault ride through immunity, etc.) of the tested devices as well as any other requirements, such as protection functions.

IV. Tested Devices and Test Sequences

The brief specifications of tested inverters are given in Table 2. Each inverter is by different manufacturer, and all the models are currently commercially available in the Czech Republic as well as in Europe. All of them were set to default settings for the Czech Republic, which is predefined by a manufacturer with firmware updated to most recent version.

Table 2. Inverters specifications

Mark	Nominal power P_n [W]	Number of phases [-]	Type
A	10000	3	hybrid
B	6000	3	hybrid
C	6000	3	hybrid

The presentation of the measured data in the next section is designed so that every part of the test sequence essential for inverter's properties verification purposes is evident. For this reason and to prevent data redundancy, complete test sequences will not be listed in this section. These basically consisted of keeping all of the quantities constant during synchronization of the inverter and ramping up the available power, following with changing the quantity (voltage or frequency) needed to verify the specific requirement. During the changes of one of the quantities, all other quantities were kept constant (for example, during frequency changes, the voltage and available power of the DC power supply were constant). Inverters B and C were tested with a power corresponding to 100% P_N, while inverter A was tested with a power corresponding to 80% P_N. This is because we observed some problems with the MPP tracker of the inverter A when working at 100% power resulting in repeated disconnections of the inverter from the grid.

V. Results

The test results are sorted according to the relevant requirements described in section II. The results for all three inverters are plotted in a single graph. Active and reactive power are converted into the per unit system in relation to nominal power of each inverter for easy comparison. Quantities were measured using power analyzer and recorded every 50ms.

A. Automatic Connection and Starting to Generate Electrical Power

Course of active power since the voltage was applied at the inverters terminals is shown in Fig. 6. Inverter A does not respect both the requirement for automatic connection and starting to generate electrical power. It synchronizes approximately 30s after applying the voltage and increased his

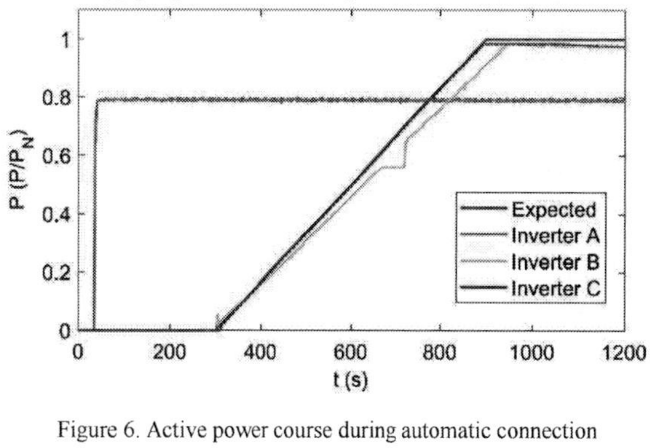

Figure 6. Active power course during automatic connection

power output rapidly. Both invertor B and C respects 5 min delay before synchronization. After that, the invertor B increase power output with slightly slower gradient (10,66 %P_N/min), while inverter C maintain exactly 10% P_N/min.

B. Rate of Change of Frequency (ROCOF) Immunity

To verify ROCOF immunity, three rapid frequency changes were made at the ROCOF rate of 2 Hz/s. The first change was increase the frequency by 1 Hz upwards, followed by 2 Hz downwards and finally the frequency was returned to the nominal value. Detail of active power inverter response is in Fig. 7. It is obvious that all of the tested inverters remained connected and continued to supply electricity. It is also possible to observe the response of inverters B and C to the increase in frequency as a decrease in active power due to activation of the function for active power response to overfrequency.

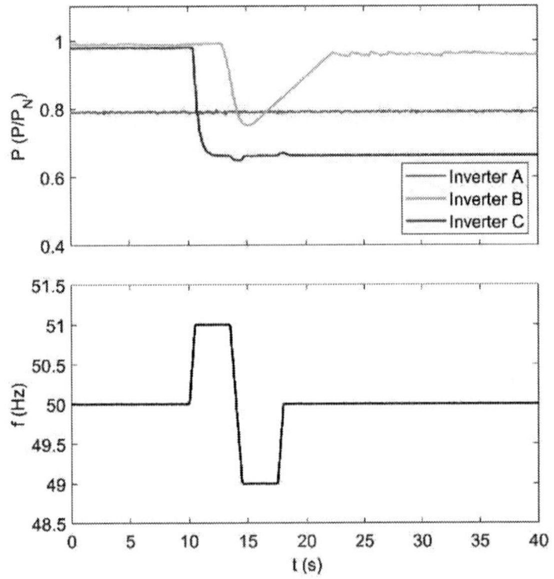

Figure 7. Active power response to rapid frequency changes

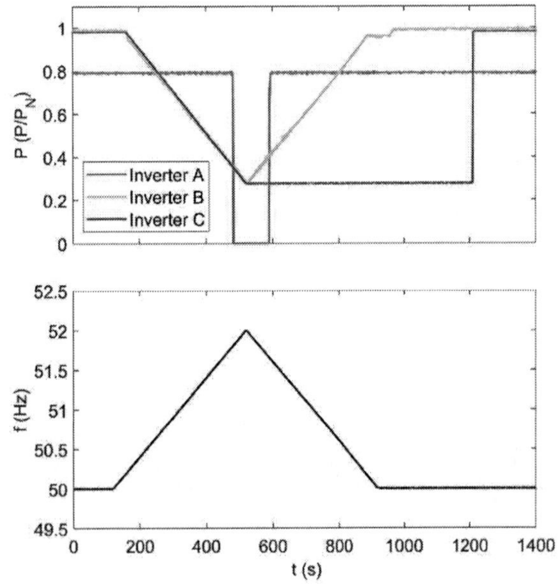

Figure 8. Active power response to overfrequency

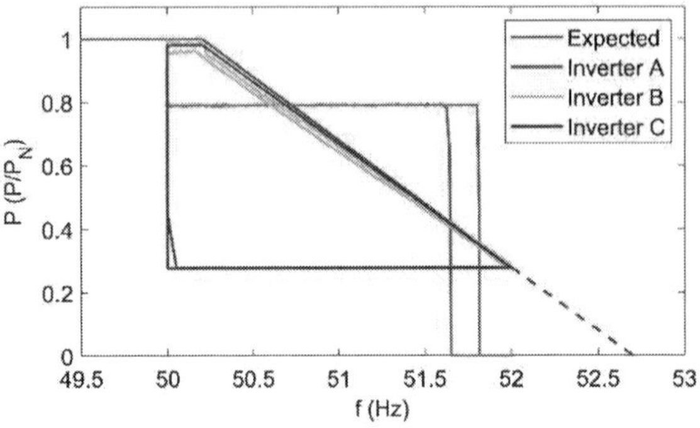

Figure 9. Active power response to overfrequency in P-f area

C. Power Response to Overfrequency

To verify the overfrequency response, a sequence was designed that included a gradual increase in frequency from nominal value to 52 Hz and back. The frequency change was performed continuously with 5 mHz/s ramp rate. Measured values are presented in Fig. 8. They were averaged over 1s and plotted into P-F area (Fig. 9) together with required control characteristic for better comparison. Inverter A does not react to frequency changes in any way until its frequency protection tripped at 51.8 Hz. In addition to not responding, tripping is also incorrect behavior, because frequency protection should trip at 52 Hz according to the Czech network code [3]. Power response to overfrequency is active for the remaining two inverters. Inverter B has correct frequency threshold and droop, but frequency threshold for deactivating is not set. Inverter C fulfill almost all of the requirements, but it is still deviating. Instead of increasing of active power gradually after frequency threshold for deactivation is reached, it is waiting for 300s and then increase its power output rapidly.

D. Voltage Support by Reactive Power and Voltage Related Active Power Reduction

Voltage support was verified using the testing sequence with gradual increasing in voltage from nominal value to 258 V, following by decreasing to 202 V and then increasing again until the voltage reached nominal value. Response of active and reactive power can be observed in Fig. 12. Measured values were averaged over 1 s and plotted into Q-U (Fig. 10) and P-U area (Fig. 11) respectively. It is clear, that voltage support by reactive power and active power reduction of the inverter A is not active. Furthermore, it is not able to operate and supply electricity to the grid when the voltage is low (approximately below 215 V). During low voltage period is connects and disconnects repeatedly. This could have negative impact on power quality. Values falling

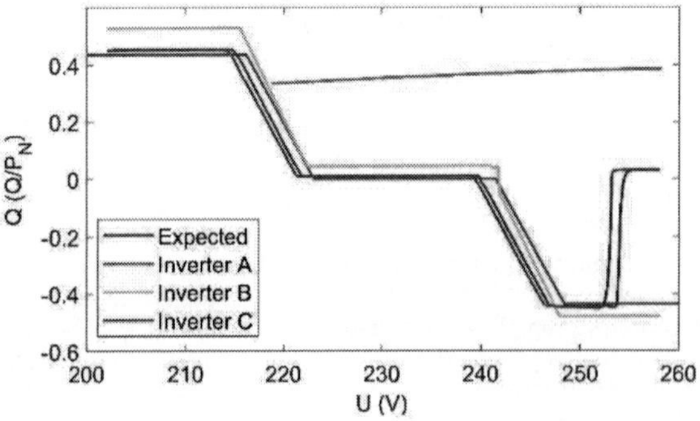

Figure 10. Reactive power response to voltage changes in Q-U area

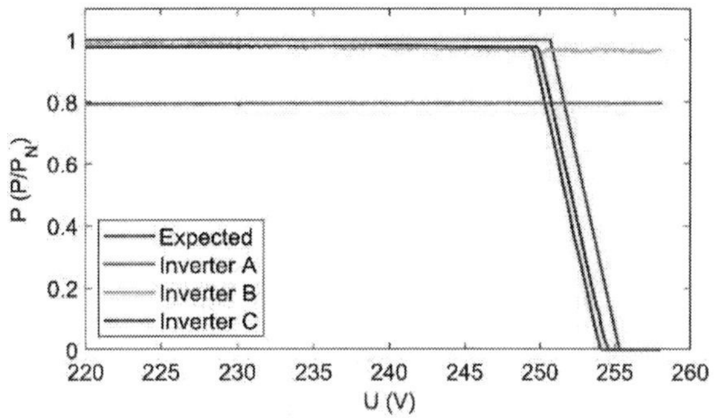

Figure 11. Active power response to voltage changes in P-U area

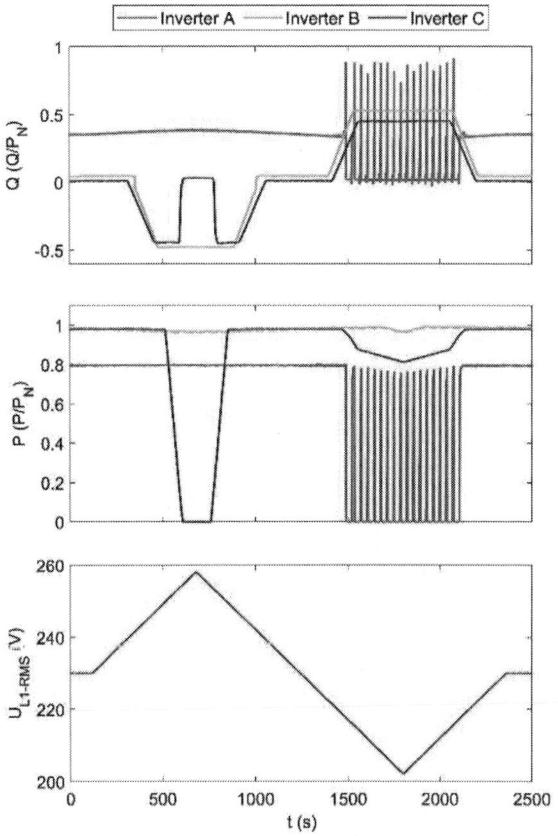

Figure 12. Active and reactive power response to voltage changes

within this period were excluded from processing in the Q-U and P-U areas (Fig. 10, Fig. 11) in order to maintain appropriate readability. Last fact to point out is, that inverter delivered reactive power to the grid all the time and therefore operated with lower power factor, approximately 0.9. Inverter B activated only voltage support by reactive power, which followed the required control characteristic (Fig. 2), but maximum and minimum required reactive power was exceeded. It can be also observed, that even in the dead band of the characteristic, certain amount of reactive power was supplied to the grid. Inverter C activated both of the functions. Active power was controlled according to required control characteristic (Fig. 3), but it was shifted few volts to the left and also small hysteresis can be observed. Reactive power control respected maximum and minimum values of the characteristic, but the same as in the previous case applies (shift and hysteresis). It is also obvious, that inverter was not able to continue in reactive power consumption when active power response was activated and active power decrease to zero.

VI. Conclusion

The paper describes results obtained from inverters testing in ResLab laboratory. The main goal was to point out the serious problems with setting and parameterization of inverters operation properties according to valid legislation for this type of sources in the Czech Republic. Based on the obtained results, it was proved that none of the tested inverters meets the requirements in full extent. Found facts point to the inability of PV inverters manufacturers correctly and consistently implement the requirements defined by national and international standards and regulations. In this respect, the absence of a certain control mechanism, over time, there could be a situation where there will be a significant number of inverters without correctly set control, which can very adversely affect the operation of the electric power system. Based on our experience, we are convinced that it is necessary to clearly define the conditions for verifying the settings of inverters before their introduction to the Czech market.

Acknowledgment

This research work has been carried out in the Centre for Research and Utilization of Renewable Energy (CVVOZE). Authors gratefully acknowledge financial support from the Technology Agency of the Czech Republic (project No. TK04020103).

References

[1] International Energy Agency, "Global Energy Review 2021," Paris, France, pp. 22–24, April 2021.

[2] Erzeugungsanlagen am Niederspannungsnetz – Technische Mindestanforderungen für Anschluss und Parallelbetrieb von Erzeugungsanlagen am Niederspannungsnetz, VDE-AR-N 4105:2018-11, 2018.

[3] Energy Regulatory Office, "Rules for operation of distribution power system Annex 4: Rules for parallel operation of generators and accumulation devices with the distribution system operator's power system", June 2021.

[4] HETS Grid Code, IPTO, 2022. [Online]. Available: https://www.admie.gr/en/market/regulatory-framework/esmie-operation-code

[5] IEEE, "IEEE Standard for Interconnection and Interoperability of Distributed Energy Resources with Associated Electric Power Systems Interfaces," IEEE Std 1547-2018 (Revision of IEEE Std 1547-2003) , pp.1-138, 6 April 2018, doi: 10.1109/IEEESTD.2018.8332112.

[6] European Commision, "COMMISSION REGULATION (EU) 2016/631 of 14 April 2016 establishing a network code on requirements for grid connection of generators." [Online]. Available: https://www.entsoe.eu/network_codes/rfg/

[7] European Commission, Directorate-General for Energy, Ali, M., Makki, E., Cazal, C., et al., "Implementation of the network code on requirements for grid connection of generators : final report," 2021. [Online]. Available: https://data.europa.eu/doi/10.2833/235293

[8] EN 50549-1 - Requirements for generating plants to be connected in parallel with distribution networks - Part 1: Connection to a LV distribution network - Generating plants up to and including Type B, 2019.

[9] EN 50549-2 - Requirements for generating plants to be connected in parallel with distribution networks - Part 2: Connection to a MV distribution network - Generating plants up to and including Type B, 2019.

[10] EN 50160 - Voltage characteristics of electricity supplied by public electricity networks, 2010.

On Influence of Smart Grid Technologies on the Safe Operation of NPPs

Štefan Čerba, Branislav Vrban, Jakub Lüley, Vendula Filová, Vladimír Nečas
Institute of Nuclear and Physical Engineering
FEI STU
Bratislava, Slovakia
stefan.cerba@stuba.sk, branislav.vrban@stuba.sk, Jakub.luley@stuba.sk, vendula.filova@stuba.sk, vladimir.necas@stuba.sk

Ján Poničan
Institute of Nuclear and Physical Engineering
FEI STU
Bratislava, Slovakia
jan.ponican@stuba.sk

Abstract— **For safe and secure operation as well as accident mitigation of Nuclear Power Plants (NPPs) during normal operation and in case of emergency scenarios, offsite electricity supply is considered as the most preferred and reliable source. The nuclear safety standard used worldwide and those adopted in Slovakia requires that electricity from the grid should be derived to the NPP via two physically independent circuits in order to minimize the likelihood of their simultaneous failure. These circuits shall be also designed to be available in sufficient time following a loss of all options for onsite power supply, to assure that specified acceptable fuel design limits and design conditions of the reactor coolant pressure boundary are not exceeded. To meet these requirements the electric grid has historically evolved along a consistent physical configuration and mode of operation. The current large and centralized grid consists of high voltage power stations generating electric power, subsequently distributed via a network of high-voltage transmission lines and substations, lower voltage distribution lines and substations, and finally through low voltage local distribution lines to supply the electrical loads of the end users. The current fleet of NPPs has been designed, built, licensed, and operated under this traditional configuration of the electric grid. As the modern electric grid evolves incorporating smart grid innovations, distributed resources, renewable energy sources, high speed sensing and controls, microgrids, large-scale energy storage, and other advanced technologies, NPPs must also adapt to these changes. It naturally brings challenges also to the new state-of-the-art reactor technologies, such as GEN IV reactors or small modular reactors, but most importantly to the current fleet of reactors in operation, which has only limited maneuverability. This paper presents the results of a comprehensive literature survey conducted on Smart Grid technologies, renewable energy integration, grid reliability standards, and electric grid regulations and guidance. It summarizes the measures developed worldwide and tries to adopt some of them for the conditions of Slovakia.**

Keywords — smart grid, nuclear power, small modular reactors, hybrid energy systems

I. INTRODUCTION

For a nuclear newcomer country, the introduction of nuclear power plants (hereinafter NPPs) in its electric grid is a huge commitment. It requires to develop and build the whole infrastructure necessary to connect the power generated in a NPP in an economically feasible and safe manner. One of the most significant parts of such an infrastructure is the electric grid and the connection of the NPP to it. Nowadays most countries, even those without nuclear experience, already have an electric grid, but connecting a large NPP to it always brings new technical challenges. If the built NPP should operate in a safe, secure, reliable, and effective manner, the same must be true for the electric grid. Interconnected electric grids can encompass several countries and are often seen as the largest machines in the world [1]. They are formed by thousands of kilometres of transmission and distribution lines, hundreds or thousands of suppliers and uncountable final consumer stations and loads.

Compared to other types of power plants, such as hydro, renewable or coal fired plants, NPPs are usually very large and extremely sensitive installations. They are producers and consumers at the same time, and they must always maintain their connection to the electric grid. Due to their relatively large power, they supply a huge amount of electricity to the grid, but at the same time, thy require stable electricity supply for the reliable operation of their safety systems. It is especially true for start-up, power climbing, operation, and shutdown. This stable source of electricity is often referred as "off-site power supply" [1].

A balanced, maintained, modern but reliable electric grid is the key for implementing new nuclear power plants. The grid also plays an important role after the reactor has been shut down, since most safety systems ensuring cooling the reactor core require electric supply. Here instabilities and interruptions could have significant economic, social and environmental consequences.

The connection of the electric grid with the NPP forms a complex controlled and dynamic system that depends on the following:

- The size and interconnection of the grid
- Nuclear safety requirements
- Balance between consumption and supply
- Nature of the produced electricity

As electricity is difficult to store, the grid must ensure matching the demand for electricity with the actual supply and therefore the grid always requires continuous control and balancing actions using special systems. Due to its nature and its low maneuverability due to nuclear safety limits or material constrains, NPPs are usually operated in steady-state mode as baseload sources. The integration of large NPPs into an electric grid brings nuclear safety requirements that impose additional requirements on the grid design, operation, and stability [1]. The steady-state operation, when the NPP produces electricity, is not problematic. However, when the nuclear reactor is shut down after operation for refueling or maintenance, to keep the safety systems in operation and ensure cooling, it still requires electric supply from the grid. In addition, the electricity supply is also required for the NPP to be able to restart when required.

II. CHANGING THE PHILOSOPHY OF ELECTRIC GRIDS

The whole philosophy of the design and operation of the electric grid has been changing. The grid, which was primarily designed to transfer electricity from centralized stations to suppliers is now implementing new technologies and therefore also economically more reliable operation scenarios come to the fore. Over the last decades a concept of distributed energy sources (DER) has been getting more and more popular. In addition to distributed energy sources the modernizations of the grid also include smart grid solutions with high-speed sensing and control units and energy storage systems [2]. The grid for which the majority of currently operated NPPs was designed and built, and therefore the NPPs will have to adapt to these changes while keeping their economic effectiveness and safety standards.

Before going into details, how these changes influence the NPPs and their operation, let's think about what a smart grid (SG) is. It comes as no surprise that there is no exact definition of the Smart Grid. It could be an electricity network based on digital technologies used to supply electricity to consumers via two-way digital communication channels. SG could be also a complex machine with a variety of technology options at hand to improve its functionality. As a result of broad definitions, there are different architectures, technologies called as smart grids components. Looking from the perspective of the whole electric power delivery system, which has often been cited as the greatest and most complex machine ever built, the smart grid can be defined as an innovation that has the potential to revolutionize the transmission, distribution and conservation of energy [3].

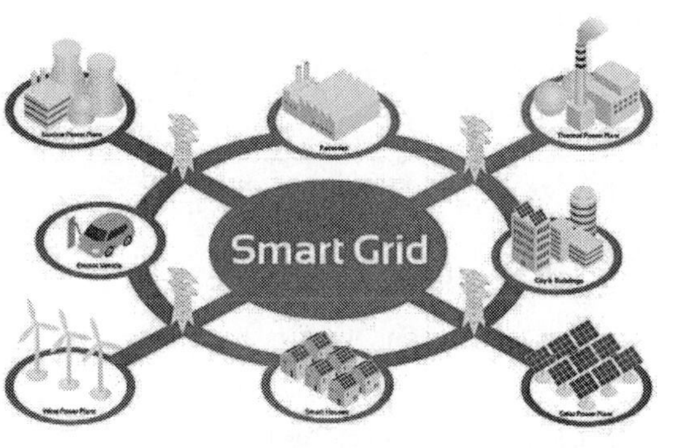

Figure. 1 – Schematic view of a smart grid combining various power sources and consumers [4]

Regarding the impacts that smart grid technologies may have on the electrical transmission grid as a source of off-site power, the following parameters are seen as the most crucial ones:

1. Power quality
2. Grid reliability and stability
3. Protection systems
4. NPP - grid interface

In terns of power quality, the most important issues are related to the transmission system voltage - VAR control, frequency regulation and harmonics concerns. Based on [2] the basic requirement for the offsite power systems in terms of power quality should be defined as follows:

"An onsite electric power system and an offsite electric power system shall be provided to permit functioning of structures, systems and components important to safety."

The basic requirement for the reliability and stability of the grid can be defined as [2]:

"Electric power from the transmission network for the onsite electric distribution system shall be supplied by two physically independent circuits. Each of these circuits should be designed to be available in sufficient time following a loss of all onsite alternating current power supplies."

The traditional operating configuration of the electrical transmission grid consists of large, centrally located, high-voltage power stations generating electricity subsequently transmitted from these sources to the suppliers. Power transmitted in one direction, from the source through the grid of transmission lines and substations, lower voltage distribution lines and substations, and finally through low-voltage local distribution lines to supply the electrical loads of the end users. Electrical protection systems are used to take advantage of this fact and to coordinate the settings of overcurrent devices, establish directional overcurrent protection schemes, transmission line distance relaying schemes, and pilot wire protection systems [2]. The NPP – grid interface has always been considered as a potential area of vulnerability, due to different approaches of maintenance, operations, regulation, requirements on training and the overall philosophy of the system. The integration of the smart grid will require changes in the planning, design and operation of power systems [2].

III. VULNABIRITY OF ELECTRICAL SYSTEMS DUE TO SMART GRID TECHNOLOGIES

As more and more new smart grid technologies are introduced to the traditional energy grid, vulnerabilities can be identified. The major ones accompanied by these technologies are the distributed energy resources, especially the highly variable renewable generation sources such as solar sources and wind turbines, energy storage systems, microgrids and sensing devices. Among the concerns related to distributed energy sources, renewable energy sources are the most crucial [2]. Their output strongly depends on local environmental conditions, such as temperature, wind speed and direction or solar insolation. Since these conditions are always changing, they are difficult to predict, planning the electric output of energy sources relying on these conditions is also difficult for a longer time. This variability can be problematic to grid operation if it is not managed, and can lead to voltage and frequency variations, as well as power quality issues. There is no doubt about multiple benefits resulting from accommodating energy storage systems in the smart grids, but the reliance on theses systems may also lead to vulnerabilities of electrical systems. A failure of a large energy storage system may cause large disturbance in the grid, effect the inertia response and may also have fatal effects on the grid operation [2]. Fig. 2 presents several examples of failures of energy storage systems, their impacts and likelihoods.

Grid applications	Storage System Capacity	Duration (and frequency) required to be effective	Potential immediate impact caused by loss or failure of storage system	Likelihood of impact based on a single loss or failure	Other mitigation means
Electric energy time-shift (arbitrage)	Centralized	A few hours (daily)	Economic losses: A curtailment of generation (renewable or conventional generation)	Very likely: Generation has to be curtailed without ESSs	Other types of ESS.
Regulation	Centralized	Half a minute to a few minutes (constantly)	Power quality and economic losses: Increased frequency deviation and/or increased wear-and-tear of conventional generator;	Likely: May be compensated for partially by online conventional generators.	Increasing spinning reserves
Black start	Centralized	A few minutes at the beginning of a recovery (upon a blackout)	Economic losses: Delayed recovery from blackouts	Very likely: The recovery has to be delayed if the ESS fails	Generators with black start capability
Frequency response	Centralized	Seconds to less than a minute (upon a large disturbance)	Stability: Degraded frequency responses and frequency may become unacceptable	Very likely: Difficult to mitigate using conventional generators that are not responsive;	Other types of ESSs and/or gas turbines with quick ramping capabilities
Load following	Centralized	Several minutes (constantly)	Power quality and economic losses: Increased frequency deviation and/or increased wear-and-tear of conventional generator;	Very likely: Can be compensated for partially by online conventional generators.	Spinning reserves, or redundant ESSs
Renewable integration (capacity firming or ramping control)	Centralized or Distributed	Several minutes (every few minutes)	Power quality and economic losses: Increased frequency and voltage fluctuations, curtailment of renewable generation;	Very likely: May have to curtail renewables; Frequency and voltage responses may be partially improved using spinning reserves;	Spinning reserves, or redundant ESSs
Generation, transmission, or distribution upgrade deferral	Centralized or Distributed	A few hours to a few days (yearly)	Power quality and loss of load: Voltage fluctuation, load curtailment	Very likely: ESSs are designed specifically for peaking conditions.	Distributed generators, or mobile ESSs.
Customer side of the meter storage for enhancing power quality	Distributed	Seconds to minutes (Upon power outage/interruption)	Economic losses and reduced reliability Loss of load	Very likely: No response time for alternative options.	Other types of ESSs

Figure. 2 – Potential grid impacts caused by loss of energy storage [2]

As it can be seen, the RES and energy storage system may have strong impact on the energy grid, but the major vulnerability is related to the possibility of propagating cascading instability, originated from microgrids, beyond the distribution systems.

The vulnerabilities introduced by DERs and microgrids may be more harmful to the operation of NPPs when they are physically and electrically near the NPPs, e.g., a large frequency dip may simply be a localized regional issue, but such a transient can nevertheless result in the trip of a nuclear plant nearby [2].

Finally, the sensing and intelligent devices as well as communication platforms used to manage the demand and supply of electricity in smart grids in an effective manner may also influence the vulnerabilities of electrical systems. In the times of cyber-attacks, these components can be entry points to the network for direct threats or attacks, which may lead to a breach of network functionality. Due to the size and massiveness of the smart grids, monitoring and detection of potential attacks may be very difficult or even impossible.

IV. Relation of NPP operation guidelines and smart grid technologies

According to [2] the General design criteria and rules for judging the influence of smart grid components on NPP operation can be divided to the following categories:

1. Electric power systems
2. Inspection and testing of electric power systems
3. Reactor coolant makeup, residual heat removal, emergency core cooling, containment heat removal, containment atmosphere cleanup and cooling water
4. Loss of all alternating current power
5. Monitoring the effectiveness of maintenance at NPP
6. Stability of offsite power systems
7. Station blackout
8. Monitoring the effectiveness of maintenance at NPPs

The summary of investigated impacts of smart grid technologies on meeting the General design criteria and listed regulation rules is shown in Table 1.

Table. 1 – Impact of smart grid soulutions on meeting NPP operation rules [2]

Rule	Impact of smart grid on NPP meeting the rule
1;	• Reduction in the reliability of the electric grid than that assumed in the design of the power plant systems can result in failure to meet this requirement. • Guidance on assessment of the reliability of the grid with implementation of the smart grid.
2;	• Inspection procedures may need to be adjusted to consider implementation of the smart grid.
3;	• These requirements are expected to remain unchanged.
4;	• This rule assumes recovery of the offsite power. Currently, this rule is being revised considering lessons learned from Fukushima accident. • This aspect of the rule may need to be reviewed.
5;	• Grid-risk sensitive maintenance activities may need to be defined considering the smart grid implementation.
6;	• Examination of the capability to withstand the loss of the largest unit. • Grid stability analysis should be defined considering the smart grid technology and its effect.
7;	• Assignment to offsite power design characteristic groups may be different. • Restoration of offsite power may need to be discussed considering implementation of smart grid.
8;	• Define guidance for consideration of SSCs that may need to consider and monitored under MR. • Define risk consideration for maintenance of SSCs incorporated as part of smart grid implementation.

V. Summary of reccomendations

Based on the status of smart grid implementation and research activities planned for the forthcoming years also positive effects of smart grids on NPPs can be identified:

1. Improved resilience, efficiency, and cost effectiveness of the grid
2. Reduced grid losses and improved volt/VAR support
3. Reduction of power disruption in extreme weather conditions
4. Real-time monitoring and visualization of system status
5. System analysis using analytical software tools
6. Improved interfaces and decision support
7. Enhancement of distributed and renewable energy resource integration
8. Better capability to meet customer needs
9. Improved energy security from increased fuel diversity

Along with these benefits we should not forget about potential adverse impacts that may require some changes or actions for the NPPs that use the grid as their source of off-site power. The potential adverse impacts of smart grid technologies on the reliability of NPP offsite power supply are summarized in Table 2. Following the findings of chapter III, the identified adverse impacts of smart grid solutions on NPPs are divided into 4 major categories, i.e. distributed energy sources, energy storage systems, microgrids and advanced sensing, measuring and control tools.

Table. 2 – Potential impact of smart grids on the reliability of NPP offsite power supply [2]

Smart grid technology	**Adverse impacts**
Distributed energy resources	• Negative impact on grid performance due to power output variations. • Reduced system inertia due to RES.
Energy storage systems	• Grid vulnerability due to loss of a single energy storage system. • Grid operation heavily dependent on storage impacted by supply chain issues. • Storage equipment vulnerable to physical and cyber-attacks. • Reduced system inertia with inverter-based storage systems.
Microgrids	• Cascading instability of microgrids may be propagated beyond the distribution systems • Transmission and distribution networks may be much more coupled due to microgrids. • Most relevant for NPPs directly connected to the transmission network.
Advanced sensing, measuring and control tools	• Makes grid operation more complex requiring a much deeper understanding of the interface between grid and NPP operators. • Two-way communications envisioned may imply different load follow characteristics for NPPs. • Self-healing characteristics should not have any adverse effects for NPPs. • IIDS systems provide too fast changes to NPP decision making.

VI. NUCLEAR-RENEWABLE HYBRID SYSTEMS

Currently, around 10% of the world's electricity is generated from nuclear power plants. The 2553 TWh of electricity in 2020 was produced from 440 reactors [5]. This makes an average power of a single reactor used for electricity production around 660 MWe. Considering an availability factor of nuclear reactors at least 80 %, it corresponds to an average electric power of 827 MWe per unit. In 2021 there were 55 reactors in construction worldwide, most of them with installed capacity of 1000 MWe or more. Slovakia is an exception, since the 2 units of the Mochovce NPP being constructed are VVER-440 units with 440 MWe installed capacity [6]. Considering the size of Slovakia and the options for backup power, the 440 MWe installed capacity can still be seen as high.

In general, high installed capacity of a nuclear reactors requires strong off-site power supply, which more emphasises the influence of smart grid technologies on NPP operation. However. there are possible ways to overcome some of these challenges by introducing Small Modular Reactor (SMR). There are a few initiatives to integrate nuclear and renewable technologies to provide high-performance energy systems with maximum utilization of nuclear power plant resources as well as to achieve balanced energy generation to meet load profiles and demands. These are the Nuclear- Renewable Hybrid Energy System (N-R HES) [7].

The nuclear-renewable integration can be carried out in two ways: large scale and small scale nuclear-renewable integration. If the RES based power generation systems are installed around the traditional large NPP, it is called large scale nuclear-renewable integration. When the small-scale nuclear power, such as SMR or MMR, are moved to the site of RES and integrated with RES based power generation system to support the local electricity demand, it is called small scale nuclear-renewable integration. A grid-connected N-R HES makes a worthy solution for large industry, office, and university. It may reduce the dependency on the grid and lower the energy cost. Research simulations estimate the criteria and help to choose the appropriate energy system depending on electricity load profile [7]. There are mainly four types of coupling method in nuclear-renewable integration:

1. Loosely Coupled
2. Multiple Tightly Coupled Products
3. Multiple Tightly Coupled Resources
4. Tightly Coupled

The schemes of the RES and nuclear coupling options are shown in Fig. 3 – 6.

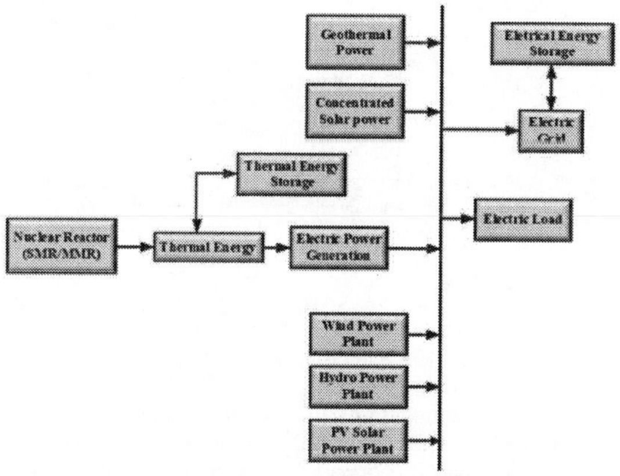

Figure. 3 – Losely Coupled N-R HES [7]

In the first case, the RESs are combined with an NPP and the main consideration is to produce electricity. The RES of various size and the small modular nuclear reactor generate electric power separately, but they are linked together at the level of electricity distribution. The combined system may supply the surplus electricity to the electric grid and the N-R HES can be also to the electric grid [7].

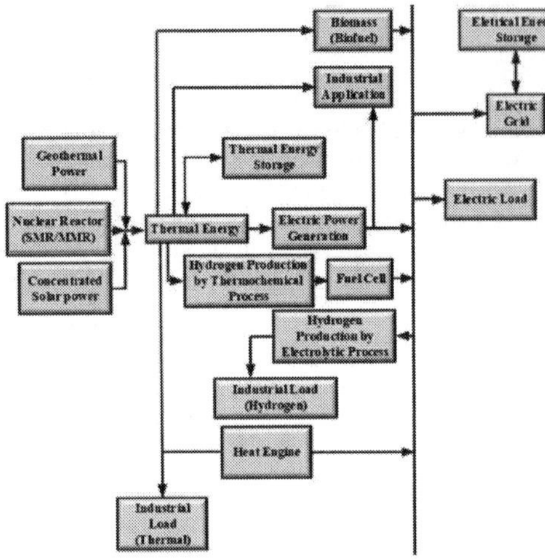

Figure. 4 – Multiple Tightly Coupled Products N-R HES [7]

In case of option 2 shown in Figure 4, the small size modular nuclear reactor is integrated with RES, which is also yielded by the nuclear power inside the N-R HES. Here no external RES are combined with a nuclear reactor. The heat and electric power from the nuclear reactor are used for biomass or hydrogen production for the RES. The electricity generated from the RES is used to supply the electric power of the NPP. In N-R HES, the hydrogen and biomass are considered as RES. [7]

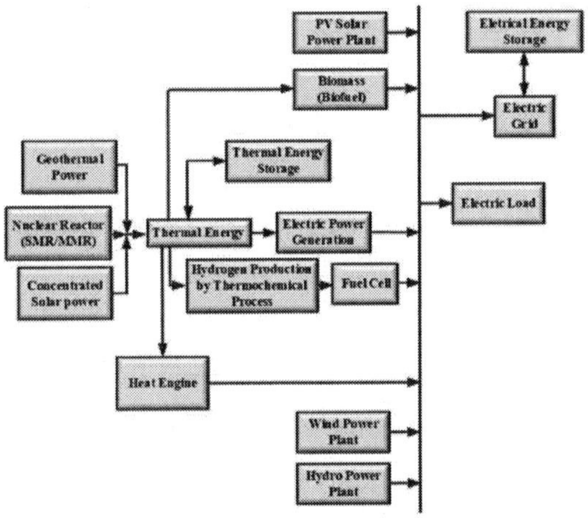

Figure. 5 – Multiple Tightly Coupled Resources N-R HES [7]

The coupling shown in Figure 5 offers two level energy integration, i.e. thermal and electric. In this scheme there are geothermal power sources or concentrated solar power sources, which produces a considerable amount of heat. This thermal power together with the nuclear reactor´s thermal power is used to generate electricity. In the electric level coupling, the electricity generated from a nuclear reactor is linked with the electricity produced from other RES (wind, solar, hydro, biomass, etc.). An electric grid is also connected to this hybrid system to take surplus electric energy to the grid or to supply electricity to the N-H RES, if no sufficient electric output is available from the RES. However, the focus of this coupling method is to generate electricity. [7] In the last coupling method, all forms of energy mixes are coupled together to achieve maximum utilization of the available energy in the hybrid system. The electricity is produced by adding all thermal energy together. The available thermal and electric energy is used in all possible applications, such as seawater desalination plant, fuel cell, district heating, heat engine, calcination, etc. [7]

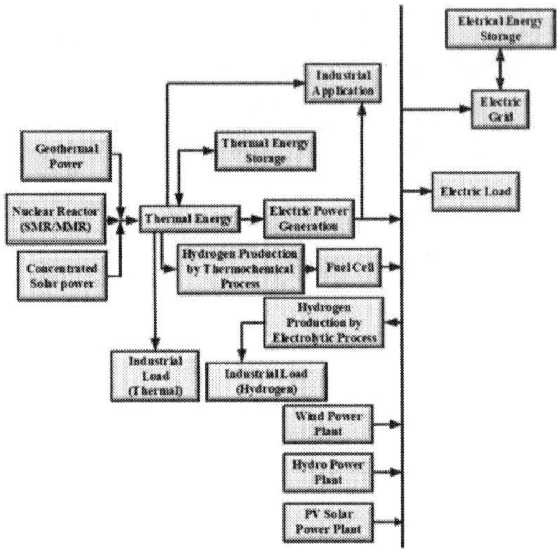

Figure. 6 – Tightly Coupled N-R HES [7]

VII. Conclusion

The nuclear safety standards used worldwide require that electricity from the grid should be derived to the NPP via two physically independent circuits. To meet these requirements the electric grid has historically evolved along a consistent physical configuration and mode of operation. As the modern grid evolves incorporating smart grid innovations, distributed resources, renewable energy sources, high speed sensing and control units, microgrids or large-scale energy storage systems, NPPs must also adapt to these changes. It naturally brings challenges also to the new reactor technologies, such as GEN IV reactors or small modular reactors, but most importantly to the current fleet of reactors in operation, which has only limited manoeuvrability.

As smart grid components are being introduced to the grid, concerns related to that the increased variability of the grid are arriving as well. The low inertia of renewable energy sources, the non-linear characteristics of the inverters connecting renewable energy sources and large-scale energy storage to the grid may lead to a variety of power quality concerns. The integration of wind and photovoltaic sources into the grid brings the greatest concerns. As the penetration level of these renewable sources increases, the amount of dispatchable power that grid operators will need to maintain may also increase. Vulnerabilities introduced by distributed energy resources, energy storage systems, and microgrids have also been identified. The potential issues with them are that all may cause large scale power outages. The occurrences of such large-scale outages are expected to be rare, but the consequences would be high. Additional vulnerabilities can be seen in cyber security. The used smart sensing and control units may be entry points to the network for direct threats or cyber-attacks.

It is clear, that the NPPs will have to adapt to all these changes in the electric grid. On one hand, it may be very difficult for them, since the current fleet has been designed to centralized grids and have high output power directly requiring high off-site backup power for their safety systems. A solution for the future could be the introduction of small modular reactors either as standalone units or as part of a hybrid system also including RES. The heat generated from this hybrid system used for the industrial purpose, whereas the generated electricity from can be used used to support the electric demand. The surplus energy produced from the system can be always supplied to the grid, or if backup power is required, the grid may support the hybrid system.

ACKNOWLEDGMENT

This publication was created thanks to support under the Operational Program Integrated Infrastructure for the project: International Center of Excellence for Research on Intelligent and Secure Information and Communication Technologies and Systems – 2. stage, ITMS code: 313021W404, co-financed by the European Regional Development Fund, by the Scientific Grant Agency of the Ministry of Education of Slovak Republic No. VEGA 1/0615/21 and partially by the ESFR-SIMPLE project funded by the European Union's Horizon Europe research and innovation program.

REFERENCES

[1] IAEA, “Interfacing Nuclear Power Plants with the Electric Grid: the Need for Reliability and Complexity.” Internet: https://www.iaea.org/sites/default/files/gc/gc53inf-3-att5_en.pdf [March. 10, 2022].

[2] M. Villaran, “A Literature Review: Smart Grids Impacts on Nuclear Power Plants – Task4: Project final report”, Brookhaven National Laboratory, US DoE, 2016.

[3] M. Shabanzadeh and M.P. Mofhaddam, “What is the Smart Grid? Definitions, Perspectives and Ultimate Goals,” *International Power Systems Conference*, 2013, pp. 1-4.

[4] European Union Agency for Cybersecurity, “Smart Grid Security Certification in Europe: Challenges and Reccomendations.” Internet: https://www.enisa.europa.eu/news/enisa-news/smart-grid-security-certification-in-europe-challenges-and-recommendations [Apr. 2, 2022].

[5] World Nuclear Org, “Nuclear Power in the World.” Internet: https://world-nuclear.org/information-library/current-and-future-generation/nuclear-power-in-the-world-today.aspx [Apr. 5, 2022].

[6] Slovenské elektrárne a.s. „Atómové elektrárne.“ Internet: https://www.seas.sk/o-nas/nase-elektrarne/atomove-elektrarne/ [Apr. 5, 2022].

[7] IEEE Smartgrid, “Grid-Connected Nuclear-Renewable Micro Hybrid Energy Systems.” Internet: https://smartgrid.ieee.org/bulletins/august-2019/grid-connected-nuclear-renewable-micro-hybrid-energy-system [Apr. 5, 2022].

Phases of the Smart Low-Voltage Distribution Network Implementation by System Theory Approach

Péter Holcsik
Óbuda University,
Automatizálási és Energiarendszerek Intézet
1034 Budapest, Bécsi út 94-96. BC 323 Hungary
holcsik.peter@kvk.uni-obuda.hu

Norbert Nieberl
MVM Émász Áramhálózati Kft.
3525 Miskolc, Dózsa György út 13.
norbert.nieberl@mvmemasz.hu

Judith Pálfi
Óbuda University,
Automatizálási és Energiarendszerek Intézet
1034 Budapest, Bécsi út 94-96. BC 323 Hungary
palfi.judith@uni-obuda.hu

Abstract — **The continuously increasing expectations of less outages, the growing number of household-sized power plants and the available technological and regulatory options require a rethink of the low-voltage distribution network strategy. The aim of this article is to summarize the answers to the challenges ahead and the concept of LV operation management system is also presented as an integral part of it.**

Keywords: smart low-voltage network, smart modem, SCADA.

I. THE CURRENT CHALLANGES OF THE LOW-VOLTAGE DISTRIBUTION NETWORK

The continuously increasing expectations of less outages, the growing number of household-sized power plants (abbreviation of its Hungarian name: HMKE) and the available technological and regulatory options require a rethink of the low-voltage distribution network (LV) strategy. This new approach can be called SMART philosophy, which is now existing on high-voltage transmission and medium-voltage (MV) distribution grids thanks to a larger amount of remote control and monitoring devices. Although these devices were available for the LV, their reasonableness hasn't been justified so far.

A. Challenges caused by household sized PV generators

Today, the installed household sized power plants (HMKE) are typically photovoltaic plants (PV) in Hungary. [14] [15] [23] These plants are connected to the LV. The designers of LV network have determined the parameters of the elements by calculating with a one-way load-flow. The natural part of it is the voltage decrease from the supply point to the consumers. The longer an overhead line, the greater the voltage decrease (in case of cable networks the problem typically does not occur due to the better R/X ratio). However, by implementing the HMKE PVs, new load-flows appeared that LV designers have not been able to calculate in the past. [8] [9]

Due to these new load-flow directions the voltage increased during the spring-summer produce period. This increased voltage rate can activate the protection of the inverters and the production stops. This is to ensure that consumer equipments do not fail.

It is also typical that PVs are implemented along with the heat pump heating systems that consume a significant amount of electricity. As a result, on a cold, cloudy winter day the consumption will increase more than previously planned.

In case of the lack of production, the voltage can decrease below standard level (Figure 1).

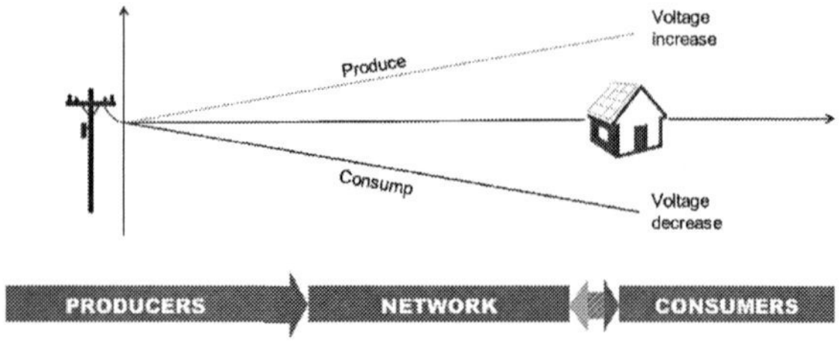

Figure 1. Illustrates the voltage decrease and increase problems

Distribution system operators (DSOs) face huge challenges due to these voltage deviations. [6] [7] The concept presented in this article can be a system level solution to these problems.

B. Challanges of improving quality of supply

Reducing consumer outages is a key goal of all DSOs and the main pillar of their investment strategy. During the last decades, implementation of various remote monitoring and control devices into the MV distribution network was an important way of improving quality indicators (SAIFI, SAIDI). Such devices are the indoor and outdoor distribution network telemechanics and short-circuit signaling devices. It is typical of these devices that a higher SAIDI, SAIFI improvement can be achieved with a lower investment intensity compared to conventional grid developments. However, other features of that solution are that their effectiveness is decreasing with the amount installed. (pls. see on Figure 2).

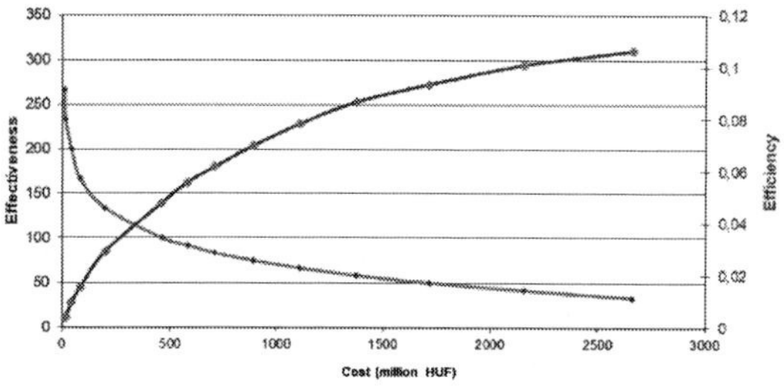

Figure 2. Indoor and outdoor MV telemechanic effects on the SAIDI [1]

In absence of remote monitoring on LV, the start time of outages is the time of the consumer's notification. For this reason, remote monitoring devices will only result in a direct SAIDI-SAIFI reduction in special cases. [10]

However, due to today's growing expectations, the DSOs need to be aware of faults before consumer notifications. This is because in the event of a blackout, the number of calls to call centers increases significantly. That overload leads to a significant additional waiting time, which is not an optimal solution for either customers or call centers.

In addition, today's devices are becoming more and more sensitive to the disappearance of voltage. In many cases, these devices feed critical processes for the continuity of business. These processes do not, or hardly tolerate the lengthy procedure of consumer notifications, dialing, call center log in, ticket recording for faults.

Given that the above mentioned MV automation devices have already been installed in large volumes, it is necessary to examine further innovative technology solutions.

II. Sensors and Actors in the First Phase

Nowadays, consumer fault reporting is the only tool for LV fault detection. Consumer notifications should continue to provide information to network operators but those are expected to gradually lose their relevance due to the increasing number of sensor and interloper/switch devices installed on the network.

An example of a device installed on the network that implements fault detection is a so-called smart modem installed on the LV bus of the MV/LV transformer stations. These devices detect, through voltage sensors connected to the outgoing phases of LV circuits, if voltage falls below a certain level, for example the fuse has disconnected the circuit. In case the voltage sensor gives a signal, the smart modem sends an automatic error message to the LV operation controller, who can send the electricians to the site.

The detection of a LV group faults allows the distribution system operator to be notified of a fault on the LV network not by customer feedback but by a fully automated system. This will reduce the duration of power outages for users (The DSO will be aware of the fault sooner, will send the electricians sooner, remediation will take place sooner.)

Thanks to its modular setup, smart modems can serve many other functions in addition to remote signaling of LV faults:

• As a modem, it allows data to be queried by receiving and transfering data from connected sensors and switches by implementing real-time communication, which enhances transparency.

• The devices can be equipped with security functions: when the door is opened, the modem sends a signal to the dispatcher.

• The device can also be equipped with environmental sensors, so that they can be used to measure light [22], temperature, humidity, air quality, etc. and to transfer the measured data. Environmental data can be used for other purposes in addition to operating the network.

• In addition to LV fault detection and measurement of environmental parameters, measuring bus-bar voltages and currents for more accurate network design and asset management can be an important function. Measuring oil temperature of the transformer can be an important monitoring tool and provides information for equipment diagnostics.

With the help of this large amount of data, a more accurate network design and asset strategy can be set up. If the measurement parameters are selected correctly this can also replace the mandatory voltage quality measurement for DSO. Measuring voltages and currents, as presented below, can be an important tool in determining the flexibility service needs.

III. Sensors and Interlopers in the Next Phases

Procurement of the smart modem has started at almost all Hungarian DSOs, however, the measurement at MV/LV transformers alone does not give a complete picture of the entire LV grids. The additional sensors and actors devices available today are presented below. Pilot data is already available for most devices, meaning that their technology maturity is typically suitable for a wider application.

A. Endpoint and branch node sensors

In addition to the indications of voltage sensors located at the outgoing phases of the MV/LV transformer stations, voltage measuring at any point in the circuit such as branch nodes or endpoints, can also be useful and necessary information for LV operation control. This also refers to the mobile voltage quality measuring devices, which will be replaced over time by an increasing number of constant voltage quality measurements.

B. Remote controlable on load MV/LV transformers

The equipment closest to the MV operation control level are the MV/LV transformers. The "smart" versions of these devices have current and voltage measurements, remote controlling and the ability to intervene and can be controlled under load (OLTC-On Load Tap Changer). By keeping the voltage of the LV bus at a certain level, OLTC devices can prevent voltage problems of end users which can also be caused by PV' on MV.

One possible definition for the border of the LV operation control is the following: From the low voltage tap of the transformer of MV/LV station to the ownership point or to the point specified in the operating agreement. (With the introduction of this definition, the process for reporting work on connection cables and measuring points to the LV operators must be carefully defined.) It follows from this definition that the MV/LV transformer falls within the scope of the MV operation control. However, the intervention of the OLTC affects the LV buses and circuits, so their signals and intervention are relevant to the LV operation control as well. The challenge ahead is to decide whether the operation of the MV/LV transformers, like the operation control of the TSO/132 kV or 132 kV/MV transformers, should be under the scope of the higher or lower level operation control. [13]

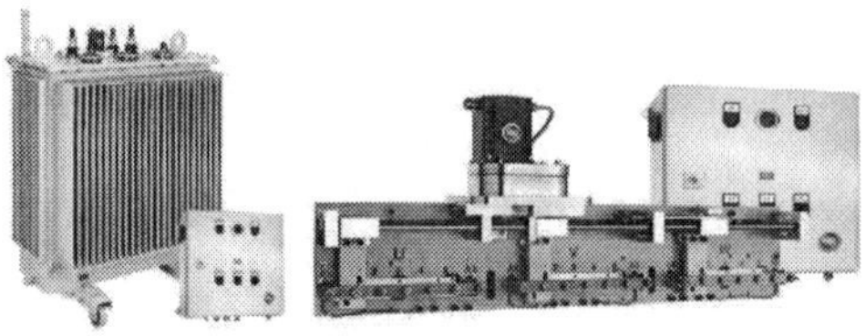

Figure 3. MV/LV transformer with on-load controller (left) and on-load tap changer and control (right) manufactured by Maschinenfabrik Reinhausen (MR)

C. *Smart Switchboard*

It doesn't seem economically feasible today, but we have to reckon with the presence of smart LV switchboard (SSB) in the future. SSB solutions can replace LV switchboards in the MV/LV stations and also on the LV lines.

Definition of the SSB is the following: a modern low voltage switchboard (compliant with the relevant standards) which, under the maintenance and operating conditions specified by the manufacturer, breaks overcurrent or short circuit current throughout the life of the equipment without partial or complete replacement of any unit (e.g. fuse). The detection of overcurrent is performed using a combination of a local measuring and evaluation system. Operation is based on a pre-defined current-time curve. The equipment is a remote signaling device, which transmits the presence of voltages (yes/no) at each measuring points, the effective and reactive current components, the voltage and power data calculated therefrom, using wireless communication (ensuring all relevant safety protocol standards) to the LV systems operation control. The device can be remotely controlled therefore suitable for remote switching. In order to ensure the safe execution of the planned works, the remote operation can be blocked on site and has a visible switch-off point and grounding option. The device also has automatic reclosing so it is suitable for performing one or more reclosing cycles based on the pre-set parameters.

As per the definition, in order to implement the SSB concept, it is necessary to replace the fuses with circuit breakers. [18] [19] Figure 4 shows an example for that:

Figure 4. One possible implementation of SSB concept [2]

Thanks to the measurement, signaling and protection tools of the SSB device, it can be suitable for supporting the concept of the curved or looped or weekly looped LV network, as well as for the implementation of switching possibilities on it.

D. *Line Voltage Regulators (LVRs)*

In addition to the nodes in the LV network, it may be necessary to use line voltage regulators depending on the network parameters and flows. Put simply, these devices are basically two inverters inverted into each other. For this reason, it is suitable for maintaining a pre-set voltage value on the secondary side, regardless of the voltage displayed on the primary side (within the value range) Using further simplifications, it can be stated that the device can fix the voltage at a certain point in the circuit.

Figure 5. Pilot line voltage regulator established on LV network [3]

The advantage of LVR is the quick installation and low investment cost compared to conventional developments. The disadvantage is that it does not improve the shock protection conditions of a directly earthed 3-phase supply (the conditions for short circuit current based circuit fuse blowing) and it preserves old, typically outdated, long networks with small cross-section and high losses.

E. *Monitoring by smart meter data*

According to the provisions of Act. No. LXXXVI of 2007 on electricity and Governmental Decree, the DSO is obliged to install a smart meter for 3 x 80 A LV users and below:

- If the capacity of the connection point - in case of connection of a new customer or expansion of the connected capacity reaches or exceeds 3 x 32 A, in case of direct measurement, if a household sized small power plant is connected to the grid or the calibration time of the metering device expires,
- If the electricity consumption of the connection point has reached or exceeded 5000 kWh / year in the previous year's billing period, and no indirect metering has been established at the connection point the smart meter shall be installed according to the time schedule specified by law.

High penetration of smart meters is a huge potential for LV operation control. If these devices are able to measure and signal the loss of voltage and the value of voltages and currents in real-time, a large amount of information about the network will be available to them. In this way, as already presented at smart modems, not only can the duration of power outages shortened but also the data need of the network asset strategy and the flexibility services can be served.

F. *Storages on distribution network*

In addition to the above, based on the Hungarian Electricity Act, the distributors also have the possibility to establish and operate electricity storages with a capacity of no more than 0,5 MW as part of the distribution network (excluding managing line capacity bottlenecks), in accordance with the principle of least cost in order to ensure the safe and reliable operation of the distribution network. The storage can be an effective solution for voltage management, peak management in some cases and it is also suitable for making reserves. Their economical efficiency needs to be examined. [5]

G. *Flexibility services*

Different flexibility services can provide a solution to LV network problems until the completion of conventional network development investments or even replace them in some cases. In practice, these services provide control of effective and/or reactive power of the production and consumption equipment connected to the network for the purpose of voltage and overload management. Inverters for small household size PVs, chargers and batteries of electric vehicles, as well as other customer storage equipment and load-schedulable devices (e.g. boilers) have a significant flexibility potential. The control of each device can be central, decentralized and autonomous. Central control logic is able to take system-level considerations, while decentralized control aims at the optimal operation of a smaller network area. In case of autonomous control, the controlled device has no (or only one-way) communication with the network operators, so the control takes place on local parameters. On the current low voltage network, autonomous control can be implemented. The conditions for the use of flexibility services through central or decentralized regulation are currently not available due to the current practice of passive network operation, in the absence of adequate online monitoring and control tools. Although with targeted developments, the distribution network can be made suitable for the use of flexibility services and thus for the implementation of more weather-dependent renewable capacity. [11]

H. *Other devices of monitoring*

It is important and necessary for the future LV system operation control, that monitoring data of the operating mobile aggregates shall be available, and properly displayed.

This system must be able to process the part collected from its own sensors, partly from external sources meteorological data. These data will be the base for load estimation of the weather-dependent renewables, thus also one of the base of the whole network load-flow forecasts.

Until this point, equipment owned by DSOs have been presented. However, there are several devices owned by customers which are connected to the low voltage distribution network and necessary to have online information about them. They are, for example, inverters of small household sized power plants or EV chargers that are measuring many electrical parameters.

IV. Expected features of the display

In the current, first phase, the primary task of the system is data collection and data transmission via interfaces. For this reason, the display at this stage only provides information on the operation of the system. Display shall support reporting function, evaluation of the experiences of the first phase, and evaluation of initial errors.

In addition to all this, the system in the additional phases can be upgraded, with a possible first step of adding network data(GIS) to the system.

If that happens, an online scheme can be available for the dispatcher. If the online scheme is available, in addition to receiving remote monitoring data, possibility of remote intervention, temporary connections, connection of mobile aggregators, planned and unplanned works and their network effect can also be displayed online. This would mean the implementation of LV operation control.

These features are typically not yet available, but a smaller subset of them are available today at the workforce management system.

V. Communication

LV operation control requires complex human-human, human-machine and machine-to-machine communication, which can be provided by the combined existence of an Internet and/or data connection, URH connection and telephone connection.

One of the basic requirements for LV operation control system is that it should be open to diverse and dynamically changing communication technologies. Two-way communication is important not only for the control of the devices, but also for the creation of the possibility of remote parameterization by the distributor in the case of automation and autonomous devices. In terms of communication technology, a distinction should be made between long-range communication with a data center and last-mile communication between data collectors, gateways and on-site devices. The long-range communication can be done with wired technology using optical fiber or wireless technologies (e.g. GPRS, 4G, LTE-450, LoRa, etc..) For last-mile communication it is advisable to use cheaper short-range wireless technologies (e.g. ZigBee, WiMAX, WIFI, etc.) For the long-range communication only GSM 4G VPN has adequate coverage and reliability. (Wired technologies can be excluded due to the investment costs involved and their uncertain operation during breakdowns.) [12]

It should be noted, however, that the spread of the real-time 5G network may open up new possibilities, such as real-time coordination of substation and distribution protections. Today, however, current trends do not allow us to predict when the reliability of this technology will reach a level where it can be used in electricity systems.

VI. Low voltage operation control

Operation control is defined as the control of the operational status of the network and the supervision of the system. It is responsible for the operation control of the system in such a way that the service is continuously provided to customers at standard parameters. **Chyba! Nenašiel sa žiaden zdroj odkazov.**

It arises from the combination of this definition that the basic condition of operation control is the availability of switching status, electrical parameters, as well as information in real time. In addition, a device or system must be available on which the operation control can display network interventions in real time. So, a real-time and modifiable scheme is needed.

The overall operation control activity includes, in addition to operation control, preparation of operation, including the determination of flexibility needs and the sub-process of system assessment. Thus, when we talk about the control of LV distribution networks, we need to look at these three basic tasks, in the context of the technical challenges, technological and legal possibilities. Below, we present a possible system design that may be able to handle the tasks detailed so far. Given the multiplicity and complexity of system connections and the fact that there is no ready-made solution on the market, development needs to be gradual.

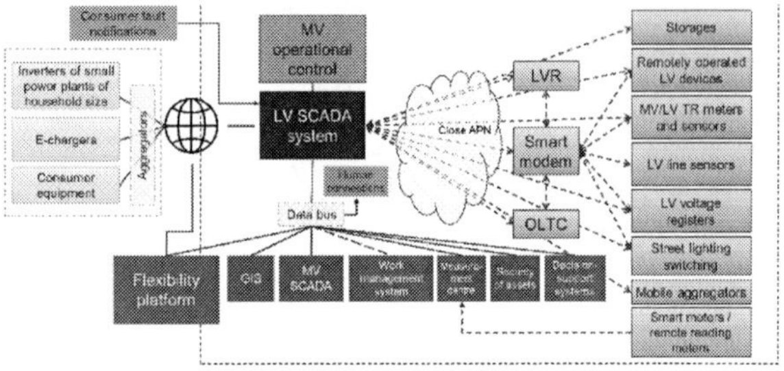

Figure 6. Possible designt of LV control management system

The development should start with the integration of sensor functions, and the LV network information obtained in this way can be used for the gradual improvement of work and asset management. After data-analysis, the intervention tools that have become marketable in the meantime, should be installed in certain focus areas. In parallel, the development and optimization of appropriate interfaces, algorithms and logic rules can begin. The establishment of logical rules requires close power company cooperation in order to align the operation strategy of the voltage-based distribution network with the schedule-based system-level operation.

Given that these solutions require a significant amount of data to be fed into the operation control system - much more than the current transmission or distribution system - it is necessary to think about pre-processing or automatic evaluation. Online load-flow calculation is also needed to plan the necessary interventions. These may require the integration of a self-learning system. [16] [21]

In order to display the real-time schema, the LV operational management system must be connected to the geographic information system (GIS), in which all network elements must be mapped in accordance with the current paths. To run simulations, the system must be able to call the network modeling software and provide the necessary data (e.g. meteorological data to calculate the energy produced by the PVs). [20] It is important to establish a proper connection with the MV operation control management system to manage the interaction between the different voltage levels, and for safety reasons. Distribution flexibility needs determined on the basis of planned reconstruction works, unplanned outages and simulation data must be handed over to the flexibility platform, where tenders can be issued and system users and aggregators can submit their bids. The success or failure of the tender is an important feedback to the operation control system.

The future operation control system must be able to handle the challenges of work organization, e.g. that different departments may be responsible for LV work management at different times of the day. Furthermore, it should be also possible to manage the system not only from fixed workstations, but also to allow the responsible colleagues to perform their tasks remotely. [17]

VII. The relation between the LV operation control and the workforce management system

Nowadays, the LV dispatchers are workforce managers assigning resources to tasks. These tasks can be planned works, like operation, maintenance, switching, etc. or unplanned works, such as troubleshooting. However, assigning resources to troubleshooting does not mean disposition over grid state.

However, the large number of LV faults and tasks requiring network intervention make it necessary that the LV operational control system which is expected to be implemented in the near future, is developed in conjunction with the workforce management system. One version of this if the current workforce management systems will be extended by operation control functions or another solution can be if the workforce management systems are given more information from the future operation control systems.

As a consequence, there are basically two well-defined directions for DSOs: to upgrade their current workforce management systems and make them suitable for LV operation control tasks, or to develop a stand-alone LV operation control system.

In the first solution, the system remains as a data platform that receives and transmits on-site signals. In the second case, current operation control functions can be removed from workforce management system, and a stand-alone operation control system will be established.

Given that this is not primarily a technological, but corporate strategy, the authors of this article should not aim to take a position between the two methods. The question is still open, it is still up to the power companies to decide.

VIII. Summary

A summary and simplified schematic of the concept of smart LV presented in this paper, is illustrated in Figure 7:

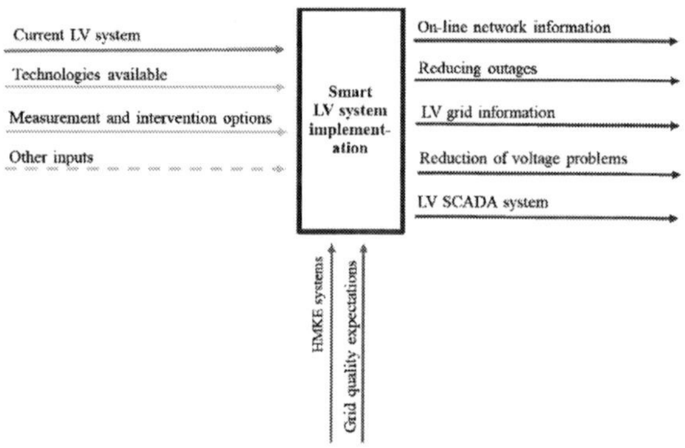

Figure 7. System theory approach of the smart LV system implementation

Approaching the smart LV model (Figure 7) from a systems theory view, it can be stated that it is active, Multi Input, Multi Output (MIMO) The development can be described by a multi-excitation, complex decision model, because in addition to the input parameters, it takes into account the additional and other excitations.

On the input side of the smart LV power system control model the following excitations are available: the current LV system, technologies available, measurement and intervention options and the other inputs that can't be seen yet.

As with any systems, the new smart LV operation control system is also affected by additional factors, which should be also considered during the design of the system. Additional factors can be for example, sporadically installed PV systems in the future and the grid quality expectations.

The output of the smart LV system will be the following: on-line network information, reduced number and duration of outages, on-line LV grid information, reduction of voltage problems and the basis of the LV SCADA system.

REFERENCES

[1] A. Csank, *Implementation of remote operation on the ELMŰ-ÉMÁSZ distribution network to improve the MKEH indicators*, 55. Elektrotechnikai Vándorgyűlés, Eger, 2008, pp. 1-15.

[2] SCHNEIDER ELECTRIC, "Compact NSX, Low-voltage cast-housing circuit breakers and disconnectors." Internet: https://download.schneider-electric.com/files?p_enDocType=Catalog&p_File_Name=Compact+NSX+LQ.pdf&p_Doc_Ref=SE100,%20201 [Feb. 20, 2022].

[3] I. Táczi, "The role of smart devices in solving observability problems in low-voltage distribution networks," *67. Elektrotechnikai Vándorgyűlés*, 2021, pp. 17-18.

[4] MVM Hálózat, Regulation of distribution, 2019, pp. 42-48.

[5] I. Diahovchenko et al., "Progress and Challenges in Smart Grids: Distributed Generation, smart Metering, Energy Storage and Smart Loads." *2020 Iranian Journal of Science and Technology, Transaction of Electrical Engineering*, vol. 44, pp. 1319-1333 2020.

[6] M. Kolcun et al., "Active and Reactive Power Losses in Distribution Transformers." *Acta Polytechnica Hungarica*, vol. 17, pp. 161-174, 2020.

[7] G. Morva et al., "Analysis of the impact of nonlinear distortion in voltage and current curves on the errors of electric energy metering devices," *2017 IEEE First Ukraine Conference on Electrical and Computer Engineering (UKRCON)*, 2017, pp. 528-533.

[8] Z. Čonka et al., "Impact of renewable energy sources on stability of EWIS transmision system," *14th International Conference on Environment and Electrical*, 2014, pp. 75-79.

[9] P. Hocko et al., "Influence of photovoltaic power plants on the power system to the prediction for year 2020," *14th International Conference on Environment and Electrical Engineering*, 2014, pp. 116-121.

[10] A. Koncz; Z.C. Johanyák, L. Pokorádi, "Fuzzy approaches in failure mode and effect analysis." *International Journal of Artificial Intelligence*, vol. 19, pp. 56-76, 2021.

[11] L. Pokorádi, S Kocak, E. Tóth-Laufer, "Fuzzy Failure Modes and Effects Analysis Using Summative Defuzzification Methods." Acta Polytechnica Hungarica, vol. 18, pp. 111-126, 2021.

[12] S. Szén, E. Rácz, "Use of the Power Line Communication System (PLC) at Low Voltage (0.4 kV) Noisy Electrical Networks – Introducing a New Concept at Power Quality," *International Conference on Renewable Energies and Power Quality (ICREPQ'12)*, 2012, pp. 1735-1740.

[13] Introducing a New Concept at Power Quality, Renewable Energy & Power Quality Journal, vol. 10, pp. 1735, 2012.

[14] G. Morva, I. Szén, "Reconstruction of the Hungarian transmission networks," *5th International Scientific Symposium on Electric Power Engineering (Elektroenergetika)*, 2009, pp. 30-31.

[15] B. Hörömpöli, E. Rácz, "Investigation of state-of-the art solar cell using color filters," *10th International Scientific Symposium on Electrical Power Engineering (Elektroenergetika)*, 2019, pp. 535-541.

[16] B. Hörömpöli, E. Rácz, "Statistical Analysis of Power Measurements made on Mono- and Polycrystalline Solar Cells" *SAMI 2018 IEEE 16th World Symposium on Applied Machine Intelligence and Informatics*, 2018, pp. 000149-000154.

[17] M. Jocic, E. Pap, A. Szakál, Djordje Obradovic; Zora Konjovic, "Managing Big Data Using Fuzzy Sets by Directed Graph Node Similarity." *ACTA Polytechnica Hungarica*, vol. 14, pp. 183-200, 2017.

[18] I.J. Rudas, J. Gáti, A. Szakál, K. Némethy, "From the Smart Hands to Tele – Operations," *ACTA Polytechnica Hungarica*, vol. 13, pp. 43-60, 2016.

[19] Z. Mitrik, P. Kádár, "Test result utilization in the development process of high voltage SF6 circuit-breaker," *ACTA Polytechnica Hungarica*, vol. 17, pp. 111-124, 2020.

[20] Z. Mitrik, P. Kádár, "System theory approach of SF6 high voltage circuit breaker development process," *2018 IEEE 16th World Symposium on Applied Machine Intelligence and Informatics (SAMI)*, 2018, pp. 137-142.

[21] M. Zsolt, "Establishment of a New Reliability Model of the Hungarian Energy Transmission Network," *Proceedings of the 9th International Scientific Symposium on Electrical Power Engineering (Elektroenergetika)*, 2017, pp. 551-555.

[22] G. Sebestyén, J. Kopják, "Battery Life Prediction Model of Sensor Nodes using Merged Data Collecting methods," *IEEE 20th Jubilee World Symposium on Applied Machine Intelligence and Informatics SAMI*, 2022, pp. 000027–000030.

[23] R. Istók, "Conducted and Radiated Emissions of LED Lamps and Intelligent Lighting Systems," *The 9th International Scientific Symposium (Elektroenergetika)*, 2017, pp. 259-262.

[24] R. Istók, P. Kádár, L. Reizer „Solar Carport with Energy Storage Capabilities," *IEEE 4th International Conference and Workshop in Óbuda on Electrical and Power Engineering*, 2021, pp. 89-94.

F: Works of PhD Students

Research of the Influence of Humidity and Contamination on Glass and Silicon Rubber Insulating Materials by Leakage Current Measurement

Ľuboš Šárpataky, Bystrík Dolník
dept. of Electric Power Engineering
Technical University of Košice
Košice, Slovakia
lubos.sarpataky@tuke.sk, bystrik.dolnik@tuke.sk

Abstract—**The reliability of transmission and distribution of electricity is highly dependent on the reliability of insulators. The insulator is the main insulating element of the transmission and distribution lines. It must operate without failure throughout its service life. However, there are situations where its reliability is compromised. Moisture and pollution are the main factors that reduce the electrical strength of insulators. This study is devoted to the formation and research of a layer of contamination formed on the surface of external insulation made of glass and silicone rubber. After mixing solutions representing different contamination levels, these solutions were applied to the insulation surfaces. Leakage current was measured on contaminated surfaces at various humidity from 40 % to 90 %. The applied layer was washed from the insulation materials to calculate the equivalent salt deposit density (ESDD) and non-soluble deposit density (NSDD) according to IEC/TS 60815-1: 2008 and determine the contamination level for the used insulation material. The measurement results showed that the effect of pollution and humidity on the magnitude of the leakage current is significant. The increase in leakage current is the most pronounced between 60 % and 70 % of relative humidity. The amount of pollution that settles on the insulator surface depends on the hydrophobicity of the material.**

Keywords: contamination; humidity; ESDD; NSDD; leakage current; glass; silicon rubber;

I. Introduction

Insulators are most at risk when exposed to various stresses. Electrical stress constantly affects the functionality of the insulator due to the electric and magnetic fields that the electric energy constantly generates and in situations such as short circuits, overvoltage, or voltage pulse. Another effect that affects insulators is the weather. There are different weather conditions in different environments, and the selection of insulators must be adapted to these conditions. The insulator selection also applies to pollution, which can reduce the surge voltage of the insulator by tens of percent [1-4].

Insulators contamination, including long-term accumulation of dust or sand particles, moisture, rain, ambient moisture, and dew, can form a conductive layer. Insulators contamination can lead to surface discharges and leakage paths, which can negatively affect the life of the insulator or lead to a flashover, thereby interrupting the power supply. Contamination settles on the insulation surface thanks to the wind. The polluting particles are fed to the insulator, where its velocity is reduced when the particle hits, and this causes the pollution to accumulate on the insulator. Constantly depositing particles on the insulator form an ever-thicker pollution layer. Over time, a continuous layer of dirt is formed. Moisture, rain, fog, and dew moisten the pollution layer, dissolve the substances in the accumulated pollution, and form a conductive layer that decreases electric strength. The decrease in the electric strength of the insulation can reach 20-25 % [2], [5-9].

Chemical reactions between water molecules and insulation (hydrolysis) can lead to degradation or destruction of the insulation. Moisture also impairs the mechanical properties of the insulation. The electrical properties change significantly due to moisture absorption (the dielectric strength of the material changes significantly, dielectric losses increase, and the distribution of electrical stress on the insulation changes). The moisture absorbed leads to sparking on the insulation surface, to failure of the insulation surface due to partial discharges followed by erosion of the insulation surface. Synergistic effects can occur between electric field and humidity, leading to the formation of water trees on the surface of the insulator on some polymers. The dynamics of condensation on the surface depend on the hydrophobicity of the material used for insulation and on the action of steam or liquid on the insulation. Condensation also depends on temperature and its changes [10-13].

Leakage current is recognized generally as one of the main parameters for measuring the condition of insulators. It has also been shown that the probability of a flashover is very high as the leakage current of the insulator approaches a certain threshold value. This value was defined based on experimental work as the amplitude of the leakage current of the half-period through which the current passed immediately before the flashover (I_{max}) [14]:

$$I_{max} = (S_{CD}/15.32)^2 A \tag{1}$$

where S_{CD} is specific creepage distance.

A lot of research is done to measure leakage current. Research clearly shows an increase in leakage current as contamination levels increase. However, the leakage current magnitude depends not only on the contamination level but also on the insulator shape, the insulating material used, and the composition of the contamination layer [15]. As the thickness of the contaminant increases, the intensity of the partial discharges also increases and can be measured as pulses on the leakage current waveforms [16].

In addition to leakage current measurement, harmonic analysis of the individual components of the leakage current is also used. They experimentally investigated the leakage current and performed its analysis. The leakage current measured was focused on the peak value, harmonic components, and the harmonic signal index of the leakage current. These characteristics were evaluated to identify and predict the flashover over the insulator's surface. The leakage current waveform was used to monitor the insulators [17], [18].

In this paper, two different insulating materials (glass and silicon rubber) are contaminated by solutions that represent different pollution levels. Firstly, we compare how different materials absorb pollutants and calculate equivalent salt deposit density (ESDD) and non-soluble deposit density (NSDD). Measurement and calculation were performed according to IEC/TS 60815-1:2008 standard. Secondly, leakage current measurement is done on clean samples and every pollution level at relative humidity from 40 % to 90 %.

II. Materials and Methods

Measurement of surface leakage current on glass insulators and polymer surge arresters was performed in a laboratory at the Technical University in Košice. Measurements were performed under different relative humidity from 40 % to 90 % in a closed chamber. Four levels of contamination (L1, L2, L3, L4) were prepared and applied to the surface of the samples. Solutions represented four levels of contamination were created. Each solution consists of a specified amount of salt, kaolin, and distilled water. The kaolin is needed to create an even distribution of the contaminant layer on the surface of the insulator. We added 40 g of kaolin for each liter of water. Table 1 lists the amount of salt per liter of distilled water used for each degree of contamination and the conductivity of the solution.

Table 1. The amount of salt in the solution and the corresponding conductivity of the solution represents different pollution levels.

Pollution level	*NaCl amount (g/l)*	*Solution conductivity (mS/cm)*
L1	1	2.88
L2	6	12.7
L3	10	21.7
L4	16	27.7

The measuring electrodes were formed by a conductive copper tape applied to the surface of the glass insulator and silicon rubber insulating material. The distance between the measuring electrodes is 4 cm. Measuring electrodes on the insulation surface are shown in Fig. 1.

Used glass insulator U160BS according to the IEC 60305 standard is shown in Fig. 1 **a)**. The diameter (*D*) of the insulator is 280 mm, spacing (*H*) is 146 mm and ball and socket coupling (*d*) are 20 mm. Glass cap-and-pin insulators are suitable for "very light" to "medium" polluted areas where a long creepage distance or an aerodynamically effective profile is not required.

ABB company surge arrester POLIM-S, according to the IEC 60099-4 standard, is a metal-oxide surge arrester without a spark gap. Direct molded silicone housing in a patented loop is designed for indoor and outdoor installations. The diameter of the shed used for measurement is 4 cm, and measuring electrodes are shown in Fig. 1 **b)**. POLIM-S surge arresters are suitable for use in all environmental conditions.

Fig. 1. Measuring electrodes used on: **(a)** glass insulator; **(b)** silicon rubber

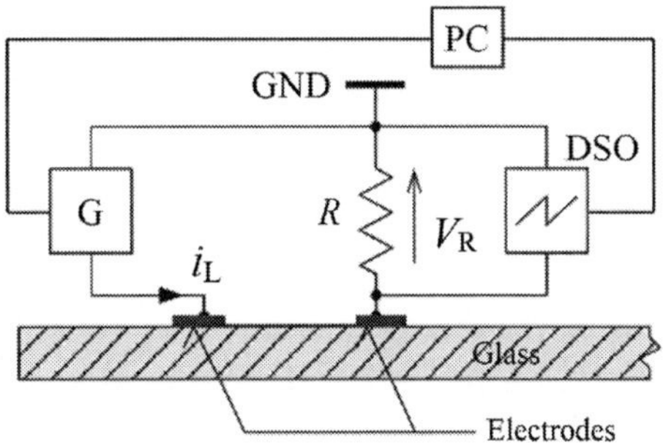

Fig. 2. Measurement diagram for leakage current measurement.

For leakage current measurement, a function/arbitrary waveform generator Agilent 33210A (G) and an oscilloscope Agilent DSO7104A (DSO) were used. To prevent electromagnetic interference from the supply network, two different frequencies of the measuring voltage were used. The first frequency, close to the industrial frequency, is 113 Hz, and the second frequency is 1000 Hz. Leakage current (i_L) was calculated as the average of 128 periods of measuring voltage. The measurement diagram is in Fig. 2.

III. Measurement of ESDD and NSDD

The research included the measurement of ESDD and NSDD for various insulation materials. These parameters indicate the number of pollutants that insulation materials can absorb. Since we have applied the same solutions to the glass and silicone rubber insulating materials that represent the contamination levels, we can point out the ability of the material to repel the contamination. Measurement and calculations were done according to IEC/TS 60815-1:2008 standard.

Firstly, the insulation materials were immersed in the solution and dried for 24 hours. After the layer of pollution on the surface of the insulating material was created, the washing process began. With the selected amount of water, the surface of the insulator was thoroughly washed. The amount of water is taken into account in the calculations, as well as the size of the surface from which the pollution was washed away. The conductivity of the collected solution from the insulator surface was measured using a Toledo S230 conductivity meter. The conductivity meter did not take into account the temperature of the solution. The temperature of the solution was recorded, and then the measured conductivity was recalculated using the factor depending on temperature *b*:

$$b = -3.2 \cdot 10^{-8}\theta^3 + 1.032 \cdot 10^{-5}\theta^2 - 8.272 \cdot 10^{-4}\theta + 3.544 \cdot 10^{-2} \quad (2)$$

where θ is the temperature of the solution.

After calculations of factor b, conductivity correction can be applied:

$$\sigma_{20} = \sigma_{\theta}(1 - b(\theta - 20)) \tag{3}$$

where σ_{20} is volume conductivity at a temperature of 20°C (S/m), σ_{θ} is volume conductivity at temperature θ (S/m). Salinity S_a is calculated:

$$S_a = (5.7\sigma_{20})^{1.03} \tag{4}$$

When we have salinity we can calculate ESDD where the amount of water V (cm^3) and surface area A (cm^2) are taken into account:

$$ESDD = S_a \cdot V/A \tag{5}$$

Measured and evaluated data are in table 2.

Table 2 Calculated ESDD parameters for glass insulating material.

Glass	**L1**	**L2**	**L3**	**L4**
Conductivity at 20°C (S/m)	0.0052	0.0136	0.0196	0.0241
Conductivity at θ °C (S/m)	0.0052	0.0143	0.0203	0.0250
Salinity (kg/m^3)	0.0269	0.0719	0.1045	0.1295
ESDD (mg/cm^2)	0.0244	0.0489	0.0712	0.0882
The standard deviation of conductivity at θ °C (S/m)	$37.5 \cdot 10^{-6}$	$5.7 \cdot 10^{-6}$	$37.9 \cdot 10^{-6}$	$54.3 \cdot 10^{-6}$

Table 3 Calculated ESDD parameters for silicon rubber insulating material.

Silicon rubber	**L1**	**L2**	**L3**	**L4**
Conductivity at 20°C (S/m)	0.0021	0.0043	0.0115	0.0125
Conductivity at θ °C (S/m)	0,0017	0.0055	0.0119	0.0129
Salinity (kg/m^3)	0.0103	0.0220	0.0605	0.0658
ESDD (mg/cm^2)	0.0250	0.0415	0.1102	0.1198
The standard deviation of conductivity at θ °C (S/m)	$11.4 \cdot 10^{-6}$	$10.9 \cdot 10^{-6}$	$21.5 \cdot 10^{-6}$	$31.3 \cdot 10^{-6}$

After the conductivity measurement, the solution was filtered through the filter. Filter was weighted before the filtering process and after the filtering process. The filter went through the process of drying for 24 hours.

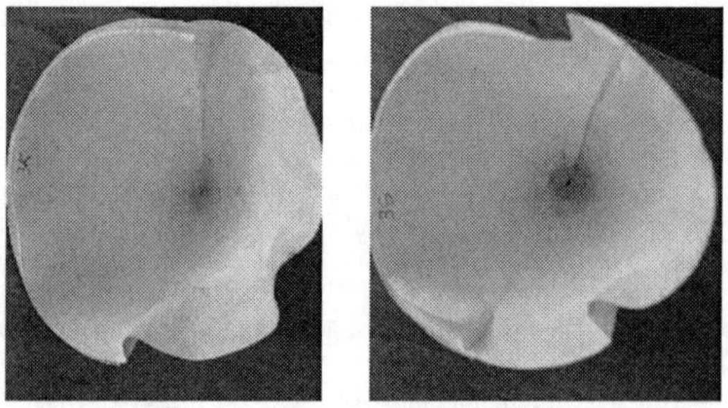

a) **b)**

Fig. 3. Amount of non-soluble pollution taken from the surface of the insulating material (L3): **(a)** silicon rubber; **(b)** glass

An example of a measured filter after the drying process is in Fig. 4.

The NSDD is then calculated from the values of the weighed filters:

$$NSDD = 1000(W_i - W_f)/A \tag{6}$$

where W_i is the weight of the filter with pollutants and W_f is the weight of the filter before the filtring process. Measured and calculated data are in tables four and five.

Table 4 Calculated NSDD parameters for glass insulating material.

Glass	**L1**	**L2**	**L3**	**L4**
W_f (g)	2.0603	2.1041	2.0322	2.0193
W_i (g)	2.3945	2.2358	2.2846	2.2354
NSDD (mg/cm^2)	0.7585	0.2989	0.5729	0.4905
The standard deviation of W_f (g)	$0.563 \cdot 10^{-3}$	$0.221 \cdot 10^{-3}$	$0.327 \cdot 10^{-3}$	$0.228 \cdot 10^{-3}$
The standard deviation of W_i (g)	$0.383 \cdot 10^{-3}$	$0.135 \cdot 10^{-3}$	$0.601 \cdot 10^{-3}$	$0.313 \cdot 10^{-3}$

Table 5 Calculated NSDD parameters for silicon rubber insulating material.

Silicon rubber	**L1**	**L2**	**L3**	**L4**
W_f (g)	2.0822	2.0395	2.0472	2.0565
W_i (g)	2.0865	2.0430	2.0556	2.0724
NSDD (mg/cm^2)	0.0257	0.0217	0.0515	0.0961
The standard deviation of W_f (g)	$0.084 \cdot 10^{-3}$	$0.079 \cdot 10^{-3}$	$0.280 \cdot 10^{-3}$	$0.499 \cdot 10^{-3}$
The standard deviation of W_i (g)	$0.134 \cdot 10^{-3}$	$0.230 \cdot 10^{-3}$	$0.811 \cdot 10^{-3}$	$0.327 \cdot 10^{-3}$

From measured values of ESDD and NSDD we can estimate the contamination level. Measured data are shown in Figures 4 and 5.

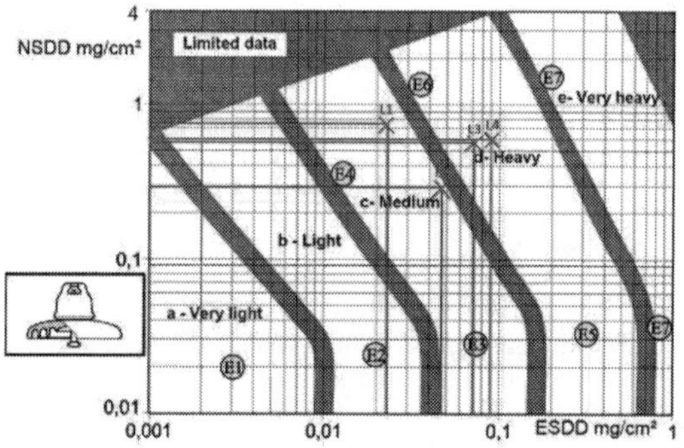

Fig. 4 Type A site pollution severity – Relation between ESDD/NSDD for glass insulator

Our mixed solutions create contamination on the glass insulator (Fig. 4) which represents a medium pollution level for L1 and L2 and a heavy pollution level for L3 and L4. Glass insulator has a high level of non-soluble deposit density. Silicon rubber insulating material has significantly lower NSDD, and ESDD values were comparable with glass insulators.

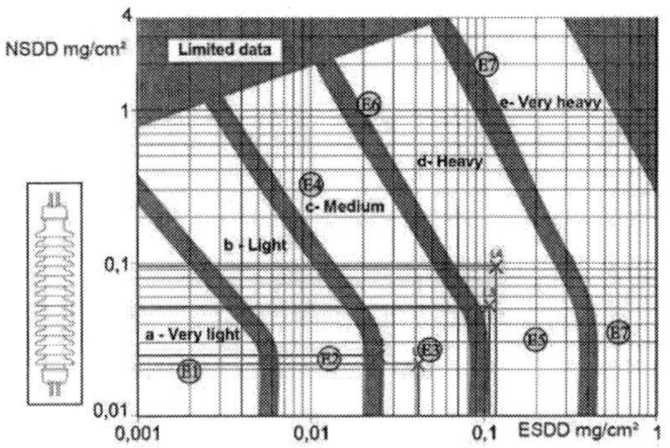

Fig. 5 Type A site pollution severity – Relation between ESDD/NSDD for silicon rubber insulation

Contamination on the silicon rubber insulating material (Fig. 5) was between light and medium pollution for L1, medium for L2, and heavy for L3 and L4. The standard deviation of all measured quantities was low, so we can consider this measurement accurate.

IV. Measurement of Leakage Current

Measured data of leakage current at frequencies 113 Hz and 1 kHz are in Tables 6 to 9. The measurement voltage was 5 V.

Measured data in Tables 6 to 9 show the increasing trend of leakage current with increasing relative humidity and increasing pollution level. Data from Tables 6 to 9 are processed in graphs in Fig. 6 and 7.

Table 6. Leakage current values measured at a frequency of 113 Hz on the glass insulator

	Leakage current (μA)			
RH (%)	**L1**	**L2**	**L3**	**L4**
40	0.0382	0.0616	0.0560	0.0566
60	0.0685	0.618	0.877	1.137
70	0.543	7.266	17.10	21.94
80	2.725	26.62	36.65	41.43
90	12.75	31.18	34.71	37.91

Table 7. Leakage current values measured at a frequency of 1 kHz on the glass insulator

	Leakage current (μA)			
RH (%)	**L1**	**L2**	**L3**	**L4**
40	0.268	0.334	0.282	0.284
60	0.302	0.870	1.094	1.507
70	0.914	10.32	19.10	27.01
80	2.983	28.34	37.84	40.94
90	14.13	32.34	36.42	38.73

Table 8. Leakage current values measured at a frequency of 113 Hz on the silicon rubber insulating material

	Leakage current (μA)			
RH (%)	**L1**	**L2**	**L3**	**L4**
40	0.0154	0.0153	0.0151	0.0161
60	0.0181	0.0187	0.0166	0.0152
70	0.0161	1.908	1.526	0.609
80	0.0159	7.947	19.34	28.47
90	0.0190	13.61	25.00	34.89

Table 9. Leakage current values measured at a frequency of 1 kHz on the silicon rubber insulating material

	Leakage current (μA)			
RH (%)	**L1**	**L2**	**L3**	**L4**
40	0.129	0.128	0.126	0.134
60	0.131	0.135	0.129	0.135
70	0.158	2.177	1.729	0.865
80	0.131	8.196	21.12	31.48
90	0.131	14.46	26.60	34.03

The rapid increase in leakage current was measured between relative humidity of 60 % and 70 %. The leakage current increase continues to the relative humidity of 80 %, and then the leakage current increase slows down between relative humidity of 80 % and 90 %. The most significant difference between pollution levels was between the first and second pollution levels. The difference between the second, third, and fourth pollution levels was not so significant but noticeable.

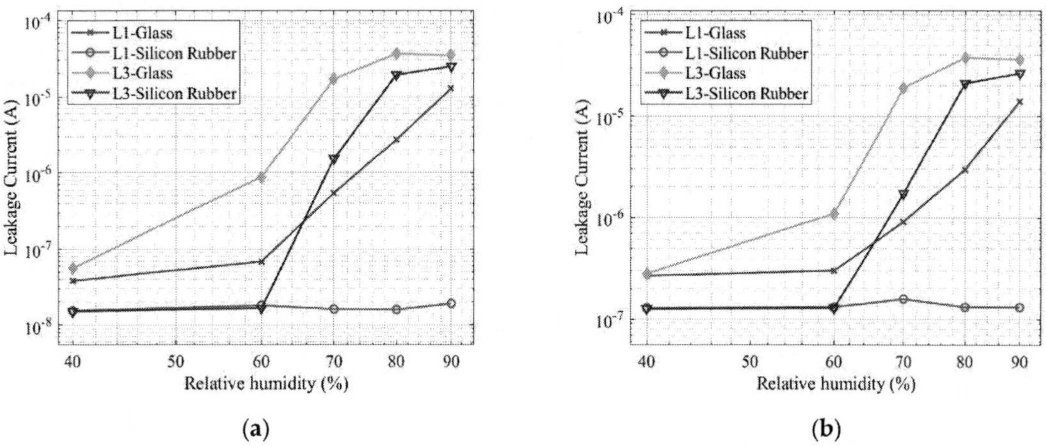

Fig. 6 Comparision of leakage current: (**a**) pollution levels L1 and L3 on glass and silicon rubber at frequency 113Hz; (**b**) pollution levels L1 and L3 on glass and silicon rubber at frequency 1 kHz

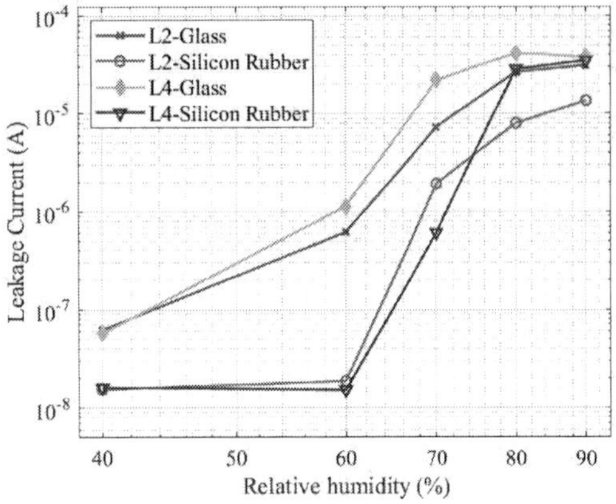

(a)

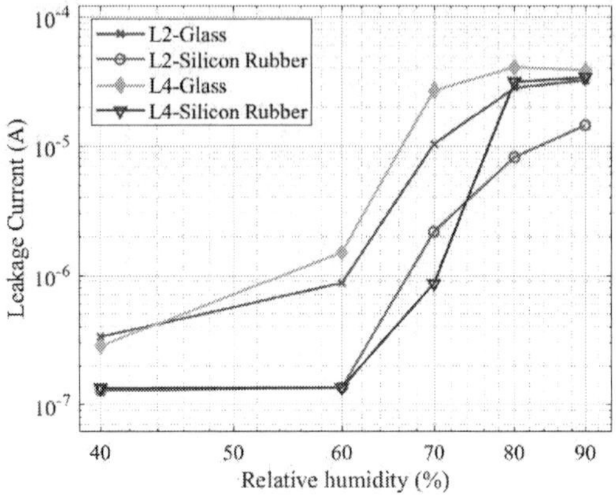

(b)

Fig. 7 Comparision of leakage current: (**a**) pollution levels L2 and L4 on glass and silicon rubber at frequency 113Hz; (**b**) pollution levels L2 and L4 on glass and silicon rubber at frequency 1 kHz

Fig. 6 proves that silicon rubber insulating material better withstands contamination and also moisture. The first pollution level (L1) is almost constant in the case of silicone rubber and started to increase at a relative humidity of 60 % in the case of glass. Silicon rubber insulating material contaminated by the third pollution level (L3) was up to a relative humidity of 60 % lower than in the case of the first contamination level (L1) on the glass insulator. There was a sharp increase between 60 % and 70 %, but the leakage current value remained lower for the silicone insulator than for the glass insulator.

A comparison of the second a fourth pollution levels in Fig. 7 proves an increasing trend of leakage current due to contamination severity. The only exception was in the case of silicon rubber insulating material at a relative humidity of 70 % where leakage current was lower for the fourth pollution level.

Conclusion

Experimental measurements in laboratory conditions were performed to investigate the pollution level on glass and silicon rubber insulating materials. Four levels of artificial contamination were prepared and applied to the insulators. Equivalent salt deposit density (ESDD) and non-soluble deposit density (NSDD) were measured and calculated to investigate how different insulating materials accept contamination on their surface. Measurement results show that hydrophobicity of silicon rubber insulating material helps to repel the water droplets after immersion that causing the minimum non-soluble particles of the contamination solution to settle on the surface. Soluble particles stay on the surface of both insulating materials and the difference between ESDD values was minimal. The ESDD related to conductivity was almost the same. That can make it difficult to visually inspect the surface for contamination of silicon rubber. NSDD was significantly higher on glass insulating material. Measurement point out the need to take pollution samples directly from the surface of the used insulators, especially if different insulating materials are used.

The leakage current across the surface of the glass and silicon rubber insulating material was measured for individual frequencies of 113 Hz and 1 kHz. The measurement results confirmed that with increasing relative humidity and increasing levels of contamination, the conductivity of the insulator surface also increases, and so does the leakage current. The most significant difference was between the first and second pollution levels. The rapid increase was between relative humidity of 60 % and 70 %. Measurement results confirm that silicon rubber insulating material has better insulating properties than glass. Up to a relative humidity of 60%, the heaviest contamination (L4) on the silicone rubber was less conductive than the first contamination level on the glass insulator. Silicone rubber achieved increased leakage current up to relative humidity of 80 % and 90 %. At high relative humidities, differences between leakage current values were much tighter. This points to the fact that the most dangerous conditions are foggy weather when the relative humidity is very high. Pollution at low relative humidity is not dangerous. Drizzles can also be dangerous if the pollution layer does not wash off the surface of the insulating material.

The next step of the experimental measurement of polluted insulators will focus on the harmonic analysis. We will investigate whether the individual harmonic component is influenced by the pollution level.

Acknowledgment

This research was funded by the Scientific Grant Agency of the Ministry of Education, Science, Research and Sport and of the Slovak Academy of Sciences under contract No. 1/0340/18; the Slovak Research and Development Agency under the contract No. APVV-15-0438; the Cultural and Educational Grant Agency of the Ministry of Education. Science. Research and Sport of the Slovak Republic (KEGA) under project No. 008TUKE-4/2019.

References

[1] M. Farzaneh, W. A. Chisholm. "Insulators for icing and polluted environments," in *Piscataway, NJ : Hoboken, NJ: IEEE Press ; J. Wiley*, 2009, pp. 706. ISBN: 978-0-47-028234-2.

[2] W. L. Vosloo, R. E. Macey, C. de Tourreil. "The practical guide to outdoor high voltage insulators," in *Johannesburg: Crown Publications*, 2004, pp. 220. ISBN: 978-0-62-031074-1.

[3] A. Haddad, D. F. Warne. "*Advances in high voltage engineering*," in Institution of Electrical Engineers, London, 2004, pp. 669. ISBN: 978-0-85296-156-2.

[4] E. Kuffel, W. S. Zaengl, J. Kuffel. "High voltage engineering: fundamentals," in *Elsevier*, 2nd ed. Oxford ; Boston: Butterworth-Heinemann, 2000, pp. 539. ISBN: 978-0-7506-3634-6.

[5] N. A. Othman, M. A. M. Piah, Z. Adzis. "Contamination effects on charge distribution measurement of high voltage glass insulator string," in *Measurement*, vol. 105, pp. 34-40, June. 2017. DOI: 10.1016/j.measurement.2017.03.044.

[6] R. Castillo-Sierra, O. Oviedo-Trespalacios, J. E. Candelo-Becerra, J. D. Soto, M. Calle. "A novel method for prediction of washing cycles of electrical insulators in high pollution environments," in *International Journal of Electrical Power & Energy Systems*, vol. 130, pp. 107026, Sep. 2021. DOI: 10.1016/j.ijepes.2021.107026.

[7] "Selection and dimensioning of high-voltage insulators intended for use in polluted conditions — Part 1: Definitions, information and general principles," DD IEC/TS 60815-1:2008, 2008.

[8] M. Sivaraman, J. Sivadasan. "Comparison of pollution flashover performance of various insulators," in *2014 International Conference on Circuits, Power and Computing Technologies [ICCPCT-2014]*, Nagercoil, Tamil Nadu, India, Mar. 2014, pp. 148-152. DOI: 10.1109/ICCPCT.2014.7054774.

[9] J. Araya, J. Montana, R. Schurch. "Electric Field Distribution and Leakage Currents in Glass Insulator Under Different Altitudes and Pollutions Conditions using FEM Simulations," in *IEEE Latin America Transactions*, 19(8), pp. 1278-1285, Aug. 2021, DOI: 10.1109/TLA.2021.9475858.

[10] M. Váry, E. Firický. "Electrical insulation system ageing caused by electrical, multifactorial and environmental stresses," *Posterus*. Internet: http://www.posterus.sk/?p=15765, June 12, 2013 [Nov. 09, 2021].

[11] M. R. Ahmadi-veshki, M. Mirzaie, R. Sobhani, "Reliability assessment of aged SiR insulators under humidity and pollution conditions," in *International Journal of Electrical Power & Energy Systems*, vol. 117, pp. 105679, May. 2020. DOI: 10.1016/j.ijepes.2019.105679.

[12] S. Deb, R. Ghosh, S. Dutta, S. Dalai, B. Chatterjee. "Effect of humidity on leakage current of a contaminated 11 kV Porcelain Pin Insulator," in *2017 6th International Conference on Computer Applications In Electrical Engineering-Recent Advances (CERA)*, Roorkee, Oct. 2017, pp. 215-219. DOI: 10.1109/CERA.2017.8343329.

[13] M. S. B. Abd Rahman, M. Izadi, M. Z. A. Ab Kadir. "Influence of air humidity and contamination on electrical field of polymer insulator," in *2014 IEEE International Conference on Power and Energy (PECon)*, Kuching, Sarawak, Malaysia, Dec. 2014, pp. 113-118. DOI: 10.1109/PECON.2014.7062424.

[14] M.P. Verma, H. Niklasch, W. Heise, G. F. Luxa, H. Lipken, H. Schreiber. "The criterion for the pollution flashover and its application to insulation dimensioning and control," in *Cigre Report*, 1978.

[15] I. Ramirez, R. Hernandez, G. Montoya. "Measurement of leakage current for monitoring the performance of outdoor insulators in polluted environments," in *IEEE Electrical Insulation Magazine*, 28(4), pp. 29-34, 2012. DOI: 10.1109/MEI.2012.6232007.

[16] D. Maadjoudj, A. Mekhaldi, M. Teguar. "Wavelet transform of the leakage current to monitoring the insulator surface under non-uniform desert pollution," in *2018 International Conference on Electrical Sciences and Technologies in Maghreb (CISTEM)*, Algiers, Oct. 2018, pp. 1-5. DOI: 10.1109/CISTEM.2018.8613424.

[17] A. A. Salem, R. Abd-Rahman, S. A. Al-Gailani, Z. Salam, M. S. Kamarudin, H. Zainuddin, M. F. M. Yousof. "Risk Assessment of Polluted Glass Insulator Using Leakage Current Index Under Different Operating Conditions," in *IEEE Access*, vol. 8, pp. 175827-175839, 2020. DOI: 10.1109/ACCESS.2020.3026136.

[18] D. Fauziah, H. Alfiadi, Rachmawati, Suwamo. "The effect of coating on leakage current characteristic of coast field aged ceramic insulator," in *2017 4th International Conference on Electrical Engineering, Computer Science and Informatics (EECSI)*, Yogyakarta, Sep. 2017, pp. 1-7. DOI: 10.1109/EECSI.2017.8239140.

The Use of WAMS in Solving Major Faults

Rastislav Šmidovič
Slovak Dispatching Centre, Žilina, Slovakia
Slovenská elektrizačná prenosová sústava a.s.
rastislav.smidovic@sepsas.sk

Vladimír Kohan, Zsolt Čonka, Michal Kolcun, Peter Havran, Róbert Štefko
Department of Electric Power Engineering
Technical university of Košice, Slovakia
vladimir.kohan@tuke.sk, zsolt.conka@tuke.sk, michal.kolcun@tuke.sk, peter.havran@tuke.sk, robert.stefko@tuke.sk

Abstract—**The article deals with the issue of large rounding angles in Power System of Slovakia. PMUs are installed in the transmission system, with which we can monitor differences in voltage angles when switching transmission lines. Although there are no shifts in standard operation that would cause rounding problems, this can happen in the event of major outages. This case is documented on a real outage in ES Veľké Kapušany.**

Keywords: reconfiguration; synchronization; phase; restoration grid topology; WAMS

I. Introduction

When the line is switched, circular conditions such as the magnitude of the voltage and its phase shift are checked.

A. Initial Fault Condition

In 3. December 2020, a major breakdown occurred in substation V. Kapušany, when all outlets connected to substation V. Kapušany were switched off due to differential protection. A detailed wiring diagram of electric power station V. Kapušany is shown in Figure 1. The V. Kapušany substation is a border electric power station where there is no 400/110 kV transformation, but where a very important international line is connected to Ukraine.

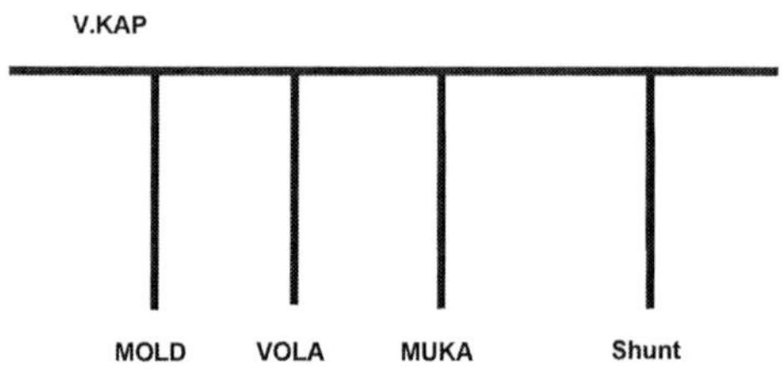

Figure 1. Connection of substation V. Kapušany

Figure 2 shows the connection of part of the TS (Transmission System) and TS MAVIR system around the V. Kapušany substation. During the failure, the Hungarian-Ukrainian line Szabolcsbek-Mukachevo was switch off, according the outage planning.

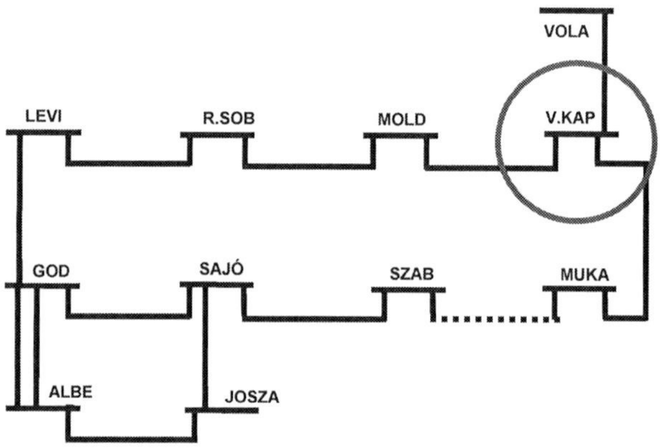

Figure 2. Connection of part of TS SR and TS MAVIR

In the time 9:50 switch off the all lines in the substation V. Kapušany and it case of a large increase amount of transmission power flow especially on the international overhead line LEVI–GOD. Priebeh výkonu na tomto vedení je zobrazené na obrázku 3.

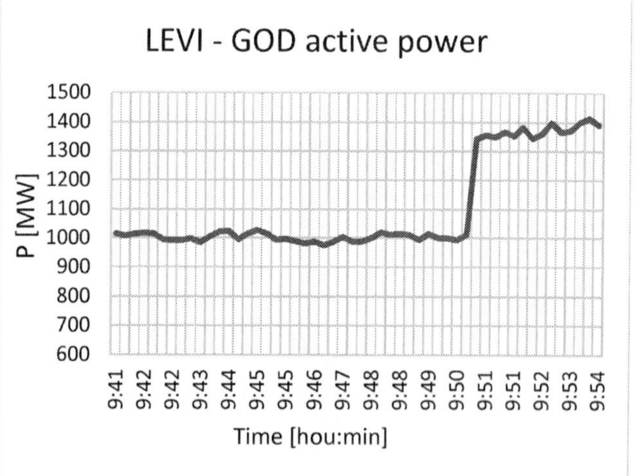

Figure 3. Active power flow on the overhead line LEVI - GOD

As of this point, the security N-1 criterion is not met and in the case of large transmissions on the LEVI - GOD line, it is possible to use reconfiguration in the GOD substation and change the connection according to Figure 4.

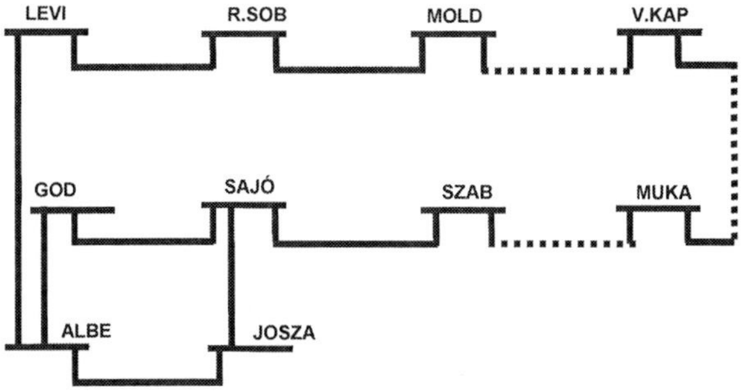

Figure 4. Connection of part of TS SR and TS MAVIR during reconfiguration in GOD substation.

It was help to reduce the active power flow on tie-line LEVI – GOD and the security N-1 criterion was fulfilled again, as shown at the Fig. 5.

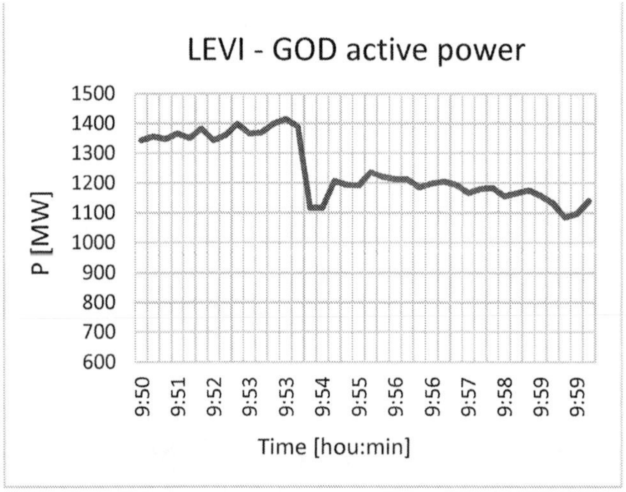

Figure 5. Active power flow on the overhead line LEVI - GOD

II. Utilization of WAMS During Disturbances

A. Current State

The implementation of WAMS in SEPS started in 2011 with a pilot project in which 4 internal lines were monitored with 8 PMUs. In 2020, totally 22 PMUs were installed. One on each interconnector in each crossborder substation (V404, V270, V424, V280, V497, V448, V449, V440, V478, V477, V052) and on both ends of 7 internal lines (V427, V492, V407, V439, V481, V480, V447) Fig. 6.

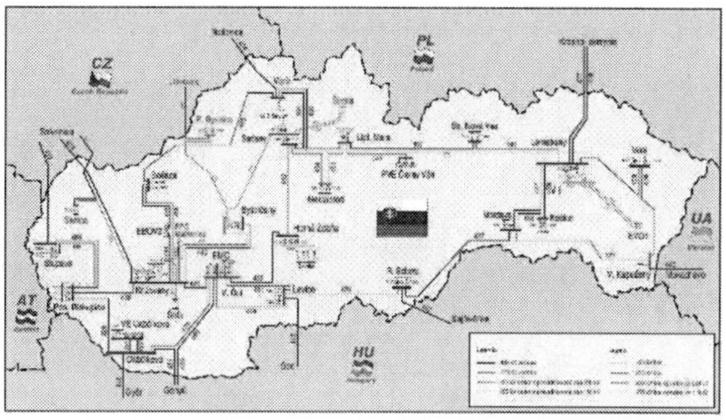

Figure 6. Lines of Power system of Slovakia equipped with PMUs

For measurement Elcom ENA460 PMUs are used. Their output values are Û, Î, P, Q, f, df/dt with a total vector error (TVE) <0,5%. It samples the electric wave with a sampling rate 12 kSa/s which are used to calculate the vectors. Output values are sent to PDC every 20ms i.e. 50/s. Communication is mediation by standard IEEE C37.118-2005. Each PMU device needs 6 phasors (3 voltages and 3 current).

B. Visual Monitoring of Parameters

SEPS has purchased several licenses for monitoring electrical quantities in the electricity system in real-time views. Such as voltages and currents (Positive, Negative or Zero sequence and etc.) or frequency. Using these parameters and the fast-sampling frequency, it is possible to visually observe the operation of the TSO in dynamic dimensions. It is also possible for the dispatchers to react earlier to adverse effects and to operate the power system with higher safety and reliability.

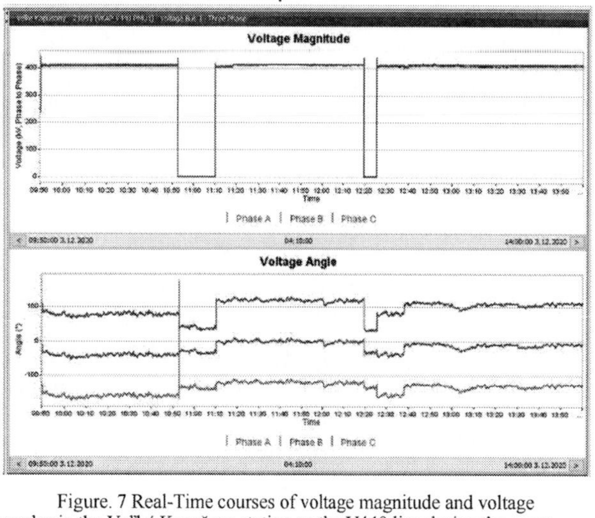

Figure. 7 Real-Time courses of voltage magnitude and voltage angles in the Veľké Kapušany station on the V440 line during the event 03.12.2020

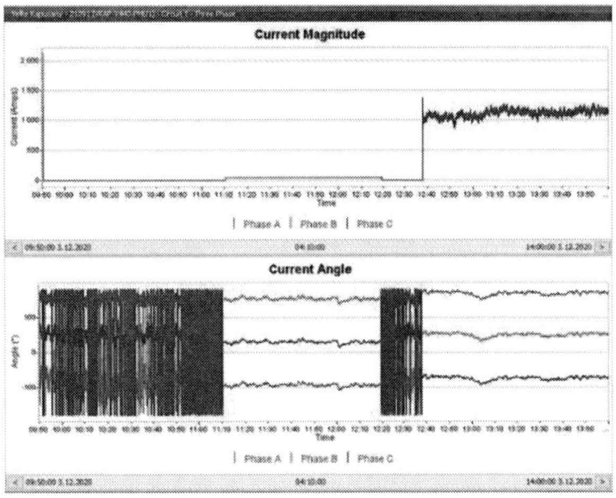

Figure 8 Real-Time courses of current magnitude and current angles in the V. Kapusany station on the V440 line during the event 03.12.2020

For an illustrative example of the size of the voltage angle, it is possible to use a polar graph, which shows the phase shifts of individual measurements throughout in the Power System of Slovakia. This display in a given operation is the best way to get an overview of the situation in the Power System of Slovakia as soon as possible (Figure 9).

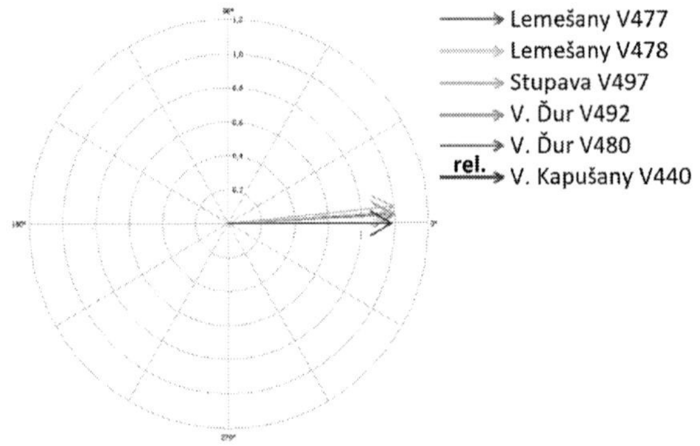

Figure 9 On-line visualization of angle differences in a polar chart

Figure 7 and Figure 8 show the monitoring of the power system by means of PMU units at a given issue (inability to switch on the line due to large angles). With the expanded WAMS systems, SEPS also participated in the implementation of PMU units in individual sub-stations. While the construction and installation of new PMU units is still in the test formation, but in the near future it will be integrated as a supplement to the SCADA system for dispatch control.

III. Real-Time Measurements During the Disturbances

Data was downloaded from the PMU storage and subsequently analyzed to understand the unexpected event being analyzed. After several unsuccessful attempts to turn on the V440 line, it was necessary to solve this problem at the international level (with MAVIR TSO Hungary). Using this data and a retrospective view of the history of measurements from PMU units, it was possible to record the event and then further analyze it and better understand it.

In Fig. 10 is displayed phase angle difference after breakdown in substation V. Kapušany. The phase angle difference between two busbar in substations Moldava and V. Kapusany is about 35 °.

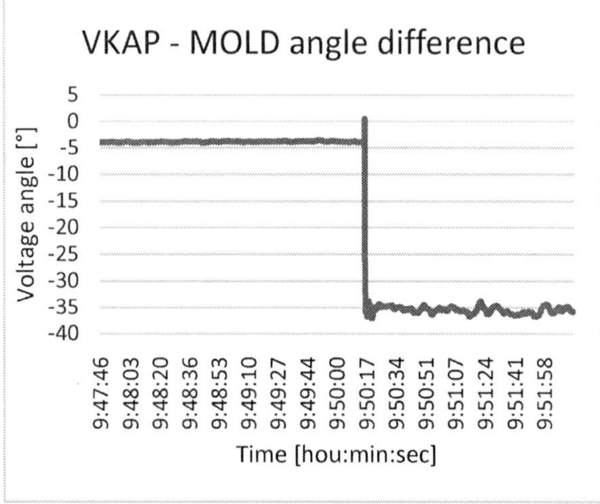

Figure 10. The phase angle difference after the breakdown between the substations V. Kapusany and Moldava

In Fig. 11 is displayed the phase angle difference between two busbar between two busbar in substations Moldava and V. Kapusany after the reconfiguration in substation GOD. The phase angle difference increase by the value about 4°. This reconfiguration was was necessary for keeping the N-1 violation and made the phase angle difference worse.

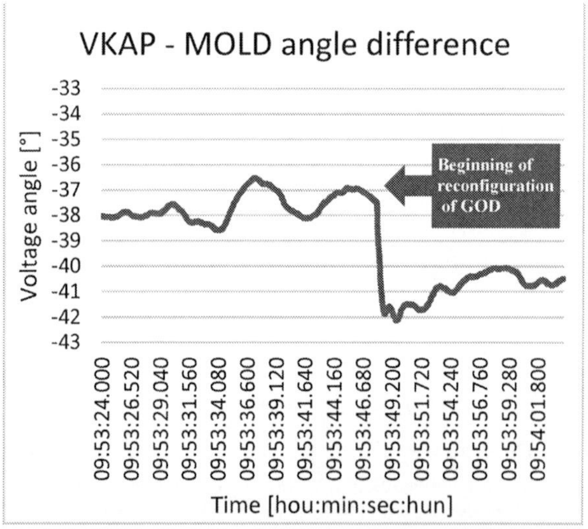

Figure 11. The phase angle difference afterv the start of reconfiguration in substation God (during shut down in substation V. Kapušany)

The phase angle difference increases by value about 2°. It is possible to see that the phase angle difference before the swiching of the circuit breaker is bigger than 30°. The critical condition for safe synchronization is angle less than 30°. To decrease the phase angle difference some pump-storage generation was put into operation but without any significant effect. For successful synchronization it was necessary to wait for the situation when the angle difference on the line dropped below 30°.

IV. Synchrozation of Paralel Lines/Busbars

High voltage diversions between busbar and terminal could cause event similar to a short-circuit followed by activation of protections. Calculation of the maximum phase angle for the synchronization is important for the protection settings and applies specific involvement of the power system. In practice, when the previous synchronization reaches the angle which value is close to the calculated critical value, the dispatcher has to perform actions to reduce this angle [3]. The conditions for synchronization (absolute voltage difference $|\Delta U|$ and angle (φ) are checked at every switching-on the line or transformer to operation.

At present, the synchronization in transmission system recommends the maximum phase value $\Delta\varphi = 30°$, and the difference between the voltage $\Delta U = 10\%$ (valid for 400 kV level). Non-fulfilment of these terms could result into the inrush current, which can result in element disconnection. If the voltage difference or the difference angle phase are greater than these default values, synchronization is not permitted and there is a need to arrange the following conditions:

- Reduction of the voltage difference - is carried out on those facilities, which are close to synchronized node (taps regulation on transformers, excitation regulation of sources, secondary voltage control in the pilot nodes).
- Reduction of the angle difference - is carried out by adjusting the network topology on the side of the transmission or distribution system.

During the disturbance operators try to do some remedial action, like to switch on the pump storage, switch off some internal lines or to choosed diferent sequence of switching internal lines. All actions were without significant effect to reduce a voltage angle and was not possible to switch on the tie-line V. Kapušany – Mukačevo outage.

It was necessary to wait for the scheduled program for power exchange on the profile between Slovakia and Hungary. At time 12:36, it was agreed with the MAVIR dispatcher that the reconfiguration at the GOD substation would be canceled. This step was very dangerous, because there was a large increase of power flow on the LEVI - GOD overhead line and a subsequent breach of the security N-1 criterion. The operators observed the size of the angle and waited for the decrease of this parameter and gave an instruction to rounding of the V. Kapušany - Mukachevo overhead line into operation. This was achieved at 12:38. The curve of active power on the overhead line LEVI - GOD is shown in Figure. 12.

The angle difference curve on the Levice – GOD overhead line are shown in Figure no. 12.

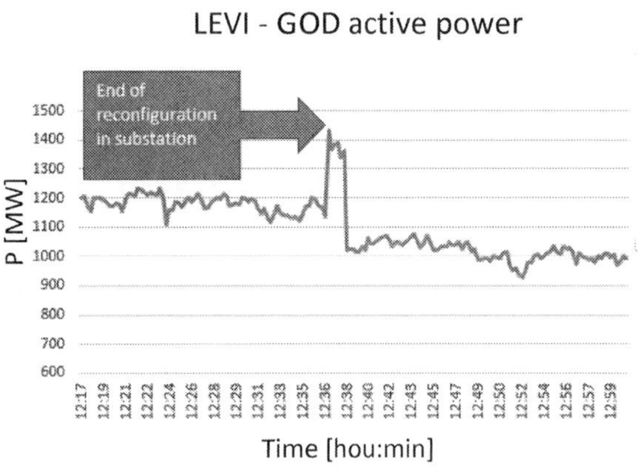

Figure 12. Active power curves on the LEVI – GOD overhead line

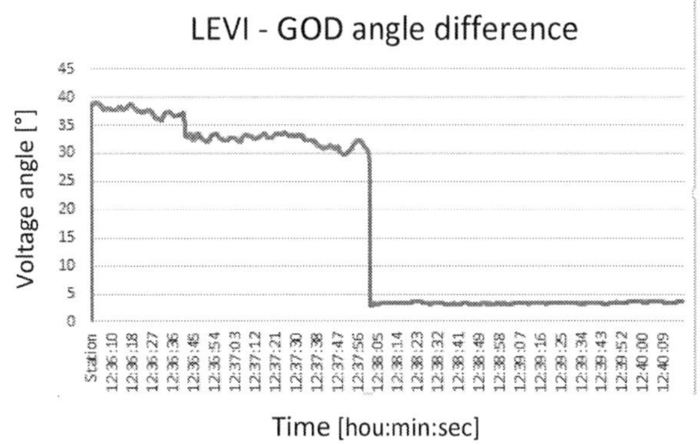

Figure 13. Voltage angle curve on the LEVI - GOD overhead line during commissioning

Figure 13 shows the angle curves before being switched on. It is possible to see that, the rounding conditions were met at the limit values and not turning on this overhead line would endanger the security of the entire power transmission system.

V. Conclusions

In the event of large network outages or disturbances, there may be situations where, even though these outages have already been resolved and the equipment has been repaired, it is not possible to put them into operation due to unsatisfied rounding conditions.

The WAMS technology seems to be promising tool for future analyses of reconfiguration activation. Implementation of WAMS to SCADA system, network analyses and awareness system would have added value to system operation and dispatcher's decision in critical situations.

For those interested in this issue, we recommend the article [4], which deals with the use of the WAMS tool in network reconfigurations, and the work [5], which deals with the use of WAMS in the detection of oscillations in the system.

Acknowledgment

This work was supported by the Slovak Research and Development Agency under the contracts No. APVV-19-0576 and APVV-21-0312 and by the Ministry of Education, Science, Research and Sport of the Slovak Republic and the Slovak Academy of Sciences under the contract VEGA 1/0757/21.

REFERENCES

[1] M. Jedinák, S. Prieložný, R. Šmidovič. "Topology changes in the transmission system of the Slovak Republic," in *10th International Conference, Control of power systems 2012* Tatranské Matliare, Slovak Republic, 2012, pp. 1-9.

[2] R. Šmidovič, M. Novák. "Kruhovanie elektrických sietí," in *Elektroenergetika*, Stará Lesná, 2005, pp. 442-448. ISBN: 978-80-553-0399-4.

[3] J. Altus, M. Jedinák, S. Prieložný, R. Šmidovič. "Reconfiguration in the Transmission System of the Slovak Republic," in *Journal of Energy and Power Engineering,* 2013, pp. 577-582.

[4] M. Grega, M. Jedinák, R. Šmidovič. "Synchronization Options during Restoration of Base Case Topology in the Transmission System by using WAMS system," in *Cigre*, 2016, IFAC.

[5] M. Grega. "Oscilácie v elektrizačnej sústave – detekcia, analýza a hodnotenie," Disertattion work, 2021, STU Bratislava.

Case Study of the Design of Renewable Energy Sources for Microgrid Systems

Róbert Štefko, Ľuboš Šárpataky, Miloš Šárpataky, Ľubomír Beňa, Zsolt Čonka, Jaroslav Džmura, Michal Kolcun and Marek Pavlík.
Department of electric power engineering
Technical university of Košice
Košice, Slovakia
robert.stefko@tuke.sk, lubos.sarpataky@tuke.sk, milos.sarpataky@tuke.sk, lubomir.bena@tuke.sk, zsolt.conka@tuke.sk, jaroslav.dzmura@tuke.sk, michal.kolcun@tuke.sk and marek.pavlik@tuke.sk

Abstract—**Advances in microgrid research are advancing significantly, and currently a few microgrid systems have already been created worldwide and their progress in construction is improving every year. For the design of power sources for a microgrid, it is necessary to know in as much detail as possible the dynamic load variations, environmental weather effects, technical and quality parameters for such a system. However, considerable care must be taken in the design to select an appropriate mix of power sources, and considerable consideration must be given to the site conditions of the microgrid under construction, as such a concept is not always contemplated. It is necessary to draw attention to this issue when designing microgrids in the future. The article deals with the selection of a suitable location for the creation of a microgrid and evaluates the conditions in Slovakia. The article next address the design of a universal mix of prospective sources that can be considered in Slovakia.**

Keywords: microgrid; design; power source; weather effects; environment; construction; research

I. Introduction

The key feature of the microgrid system is the smooth island operation during the transition from the distribution network and the ability to auto control in the island mode. In the event of a failure in the distribution network, local microgrids can increase the reliability of the energy system by switching to island mode, because local renewable energy sources (RES) continue supplying the microgrid [1]. Alternatively, we can consider the opposite case if a failure occurs in one of the microgrids, while the distribution network will be disconnected, only the affected part of the microgrid will be disconnected, where local fault management and local stabilization for the failing microgrid will take place. Precisely because of the maintenance of island operation, the selection of suitable local sources is crucial for the stable maintenance of operational and quality parameters. Currently, some of the most popular sources found in microgrid systems obtain solar systems (PV), battery storage systems (BESS), fuel cells, biogas and biomass plants, thermal power, hydropower, non-traditional electricity storage systems, and wind farms. Instability and dependence on weather conditions are fundamental problems in the use of some of the sources, for some of these sources diesel generators (DG), gas turbines or fuel cells are in practice used as backup sources. For this reason, efforts are still entailed to research new energy sources that are less weather-dependent and operate more efficiently than current ones, recent research has shown that non-traditional energy sources and electricity storage systems are promising. Naturally, there is still a need to develop battery systems for their efficient use in the energy sector, not forgetting the recycling of these battery systems and their reuse.

II. Microgrid

Research shows microgrid systems will represent a key part of building smart grids, and this major part also provides solutions for powering remote communities, industrial facilities, military bases that are usually connected to the grid so that they can disconnect and operate independently [2]. The on-grid operation mode obtains a battery-free storage system connected to the public distribution power grid, but the off-grid mode contains a battery storage system with backup power sources. All the energy produced, which cannot be used in the microgrid in on-grid mode is provided to the power grid, otherwise, the least used power sources will be disconnected to avoid consuming fuel. The stand-alone operating mode can be disconnected from the grid to operate autonomously, ensuring continuity of power supply in the case of an outage. This approach is common in locations where the grid can be unreliable, like storm-prone or mountainous areas [2] - [8]. A microgrid can be defined as a small local distribution network that supplies consumers with electricity, which generates electricity using distributed energy resources. Such grids need to satisfy the assumption of self-sufficiency in electricity generation, i.e., such grids need to contain an appropriate mix of energy sources according to the geographical location to provide sufficient generation for the daily load pattern [4], [5]. The European Union research project defines microgrids in its terms as follows: „Microgrids include low-voltage distribution systems with distributed energy resources, energy storage devices, energy storage systems and flexible loads. Such systems may operate either connected or disconnected from the main grid“ [9]. Depending on the use and location of the distributed generation resources and how the microgrid is managed, they can be divided into the following categories:

1. Institutional or Campus Microgrids

The focus of institutional or campus microgrids aim to accumulate generation resources at a site with multiple loads located near a similar geographic location so that the system can easily monitor and control [9]. The deployment of combined cold, heat, and power generation with multiple loads located on campus, or at an institution, has obtained a successful model to date and typically includes the largest microgrids to date, ranging in capacity from 4 to over 40 MW [11].

2. Military Microgrids

Cost-effective energy security to provide reliable power and fuel for critical equipment essential for their use to protect and secure critical operations during extended outages of the local power grid or during operational field deployments remain a leading driver for military microgrids [11].

3. Residential, Commercial, and Industrial Microgrids

In many industrial and manufacturing processes, short-term interruptions in electricity supply can cause acute economic losses to industries. Therefore, industries build microgrids according to their requirements with the necessary standards to avoid power interruption and eliminate economic losses [9].

The schematic of the microgrid system shown in Fig. 1 has already been practically tested on island operation and is currently working in a few places. One such system supplies electricity to the remote north of the Arctic, with approximately 160 members of Lake Colville located 50 km north of the Arctic Circle in Northern Canada. The peak load is 150 kW, and the fundamental load is 30 kW, 136 kW solar panels were installed locally, generating approximately 112 MWh per year representing about half of the community's total annual consumption, together with a 2 × 100 kW and 1 × 150 kW diesel generators complemented by a 200 kWh battery energy storage system (BESS) [2]. As shown Fig. 1. this initial system for island-only microgrid operation has been supplemented with grid connectivity for broader deployment options and uses a phasor measurement unit (PMU) for power plant monitoring needs.

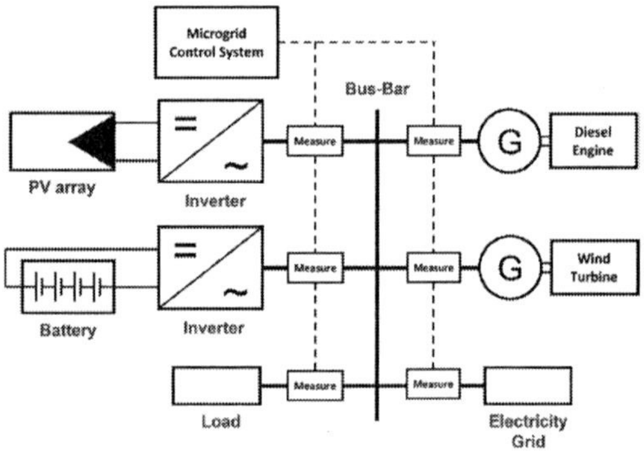

Fig. 1 Schematic of the microgrid system [2]

III. Weather Conditions on the Territory of the Slovak Republic

The territory of Slovakia has a diversity of renewable energy sources, and this chapter will be devoted to their evaluation and their future use for the construction of microgrids. Fig. 2 shows the potential sites for the place of the solar system, where the most excellent prospect is for the southern part of Slovakia, where the annual average global tilted irradiation at an optimum angle is above 1500 kWh/m^2 per year. Equally a positive for this location is the vast Danube Lowland, which does not influence the reduction of solar irradiation time, unlike the mountains in the northern areas. Which favors the location for the deployment of solar systems for areas from Komárno to Nitra. The potential place is moreover in the vicinity of Poprad, which is a mountainous area with an increased occurrence of storms and clouds. Nevertheless, the valley offers the most suitable location among all mountain areas for the northern part of Slovakia with annual average global tilted irradiation at an optimum angle (GTI opta) is about 1400 kWh/m^2 per year. The least promising place for deployment of solar systems is in the north of Slovakia around Žilina, Martin and Sabinov (GTI opta is about 1300 kWh/m^2 per year). When comparing Fig. 2 for potentially suitable locations in Slovakia with the already deployed solar systems from Fig. 3, it is possible to see a significantly higher number of DER in central Slovakia, which is also confirmed by the overview in Tab. 1 [12].

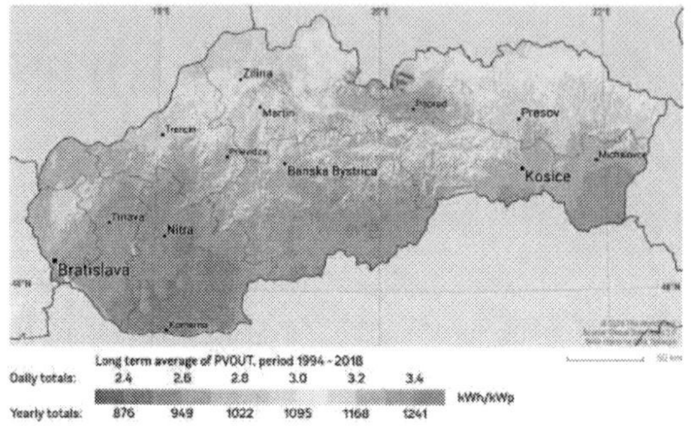

Fig. 2 Solar power potentials of the Slovak Republic [9]

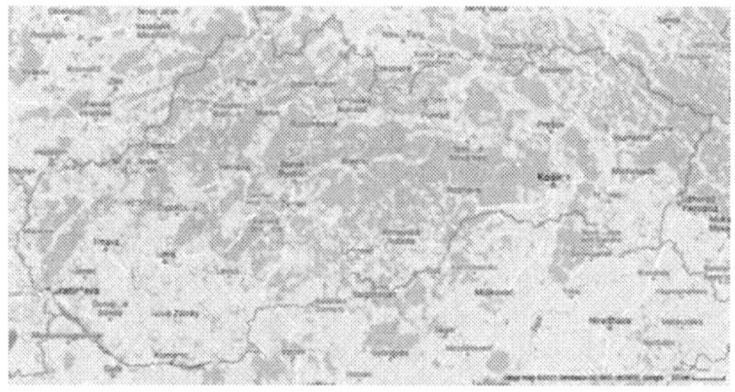

Fig. 3 Deployed solar systems in Slovakia [13]

Tab. 1 Division of power plants in Slovakia [2][14]

Energy Source	Division of the territory of the Slovak Republic		
	Western Slovakia	***Middle Slovakia***	***Eastern Slovakia***
Solar	104.28 MW	240.16 MW	133.9 MW
Hydro	1166.2 MW	1289.98 MW	111.19 MW
Biomass	43.67 MW	56.78 MW	52.39MW
Biofuel	44.41 MW	33.65 MW	24.82 MW
Wind	3.14 MW	0 MW	0 MW
Nuclear	1950 MW	0 MW	0 MW
Other RES	4.66 MW	0 MW	0 MW

The article only considered irradiance comparison, where it is assumed that for mountainous areas, especially in the winter months, there will be significantly lower production relative to the lowlands due to increased cloud cover and slower passage of clouds.

When comparing **Chyba! Nenašiel sa žiaden zdroj odkazov.**, which shows the river map of Slovakia, with Fig. 5, which shows the currently located hydropower plants, the comparison shows Slovakia still has the potential for the completion of additional hydropower plants, for example on the Laborec, Ondava, or Topľa River. The comparison of the even distribution of energy resources in Tab. 1 shows a significant scarcity of resources in eastern Slovakia. This will lead to the problematic design of microgrids due to a much smaller number of power plants and thus a larger initial investment.

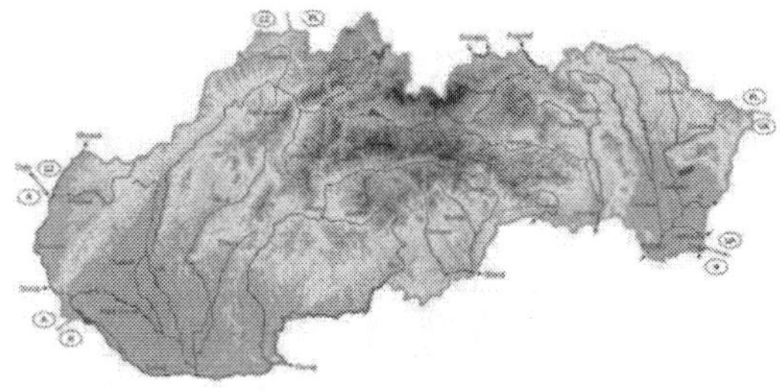

Fig. 4 River map of Slovakia [15]

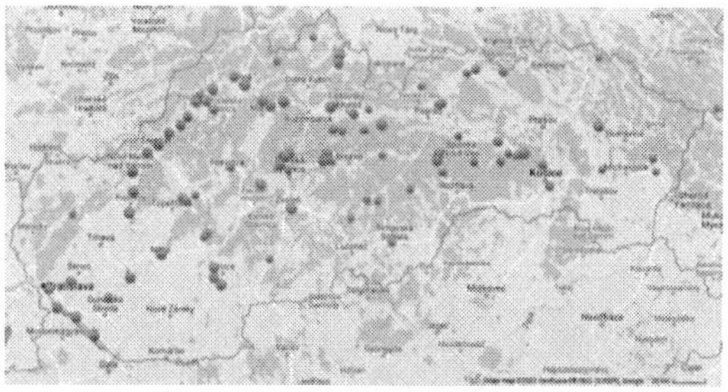

Fig. 5 Deployed Hydropower plant in Slovakia [13]

Renewable energy sources, such as biomass and biogas or biofuel, have seen a significant increase, especially in western Slovakia, which contains a significant number of small sources, while the power distribution of these sources is even in Slovakia as shown in Fig. 6. However, with support for the development of this technology, the output from these sources could be higher in the future. For this reason, further developments are needed in this area as well. The primary disadvantage of this type of renewable energy source in the efficient use of biomass attends the need for a vast area for growing suitable fuel, while biofuel requires a constant supply of biological waste, which is provided mainly by animal cooperatives, whose incidence in Slovakia is declining in recent years.

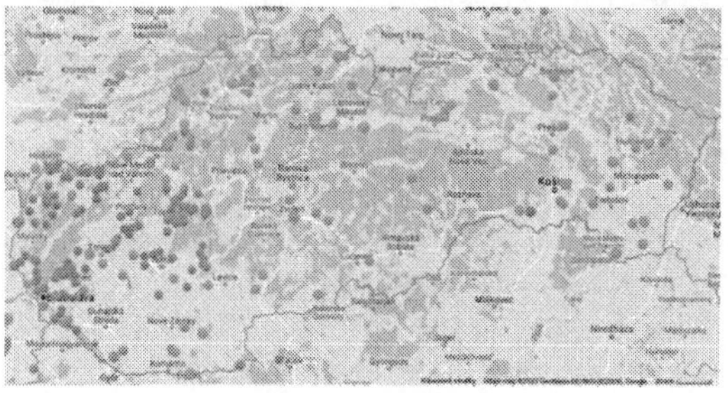

Fig. 6 Deployed Biomass and Biofuel power plants in Slovakia [13]

As shown in Fig. 7 for wind potential in Slovakia, the wind speed for vacancies that do not fall into the protected area is based on an average speed of 2.07 - 3.5 m/s, which clearly indicates the territory of Slovakia does not approach the type of this renewable source. The unique alternative for the use of wind energy is microturbines, which would be installed on the roofs of buildings, for which the required wind speed is exceedingly lower than for large turbines. However, the disadvantage of this type of source is the generation of terrible noise, which is caused by the rotation of the turbine blades, while maintaining the specified distances for the location of wind parks from the airports, the number of suitable places is considerably reduced.

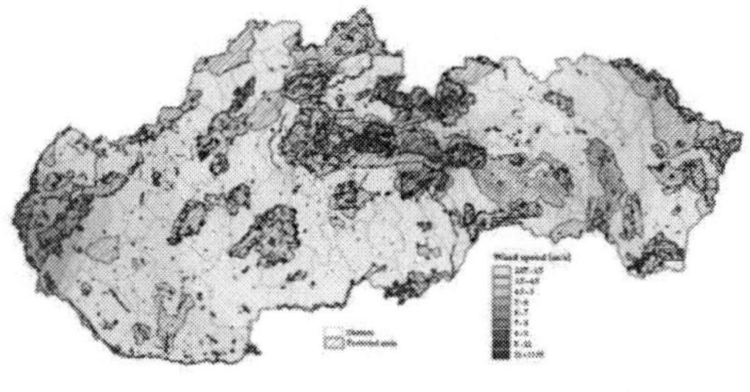

Fig. 7 Wind potential and protected areas in Slovakia [16]

IV. Calculating of the Size of the Power Source Mix for the Microgrid

The calculation of the energy mix of resources must consider the location of the microgrid, the size of the peak load (PL), the size of the minimum load and the quality parameters that will be provided by the locally distributed resources. As an example, we can consider the Off-Grid microgrid from Chapter 2, where the peak load was 150 kW and the basic (minimum) load 30 kW. In this proposal, it is necessary to consider the fact that such an Off-Grid microgrid was designed for 150 people, and after recalculation to a vaster population, the plan will be more complex, but it is appropriate to start including local power plants, which will perform the decrease investment costs. The charging of the battery system (BESSCHAR) should be unforgotten in the calculation. We consider the possibility of charging only in times of power surplus under the condition of minimum fuel costs for the diesel generator.

$$P_{PL} = P_{PV}/X + P_{DG}/Y + P_{BESS}/Z \quad (1)$$

Or

$$3.8{*}P_{PL} < P_{PV} + P_{DG} + P_{BESS} \quad (2)$$

Where P_{PL} is peak power load, P_{PV} is photovoltaic power source, P_{DG} is power of diesel generator, P_{BESS} is power of battery energy storage system and X, Y and Z are proportion constants for determining the output of individual local sources.

Formula (1) describes, in general, the condition for dimensioning the performance of local resources. While relation (1) can also be expressed according to formula (2).

We consider the possibility of charging the battery system only at the time of minimum load, under the condition of minimizing the fuel costs incurred for diesel generators or biomass and biofuel power plants. We would consider charging the battery system with only one-sixth of the charging power at the time of charging described by equation (3).

$$P_{LOAD} + P_{BESSCHAR}/6 < P_{PV}/X + P_{DG}/Y + P_{BESS}/Z \quad (3)$$

Where P_{LOAD} is basic or minimum load, P_{PV} is photovoltaic power source, P_{DG} is power of diesel generator, P_{BESS} is power of battery energy storage system, $P_{BESSCHAR}$ is charging battery storage system and and X, Y and Z are proportion constants for determining the output of individual local sources.

When designing and using a battery system, we strive for the shallowest possible cycles, thus achieving extended battery system life, and thus the total battery capacity should not fall below half of the battery system capacity. Further, the size that such a microgrid will manage with the help of distributed energy sources, the size, and type of which will ensure a major influence to establish sufficient power, will be decisive in the design of resources. One of the core parts of research in the world is, in addition to the organization of resources, the determination of a suitable amount whose stability and quality limits will be sustainable in island operations. To speed up and facilitate the design and construction process, existing power plants should also be considered. This fact will equally affect the initial investment in the construction of the microgrid, which will not be completely dependent only on the newly installed supplies but can additionally utilize older resources.

Tab. 2 The possibilities of using individual resources show the use of individual resources for the built microgrid, while it will be affected not only by the consumption but also by the weather, fuel reserves, and the season. The individual results show twelve possibilities of the power distribution of the considered resources for the proposed microgrid while the individual share constants for equation (1) must be equal to the peak power load after the calculation. In other words, Tab. 2 shows the utilization of individual sources in different ratios to determine their power optimum. The results show a high dependence on the solar system for this reason other renewable sources need to be added. The addition of other renewable sources will improve reliability.

Tab. 2 The possibilities of using individual resources

No.	*X*	*Y*	*Z*
1.	1	50	25
2.	1	42	30
3.	1	40	32
4.	1	35	40
5.	1	30	60
6.	1	28	80
7.	1	26	130
8.	1	25	200
9.	2	5	16
10.	3	70	2
11.	6	4	5
12.	9	10	2

One of the promising areas in Slovakia for the creation of a microgrid keeps the region in the Trenčín Region, which could consist of four municipalities Dubnica nad Váhom, Nemšová, Nová Dubnica and Trenčianska Teplá as shown in Fig. 10. The positivity of this location is indicated by the absence of a solar system or a diesel generator, whose functionality we try to constrain according to the individual constants of Tab. 2.

The other sources available in the region do not need to be built anymore, so it is sufficient if we add a solar system to individual houses. Where the installed current power of local sources reaches 46 kW for the solar system, 16.5 MW hydroelectric power plant, and 10.878 MW for biomass. Although this area does not have a medium solar system, considering its geographical location, it is still a relatively lucrative place to install this resource, while also providing diversification of local resources for island operations. The solar potential of the region reaches 3.805 kWh/m2 per day at an optimum angle and the annual average global tilted irradiation at optimum angle is about 1390 kWh/m2 per year with a system installed capacity of 46kWp. The estimated production would be about 50 MWh at an average annual irradiance of 1384.6 kWh/m2 per year. Fig. 8 shows the average monthly output for each month [12].

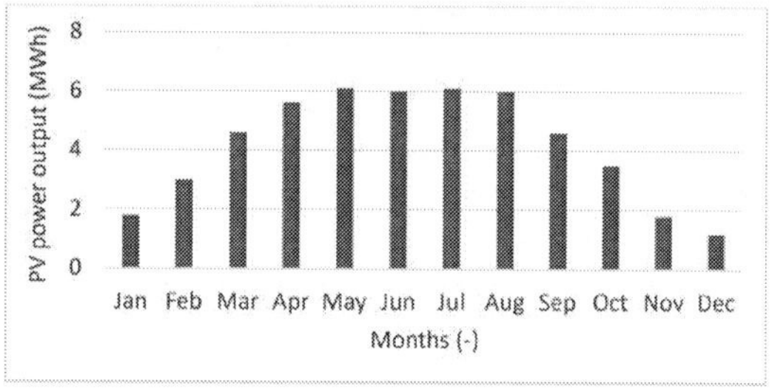

Fig. 8 Considered area for the creation of a microgrid in western Slovakia [12]

We can supplement the solar system with a battery system to compensate for dynamic changes in weather or load and gain storage options for excess energy.

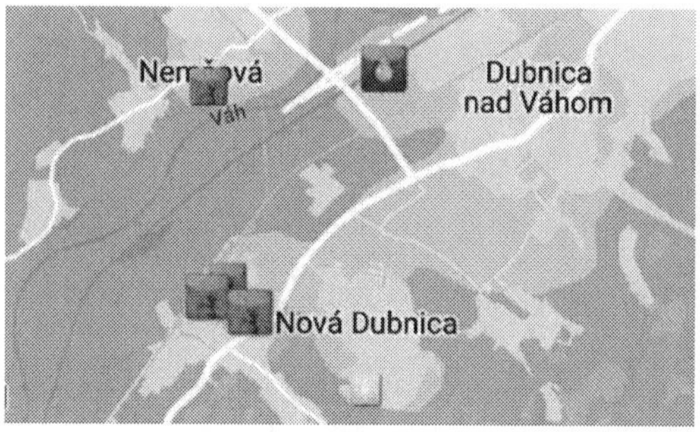

Fig. 9 Considered area for the creation of a microgrid in western Slovakia [13]

V. Battery Storage Systems

One of the most widespread ways of conserving electric power is battery storage systems, the primary disadvantage of which remains the possibility of recycling and, from the point of view of the design of dimensioning sufficient capacity, which, despite the constantly declining price of electromobility, represents a massive investment. Since 2019, Slovak energy legislation has allowed the installation of so-called local sources of electricity up to 500 kWp or up to the maximum amount of reserved capacity, whichever is lower. Another task is that all the electricity produced must be consumed within its own waste, i.e., at the point of consumption [17].

In the territory of Slovakia, there are, moreover, several applications of battery systems in more superior capacity in cooperation with renewable energy sources, but most often in connection with solar systems as shown in Fig. 10. However, this is not the most suitable option for residential use, as we can utilize this excess energy more efficiently in conjunction with a heat pump, which will significantly reduce energy costs in the industrial sector. It is such a system that is beginning to expand in the industrial sector also due to rising energy prices, like the plant in Humenné, which operates a solar system with an output of 500 kW and a 432 kWh battery system. Smart battery management is designed to ensure that sustainable energy is used as efficiently as possible and without unnecessary losses. Implemented artificial intelligence charges and discharges the battery according to the needs of the plant, thus saving funds associated with the supply of energy and thus accelerating the return on investment [17] - [19].

Fig. 10 Map of the location of large-capacity battery systems in Slovakia

The same large 432 kWh smart battery system is in the headquarters of the transport company in Senec. It is one of the first large-capacity battery systems installed in Slovakia [20].

The East Slovak distribution company was the only one to use and design the battery system sensitively and unconventionally to increase reliability in the region in which this battery system was located, while the proposal also considered the impact on fauna and flora. The construction additionally included the replacement of overhead lines with cable ones, which equally contributed to increased reliability, as it is a mountainous and prone area to weather conditions. In comparison, the cost of the air duct is three times lower than that of the cable duct. The battery system has a capacity of more than 1200 kWh and can supply power to the system in the range of up to 600 kW, lasting at least two hours, which should cover the duration of peak load [21].

The company Zapadoslovenská energetika plans to build the largest battery storage facility in Slovakia so far. It plans to place this system in the village of Nižná in the district of Piešťany for an installed capacity of 35 MW and a capacity of 35 MWh with construction from 2024. The estimated investment will be in the range of 24-32 million euros [22].

VI. Non-Traditional Energy Sources Using Gravity

All the sources considered in the design of the power supply of the micro-network have their advantages and disadvantage, for this reason. It is necessary to considerably develop electricity sources and streamline already used energy sources. However, research is advancing and proposing alternative options for electricity generation and storage, which moves the application of microgrids and smart grids closer to a significant deployment.

The ancient's way of conserving electricity is by means of pumped-storage hydroelectricity, where water can store excess electricity by gravity. This method is still used today, but it conducts a certain limitation in construction due to the large height difference of tanks. Looking at the same use of gravity and water, a modern system was designed by the startup, which no longer depends on the height difference as shown in Fig. 11. This modern system was based on the hydraulic lifting of a large massif using several water pumps. The rock gains potential energy through water and can release it when discharged back through the turbine. The system has 80 % efficiency and its usable capacity ranges from 1-8 GWh [23].

Fig. 11 New type of pumped-storage hydroelectricity [23]

Another way of exploiting gravity was developed by Energy Vault, a Swiss-American startup that reclaims 35-ton concrete bricks to construct a tower. At a time of excess electricity, it accumulates the potential energy of the massif. At a time of shortage, he operates cranes to lower the separate bricks to the ground, thus supplying electricity [24].

Acknowledgment

This research was funded by the Slovak Research and Development Agency [APVV-19-0576] and [APVV-21-0312] and the Ministry of Education, Science, Research and Sport of the Slovak Republic and the Slovak Academy of Sciences [VEGA 1/0757/21].

References

[1] R. Štefko, Z. Čonka, J. Kurimský, M. Kolcun. "Problems of protection of industrial networks with a high share of renewable energy sources," in *2020 IEEE 3rd International Conference and Workshop in Óbuda on Electrical and Power Engineering (CANDO-EPE)*, 2020, pp. 79-84. DOI: 10.1109/CANDO-EPE51100.2020.9337771.

[2] R. Štefko, Z. Čonka, M. Kolcun. "Case Study of Power Plants in the Slovak Republic and Construction of Microgrid and Smart Grid," in *Applied Sciences*. 2021, vol. 11, pp. 1-22. DOI: 10.3390/app11115252.

[3] C. Dou, B. Liu. "Multi-Agent Based Hierarchical Hybrid Control for Smart Microgrid," in *IEEE Transactions on Smart Grid*, 4(2), pp. 771-778, June 2013. ISSN: 1949-3061, DOI: 10.1109/TSG.2012.2230197.

[4] E. Entchev, L. Yang, M. Ghorab, E.J. Lee. "Simulation of hybrid renewable microgeneration systems in load sharing applications," in *Energy*, vol. 50, pp. 252-261, 2013. ISSN: 0360-5442, DOI: 10.1016/j.energy.2012.11.046.

[5] A.A. Moghaddam, A. Seifi, T. Niknam, M.R.A. Pahlavani. " Multi-objective operation management of a renewable MG (micro-grid) with back-up micro-turbine/fuel cell/battery hybrid power source," in Energy, vol. 36, no. 11, pp. 6490-6507, 2011, issn: 0360-5442, doi: 10.1016/j.energy.2011.09.017.

[6] A. Milo, H. Gaztanaga, I. Etxeberria-Otadui, E. Bilbao, P. Rodriguez. "Optimization of an experimental hybrid microgrid operation: Reliability and economic issues," in *2009 IEEE Bucharest PowerTech*, 2009, pp. 1-6. ISBN: 978-1-4244-2234-0, DOI: 10.1109/PTC.2009.5282176.

[7] Z. Dongmei, Z. Nan, L. Yanhua. "Micro-grid connected/islanding operation based on wind and PV hybrid power system," in *IEEE PES Innovative Smart Grid Technologies*, 2012, pp. 1-6. ISSN: 2378-8542, ISBN: 978-1-4673-1220-2, DOI: 10.1109/ISGT-Asia.2012.6303168.

[8] T.C. Ou. "A novel unsymmetrical faults analysis for microgrid distribution systems," in *International Journal of Electrical Power & Energy Systems*, 43(1), 2012, pp. 1017-1024. ISSN: 1017-1024, DOI: 10.1016/j.ijepes.2012.05.012.

[9] Solargis. "Global Solar Atlas 2.0." Internet: www.solargis.com/maps-and-gis-data/download/slovakia, 2019 [Mar. 4, 2021].

[10] G.B. Narejo, B. Acharya, R.S.S. Singh, F. Newagy. "Microgrids Design, Challenges, and Prospects," in *CRC Press*, Boca Raton, London, New York, 2022, pp. 301. ISBN: 978-0-367-48795-9.

[11] A. Hirsch, Y. Parag, J. Guerrero. "Microgrids: A review of technologies, key drivers, and outstanding issues," in *Elsevier: Renewable and Sustainable Energy Reviews*, 2018, vol. 90, pp. 402-411. ISSN: 1364-0321, DOI: 10.1016/j.rser.2018.03.040.

[12] Solargis. "Global Solar Atlas." Internet: www.globalsolaratlas.info/map, 2022 [May 26, 2022].

[13] Repowermap. "Renewable energies and energy efficiency in your neighbourhood." Internet: www.repowermap.org, 2022 [Feb. 15, 2022].

[14] Ministry of Economy of the Slovak Republic. "Fotovoltaika." Internet: www.enviroportal.sk/fotovoltika, 2022 [Apr. 5, 2022].

[15] Reny travel. "Slovak rivers." Internet: http://reny.sk/slovensko-rieky.htm, 2022 [Feb. 15, 2022].

[16] J. Hofierka. "Potential for renewable energy in Slovakia." Internet: www.web.science.upjs.sk/hofierka/projekty/vega_results.html, 2008 [Apr. 6, 2022].

[17] R. Potočár. "Energiu zo slnka ukladá smart batéria. Závod v Humennom využíva lokálny zdroj „na prenájom"," Internet: www.energie-portal.sk/Dokument/energiu-zo-slnka-uklada-smart-bateria-zavod-v-humennom-vyuziva-lokalny-zdroj-na-prenajom-106508.aspx, Oct. 21, 2020 [Mar. 31, 2022].

[18] E. Almašiová. "Slovenské elektrárne purchased 3,24 MWh of battery storage systems brAIn," Internet: www.fuergy.com/blog/slovenske-elektrarne-purchased-battery-storage-brain, Mar. 10, 2022 [Mar. 31, 2022].

[19] R. Istók, P. Kádár, L. Reizer. "Solar Carport with Energy Storage Capabilities," in *2021 IEEE 4th International Conference and Workshop Óbuda on Electrical and Power Engineering (CANDO-EPE)*, 2021, pp. 89-94. ISBN: 978-1-6654-2023-5, DOI: 10.1109/CANDO-EPE54223.2021.9667901.

[20] E. Almašiová. "We have implemented the biggest battery storage in Slovakia!" Internet: www.fuergy.com/blog/the-biggest-battery-storage-in-slovakia, Sep. 16, 2020 [Mar. 31, 2022].

[21] M. Rojko. "Východoslovenská distribučná inštaluje batériové úložisko pre Spišskú Maguru." Internet: www.fuergy.com/blog/the-biggest-battery-storage-in-slovakia, Mar. 9, 2021 [Mar. 31, 2022].

[22] M. Šimovič. "Elektrárne predstavujú svoju budúcnosť. Po celom Slovensku nainštalujú batériové úložiská." Internet: www.fontech.startitup.sk/elektrarne-predstavuju-svoju-buducnost-po-celom-slovensku-nainstaluju-bateriove-uloziska/, Mar. 11, 2022 [Mar. 31, 2022].

[23] Gravity Storage. "A new solution for large scale energy storage," Internet: https://heindl-energy.com/, 2021 [Apr. 7, 2022).

[24] J. ST. John. "Energy Vault Lands $110M From SoftBank's Vision Fund for Gravity Storage." Internet: www.greentechmedia.com/articles/read/energy-vault-lands-110m-from-softbanks-vision-fund-for-gravity-energy-stora, Aug. 15, 2019 [Apr. 7, 2022].

Comparative study of dielectric properties of mineral oil, synthetic ester and natural ester

Miloš Šárpataky
Department of Electric Power Engineering
FEI TUKE
Košice, Slovakia
milos.sarpataky@tuke.sk

Michal Rajňák
Department of Electric Power Engineering
FEI TUKE and UEF SAV
Košice, Slovakia
michal.rajnak@tuke.sk

Juraj Kurimský
Department of Electric Power Engineering
FEI TUKE
Košice, Slovakia
juraj.kurimsky@tuke.sk

Michal Krbal
Department of Electrical Power Engineering
FEEC VUT
Brno, Czech Republic
krbal@vut.cz

Luděk Pelikán
Department of Electrical Power Engineering
FEEC VUT
Brno, Czech Republic
Ludek.Pelikan@vut.cz

Abstract — **The current research of insulating oils for high-voltage applications focuses mainly on biodegradable oils that are in line with sustainable development. Mineral oils may in the future be replaced by more environmentally friendly materials with similar properties. One of the probable options as substitutes in the future will be synthetic or natural esters. This paper presents a comparative experimental study of mineral oil, natural ester, and synthetic ester and their dielectric properties, specifically AC breakdown voltage, lightning breakdown voltage, and thermal dependence of dissipation factor, volume resistivity, and relative permittivity. Measurements of AC and lightning breakdown voltage were performed according to IEC 60156 with a horizontally positioned sphere–sphere electrode system. Synthetic ester reached the mean value of AC breakdown voltage 60.2 kV, lightning breakdown voltage with positive polarity 144.3 kV, and with negative polarity 148.6 kV. Natural ester showed a mean value of AC breakdown voltage 58.4 kV, the lightning breakdown voltage of positive and negative polarity was 149.5 kV and 151.5 kV, respectively. Measurement of dissipation factor, volume resistivity, and relative permittivity was executed in accordance with IEC 60247 2004. The dissipation factor moved from 0.0031 at 20°C to 0.3119 at 140°C for synthetic ester and from 0.0011 at 20°C to 0.0331 at 140°C for natural ester. These results are compared with the dielectric properties of mineral oil. The challenging problem of potential oil substitution in power transformers is discussed.**

Keywords: breakdown voltage, natural ester, synthetic ester, mineral oil, insulating oils

I. Introduction

Insulating oils are used mainly in power transformers but also in other high voltage equipment such as cables, circuit breakers, reactors, power capacitors, tap changers and bushings [1], [2]. These insulating oils are expected to meet the requirements for dielectric properties and efficient heat transfer. In recent years, more and more emphasis has been placed on environmental aspects, bringing biodegradable oils to the forefront, especially natural esters (NE) and synthetic esters (SE), which can eventually replace the currently used insulating oils, especially mineral oils (MO) [3]. Biodegradability of MO is around 30 % while oils with similar dielectric and heat transfer properties reached biodegradability up to 80 % for SE and up to 99 % for NE [4], [5]. One of the properties that speak for a synthetic and natural ester is safety connected with high fire and flash point and with potential leakage that may affect the surrounding environment [6], [7]. Various studies have compared the dielectric properties of MO and esters. In [8] there are comparisons of AC breakdown voltage (AC-BDV) and lightning breakdown voltage (LI-BDV) of MO, NE and SE with different level of moisture. Results showed that AC-BDV of MO is more affected by moisture than both esters. The difference moved from 33 % (around 10 %RH (relative humidity)) to around 85 % (around 25 %RH). Slightly better results showed MO in LI-BDV measurements where MO has a higher BDV than both esters at almost all levels of moisture.

In [9] Beroual et al. compared MO Nynas AB, NE Envirotemp™ FR3™ and SE MIDEL 7131. The value of the average AC-BDV of SE was 76.8 % higher than MO and also NE reached higher values than MO, concretely AC-BDV was around 69 %. The average value of DC breakdown voltage (DC-BDV) for MO was 52.9 kV, NE reached 12.8 % higher values and SE has higher values by 19.5 %.

On the other hand, in [10] MO showed better dielectric properties by reaching higher values of LI-BDV of positive and negative polarity. Experimental results compared LI-BDV of MO, NE and SE with increasing gap distance between electrodes. Positive LI-BDV of MO increased with distance more markedly than both esters. Negative LI-BDV showed the opposite behavior when the difference between MO and esters decreased with distance incline, however, LI-BDV values of MO were higher at all gap distances.

Moreover, many other studies deal with the comparison of different insulating liquids with special emphasis on environmentally friendly fluids [6], [7], [11], [12].

II. Experiment

A. Examined insulating oils

Insulating oils used in this study were NE MIDEL eN 1204 (rapeseed-based oil) SE MIDEL 7131 and MO MOL TO 40A. The physicochemical properties of examined oils are in Table 1.

Table 1. Physicochemical properties of examined oils [1], [13], [14]

Properties	*MIDEL 7131*	*MIDEL eN 1204*	*MOL TO 40A*
Density at 20°C (g/cm^3)	0.97	0.92	0,865 (15°C)
Kinematic viscosity at 40°C (mm^2/s)	29	8	9,3
Pour temperature (°C)	-56	-31	-45
Flash point (°C)	260	>315	155
Fire point (°C)	316	>350	-
Total acid number (mg KOH/g)	<0.03 mg	0.04	-

B. Experimental setup

Measurement of AC-BDV was performed according to IEC60156 by AC test set Haefely Hipotronics. The circuit diagram of the AC-BDV test is in Figure 1. Resistor (5 MΩ) was used for current limitation to prevent degradation of samples. Test cell of volume 350 ml has 12.5 mm diameter electrodes horizontally positioned with electrode gap 2.50±0.05 mm. The voltage rise rate was 2 kV/s until breakdown occurs. The time between two consecutive measurements was 2 min. Each sample was tested 30 times.

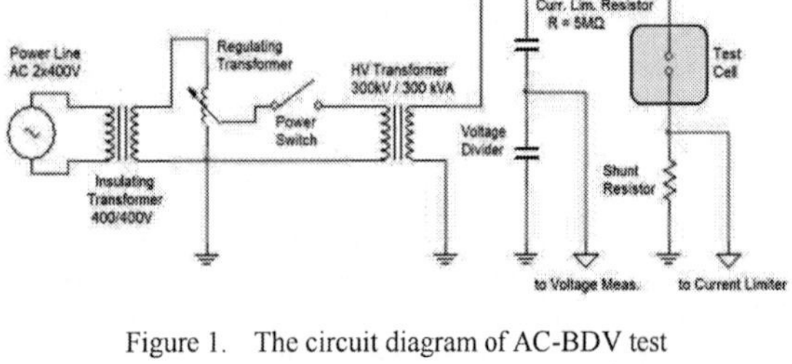

Figure 1. The circuit diagram of AC-BDV test

Measurement of LI-BDV were performed with the same test cell of volume 350 ml and 12.5 mm diameter electrodes horizontally positioned with electrode gap 2.50±0.05 mm. The LI voltage was supplied by a 1 MV – 100 kJ – 1.2/50 μs SGΔA Impulse Voltage Test System made by Heafely Test AG. Resistor (50 kΩ) was used for current limitation to prevent degradation of samples. Measurement was determined by the up-and-down method according to IEC 60060 ed3 2010-09 standard. Firstly, the proper (start) value of LI-BDV was determined. The experiment consisted of increasing the voltage value by 5 kV if breakdown did not occur and decreasing by 5 kV if the LI-BDV occurred. The breakdown voltage $U_{50\%}$ was calculated as the average of 30 consecutive measurements. Both positive and negative polarity of LI voltage was applied. The circuit diagram of the LI-BDV test is in Figure 2.

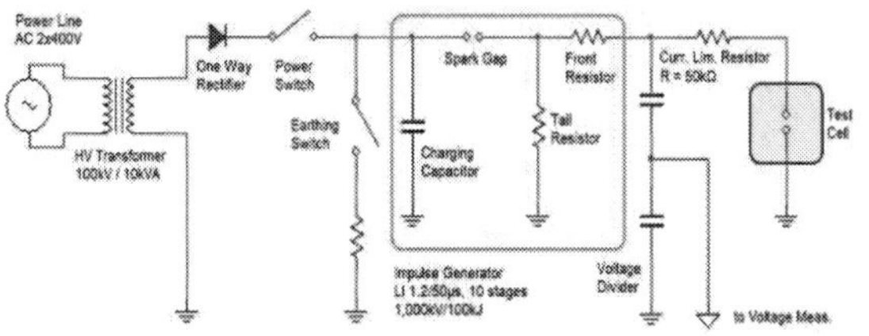

Figure 2. Circuit diagram of LI-BDV test

Measurement of dissipation factor, volume resistivity, and relative permittivity was executed in accordance with IEC 60247 2004. Test cell HAEFELY TEST AG 2903 H was filled with a sample of volume 40 ml. Tettex 2830 a Precision Oil and Solid Dielectric Analyzer was used for the measurements.. Dissipation factor, volume resistivity, and relative permittivity were measured at temperatures from 25°C to 140°C to determine the temperature dependence.

III. Experimental Results

A. AC breakdown voltages

The mean value of measurements was used to compare different insulating fluids calculated by (1).

$$\overline{U} = \frac{1}{N}\sum_{i=1}^{N} U_i \quad (1)$$

where $\overline{U}$, N, i, and U_i are the mean breakdown voltage, total measurement number, index of measurement, and breakdown voltage, respectively.

Figure 3 represents the distribution of AC BDV of examined insulating oils. The AC BDV maximal value of SE, NE, and MO is 62.3 kV, 61 kV, and 49.2 kV, respectively. The difference of maximal values between SE and MO is 26.6 % and between SE and NE 2.1 %.

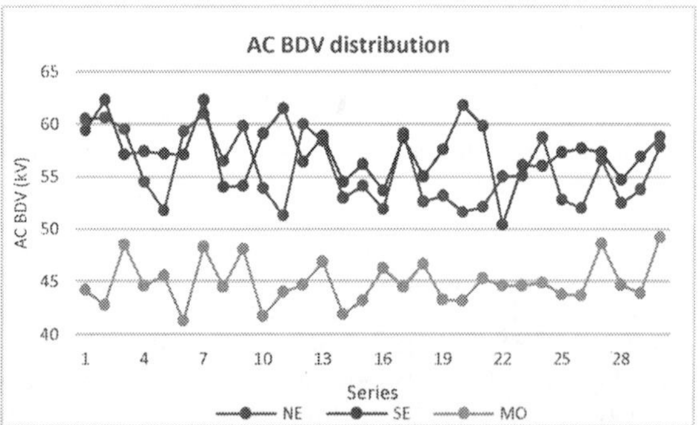

Figure 3. The distribution of AC BDV of insulating oils

The mean values of insulating oils are presented in Figure 4. The difference between SE and NE/MO is 3 % and 28.8 %, respectively. The AC BDV value of NE is 25 % higher than MO. According to the mean and maximal values of AC

BDV, the best dielectric performance among the NE, MO and SE is observed for SE.

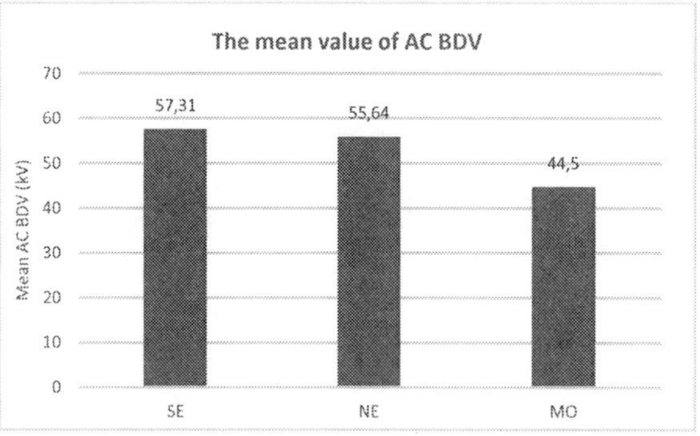

Figure 4. The mean values of AC BDV of examined insulating oils

B. Lightning breakdown voltage

The mean value of LI BDV is represented as $U_{50\%}$ (50 % probability of breakdown to occur) and computed in the same way as AC BDV.

Figures 5 and 6 show the distribution of LI BDV of positive and negative polarity. The positive LI BDV maximum showed the value at the level of 200 kV for NE. The maximum for MO and SE is lower than that for NE by 11 % and 8 %, respectively. The maximal LI BDV with negative polarity was reached by MO at the voltage level of 185 kV. Natural and synthetic esters showed conformity of values with a maximum of 180 kV.

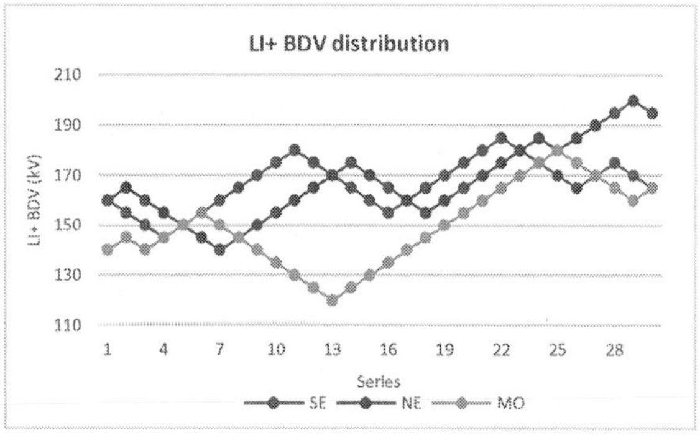

Figure 5. Distribution of positive polarity LI BDV

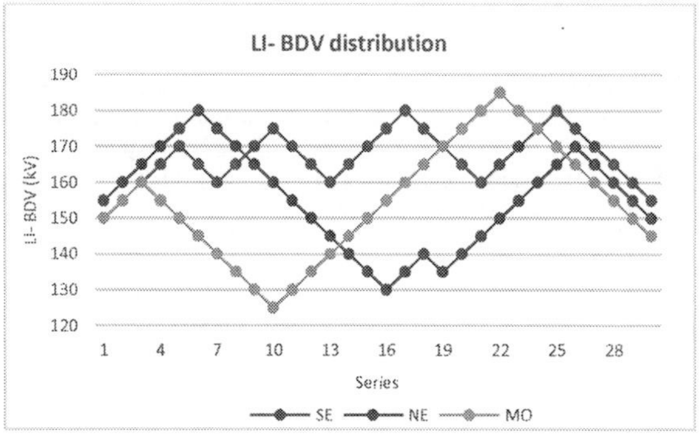

Figure 6. Distribution of negative polarity LI BDV

$U_{50\%}$ values of LI BDV of both polarities are in Figure 7. The difference of positive polarity LI BDV between SE and NE is less than 1 %, however, MO reached values 12 % lower than NE and the average level of BDV was around 150 kV. The negative polarity measurement showed that SE has significantly higher values of LI BDV than both NE and MO. Difference between SE and NE/MO reached a value of around 8 %. To sum up, the best dielectric performance among samples according to LI BDV has SE with the values between 166 kV and 167 kV for both polarities.

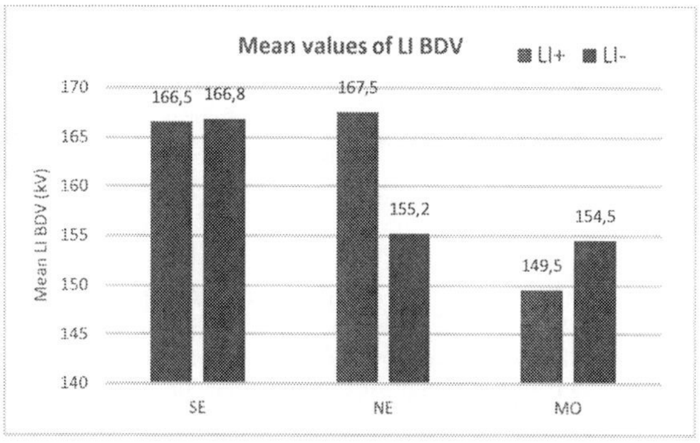

Figure 7. Distribution of negative (LI-) and positive (LI+) polarity LI BDV

C. Volume resistivity

The volume resistivity (also called DC resistivity) of the material shows how strongly this material opposes the flow of electric current through the volume of the cubic specimen. With the higher value of the volume resistivity, the electrical conductivity of the material decreases [15].

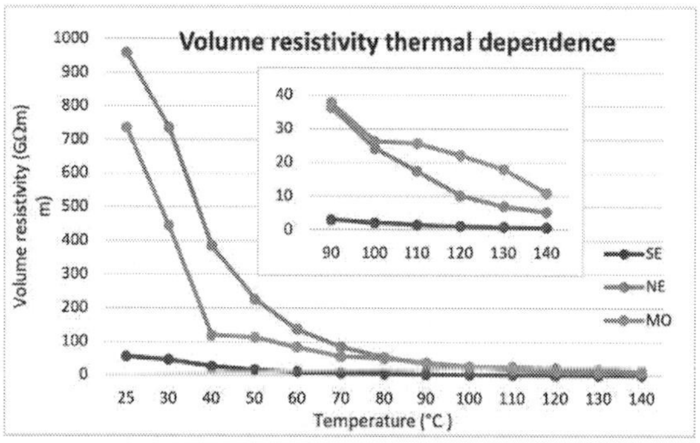

Figure 8. Thermal dependence of volume resistivity of insulating oils

In Figure 8 there are temperature-dependent resistivity differences between the examined insulating oils. It can be concluded that SE has the lowest volume resistivity among the samples throughout the course. The highest differences are at low temperatures from 25°C to 90°C. Particularly, at 25°C the resistivity of NE was 7 times higher than that of SE. The resistivity difference between MO and NE changed at around 90°C when the volume resistivity of MO started to increase up to around 250 % at temperature 130°C. However, when averaging the resistivity values over the whole temperature range, the highest volume resistivity belongs to NE.

D. Dissipation factor

The dissipation factor represents a dielectric loss mainly connected with the presence of impurities, oxidation derivates, and content of moisture in liquid. The dissipation factor defines the quality and level of degradation of materials [16].

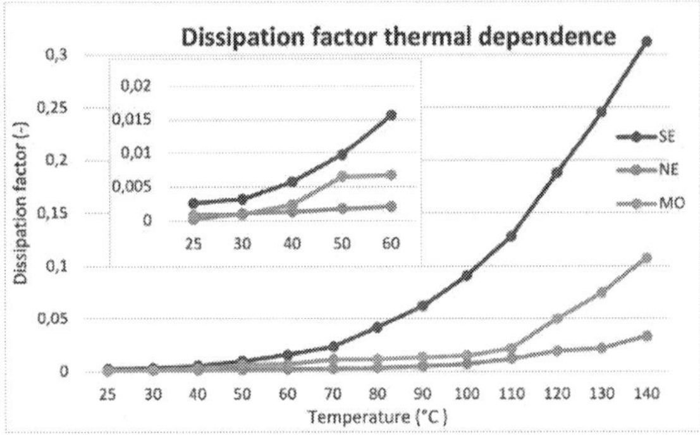

Figure 9. Thermal dependence of dissipation factor of insulating oils

Natural ester has the most stable values of dissipation factor among samples and also the lowest average number that makes NE the most suitable among samples according to this specification. Dissipation factor at 90°C (IEC 60247 2004) in SE is 4,6 times higher than that of MO and around 9 times higher than in NE. These differences between SE and NE/MO increase with higher temperature up to 140°C when SE has 9,5 times higher value than NE. The difference between NE and MO is not that significant below the temperature 120°C. Above this temperature, the difference started to rise and moved from 161 % at 120°C to 225 % at 140°C (MO has higher values).

E. Relative permitivity

The relative permittivity, also called the dielectric constant, determines the degree of polarization in response to an applied electric field [17].

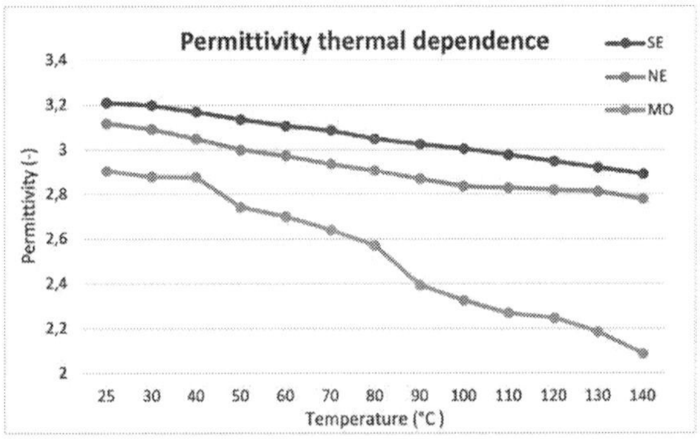

Figure 10. Thermal dependence of relative permittivity of insulating oils

The highest value of relative permittivity among the samples has SE, as can be seen in Figure 10. The difference between SE and NE throughout the course is almost stable around 5 %. The highest decrease of relative permittivity in dependence on temperature has MO and the difference between MO and NE moved from around 6 % at 40°C to around 33 % at the highest examined temperature 140°C.

IV. Discussions and conclusions

Different insulating oils reached different values of examined properties according to their structure. The main difference can be seen in the breakdown voltage where esters showed better dielectric strength. The behavior of oils at various temperatures represented by dielectric properties, concretely dissipation factor, volume resistivity, and relative permittivity showed that SE had worse dielectric performance in comparison with MO and NE according to this experiment.

SE reached the highest values among examined samples in breakdown measurements. Mean AC BDV of SE was around 3 % higher than in NE and by around 28 % higher than in MO. LI BDV of SE with positive polarity was less than 1 % lower than in NE and more than 11 % higher than in MO. Negative polarity testing showed that SE has higher values of LI than NE and MO and the difference is around 7 % and 8 %, respectively. However, the values of dielectric properties dependent on temperature showed aggravated results, as compared with NE and MO. Volume resistivity of SE was lower than that of NE and MO throughout the course at all temperatures. SE had a 17 times higher number of volume resistivity than NE at 25°C. The dissipation factor of SE was higher compared to NE and MO at all temperatures and also the relative permittivity had the highest number among samples throughout the course.

AC BDV of NE was 3 % lower than that of SE and 25 % higher than that of MO. Negative LI BDV in NE was 7 % lower than the one found for SE, but negligible difference in LI BDV is observed for NE and MO. However, the LI BDV of positive polarity showed the highest number for NE , and the difference between NE and SE/MO was around 1 % and 12 %, respectively. Dielectric performance of NE at different temperatures was in all cases better than in SE. In comparison with MO, the volume resistivity was higher up to temperature 80°C and at higher temperatures, the resistivity reached lower values than MO. The dissipation factor of NE was the lowest among the samples mainly at higher temperatures. The relative permittivity of NE was lower than the one of SE and higher than that of MO at all examined temperatures.

MO reached the lowest values of AC and LI BDV among all samples. However, dielectric performance according to properties examined with dependence on temperature was better than in SE. MO reached the highest values of volume resistivity at temperatures higher than 80°C. The dissipation factor of MO was comparable to NE to temperature 120°C. The relative permittivity of MO compared to NE and SE was lower at all temperatures and the difference moved from 6 % at 40°C compared to NE to around 38 % at 140°C compared to SE.

To sum up, according to experimental data SE showed the best dielectric strength among these samples with the highest mean values of AC and LI BDV. However, its dielectric performance was lower than that of other samples at all temperatures. MO with the lowest dielectric strength among samples reached excellent dielectric performance mainly at higher temperatures. The most complex results among samples were reached by NE. Dielectric strength of NE is slightly lower than in SE but differences of AC and LI breakdown voltages were not that significant. Moreover, measurement of positive LI BDV showed that NE reached the highest mean value among all samples. Dielectric performance of NE, according to dissipation factor and volume resistivity, showed better results than MO and SE. Value of relative permittivity of NE moved between SE and MO at all examined temperatures.

Acknowledgment

This research was funded by the Slovak Academy of Sciences and Ministry of Education in the framework of projects VEGA 2/0011/20, VEGA 1/0154/21, and Slovak Research and Development Agency under the contract No. APVV-18-0160.

References

[1] H. Duzkaya and A. Beroual, 'Statistical Analysis of AC Dielectric Strength of Natural Ester-Based ZnO Nanofluids', *Energies*, vol. 14, Dec. 2020, doi: 10.3390/en14010099.

[2] I. Fofana, '50 years in the development of insulating liquids', *IEEE Electr. Insul. Mag.*, vol. 29, no. 5, pp. 13–25, Sep. 2013, doi: 10.1109/MEI.2013.6585853.

[3] M. Šárpataky, J. Kurimský, and M. Rajňák, 'Dielectric Fluids for Power Transformers with Special Emphasis on Biodegradable Nanofluids', *Nanomaterials*, vol. 11, no. 11, Art. no. 11, Nov. 2021, doi: 10.3390/nano11112885.

[4] R. Asano and S. A. Page, 'Reducing Environmental Impact and Improving Safety and Performance of Power Transformers With Natural Ester Dielectric Insulating Fluids', *IEEE Trans. Ind. Appl.*, vol. 50, no. 1, pp. 134–141, Jan. 2014, doi: 10.1109/TIA.2013.2269532.

[5] J. Tokunaga, M. Nikaido, H. Koide, and T. Hikosaka, 'Palm fatty acid ester as biodegradable dielectric fluid in transformers: a review', *IEEE Electr. Insul. Mag.*, vol. 35, pp. 34–46, Mar. 2019, doi: 10.1109/MEI.2019.8636104.

[6] M. Rafiq, M. Shafique, A. Azam, M. Ateeq, I. A. Khan, and A. Hussain, 'Sustainable, Renewable and Environmental-Friendly Insulation Systems for High Voltages Applications', *Molecules*, vol. 25, no. 17, Art. no. 17, Jan. 2020, doi: 10.3390/molecules25173901.

[7] P. Rozga, 'Properties of new environmentally friendly biodegradable insulating fluids for power transformers', *Eur. Sci. J. ESJ*, Jul. 2013, doi: 10.19044/esj.2013.v9n21p%p.

[8] D. Martin and Z. Wang, 'An Overview of the Suitability of Ester Oil Dielectrics for use in Large ...', *yumpu.com*. https://www.yumpu.com/en/document/read/34521632/an-overview-of-the-suitability-of-ester-oil-dielectrics-for-use-in-large- (accessed Mar. 02, 2022).

[9] A. Beroual, U. Khaled, P. S. Mbolo Noah, and H. Sitorus, 'Comparative Study of Breakdown Voltage of Mineral, Synthetic and Natural Oils and Based Mineral Oil Mixtures under AC and DC Voltages', *Energies*, vol. 10, no. 4, Art. no. 4, Apr. 2017, doi: 10.3390/en10040511.

[10] M. Nguyen Ngoc, O. Lesaint, N. Bonifaci, A. Denat, and M. Hassanzadeh, 'A comparison of breakdown properties of natural and synthetic esters at high voltage', pp. 1–4, Oct. 2010, doi: 10.1109/CEIDP.2010.5724062.

[11] M. Rafiq, Y. Z. Lv, Y. Zhou, K. B. Ma, W. Wang, C. R. Li and Q. Wang, 'Use of vegetable oils as transformer oils – a review', Renew. *Sustain. Energy Rev.*, vol. 52, pp. 308–324, Dec. 2015, doi: 10.1016/j.rser.2015.07.032.

[12] I. Fernández, A. Ortiz, F. Delgado, C. Renedo, and S. Pérez, 'Comparative evaluation of alternative fluids for power transformers', *Electr. Power Syst. Res.*, vol. 98, pp. 58–69, May 2013, doi: 10.1016/j.epsr.2013.01.007.

[13] U. Khaled and A. Beroual, 'AC Dielectric Strength of Synthetic Ester-Based Fe 3 O 4 , Al 2 O 3 and SiO 2 Nanofluids - Conformity with Normal and Weibull Distributions', *IEEE Trans. Dielectr. Electr. Insul.*, vol. 26, pp. 625–633, Apr. 2019, doi: 10.1109/TDEI.2018.007759.

[14] 'MOL TO 40A - MOL Hungary'. https://mollubricants.com/en/lubricants-autochemicals/401-industry/industrial-lubricants/insulating-oils/5364-mol-to-40a (accessed Mar. 18, 2022).

[15] 'Volume Resistivity: Electrical Resistivity of Plastic'. https://omnexus.specialchem.com/polymer-properties/properties/volume-resistivity (accessed Mar. 21, 2022).

[16] D. Gnanasekaran and V. P. Chavidi, 'Properties of Vegetable Fluids: A Green Insulator for Power Sector', in *Vegetable Oil based Bio-lubricants and Transformer Fluids*, Singapore: Springer Singapore, 2018, pp. 125–155. doi: 10.1007/978-981-10-4870-8_7.

[17] P. Havran, R. Cimbala, J. Kurimský, B. Dolník, I. Kolcunová, D. Medveď, J. Király, V. Kohan and Ľ. Šárpataky, 'Dielectric Properties of Electrical Insulating Liquids for High Voltage Electric Devices in a Time-Varying Electric Field', *Energies*, vol. 15, no. 1, Art. no. 1, Jan. 2022, doi: 10.3390/en15010391.

Testing of the Faulty Phase Earthing System Using Real-Time Simulator

Viktor Jurák, David Topolánek, Jaroslava Orságová
Brno University of Technology, Faculty of Electrical Engineering and Communication, Department of Electrical Power Engineering,
Technická 12, 616 00 Brno, Czech Republic
Viktor.Jurak@vut.cz

Abstract — **To extinguish arc of a single-phase fault in resistor earthed distribution networks, an automatic reclosing system is usually used. However, this operation results in a short-term power supply interruption. As an alternative to the automatic reclosing function, it is possible to apply a faulty phase earthing (FPE) system. The main objective of this paper is to test the faulty phase earthing system using the real-time digital simulator (RTDS) environment, where behaviour of a real distribution network will be numerically simulated. One of the advantages of the real-time testing is the possibility to bring the output of the protection device (i.e. the input for the faulty phase earthing or triggering the circuit breaker operation) back to the simulation and, therefore, test the hardware in the loop. As a test object, the ABB feeder protection is used. This device is programmed according to the defined requirements of the FPE algorithm. The aim is to prove proper functionality of the algorithm and its correct implementation into the protection terminal settings, which is crucial for securing high reliability of the FPE system prior its installation to real resistor earthed network. The paper is focused on verification of the FPE system prior its real application in pilot test as alternative to autoreclosing system in real resistor earthed distribution network.**

Keywords: faulty phase earthing; real-time simulation; IEC61850; hardware-in-the-loop simulation; resistor earthed network; single-phase fault; protective relay

I. INTRODUCTION

The electricity system is undergoing a great number of changes recently. One of the recent trends is massive cabling. Medium voltage (MV) distribution networks in the Czech Republic are generally operated as compensated (resonant earthed). The increase in the number and length of cables is then reflected in an increase in capacitive current. Until recently, a regulation was in force in the Czech Republic, which limits maximum extent of a compensated network in terms of capacitive current without detailed verification of touch or step voltages. However, it can be assumed that distribution system operators (DSOs) will still operate cable and mixed networks as resistance-earthed for the time being.

In resistor-earthed networks, in contrast to compensated networks, there is no self-extinction of single-phase temporary faults. In the case of a single-phase temporary fault, the use of auto reclosing (AR) is necessary, which leads to a short-term interruption of the power supply to the consumers. The solution to reduce this negative effect is the use of the faulty phase earthing (FPE) as an alternative to AR. In this case, the FPE is first used, and only if this attempt is unsuccessful then full AR is performed. With FPE, the negative consequences on power quality and continuity according to [1] are significantly less than in the case of full AR application. This approach is most suitable especially for mixed networks. Since this type of network is characterized by both a high capacitive current, which makes it impossible to operate the network as compensated and a high number of single-phase temporary faults. On the other hand, for a purely cable network, this approach is inappropriate, as the incidence of single-phase faults is not as high and, in addition, there is no assumption of temporary faults on the cable, which disappear after the fault current is reduced.

The issue of faulty phase earthing is already well known in isolated or compensated networks, where the FPE reduces the residual current and thus helps to extinguish faults that the arc suppression coil (ASC) is unable to eliminate. This issue is discussed in [2-3]. The application of this system to resistor-earthed networks and an evaluation of the advantages and disadvantages is summarized in [1].

For the application of the system in practice, it is necessary to design the FPE as an alternative to AR algorithm so that it could be implemented in serially produced IEDs and, above all, it is necessary to test the implemented algorithm. It is performed by testing that it is necessary to verify whether the implemented algorithm works correctly, since dangerous conditions could occur in case of a wrong operation. One of these undesirable conditions is, for example, the earthing of multiple phases. This faulty operation would lead to the creation of a multi-phase short circuit right into the substation, resulting in disruption of the distribution network.

Therefore, RTDS was used for the testing. Using it, the configured IED (test object) can be connected into a closed loop (HIL) and integrated into the simulation being performed, in which both the power part and the control part are simulated. The test object is then exposed to conditions that are almost identical to real operation. A further advantage is the possibility of including a process communication according to the IEC 61850 standard, which allows receiving control commands from the IED using GOOSE. The testing of IEDs using RTDS via HIL is described e.g. in [4-5]. A similar system to the one presented in this paper is described in [6], but in this paper the algorithm is completely different due to the different type of considered network.

This research work has been carried out in the Centre for Research and Utilization of Renewable Energy (CVVOZE). Authors gratefully acknowledge financial support from the Ministry of Education, Youth and Sports of the Czech Republic under BUT specific research programme (project No. FEKT-S-20-6449)

This research work has been carried out in the Centre for Research and Utilization of Renewable Energy (CVVOZE). Authors gratefully acknowledge financial support from the Technology Agency of the Czech Republic (project No. TK 01030094))

II. Algorithm Design and Implementation

The sequence of the FPE function as an alternative to AR is presented in [1] and illustrated in Figure 1. After the trip signal from the overcurrent protection function PHLPTOC is activated, it is evaluated whether the fault is single-phase or multi-phase. If it is a single-phase fault, the faulted phase is earthed for 300 ms. If the fault remains (the start signal of the PHLPTOC protection function has logical value 1), the sequence continues with a three-phase switch OFF of the feeder circuit breaker for 300 ms (AR sequence). If the insulation level of the feeder is not restored during this clearing period and the fault persists, the start signal of the overcurrent leads to the final tripping of the faulty feeder. If the fault is other than single-phase, the sequence is similar, but the FPE cycle is omitted and the AR is performed directly.

This sequence has been implemented in the multifunctional IED REX 640 from the ABB company. The configuration of this device is done via the PCM 600 software. Since the FPE as an alternative to AR sequence is not yet industrially applied, its configuration was a challenging task. For the configuration, it was necessary to use only those function blocks that are available in sufficient amounts in the given IED configuration.

The core of the sequence is the PTGAPC1 function block. It is eight independent pulse timers. The output of the pulse timer takes logic 1 immediately after the rising edge on the input, and resets in a set time, regardless of the value of the input signal during this time set. The output of the timer is connected to the input of the next timer through the logic function NOT (negation). This creates a cascade sequence where each following pulse starts with the falling edge of the previous one.

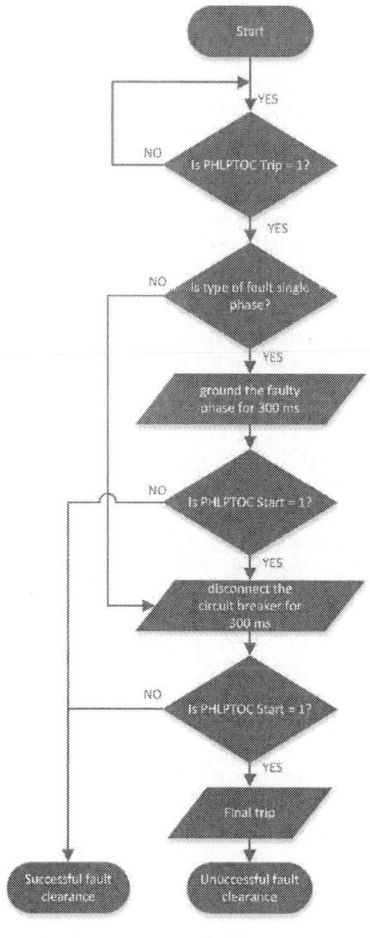

Figure 1. Flow chart of the FPE algorithm as an alternative to the AR

This core of the sequence is illustrated in Figure 2. If the input IN1 of the PTGAPC1 is not blocked, the sequence is triggered by the coming of the rising edge from the trip output of the low stage of overcurrent protection (PHLPTOC). At the same time, output *Q1* of the PTGAPC1 block, which is the driving signal to turn on the earthing breaker, is also activated. When the set time is over, output *Q2* is activated, which is the driving signal to turn off the earthing breaker. The earthing breaker consists of three separate single-phase breakers that can be switched individually.

Input *IN1* of PTGAPC1 (sequence start) is blocked by two signals. The first is a blocking signal that is active after the sequence has started (Figure 2 above). This signal prevents the repetition of the sequence while the sequence is already running. The second blocking signal is a logic value that indicates the inexistence of a multi-phase fault. To generate this signal, it is necessary to know at which phase the fault occurred. Since the IED configuration is missing this output of the overcurrent function block, it is necessary to deal with this problem alternatively.

In the REX640 IED, there are two ways to identify the faulty phase. The first is by using IEC 61850 GOOSE communication. Despite the fact that the start and trip signal of the protection function for each phase is not available in the IED configuration (using function blocks), it is possible to send it to another device and work with it there. However, this solution would require the use of two IEDs. The second option is to use the LEDPTRC function block, which is primarily used to indicate the faulted phase on the human-machine interface (HMI) LED indicators and is not primarily intended for tripping. In the configuration presented in this paper, the second way was used.

Switching on the earthing breaker is done only for the faulty phase. Switching on of the breaker in each individual phase is conditioned by active output *Q1* of the PTGAPC1 function block, inactive interlock signal (in case of multi-phase faults) and active fault signal in the corresponding phase. If the fault does not persist after the earthing breaker is switched off (after *PTGAPC1_Q2* is activated), the start signal of the overcurrent protection function (variable *PHLPTOC_St*) is not active and the tripping of the three-phase feeder breaker

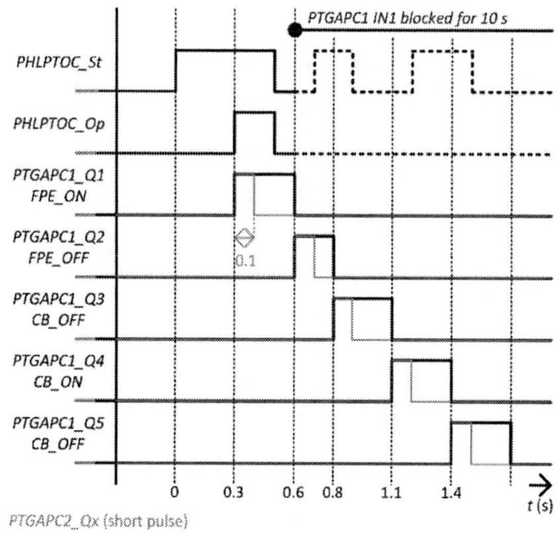

Figure 2. The core of the FPE as an alternative to OZ function

is blocked. Otherwise, a three-phase trip of the feeder breaker (*PTGAPC1_Q3*) occurs and after 300 ms it is switched on again (*PTGAPC1_Q4*). If the *PHLPTOC_St* signal is inactive after the breaker is switched on, the next trip is blocked. Otherwise, permanent tripping of the breaker occurs.

If an interlock signal indicating a multi-phase fault is active, the FPE sequence is skipped and the reclosing is activated directly.

Since it is not appropriate to have the signals for switching on and off the breakers active for the entire duration of the state defined by the PTGAPC1, the outputs of this block are connected to the inputs of the PTGAPC2 block, which is used to generate short pulses (green pulses in Figure2.). These pulses always last 100 ms. The final outputs for controlling the breakers are routed to the MVGAPC function block. This function block allows the creation of the events that can be transmitted via IEC 61850 GOOSE communication.

III. Testing the Implemented Algorithm Using a Real-time Simulator

The configured IED was tested using a real-time simulator RTDS. This method of testing allows performing HIL tests when the IED under test is connected to the simulator in a closed loop. The output of the simulation is the currents and voltages that are input to the IED. The IED reacts based on these values and sends signals back to the simulator where they are reflected by a change in the simulation. The internal schematic of the simulation and the external schematic of the connection of the IED to the RTDS is shown in Figure 3.

A simple HV/MV substation with a three winding transformer and one feeder is simulated. The simulated network starts with a 110 kV line feeder with an initial short-circuit inrush current Ik" = 10 kA and a ratio R/X = 0.1. The 110/22/6.3 kV transformer has a nominal capacity of 25 MVA and its primary side is directly earthed. The neutral node of the secondary side of the transformer is earthed through a 1 kA resistor. This current value corresponds to a resistance value of 12 Ω. The secondary side of the transformer is connected to an 15 km long overhead line with AlFe 110/22 conductors in side-by-side arrangement. At the end of the line there is a 5 MW load in a delta connection. A fault is located in the middle of the line, for which its resistance and which phase it is in can be freely set.

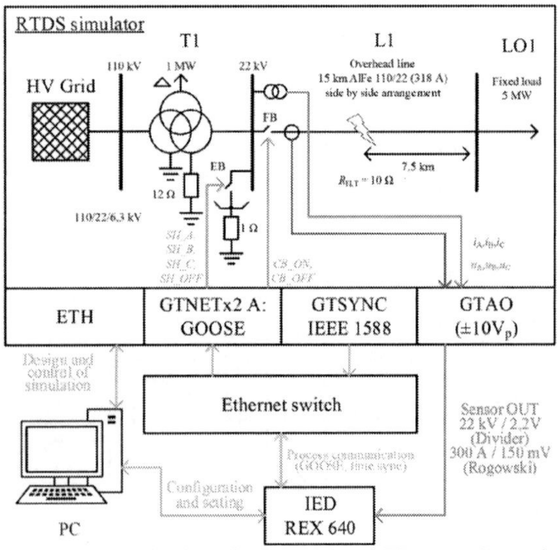

Figure 3. Schematic of simulated network and IED connection to the RTDS

The considered resistance of the fault is 10 Ω. In the substation, a three-phase circuit breaker is located at the feeder. In Figure 3. the feeder breaker is labeled "FB". The earthing breaker is connected to the busbars in the 22 kV substation. This breaker is capable of switching individual phases to a common node, which is connected to the substation earthing system through a 1 Ω resistor. Earthing breaker is labeled "EB" in Figure 3.

Voltage metering is installed on the busbars of the 22 kV substation. Current measurement is located at the feeder behind the circuit breaker. This measurement is output to the analog output card GTAO of the simulator. From this card, wires are routed to the sensor inputs of the REX 640 IED. The scaling factor for the voltage is 10000:1, so 22kV in the simulation corresponds to 2.2 V at the output of the GTAO card. For current, the nominal value of 300 A in the simulation corresponds to 150 mV at the output of the GTAO card. However, the currents need to be first derivated in the simulation, since the sensor inputs of the protection do not allow any other type of signal than the type of signal coming from the Rogowski coil.

The IED sends commands for switching on/off of the feeder breaker and for switching on the earthing breaker for each individual phase and for central switching off the earthing breaker via IEC 61850 GOOSE communication. These signals then directly trigger the breaker elements in the simulation. All breakers have a set on/off delay of 50 ms.

The simulation is controlled from a PC connected to the simulator. The same PC is also used to configure and parameterize the IED. The IED, the Ethernet switch and the simulator are synchronized with each other, via IEEE 1588. The time base is the internal clock of the Ethernet switch. The simulation is configured in RSCAD software using the "Draft" module and controlled using the same program in the "Runtime" module. An example of the "Runtime" module is shown in Figure 4.

The simulation results can be seen in Figure 5, where the binary signals are plotted, and in Figure 6, where the RMS values of the voltages and currents are plotted. These signals are recorded directly in RTDS in the "Draft" window of the RSCAD software.

Signals with the prefix "G_" are received GOOSE messages which are sent by the IED. The signals *G_SH_A*, *G_SH_B* and *G_SH_C* are commands to turn on the earthing breaker in a selected phase A, B or C. The signal *G_SH_OFF* is a central command to turn off the earthing breaker. The signals *G_CB_ON* and *G_CB_OFF* are commands to close and open the feeder breaker. The signals *G_PHLPTOC_Str_general* and G_PHLPTOC_Op_general are

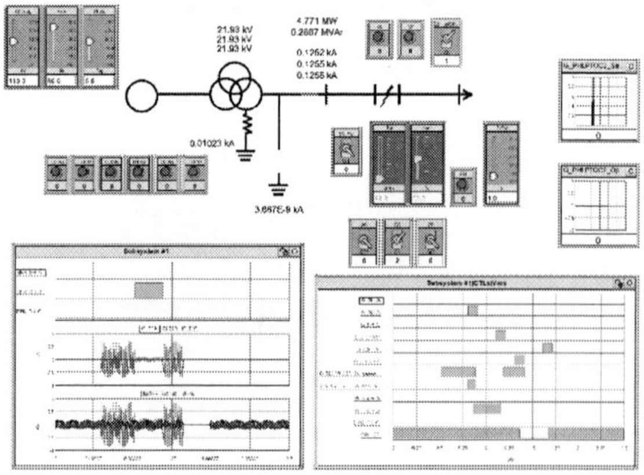
Figure 4. Sample of the simulation control panel (module "Runtime")

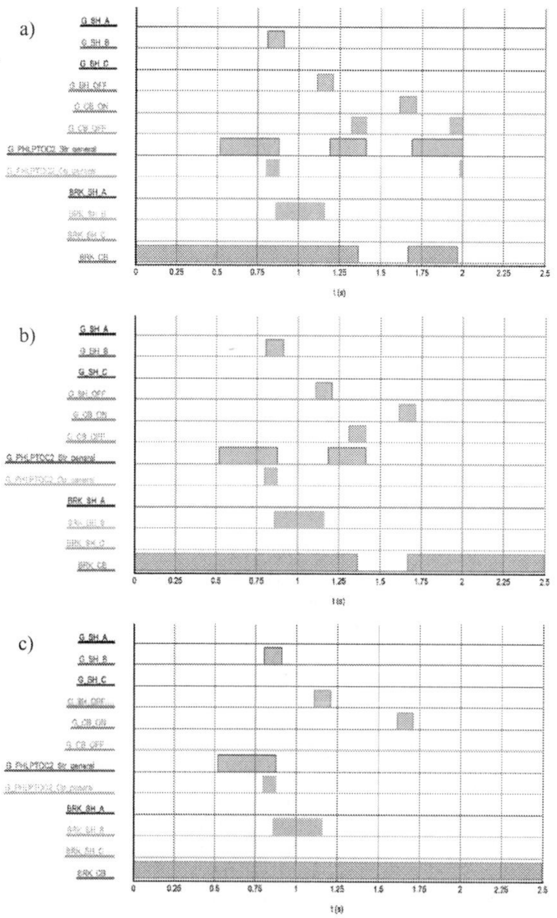

Figure 5. Simulation results a) the fault is cleared after FPE, b) the fault is cleared after AR, c) the fault is permanent

the start and trip signals from the overcurrent protection function that triggers the FPE as an alternative to the AR sequence. These signals are used in the simulation for control purposes only. The signals *BRK_SH_A*, *BRK_SH_B* and *BRK_SH_C* are the states of the individual earthing breakers and the signal *BRK_CB* is the state of the pin switch. For the states, a logic 1 means that the element is switched on.

Figure 5. shows the correct function of the configured sequence. Figure is divided into three parts. In part a), the fault has a permanent character. In part b), the fault is cleared by full AR. In part c), the fault is cleared during FPE (ideal case). A single-phase fault in phase B was simulated.

From Figure 5. a) it can be seen that the start of the overcurrent function occurred immediately after the application of the fault (t ~ 0.5 s). After a set time of 300 ms, the trip signal was activated (t ~ 0.8 s), along with which the earthinging of the faulty phase B occurred, which lasted until the earthing breaker was turned off (t ~ 1.1 s). Thus, the algorithm correctly recognized the faulty phase. Subsequently, the fault continued and so after a further 200 ms (t ~ 1.3 s), a command to trip the feeder breaker was sent. This breaker was switched back on. after 300 ms (t ~ 1.6 s). During the voltage-free interval, the fault did not clear, so after another 300 ms (t ~ 1.9 s) the feeder breaker was turned off permanently. This is the worst case where the fault is permanent.

Figure 5. b) is similar to the previous one, but now the fault has been cleared during the voltage less pause during AR.

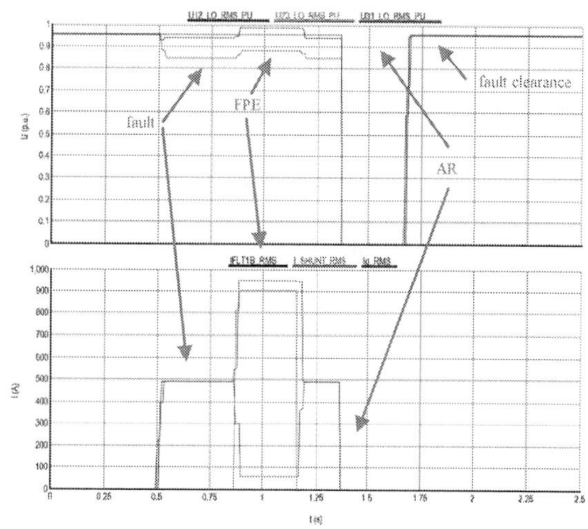

Figure 6. The phase-to-phase voltages at the end of the line and the current flowing through the fault, the shunt and the neutral resistor.

From this figure, it can be seen that the next trippings of the feeder are blocked since the start signal of the overcurrent protection is no longer active after AR. However, even this case is not beneficial, as the distribution of electrical power to customers was interrupted.

Figure 5. c) is again in the similar format as the previous two figures, but now the fault has been cleared after the FPE. It can be seen from the figure that once again the tripping of the feeder breaker is blocked when the start signal of overcurrent protection is inactive. This case is the most beneficial, since it was not necessary to interrupt the power supply to clear the fault.

The other scenarios where the fault is placed in a different phase were also verified in the simulation. The algorithm correctly identified the phase each time. Tests were performed for two-phase and three-phase fault conditions. The configured algorithm in the IED recognized this state every time, and the FPE sequence was not applied at all and the AR occurred directly.

The advantage of the FPE as an alternative to AR can be seen in Figure 6. Despite the fact that this figure depicts an unbeneficial condition where the fault is cleared only by AR, the benefit of using FPE is evident. At the top, the curves *U12_LO_RMS_PU*, *U12_LO_RMS_PU*, *U12_LO_RMS_PU* are plotted, which are the RMS values of the L-L voltages at the point of connected load (at the end of the line) in per-units (referenced to the nominal value of the line to line voltage U_n = 22 kV). The curves show that compared to the fault condition, the voltage between phases A and B (black curve) and further between phases B and C (red curve), has slightly increased, while the voltage between phases C and A (blue curve) remained unaffected.

In the second part of Figure 6, the RMS values of the currents are plotted. The variable *IFLT1B_RMS* (black curve) is the current passing through the fault. It can be seen that this current takes on a value of approximately 500 A during a fault in the simulated network. During the FPE, this current rapidly decreases to a value of approximately 60 A, which can ensure self-extinguishing of the fault. The variable *I_SHUNT_RMS* (red waveform) is the current that passes through the earthing breaker. This current takes a value of approximately 900 A when the FPE is applied. The earthing breaker both switches this current on and off. It is therefore necessary to take into account the increased stress of this switching element when dimensioning the earthing breaker. The last curve is the *Ig_RMS* (blue curve), which is the current that passes through the neutral node of the secondary side of the HV/LV transformer. This curve follows the fault current curve, but when FPE is applied, it increases up to a value of about 950 A.

IV. Conclusion

In this paper, an implementation of the FPE algorithm as an alternative to AR was proposed and the configuration into a commercially available IED was described. The paper also summarized the testing of the configured IED using a real-time simulator, which was successful. The paper shows the benefits of real-time testing, which greatly facilitates the verification of the prototype configuration.

Regarding the FPE itself as an alternative to AR, here the benefit in case of transient fault is evident, as the use of the proposed sequence allows the clearing of the temporary faults without power supply interruption. This leads to an improvement of the continuity of supply indicators, which can lead to financial savings for the distributor. On the other hand, the disadvantage is the increase in the tripping time of a permanent fault in case that FPE does not clear the fault. Here the tripping time is extended by the FPE sequence.

Acknowledgment

This research work has been carried out in the Centre for Research and Utilization of Renewable Energy (CVVOZE). Authors gratefully acknowledge financial support from the Ministry of Education, Youth and Sports of the Czech Republic under BUT specific research programme [project No. FEKT-S-20-6449]

This research work has been carried out in the Centre for Research and Utilization of Renewable Energy (CVVOZE). Authors gratefully acknowledge financial support from the Technology Agency of the Czech Republic [project No. TK 01030094].

References

[1] D. Topolanek, M. Ptacek, V. Krcal, J. Drapela and P. Toman, "FAULTY PHASE EARTHING AS AN ALTERNATIVE TO AUTORECLOSING IN RESISTOR EARTHED NETWORKS," CIRED 2021 - The 26th International Conference and Exhibition on Electricity Distribution, 2021, pp. 1293-1297, doi: 10.1049/icp.2021.2126.

[2] D. Topolánek, J. Orságová, J. Dvořák and P. Toman, "The method of the additional earthing of the affected phase during an earth fault and its influence on MV network safety," 2011 IEEE Trondheim PowerTech, 2011, pp. 1-8, doi: 10.1109/PTC.2011.6019222.

[3] A. Nikander, "Development of Modern Phase Earthing System for Improving Quality of Supply in the MV Network," in IEEE Transactions on Power Delivery, vol. 32, no. 3, pp. 1228-1235, June 2017, doi: 10.1109/TPWRD.2015.2506820.

[4] A. Umair, A. Nikander and P. Järventausta, "Simulation environment for centralized protection and control applying dSPACE and RTDS with IEC 61850 9–2 communication," 2016 IEEE PES Innovative Smart Grid Technologies Conference Europe (ISGT-Europe), 2016, pp. 1-5, doi: 10.1109/ISGTEurope.2016.7856231.

[5] A. S. Makhzani, M. Zarghami, B. Falahati and M. Vaziri, "Hardware-in-the-loop testing of protection relays in distribution feeders with high penetration of DGs," 2017 North American Power Symposium (NAPS), 2017, pp. 1-6, doi: 10.1109/NAPS.2017.8107191.

[6] A. Nikander, "Development and Testing of New Equipment for Faulty Phase Earthing by Applying RTDS," in IEEE Transactions on Power Delivery, vol. 32, no. 3, pp. 1295-1302, June 2017, doi: 10.1109/TPWRD.2015.2505779.

Dynamic Changes of Photovoltaic Power Plants Power Supply

János Kurcz
Faculty of Electrical Engineering and Information Technology STU
Institute of Power and Applied Electrical Engineering
Bratislava, Slovak republic
janos.kurcz@stuba.sk

Ján Poničan
Faculty of Electrical Engineering and Information Technology STU
Institute of Power and Applied Electrical Engineering
Bratislava, Slovak republic
jan.ponican@stuba.sk

Michal Chudý
Faculty of Electrical Engineering and Information Technology STU
Institute of Power and Applied Electrical Engineering
Bratislava, Slovak republic
michal.chudy@stuba.sk

Vladimír Šály
Faculty of Electrical Engineering and Information Technology STU
Institute of Power and Applied Electrical Engineering
Bratislava, Slovak republic
vladimir.saly@stuba.sk

František Janíček
Faculty of Electrical Engineering and Information Technology STU
Institute of Power and Applied Electrical Engineering
Bratislava, Slovak republic
frantisek.janicek@stuba.sk

Marek Gebura
Slovak Academy of Sciences
Institute of Materials and Machine Mechanics
Bratislava, Slovak republic
marek.gebura@savba.sk

Abstract—**The aim of this work is to analyse the dynamics of the power production curve of Photovoltaic Power Plant (PVPP). The paper is focused on selected results of experimental activities related to photovoltaic production. Moreover, the theoretical background of inclusion of PVPP to the daily load diagram of a Slovak TSO (transmission system operator) is discussed. In the theoretical part, technical and environmental effects of the PVPP dynamic power production and corresponding mathematics and statistics is listed. Dynamic changes in PVPP production are statistically compared utilising data taken from selected PVPPs various technology and installed power. Dynamic changes of the PVPP power production can be used to optimise the analysed power system. These results are important to properly set regulation system settings of other Smart Grids. Technical solutions of virtual power plants, power regulation and others are verified and correlated with experimental data of Smart Grid systems involving PVPP.**

Keywords: power output, dynamical changes of power output, photovoltaic power plant, Smart Grid, power regulation

I. INTRODUCTION

Electricity production is bound to be the primary energy source. For electricity production humankind primarily uses thermal power plants which burn fossil fuels (72.9% of electricity production in the world [1]). The maximal efficiency of thermal power plants (efficiency of transformation of primary energy source to electricity) is about 41.6% [2]. It means that nearly 60% of the energy of primary fossil fuels is not used and it is wasted by nature in these power plants.

Renewable energy sources (RES) produce electricity constrained by natural conditions. They produce electricity from sources which are constantly renewing naturally or by human activity. In this way; these sources are inexhaustible.

RES are at the moment not capable of replacing all fossil power plants. There are two primary reasons: The first one is because most RES power plants have no storage of primary energy sources. They produce electricity only under suitable conditions. This makes them hardly adaptable to the power load of the electric system. The second reason why they cannot replace traditional power plants so easily is that the technology of these systems is more expensive. The production level of RES systems is increasing, because of response to investors' demand. By increasing the demand in the market, the price of these technologies should decrease. The increase in electricity price has a positive effect on RES due to better and sooner return on investment. The operation of these sources should not be affected by fluctuation of primary sources (fossil fuels) price, too [3].

Not counting hydro power, the most prevalent RES in Slovakia is power produced from biomass and biogas. Photovoltaic power plants (PVPPs) are only the third largest renewable electricity source [4]. Power plants which produce electricity from biomass and biogas have continuous production curves and they do not affect system energy imbalance. In contrast, PVPPs are highly dependent on weather conditions and their supplied power changes sharply. It depends not only on weather conditions, but on power electronics, too.

Because of early economic payback, the operator of residential PVPP is trying to exploit the power from PVPP in the place of production. This mean, operator is trying to adapt its own consumption to PVPP production [5]. If this adaptation to consumption is intelligently managed and it is automatized, then we are moving to the vision of Smart Grids. Definition of

Smart Grids is different in different literature, but we have to deliberate, that these systems must be a superstructure of existing systems. We have to adjust measurement and control of these existing systems [6].

If we want to merge control of different devices in Smart Grid systems, we should recognize these individual devices. For this reason, we are focusing on the dynamics of power production from PVPPs.

II. Operating of RES

Operator of PVPP is trying to produce as much power as possible from PVPPs, because the reward stands for produced MWh. However, power producers are by law not permitted to trade the produced energy and they are usually not responsible for the power plant energy imbalance. The repurchaser (energy supplier/trader) is responsible for the PVPP (or other RES power plant) energy imbalance and is trading the produced energy.

Impact of RES to the power system can be divided into two major categories: local impacts and impact to the whole power system. Impact to voltage, short circuit power, supply of harmonics current, impact to mass remote control, flicker and others are local impacts. Operator of distribution system have to follow the rules of electricity quality, which is described in norm *STN EN 50 160 Voltage characteristics of electricity supplied by public electricity networks.* The distribution system operator issues document *Technical conditions of the distribution system operator (valid from 01-05-2019)* for this reason. In this document, all technical requirements and standards to connect new power sources are mentioned

Impact to system stability, insertion of RES power production to daily load diagram and energy imbalance caused by RES are impacting the whole system. Overload of transmission lines caused by RES is another issue, which is closely related to RES development in the power system. This overloading can cause non-compliance of criterion N-1 of the power system [7].

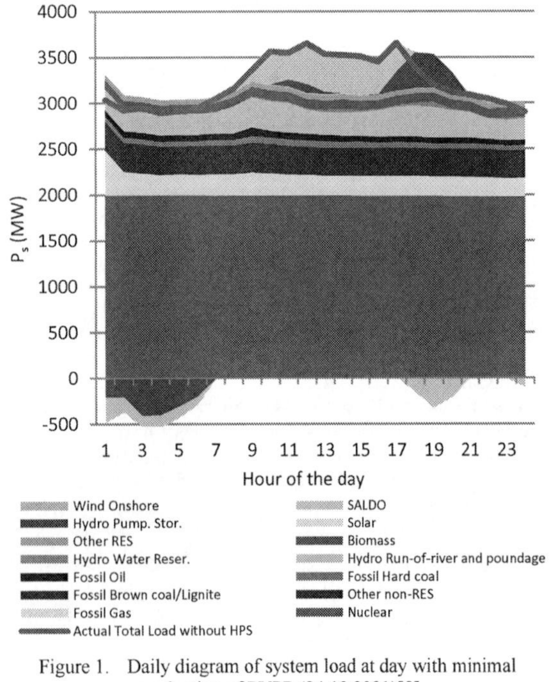

Figure 1. Daily diagram of system load at day with minimal production of PVPP (24.12.2021)[8]

Figs. 1 and 2 show which part of electricity consumption is covered by RES (especially PVPPs) during the day with maximal production of PVPPs and during the day with minimal production of PVPPs.

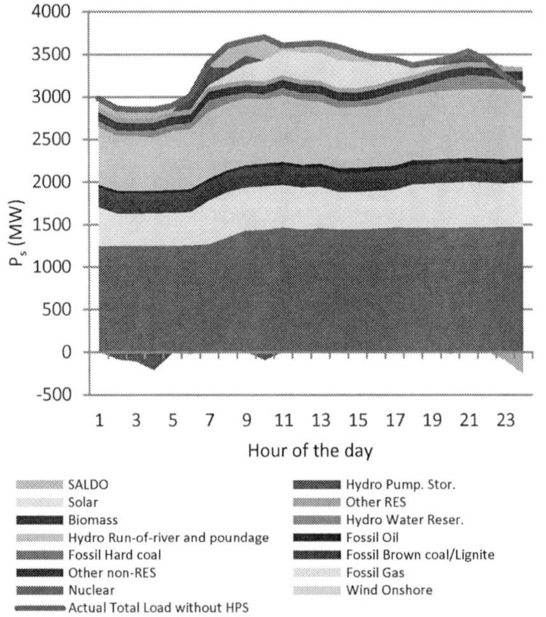

Figure 2. Daily diagram of system load at day with minimal production of PVPP (26.05.2021)[8]

Rentability of investments to RES were not profitable in the past. Therefore, electricity produced by RES was subsidised. Some subsidies are still active. Electricity market is run by 1 hour intervals, but energy imbalance is evaluated in 15 min intervals. This results in a higher cost of energy imbalance for a purchaser of this electricity as even very precise hourly predictions cannot eliminate energy imbalance in 15 min time intervals.

III. Theory

Modern inverters and DC-DC converters for PV systems use MPPT (maximum power point tracker) technology. This technology consists of an electronic circuit, which adapts input voltage from photovoltaic panels. This circuit looks for an optimal operating point at the VA curve in order to gain maximum power (P_m.) This power in W is determined as follows:

$$P_{mp} = U_{mp} I_{mp} \quad [W] \tag{1}$$

where U_{mp} is the voltage in V and I_{mp} is the current in A at maximum power P_{mp}. The current at each point of the VA characteristic can be determined from the following equation:

$$I = I_L - I_0 \left[exp \frac{e(U+IR_S)}{mkT} - 1 \right] - \frac{U+IR_S}{R_P} \quad [A] \tag{2}$$

where I_0 is the saturation current in A of PN junction dependent from temperature, e is the electron charge in C, U is the belonging voltage in V on VA characteristic, m is the ideality factor (-) of the semiconductor element, k expresses Boltzmann's constant in m^{2}kgs^{-2}K^{-1}, T is the actual temperature in K of PN junction and R_s with R_p are parasitic resistances (serial and parallel) in Ω, which depend on type of solar cell.

These resistances are nonlinear elements and they vary on temperature. The current I_L is produced by light in A:

$$I_L = eA_c \int_{E_G}^{\infty} S(E)\eta_{qe}(E)dE \quad [A] \tag{3}$$

where $S(E)$ is the number of photons with energy E ($cm^{-2}s^{-1}$), affecting PN junction with external quantum efficiency η_{qe} (-) on the area of A_c in cm^2. Integral is limited from the bottom by the energy of E_G in J, which is the wildness of the forbidden zone of semiconductor.

From the equations (2) and (3) one can see that the magnitude of the current and the magnitude of the power from each cell of the photovoltaic panel depends on solar irradiation intensity and temperature Other parameters are constants or they depend on used material of photovoltaic cells and their structure [9].

The curve of power delivered to the network is also affected by the efficiency of the power inverter. These losses are produced by current conversion (DC to AC) and they increase with the square of current. The supplied power is even reduced by the inverter's own power consumption, while not depending on supplied power. It can be various under different operating states of the inverter (daytime operation, nightly operating, disconnected PV, disconnected grid and combinations of these operating states).

In the datasheet of the inverter of PVPP no. 1, just two distinct values are stated: the maximum efficiency of the inverter, which is 98%, and the night consumption of 0.5 W. Measurements have shown that average consumption is 5.9 W during night operation. In the datasheet of inverters used for PVPP no.2, the dependence of efficiency on output power (P_{AC}) is shown in detail (Fig. 3).

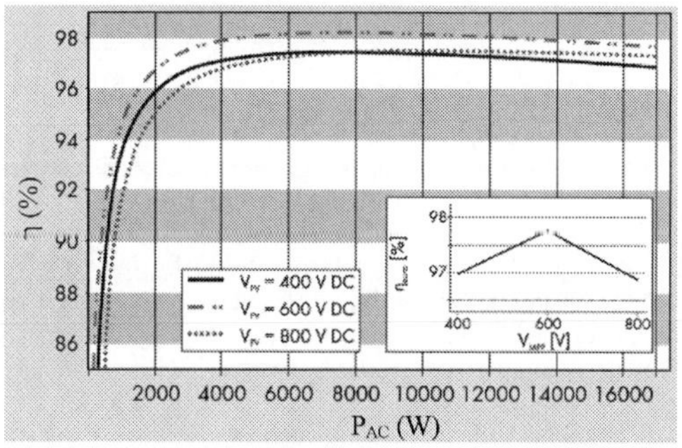

Figure 3. Efficiency dependence on produced power of inverters used for PVPP no. 2

Both types of inverters (Table 1.) have an installed power of 17 kW and the efficiency is around 98% for output power higher than 12% of the installed power (2000 W) at V_{FV} = 600 V. Errors caused by non equivalency of 2 types of inverters are negligible for the analysis below.

IV. Dynamical Curve of Produced Power by PV

Power production of PVPP is possible to be predicted in 15 minutes or hour intervals using meteorological data. But the supplied power of PVPP changes more rapidly (every second, or even less) due to the changes in solar behaviour. Rate of this power change depends on the current produced by the PV cell and also on the reaction time of MPPT of the inverter (frequency of maximum power point evaluation). Measured data in this work are produced by photovoltaic power plants at the Institute of materials and machine mechanics of Slovak Academy of Sciences. The data acquisition rate here is in second resolution.

We will consider the relative percentage performance p for clarity in %:

$$p = \frac{P_t}{P_{inst}} \cdot 100 \quad [\%] \tag{4}$$

where P_t in W is supplied power in the time period of t in seconds and P_{inst} is the installed power of PVPP in W. Similarly, the second power change (related to installed power) will be considered as follows:

$$\Delta p_t = \frac{P_t - P_{t-1}}{P_{inst}} \cdot 100 \quad [\%] \tag{5}$$

It means that positive change will increase power and negative change will mean power reduction. Two indicators of dynamic changes will be used. The first one is the average of absolute values of power changes per second Δp_Φ (related to installed power) in %:

$$\Delta p_\Phi = \frac{\frac{\sum_{t=2}^{n}|P_t - P_{t-1}|}{n-1}}{P_{inst}} \cdot 100 \quad [\%] \tag{6}$$

The second indicator for dynamic power changes analysis will be square root of summed squared changes per day (related to installed power) in %:

$$s_P = \frac{\sqrt{\frac{\sum_{t=2}^{n}(P_t - P_{t-1})^2}{n-1}}}{P_{inst}} \cdot 100 \quad [\%] \tag{7}$$

In both cases n (-) is the number of measured data on a given day [10].

Table 1. Parametres of analysed PVPPs

Analysed PVPPs	PVPPs	
	PVPP no.1	*PVPP no.2*
Installed power Pinst [kW]	16.2	800
Technology of PV panels	Polycrystal	Polycrystal
Azimut [°]	210	180
Slope [°]	42	35
Technology of inverters	One 17 kW inverter	47 psc of 17 kW inverters

For analysis below, measurements from two different PVPPs placed relatively close to each other (10.8 km) were used. Parameters of PVPPs are shown in Tab. 1. The weather conditions were analysed as being very similar for relevance of comparison.

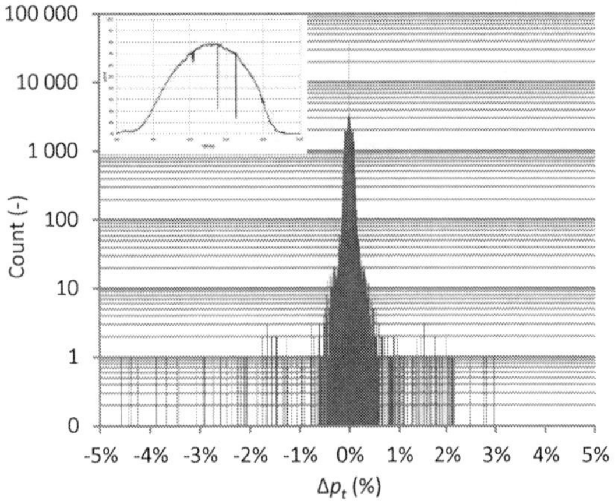

Figure 4. Statistical frequency of power changes in 1 second period—sunny day (PVPP no. 1)

In Figs. 4, 5, 6 and 7 statistical expressions of changes in frequency of PVPPs are depicted. These power changes are expressed as Δp_t and they are divided into intervals of 0,01% from installed power (interval of 0—0.01%, 0.01%—0.02%, 0.02—0.03% and so on).

In Figs. 4 and 5 it is visible that the power change between two measurements in a row was not greater than ±5% (only 1 measured point is higher in Fig. 5) of installed power during the sunny day. (The course of the power output of that day is displayed on the top of graph's). The results and indicators are shown in Tab. 2.

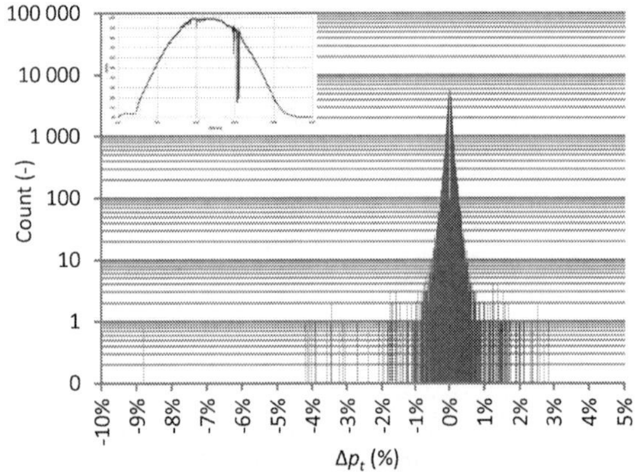

Figure 5. Statistical frequency of power changes in 1 second period—sunny day (PVPP no. 2)

The statistical expression of frequency of changes for both PVPPs are almost the same during sunny days. On the other hand, the power changes during a partly cloudy day were significantly higher and more frequent (Fig. 6 and 7).

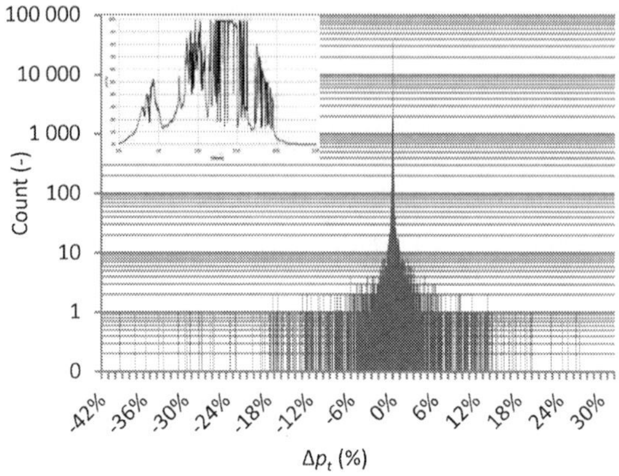

Figure 6. Statistical frequency of power changes in 1 second period—partly cloudy (PVPP no. 1)

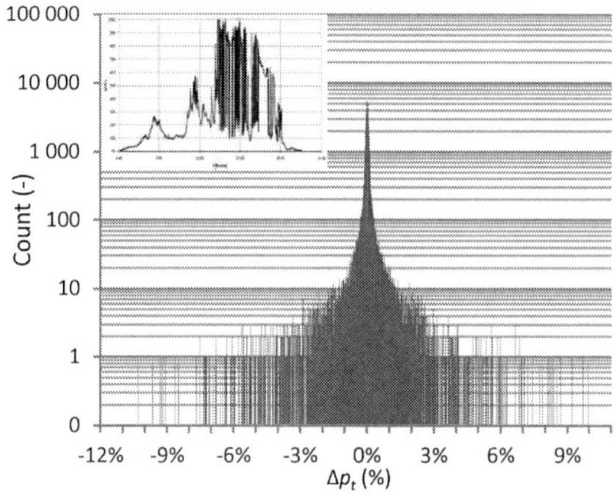

Figure 7. Statistical frequency of power changes in 1 second period—partly cloudy (PVPP no. 2)

The larger PVPP needs a larger cloud for the panels to be totally overlapped. Similarly, at the same wind speed (rather same cloud movement speed), the total overlap of large PVPP is slower than total overlap of smaller PVPP. It means that larger PVPPs (extending in a larger area) will lower absolute power changes and thus should achieve better results of Δp_{Φ} and s_p (mainly s_p). This hypothesis is confirmed by comparison of the statistical expression of changes in frequency of both PVPPs on partly cloudy day (Tab. 2).

Table 2. Results of analysis and value of dynamic changes indicators

Indicators	Sunny day		Partly cloudy day	
	PVPP no.1	*PVPP no.2*	*PVPP no.1*	*PVPP no.2*
Maximal positive change [%]	3.0	2.8	26.8	11.4
Maximal negative change [%]	-4.6	-8.8	-41.9	-10.3
Amount of changes upper than +10%	0	0	99	2
Amount of changes under -10%	0	0	132	1
Average of absolute values of power changes Δp_{Φ}	0.0347	0.0664	0.157	0.15
Square root of summed squared changes per day s_p	0.095	0.1277	1.025	0.5

V. PVPP AS A DEVICE IN SYSTEM OF SMART GRID

Operator of the transmission system orders in a way that each power plant with installed power higher than 50 MW has to provide supporting services in the *Operation rules* document [11]. It means, if more and more small power plants (with installed power under 50 MW) will be in place, and large fossil power plants will be closed, the regulation power of the power system will decrease. It will increase the demand for regulation power by installing new small non predictable or hardly predictable power sources (such as PVPPs and wind power plants). It is necessary to find a solution for this issue. One of the solutions could be joint regulation of various power plants within virtual power plants. If the aggregator will have more virtual power plants joined by an operation and optimization system, it could reach the required minimal capacities in *Operation rules* of transmission system operator [11]. Using an appropriate combination of hardly predictable power sources and systems with the possibility of accumulation, the system of a virtual power plant can be not only predictable, but also it can become an interesting player in the energy market. It is necessary to have at least one accumulation system with power regulation in the system of a virtual power plant. This accumulation system can be represented by various technologies: directly accumulated electric energy (batteries, bladder vanadium flow batteries and others) or power accumulation in primary sources (biogas, biomass and others). The regulation of supplied power from PVPP is technically possible, too. But it is not profitable due to the actual supporting system of RES in Slovakia.

VI. CONCLUSION

Statistical analysis mentioned above points out to dynamic changes of PVPP power supply during partly cloudy days. Those days are the most difficult for prediction. It was shown that PVPP with higher installed capacity showed lower dynamic output changes. For a small PVPP up to 10 kWp (residential PVPP), the second dynamic changes can be more than 30% during partly cloudy days. Although, the power consumption of households might change, too. As a result, the 15-minute integrated production mathematically covers the consumption, but on the other hand in some moments the consumption can exceed the production of PVPP or vice versa. However, the production and the consumption on electricity meter are the same during the 15-minute period.

PVPPs with installed power of 100 kW and more (some MW) have second dynamic power change lower than small PVPPs (of approx 11%). Although, this does not mean that their performance does not suddenly change in the event of a cloud crossing.

This comparison could be beneficial for virtual power plant operators, who have one or more PVPPs in their balance responsible party. It can be considered as a valid input for the virtual power plant operation.

One of the ways to become independent from importing fossil fuels is deployment of power plants which have primary sources in the inland. One of the benefits of RES power plants is that most of them have primary energy sources at the place or close to the place of consumption. They are independent of their primary energy source, but on the other hand they are dependent on weather and other hardly predictive conditions. Therefore, issues with supplied power fluctuation and the energy storage are required to be solved. Virtual power plants can bring solutions to these challenges and make the RES power plants more competitive against traditional power plants on the energy market.

ACKNOWLEDGEMENT

This article was supported by the Agency for support for research and development on the basis of the contract no. APVV-20-0157, APVV-19-0049, APVV-18-0029 and the Operational program Integrated infrastructure for the project International centre of excellence for the research of intelligent and safe information-communication technologies and systems – II. phase, code ITMS: 313021W404, partially financed from the sources of the European fond for regional development.

REFERENCES

[1] REN21 Secretariat. (2021) "Renewables 2021 Global Status Report." *REN21*, [On-line], pp. 37. ISBN 978-3-948393-03-8, Available: www.ren21.net/wp-content/uploads/2019/05/GSR2021_Full_Report.pdf [Feb. 4, 2022].

[2] POWER editorial staff. (2017) "Who Has the World's Most Efficient Coal Power Plant Fleet?" *POWER*. [On-line], Available: www.powermag.com/who-has-the-worlds-most-efficient-coal-power-plant-fleet/ [Feb. 4, 2022].

[3] F. Janíček, M. Búran, A. Cerman, T. Hanus, J. Kubica, M. Szabová. "Obnoviteľné zdroje na Slovensku a ich vplyv na elektrizačnú sústavu," in Spektrum STU, 1st ed., Bratislava, 2019, pp. 9-14, pp. 109-112. ISBN: 978-80-227-4966-4.

[4] Slovenská elektrizačná prenosová sústava. (2021). "Individuálna a konsolidovaná výročná správa 2020," *SEPS*, pp. 24-26, [On-line]. Available: https://www.sepsas.sk/media/5027/seps-sk-2020-final.pdf [Feb. 22, 2022].

[5] F. Janíček, J. Poničan, M. Sadloň. "Modified discounted payback period of various small-scale photovoltaic installations," in *10th International Scientific Symposium on Electrical Power Engineering, Elektroenergetika*, 2019, pp. 468-473.

[6] Ministry of economic of Slovak republic. (2007) "Stratégia vyššieho využitia obnoviteľných zdrojov energie v SR." [On-line], Available: www.economy.gov.sk/uploads/files/MuZlb3Ut.pdf [Feb. 22, 2022].

[7] M. Koníček, A. Beláň, M. Ivanič, "Impact of Distributed Sources on Safety Operation of Power System," in *8th International Scientific Symposium on Electrical Power Engineering, Elektroenergetika*, 2015, pp. 492-495.

[8] SEPS. "Data from Damas Energy information system for market control of transmission system of system operator, march of 2020." Internet: https://dae.sepsas.sk/public/?VIEW_CODE=SYS_DE-SS_PUB-PUB-DATAFLOW-PRODUCTION_CONSUMPTION_VIEW [Mar. 8, 2022].

[9] E. Lorenzo, G. L. Arajújo, A. Cuevas, M. A. Egido, J. C. Miñano, R. Zilles, "Solar electricity engineering of photovoltaic system," in *Artes Gráficas Gala*, S. L. Spain, 1st ed., 1994, pp. 69-80. ISBN 84-86505-55-0.

[10] I. Tomiš, F. Janíček, M. Kolcun, "Improving Photovoltaic Power Production Profile with Regard to power Output Variability," in *8th International Scientific Symposium on Electrical Power Engineering, Elektroenergetika*, 2015, pp. 448-450.

[11] Slovenská elektrizačná prenosová sústava. (2021) "Technické podmienky prístupu a pripojenia, Pravidlá prevádzkovania prenosovej sústavy, Dokument B," 22nd ed., pp. 5, [On-line]. Available: www.sepsas.sk/media/5329/dokument-b-tp-januar-2022.pdf [Feb. 8, 2022].

Impacts of EU Taxonomy Implementation on the Energy Sector

Adam Kubín, Tomáš Králík, Jiří Vašíček
Department of Economics, Management and Humanities
Czech Technical University
Prague, Czech Republic
kubinada@fel.cvut.cz, tomas.kralik@fel.cvut.cz, vasicek@fel.cvut.cz

Abstract — **This paper analyzes the implications of the EU Taxonomy, as the European Union (EU) instrument to mitigate climate change and reduce import dependency, for the European energy sector in the context of resource adequacy. Resource adequacy is assessed in the medium-term horizon using the case example of the Czech Republic (CZ) as a country facing a significant energy transition, as its resource mix is still largely made up of coal. The analysis is based on the country's energy balance and the available outlooks published in the energy concepts and strategic documents of the CZ. These outlooks are quantitatively analyzed and critically assessed. According to the current set of conditions for new resources resulting from the EU Taxonomy, there is a risk that in the medium term (2040) there will be no new resources, technologies and fuels available in the CZ that would comply with the conditions of the EU Taxonomy and at the same time cover the country's energy needs. This may lead to the situation where Czech Republic's electricity system will be inadequate in terms of generation and installed capacity sooner than would be the case if the implications of the EU Taxonomy were not considered. As a result of the EU Taxonomy, other European countries on the threshold of energy transition may face a similar problem and their energy security may be compromised. The results show that EU Taxonomy can have a significant impact on future development of European energy sector in the context of energy mix.**

Keywords: EU Taxonomy; energy mix; resource adequacy; Green Deal

I. INTRODUCTION

Europe is currently undergoing a significant geopolitical and energy transformation against the backdrop of global climate change. The European Union (EU) is the current world leader in climate, energy and environmental policy. The EU has committed to becoming climate neutral by 2050 through the Green Deal [1]. The reason for this transformation is not only to mitigate climate change, but also to reduce the EU's import dependence on fossil fuels from other continents, which is proving more important than ever in the current context of the war in Ukraine and the disruption of geopolitical relations between Russia and the EU. Indeed, energy insecurity and import dependence can have a negative impact on energy security and on the economy of the importing continent.

The current tense situation and fears of supply stoppages are resulting in a very significant increase in the prices of energy commodities. According to statistics [2], payments to Russia for natural gas formerly amounted to EUR 190 million per day. At the beginning of March, these payments amounted to EUR 610 million per day. This is about a 320% increase. According to a report by the International Energy Agency [3], Russian gas supplies to the EU in 2021 amounted to almost 155 bil. m^3, representing around 45% of the EU's total gas imports and almost 40% of its total consumption. According to [4], Norway is the second largest importer of natural gas into the EU, followed by Algeria and Qatar. The latter two countries cannot be considered politically stable regions, like Russia. It is therefore questionable whether the EU should rely on this source of energy and whether it is able to manage the planned transition to carbon neutrality in this situation. Given that this is a long-term transition, it can be assumed that the final 2050 target will be maintained, with likely adjustments to sub-targets.

One of the tools to feed EU climate policy, mitigate climate change and reduce import dependency by 2050 is the EU classification system or EU Taxonomy. As a scientifically based tool, the EU Taxonomy classifies economic activities in terms of their sustainability. The purpose is to guide investments in sustainable projects and activities to help the energy transition towards a green and self-sufficient energy. The EU Taxonomy does not only cover activities in the energy sector, but also in sectors such as transport, forestry, manufacturing and construction. The legislative framework of the EU Taxonomy is anchored in the EU Taxonomy Regulation 2020/852, which entered into force on 12 June 2020 [5]. This regulation defines the terminology and at the same time establishes the framework for the EU Taxonomy by setting out conditions that each activity must meet to qualify as environmentally sustainable. The Regulation also defines environmental objectives for which the Regulation obliges European Commission (EC) to establish technical criteria through Delegated Acts. The Acts define the technical screening criteria (minimum requirements) by which certain economic activities qualify as significantly contributing to these objectives and by which it can be determined whether these economic activities do not significantly harm any of the relevant environmental objectives.

The EU Taxonomy thus does not in principle prohibit or mandate anything. It "only" cuts off non-green activities or technologies from EU funds, making it more difficult to obtain financing from banks and other financial regulated institutions, which is a common way of co-financing larger projects. The investor can pay for the investment from his own funds or loans from the state, but the cost of own capital is usually higher than that of foreign capital, which leads to complications with the return on projects and pressure on the increase in prices of the commodity produced and consequently lower competitiveness. The technology requirements imposed by the screening criteria can also mean significantly higher investment costs for some projects, which discredits the purpose of the project. The EU

Taxonomy therefore does not theoretically restrict Member States in the energy mix they choose. However, at a practical level, it is certain that rational investors will not put their money into activities outside the green list.

The topic of the EU Taxonomy and its effects (impacts) on different sectors is addressed in many publications. Publication [6] focuses on the impact of the EU Taxonomy on the greenwashing in the financial sector, which according to the authors is owing to it reduced. Study [7] deals with a general assessment of the financial impact of EU Taxonomy. According to the authors, the EU Taxonomy as a tool will help direct financial resources towards sustainable economic activities and help fill investment gaps in relevant sectors. The impact of the EU Taxonomy on the energy sector only is addressed in paper [8]. The author is of the opinion that the EU Taxonomy will have a very significant impact on investment decisions in the energy sector. This statement is confirmed in general terms, i.e., without taking into account the specific sector, by [9] and [10].

A yet unexplored issue is the assessment of the impact of the key technical screening criteria on the energy sector, specifically the electricity sector and marginally the district heating sector (DHS) in terms of resource adequacy. This article is focused on the analysis of generation adequacy and installed capacity adequacy in the electricity sector in the years 2025, 2030, 2035 and 2040 for the case example of the Czech Republic (CZ), as a typical representative of a country that is still dependent on fossil fuels, especially coal, and which is facing a massive energy transition (decarbonization) in the coming years. The analysis conducted for the period until 2040 for the case example of the CZ is based on the energy balance of the country and the available outlooks published in the energy concepts and strategic documents of the CZ. These outlooks are quantitatively analyzed and critically assessed. The DHS in the CZ is only marginally addressed in the discussion, as by its nature there are no conceptual strategies for its development.

II. Methodology

This chapter describes the methodology for assessing the impacts of the EU Taxonomy on the resource adequacy of the energy mix.

A. *Identification of key Screening Criteria*

Firstly, key screening criteria from the Delegated Act [11] and the Complementary Delegated Act [12] related to economic activities in the energy sector are identified to analyze the impact of the EU Taxonomy on the future development of the energy mix. The first Delegated Act, which entered into force in 2021 and became effective on 1 January 2022, defines technical screening criteria for economic activities that contribute to climate change mitigation and adaptation objectives. For the remaining objectives complementary acts will be published in the future. The last and important change from an energy perspective was the approval of the Complementary Delegated Act on 2 February 2022, where the criteria related to nuclear and natural gas energy generation were published. If the Act goes through the approval process, it will enter into force on 1 January 2023. Currently, there is no indication that approval will not be forthcoming.

B. *Energy Mix Outlook*

Subsequently, an analysis of the current state of the CZ generation mix is made, both in terms of the installed capacity of individual types of resources and in terms of the energy produced by them. The analyzed data is taken from the to Energy regulatory Office (ERO) Annual Report on the Operation of the Electricity System (ES) for 2020 [13], the Preparation of Operation for 2022 issued by the transmission system operator ČEPS [14] and the ENTSO-E Transparency Platform [15].

The development of the resource mix is a part of the national strategic concept documents, which, however, due to the frequency of their update, may be outdated and thus not reflect current legislative and geopolitical developments. Since the State Energy Concept is outdated (2015), the document from the ČEPS: Resource Adequacy Assessment of the Electrical Grid of the Czech Republic until 2040 (MAF CZ) [16] is used for the purposes of the resource mix outlook of the CZ. This annually published document analyzes the impacts of different scenarios for the development of the resource mix in the CZ in the medium term and highlights potential resource shortages in the grid. The most recently published document respects the current EU legislation, however, due to the novelty of some of the EU legislation documents, their influence will be considered in the next edition of MAF CZ. That is the case of newly presented strategy REPowerEU and the issued Complementary Delegated Act. Because MAF CZ does not fully consider the impact of the EU Taxonomy, in addition to the outlooks presented in the MAF CZ, the authors' modified outlook in this paper is also presented, respecting the impact of the EU Taxonomy on the shape of the resource mix of the CZ in the following years. Based on this adjusted outlook of the resource mix, an assessment of the resource adequacy of the ES is made.

For the authors' impact analysis of the EU Taxonomy on a future development of the energy mix in this paper, a MAF CZ outlook of the resource mix corresponding to a conservative scenario is considered as a reference. Given the geopolitical situation, it seems most likely. The conservative scenario in the MAF CZ expects the decline of coal resources by 2038, which is different from the progressive scenario, which foresees a decline already in 2033. Part of these resources is replaced by gas resources. In the DHS, a transformation to natural gas is expected by the end of 2030. In terms of nuclear power development, a new nuclear unit at Dukovany (1 140 MW) is expected by the MAF CZ to be commissioned from 2036. The development of renewable energy sources (RES) is realistically predicted in relation to the financial resources of the Modernization Fund and other programs designed to support RES development. The outlook in the MAF CZ also considers the development of the generation categories: battery storage and combined heat and power plant, which, however, do not correspond to the standard division of the resource mix. Battery storage is not considered for the purposes of the resource adequacy assessment in this paper, as it is assumed that the installed capacity of battery storage will be used primarily for the purpose of providing power balance services, the volumes of which could be increased due to the planned increase in installed capacity of intermittent RES. The development of the installed capacity of the combined heat and power plant category is also not considered separately, but as part of the coal and gas categories. It should also be noted that in terms of emissions produced by the energy sector, even the MAF CZ conservative scenario is in the line with the 2030 targets presented in the Fit for 55 package. According to the results of the ČEPS, considering the conservative scenario, the resource adequacy should not be breached until 2040, when the operation of such a system would result in an increased number of supply outages and grid instabilities that would exceed the permitted limits of adequacy assessment indicators.

C. *Resource Adequacy Assessment*

Based on the expected development of the energy mix of the ES of the CZ, an assessment of resource adequacy is made. Resource adequacy is examined in terms of generation adequacy, i.e., a comparison of the annual balance of net generation and consumption, and installed capacity adequacy. The annual energy production of individual resource categories is based on the outlook presented in the MAF CZ and on the authors' own analysis of the time of utilization of the installed capacity of these resource categories.

While [17], [18] or [19] use simulation or probabilistic models to assess the installed capacity adequacy, in this paper, for given years the load coverage by the deployed capacities is assessed deterministically. Thus, this is not an ENTSO-E compliant methodology and there are no outputs in the form of adequacy assessment indicators. The next section of this chapter is devoted to a description of the methodology for deterministic installed capacity adequacy assessment.

The assessment of installed capacity adequacy evaluates the ability of the available generation mix to meet the demands of the electricity system under both favorable and unfavorable conditions. The main objective of this analysis is to highlight the worst-case scenarios in terms of installed capacity adequacy that may commonly occur on the system and how resources within the ES can cope with them. These are not even exceptional conditions resulting from an unplanned outage of a major resources, but rather conditions that, based on available data, may commonly occur in the future period in view of expected growing consumption and increasing intermittent generation.

Information of the current operating regime is used to estimate the future expected deployment of each resource group. Historical data available on the ENTSO-E Transparency Platform website for 2019, 2020 and 2021 on the deployment of each category of resources is used. The expected size of the system load is proportionally adjusted based on the information concerning the total consumption in the analyzed time series according to the outlook available in the MAF CZ and the 2021 consumption curve available on the ČEPS website [20]. Installed capacity adequacy is examined for the following cases:

- Day with the occurrence of a maximum hourly load of the ES (winter period), with minimum and maximum production of intermittent RES
- Day with the occurrence of a maximum hourly load in the summer period during the outage of one nuclear unit, with minimum and maximum production of intermittent RES

Based on the historical deployment of each category of generation resources, the years 2025, 2030, 2035 and 2040 are analyzed with respect to the assumptions of maintaining the same operating modes in the expected resource mix. Individual generation resources are considered in the following categories: nuclear, coal (brown, lignite, hard, biomass), gas, solar, wind, hydro (run-of-river and poundage and water reservoir), hydro pumped storage, other RES (biogas, biomass). The following reasons for the unavailability of installed capacity are considered when estimating the deployment of each category of resources:

- Procurement of balancing reserves (power required to ensure the security of electricity supply)
- Maintenance and repairs (shutdowns of combustion resources and nuclear power plants for refueling)
- Unavailable capacity - reduction in the available capacity of resources for various reasons (unintentional unavailability, e.g., for heat supply; constraints due to availability of primary energy like solar, wind or hydro and intentional unavailability like retrofits or greening)

If the deployed capacity of the resources is higher than the load, it means that there is enough capacity in the ES to meet the electricity demand. Otherwise, the lack of capacity indicates the need to import electricity.

III. Key Screening Criteria for Deployment of Energy Sector in Terms of Resource Adequacy

The technical screening criteria refer to thresholds or performance levels (minimum requirements) that an activity should achieve to qualify as a significant contributor to one of the climate objectives while avoiding significant harm to the other environmental objectives. For activities related to energy production, the key screening criteria are usually emission limits, safety requirements and time constraints. Mentioned Acts, or their annexes, are structured according to the economic sectors and economic activities concerned, for which technical criteria are defined. It is a list of activities and to each of them are defined the technical conditions for their classification as sustainable.

Energy generation activities and conditions of performing these activities are crucial regarding the development of the Czech Republic's energy mix. The identified screening criteria of each energy production activity relating to the climate objectives: climate change mitigation and climate change adaptation are used in this paper to assess the feasibility of activities, i.e., their possible inclusion in the future energy mix. The structure of the evaluation of the technical screening criteria is maintained in this article as in the Acts, i.e., for each energy generation activity the key limiting screening criteria are highlighted in relation to the future development of the energy mix.

From the analysis of the significant limiting screening criteria for energy generation activities, it can be concluded that in terms of the electricity and heating sectors, there should generally be no problem with the construction of RES in the future, as they can be considered sustainable in the long term and do not impose unreasonable technical requirements. The only significant problem can be identified in the construction of biomass-fired power and heating plants, because using biomass must be environmentally sustainable, i.e., it must only be grown on a specific site and its import distance is limited. Thus, the EU Taxonomy should not significantly affect the planned construction of new solar, wind and hydro power plants; energy can also be obtained from geothermal energy, renewable non-fossil gaseous and liquid fuels (e.g., hydrogen or biomethane) and bioenergy (e.g., biomass or biogas). Thermal energy can also be produced from solar thermal collectors and waste heat. For these selected resources it should not be a problem to reach the limit of 100 g CO_2e/kWh of energy output for life cycle emissions (and for some 270 g CO_2e/kWh of direct emissions), as confirmed by the results of a study published by the United Nations Economic Commission for Europe [21].

The situation is different for nuclear and natural gas energy generation, which can be classified as temporarily green, i.e., they will only be supported up to a certain point in time and then they will no longer appear on the list of sustainable investments. The transitional nature of nuclear power is that the expansion of existing nuclear power plants, i.e., increasing their installed capacity, will require operating authorization (construction permit) until 2040. A construction permit for this expansion can only be granted if a Member State presents plans for the construction of a deep repository for radioactive waste that will be in operational by 2050. For new nuclear power plants, the deadline for obtaining an operating authorization is set for 2045 under the same conditions. Therefore, the prospect of building new nuclear power plants or expanding them to supply electricity and heat cannot be envisaged if a country does not plan or has not already built a nuclear waste repository. Nor should there be a problem with a life cycle emissions threshold for nuclear power. According to [21], nuclear power generation has the lowest life cycle emissions compared to other technologies, with only 5.1 to 6.4 g CO_2e/kWh.

The transitional nature can also be identified in the case of natural gas-fired power generation, which is also severely constrained by strict technical conditions. It can be identified as sustainable under the current criteria if either the cross-cutting threshold of 100 g CO_2e/kWh of energy output for life cycle emissions is met or if the whole set of criteria is met simultaneously. In view of the results of the study [21], meeting the life cycle emissions threshold is not very realistic, as life cycle emissions for a gas-fired power plant were calculated to be in the range of 403 to 513 g CO_2e/kWh. In the case of a waste heat recovery plant, the emission results would certainly be better, but would most likely not exceed the threshold. The transient nature of the activity can be seen from the second option for classifying the activity as sustainable, which is the recognition of natural gas-fired generation only until the end of 2030. Moreover, this condition is subject to the cumulative fulfillment of the sub-criteria. The first criterion is the limit on direct emissions, which must not exceed 270 g CO_2e/kWh. In view of [21], which states that over 80% of the emissions produced during the life cycle of a gas resource are direct, this means that new gas resources will have to have carbon capture and storage technology installed. In terms of the planned construction of new gas resources, this means a significant increase in financial investment costs. Furthermore, it is also the case that green status will only be granted to gas resources that are built as a direct replacement for fossil fuel burning resources, and only if they cannot be replaced by RES. An essential condition for the planned construction of new gas-fired resources is that they must burn 100% low-emission fuel from 1 January 2036. Biomethane and primarily hydrogen are under consideration. If Member States fail to meet these conditions, no new gas-fired power plants can be expected. This requirement is a challenge to local market opportunities and the availability of low-emission fuel technology. However, this may also be a problem from a pan-European perspective, where the demand for production facilities capable of burning these gases (e.g., gas turbines) will not be sufficiently met by 2036. This poses a significant problem for the electricity and heating sector as new gas-fired resources will not be able to be developed without the provision of sufficient quantities of emission-low fuels and technologies capable of burning these fuels. It should also be mentioned that there are still no concepts or plans at present to obtain the necessary quantities of these gases by the end of 2035 or to address specific technical problems. The following section is devoted to the application of the practical implications of the EU Taxonomy to the future energy mix of the CZ, as a state which energy transition is based on nuclear and natural gas.

IV. Energy Mix Outlook of the Czech Republic Depending on Key Screening Criteria

This chapter is devoted to the description of the current state of the generation mix of the ES of the CZ and the energy balance of the DHS of the CZ. Furthermore, the considered scenario of the development of the generation base of the ES in relation to the presented impacts of the EU Taxonomy and the projected development of the consumption of the CZ is presented.

A. Current State of Generation Mix

The current state of the resource base (structure of the resource mix) of the CZ is shown in Fig. 1. The figure shows that coal-fired resources still have the largest share in terms of installed capacity (45%), followed by nuclear (20%), solar (11%), gas-fired resources (9%), pumped storage and hydro power plants (6% and 5%) and wind power plants (2%). The Other RES category, which is mainly made up of biogas-fired resources, accounts for the remaining 2%. The total capacity of all resources in the electricity grid of the CZ is 20 333 MW.

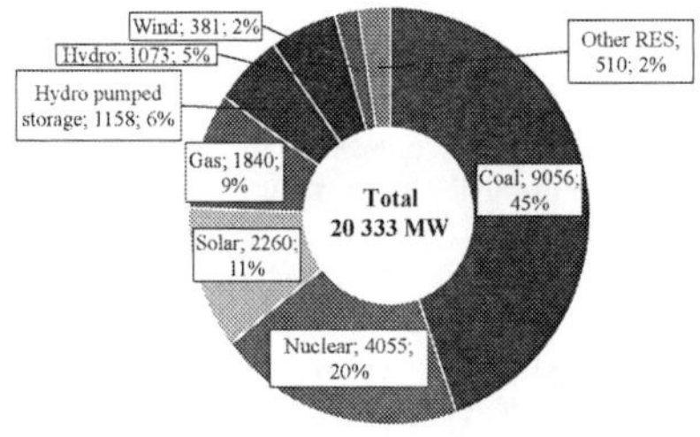

Figure 1. Overview of the net installed capacity of individual types of resources within the electricity system of the Czech Republic in 2021 [MW], Data source: ERO, ENTSO-E Transparency Platform, ČEPS - Preparation of operation for 2022; authors' own processing

Fig. 2 shows the share of resources in the net electricity generation in the ES of the CZ in 2020. The data come from the ERO Annual Report on the Operation of the Electricity System for 2020 and from the analysis of generation time series data from the ENTSO-E Transparency Platform. The figure shows that nuclear power plants contribute significantly more to generation than installed capacity due to their baseload use and high annual installed capacity utilization (over 7 000 hours). On the other hand, solar plants contribute significantly less to generation compared to installed capacity, due to the seasonality of generation (annual installed capacity utilization of around 1 000 hours). The annual net electricity consumption in the CZ, including losses and consumption for hydro pumped storage pumping, is around 66 TWh. The annual balance of production/consumption in the CZ is thus currently export-oriented.

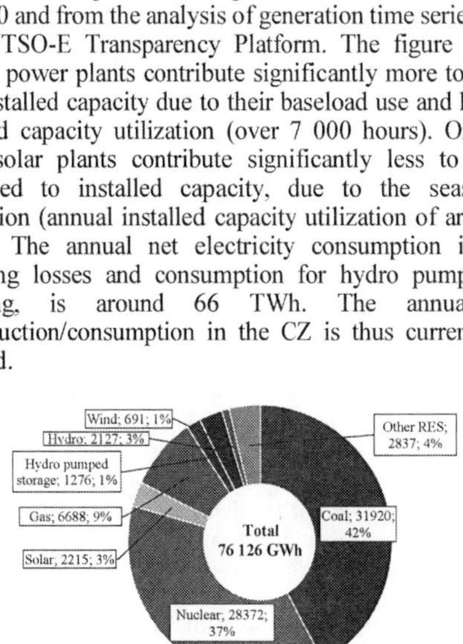

Figure 2. Overview of net production of individual types of resources within the electricity system of the Czech Republic in 2020 [GWh], Data source: ERO, ENTSO-E Transparency Platform, authors' own processing

B. Energy Mix Outlook of the Czech Republic

Following Tab. 1 presents the outlook for the resource mix of the CZ based on the available predictions in the MAF CZ and on the authors' own analysis of the impact of the EU Taxonomy on the development of the resource mix (see the number in brackets where values differ from the MAF CZ outlook). Despite the shutdown of coal-fired power plants, there is an overall increase in installed capacity until 2035, which is mainly due to an increase in the installed capacity of solar and wind power plants. The increase in installed capacity of RES is fully in line with the EU Taxonomy and therefore no changes have been made compared to MAF CZ outlook. Decrease of installed capacity of category Other RES in 2030 is due to the transformation of biogas plants into biomethane plants. The changes in the outlook concern only gas resources, specifically in the 2030, 2035 and 2040 time series. In 2025, the increase in installed capacity of gas-fired resources is left as presented by the MAF CZ, as these values are based on the questionnaire survey, and it is very likely that these new gas-fired capacities already have construction permit and thus will not be subject to the EU Taxonomy. However, the MAF CZ foresees a further increase in the installed capacity of gas-fired power plants in future years, both due to the decarbonization of the DHS (additional approx. 1 150 MW) and to compensate for the loss of installed capacity in the coal-fired power plant category (additional approx. 600 MW). However, due to a number of technical and economic constraints as outlined in previous chapter, it can be assumed that new gas-fired plants will not be built. Therefore, in this paper authors consider a lower installed capacity of gas sources by the mentioned approx. 600 MW (which should replace coal sources) compared to the MAF CZ's outlook. The decrease in the MAF CZ's outlook between 2030 and 2040 is due to the decline in the installed capacity of thermal power plants. The assumption of the transition of the DHS to gas by 2030 is maintained due to the very limited possibilities of local heating systems. The renewal of gas capacities could occur after 2040, when according to the Hydrogen Strategy [22] the gradual use of hydrogen in the energy sector is expected.

Table 1. Considered outlook of the electricity system resource mix in the years 2025, 2030, 2035 and 2040 based on the values in the MAF CZ and on the basis of the EU Taxonomy impact analysis [MW]

Installed capacity [MW]	**2025**	**2030**	**2035**	**2040**
Solar	4 268	8 133	9 317	10 005
Wind	529	742	942	1 141
Other RES	555	451	538	648
Hydro	1 073	1 073	1 073	1 073
Hydro pumped storage	1 158	1 158	1 158	1 158
Gas	2 802	4 545 (3 929)	4 277 (3 661)	3 975 (3 359)
Coal	6 238	2 851	2 382	0
Nuclear	4 047	4 047	4 047	5 187
Total	**20 670**	**23 000 (22 384)**	**23 734 (23 118)**	**23 187 (22 571)**

One could speculate on an increase in installed capacity for nuclear, as the CZ does not have and does not plan to build a nuclear storage facility, which would mean a failure to meet the taxonomic condition. However, in view of all the available information, the completion of the Dukovany unit should be fully paid for by the state. The installed capacity outlook of the nuclear power plants in the CZ in this article is the same as in MAF CZ.

C. *Electricity Consumption Outlook of the Czech Republic*

Based on the available forecasts in the MAF CZ, the electricity consumption outlook is considered as shown in Tab. 2. This is the net electricity consumption including losses in the grid and consumption for hydro pumped storage pumping. The consumption outlook corresponds to a conservative scenario, i.e., medium electrification. Consumption is on an upward trend. This is due to the expected growth in the number of electric vehicles and heat pumps. From the projected load outlook, consumption peaks are also determined to assess the installed capacity adequacy of the CZ.

Table 2. Considered development of net electricity consumption and development of winter and summer maximum loads for the years 2025, 2030, 2035 and 2040 based on the conservative scenario in the MAF CZ [TWh/year] and consumption curve in 2021 [MW], modified

Year	**2025**	**2030**	**2035**	**2040**
Consumption [TWh]	73.0	76.6	81.2	83.7
Winter max. load [MW]	11 991	12 578	13 333	13 752
Summer max. load [MW]	9 123	9 569	10 144	10 463

D. *Energy Balance of District Heating Sector in the Czech Republic*

Following Fig. 3 shows the current annual energy balance of the DHS in the CZ, or the share of fuels in the production of thermal energy. The data is from the ERO Annual Report on the Operation of Heating Systems for 2020 [23].

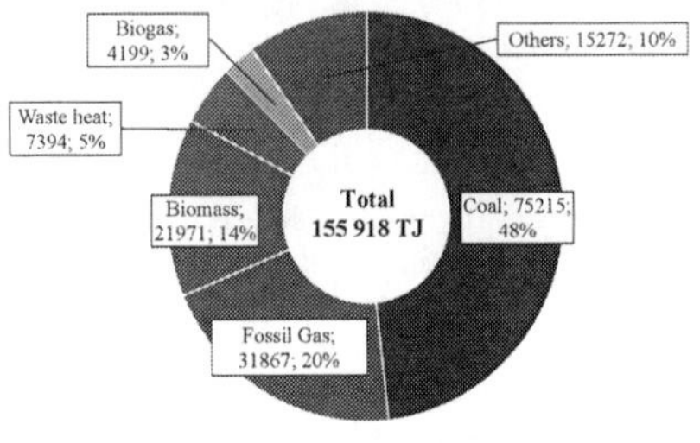

Figure 3. Overview of the share of fuels in the production of thermal energy in the district heating sector of the Czech Republic in 2020 [TJ], Data source: ERO, authors' own processing

The total thermal energy production in 2020 was 155.9 PJ. Most of the thermal energy is still generated by coal (48%), followed by natural gas (20%) and biomass (14%). The net supply of thermal energy in 2020 was about 85.9 PJ. The rest of the energy was essentially self-consumption by heat and power plants and losses. The total installed thermal capacity of the heat production plants is about 40 GW_t. However, a comprehensive concept for these resources is missing for the Czech DHS. This is because district heat generation is a local monopoly and therefore significantly different from electricity generation and distribution. For this reason, the impact of the EU Taxonomy on the Czech DHS is only marginally addressed in the discussion using annual heat energy balances.

V. Results

This chapter presents and describes the results obtained from the analysis of the generation and installed capacity adequacy of the ES of the CZ affected by the consequences of the EU Taxonomy.

A. *Generation Adequacy*

Based on the considered outlook of the resource mix of the ES (Tab. 1) and the analysis of the time utilization of the installed capacity of individual categories, the expected net production of the considered resource mix was determined for individual years. Fig. 4 presents the results of the annual balances of net electricity production and consumption.

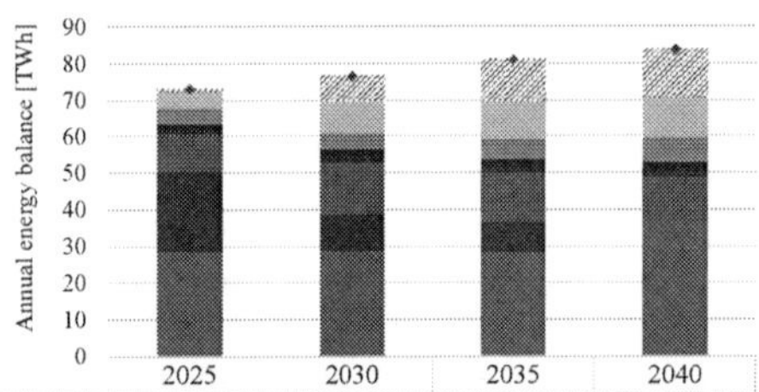

	2025	2030	2035	2040
Import	0,7	6,9	11,8	13,2
Solar	4,7	9,0	10,4	11,1
Wind	1,3	1,9	2,6	3,3
Other RES	3,0	2,4	2,9	3,5
Hydro	1,8	2,1	2,4	2,5
Hydro pumped	1,1	1,2	1,4	1,5
Gas	10,2	14,3	13,3	12,2
Coal	21,8	10,0	8,3	0,0
Nuclear	28,5	28,6	28,2	36,5
Consumption	73,0	76,6	81,2	83,7

Figure 4. Generation adequacy outlook of the electricity system of the Czech Republic, Data source: MAF CZ, authors' own calculation

It is evident that despite the increase in installed capacity, the generation adequacy of the ES of the CZ will not be ensured in the future. This is because the installed capacity of coal-fired power plants with an annual utilization of around 3 500 hours is replaced by RES, which have an annual utilization of around 1 000 hours for solar and 2 000 hours for wind in the conditions of the CZ.

This means that the CZ could become an importing country in the future with imports of around 13,2 TWh in 2040.

B. Installed capacity adequacy

Based on the considered outlook of the resource mix of the ES the installed capacity adequacy assessment was performed for individual years. The results are summarized in Tab. 3.

Table 3. Installed capacity adequacy assessment of the electricity system of the Czech Republic

Year		2025	2030	2035	2040
Summer day with max. load	max. RES generation	OK	not OK	not OK	not OK
	min. RES generation	not OK	not OK	not OK	not OK
Winter day with max. load	max. RES generation	OK	not OK	not OK	not OK
	min. RES generation	not OK	not OK	not OK	not OK

The results show that under unfavorable conditions for the ES (high load), the installed capacity adequacy at the minimum RES production can be disturbed already in 2025. Fig. 5 depicts resource deployment and daily load in 2025 in a summer day with maximum and minimum generation from RES. If the generation from RES is high, the load is sufficiently well covered. On the other hand, imported power is needed in case of lower RES production. In the next analyzed years, the situation worsens as coal-fired power plants are phase out and even higher RES generation does not solve the inadequacy. Although coal-fired sources are being replaced be new resources (the total installed capacity of the ES is increasing), these resources are not able to adequately replace them due to the nature of their production. This situation is depicted in Fig. 6.

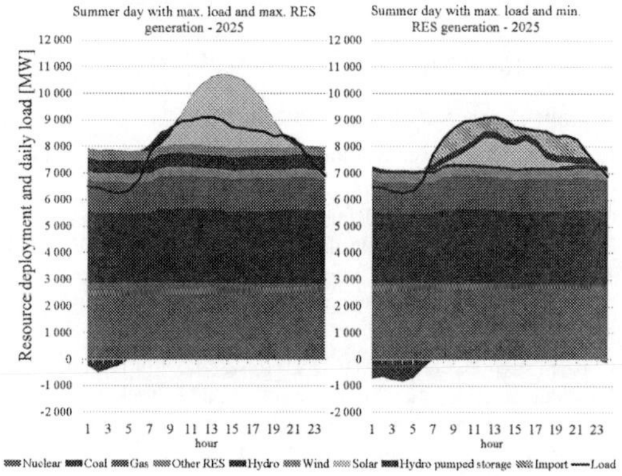

Figure 5. Installed capacity adequacy in 2025, Data source: authors' own processing

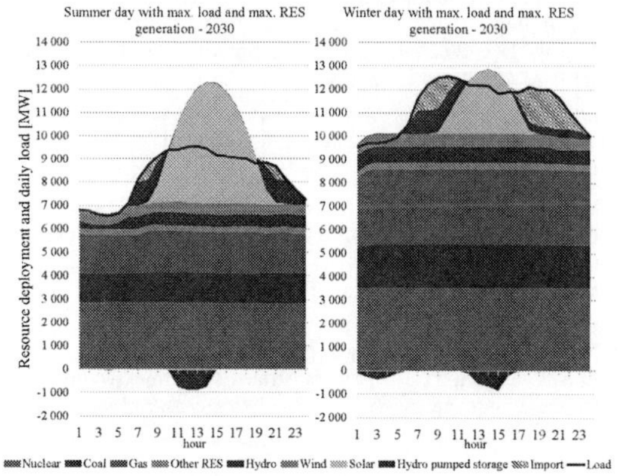

Figure 6. Installed capacity adequacy in 2030, Data source: authors' own processing

Installed capacity inadequacy will not improve even with the commissioning of a new nuclear unit, as shown in Fig. 7, which shows the state of resource deployment and load in 2040. In addition to increased nuclear capacity, Fig. 7 also shows zero coal generation and very high-RES generation. However, storage elements in the system would be needed to use it in the time

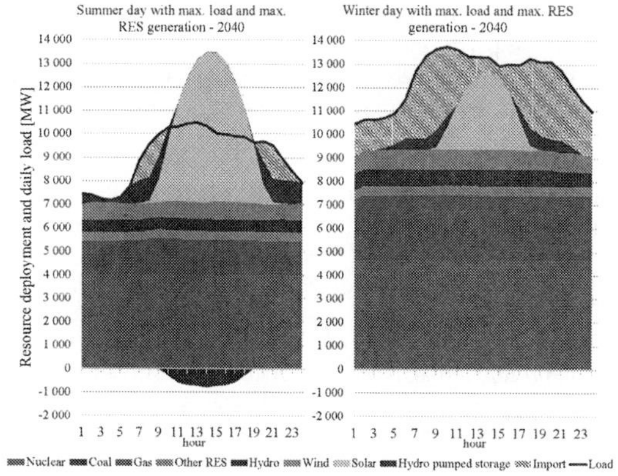

required.

Figure 7. Installed capacity adequacy in 2040, Data source: authors' own processing

VI. Discussion and Conclussions

According to the current set of conditions for new resources resulting from the EU Taxonomy, there is a risk that in the medium term there will be no new resources, technologies and fuels available in the CZ that would comply with the conditions of the EU Taxonomy and at the same time cover the country's energy needs. This may lead to the situation where the ES of the CZ will be resource inadequate sooner than would be the case if the implications of the EU Taxonomy were not considered. In the case of generation adequacy, the CZ can be expected to be an importing country from 2025 onwards, which will remain unchanged in the monitored outlook. In case of installed capacity adequacy, the CZ is not prepared for future increases in total load. Higher RES generation may help avert unfavorable conditions for the system as early as 2025, while from 2030 the CZ will have to rely on electricity imports from abroad in such situations (this require a possibility of electricity imports in the CEE region!). The EU Taxonomy contributes to the power inadequacy mainly by the fact that controllable sources comparable in technology and generation character are not built to replace decommissioned coal-fired sources. The case of the CZ should be a warning to other EU countries on the threshold of energy transition, e.g., Poland, Bulgaria or Estonia. These states may face a similar problem as a result of the EU Taxonomy and their energy security may be compromised. The results show that EU Taxonomy can have a significant impact on future development of European energy sector in the context of energy mix.

However, the EU Taxonomy does affect not only the future development of the electricity sector but also the heating sector. The situation here is more complex, as most district heating systems are local monopolies. It is therefore not possible to secure supplies of thermal energy from other systems or even to import it from foreign countries. As a result, given the local marginal conditions of the systems, there is simply no alternative but to decarbonize the DHS temporarily with new natural gas-fired resources. This is because the investor has no other option, unlike investments in the electricity sector, where decentralized RES can be built on a suitable site and the electricity simply transmitted to the customer. This cannot be done so easily with a heat transfer medium such as water or steam. The use of RES in the heating sector is limited by its lower energy density. The ownership structure of the DHS, specifically the large representation of public institutions in the ownership of DHS, also plays a role in the decision-making of investors. It is also interesting to look at the whole issue from the perspective of the energy balance of the DHS and the available outlooks in the National Action Climate Plan (NCEP) [24].

Significant problems can be seen in the replacement of about 75.2 PJ of energy from coal. This energy should be replaced by natural gas and partly by biomass or biogas by 2030. There is a great deal of uncertainty about the potential of biomass and biogas for the heating sector. However, according to the outlook, an additional 14.7 PJ of coal could be replaced by biomass and an additional 9.3 PJ by biogas in 2030. Other technologies such as heat pumps and solar collectors could contribute a further 18.6 PJ. This means that the remaining 32.6 PJ of thermal energy would have to be replaced by new gas resources. If the taxonomic conditions were to be respected, this would mean that this energy would have to be replaced by low-carbon gases, hydrogen and biomethane after 2035. In terms of biomethane, a significant increase in production is expected, but also in use in other sectors. According to the NCEP data and maintaining the trend, biomethane could thus cover an additional 2.4 PJ in 2036. However, the biggest uncertainty about the development is in the hydrogen sector. According to the Hydrogen Strategy, hydrogen should be used more massively after 2040. For 2036, only about 1 PJ of hydrogen energy for the heating sector can be counted on. It is obvious from the energy balance that the production of renewable gases in the CZ will have to be accelerated and increased or the consumption must be reduced due to savings in the distribution system and final consumption. The question is whether we are able and ready to produce hydrogen on a larger scale. Constraints may be the technical immaturity of large-scale production and storage technology, the hydrogen distribution system and geographical conditions. However, given the current geopolitical tensions, the problem of natural gas substitution may come much sooner than in 2035. The very first plans (discussed as REPowerEU) are aiming at independency on Russian fuels by 2030. This would mean a significant increase in hydrogen and biomethane production and utilization. However, such plans will put another stress factor on generation and installed capacity adequacy.

ACKNOWLEDGMENT

This work was supported by the by Czech Technical University in Prague within the project [SGS21/118/OHK5/2T/13].

REFERENCES

[1] A European Green Deal. "European Commission - European Commission." https://ec.europa.eu/info/strategy/priorities-2019-2024/european-green-deal_en [Mar. 16, 2022].

[2] L. Elliott, West feeling impact of Russia sanctions too as oil and gas prices soar. "The Guardian". [Online]. Available: https://www.theguardian.com/business/2022/mar/07/west-impact-russia-sanctions-oil-gas [Mar. 13, 2022].

[3] IEA. "How Europe can cut natural gas imports from Russia significantly within a year - News." [Online]. Available: https://www.iea.org/news/how-europe-can-cut-natural-gas-imports-from-russia-significantly-within-a-year [Mar. 13, 2022].

[4] Eurostat. "Shedding light on energy on the EU: From where do we import energy?, Shedding light on energy on the EU" [Online]. Available: https://ec.europa.eu/eurostat/cache/infographs/energy/bloc-2c.html [Mar. 13, 2022].

[5] Regulation (EU) 2020/852 of the European Parliament and of the Council of June 2020 on the establishment of a framework to facilitate sustainable investment, and amending Regulation (EU) 2019/2088, vol. 198. 2020. [Online]. Available: http://data.europa.eu/eli/reg/2020/852/oj/ces, [Mar. 13, 2022].

[6] S. Lentfer, L. Mavon, and S. Stenberg. "The Impact of the EU Taxonomy on Greenwashing: With a Case on the Swedish Sustainable Finance Sector". 2021 [Online]. Available: http://urn.kb.se/resolve?urn=urn:nbn:se:hj :diva-52777, [Mar. 16, 2022].

[7] L. Allesi, S. Battiston, A. Melo, and Roncoroni, A. "The EU Sustainability Taxonomy: a financial impact assessment. European Commission". 2019. [Online]. Available: https://ec. europa. eu/jrc/en/publication/eusustainability-taxonomy-financial-impact-assessment.

[8] A. Piebalgs and C. Jones. "The importance of the EU taxonomy: the example of electricity storage" *European University Institute*, 2021. doi: 10.2870/886395.

[9] F. Schuetze and J. Stede, "EU Sustainable Finance Taxonomy – What Is Its Role on the Road towards Climate Neutrality?", *Social Science Research Network, Rochester, NY, SSRN Scholarly Paper* ID 3749900, Dec. 2020. doi: 10.2139/ssrn.3749900.

[10] F. Schütze, J. Stede, M. Blauert, and K. Erdmann, "EU taxonomy increasing transparency of sustainable investments", *DIW Weekly Report*, vol. 10, no. 51, pp. 485–492, 2020, doi: 10.18723/diw_dwr:2020-51-1.

[11] Eur-lex. "Commision Delegated Regulation (EU) supplementing Regulation (EU) 2020/852. 202AD." [Online]. Available: https://eur-lex.europa.eu/legal-content/EN/TXT/?uri=PI_COM%3AC%282021%292800 [Mar. 13, 2022].

[12] European Commission. "EU taxonomy: Commission presents Complementary Climate Delegated Act to accelerate decarbonisation", *European Commission - European Commission*. [Online]. Available: https://ec.europa.eu/info/publications/220202-sustainable-finance-taxonomy-complementary-climate-delegated-act_en [Feb., 2022].

[13] ERÚ "Zprávy o provozu elektrizační soustavy." https://www.eru.cz/zpravy-o-provozu-elektrizacni-soustavy [Mar. 13, 2022].

[14] ČEPS, a.s., "Roční příprava provozu (2022)". 2022. [Online]. Available: https://www.ceps.cz/cs/priprava-provozu, [Mar. 13, 2022].

[15] ENTSOe. "Transparency Platform". https://transparency.entsoe.eu/dashboard/show [Mar. 16, 2022].

[16] ČEPS, a.s., "Resource Adequacy Assessment of the Electrical Grid of the Czech Republic until 2040, (MAF CZ 2021)". [Online]. Available: https://www.ceps.cz/cs/tiskove-zpravy/novinka/ceps-vydava-maf-cz-2021-analyzuje-dopady-ruznych-smeru-vyvoje-energetickeho-mixu-do-roku-2040, [Feb. 15, 2022].

[17] C. K. Simoglou and P. N. Biskas, "Assessment of the impact of the National Energy and Climate Plan on the Greek power system resource adequacy and operation", *Electric Power Systems Research*, vol. 194, p. 107113, May 2021, doi: 10.1016/j.epsr.2021.107113.

[18] W. Cole, D. Greer, J. Ho, R. Margolis. "Considerations for maintaining resource adequacy of electricity systems with high penetrations of PV and storage", *Applied Energy*, vol. 279, p. 115795, Dec. 2020, doi: 10.1016/j.apenergy.2020.115795.

[19] V. Guerrero-Mestre, M. Poncela, G. Fulli, and J. Contreras, "A probabilistic analysis of power generation adequacy towards a climate-neutral Europe", *Energy Reports*, vol. 6, pp. 3316–3333, Nov. 2020, doi: 10.1016/j.egyr.2020.11.243.

[20] ČEPS, a.s.; "Data" https://www.ceps.cz/cs/data, Mar., 2022.

[21] UNECE, "Life Cycle Assessment of Electricity Generation Options". 2021. [Online]. Available: https://unece.org/sites/default/files/2021-10/LCA-2.pdf [Mar. 14, 2022].

[22] MPO. "Vodíková strategie ČR schválena vládou". https://www.mpo.cz/cz/prumysl/strategicke-projekty/vodikova-strategie-cr-schvalena-vladou--262590/ (accessed Mar. 23, 2022).

[23] ERÚ. *"Výroční zpráva o provozu TS za rok 2020"*. https://www.eru.cz/teplo [Mar. 13, 2022].

[24] MPO. "Vnitrostátní plán České republiky v oblasti energetiky a klimatu". https://www.mpo.cz/cz/energetika/strategicke-a-koncepcni-dokumenty/vnitrostatni-plan-ceske-republiky-v-oblasti-energetiky-a-klimatu--252016/ [Mar. 23, 2022].

The Compliance of Green Hydrogen Targets

Green hydrogen targets feasibility within European climate strategies

Šárka Jablonská
Department of Economics, Management and Humanities
Czech Technical University in Prague
Prague, Czech Republic
jablosar@fel.cvut.cz

Abstract— **European countries face a significant challenge on their way to the goals set by "Clean Energy for All Europeans". To achieve European clean policy targets, hydrogen, has become a significant part of national strategies. According to Fit-for-55 Package, the European Commission wants 50 % of hydrogen consumed by the industry to be green by 2030. In 2022 the chemical industry is the only producer and consumer of hydrogen. However, that hydrogen is produced from fossil fuels, mainly natural gas. Due to the current Ukraine-Russian crisis, it is also very likely that new and more ambitious targets will be set by European Union to reduce the dependence on Russian gas. As a result, the importance of green hydrogen might increase even more. Due to increasing gas prices, the green hydrogen might even soon become economically competitive. Naturally, the targets differ throughout countries based on their geographical, economic, and other conditions. This paper aims to compare whether the green hydrogen outlooks by 2030 in four selected Central-European countries (Austria, Czech Republic, Germany, Slovakia) comply with their feasibility based on renewable resources potential.**

Keywords: green hydrogen, European hydrogen strategies, renewable resources

I. INTRODUCTION

Currently, most of the hydrogen production (96 %) [1] comes from fossil fuels (natural gas) throughout pyrolysis which is not a zero-emission process even if Carbon Capture and Storage (CCS) technologies are considered. Such fossil fuel-based hydrogen is called grey hydrogen for its high amount of emissions. For 1 ton of grey hydrogen, it is estimated that 9.3 tons of CO_2 are emitted. [1] In the case of CCS, the emissions are captured and stored, and the hydrogen is called blue hydrogen. However, blue hydrogen production does not appear to be economically competitive and other alternatives are to be searched. Moreover, some experts question blue hydrogen emissions due to natural gas used to power CCS equipment. The CO_2 emissions for producing 1 ton of blue hydrogen are 4 tons on average. [2]

Table 1: Hydrogen emissions based on its color

Hydrogen color	Average emissions $t\ CO_2/t\ H_2$
Grey	9.3
Blue	4
Green	0

That is why a lot of interest is focused on green hydrogen, even though its research is still in an early stage. Due to the current Ukraine-Russian crisis, it is also very likely that new and more ambitious targets will be set by European Union to reduce the dependence on Russian gas. Nevertheless, increasing fossil fuel prices favor green hydrogen competitiveness. As a result, the importance of green hydrogen might increase even more. So far, there has been defined only one major procedure for green hydrogen generation which is electrolysis. To be considered green, all electricity used for the process must also come from renewable resources such as solar, wind, and hydro. To produce 1 kg of hydrogen approximately 50 kWh of electricity and 9 l of water are consumed. [3] The energy contained in 1 kg of hydrogen is 33,33 kWh. [4] That means that the process efficiency is approximately 66 %. However, in the actual conversion of renewable energy into electricity, the efficiency is already in the range of 20-50% (lower for PV, higher for WP). [5] [6] This means that the efficiency of the whole process is only 13-33%.

Table 2:Green hydrogen balance numbers

Parameter	Value
Water consumption	9 l/kg
Electricity consumption	50 kWh/kg
Hydrogen consumption	33,33 kWh/kg

II. DATA AND METHODS

The study assesses the feasibility of green hydrogen production from renewable resources based on low and high scenario assumptions according to a study prepared for the European Union [7]. No other renewable resources than solar and wind were considered. To state a green hydrogen outlook, national documents such as National Energy and Climate Plans (NECP) and Hydrogen Strategies were analyzed. [8]-[11]

Historical reports of energy balance [12]-[16] were analyzed as well, in order to state the expected production of renewable resources. First, the dependence between the annual production and installed power was estimated. This value was stated as an average from four previous years 2018-2021. For instance, in the Czech Republic, the annual solar production overcomes its installed capacity by 1 078 times on average. [17] Both, solar and wind sources are intermittent, the production varies according to the season and annual production is not sufficient information. Therefore, an annual distribution of renewable production was set, separately for wind and solar. The results for CZ and DE can be seen in Figure 1. [17] [18] The annual distribution of production was important to state the installed capacities of electrolyzers. If this assumption was not taken into account, the electrolyzer capacity would have been stated as an average value which would not be so precise. Afterward, hydrogen production was stated as well as water consumption, which is an important aspect to be considered when implementing green hydrogen.

III. Current Hydrogen Strategies

To state a green hydrogen outlook by 2030, an analysis of the latest National Hydrogen Strategies was an inevitable part of the paper. The methodology applied in National Strategies appeared to be the same in all countries in the selected region – to replace currently consumed grey hydrogen with a low carbon variant – blue or green hydrogen. The 2030 strategies are mainly focused on industry and transport. The reason is to produce and consume hydrogen on a local level since hydrogen transport is not fully developed yet. Also, the water resource is usually present which is an important aspect for electrolysis. However, the ambitions vary throughout countries due to different energy mixes and the potential of renewable resources. [13]

A. Germany

Germany's total natural gas demand is approximately 920 TWh. [20] Regarding the volume of hydrogen, Germany appears to be the most ambitious country in the selected region. Currently, the German industry uses approximately 55 TWh (1.7 million tons) of hydrogen per year. This hydrogen demand is expected to be doubled by 2030. However, 95 % of current demand is covered by grey hydrogen. According to its newest Hydrogen Strategy, the target is to increase the current level of installed electrolyzers from 750 MW to 5 GW by 2030 with a total production of 14 TWh of green hydrogen per year. [11] [15] [16]

B. Austria

Austria's 2020 natural gas consumption was 90 TWh. [21] Austrian NECP considers renewable hydrogen as a key technology for sector integration and coupling and comprises the concrete target of a renewable electricity-based hydrogen consumption of 1.1 TWh (33 000 tons) by 2030. This target complies with Austrian ambition to cover all electricity consumption by renewable resources by 2030. Austria expects significant development of hydrogen demand in industry, followed by transport. Major growth in household consumption is not expected within this period. [9] [14]

C. Czech Republic

Czech natural gas consumption is approximately 93 TWh. [22] In the Czech Republic, currently, the largest and practically only producer and consumer of (grey) hydrogen is the chemical industry. Within the 2030 hydrogen targets, there are two steps to convert to a cleaner solution. By 2025 the total consumption is expected to be 7 000 tons of low carbon hydrogen. The first step is to replace grey hydrogen by blue hydrogen in the chemical industry (2 000 t/y) and implement 5 000 tons of hydrogen in transport. However, by 2030 the hydrogen applications should increase significantly (households, other industries, transport) with a total consumption of 95 000 tons per year. In the last strategy, green hydrogen was not planned before 2030. [10] [12]

D. Slovakia

For Slovakia, natural gas covers about 50 TWh of its energy demand. [23] Hydrogen consumption in Slovakia concerns mainly the chemical industry with a yearly grey hydrogen consumption of 200 000 t/y (6.7 TWh). By 2030, Slovakia does not expect any significant changes in this consumption and assumes to dedicate 1 % of hydrogen consumption to be covered by green hydrogen. Even though Slovakia realizes that it is desirable to find alternatives to fossil-based hydrogen, its latest hydrogen strategy does not specify any other concrete way to do so. [8] [13]

Table 3: 2030 expected hydrogen demand

Country	**2030 Expected H_2 demand**	
	t/y	*TWh/y*
AT	1 700 000	56.7
CZ	33 003	1.1
DE	95 000	3.2
SK	200 000	6.7

IV. More Ambitious Scenarios

The binding targets of the EU's ambitious climate and energy policy have been changing and tightening up very dynamically in recent years. For instance, in 2014 EU set a target to decrease greenhouse gases emissions at least by 40 % by 2030 compared to 1990. In 2020 the Commission decided to set a more ambitious plan – a 55% GHG reduction by 2030. The uncertain political situation in Europe following the Ukrainian-Russian war conflict, which began in 2022, has exposed the EU to the need to reconsider the current energy plans for the future of primary fuels in energy production, especially given the growing threat of restricting Russian gas supplies. As natural gas plays a key role in the planned energy transformation and the imminent loss of supply without a change in the original EU´s energy strategy could result in Europe-wide and long-term energy shortages. Most of those strategies emphasized natural gas as a transition bridge towards clean energy. Natural gas was also an important commodity in hydrogen scenarios, as it was supposed to be the main source of blue hydrogen.

A. Blue Hydrogen Replacement

Due to efforts to reduce dependency on Russian natural gas imports, it is very presumable, that blue hydrogen utilization will be reduced. The most pessimistic scenarios might even completely abandon it. Replacing gray hydrogen with green hydrogen is a technologically undemanding process in the case of existing technologies and end-consumer applications. On the contrary, green hydrogen production is a complex question of research, since there are many challenges and barriers. For green hydrogen to become economically competitive, the prices of renewable electricity must become more competitive as well. Moreover, there must be enough renewables to cover the hydrogen demand.

B. The Technical Potential for Renewable Resources

So far, the only possible way to produce green hydrogen is electrolysis with all the electricity originating from renewable resources. In Central Europe, the conditions for renewables might not be as pleasant as in southern or northern parts. The conditions are not ideal neither for solar nor for wind power. Hydropotential is often already exhausted, mostly for electricity generation. Austria, on the contrary, has high shares of hydropower. [24] Renewable resources are mostly considered intermittent, which makes it harder to guarantee the supply when there is a demand. Moreover, the production changes during the year. To secure a green electricity supply, it is important to find complementary renewable resources such as photovoltaics together with wind power, as can be seen in Figure 1. The figure shows solar and wind production distribution within one year (2020) in the Czech

Republic and Germany. [17] [18] Thanks to the close geographical location, the results are very similar. The complementary trend of the two resources is obvious. Whereas the solar power plants produce the most electricity from April to August, the wind power plants are on minimal production.

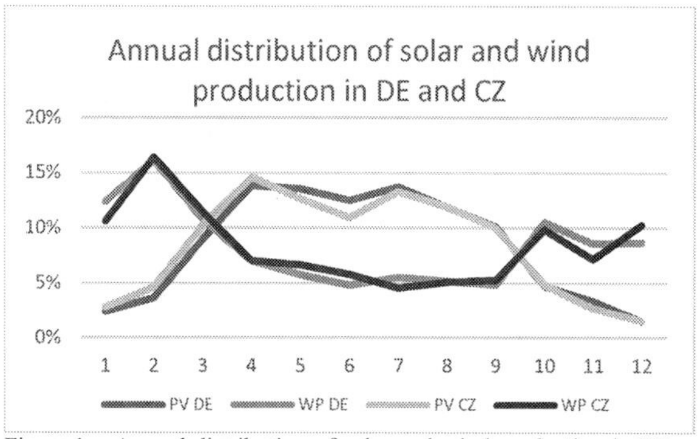

Figure 1. Annual distribution of solar and wind production in DE and CZ

Based on the additional renewable production potential [7], electrolyzers' capacities were established. In the European Opportunities for Hydrogen Energy Technologies [7], two scenarios for renewable potential were set. The two scenarios were set based on different levels of ambition linked to the national context in each state. In Figures 2 and 3, two examples are presented. If low scenarios were implemented, the need for installed electrolyzers capacity does not vary so significantly as for high scenarios. Furthermore, the annual installed capacities results differ for the two presented countries. Installed production capacities of solar and wind power plants in Germany show much more identical values than in the case of Austria. The installed solar capacities in Germany are much higher compared to installed wind capacities which causes the production of these sources to be much closer to each other. Whereas the installed production capacity of wind farms in Austria significantly exceeds the capacity of solar power plants, this results in this specific production curve which reminds the shape of the wind annual distribution curve as seen in Figure 1.

The problem is, that in Central European conditions, the wind and solar power plants will most likely not be constructed in the same area. Therefore, it is very likely that one electrolyzer will produce hydrogen either from solar or wind power. In both cases, the lack or overflow of energy must be considered and solved in order to guarantee the supply. Looking from another perspective, hydrogen could be the solution to energy overproduction, for example in summer for solar power plants, and could play the role of a seasonal accumulation.

Table 4:Installed capacities of renewables for hydrogen production

Country	Low scenario			High scenario		
	Solar	Wind	Total	Solar	Wind	Total
	MW	*MW*	*MW*	*MW*	*MW*	*MW*
AT	450	930	1 380	1 580	3 300	4 880
CZ	130	280	410	520	1 140	1 660
DE	6 800	3 310	10 110	31 360	15 280	46 640
SK	110	220	330	310	620	930

To get a flatter curve for the required installed capacity of the electrolyzer, it would be necessary to optimize the installed solar and wind capacities. However, this was not part of this research.

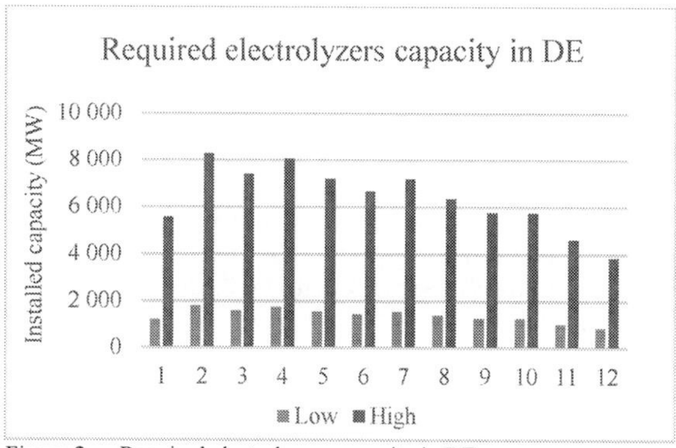

Figure 2. Required electrolyzers capacity in DE

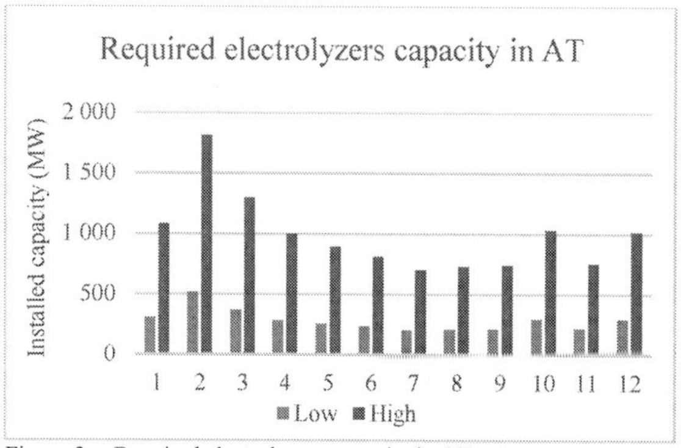

Figure 3. Required electrolyzers capacity in AT

V. Low and High Scenario Results

Based on renewable potentials and annual distributions of productions the desired electrolyzers capacities have been analyzed. The results for the low and high scenarios are shown in Tables 4-8 and Figures 4 and 5. It can be seen that Germany has the highest potential for green hydrogen production with a total production of 251 000 t/y, followed by Austria. Czech and Slovak potential is very similar in the case of low scenarios.

The highest spread between scenarios can be observed in Germany with a difference of more than 350 %, the lowest in Slovakia. Based on the results, Slovakia appears to have the lowest potential for green hydrogen in the defined region. However, Slovakia is also the smallest country in the selected region.

Table 5: Renewables annual production based on Table 4

Country	Low scenario			High scenario		
	Solar	Wind	Total	Solar	Wind	Total
	GWh/y	*GWh/y*	*GWh/y*	*GWh/y*	*GWh/y*	*GWh/y*
AT	500	2 023	2 523	1 755	7 103	8 858
CZ	141	545	687	566	220	786
DE	6 311	6 232	12 543	29 106	28 769	57 875
SK	121	471	592	341	1 328	1 669

Naturally, the absolute values of hydrogen technical potential differ throughout selected countries. Therefore, the results were transformed into relative values by dividing the hydrogen production by country area in km^2, so that they can be compared to each other. [25] The results are presented in Tables 6 and 7. Even in relative values, Germany can still be considered the country with the highest hydrogen potential, followed by Austria. However, the hydrogen potential for Austria and Germany appears to be much closer, especially for the low scenario. The Czech Republic and Slovakia are also much closer.

Table 6: Low scenario results

Country	Low scenario results				
	Electrolysis capacity	*Electricity consump.*	*Hydrogen production*		
	MW	*MWh/y*	*MWh/y*	*t/y*	*kg/y/km²*
AT	550	2 523 080	1 681 885	50 462	612
CZ	150	686 692	457 749	13 734	178
DE	1900	12 543 413	8 361 439	250 868	703
SK	120	592 130	394 714	11 843	246

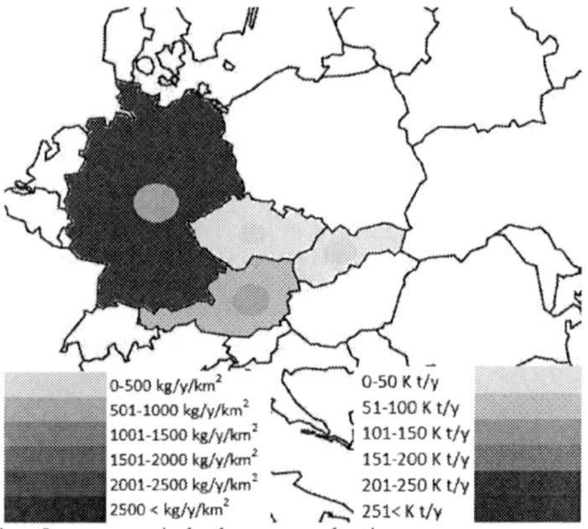

Figure 4. Low scenario hydrogen production

Table 7: High scenario results

Country	High scenario results				
	Electrolysis capacity	*Electricity consump.*	*Hydrogen production*		
	MW	*MWh/y*	*MWh/y*	*t/y*	*kg/y/km²*
AT	1 960	8 857 859	5 904 649	177 157	2 147
CZ	600	2 785 714	1 856 957	55 714	722
DE	8 300	57 875 621	38 579 889	1 157 512	3 242
SK	320	1 668 730	1 112 376	33 374	694

The results of this research show the potential for green hydrogen production in the selected region and compare it with the expected consumption. Already within the region, the hydrogen potential varies quite considerably, even in relative terms. The highest production potential can be observed in Germany with 1.2 million t/y, followed by Austria with 180,000 t/y. However, Germany's expected consumption is 1.7 million t/y, 0.5 million t/y of hydrogen is missing, while Austria's expected hydrogen consumption is lower at only 33,000 t/y. Hydrogen shortages are also expected in the Czech Republic and SK, where the shortage is more than 50%. Moreover, according to the latest Czech Energy strategies, the production of green hydrogen was not expected sooner than 2030.

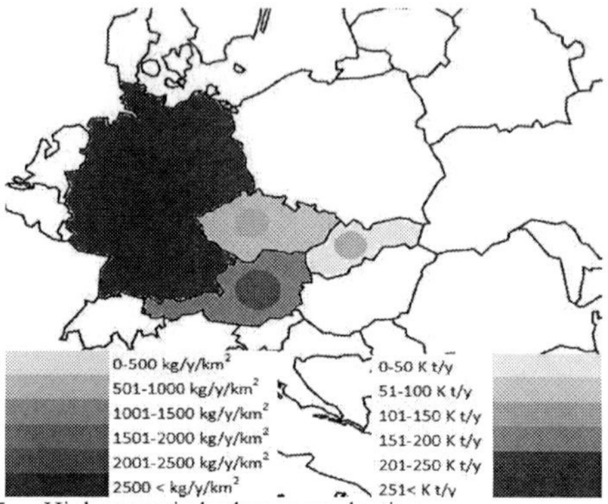

Figure 5. High scenario hydrogen production

Table 8: Water consumption for electrolysis

Country	Water consumption			
	Low scenario		High scenario	
	hl/y	*hl/y/km²*	*hl/y*	*hl/y/km²*
AT	4 542	5.5	15 944	19.3
CZ	1 236	1.6	5 014	6.5
DE	22 578	6.3	104 176	29.2
SK	1 066	2.2	3 004	6.2

Another important results involve water consumption as an essential commodity for electrolysis. When building a new electrolyser, this must be thought of. Naturally, the level of water consumption complies with the hydrogen production as can be seen in Table 8. Corresponding to that, Germany has the highest water consumption with approximately 6-30 hl/y/km^2, followed by Austria with 6-20 hl/y/km^2. The Czech and Slovak relative water consumptions are more or less the same.

VI. Conclusion and Discussion

The results show that for the selected region (Austria, Czech Republic, Germany, Slovakia) there is a high amount of technical potential for additional renewable resources, which could be dedicated to electrolysis. The strategies consider solar and wind power as these sources can be considered complements and hydropotentials are usually exhausted for electricity generation. It is natural, that throughout countries, the potential varies. The highest potential can be seen in Germany, however, not even the production in the high scenario would cover its expected demand. Austria appears to be the only country with a higher production compared to demand. Furthermore, Austria has a high amount of water supply, which is necessary for the electrolysis process.

When the results were adjusted by the state's surface areas, there were not such significant differences in the selected region. The Czech Republic and Slovakia have almost the same relative hydrogen potential (200-700 kg/y/km^2). Also, the differences between Austria and Germany lowered significantly. That means that the differences were caused by a different surface area.

Differences in supply and demand across countries open a question of the need for hydrogen transport and market, which are currently not prepared neither physically in the form of pipelines nor legislatively. Furthermore, an efficient system for "Guarantee of origin" needs to be developed. The whole Engineering, Procurement, and Installation process is time-consuming and can take on the order of years. That is why no significant changes are expected by 2025 which complies with the set targets. Another question is government support in order to find the right motivation for possible investors to be able to use the technical potential for renewables at least by 2030.

None of the selected countries' national water supplies should be endangered by such an additional level of water consumption.

However, it is still necessary to guarantee the water sources at the same place where the electrolyser will be. In the case of Germany, defining water resources for green hydrogen production will be appropriate, especially within the high scenario. In this case, the relative results in hl/y/km^2 might not be the best indicator, since it is important to consider water resources on a local level on a larger scale.

The potential might increase if biomass was considered as a possible source for electrolysis electricity supply, not only in the case of Germany but for the rest of the countries as well. There have already been a few studies on that topic. [26] Unfortunately, biomass has not been classified as a resource to produce green hydrogen by Taxonomy so far. [27] Furthermore, in the selected region, biomass will play a key role as a primary substitute for coal.

Further research following this article could concern the optimization of solar and wind installed power. Such research would not change the results of this article significantly but possibly could affect costs input for green hydrogen production.

Europe is currently facing considerable uncertainty in its energy supply. It is therefore possible that the REPowerEU [28] targets will increase the importance of green hydrogen and the pressure will be on to deliver the outcomes of the scenarios discussed in this paper as soon as possible. The results could also be used to identify hydrogen corridors within the European Hydrogen Backbone initiative.

Acknowledgment

This work was supported by the Student Grant Competition of Czech Technical University in Prague [SGS21/117/OHK5/2T/13].

References

[1] European Parliament, "EU hydrogen policy," European Union, 2021.

[2] C. Naschert, „Blue hydrogen under microscope over emissions, role as transition fuel," 29 September 2021. [Online]. Available: https://www.spglobal.com/marketintelligence/en/news-insights/latest-news-headlines/blue-hydrogen-under-microscope-over-emissions-role-as-transition-fuel-66572882. [March 2022].

[3] IRENA, „Green Hydrogen Cost Reduction: Scaling up Electrolysers to Meet the 1.5°C Climate Goal," *International Renewable Energz Agencz*, Abu Dhabi, 2020.

[4] Clean Cities, „Alternative Fuels Data Center: Fuel Properties Comparison," *U.S. Department of Energy*, 2014.

[5] "All you need to know about wind power," [Online]. Available: https://www.edfenergy.com/for-home/energywise/all-you-need-to-know-about-wind-power. [Accessed 30 May 2022].

[6] A. Vourvoulias, "How Efficient Solar Panels Are?," 19 April 2022. [Online]. Available: https://www.greenmatch.co.uk/blog/2014/11/how-efficient-are-solar-panels. [May 30, 2022].

[7] Trinomics B.V., "Opportunities for Hydrogen Energy Technologies Considering the National Energy & Climate plans," *Fuel Cells and Hydrogen*, Rotterdam, 2020.

[8] Slovak Ministry of Economy, „Integrated National Energy and Climate Plan for 2021-2030," *Slovak Ministry of Economy*, Bratislava, 2019.

[9] Federal Ministry, Republic of Austria, Sustainability and Tourism, „Integrated National Energy and Climate Plan for Austria 2021-2030," *Federal Ministry*, Republic of Austria, 2019.

[10] Ministry of Industry and Trade, „National Energy and Climate Plan of the Czech Republic," Ministry of Industry and Trade, 2019.

[11] Federal Ministry for Economic Affairs and Energy, „Integrated National Energy and Climate Plan," *Federal Ministry for Economic Affairs and Energy*, 2019.

[12] Ministry of Industry and Trade, „The Czech Republic's Hydrogen Strategy," *Ministry of Industry and Trade*, Prague, 2021.

[13] J. Sinay, J. Weiterschütz, P. Blaškovitš a R. Sulík, „Národná vodíková stratégia, "Pripravení na budúcnosť"," *Úrad vlády Slovenskej republiky*, 2021.

[14] M. Belaunde, „Hydrogen in Austria," Federal Ministry of *Austria for Climate Action, Environment, Energy, Mobility, Innovation and Technology*, 2021.

[15] Federal Ministry for Economic Affairs and Energy, „The National Hydrogen Strategy," *Federal Ministry for Economic Affairs and Energy, Public Relations Division*, Berlin, 2020.

[16] Deutsche Windguard, „Status of Wind Energy Development in Germany," *Deutsche Windguard*, 2020.

[17] Ministry of Industry and Trade, „Souhrnná energetická bilance České republiky," Ministry of Industry and Trade, 2021.

[18] Bundesverband WindEnergie, "German wind energy in numbers," 2022. [Online]. Available: https://www.wind-energie.de/english/statistics/statistics-germany/. [Mar. 2022].

[19] Windfakten, „Windenergie in Österreich," 2022. [Online]. Available: https://www.wind-energie.de/english/statistics/statistics-germany/. [Mar. 2022].

[20] N. Sönnichsen, "Natural gas consumption in Germany from 2005 to 2020," 26 August 2021. [Online]. Available: https://www.statista.com/statistics/703657/natural-gas-consumption-germany/. [Mar. 2022].

[21] N. Sönnichsen, "Natural gas consumption in Austria from 2005 to 2020," August 2021. [Online]. Available: https://www.statista.com/statistics/703616/natural-gas-consumption-austria/. [Mar. 2022].

[22] N. Sönnichsen, „Natural gas consumption in Czechia from 2005 to 2020," 26 August 2021. [Online]. Available: https://www.statista.com/statistics/703643/natural-gas-consumption-czechia/. [Mar. 2022].

[23] CEIC, "Slovakia Natural Gas: Consumption," 2021. [Online]. Available: https://www.ceicdata.com/en/indicator/slovakia/natural-gas-consumption. [Mar. 2022].

[24] A. Gurung a A. Bosdorf, „Rethinking Pumped Storage Hydropower in the European Alps: A Call for New Integrated Assessment Tools to Support the Energy Transition," *MRD Journal*, p. 225, 2016.

[25] The World Bank, Data, "Land area (sq. km) - Germany, Austria, Czech Republic, Slovak Republic," 2020. [Online]. Available: https://data.worldbank.org/indicator/AG.LND.TOTL.K2?end=2020&locations=DE-AT-CZ-SK&start=2009. [Mar. 2022].

[26] K. Sang-Hyoun, K. Gopalakrishnan, C. Wei-Hsin a K. K. Samir, „Renewable hydrogen production from biomass and wastes (ReBioH2-2020)," *Bioresource Technology*, 2021.

[27] R. Oyarzabal and P. Mertenskötter, "New Definitions for Blue and Green Hydrogen: The European Commission's Package on Hydrogen and Decarbonized Gas Markets," January 2022. [Online]. Available: https://www.insideenergyandenvironment.com/2022/01/new-definitions-for-blue-and-green-hydrogen-the-european-commissions-package-on-hydrogen-and-decarbonized-gas-markets/. [Mar. 2022].

[28] European Commission, "REPowerEU: affordable, secure and sustainable energy for Europe," 18 May 2022. [Online]. Available: https://ec.europa.eu/info/strategy/priorities-2019-2024/european-green-deal/repowereu-affordable-secure-and-sustainable-energy-europe_en. [May. 2022].

Application of Digital Twin Technology in Distribution Networks

Matěj Vrtal, Petr Toman
Brno University of Technology
Faculty of Electrical Engineering and Communication
Department of Electrical Power Engineering
Matej.Vrtal@vut.cz, Toman@vut.cz

Abstract — **Meanwhile the demand for managing and commissioning of a wide range of equipment increases, many technology industries around the world are facing significant shortages of key components and parts for their operations. Part of the solution to both of these problems lies in a relatively new trend today - the deployment of so-called digital twins to simulate manufacturing processes. The principle of the digital twin function is based on the creation of a virtual model of a specific physical device or process. The potential applications of digital twin technology in the energy sector are similar to those of other technology sectors. By creating a reliable numerical model of any part of the distribution network, we can achieve the perfect tool for finding weak points in the network, analyzing and optimizing control processes, or simulating fault conditions for different scenarios. Although the number of articles on digital twins is increasing every year, only a small number of publications are devoted to the use of this technology in the energy industry. This paper analyses the possible applications of digital twin technology in distribution networks, describes the key issues that need to be addressed when building a specific model of a distribution network digital twin, and discusses the potential of optimization processes for improving the reliability of distribution networks.**

Keywords: digital twin, power grid, distribution network, smart grid, smart city, simulation, optimization

I. Introduction

The term digital twin (DT) first appeared in 2003 at the Management Forum of the American Manufacturing Engineering Association. Originally, this technology was intended to serve the US Department of Defense for the maintenance and support of aerospace vehicles [1]. In 2012, the concept of DT was defined as a multiphysics, multiscale probabilistic simulation that timely reflects the state of a corresponding twin based on historical data, real-time sensor data, and a physical model by the National Aeronautics and Space Administration (NASA) [2].

DTs are used not only to improve existing processes but also to better understand their real state, simulations, and non-standard events or economic analyses. Essentially, these are real objects and processes converted into digital form. The DT model must contain all the parameters of the real object or process. Using DT technology, we can run the entire physical model cycle, including design, development, manufacturing, service, maintenance, and even scrap recovery [1]. If desired, the DT environment may also include an interactive interface to provide users with simulations. An example of such an interface may also be a mapping environment.

Although there is still no clear definition of DT and no clear idea of what structure DT should contain, the technology is being applied in many different technology sectors. There are also several references to DT in scientific publications in the context of power grids [1],[3],[4]. One of the sectors that could benefit from the incorporation of DT is optimization in distribution networks. Along with the evolution of the electricity sector, the implementation of distributed energy resources (DER), and smart technologies, society is increasingly dependent on critical infrastructures of data and distribution networks. When the power supply is interrupted, many production processes and safety technologies are affected. In addition, with each outage comes the cost of interrupting of the power supply to individual customers. According to [5], these costs have more than doubled since 1990. It can be assumed that, together with the implementation of new technologies and digitization, these costs will continue to rise, and this topic is very topical.

A clear understanding of the costs of power interruption is essential for effective investment in the development of distribution networks. The reliability of distribution networks is therefore a factor that plays an important role for both parties involved - the distributor and the customer. For these reasons, this paper first deals with the description of DT itself and then analyses the possibilities of applying this technology in the optimization of distribution networks in terms of reliability.

II. Digital Twin Technology

The DT is important not only for the exact reproduction of the physical world but also for the prediction of possible future events. DT technology provides effective decision making through the same mechanism of operation as in the physical world. DT also serve for the verification of certain environments or states that cannot be artificially simulated in the physical world.

Today, the potential of DT technology is enormous. It can break through the limitations of physical conditions and be used in many scenarios where there is no legal entity available for monitoring, where the cost of the risk of monitoring is high. DT can not only simulate the behavior of objects in the physical world, but also monitor changes, reflect the operational state, and evaluate the state of the physical world.

In addition, neural networks and other algorithms can be used to predict the future trend of system development and upgrade the entire system. The application of DT technology in various fields of production is expected to bring great changes to economic development.

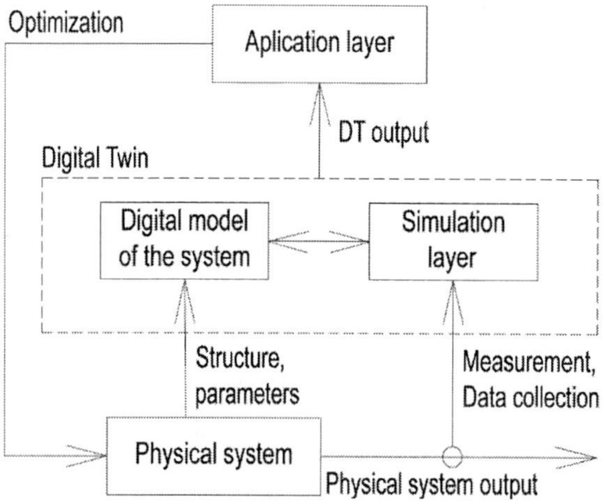

Figure 1. Overview diagram of DT functionality

The overview diagram of the DT functionality can be seen in Figure 1. First, a digital model of the physical system is created. Depending on the type of system, this can be achieved manually or by automated data transfer. The key to the correct functioning of a digital model is the precise knowledge of all parameters that affect the function of a physical system. Once the model is created, various simulations can be implemented in the model. Even non-standard operating conditions can be simulated. The simulation results can be compared with the real data obtained by measuring the output of the physical system. Based on these results and DT algorithms, optimization solutions can be designed for the simulated physical system.

A. Applications of DT Technology

It has already been mentioned that the term DT was firstly used in the context of aerospace engineering. Today, DT technology is used in various engineering sectors of the world. A 2019 study in [6] conducted to analyze DT articles shows that two thirds of relevant articles published on the Web of Science (WoS) deal with the application of DT in manufacturing (54 articles in total), the remaining third is related to the following industries: aerospace (5), energy (4), automotive (3), marine (3), petroleum (2), agriculture (2), healthcare (2), public sector (1), and mining (1). The number of publications is given in brackets [6].

B. The Purpose of DT Applications

In general, applications of DT technology can also be classified according to the following three purposes [6]:

- simulation,
- monitoring and
- control.

The specific application may then involve a combination of these three purposes. In the case of simulations, the application model contains a layer in which the behavior of physical objects in a virtual space can be reproduced. This allows processes to be planned or optimized without the need to rely on a physical object. The monitoring purpose includes all applications aimed at displaying the current state and its interpretation of the physical object. The purpose of control includes applications where DT directly affects products or means of production [6].

III. Application of DT Technology in Distribution Networks

The DT of the distribution network provides a seamless connection between distribution planning models and external information systems. This part of the paper outlines the possibilities of using DT technology in distribution networks.

A. Creation of a DT Model

Due to the breakdown of DTs mentioned above according to their application and use, the most frequently used DTs in power networks are applications for simulation and monitoring of the network operation. The most often is a combination of both.

To create a DT of a distribution network, it is necessary to obtain a plausible image of the topology of the real network and include all operational parameters. To create a network topology image, a list of individual nodes in the network, data on lines, transformers, and individual switching elements is needed. The coordinates of the individual nodes are also necessary to display the network in the map base. A suitable way to obtain these data is to directly export them from the Geographic Information Systems (GIS) of the Distribution System Operators (DSO).

Data on network operating parameters include data on the current status of switching elements and power flows, production, and consumption. These data can be obtained from load measurements of the distribution network (winter measurements, flow measurements, etc.) or estimation methods.

It is important to remember that the creation of a DT distribution network involves working with very sensitive data that can be easily misused. If it is not necessary to work with real data, it is advisable for general simulations to use the possibility of anonymizing the data used or creating an unrealistic random network. The problem of creating an anonymous model of a distribution network is discussed in more detail in [7].

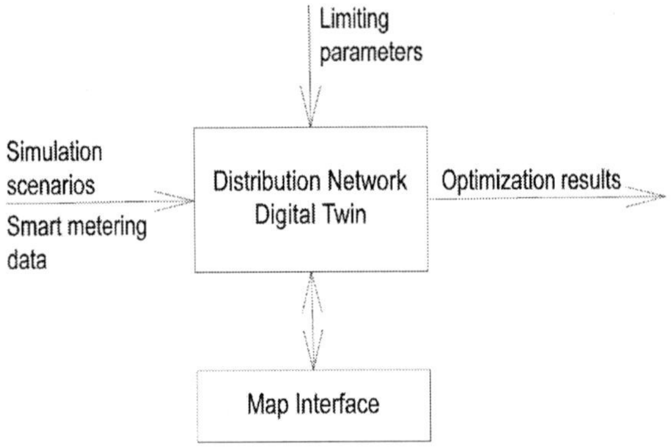

Figure 2. Example of a DT optimization method regarding the reliability of the distribution network

Figure 2 shows an example of a distribution network optimization principle using DT. Once a distribution network model is built, it can be used to simulate operational and emergency conditions. At this point, a realistic picture of the physical distribution network is created that can also be displayed to users in a map interface. The way in which the DT model can be extended with the map environment is described in [8]. The functionality of the model must then be verified, for example, by calculating the power flow of the network. Due to its plausibility, the model can be used for simulations of different scenarios or for optimization processes. This paper focuses on optimizing the reliability of the distribution system. The essence of successful optimization is the correct setting of the constraining parameters of the simulation. By limiting parameters in this case, for example, the maximum value of the Customer Average Interruption Duration Index (CAIDI).

The output of the simulation in DT can be the proposal of an appropriate optimization of a specific part of the network that the simulator evaluates as insufficient. In the following part of the paper, the possibilities of the proposed optimization methods to improve the operational reliability indices for different parts of the distribution system are presented.

B. Optimization Methods

Distribution Network Optimization is essentially a balancing act between operating expenses and working capital as opposed to Service Level requirements or Service Level versus Costs. As part of optimization solutions, weak points in the system must be identified. Then, for every weak point in the system, an optimization solution can be determined. Many optimization methods are used in distribution networks, such as decreasing the fault rate of conductors, optimizing the number of switches and their location, increasing the level of interconnection, and distributed automation. Research on the optimization of distribution systems can be broadly divided into two categories [9]:

- planning (integration of distributed energy resources, power quality mitigation),
- operation (the daily management and operation in the use of network analysis and optimization).

This paper will focus on one of the optimization options in distribution networks, improving distribution reliability and its implementation using digital twin technology. Basically, the reliability of the electrical supply can be improved in the following ways [5]:

- reducing the frequency of faults,
- reducing the duration of a fault,
- reducing the magnitude of interruptions.

In practice, we encounter traditional technologies based on the above principles to increase the reliability of distribution networks. Some of the most commonly used practices include the relocation of existing overhead lines to cable lines, the replacement of non-insulated overhead lines with insulated conductors, underground cabling, switches, primary substations, and ground fault current compensation [10]. When deciding between the use of each method, the effect on the System Average Interruption Frequency Index (SAIFI) and the CAIDI supply reliability indicators is taken into account.

According to the analysis presented in [11], overhead line cabling and ground fault current compensation is an effective means of reducing the SAIFI indicator. Then, according to the same study, reserve connections and distribution automation are particularly suitable for reducing CAIDI.

IV. Reliability Indicator Improvement Methods

According to [5], the following methods can be used to improve the reliability of distribution networks:

- underground (UG) cables and insulated overhead lines (OHL),
- new substations and reserve connections,
- earth fault current compensation,
- distributed automation,
- meshed operation of MV networks,
- control and monitoring of LV networks.

A. Underground Cables and Insulated Overhead Lines

One way to improve reliability indicators is to replace existing lines with underground cables and insulated overhead lines. Research carried out in [10] shows that the failure rate of MV overhead lines is typically three to five times higher than that of underground MV cables. Using modern XLPE cables, the DSO can achieve even lower failure rates. The main advantage of underground cables compared to overhead lines is their resistance to weather conditions.

B. New Substations and Reserve Connections

Distribution network topology has a significant impact on the reliability of electricity distribution. Three simplified network topology configurations are used in distribution networks: radial, loop, and network. Loop and network configurations are also called meshed systems [5]. In radial topology, customers are supplied from a single path, in circuit topology, from two paths, and in network systems, customers are supplied from multiple paths. Most meshed systems are meshed by design only; the network is normally operated radially. The main advantage of meshed networks is that the repair time does not determine the duration of the interruption. This is particularly important in urban networks, where the repair of a fault in underground lines takes a long time [10].

C. Compensating the Earth Fault Current

The most common type of fault in electrical distribution networks is a ground fault. According to various sources, there are approximately 50-80 % faults of this type in Central Europe [12]. In MV networks, a Petersen coil can be placed between the neutral conductor of the transformer and the ground, and its inductive reactance is set to the value of the capacitive reactance of the distribution network. In this way, the fault current is compensated and the distribution network can continue to operate. The advantage of operating the network with a Petersen coil is that it avoids interruptions in the supply of electricity to customers.

D. Distributed Automation

Once the power supply is interrupted, it is necessary to look for ways to reduce the interruption time. One way to reduce this time is to implement distributed automation in distribution networks [13]. One of the advantages of this optimization method is that it can be installed in the existing distribution network without the need for extensive reconstruction of the existing network, which in addition to the cost also reduces the installation time. The impact of distributed automation has been studied in detail, for example in [14].

E. Control And Monitoring of LV Networks

In addition to remote reading, smart meters can also be used to monitor LV networks. The widespread installation of smart meters for end customers allows for easier detection of fault locations of LV networks and thus is also one of the ways to reduce the time of power outages.

V. CONCLUSION

DT is a term that is often found in modern technologies. Although the term was not invented in the context of power grids, it is a technology that has been developed in almost all sectors, and the energy industry is not an exception. In this paper, a search for publications on DT technology research has been conducted, and the principles of this technology have been presented in different sectors of the technology. In the context of this research, a procedure for the creation of a DT distribution system has been outlined. The possibilities of using DT for optimizations in distribution networks are almost limitless.

With the increasing cost of interrupting the power supply to individual customers in recent years, the need to improve the reliability of distribution networks has increased. These costs are expected to continue to increase, and DT technology, in particular, has potential in the area of distribution network optimization.

To this end, the paper also describes the indicators of supply reliability in distribution networks and summarizes commonly used practices to improve these indicators.

ACKNOWLEDGMENT

This research work has been carried out in the Centre for Research and Utilization of Renewable Energy (CVVOZE). Authors gratefully acknowledge financial support from the Technology Agency of the Czech Republic (project No. TK02030013).

REFERENCES

[1] H. Pan, Z. Dou, Y. Cai, W. Li, X. Lei, and D. Han. "Digital twin and its application in power system," in *2020 5th International Conference on Power and Renewable Energy (ICPRE)*, pp. 21–26, 2020.

[2] E. Glaessgen and D. Stargel. "The digital twin paradigm for future nasa and us air force vehicles," in *53rd AIAA/ASME/ASCE/AHS/ASC structures, structural dynamics and materials conference 20th AIAA/AS-ME/AHS adaptive structures conference 14th AIAA*, p. 1818, 2012.

[3] T. Liu, H. Yu, H. Yin, Z. Zhang, Z. Sui, D. Zhu, L. Gao, and Z. Li. "Research and application of digital twin technology in power grid development business," in *2021 6th Asia Conference on Power and Electrical Engineering (ACPEE)*, pp. 383–387, 2021.

[4] C. Li and D. Yang. "Construction of power grid digital twin model based on gan," in *2021 China Automation Congress (CAC)*, pp. 7767–7771, 2021.

[5] O. Siirto, M. Hyvärinen, M. Loukkalahti, A. Hämäläinen, and M. Lehtonen. "Improving reliability in an urban network," *Electric Power Systems Research*, vol. 120, pp. 47–55, 2015. Smart Grids: World's Actual Implementations.

[6] M. Enders and N. Hoßbach, "Dimensions of digital twin applications - a literature review," Aug. 2019.

[7] M. Vrtal, J. Benedikt, D. Topolanek, R. Fujdiak, P. Blazek, P. Toman and J. Misurec. "Power grid and data network simulator," *22nd International Scientific Conference on Electric Power Engineering (EPE)*, unpublished.

[8] J. Benedikt, M. Vrtal, D. Topolanek, R. Fujdiak, P. Blazek, P. Toman, and J. Misurec. "Virtualization Platform for Urban Infrastructure," *22nd International Scientific Conference on Electric Power Engineering (EPE)*, unpublished.

[9] H. Liao. "Review on distribution network optimization under uncertainty," *Energies*, vol. 12, no. 17, p. 3369, 2019.

[10] H. Ying. "Study and analysis of distribution equipment reliability data," *Elforsk Report* 10: 33, 2010.

[11] M. Gonzalez. "Improvement of saidi and saifi reliability indices using ashunt circuit-breaker in ungrounded mv networks," in *22nd International Conference and Exhibition on Electricity Distribution (CIRED 2013)*, pp. 1–4, 2013.

[12] S. Hanninen. "Single phase earth faults in high impedance grounded networks: characteristics, indication and location." *VTT Technical Research Centre of Finland*, 2001.

[13] O. Siirto. "Distribution Automation and Self-Healing Urban Medium Voltage Networks." *Doctoral thesis, Aalto University, School of Electrical Engineering*, 2016.

[14] K. Kamps, F. Mohrke, M. Zdrallek, P. Awater, and M. Schwan. "Reliability of decentralized network automation systems and impacts ondistribution network reliability," in *2020 International Conference on Probabilistic Methods Applied to Power Systems (PMAPS)*, pp. 1–6, 2020.

Fault Detection in Mesh Network Utilizing Measured Data from MV/LV Transformer Stations

Vit Krcal, David Topolanek
Brno University of Technology,
Faculty of Electrical Engineering and Communication, Dept. of Electrical Power Engineering
Brno, Czech Republic
vit.krcal@vut.cz, david.topolanek@vut.cz

Abstract — **This paper aims at fault detection in a low voltage meshed network based on current monitoring at distribution transformer stations. For indication of fuse (or line) interruption a simple algorithm has been designed. The algorithm is based on evaluating current changes at low voltage feeders and comparing them to threshold values derived from normal operation state. For correct setting of threshold values of current changes, sensitivity analysis of load behavior is carried out. The network sensitivity to load changes is presented using both normalized load profile data and real measured data from several distribution stations. The algorithm performance is then tested on a numerical model considering interruption of each individual line section (fuse breaking). The simulations are carried out with a number of different threshold settings using extensive meshed network model. The study is supplemented with topological categorization of line sections to specify the position of the interrupted line within the network.**

Keywords: meshed network; distributed measurement; fault detection; current changes; low voltage

I. Introduction

Nowadays, the integration of smart grid components in distribution networks is thriving and various measurement units are being deployed in distribution systems, even on the low voltage (LV) level. Data measured on such devices should be used for enhanced monitoring functionalities to provide better insight of the network status to the distribution system operator (DSO) and enable e.g., to optimize the operation. These functionalities include detection and localization of faults and abnormal events, which is beneficial especially in meshed networks.

The operation of LV distribution network (DN) in meshed topology is not a common practice, and therefore not many studies dealt with this kind of problems. Due to the lack of test DNs of meshed topology a model of large real-based LV meshed network has been created [1] and is used for simulations in this paper. This extensive meshed network (EMN) consists of 1498 underground cable lines (1424 on LV level), 1268 nodes 79 distribution transformers (DT) and 55 manually controlled bus couplers. The overview network topology can be seen in Figure 2.

Previous work [2] has revealed that distributed voltage monitoring in such complex meshed network cannot reliably indicate line interruptions even in the case of cascading failure scenarios. Widely distributed monitoring is not considered in this stage of the study also due to large investment cost. It has been indicated that monitoring current changes at LV feeders in distribution transformer stations (DTS) is more realistic and has bigger potential of fault detection. As the DSOs in Czech Republic plan to equip most of the DTS with measuring devices, monitoring at all DTS can be considered in this paper.

II. Problem Definition

The goal of the study is to use measured currents at LV feeders I_{fe} in a meshed distribution system, which is depicted in Figure 1, for sensing abnormalities such as line interruptions. For the simulation purposes, the large network (EMN) is used (shown in Figure 2), of which several LV feeders are already equipped with monitoring. These real-based data will serve for analysis and will enable correct parametrization of proposed algorithm.

A. Placement of Current Monitors

Considered current monitoring at all LV feeders presents possibilities for abnormalities and faults indication. One advantage of monitoring all feeders is that the whole network is covered roughly uniformly and it is possible to use the same indication mechanisms in all parts of the network, at all MV/LV stations. The other advantage rests in having the information about the total power supply and its distribution among the individual transformers and feeders. This could help for differentiation between load changes and changes of topology (caused by fault).

B. Type of Faults

Because of the nature of meshed distribution systems, a fault does not necessarily mean supply interruption or relay tripping. Both the short-circuit (SC) faults and overloading may lead to fuse breaking in any part of the network and following power flow redistributions in the unmonitored meshed system pose a danger to the network in form of cascading failure. Considering fuses are installed at all cables in all electrical nodes, any SC fault should cause the fuse braking which will effectively interrupt the protected line. Therefore, considering that most of the faults in LV network transform into breaking of fuses, the line interruption fault (LIF) has been defined and is used for simulation.

III. Proposed Algorithm

A simple deterministic algorithm for LIF indication was designed based on data from all the LV feeders in the meshed network. The algorithm is supplemented with topology analysis and with analysis of currents measured (and simulated) during normal operation.

A. Differentiation of Lines Topology

The logic for indication and localization is based on categorization of lines according to their role in the network. In the LV mesh network, line sections can be divided into three main categories: feeders (type 1), ring lines (2) and radial lines (3). Feeders are directly connected to a LV bus of a transformer, ring lines connect nodes powered from several feeders and radial lines mostly serve for supplying consumers. The feeders are further divided into 3 sub categories reflecting special cases of their connection. Type 11 is defined as a line interconnecting two LV feeders, type 13 represents a radial feeder supplying only one node. The rest of the feeders are considered as common feeder (type 1). The topology line differentiation is shown in Figure 1. The differentiation for EMN is then shown in Figure 2, where color assignment reflects the one in Figure 1. The categorization algorithm follows simple deterministic rules:

- types 1 and 13 are connected to transformer LV busbar from one end,
- type 11 interconnects 2 transformer LV busbars,
- interruption of types 3 and 13 leads to isolation of a node,
- the rest of lines is of type 2.

It is clear that faults occurring at different types of lines will cause different response and will invoke different redistribution of currents at LV feeders. The interruption of feeders (types 1, 11, 13) will cause a drop of the current to zero level and should be recognized easily. The interruption of line of type 3 will cause a supply interruption, whereas the breaking of line type 2 will only cause power flow redistribution. More detailed fault localization method and differentiation will be a subject to following study.

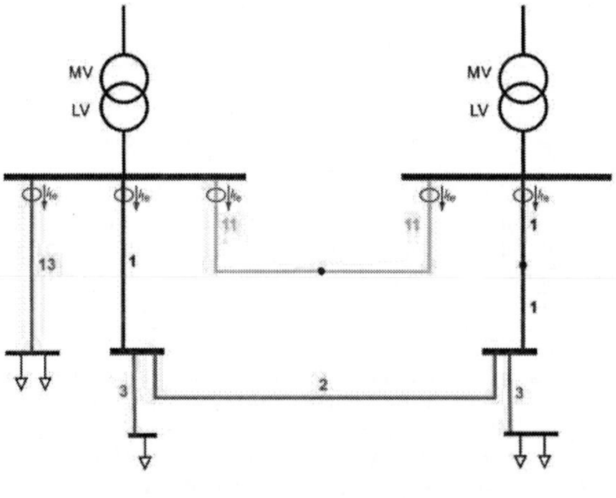

Figure 1. Line sections topology differentiation

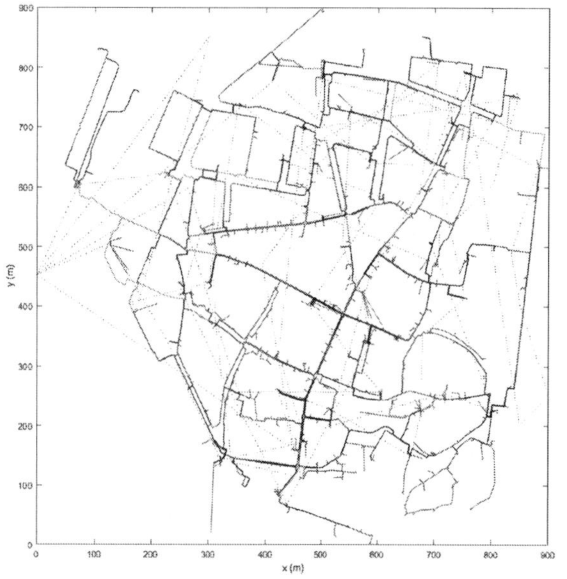

Figure 2. Line topology differentiation in EMN

B. Indication Algorithm

Based on calculation of current changes at all LV feeders a simple algorithm has been designed to reveal an abnormality caused by line interruption, fuse breaking, reconfiguration, or major load change. To differentiate it from common operation, each feeder is associated with different threshold values derived from normal operation data. Further analysis of LV feeder currents combined with line topology analysis may lead to fault localization.

The principle of algorithm is depicted in flowchart in Figure 3. In the first step, the currents measured at all LV feeders I_{fe} (see Figure 1) are transferred to SCADA every second. From these values changes ΔI_{fe} are calculated as a difference between two consecutive samples, or with predefined aggregation interval to mitigate impacts of stochastic load switching. These changes are then compared with set measurement uncertainty ΔI_{meas} to suppress measurement errors. If the change is lower than the set limit, the impulse is marked as unmeasurable, otherwise it is further compared with defined threshold ΔI_{thr}. For each LV feeder the threshold value is set individually depending on evaluation of normal operation state, which may be derived from historical data recorded at the feeder. If the ΔI_{fe} is lower than ΔI_{thr}, the impulse is marked as measurable but insensitive and the possible abnormality is not revealed by this feeder's data. In the opposite case, the feeder's current data are stored into an indication table, which is further analyzed.

Based on the number of feeders stored in the indication table the event is further assessed and possible LIF is localized. Empty table means that no feeder recorded sufficiently high current change to indicate abnormal behavior. Non zero number of items in the table detects abnormality and using the stored data supplied with topology analysis data enables closer problem localization.

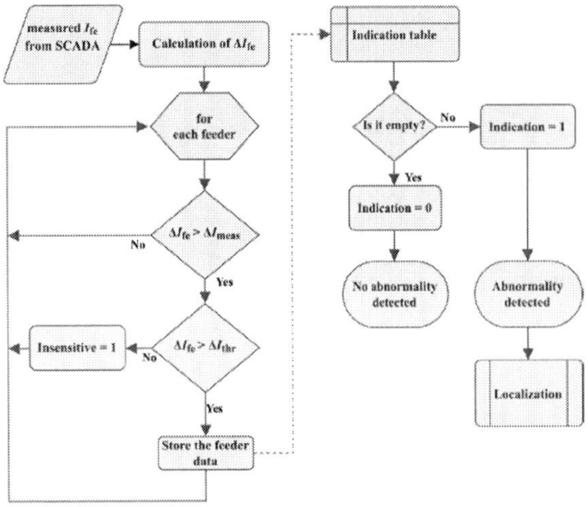

Figure 3. Flowchart of proposed indication algorithm

C. Sensitivity Analysis – Normal Operation Analysis

The key and most important part of the indication algorithm is the threshold setting for each feeder. To enable correct setting of this parameter, an analysis of network behavior must be carried out. In this paper, analysis of the changes ΔI_{fe} caused by regular network operation was carried out in two forms – the first one is calculated from real measured data, the other is derived from normalized load profile (NLP) applied to the EMN numerical model.

As mentioned in the introduction, in the EMN, there are only a few feeders equipped with current measurement as of now. For analysis purposes, the DSO provided measured data from those feeders for each minute for one-year operation. It is important to mention that these values represent instant values, sent to SCADA each second, and are not in any way aggregated. That means, that only 1/60 = 1,66% of measured data are available for analysis.

However, a sample of one-day data recorded at one feeder each second was provided to enable closer look at measured data. In Figure 4 values of current magnitudes measured at one feeder are depicted. In the middle plot, the difference between instant values (read every minute) and average values (calculated as mean average within the minute) are compared. From the bottom plot it can be seen that the difference between instant hour data (read every hour) and averaged hour data is small, when the average deviation is 10%. That enables to use instant data for initial analysis. Also, the bottom plot shows a comparison between measured current and current calculated using the EMN numerical model with NLP-defined loads. The NLP curve is offset in average by 15.2A, meaning that the actual demand at this feeder was significantly lower than modelled NLP. Despite this offset, the curves still dispose of well matching waveforms (less than 20% average deviation after offset compensation).

The Figure 5 shows changes of current ΔI_{fe} calculated as an absolute value of the difference between two consecutive samples of waveforms depicted in Figure 4. The plots show that used aggregation of measured instant data has only little effect on mitigation of peaks caused by stochastic load changes, and therefore, the correct setting of aggregation interval for changes calculation should be further analyzed on larger datasets. From the comparison of curves representing measured hour average and NLP it can be concluded that setting of the threshold value ΔI_{thr} can be derived from NLP for simulations as the measured data are not available.

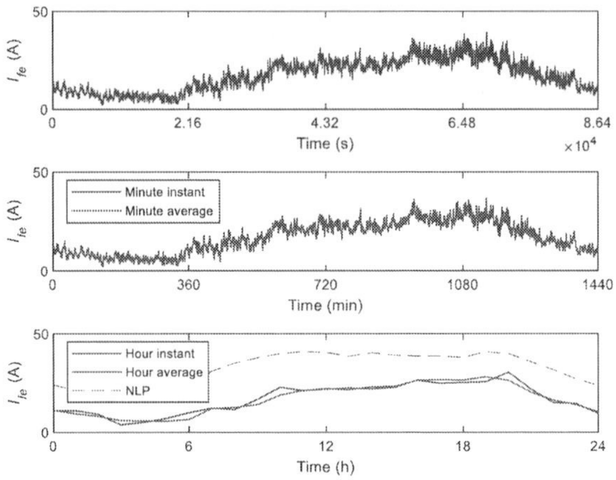

Figure 4. Current magnitude at one feeder during one-day operation

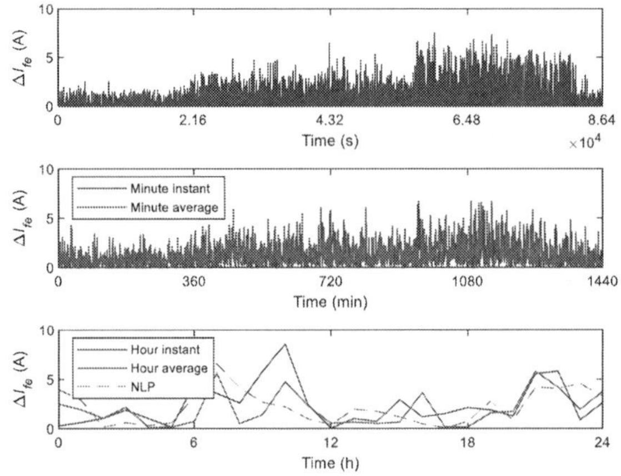

Figure 5. Current changes at one feeder during one-day operation

From all available measured data (a few transformer stations), an analysis of currents magnitudes and changes was carried out. For clarity, only one feeder data at one selected transformer (T7) are used for demonstration.

In Figure 6 magnitudes of currents I_{fe} at measured feeders at T7 are depicted in the scope of the whole year. The top plot displays minute instant measured values, the middle plot displays hourly mean average of minute instant data and the bottom plot shows waveforms of feeder currents calculated according to NLP. Waveforms of individual feeders are differentiated by color. Opposed to simulated NLP data, the measured currents waveforms include peaks and irregularities, which may be caused by sudden changes of loads, network reconfiguration or disconnecting due to maintenance work.

The Figure 7 shows changes of current ΔI_{fe} calculated as an absolute value of the difference between two consecutive samples of corresponding waveforms depicted in Figure 6. It is noticeable that the level of changes decreases with aggregation except for the 'yellow' feeder, which has the highest level of current and its oscillation. The peaks of currents waveform in Figure 6 also transfer to Figure 7 as they represent significant events in the network. For example, at the beginning of the year, a series of large peaks is apparent. This indicates changes in the network topology rather than load stochasticity behavior. For detailed analysis of irregular network behavior, dispatcher log should be examined.

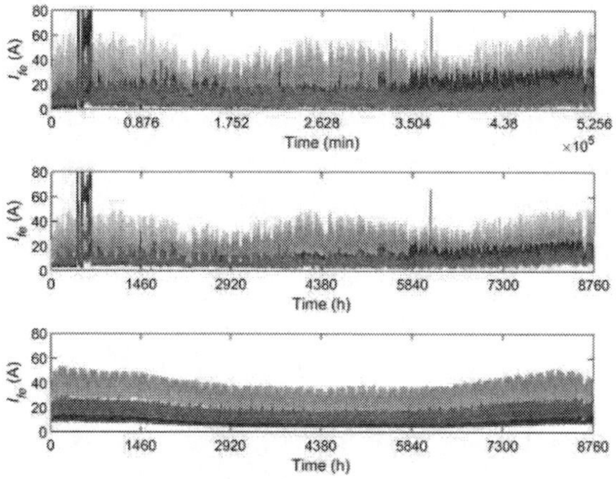

Figure 6. Current magnitudes at feeders at T7 during one-year operation

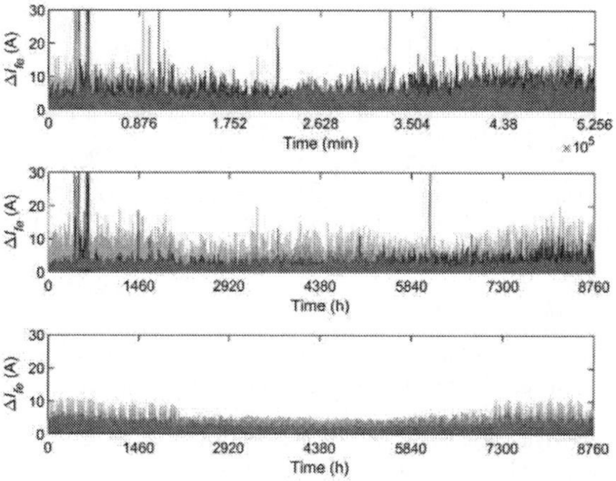

Figure 7. Current changes at feeders at T7 during one-year operation

Histograms in Figure 8 show distribution of current changes ΔI_{fe} at all feeders at T7 during one-year operation. Both histograms are depicted in logarithmic scale and in range up to 30A changes, although there are several outliers with ΔI_{fe}>30A. The distributions of both hourly and minute changes resemble normal distribution curves with majority of values lower than 15A. The NLP-derived changes ΔI_{fe} do not reach to as high values as the measured changes. Looking at Figure 7 this is mainly due to the 'yellow' feeder oscillation and also due to several peaks caused by abnormal network behavior. Otherwise, both the changes waveform and distribution are comparable. Since all the hourly, minute and second changes ΔI_{fe} not dispose of similar waveforms, levels and distribution, the thresdhold can be for derived from calculated NLP-based hourly changes ΔI^{h}_{fe}. The next step of ΔI_{thr} setting would be examination of more suited aggregation intervals and methods, using more real data to carry out the analysis and to employ artificial intelligence algorithms (machine learning, neural network).

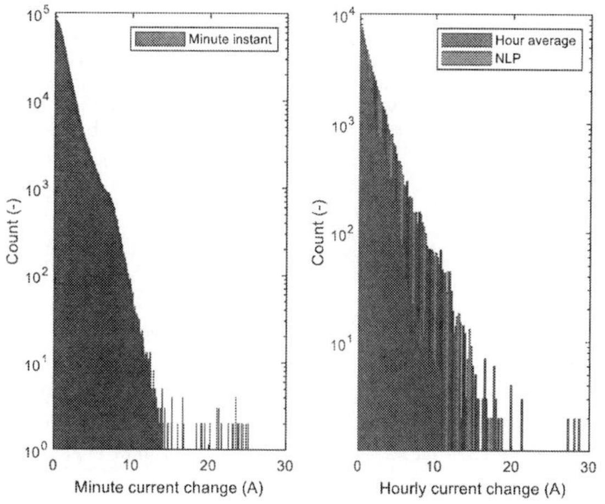

Figure 8. Histograms of current changes at all feeders at T7 during one-year operation

IV. Line Interruption Faults Simulation

The simulation was carried out using single-phase equivalent model of EMN, which is considered symmetrical in both topology and loads. Theoretically, when ignoring the MV part of the network, this approach enables to solve each phase individually and, therefore, to examine one-phase, two-phase and three-phase faults utilizing measured data associated with individual phases.

To test the algorithm performance, an interruption of each line (LIF) has been simulated separately using identical setting of loads. The loads setting was derived from normalized hourly load profile and the year consumption data. Considering only one line at fault at a time, the total of 1424 line interruptions has been simulated.

A. *Algorithm Sensitivity Setting*

Based on analysis of feeder currents in normal operation state (see section III.C) the threshold values ΔI_{thr} were set to ΔI_{fe} values derived from NLP for each ith feeder and j^{th} hour. Four different settings, denoted as SET 1 to SET 4, were used to test the algorithm performance. Thresholds in SET 1 and SET 2 vary in time and depend on the moment when the change is evaluated. SET 3 and SET 4 are defined as a constant value throughout the whole year.

In SET 1 the threshold value at each ith feeder is defined as hourly change in corresponding j^{th} hour of the year in normal operation (from NLP). To decrease the indication sensitivity and to more suppress indication of load changes, the value ΔI_{thr} for SET 2 is increased to twice the value ΔI_{thr} of SET 1. The threshold value for SET 3 represents the maximum hourly change in one-year normal operation at each ith feeder. In SET 4, the value ΔI_{thr} is again double the value ΔI_{thr} in SET 3. The setting is apparent from Table 1.

Table 1. Setting of threshold values

Setting	Threshold value	$\Delta I^{j}_{thr,i}$
SET 1	Hourly change in evaluated hour	$\Delta I^{h,j}_{fe,i}$
SET 2	2 x Hourly change in evaluated hour	$2 \cdot \Delta I^{h,j}_{fe,i}$
SET 3	Maximal hourly change throughout the year	$max\left(\left\{\Delta I^{h,j}_{fe,i}\right\}_{j=1}^{8760}\right)$
SET 4	2 x Maximal hourly change throughout the year	$2 \cdot max\left(\left\{\Delta I^{h,j}_{fe,i}\right\}_{j=1}^{8760}\right)$

As some of the LIFs have minimal impact on evaluated changes ΔI_{fe}, the measurement uncertainty ΔI_{meas} was in the simulation set as 1% of threshold value ΔI_{thr} individually at each feeder, so that for j^{th} hour at i^{th} feeder it can be stated

$$\Delta I^{j}_{meas,i} = 0.01 \cdot \Delta I^{j}_{thr,i} \quad (1)$$

B. Results

The results of LIF simulations are shown in Table 2. For each SET the lines of different topology type (according to section III.A) are evaluated separately. The column Measurable represents the number of interrupted lines, which incited at least one feeder current change ΔI_{fe} higher than defined measurement uncertainty ΔI_{meas}. Low success rate by radial topology types (13 and 3) are affected by neglectable loading of these lines before LIF occurred. Lines of type 2 are measurable in ca. 90% cases, meaning that 10% of these lines were loaded insignificantly in given simulation power distribution as well.

The success rate of LIF indication is split into two columns; one is related to only measurable cases, the other to all simulated LIFs. For dynamic settings (SET 1 and 2) the success rate is quite high even for interruptions of radial lines (type 3). Obviously, LIF at feeders (types 1, 11 and 13) have the biggest impact on ΔI_{fe} and should be easily indicated. The success rate by the type 2 is also very high, which is promising for further algorithm development. For SET 3 and 4, which use high constant values ΔI_{thr}, the sensitivity is lower and many LIFs are not indicated. Despite that, SET 3 still provides solid performance for LIF of type 2 meaning that ca. LIF at 80% of these lines cause ΔI_{fe} higher than maximum hourly change calculated in normal one-year operation. Extremely high threshold setting SET4 presents worst results.

The lowest success rates by types 3 (ranging 44.9-91.4% of measurable LIFs) in Table 2 confirm that LIFs at lines of this type are the most difficult to indicate when utilizing only feeder currents data. In the case of LIF at a radial line (type 3), the fault causes supply interruption at customer's end and the indication could be complemented with customer data via i.e. SM or power outage notification. Monitoring of feeder currents is more helpful for indicating LIFs at lines of type 2, for which the success of indication ranges from 63.2% to 98%. Since there is no other acceptable way to reveal LIFs at this line type, the monitoring of feeder currents should be further examined with the emphasis on ring topology line (type 2).

V. Conclusion

In this paper, a simple deterministic algorithm for indication of line interruptions was designed. The algorithm is based on evaluating current changes at all LV feeders in MV/LV distribution stations of a meshed network. The indication is based on comparison of evaluated changes and predefined threshold values.

Table 2. Simulation results

LIF	Interrupted lines		Measurable		Indicated		
	Top. type	*Count*	*[-]*	*[%]*	*[-]*	*[%]**	*[%]*
SET 1	1	472	456	96.6	454	99.6	96.2
	11	86	86	100	86	100	100
	13	139	71	51.1	71	100	51.1
	2	228	203	89.0	199	98.0	87.3
	3	499	394	79.0	360	91.4	72.1
SET 2	1	472	456	96.6	454	99.6	96.2
	11	86	86	100	86	100	100
	13	139	71	51.1	71	100	51.1
	2	228	203	89.0	194	95.6	85.1
	3	499	393	78.8	342	87.0	68.5
SET 3	1	472	456	96.6	448	98.2	94.9
	11	86	86	100	80	93.0	93.0
	13	139	71	51.1	71	100	51.1
	2	228	203	89.0	162	79.8	71.1
	3	499	387	77.6	264	68.2	52.9
SET 4	1	472	456	96.6	424	93.0	89.8
	11	86	86	100	70	81.4	81.4
	13	139	71	51.1	67	94.4	48.2
	2	228	204	89.5	129	63.2	56.6
	3	499	385	77.2	173	44.9	34.7

* of measurable LIF

From analysis of currents measured at several feeders in normal operation and their comparison with NLP, it can be concluded that simulated load profile values may be used as threshold values in the algorithm. The paper also presents novel topological categorization of line sections in meshed network to aid fault indication and localization.

Interruptions of each line section were simulated to test the algorithm performance. With four different settings of threshold values derived from NLP the indication success rate of the algorithm was overall high, when most of the faults were discovered. Using the topological categorization of lines, the algorithm could be further complemented with localization functionality. The following work should mostly aim at detailed specification of the threshold values, which should be derived from more complex analysis of real measured data using combination of deterministic rules and AI techniques.

Acknowledgment

This research work has been carried out in the Centre for Research and Utilization of Renewable Energy (CVVOZE) and supported by the Ministry of Education, Youth and Sports of the Czech Republic under BUT specific research programme [grant number FEKT-S-20-6449]; and Technology Agency of the Czech Republic [grant number TK02030039].

References

[1] D. Topolanek, "Meshed LV distribution network", In: *IEEE Dataport*, Mar. 2022 doi: https://dx.doi.org/10.21227/kssw-1457.

[2] V. Krčál, D. Topolánek, V. Vyčítal, J. Vaculík, "Indication of Abnormal Operation Conditions in a Mesh Network Based on Data from Distributed Measurement". In *CIRED 2021 - The 26th International Conference and Exhibition on Electricity Distribution. CIRED – Open Access Proceedings Journal*. 2021. p. 1567-1571. ISBN: 978-1-83953-591-8. ISSN: 2515-0855.

Electric Vehicle Charging Management System in the Residential Streets for Individual Houses Using a Smart Controller in a Present Grid

Rikin Tailor
Dept. of Electrical Power Engineering
Technical University of Košice
Košice, Slovak Republic
rikin.jitendrakumar.tailor@tuke.sk

Zsolt Čonka
Dept. of Electrical Power Engineering
Technical University of Košice
Košice, Slovak Republic
zsolt.conka@tuke.sk

Abstract — **Electric Vehicle (EVs) is rapidly substituting combustion engine vehicles. It is inevitable to develop Electric Vehicle charging infrastructure. It is certain that we are looking for smart grid architecture for the future. However, it is impossible to do immediate changes in the present grid. Electric vehicles are not just served for transport but also for the energy storage system. The present grid is designed to serve a load that we have been utilizing for quite a long time and according to that energy-carrying capacity is also in that order. So, if we want to charge an electric vehicle in the present grid then a proper management system is mandatory. In this paper, we are presenting the smart management system for the Electric Vehicle charging infrastructure for the individual home in residential streets. It is presented with the algorithm and case study. In this way, it is possible to maintain electrical stability in the residential grid during the charging of EVs.**

Keywords: Electric Vehicle (EVs), Electric vehicle charging infrastructure, smart control

I. INTRODUCTION

The transportation systems highly dependent on fossil fuels for many decades and it is time to change in it. The most important reason behind the change in the transportation system is climate change. Almost 24% of CO2 emissions are produced by transportation [1]. The transportation sector base is changing from combustion engines to electrical chargeable batteries and electric motors.

The rising question is whether these EVs have adequate development to replace the normal cars? Well, EVs are at the peak research and development stage. As it is electric, starting torque is high, and it is silent so no noise pollution and no carbon emissions. More research is going on EV's long-distance capacity and fast charging systems. EVs are not just used for transportation but also for electricity storage. So, it can work as a grid to a vehicle (G2V) and vehicle to grid (V2G) technology.

However, as the transportation sector is moving towards electrification that much of the burden it is creating on the electrical utilities. It is certainly possible that it will create a big negative impact on the distribution grid. Consequently, an extensive study must be considered to minimize this negative effect on the power system network [2, 3, 4]. According to the demand-supply study conducted in the USA for the electric vehicle charging systems, 73% of USA vehicle demand can be attained by the HV power system [5]. It is possible that high voltage generation can meet the demand if we plan it reasonable way and do predictable expansion of the power plant. However, more research claims that the distribution system is easy to get affected by the random charging cycle of EVs [6]. As distribution system is located where charging stations are built. It is also certainly true that it is not possible to change the existing infrastructure in somewhat near time. Because it takes many years and billions of euros investment to do it. As mentioned earlier EVs strategies are for G2V and V2G technology. However, uncontrolled bi-directional flow of power for either charging (i.e. G2V) or discharging (i.e. V2G) can cause problems for the distribution systems, such as voltage fluctuations, transformer overloading, harmonic injection, etc. Harmonic has always been the biggest challenge in power systems. This harmonic can cause electrical power signal distortion and lower the power factor. The adverse effect of EVs on the distribution networks is also presented by Oliver Marcincin. in his paper [7]. To attain a smooth charging process and stabilize the whole grid, three different categories are introduced: Uncontrollable, controllable, and autonomous [8, 9].

Now, electric vehicle charging infrastructure must develop but where? Location plays a crucial role when we plan to do any infrastructure changes. There are three-four different methods to find the location of the electric vehicle charging infrastructures such as linear programming, multi objective optimization, and genetic algorithms. Author [10] have used genetic algorithm techniques to plan electric vehicle charging infrastructure location.

How much charging infrastructure is needed is another one of the most essential questions at the current time? Simon Arpad and his colleagues have given a review on the same subject. They have figured out that home charging is the most important charging infrastructure in comparison to working base charging and public charging. Public charging infrastructure is a substitute in the case of the countries or regions where home charging availability is low [11]. According to Hardman et al. [12], 50-80% of all charging events occur at home. A work charging infrastructure stand exactly after the home with 15-25% of charging incident and near about 10% of all charging events occur at the public charging station. Oliver also provided information in his paper about how the auto companies are working together to develop the charging station in entire Europe. Adding to that he has demonstrated a small charging station cycle in Ostrava [7].

Another challenge in charging EVs is congestion on the distribution side. To mitigate this issue, Subhasis Deb has tried to use the random forest method to generate EVs SoC after PEV's journey [13]. The authors in [14] come up with two methods of smart charging strategies for PEV within existing distribution systems infrastructure. Both strategies' main essence is to be minimizing the total daily cost and maintain the peak-to-average ratio. A comparative analysis gave results that one strategy is beneficial in terms of the economy and the other in technical perspectives.

Similarly, Morsy Nour has approached the autonomous method in his paper by considering a fuzzy logic controller. It is two input and one output method, where inputs are the SoC of the vehicle battery and voltage at the point of connection [9]. The aim of the controller is to charge the EV while managing the voltage and power limit to acceptable limits.

Li Xia and colleagues also have proposed a local measurement base EV charging algorithm. Instead following the centralized approach. they went with the idea in charging protocols are made individually in each household. Like voltage level checked at each controller to a made decision. It does not require any communication platform to exchange the data which helps in minimizing the cost and is possible to implement in the current grid. However, the system is not that efficient which can be implemented [15].

To tackle the challenge of managing large numbers of EVs in the future, an intelligent energy management system is required to minimize potential problems whilst maximizing potential benefits to participants in distributed electricity supply systems [16, 17]. To attain the intelligent management system, Sayidul M. derived the artificial neural network-based intelligent decision-making system with the use of a machine-to-machine communication platform. Data has been collected from the AMI to forecast the charging time and capacity [18]. However, this technique requires big data collection and machine learning to train the system. One good point that can be considered in this paper is machine-to-machine communication.

A communication platform is most essential in the case when we want to function the system at distribution side. To provide adequate charging facility to the residential consumer and still balancing the grid, I have come up with idea to use 2 smart controller and communication platform between them. The paper is designed in following way, section II is introducing basic idea behind the architecture. Algorithm and its explanation are presented in III sections. Section IV is case study which results are derived in section V. The conclusion and future scope are given in section VI and VII.

II. Basic understanding of smart Management system

This system is designed for one big area that shares a common distribution transformer. It is an independent and autonomous system. As mentioned earlier 2 smart controllers are considered in this system. One controller works as a central controller that is placed at the transformer side and the second is given to each EVs owner. Two algorithms perform in this system which are introduced in the next section. The home base controller is placed exactly before the EVs charger. The work of this controller is to collect data from the charger about the SOC of the vehicle and check the voltage and power level at the connection point with respect to the time. This controller also collects the data from the user about the charging end time.

The main or centralized controller performs the duty as collect the power and voltage level of the street and according to that giving the command to the home base controller whether to initiate the charging cycle or not. The system is designed in such a way that it does not create an overload in the system. Central controllers give commands in accordance with balancing the system. However, it is designed to fully utilize the system.

The system gets revised every time when new EV gets connected to a charger. The vehicle which has the lowest battery gets prioritized and more power is applied to level up the battery charging status. This system works on three inputs and one output. Three inputs are voltage-power level, SOC of EV, and time given by the user to disconnect the EV.

I have assumed that the user will behave in manner and do not cheat by entering the wrong time of disconnection of the EV.

By considering this input, the system automatically calculates the priority of different charger in street and perform accordingly. In the case of a sudden rise in electricity demand and only a few cars can get charged then it will give priority to the EV which is connected first in a daytime and following it. It is possible to increase and decrease the input power to the charger by measuring demand and availability. If the SOC of any EV reached its maximum value, then it will disconnect automatically, and the EV will get removed from the centralized controller in order to perform the task. Basic schematic diagram is shown in figure 1.

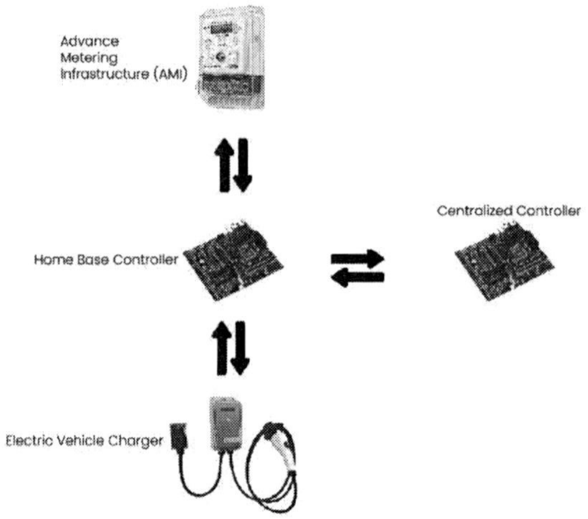

Figure 1. Basic schematic diagram of process

III. ALGORITHM

It is easier to understand the algorithm with an example. Figure 2 and figure 3 are the home base charger algorithm and centralized charger algorithm respectively. Consider EV is connected to the charger at residence. At this time, the home base controller comes into action. As the car is connected, it satisfied the first condition. Now, the user will have to enter the charging end time of EV $T_{(ce)}$ or the departure time into the application. The system confirms and processes the charging end time and stores the time of connection $T_{(c)}$. The further controller will communicate with the charger and get the actual SOC of the EV. If it is maximum (100%) then there is no need for charge and the task will stop, but if it is less than that it goes for the further process.

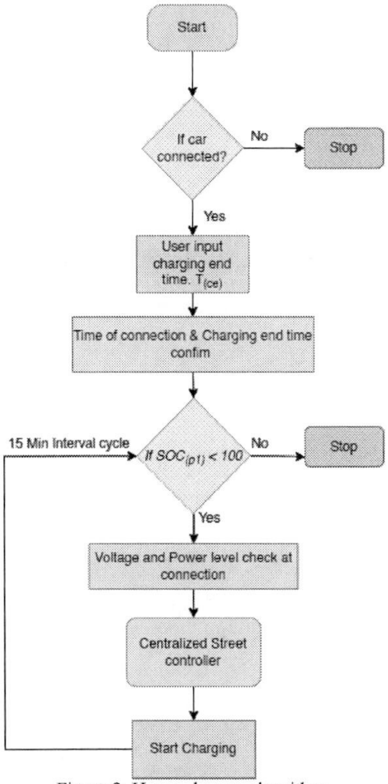

Figure 2. Home charger algorithm

The next step is a voltage and power level check at the time of connection. Once it is checked, the home base control sends all these data to the central controller. Central controllers initialize the process with the collecting line and bus data. Further, it checks the power and voltage level from collected data. If it is insufficient, then the process stops here. Otherwise, it measures the number of vehicles connected in street, $SOC_{(pi)}$ state of charge of n vehicle at the present time, and time allocation from the user side. After this step-by-step process, the controller performs the mathematical operation and estimates the charging time for each EV and enable the power flow in that charger. The system will check SOC every 15 minutes to get a status update.

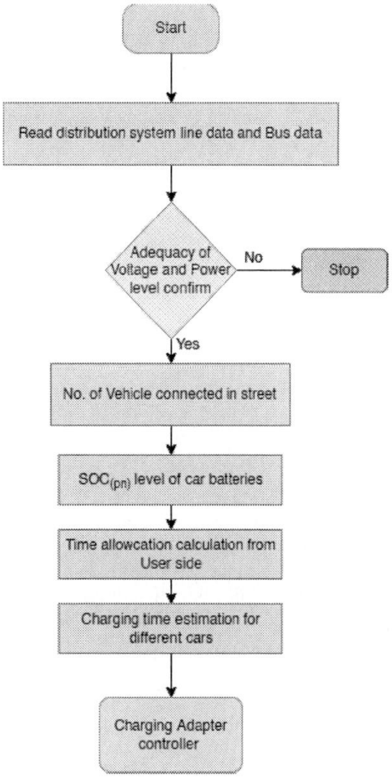

Figure 3. Centralized charger algorithm

The next step is a voltage and power level check at the time of connection. Once it is checked, the home base control sends all these data to the central controller. Central controllers initialize the process with the collecting line and bus data. Further, it checks the power and voltage level from collected data. If it is insufficient, then the process stops here. Otherwise, it measures the number of vehicles connected in street, $SOC_{(pi)}$ state of charge of n vehicle at the present time, and time allocation from the user side. After this step-by-step process, the controller performs the mathematical operation and estimates the charging time for each EV and enable the power flow in that charger. The system will check SOC every 15 minutes to get a status update

As soon as EVs battery status reached maximum value, the process will stop, and EV will get disconnected from the charger and also from the queue of the centralized controller.

Total demand of power in street for the EVs charging can be given as,

$$D_{cs}^{d} = SOC_{p1} + SOC_{p2} + SOC_{p3} + \cdots + SOC_{pi} \qquad (1)$$

where,
D_{cs}^{d} = energy demand at central system, kWh
SOC_{p1} = State of charge of EV connected at charger 1 at present time, kWh
SOC_{pi} = State of charge of EV connected at charger n at present time, kWh

- EVs battery charging equation,

$$\frac{P_A^t}{R_c} = \frac{(SOC_{max} - SOC_{(pi)}) * 100}{\mu} \qquad (2)$$

where,

P_A^t = Power available at time at the charger, kW
R_c = rate of charge, kW/h
SOC_{max} = maximum SOC capacity kWh
$SOC_{(pi)}$ = state of charge of EV at present time, kWh (i = 1, 2, 3, …, n)
μ = efficiency of charger

$$SOC_{(pi)} \leq SOC_{max} \quad (3)$$

- SOC of EVs,

$$SOC_{(pi)} = SOC_{(pi-1)} + \frac{P_A^t * \mu}{R_c * 100} \quad (4)$$

where,

$SOC_{(pi-1)}$ = state of charge of EV at time (p-1), kWh

IV. CASE STUDY

An area is consisting of 10 houses. Each house has a different load profile. Figure 4. indicates a total load of houses in street and available power for the EVs. Figure 5. is the load profile of houses. Given load, the profile is real data consider from houses in Kosice, Slovakia. The total time considers for the load profile is 24 hours with a 1-hour cycle time. A total of 10 EVs are considered for each house.

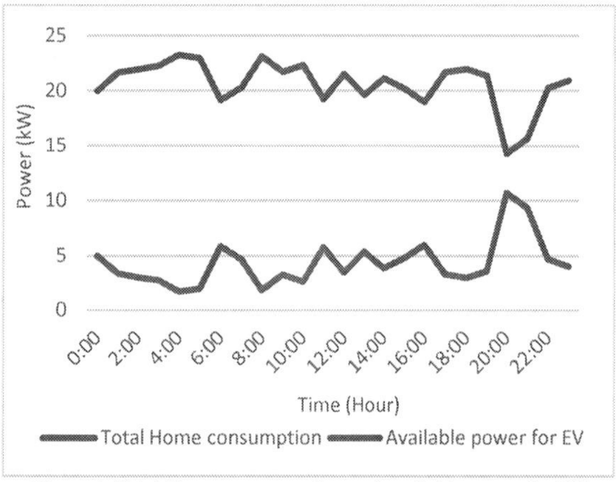

Figure 4. Total Houses consumption and available power for EV

We have considered the ideal power supplying capacity as 25 kW to the area to evaluate the true potential of the management algorithm. EVs connection time, Total EVs capacity, SOC at the time of connection, and disconnection time are given in table 1. The minimum SOC for boosting the charger first is less than 20%. It means if the battery percentage is below 20% then it will charge first until reaches to the target value. Charger efficiency is 90%. Two chargers are used: 7.6 kW & 11.5 kW.

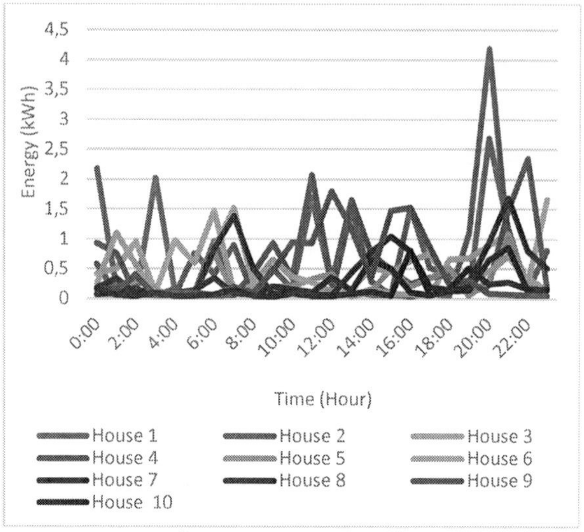

Figure 5. Houses Load Profile

Each house has its own control where they can enter disconnection time. The controller will perform autonomously recharge the EVs battery.

Table 1. EVs specification and SOC

Evs	Connection Time (24 hours)	Total capacity (kWh)	SOC (kWh)	Disconnection Time (24 hours)
V1	0:00	100	40	12:00
V2	2:00	40	10	20:00
V3	2:00	40	5	8:00
V4	4:00	37.8	20	10:00
V5	1:00	80	50	10:00
V6	3:00	60	30	14:00
V7	3:00	40	20	8:00
V8	6:00	37.8	2	12:00
V9	8:00	60	50	22:00
V10	4:00	60	0	11:00

We are going to evaluate how exactly charging will initiate for each EVs and how much SOC will have at the disconnection time.

V. RESULT

The result is very satisfactory. Figure 6. represents the EVs charging cycle during the day-nighttime. Steep rise and sudden decline in the graph indicate the Connection of EV and disconnection of EV respectively. All EVs are charged more than 25% which is clearly seen in the graph. Out of the 10 EVs, at least 6 EVs are charged more than 90%. Even the charging connection time was less for most of the EVs.

Much electricity is left even after the completion of charging. Priority given to each EVs was very accurate in that not a single EVs left behind from the charging.

EV1: Charging start at 0:00 with 40 kW battery reached to 92.64 kW

EV2: Charging start at 2:00 with 10 kW battery reached to 40 kW

EV3: Charging start at 2:00 with 5 kW battery reached to 30.99 kW

EV4: Charging start at 4:00 with 20 kW battery reached to 33.48 kW

EV5: Charging start at 1:00 with 50 kW battery reached to 80 kW

EV6: Charging start at 3:00 with 30 kW battery reached to 60 kW

EV7: Charging start at 3:00 with 20 kW battery reached to 34.95 kW

EV8: Charging start at 6:00 with 2 kW battery reached to 20.57 kW

EV9: Charging start at 8:00 with 50 kW battery reached to 60 kW

EV10: Charging start at 4:00 with 0 kW battery reached to 26.85 kW

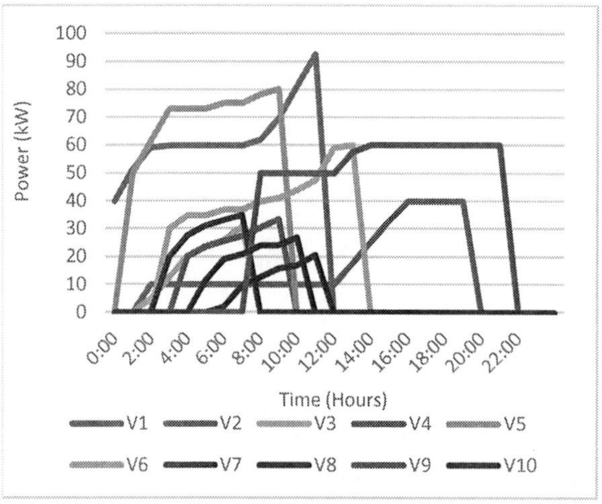

Figure 6. EVs Charging cycle

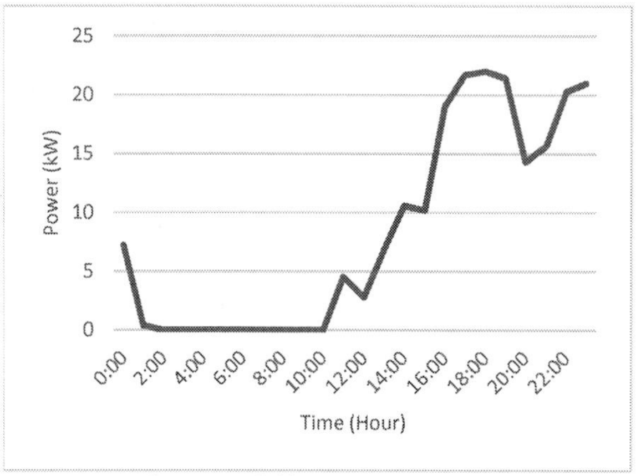

Figure 7. Available Power after EV charging

Figure 7 represents the available power after the EVs charging. It is shown in the graph that not much power has been used after the 14:00 to till the end of the day.

VI. Conclusion

EVs are coming into the market with tremendous amounts of speed. It is quite challenging for all the engineers, scientists, and utility companies to meet the electricity demand that EVs are creating. Much research is ongoing for the long-distance battery of EVs, charging infrastructure on the public-residential-work side, fast charging of EVs, etc.

In this paper, I tried to put my concept to manage EVs charging on the residential side of the present grid. I focused on the availability of power, the balance of the grid, utilization of the grid in an adequate manner, and charging of EVs in an autonomous way. I have used two different controllers to attend to this demand. One for the user and another centralized controller which can handle the charging of all EVs in a reliable manner. The results are decent and indicate that it is possible to implement such an algorithm to charge EVs.

VII. Future Scope

- Two-way power flow algorithm G2V and V2G
- More option to user if they just want to charge car at certain battery level so they can put in the input
- Much more powerful algorithms to charge EVs in more effective way.

Acknowledgment

This work was supported by the Slovak Research and Development Agency under the contracts No. APVV-19-0576 and APVV-21-0312 and by the Ministry of Education, Science, Research and Sport of the Slovak Republic and the Slovak Academy of Sciences under the contract VEGA 1/0757/21.

References

[1] S. Labatt, R.R. White. *The Financial Implications of Climate Change*. Hoboken, NJ: Wiley, 2007, pp. 288.

[2] D. Hilshey, P.D.H. Hines, P. Rezaei, J.R. Dowds, "Estimating the impact of electric vehicle smart charging on distribution transformer aging," *IEEE Trans. Smart Grid*, vol. 4, pp. 905–913, June 2013.

[3] M.K. Gray, W.G. Morsi, "Power Quality Assessment in Distribution Systems Embedded with Plug-In Hybrid and Battery Electric Vehicles," *Power Syst. IEEE Trans.*, vol. 30, pp. 663–671, March 2015.

[4] R.C. Leou, C.L. Su, C.N. Lu, "Stochastic analyses of electric vehicle charging impacts on distribution network," *IEEE Trans. Power Syst.*, vol. 29, pp. 1055–1063, May 2014.

[5] P. Richardson, D. Flynn, A. Keane, "Impact assessment of varying penetrations of electric vehicles on low voltage distribution systems," *IEEE Power and Energy Society General Meeting.*, pp. 1-6, July 2010.

[6] S.S. Raghavan, A. Khaligh, "Impact of Plug-in Hybrid Electric Vehicle Charging on a Distribution Network in a Smart Grid Environment," *IEEE PES Innovative Smart Grid Technologies*, vol. 36, pp. 1-7, Jan. 2012.

[7] O. Marcincin, Z. Medvec and P. Moldrik, "The Impact of Electric Vehicles on Distribution Network," *18th Inter. Scient. Conf. Electr. Pow. Engin.*, 2017, pp. 1-5.

[8] S. Faddel, A.T. Al-awami, "Charge Control and Operation of Electric Vehicles in Power Grids: A Review," *Energies*, vol. 14, pp. 1-21, Feb. 2018.

[9] M. Nour, S.M. Said, H. Ramadan, A. Ali, and C. Farkas, "Control of Electric Vehicles Charging without Communication Infrastructure," *2018 Twentieth International Middle East Power Systems Conference (MEPCON)*, 2018, pp. 773-778.

[10] D. Efthymiou, K. Chrysostomou, M. Morfoulaki, G. Aifantopoulou, "Electric vehicles charging infrastructure location: a genetic algorithm approach," *Eur. Transp. Res. Rev.*, vol. 9, pp. 1-9, May 2017.

[11] S. A. Funkea, F. Spreib, T. Gnanna, P. Plotza, "How much charging infrastructure do electric vehicles need? A review of the evidence and international comparison," *Transportation Research Part D: Transport and Environment*, vol. 77, pp. 224-242, Dec. 2019.

[12] S. Hardman, et al., "A review of consumer preferences of and interactions with electric vehicle charging infrastructure," *Transp. Res. Part D: Transp. Environ.*, vol. 62, pp. 508-523, July 2018.

[13] S. Deb, A.K. Goswami, R.L. Chetri and R. Roy, "Distribution system congestion management by charging coordination of plug-in electric vehicle, *2020 3rd International Conference on Energy, Power and Environment: Towards Clean Energy Technologies*, 2021, pp. 1-6.

[14] R. Mehta, D. Srinivasan, A.M. Khambadkone, J. Yang, A. Trivedi, "Smart Charging Strategies for Optimal Integration of Plug-in Electric Vehicles within Existing Distribution System Infrastructure," *IEEE Transactions on Smart Grid*, vol. 9, pp. 299-312, Jan. 2018.

[15] L. Xia, I. Mareels, T. Alpcan, M. Brazil, J. de Hoog and D.A. Thomas, "A Distributed Electric Vehicle Charging Management Algorithm Using Only Local Measurements," *ISGT 2014*, 2014, pp. 1-5.

[16] O. Vermesan, L.-C. Blystad, R. John, P. Hank, R. Bahr and A. Moscatelli, "Smart, Connected and Mobile: Architecting Future Electric Mobility Ecosystems," *Proceedings of the Conference on Design, Automation and Test in Europe EDA Consortium*, 2013, pp. 1740–1744.

[17] S. Morsalin, K. Mahmud and G. E. Town, "Smart Data-logging System for Coordination of EVs in Distributed Energy Management System," *23rd World Congress on Intelligent Transport Systems (ITS)*, Melbourne, Australia, 10-14 Oct. 2016.

[18] S. Morsalin, K. Mahmud, G. Town, "Electric Vehicle Charge Scheduling Using An Artificial Neural Network," *2016 IEEE Innovative Smart Grid Technologies – Asia (ISGT-Asia)*, 2016, pp. 276-280.

Methodology of Battery Life Prediction for Hybrid Energy Storage System in Electric Vehicles

Dávid Bodnár, Juraj Biľanský, František Ďurovský
Department of electrical engineering and mechatronics
Technical University of Kosice
Kosice, Slovakia
david.bodnar@tuke.sk, juraj.bilansky@tuke.sk, frantisek.durovsky@tuke.sk

Abstract — **Increase in electric mobility in recent years brings higher demands in battery production and performance. Battery aging still raises big concerns among the general public. One of the possible solutions to improve battery life is to reduce load on the battery by using hybrid battery storage system, which consists of battery and supercapacitor. Different topologies of the hybrid energy storage system (HESS) are discussed in this paper, including semi-active topology which was modeled in Matlab/Simulink. Methodology for battery life estimation is proposed which allows viability assessment of different sizes of modular traction batteries. Battery aging is investigated on empirical equivalent circuit model. Created model also allows testing of new energy management strategies before their implementation.**

Keywords: Hybrid battery storage system, semi-active topology, energy management system, DC/DC converter

I. INTRODUCTION

The European Union's efforts to decarbonize the economy are putting pressure on the massive use of electric vehicles in transport. Traction batteries are still a critical component of electric vehicles (EVs). Current battery technologies do not provide such energy density, and battery life that electric cars can fully compete with internal combustion engine vehicles [1]. Electric cars are currently suitable for operation in cities where they are not required to travel long distances, and they can advantageously use the kinetic energy of the vehicle during braking due to their ability to regenerate the energy of the vehicle [2]. Night charging is battery-friendly and at the same time helps to balance the consumption of electricity from the public network [3]. An interesting possibility is their ability to accumulate (supply) part of the electricity from the public network, and thus help to compensate fluctuations in energy supply and consumption in the public distribution network [4]. Despite these possibilities, electric vehicles have not yet found enough customers among ordinary users, who prefer long-range vehicles and convenient use outside cities.

The situation is different in the area of public transport. Public transport vehicles travel on permanent routes with frequent accelerations and braking. In addition, they drive in an area where the emphasis is on minimal or no emissions [5]. The deployment of electric buses is supported by states and local governments, so a wider deployment of electric vehicles can be expected in this area. The development of battery technologies is moving in several directions. The main challenge is to change the technology of the battery itself in order to increase its energy density, the ability to fast charge and discharge without damaging the battery, and increase its safety. In addition, there are other ways to extend battery life. High charging and discharging currents are not desirable for battery lifetime. However, during its operation, the EV requires increased power during acceleration as well as the ability to absorb energy quickly during braking. Such driving cycles are typical, especially for urban traffic. To protect the battery from high currents, but at the same sustain possibility to accelerate rapidly, a device capable of receiving and providing a large amount of energy in a short time is needed.

One solution is to add a supercapacitor (SC) to the traction chain [6]. Supercapacitors are also known as ultracapacitors (UC) but both terms represent the same storage technology. The SC is characterized by high specific power, which means that it can receive and supply high currents [7]. On the other hand, it has low specific energy, so it does not have much effect on extending the range of vehicle. Ensuring a sufficient range of the vehicle remains a responsibility of the battery. As a result, it is advantageous for the high power peaks to be covered by SC and the long-term load by battery.

The SC has a significantly lower internal resistance than the average battery, which will reduce the overall internal losses of the power storage and increase its efficiency [8]. However, the use of a battery-SC combination also has its drawbacks. The SC voltage varies significantly based on the state of SC charge, which requires the use of an additional DC/DC converter. The SC would be connected to DC bus via DC/DC converter. In addition, the inclusion of SC and another converter in the vehicle's traction chain increases the cost and weight of vehicle. The size of batteries and SC must therefore be optimized for the application [9].

In this paper, simulations of Ultracapacitor semi-active topology with different sizes of modular traction batteries are shown. A rule-based energy management system based on a control of SC state of charge with respect to the vehicle kinetic energy is utilized. Supercapacitor modules ensure high power peaks, hence reducing the load on the battery pack. Simulations in Matlab/Simulik were performed for a 7-ton midibus operated in city traffic. Designers of small electric buses often struggle with fitting huge battery packs into chassis to reach the desired range and power demands. If a long vehicle range is not required then a smaller battery pack can be utilized [10]. Supercapacitor modules provide power peaks, so a battery pack with lower capacity doesn't exceed maximum current limits. Therefore, vehicle mass could be reduced and it also facilitates designing for vehicle designers.

This paper is organized as follows. Section II presents possible HESS topologies, some energy management systems are described in section III. Section IV introduces the basis of battery aging and its modeling, and section V presents a model of semi-active HESS and simulations with modular traction batteries.

II. Topologies of energy storage system

There are several possible battery-SC topologies. All topologies can be divided into 3 groups: passive configuration, semi-active configuration, and fully-active configuration [1].

A. Passive configuration

Improvements in efficiency and performance are already achieved in a passive configuration, where the battery and SC are connected in parallel without the use of a DC/DC converter. The SC works as a low-pass filter, the larger is the capacity, the greater is the filtering effect. However, the SC potential remains largely unused (Fig. 1).

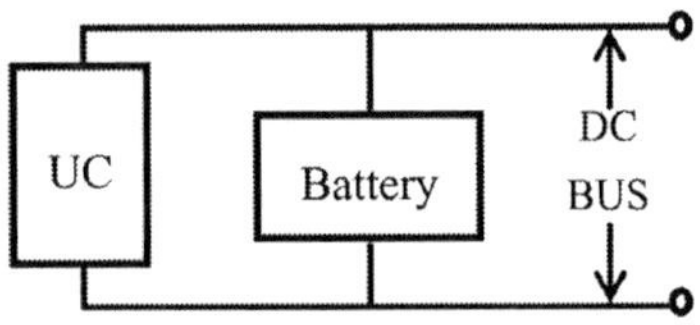

Figure 1 Passive configuration of the HESS [12]

B. Semi-active configuration

By adding a DC/DC converter, the possibility of SC operation in a wide range of voltages is achieved and a larger percentage of SC capacity can be used [13]. The use of a bidirectional DC/DC converter for the SC seems to be a good compromise between achieving the desired effect and the system costs. This is called a semi-active configuration. Due to these advantages, the mentioned solution is currently the most frequently used topology [14]. The semi-active configuration can be divided into two main topologies (Fig. 2): Ultracapacitor semi-active topology and Battery semi-active topology [15]. In the Supercapacitor semi-active topology, the battery is directly connected to the DC bus and the SC is connected in series with a DC/DC converter. The DC/DC converter controls the power flow to and from the SC. This topology is used more frequently compared to the Battery semi-active topology due to the more stable voltage on the battery.

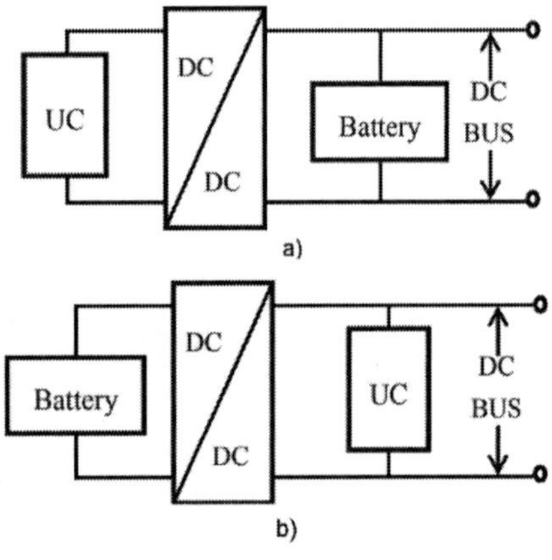

Figure 2 Semi-active configuration of the HESS,
a) Ultracapacitor semi-active topology, b) Battery semi-active topology [12]

There is also a possibility of a semi-active topology shown in Figure 3, with one DC/DC converter between both energy sources and one power diode. This allows size reduction of the converter and also cost reduction because it can be unidirectional and operates just in case the SC is fully charged [16]. However, power flow from both components can not be controlled and is decided according to the voltage of both components.

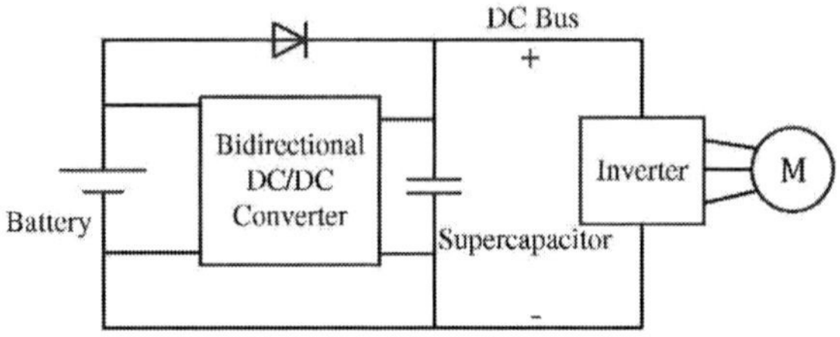

Figure 3 Semi-active topology with additional power diode [17]

C. Fully-active configuration

In fully-active configurations, both components are controlled by power converters or switches. They contain more components which usually results in higher costs and losses. On the other hand, they have better controllability. The fully-active configuration can be created by adding a second DC/DC converter. In a parallel fully-active topology, the battery and the SC are both connected to a DC bus via DC/DC converter (Fig. 4 a)). This topology can be used when the battery voltage differs from the bus voltage or if constant bus voltage is required. There are also cascaded fully-active topologies depicted in Fig. 4 b) and c). One of the components is located between two DC/DC converters, one converter provides a connection to the DC bus, and another one to the second component. This topology provides effective SC voltage regulation, which allows the system to respond to future power demand by having optimally charged SC [18].

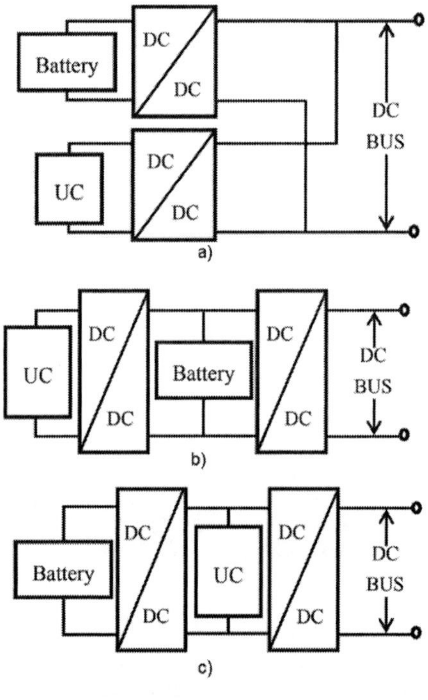

Figure 4 Fully-active configuration of the HESS [12]

In [19] a novel topology of multimode HESS and its energy management system (EMS) is proposed (Fig 5). Two energy sources are connected to each of the two ends of bidirectional DC/DC converter. A power diode is used to avoid the batteries being recharged directly. There are also two power switches to provide the power directly from the power source to the load without the DC/DC converter. That enables the converter to operate at its peak efficiency. The proposed energy management strategy provides 9 operating modes and thus it could further enhance the efficiency of the system.

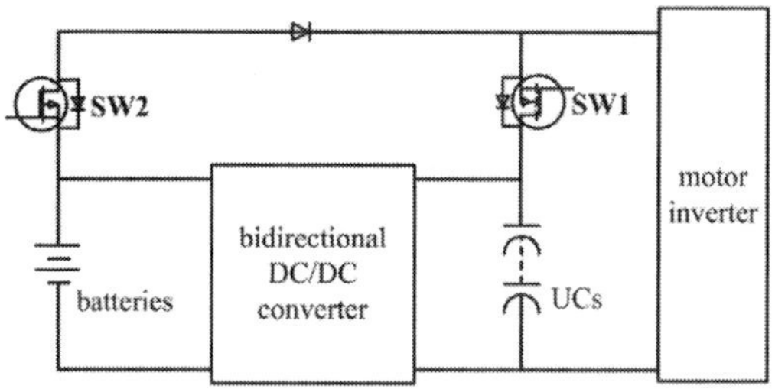

Figure 5 Fully-active configuration of the HESS with an additional control circuit [19]

One of the possible topologies of fully-active configuration is to use a pair of complementary transistors instead of a DC/DC converter [20]. Either the battery or the SC would be in operation, but not at the same time (Fig. 6). This would improve the efficiency of the ESS, but the DC bus voltage would fluctuate significantly with respect to the SC voltage, which would require a more complex motor-drive control algorithm.

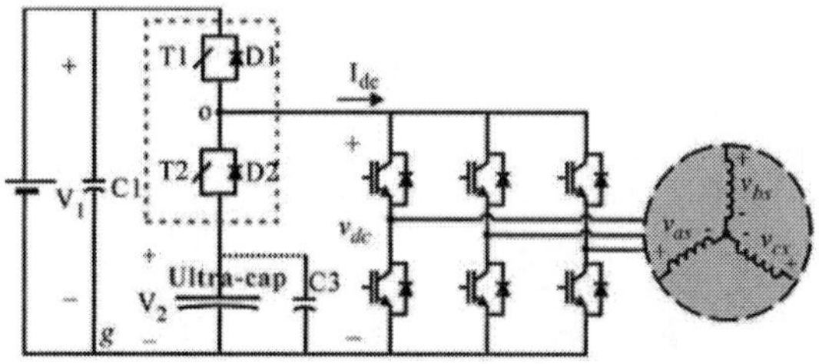

Figure 6 HESS with a pair of complementary transistors [20]

III. Energy Management System

The energy flows between the battery, SC, and motor are controlled by energy management system. In this section, only the ultracapacitor semi-active topology is discussed because it presents a good compromise between the performance and the system cost. Many different approaches have been proposed for this topology, which have been based on optimization techniques, fuzzy logic, neural networks, etc. In general, energy strategies are divided into two groups, rule-based strategies, and optimization-based strategies.

A. Rule-based strategies

One of the rule-based methods is the utilization of a low-pass filter. High-frequency current is supplied by the SC and low-frequency current by the battery. However, this type of filter-based strategy has several disadvantages. It provides a very large phase shift [21], the cutoff frequency of the filter needs to be adjusted in the design in accordance with load demands, and there are also problems with keeping the SC voltage within the desired range.

Another method is applying a minimization function [22]. This design covers a wide range of EV powers, but the SC is sometimes discharged to zero and relatively heavily loaded.

A rule-based energy management algorithm is well described in [23] and is based on the control of SC state of charge with respect to the vehicle kinetic energy. This allows high utilization of energy received during regenerative braking and provides energy for high accelerations at any time. The goal is to discharge the SC in such a way that there is anytime enough space left to absorb the energy from regenerative braking. The SC is discharged by lower power (Figure 7), so the SC is fully charged when braking at zero speed. This requirement prevents excessive discharge, and therefore it is possible to start a new driving cycle even at high acceleration requirements. As a result, in the case of a long-term load, energy is taken from the battery, and in the case of a short-term peak load, the SC also cooperates, while only the SC works in the case of recuperation braking.

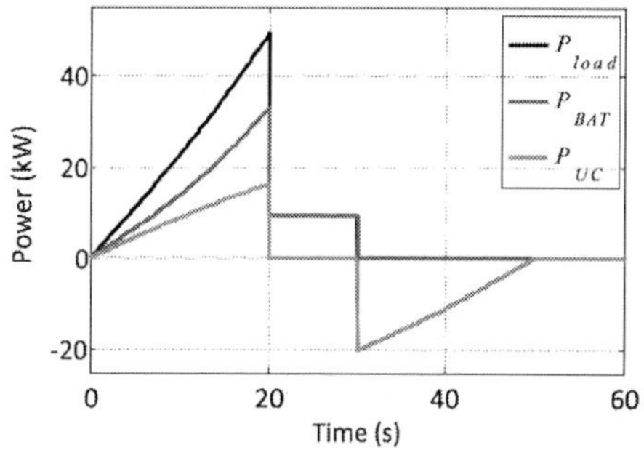

Figure 7 Load power and power distribution [16]

A good option is to use fuzzy logic [24]. However, in this design, discharging of the SC prevailed over charging, so there were situations where the SC was significantly discharged even at low speeds. Another possibility is a combination of a fuzzy logic controller with a low pass filter (Fig. 8), so the SC would perform the function of a high-pass filter, where the low-frequency load would cover the battery and the high-frequency load SC [25].

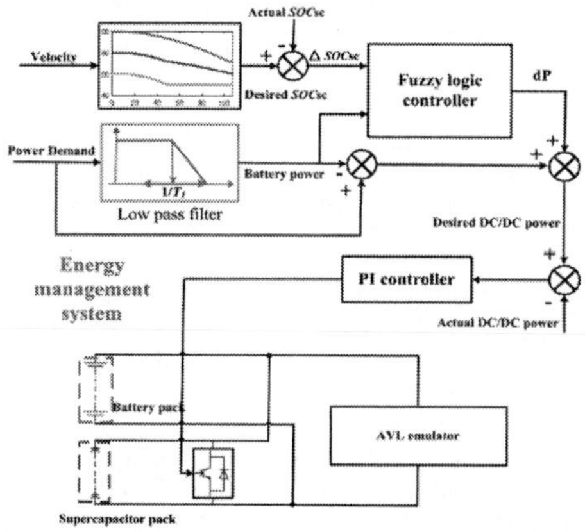

Figure 8 Fuzzy logic controller combined with a low pass filter [25]

B. Optimization-based strategies

There are also many optimization-based control strategies that utilize either optimal control theory or soft computing algorithms. The optimization-based methods include genetic algorithms [26], particle swarm optimization [27], neural networks [28], model predictive control [29], dynamic programming [30], and others. These strategies are more advanced compared to rule-based strategies but usually require high computational power and complex programming. These strategies will be discussed in another paper in detail.

IV. Methodology for Battery Aging Estimation

Battery aging is a very complex process, and many variables have a significant effect on battery degradation. Battery state of charge (SoC), temperature, C-rate (one-hour discharge current), and pulse current are the most important factors that can have an impact on battery life. Furthermore, battery aging can be divided into cycling aging and calendar aging. Moreover, the aging of a battery leads mainly to loss of capacity but also to an increase of internal resistance. Therefore, it is very demanding to accurately describe battery behavior for all possible operating conditions.

In [31], APR18650m1 LiFePO4 battery capacity fading was investigated. The cells have been cycled at different temperatures, initial state of charge, SoC deviation from the average SoC (ΔSoC), and discharge C-rates. They observed that capacity fading is directly proportional to charge processed through the battery but the capacity fading rate is highly dependent on room temperature, SoC, and ΔSoC. The temperature effect can be modeled by Arrhenius equation where the capacity fading rises dramatically above room temperatures. However, this model is valid just for temperatures above 25 °C because for lower temperatures different capacity fading mechanism occurs. For charging at lower temperatures lithium plating is present and different equations take place. It was concluded that the capacity fading is lower for a low average SoC and a small ΔSoC. Discharge C-rates did not directly affect capacity fading if battery current did not exceed the maximum current at room temperature. However, ohmic heating raises cell temperature and if is not cooling properly, could accelerate capacity fading.

In [32], T. Plattard et al. were focusing on the implementation of a weighted ampere-hour throughput model of Li-ion Nickel-Manganese-Cobalt (NMC) battery cell, taking into account the operating parameters, and modulating the impact of an exchanged ampere-hour. They were observing that the cycling aging is proportional to the number of exchanged ampere-hours and the calendar aging is proportional to the square root of time.

V. MODEL OF SEMI-ACTIVE CONFIGURATION

In this section the model of Ultracapacitor semi-active topology in Matlab/Simulink is shown with an SC power controller. Energy management system is based on the algorithm proposed in [23].

A. Model of DC/DC converter and HESS

Bidirectional back-boost converter is used as depicted in Fig. 10. Battery and DC bus are connected to the high side of the converter and SC to the low side. By switching Q2 transistor, the energy of SC is accumulating on inductance, and then after turning off Q2 transistor is the energy released to DC bus. On the other side, energy flows from DC bus to SC by switching Q1 transistor.

The power controller works on a principle of measuring SC voltage and current, and the actual SC output power is compared with desired SC power (Fig. 9). Two pulse generators generate duty cycles for both switches. Switch Q1 is enabled when DC/DC converter works in a boost mode, Q2 when it works in a back mode.

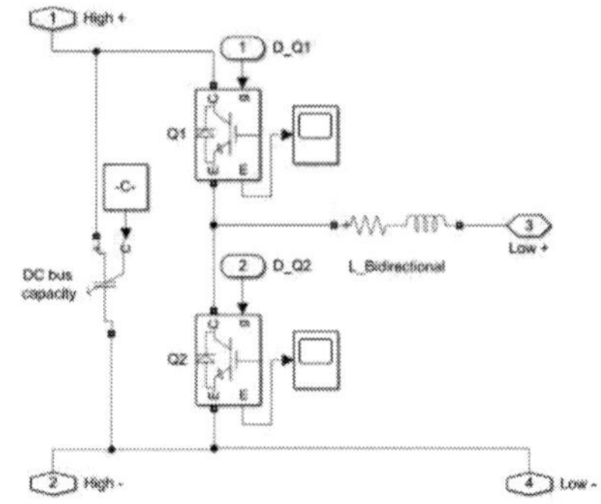

Figure 10 Model of DC/DC converter in Matlab/Simulink

B. Performance comparison of batteries with different capacities

One of the suitable applications for HESS is a city bus because of its higher weight and frequent acceleration and braking. Power demands for a 7-ton midibus were estimated based on traction resistances for a stabilized phase of Federal Test Procedure (FTP-75) driving cycle. The beginning of this driving cycle is depicted in Fig. 11. Simulations were performed for 3 different sizes of a battery pack: 90 kWh, 135 kWh, and 180 kWh. All of them were simulated as a single battery pack and as a part of HESS. Supercapacitor modules had a maximum voltage of 180V and a capacity of 60 F.

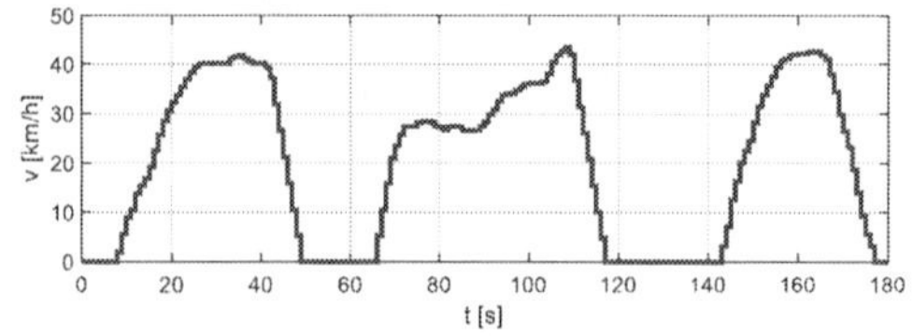

Figure 11 Simulated part of FTP-75 driving cycle

C-rate of all battery packs is shown in Fig. 12. C-rate of the smallest 90 kWh pack sometimes reaches values around 2.5 C, which means the battery pack is heavily loaded. On the other hand, 180 kWh battery is loaded about twice less.

A C-rate comparison of batteries of the same size in battery-only energy storage system (BESS) and HESS is shown in Fig. 13 and 14. Different weights of the vehicle caused by an added mass of SC and DC/DC converter (75 kg) were also included in the simulation. It is visible the battery pack is much less loaded as a part of HESS. Charge processed by a single battery cell during the whole driving cycle was also calculated to estimate battery aging. It is calculated as an integral of the absolute value

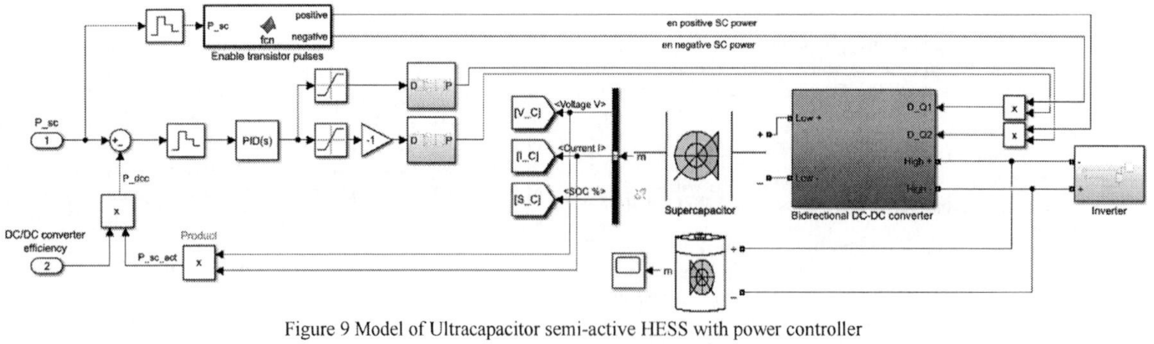

Figure 9 Model of Ultracapacitor semi-active HESS with power controller

of the cell current. For this purpose parameters of INR18650-30Q battery cells were used, which have a nominal capacity of 2950 mAh. Results are shown in Tab. 1. According to simulations, the charge processed through one cell during one driving cycle was for HESS around 47.5 % lower compared to a battery pack without SC. Including charging, processed charge savings are around 30.3 % (Tab. 2). That means cycling aging could be significantly reduced. Moreover, power peaks are eliminated, so less energy is transferred to heat by the internal resistance of the pack. Tab. 3 shows the maximum C-rate od the battery during this driving cycle. The maximum C-rate for battery in HESS was just around 38% of the maximum C-rate without SC modules.

Simulations show very promising results that could significantly improve battery lifetime. However, simulations were performed for ideal conditions, which means there were not any elevation changes, temperature 20 °C, and all energy of the vehicle was regenerated by SC until is fully charged. The maximum braking power was 80 kW, so to achieve those assumptions DC/DC converter must be also designed for this power, or a different energy management system must be used. In the case of emergency braking, braking power is even higher, so a part of energy must be dissipated in mechanical brakes.

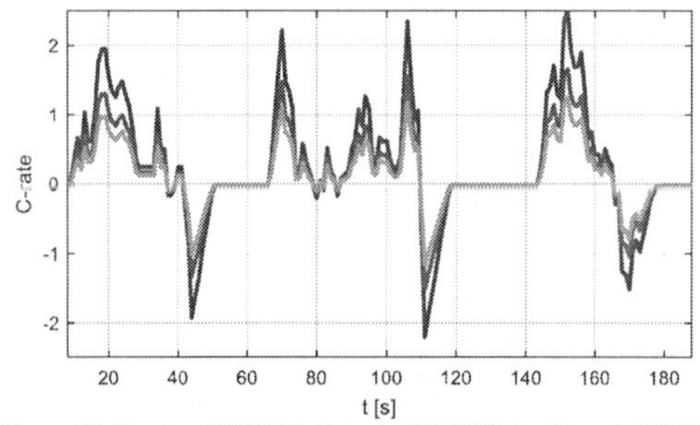

Figure 12 C-rate of 90 kWh (blue), 135 kWh (red), and 180 kWh (yellow) battery packs

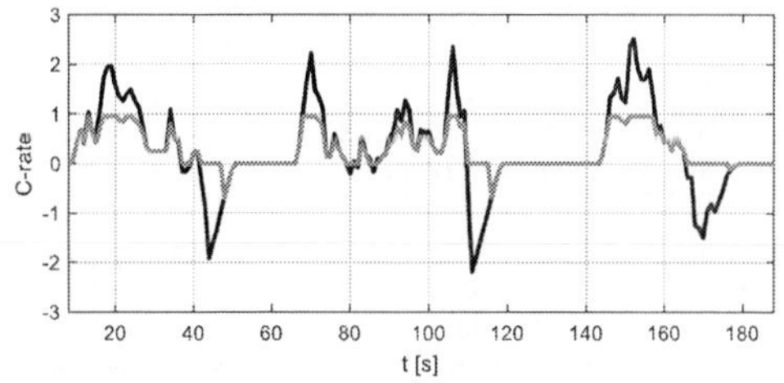

Figure 14 C-rate of 90 kWh battery pack as BESS (black) and as a part of HESS (green)

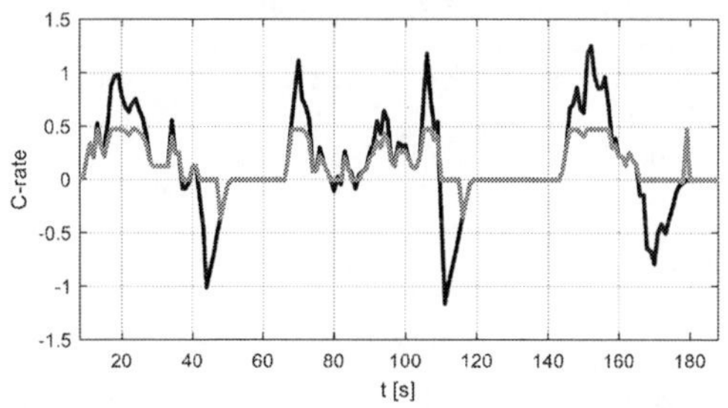

Figure 13 C-rate of 90 kWh battery pack as BESS (black) and as a part of HESS (green)

Table 1 Charge processed by single cell during one driving cycle

Battery capacity [kWh]	Charge processed on the road		
	Single component [mAh]	*HESS [mAh]*	*Savings in processed charge*
90	452.3	237.3	47.5 %
135	305.3	160.9	47.3 %
180	231.7	121.0	47.7 %

Table 2 Charge processed by a single cell during one driving cycle including charging

Battery capacity [kWh]	Charge processed with charging		
	Single component [mAh]	*HESS [mAh]*	*Savings in processed charge*
90	661.7	461.0	30.3 %
135	446.4	311.6	30.2 %
180	338.8	235.7	30.4 %

Table 3 Maximum current

Battery capacity [kWh]	Battery maximum current	
	Single component [C-rate]	*HESS [C-rate]*
90	2.51	0.958
135	1.664	0.636
180	1.251	0.476

VI. Conclusion

Ultracapacitor semi-active topology of hybrid energy storage system was modeled in this paper. The simulations have confirmed that the currents flowing through the battery can be significantly reduced by using a supercapacitor. This mainly applies to peak battery currents during the acceleration and braking of the vehicle. Methodology of cycle aging estimation was proposed, however, the battery aging was only estimated based on processed charge within one driving cycle. Results show that for a city driving cycle processed charge can be reduced by around 30% by using HESS. A more accurate battery aging model must be created in order to better estimate battery aging. The ideal way to determine the degree of battery aging is to physically discharge it on a battery cell tester, which would determine the aging rate at different currents. The results would later be used to modify the battery model and optimize the sizing of the battery-SC assembly.

The optimal sizing is important due to the high initial costs of the SC and DC-DC converter. It was shown that it is possible to reduce battery capacity if it is combined with supercapacitor modules. That allows leveraging smaller battery which fits better into midibus chassis and reduces vehicle mass without overloading battery. The building of a physical HESS and verification of simulation results will be a subject of future research.

Acknowledgment

This work was supported by Slovak Research and Development Agency on the basis of Contract no. APVV-18-0436.

References

[1] J. Deng, C. Bae, A. Denlinger, T. Miller, "Electric vehicles batteries: requirements and challenges." *Joule*, vol. 4, pp. 511-515, 2020.

[2] M.K. Yoong, Y.H. Gan, G.D. Gan, C.K. Leong, Z.Y. Phuan, B.K. Cheah, K.W. Chew, "Studies of regenerative braking in electric vehicle," *2010 IEEE Conference on Sustainable Utilization and Development in Engineering and Technology*, pp. 40-45, 2010.

[3] V.D. Kam, M. van Sark, W. van Sark. "Smart charging of electric vehicles with photovoltaic power and vehicle-to-grid technology in a microgrid; a case study." *Applied energy*, vol. 152, pp. 20-30, 2015.

[4] T.K. Miao, V.K. Ramachandaramurthy, J.Y. Yong. "Integration of electric vehicles in smart grid: A review on vehicle to grid technologies and optimization techniques." *Renewable and Sustainable Energy Reviews*, vol. 53, pp. 720-732, 2016.

[5] M. Behnke, T. Kirschstein, "The impact of path selection on GHG emissions in city logistics." *Transportation Research Part E: Logistics and Transportation Review*, vol. 106, pp. 320-336, 2017.

[6] L. Shuai, K.A. Corzine, M. Ferdowsi, "A new battery/ultracapacitor energy storage system design and its motor drive integration for hybrid electric vehicles." *IEEE Transactions on vehicular technology*, vol. 56, pp. 1516-1523, 2007.

[7] G. Lijun, R.A. Dougal, S. Liu. "Power enhancement of an actively controlled battery/ultracapacitor hybrid." *IEEE transactions on power electronics*, vol. 20, pp. 236-243, 2005.

[8] R.A. Dougal, S. Liu, and R.E. White, "Power and life extension of battery-ultracapacitor hybrids." *IEEE Transactions on components and packaging technologies*, vol. 25, pp. 120-131, 2002.

[9] S. Junyi, S. Dusmez, A. Khaligh, "Optimization of sizing and battery cycle life in battery/ultracapacitor hybrid energy storage systems for electric vehicle applications." *IEEE Transactions on industrial informatics*, vol. 10, pp. 2112-2121, 2014.

[10] S. Junyi, A. Hasanzadeh and A. Khaligh, "Optimal power split and sizing of hybrid energy storage system for electric vehicles," *2014 IEEE Transportation Electrification Conference and Expo (ITEC)*, 2014, pp. 1-6.

[11] A. Khaligh, L. Zhihao, "Battery, ultracapacitor, fuel cell, and hybrid energy storage systems for electric, hybrid electric, fuel cell, and plug-in hybrid electric vehicles: state of the art." *IEEE Transactions on Vehicular Technology*, vol. 59, pp. 2806-2814, 2010.

[12] C. Xiang, Y. Wang, S. Hu, W. Wang, "A new topology and control strategy for a hybrid battery-ultracapacitor energy storage system." *Energies*, vol. 7, pp. 2874-2896, 2014.

[13] L. Gao, R.A. Dougal, S. Liu, "Power enhancement of an actively controlled battery/ultracapacitor hybrid." *IEEE Transaction on Power Electronics*, vol. 20, pp. 236–243, Jan. 2005.

[14] L. Kouchachvili, W. Yaïci, E. Entchev, "Hybrid battery/supercapacitor energy storage system for the electric vehicles." *J. Power Sources*, vol. 374, pp. 237-248, 2018.

[15] Y. Wang, L. Wang, M. Li, Z. Chen, "A review of key issues for control and management in battery and ultra-capacitor hybrid energy storage systems." *ETransportation*, vol. 4, pp. 1-12, 2020.

[16] J. Armenta, C. Núñez, N. Visairo, I. Lázaro, "An advanced energy management system for controlling the ultracapacitor discharge and improving the electric vehicle range." *J. Power Sources*, vol. 284, pp. 452-458, 2015.

[17] Z. Song, J. Li, X. Han, L. Xu, L. Lu, M. Ouyang, H. Hofmann, "Multi-objective optimization of a semi-active battery/supercapacitor energy storage system for electric vehicles." *Applied Energy*, vol. 135, pp. 212-224, 2014.

[18] M.J. Lencwe, P. Shyama, D. Chowdhury, T.O. Olwal, "Hybrid energy storage system topology approaches for use in transport vehicles: A review." *Energy Science & Engineering*, vol. 10, pp. 1449-1477, 2022.

[19] B. Wang, J. Xu, B. Cao, X. Zhou, "A novel multimode hybrid energy storage system and its energy management strategy for electric vehicles." *Journal of Power Sources*, vol. 281, pp. 432-443, 2015.

[20] L. Shuai, K.A. Corzine, M.A. Ferdowsi, "A new battery/ultracapacitor energy storage system design and its motor drive integration for hybrid electric vehicles." *IEEE Transactions on Vehicular Technology*, vol. 56, pp. 1516-1523, 2007.

[21] S.R. Srinivasa, R. Natarajan, "Energy management techniques and topologies suitable for hybrid energy storage system powered electric vehicles: An overview." *International Transactions on Electrical Energy Systems*, vol. 31, pp. 1-30, 2021.

[22] M.-E. Choi, S.-W. Seo, "Robust energy management of a battery/supercapacitor hybrid energy storage system in an electric vehicle," *Electric Vehicle Conference (IEVC)*, 2012, pp. 1-5.

[23] J. Armenta, C. Núñez, N. Visairo, I. Lázaro, "An advanced energy management system for controlling the ultracapacitor discharge and improving the electric vehicle range." *J. Power Sources*, vol. 284, pp. 452-458, 2015.

[24] L. Rosario, P.-K. Luk, "Applying management methodology to electric vehicles with multiple energy storage systems," *International Conference on Machine Learning and Cybernetics*, 2007, pp. 4223-4230.

[25] Z. Qiao, G. Li, "Experimental study on a semi-active battery-supercapacitor hybrid energy storage system for electric vehicle application." *IEEE Transactions on Power Electronics*, vol. 35, pp. 1014-1021, 2019.

[26] W. Maciej, M. Lewandowski, "A mathematical representation of an energy management strategy for hybrid energy storage system in electric vehicle and real time optimization using a genetic algorithm." *Applied energy*, vol. 192, pp. 222-233, 2017.

[27] Z. Chen, R. Xiong, J. Cao, "Particle swarm optimization-based optimal power management of plug-in hybrid electric vehicles considering uncertain driving conditions." *Energy*, vol. 96, pp. 197-208, 2016.

[28] Z. Chen, C.C. Mi, J. Xu, X. Gong, C. You, "Energy management for a power-split plug-in hybrid electric vehicle based on dynamic programming and neural networks." *IEEE Transactions on Vehicular Technology*, vol. 63, pp. 1567-1580, 2014.

[29] L. Wang, Y. Wang, C. Liu, D. Yang, Z. Chen, "A power distribution strategy for hybrid energy storage system using adaptive model predictive control." *IEEE Transactions on Power Electronics*, vol. 35, pp. 5897-5906, 2020.

[30] Y. Wang, X. Li, L. Wang, Z. Sun, "Multiple-grained velocity prediction and energy management strategy for hybrid propulsion systems." *Journal of Energy Storage*, vol. 26, pp. 1-8, 2019.

[31] L. Long, P. Bauer, "Practical capacity fading model for Li-ion battery cells in electric vehicles." *IEEE transactions on power electronics*, vol. 28, pp. 5910-5918, 2012.

[32] T. Plattard, N. Barnel, L. Assaud, S. Franger, J.M. Duffault, "Combining a fatigue model and an incremental capacity analysis on a commercial NMC/graphite cell under constant current cycling with and without calendar aging." *Batteries*, vol. 5, pp. 1-16, 2019.

Concept of Testbed Vehicle-to-Home Technology in a Household Off-Grid System Environment

Vojtech Blazek
ENET Centre
VSB-TUO
Ostrava, Czech Republic
vojtech.blazek@vsb.cz

Pavel Kedron
ENET Centre
VSB-TUO
Ostrava, Czech Republic
pavel.kedron@vsb.cz

Zdenek Slanina
FEI, ENET Centre
VSB-TUO
Ostrava, Czech Republic
zdenek.slanina@vsb.cz

Ivo Pergl
ENET Centre
VSB-TUO
Ostrava, Czech Republic
ivo.pergl@vsb.cz

Abstract — **The article analyzes, summarizes, validates, and describes the concept of a testbed platform for Vehicle-to-Home (V2H) technologies. The testbed platform has been developed to research bidirectional inverters, control algorithms, communication between electric vehicles (EV) and off-grid or distribution networks, prediction and optimization tools, active relay protection systems and battery diagnostic. The concept developed by us was tested in the Technical University of Ostrava campus. The article focused on the impact of power quality (PQ) parameters during charging and discharging EVs in off-grid household environments. The article further evaluates the measured values according to the PQ standard EN 50160 ED.3 and outlines future research on this platform.**

Keywords: ***Vehicle-to-Home, Electric Vehicle, Power Quality***

I. INTRODUCTION

Today, the range of electric vehicles is increasing, the life of their batteries is growing, and their price is decreasing. This has led to increased electric vehicles (EV), both in individual use and company fleets. Therefore, the (EV), especially a group of EVs using traction batteries with tens of kilowatt-hours, represents an essential device in terms of logistics and energy storage and a potential standby source of electricity if the infrastructure design allows such use. Figure 1 shows the growth of electric vehicles over the years.

For this purpose, e.g., Vehicle-to-Home (V2H) systems are used. An important area of application of this approach is energy units designed as off-grid systems. An essential aspect of such solutions is the influence of the power quality parameters inside the power unit during the power transfer processes between the power unit and the electric vehicle. The maintenance of these quality parameters is necessary for both transfer directions, both when using the power from the EVs battery in V2H and when charging the electric vehicle from the power unit [1].

Researchers from the Department of Electrical Power Engineering and the ENET Research Centre have developed an experimental power unit platform operating in off-Grid mode since 2011. The experimental power unit is located on the premises of the VSB - Technical University Ostrava. Its primary purpose is to simulate household conditions using the most common household appliances. The experimental platform has undergone many changes during the research. In the past, the research team has developed several prototype testbeds, which it has further used for its research.

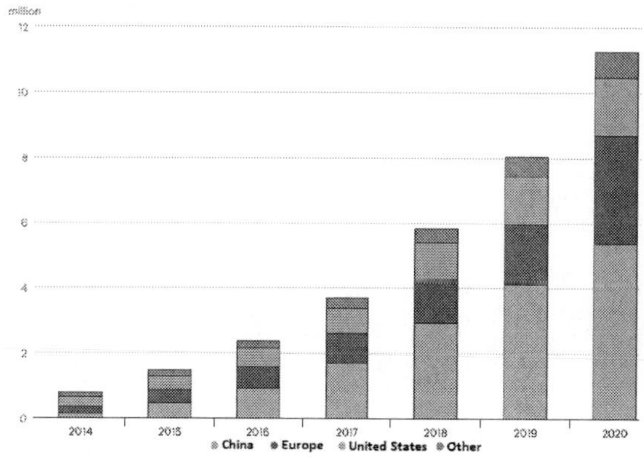

Figure 1. Growth of electric vehicles over the years [2]

In the past, the platform has been tested predictive models for wind and photovoltaic power generation, specific adaptive relays protection systems designed for AC coupling off-grid systems or battery diagnostics [3]–[9].

As a result, the research team has published highly regarded papers in many relevant journals and conferences. Furthermore, the research team has already focused several times on the effects of different appliances on power quality, not only in the distribution system but also in off-grid systems with low short-circuit power [10]–[14].

Publications and research in academia are mainly concerned with mathematical models of V2H, implementing V2H itself in only one appliance, or focusing on individual parts of the system and no longer dealing with the interconnection of individual components and control systems. Furthermore, they do not address the use of the technology itself as a potential diagnostic platform[15]–[20].

The author's motivation is to develop a platform to test V2H technologies regularly. The article outlines the concept of the testing platform and its possible future potential.

The article brings the following innovations:

- presents the V2H testbed concept
- describes the different parts of the V2H testbed
- verifies the functionality of the V2H concept
- outlines future research on the V2H testbed
- summaries of the development of V2H tested to date

The article is structured into five sections. Section 1 introduces the topic, motivation, and objectives of the authors. Section 2 describes the power unit platform and control system. Section 3 describes the experiment, Section 4 summarizes the experiment's results, and finally, Section 5 presents the experiment's conclusions.

II. Description of the V2H Testbed

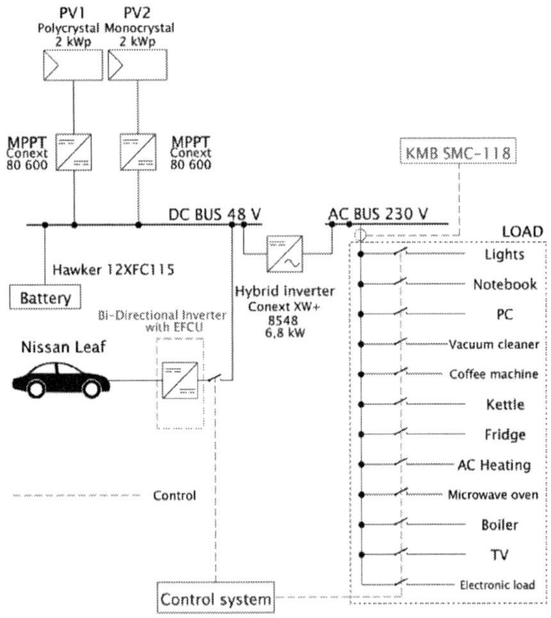

Figure 2. Electric scheme of the testbed platform in V2H DC bus configuration [21]

Figure 2 show the electric scheme of the entire V2H testbed. The testbed centre is a Hybrid inverter Conext XW + 8548 with a rated output power of 6.8 kW. The testbed's primary energy sources are two photovoltaic plants (PV) with a nominal output of a 2-kW peak. These panels are connected via MPPT inverter to DC 48 V bus. Electric energy in the V2H testbed was stored in 40 pieces of Hawker 12XFC115 batteries. These batteries are connected in four parallel groups, where each group has four batteries in series, and the nominal voltage is 48 V [21], [22]. The Nissan Leaf is connected to the bus via the V2H inverter with Energy Control Flow Unit (EFCU) [23], [24].

The test platform has been designed and implemented as a set of technical means enabling modular construction of test and test systems applicable for obtaining data describing the behaviour of devices used and usable in systems of bidirectional power transfer between the power unit and the electric vehicle battery, collection and transmission of continuous information about the parameters of the electric vehicle battery, power quality parameters in the power unit and their influence on the power transfer processes between the power unit and the electric vehicle in both directions, etc.

The platform consists of groups of technical and communication devices that provide specific functions within the test system assembly:

- Equipment for the measurement of electrical energy parameters
- Communication and signal interfaces
- Equipment for simulation/implementation of power unit loads
- Software means for data acquisition, storage and processing and parameterization of platform components
- Electric vehicles to be connected in the test

Key components of the test platform are summarized in the following points:

A. *Equipment for the transfer and conversion of electrical energy between the electric vehicle and the power unit*

- Schneider Conext XW+ hybrid inverter

Bidirectional AC-DC converter used in platform configurations in off-grid system mode.

- 48 VDC, 115 Ah stationary battery storage

Energy storage for platform configurations in off-grid mode.

- 4Q inverter Zekalabs LB-1071

Bidirectional power conversion between the stationary battery of the island system and the battery of the EV in DC charge/discharge modes in relevant platform configurations.

- V2H converter Setec Power SET-V2H-3kW

DC->AC converter used in platform configurations targeting EV battery power transfers in DC mode.

- Wallbox Schneider EVlink EVB1A22P2RI

The charging device used in platform configurations focused on tasks and measurements related to power transfers to the EV battery in AC charging mode.

B. *Electrical power parameter measurement devices*

- AC metering equipment fitted with an Eastron SDM630

Measurement of power quality parameters on the power unit side for mobile configurations: voltage, current, power factor, frequency

- AC metering equipment equipped with a KMB SMC-118 network analyzer

Measurement of power quality parameters on the power unit side for stationary configurations: voltage, current, power factor, THD I, THD U, PLT, PST, Pint, frequency

- DC metering device equipped with Carlo Gavazzi CPA3001LS1X power meter

Voltage and current measurement on the EV side in platform configurations implementing DC charging/discharging cases of the EV.

C. Communication and signal interfaces

- Communication and signal interface and EFCU test controller

A device of original design based on a single-board computer with ARM architecture, equipped with Ethernet, CANbus, RS485 communication interfaces and a 12 V digital I/O, designed to listen, simulate and provide control communication and signalling tasks for test platform configurations covering DC charging/discharging of an electric vehicle battery according to CHAdeMO standards. CHAdeMO communication (CANbus), CHAdeMO signalling (12 VDC digital I/O), CHAdeMO gateway implementation, CHAdeMO sniffer implementation, CHAdeMO control sequence implementation, DC/DC inverter control, freely programmable control functions and tasks, protocol conversion

- Vgate Scan OBD II communication interface

Collects EV traction battery diagnostic data from the EV on-board network during EV charging/discharging.

D. Equipment for simulation/implementation of power unit loads

- Stationary controlled load simulation unit

The original design's device allows loading the test platform's AC distribution circuit in the load power range 0 to 4kW with adjustable $cos\phi$in the range -0,5 to 0,5. In addition, it allows monitoring of changes in power quality parameter values depending on the set/dynamically adjustable characteristics of the controlled load.

- Stationary Real Load Control Unit

A device of original design allowing to load the AC distribution circuit of the test platform by a method of automated attachment and detachment of combinations of real loads (boiler, microwave oven, kettle, TV, air conditioning, lighting, refrigerator). Loading the AC distribution of a test island system while transferring power between the EV battery and the island system. Allows monitoring of changes in power quality parameter values depending on the load consisting of a set/dynamically adjustable combination of real appliances of the model off-grid.

- Mobile passive load unit

A device of original design implementing a 2 kW resistive load for appliance simulation. Loading the AC distribution system during power transfer between the EV battery and the platform.

E. EVs connected in platform test operation

- Nissan Leaf

The following are the implemented testbed configurations concepts, representing a comprehensive example of the investigated problem's task. During the operation of the test platform, a considerable number of configurations were used to monitor partial problems and aspects. A description of all such iterations and variations would not be fruitful; therefore, only selected configurations that best demonstrate the platform's capabilities are described below.

III. THE EXPERIMENTAL DESCRIPTION

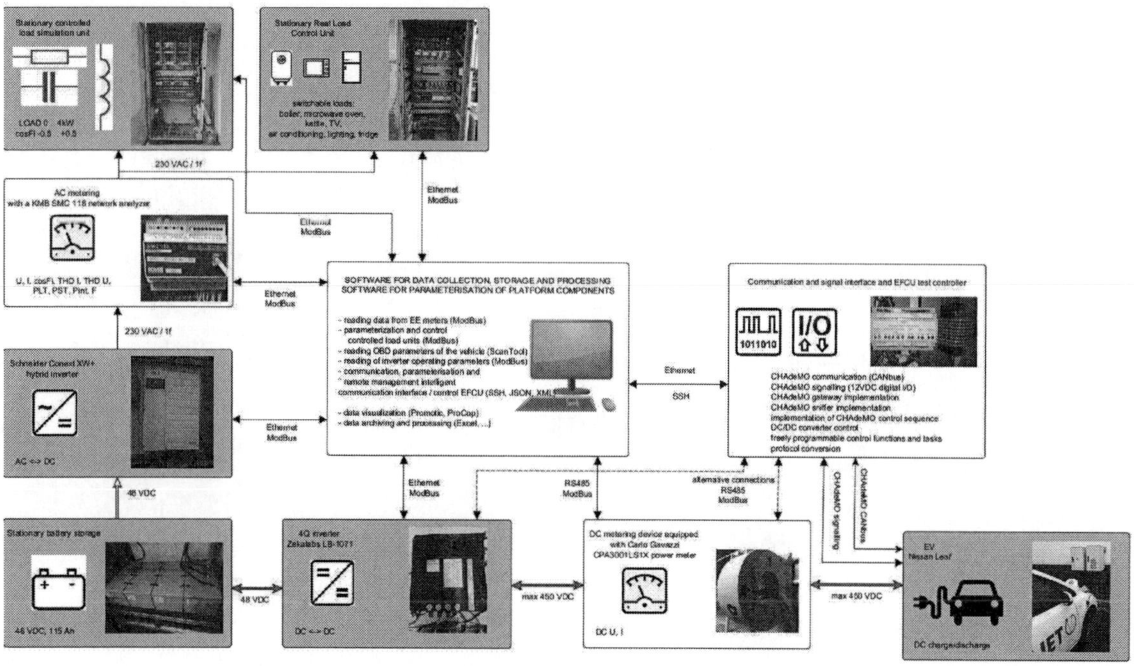

Figure 3. Schematic diagram of the platform

The experimental part deals with verifying the concept, which you can see in the Figure 3. The experiment was performed according to figure Figure 2. During the experiment, the KMB SMC-118 network analyzer was used to comprehensively evaluate the power quality in the network. The analyzer can evaluate harmonic content, symmetry, energy flows, presence of flicker, HDO signals, evaluate EMC and more. Experiments were taken from a connected Nissan Leaf EV with a fully charged battery. The experiment was taken during the platform's regular operation and is intended only to demonstrate the functionality V2H testbed. The experiment was carried out during sunny weather on July 18, 2021.

IV. Results

The following are Figure 5., Figure 6., Figure 7., Figure 8. Figure 9. shows the effect of the electricity from the EV to the off-grid on the AC side. In the experiment, we monitored the following electrical quantities: Long Term Flicker Perceptibility (Plt), Short Term Flicker Perceptibility (Pst), Voltage (U), Frequency (F), Total Harmonic Voltage Distortion (THDV), Total Harmonic Current Distortion (THDI), Active power (P). Reactive power (Q), Apparent Power (S) and Distortion Power (D).

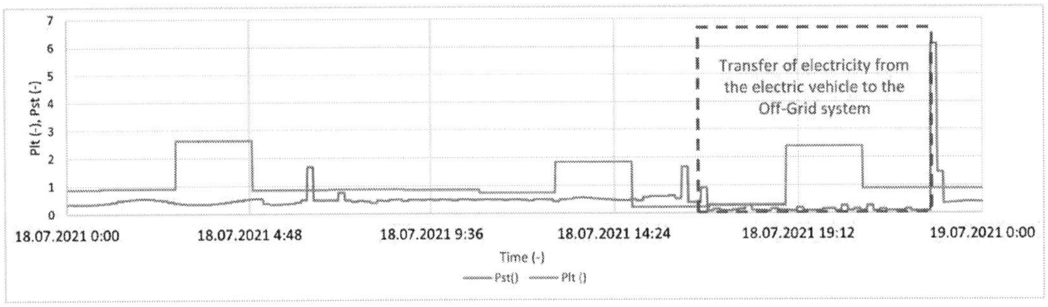

Figure 4. Long Term Flicker Perceptibility (Plt) and Short Term Flicker Perceptibility (Pst) values during the experiment

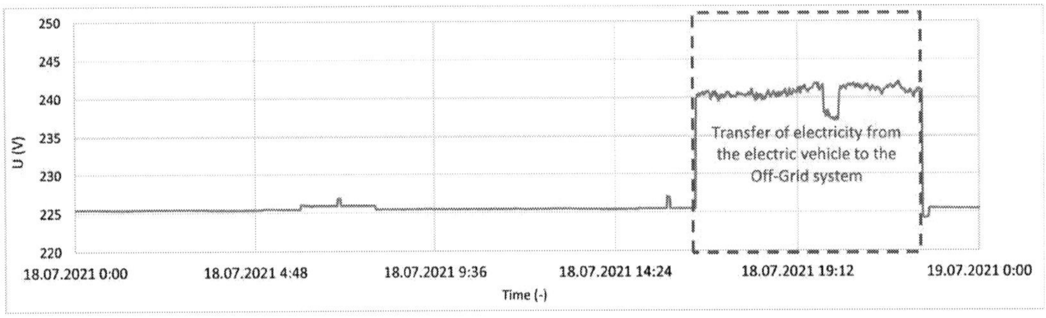

Figure 5. Voltage (U) values during the experiment

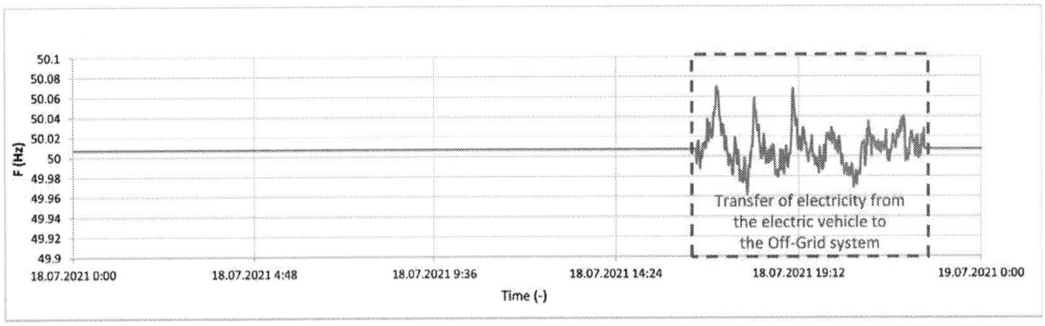

Figure 6. Frequency (F) values during the experiment

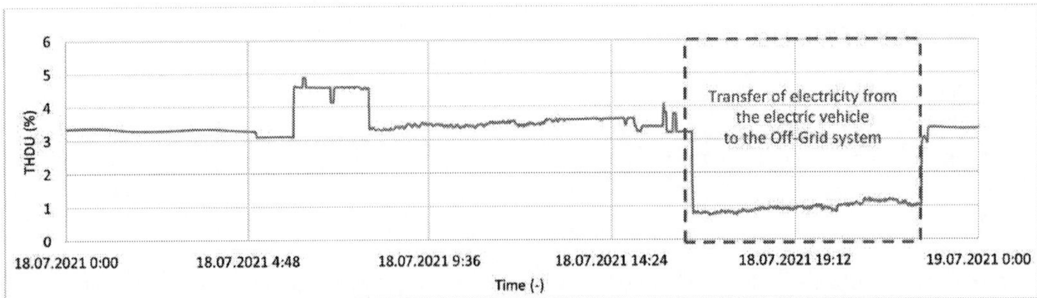

Figure 7. Total Harmonic Voltage Distortion (THDU) values during the experiment

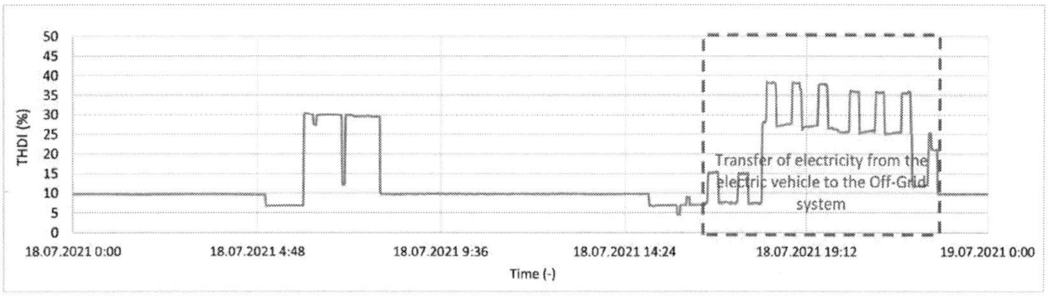

Figure 8. Total Harmonic Current Distortion (THDI) values during the experiment

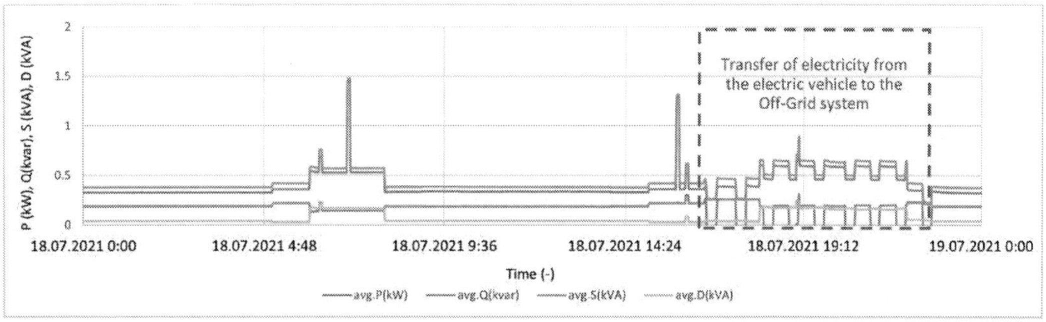

Figure 9. Active power (P). Reactive power (Q), Apparent Power (S) and Distortion Power (D) values during the experiment

V. Conclusion

The result shows that the EV affected the power quality in the AC part of the off-grid. The experiment showed that the energy supplied from the EV to the off-grid could improve but also deteriorate the power quality, even though the energy was supplied to the DC part of the off-grid system. The change in parameters is mainly caused by the semiconductor elements in the off-grid system.

The measurement results can be summarized in the following points:

- The energy transferred from the EV to the DC part affected the power quality of the AC part of the off-grid.
- During the energy transfer, the voltage on the AC side increased
- THDU decreased during power transfer
- During power transfer, the frequency was less constant
- After the end of the power transfer from the EV to the off-grid, a transient event occurred that resulted in a deterioration of the Pst parameter
- The power quality parameters did not exceed the standard CSN 50 160 ed. 3.
- During the measurements, the Pst and Plt parameter was measured to exceed the standard for public distribution networks IEC 61000-3-3:2013/AMD2:2021 (Pst Less than or equal to 1.0, Plt Less than or equal to 0.65)

The article dealt with the description and verification of the testbed as a potential future research platform for Vehicle-to-Home technology. Our future work will look at non-traditional approaches to using the EV as a battery for electrical energy storage.

Acknowledgement

This paper was supported by the following projects: This work was supported by the Doctoral grant competition VSB—Technical University of Ostrava, reg. no. CZ.02.2.69/0.0/0.0/19 073/0016945 within the Operational Programme Research, Development and Education, under project DGS/TEAM/2020-017 "Smart Control System for Energy Flow Optimization and Management in a Microgrid with V2H/V2G Technology", FV40411 Optimization of process intelligence of parking system for Smart City and project TN01000007 National Centre for Energy and by project WZ/WE-IA/2/2020 of the Bialystok University of Technology.

References

[1] IEA, "Global Electric Vehicle Outlook 2021." IEA, Feb. 2021.

[2] "Global electric vehicle stock by region." Internet: https://www.iea.org/data-and-statistics/charts/global-electric-vehicle-stock-by-region-2010-2020, [Feb. 15, 2022].

[3] L. Prokop, S. Misak, V. Snaásel, J. Platos, and P. Krömer, "Supervised learning of photovoltaic power plant output prediction models," *Neural Network World*, vol. 23, pp. 321, 2013.

[4] S. Misak, J. Stuchly, J. Vramba, T. Vantuch, and D. Seidl, "A novel approach to adaptive active relay protection system in single phase AC coupling Off-Grid systems," *Electric Power Systems Research*, vol. 131, pp. 159–167, 2016.

[5] T. Sikora, L. Prokop, J. Fulnecek, and T. Vantuch, "Industrial Uniterruptible Power Supply Batteries Monitoring Augmentation by Infrared Camera," *Proceedings of the 2019 20th International Scientific Conference on Electric Power Engineering (EPE)*, 2019, pp. 94–97.

[6] L. Prokop, S. Misak, and P. Krejci, "Analysis of Long time WPP PowerFlows Measurement," *Proceedings of the 10th International Scientific Conference Electric Power Engineering 2009*, 2009, pp. 373–378.

[7] T. Vantuch, A. G. Vidal, A. P. Ramallo-González, A. F. Skarmeta, and S. Misák, "Machine learning based electric load forecasting for short and long-term period," *2018 IEEE 4th World Forum on Internet of Things (WF-IoT)*, 2018, pp. 511–516.

[8] T. Vantuch, J. Fulneček, M. Holuša, S. Mišák, and J. Vaculík, "An Examination of Thermal Features' Relevance in the Task

of Battery-Fault Detection," *Applied Sciences*, vol. 8, pp. 1-16, 2018.

[9] S. Misak, L. Prokop, T. Sikora, T. Poplawski, K. Dasal, and B. Rusek, "Wind Power Plant Production Prediction Using Various Models," *Proceedings of the 10th International Scientific Conference Electric Power Engineering 2009*, 2009, pp. 152+.

[10] S. Misak, J. Stuchly, J. Vramba, L. Prokop, and M. Uher, "Power quality analysis in off-grid power platform," *Advances In Electrical And Electronic Engineering*, vol. 12, pp. 177–184, 2014.

[11] S. Misak, L. Prokop, and J. Dvorský, "Optimizing the mathematical model for prediction of energy production in Wind Power Plants," *Przeglad Elektrotechniczny*, vol. 87, pp. 74–78, Mar. 2011.

[12] J. Rodway, P. Musilek, S. Misak, and L. Prokop, "Prediction of PV power quality: Total harmonic distortion of current," in *2013 IEEE Electrical Power Energy Conference*, 2013, pp. 1–4.

[13] M. Petružela, V. Blažek, J. Vysoký, "Analysis of Appliance Impact on Total Harmonic Distortion in Off-Grid System," in *AETA 2018 - Recent Advances in Electrical Engineering and Related Sciences: Theory and Application*, 2020, pp. 844–849.

[14] V. Blazek *et al.*, "Error Analysis of Narrowband Power-Line Communication in the Off-Grid Electrical System," *Sensors*, vol. 22, no. 6, pp. 2265, Mar. 2022.

[15] V. Kaczmarczyk, Z. Bradac, and P. Fiedler, "A Heuristic Algorithm to Compute Multimodal Criterial Function Weights for Demand Management in Residential Areas," *Energies*, vol. 10, pp. 1-28, 2017.

[16] B. Bibak, H. Tekiner-Mogulkoc, "The parametric analysis of the electric vehicles and vehicle to grid system's role in flattening the power demand," *Sustainable Energy, Grids and Networks*, vol. 30, pp. 1-9, 2022.

[17] N. Koltsaklis, I. Panapakidis, G. Christoforidis, and J. Knápek, "Smart home energy management processes support through machine learning algorithms," *Energy Reports*, vol. 8, pp. 1–6, 2022.

[18] M. Wang and M. A. A. Abdalla, "Optimal Energy Scheduling Based on Jaya Algorithm for Integration of Vehicle-to-Home and Energy Storage System with Photovoltaic Generation in Smart Home," *Sensors*, vol. 22, pp. 1-23, 2022.

[19] H. al Attar, M. A. Hamida, M. Ghanes, and M. Taleb, "Llc dc-dc converter performances improvement for bidirectional electric vehicle charger application," *World Electric Vehicle Journal*, vol. 13, pp. 1-20, 2022.

[20] D. Borge-Diez, D. Icaza, E. Acikkalp, and H. Amaris, "Combined vehicle to building (V2B) and vehicle to home (V2H) strategy to increase electric vehicle market share," *Energy*, vol. 237, pp. 1-17, 2021.

[21] V. Blazek, M. Petruzela, J. Vysocky, L. Prokop, S. Misak, and D. Seidl, "Concept of Real-Time Communication in Off-Grid System with Vehicle-to-Home Technology," *2020 21st International Scientific Conference on Electric Power Engineering (EPE)*, 2020, pp. 1–6.

[22] V. Blazek, M. Petruzela, T. Vantuch, Z. Slanina, S. Mišák, and W. Walendziuk, "The Estimation of the Influence of Household Appliances on the Power Quality in a Microgrid System," *Energies*, vol. 13, pp. 1-19, Aug. 2020.

[23] M. Kosinka, Z. Slanina, M. Petruzela, and V. Blazek, "V2H control system software analysis and design," in *2020 20th International Conference on Control, Automation and Systems (ICCAS)*, 2020, pp. 972–977.

[24] M. Kosinka, Z. Slanina, M. Petruzela, and V. Blazek, "Control system for V2H applications," in *2020 20th International Conference on Control, Automation and Systems (ICCAS)*, 2020, pp. 916–921.

Impact of Selected Parameters on Results of Dielectric Frequency Response

Ondřej Kabot
CEET, ENET Centre
VSB - Technical University of Ostrava
Ostrava, Czech Republic
ondrej.kabot@vsb.cz

Jan Fulneček
CEET, ENET Centre
VSB – Technical University of Ostrava
Ostrava, Czech Republic
jan.fulnecek@vsb.cz

Denisa Fulnečková
Rubix Czech s.r.o.
Ostrava, Czech Republic
denisa.fulneckova@rubix.com

Abstract — **This article focuses on usage of dielectric frequency response as a diagnostic tool for the purpose of state assessment of oil insulated transformers. Main goal of it is to compare various settings, mainly voltage and stop frequency and their impact on the results. For the purpose of comparison with other methods, Karl Fischer Stoichiometric Coulometric Titration method is described and main differences between the two mentioned methods are pointed out.**

Keywords: dielectric frequency response; transformer; transformer oil; high voltage insulation system; electrical device diagnostics

I. Introduction

Dielectric response is method of diagnostics, which can be used to determine the dielectric properties of materials. This method has found its use in various industries, for example chemical and electrochemical industry. The principle of this method is using generating voltage at variating frequencies and using it to create an electrical field inside the device under test. Diagnostics is then carried out by analyzing the object response to this field, thus the method is called dielectric frequency response. Main use of this method in diagnostics of electrical devices is determination of frequency dependent complex permittivity of various materials. This method brings huge advantages, especially in diagnostics of oil transformers [1]. But it can be theoretically used for any insulation system or device like cable lines, capacitors and more. In case of transformers and capacitors, this method can be used for both to monitor the trends of the device as well as analysis of current state of it without knowledge of historical data. When it comes to cable lines, this method is only suitable for trend monitoring. Particularly in a case of HV cable/lines, it is not possible to determine the condition of insulation based just on one measurement. Therefore, other methods are more often used in a case of cable diagnostics [2][3]. Usage of this method for analysis of state of oil transformers has a big advantage because it allows the diagnostics of oil state without requirement to acquire an oil sample. Dielectric properties change by the effect of its degradation or presence of water. These changes can be detected by dielectric response.

II. Dielectric response with use of IDAX device

One of the devices that can be used for dielectric response analysis is IDAX from Megger. Values acquired by measurements are capacity of insulation system, dissipation factor TD, moisture content in oil and its conductivity. These values are shown for basic diagnostics of an oil transformer. Values of permittivity etc. can also be displayed and analyzed. Evaluation of measured variables is made by comparing them with model curves and values that are included in IDAX software. Model curve represents the ideal state of device under test in ideal working condition and measured values represent the real state of the device. The bigger is the difference between the model and measured curve, the bigger is the deterioration of devices insulation system. Because of vast database of model curves, there is no requirement for historical data when it comes to diagnostics of transformers. That also means, the bigger is the database, the more accurate results we can achieve by using this method. Figure 1 shows how different aspects such as moisture, temperature and conductivity of an oil change the shape of curve acquired by measurement [4].

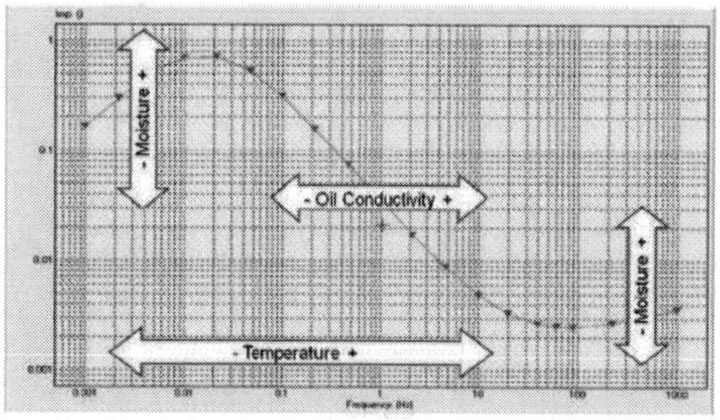

Figure 1 Influence of oil conductivity, moisture and temperature [4]

The measurement itself requires disconnection of both primary and secondary winding and following connection of electrodes on each side. To the high voltage side, the test voltage is being applied and the low voltage side is connected to the test probe. The part being tested is therefore the main insulation between high and low voltage parts of the transformer. The grounding points of the test device, amplifier (in case it is used) and device under test are all interconnected and grounded. The time requirement of measurement depends on the lowest used frequency. The highest stop frequency is 1 Hz where measurements only take couple of minutes the lowest stop frequency is 0.1 mHz and this measurement takes several hours. There are multiple criteria that are used to determine the stop frequency. One of them is temperature of the device under test. The lower the temperature, the lower should also be the stop frequency. Also, if detailed analysis is required, the lowest possible stop frequency should be used without regard to objects temperature. Voltage is another parameter that can be set. The maximal value of test voltage when only IDAX is used is 140 V_{RMS} in case amplifier is used, test voltage of 1400 V_{RMS} can be applied to the device under test. The main goal of this paper is to determine the influence of set voltage and stop frequency on the analysis results. From the principle of this method, the lower the stop frequency, the more accurate results. But the consumed time has to be also taken into account. As for the voltage, there is restriction caused by available test devices.

III. Diagnostics of oil transformer with use of IDAX device

Tested object in this case is oil transformer 22/0,4 kV.

Table 1. Transformer parameters

Type	T0294/22 ONAN	
Connection	Yzn1	
Power	100 kVA	
u_k	6,93%	
	Primary	Secondary
Nominal voltage	22 000 V	400 V
Nominal current	2,62 A	144,4 A

For diagnostics of this object, IDAX along with amplifier VAX was used. To allow comparison of both frequency and voltage influence on the results, series of measurements were taken with three different voltage levels and two different stop frequencies. The connection was done as shown on Figure 2, with only difference that the voltage output was from amplifier VAX, that was connected to the IDAX unit via external port.

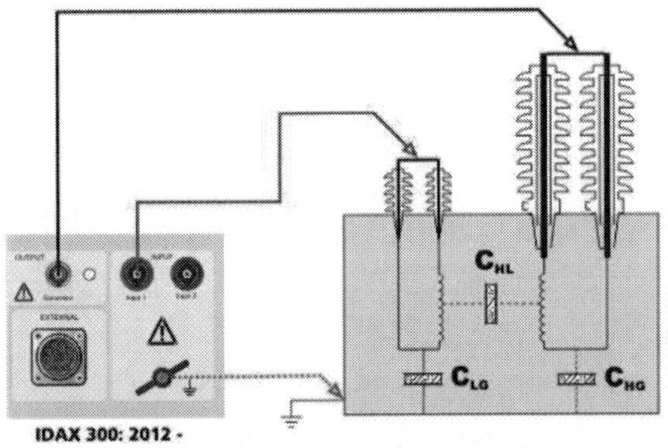

Figure 2 Connection of IDAX to transformer [4]

First the measurements with stop frequency 1 mHz were carried out and then second set of measurements with stop frequency 0,1 mHz. In both cases there were 3 different voltage levels of 14, 140 and 1400 V_{RMS}. The trends are only displayed for the measurements with the stop frequency of 0,1 mHz. It is because the curves describing the measurement with stop frequency 1 mHz are identical only missing few additional values. The differences are evaluated in Table 2.

Table 2. Values calculated by IDAX software

Test Voltage	14 V		140 V		1400 V	
Stop Frequency (mHz)	1	0,1	1	0,1	1	0,1
Moisture (% (wt/wt))	3,3	3,2	3,3	3,2	3,3	3,3
Dissipation Factor (% td)	0,708	0,73	0,70	0,72	0,705	0,717
Conductivity (pS/m)	10,10	11,80	9,44	11,3	9,1	9,6

From the values in Table 2 it is apparent, that influence of selected voltage level when stop frequency is se to 1 mHz is very low, the conductivity is only parameter that significantly changes. When the stop frequency is set to 0,1 mHz, there are also changes in dissipation factor. Also, the measured values for dissipation factor are for frequency of 50 Hz and temperature 20 °C [5].

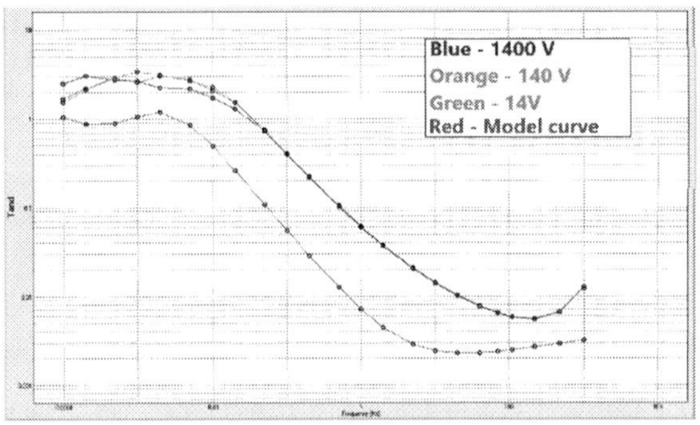

Figure 3 Dependency of dissipation factor on frequency

From curves that are shown at Figure 3 it is apparent that measurement of dissipation factor at voltage levels 14 and 140 V_{RMS} are very similar besides small nuances. Measurement done at voltage level 1400 V_{RMS} is quite different in both shape and values. It must be kept in mind the Y axis has logarithmical scale. This shows that in case of detailed analysis and with wide frequency spectrum there are significant differences in acquired values. In case the spectrum would be even wider, the presumption is that the differences would be even more significant. For correct evaluation of measurement, it is important that acquired curves have maximum in the shape of peak at lower frequencies. In case this is not the case, and the shape is not present, the stop frequency was not low enough and it is insufficient for correct assessment of devices current state [6].

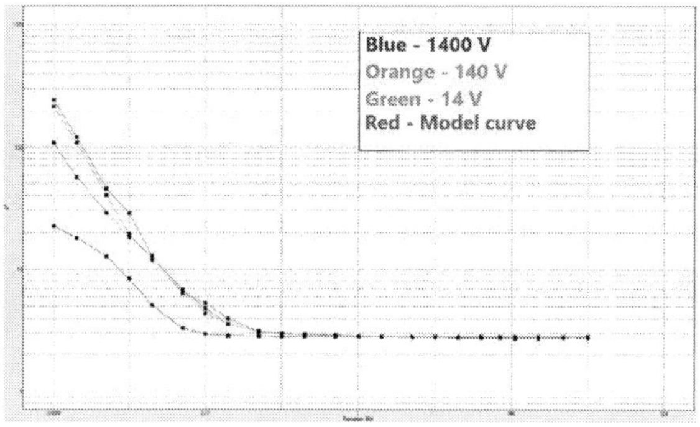

Figure 4 Dependency of real part of permittivity on frequency

The influence of testing voltage on measured permittivity value can be seen in fig. 4. The results of real part of permittivity are quite similar to those of dissipation factor. The main difference is that in this case the curves also overlap with model curve from 1000 Hz to 1 Hz. At lower frequencies not only do measured curves differ from model one, but also the curve of the highest voltage level is different from two lower ones.

IV. Coulometrical Method

Dielectric frequency response is not the only method that can be used for the purpose of assessment of the moisture in transformer oil. One of the other options how to analyze the state of oil transformer is detection of water content in oil. This can be done by coulometric method.[7]

For this purpose, Karl Fischer Stoichiometric Coulometric Titration method was used. This method can be used to determine small or even trace amounts of water in oil samples. Method is based on principle of iodometric water content determination in solution and is done by spraying the investigated substance into electrochemical container.

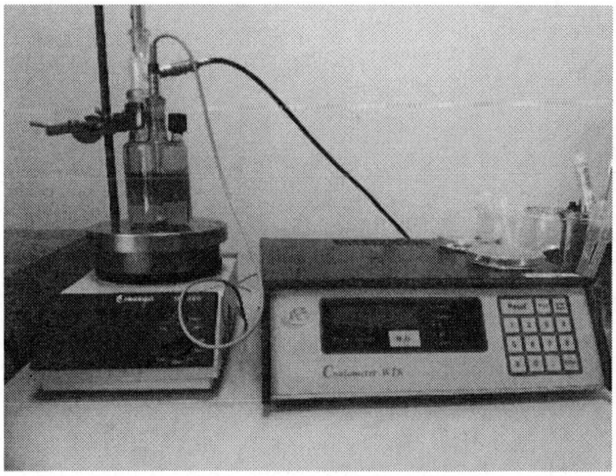

Figure 5 Coulometer

Water reacts with iodine and other substances. For 1 mol of iodine 1 mol of water is consumed this leads to transfer of charge equivalent to 1 mol of electron. By conduction of current through electrochemical titration container iodine oxidizes on generation platinum plate as elementary iodine. That allows the reaction of the water with another components of solution. Point of equivalency is then determined bipotentiometrically with use of two platinum electrodes, that are polarized by alternating current of constant value. Based on value of electrical charge we can use Faradays law to calculate iodine that was consumed by past reactions. Three oil samples were taken from the transformer and analyzed [8].

Table 3. Water content in samples of transformer oil

Number of sample	1	2	3
Weight of sample (g)	2,515	2,754	2,712
Weight of water (g)	6,48E-05	6,00E-05	6,16E-05
Water content (%)	2,58E-03	2,18E-03	2,27E-03

V. Conclusion

As it is clear from the results, for general purpose, there is no need for high voltages and very low stop frequencies as they do not affect the calculated result that much. In the case the result of the diagnostics should be detailed analysis of the tested object, with detailed examination of individual characteristics, the highest possible voltage and the lowest possible stop frequency should be used. Also, if there is a lot of interference, higher voltages are suggested to avoid distortion of measured signals. Every acquired curve shows that the biggest differences occur in the spectrum below 0,1 Hz in case of values measured at frequency 100 Hz and higher, curves are practically the same.

The main advantage of this method is that it can be carried out on a transformer that is cooled down. It is because it analyses the dielectric properties of the whole insulation system. That means it considers both the moisture content in solid part of the insulation system (usually paper) and the oil itself. On the other hand, when Coulometric method is used, there is a requirement to have the transformer heated up to operating temperature. This is because in cold state most of water content is stored in solid part of the insulation system and cannot be detected by analysis of oil. By heating up the transformer, most of the water migrates from solid part of the insulation into oil. This can also be seen in Tables 2 and 3. Coulometric method says that the water content is low but according to the IDAX, water content is quite significant. That is because the oil sample was taken from the transformer when it was cooled down. There are equations that can be used for recalculation of the water content, but some of them can be quite complicated to use and might not be as accurate.

Acknowledgment

This work was supported by the Doctoral grant competition VSB – Technical University of Ostrava, reg. no. CZ.02.2.69/0.0/0.0/19_073/0016945 within the Operational Programme Research, Development and Education, under project DGS/TEAM/2020-015 "Partial Discharge Detection in Insulation Systems".

References

[1] J. Liu, L. Zhou, Y. Luo, X. Huang and G. Wu, "Dielectric frequency response of oil-paper composite insulation with transient moisture distribution," *IEEE Transactions on Dielectrics and Electrical Insulation*, vol. 20, pp. 1380-1387, August 2013.

[2] M. Krátký, S. Mišák, P. Gajdoš, P. Lukáš, R. Bača, P. Chovanec, "A Novel Method for Detection of Covered Conductor Faults in Medium Voltage Overhead Line Systems." *IEEE Transactions on Industrial Electronics*, vol. 65, pp. 543-552, 2018.

[3] J. Liu, Y. Xu, X. Wei and X. Li, "Research of Dielectric response on insulation ageing assessment of XLPE cables," *2013 Annual Report Conference on Electrical Insulation and Dielectric Phenomena*, 2013, pp. 140-143.

[4] Megger Sweden AB, "IDAX 300/322/350 User guide", Doc, AG033586EE, V06a, 2020.

[5] D. Martin and T. Saha, "A review of the techniques used by utilities to measure the water content of transformer insulation paper," *IEEE Electrical Insulation Magazine*, vol. 33, pp. 8-16, May-June 2017.

[6] M. Ohlen and P. Werelius, "Dielectric Frequency Response and temperature dependence of power factor," *2010 IEEE International Symposium on Electrical Insulation*, 2010, pp. 1-7.

[7] M. Lyutikova, "An Improved Electrochemical Method for Moisture Determination in Mineral Oil," *IEEE Transactions on Dielectrics and Electrical Insulation*, vol. 27, pp. 2172-2178, December 2020.

[8] L. Lewand and P. Griffin, "Discrete and continuous moisture-in-oil measurements in electrical apparatus," *2008 International Symposium on Electrical Insulating Materials (ISEIM 2008)*, 2008, pp. 676-676.